Biomathematics

Volume 19

J. D. Murray

Mathematical Biology

With 292 Figures

Springer-Verlag
Berlin Heidelberg New York
London Paris Tokyo

James D. Murray, FRS
Professor of Mathematical Biology
Director, Centre for Mathematical Biology
Mathematical Institute
University of Oxford
24–29 St. Giles'
Oxford OX1 3LB
Great Britain

Cover picture: A model based on a system of reaction-diffusion equations has been suggested by the author of this book to explain how the coat markings on the leopard and other mammals are generated. In this book, he gives a whole range of animal patterning examples – from the stripes on the zebra to the eyespots on the wings of butterflies – to demonstrate the wide applicability of such models.

Mathematics Subject Classification (1980): 34C, 34D, 35B, 35K, 92A06, 92A08, 92A12, 92A15, 92A17, 92A90

ISBN 3-540-19460-6 Springer-Verlag Berlin Heidelberg New York
ISBN 0-387-19460-6 Springer-Verlag New York Berlin Heidelberg

Library of Congress Cataloging-in-Publication Data.
Murray, J. D. (James Dickson)
Mathematical biology / James D. Murray.
p. cm. – (Biomathematics; v. 19)
Bibliography: p. Includes index.
ISBN 0-387-19460-6 (U.S.: alk. Paper)
1. Biology – Mathematical models. I. Title. II. Series
QH323.5.M88 1989 574'.072'4–dc19 88-34184 CIP

© Springer-Verlag Berlin Heidelberg 1989
Printed in the United States of America

Media conversion: EDV-Beratung Mattes, Heidelberg
2141/3140-543210   Printed on acid-free paper

Preface

Mathematics has always benefited from its involvement with developing sciences. Each successive interaction revitalises and enhances the field. Biomedical science is clearly the premier science of the foreseeable future. For the continuing health of their subject mathematicians must become involved with biology. With the example of how mathematics has benefited from and influenced physics, it is clear that if mathematicians do not become involved in the biosciences they will simply not be a part of what are likely to be the most important and exciting scientific discoveries of all time.

Mathematical biology is a fast growing, well recognised, albeit not clearly defined, subject and is, to my mind, the most exciting modern application of mathematics. The increasing use of mathematics in biology is inevitable as biology becomes more quantitative. The complexity of the biological sciences makes interdisciplinary involvement essential. For the mathematician, biology opens up new and exciting branches while for the biologist mathematical modelling offers another research tool commmensurate with a new powerful laboratory technique but *only* if used appropriately and its limitations recognised. However, the use of esoteric mathematics arrogantly applied to biological problems by mathematicians who know little about the real biology, together with unsubstantiated claims as to how important such theories are, does little to promote the interdisciplinary involvement which is so essential.

Mathematical biology research, to be useful and interesting, must be relevant *biologically*. The best models show how a process works and then predict what may follow. If these are not already obvious to the biologists *and* the predictions turn out to be right, then you will have the biologists' attention. Suggestions as to what the governing mechanims are, may evolve from this. *Genuine* interdisciplinary research and the use of models can produce exciting results, many of which are described in this book.

No previous knowledge of biology is assumed of the reader. With each topic discussed I give a brief description of the biological background sufficient to understand the models studied. Although stochastic models are important, to keep the book within reasonable bounds, I deal exclusively with deterministic models. The book provides a toolkit of modelling techniques with numerous examples drawn from population ecology, reaction kinetics, biological oscillators, developmental biology, evolution, epidemiology and other areas.

The emphasis throughout the book is on the practical application of mathematical models in helping to unravel the underlying mechanisms involved in the biological processes. The book also illustrates some of the pitfalls of indiscriminate, naive or uninformed use of models. I hope the reader will acquire a practical and realistic view of biological modelling and the mathematical techniques needed to get approximate quantitative solutions and will thereby realise the importance of relating the models and results to the real biological problems under study. If the use of a model stimulates experiments – even if the model is subsequently shown to be wrong – then it has been successful. Models can provide biological insight and be very useful in summarizing, interpreting and interpolating real data. I hope the reader will also learn that (certainly at this stage) there is usually no 'right' model: producing similar temporal or spatial patterns to those experimentally observed is only a first step and does not imply the model mechanism is the one which applies. Mathematical descriptions are *not* explanations. Mathematics can never provide the complete solution to a biological problem on its own. Modern biology is certainly not at the stage where it is appropriate for mathematicians to try to construct comprehensive theories. A close collaboration with biologists is needed for realism, stimulation and help in modifying the model mechanisms to reflect the biology more accurately.

Although this book is titled *mathematical biology* it is not, and could not be, a definitive all-encompassing text. The immense breadth of the field necessitates a restricted choice of topics. Some of the models have been deliberately kept simple for pedagogical purposes. The exclusion of a particular topic – population genetics for example – in no way reflects my view as to its importance. However, I hope the range of topics discussed will show how exciting intercollaborative research can be and how significant a role mathematics can play. The main purpose of the book is to present some of the basic and, to a large extent, generally accepted theoretical frameworks for a variety of biological models. The material presented does not purport to be the latest developments in the various fields, many of which are constantly developing. The already lengthy list of references is by no means exhaustive and I apologise for the exclusion of many that should be included in a definitive list.

With the specimen models discussed and the philosophy which pervades the book the reader should be in a position to tackle the modelling of genuinely practical problems with realism. From a *mathematical* point of view, the art of good modelling relies on: (i) a sound understanding and appreciation of the biological problem; (ii) a realistic mathematical representation of the important biological phenomena; (iii) finding useful solutions, preferably quantitative; and what is crucially important (iv) a biological interpretation of the mathematical results in terms of insights and predictions. The mathematics is dictated by the biology and not vice-versa. Sometimes the mathematics can be very simple. Useful mathematical biology research is not judged by mathematical standards but by different and no less demanding ones.

The book is suitable for physical science courses at various levels. The level of mathematics needed in collaborative biomedical research varies from the very simple to the sophisticated. Selected chapters have been used for applied mathematics courses in the University of Oxford at the final year undergraduate and first year graduate levels. In the U.S.A. the material has also been used for courses for students from the second year undergraduate level through graduate level. It is also accessible to the more theoretically orientated bioscientists who have some knowledge of calculus and differential equations.

I would like to express my gratitude to the many colleagues around the world who have, over the past few years, commented on various chapters of the manuscript, made valuable suggestions and kindly provided me with photographs. I would particularly like to thank Drs. Philip Maini, David Lane and Diana Woodward and my present graduate students who read various drafts with such care, specifically Daniel Bentil, Meghan Burke, David Crawford, Michael Jenkins, Mark Lewis, Gwen Littlewort, Mary Myerscough, Katherine Rogers and Louisa Shaw.

Oxford, January 1989   *J. D. Murray*

*If the Lord Almighty had consulted me
before embarking on creation I should have
recommended something simpler*

Alphonso X (Alphonso the Wise), 1221–1284
King of Castile and Leon (attributed)

Table of Contents

1. **Continuous Population Models for Single Species**   1

   1.1   Continuous Growth Models   1
   1.2   Insect Outbreak Model: Spruce Budworm   4
   1.3   Delay Models   8
   1.4   Linear Analysis of Delay Population Models: Periodic Solutions   12
   1.5   Delay Models in Physiology: Dynamic Diseases   15
   1.6   Harvesting a Single Natural Population   24
 *1.7   Population Model with Age Distribution   29
   Exercises   33

2. **Discrete Population Models for a Single Species**   36

   2.1   Introduction: Simple Models   36
   2.2   Cobwebbing: A Graphical Procedure of Solution   38
   2.3   Discrete Logistic Model: Chaos   41
   2.4   Stability, Periodic Solutions and Bifurcations   47
   2.5   Discrete Delay Models   51
   2.6   Fishery Management Model   54
   2.7   Ecological Implications and Caveats   57
   Exercises   59

3. **Continuous Models for Interacting Populations**   63

   3.1   Predator-Prey Models: Lotka-Volterra Systems   63
   3.2   Complexity and Stability   68
   3.3   Realistic Predator-Prey Models   70
   3.4   Analysis of a Predator-Prey Model with Limit Cycle Periodic
          Behaviour: Parameter Domains of Stability   72
   3.5   Competition Models: Principle of Competitive Exclusion . . .   78
   3.6   Mutualism or Symbiosis   83
   3.7   General Models and Some General and Cautionary Remarks .   85
   3.8   Threshold Phenomena   89
   Exercises   92

---

* Denotes sections in which the mathematics is at a higher level. These sections can
be omitted without loss of continuity.

4. Discrete Growth Models for Interacting Populations 95

   4.1 Predator-Prey Models: Detailed Analysis 96
   *4.2 Synchronized Insect Emergence: 13 Year Locusts 100
   4.3 Biological Pest Control: General Remarks 106
   Exercises   107

5. Reaction Kinetics 109

   5.1 Enzyme Kinetics: Basic Enzyme Reaction 109
   5.2 Michaelis-Menten Theory: Detailed Analysis and the
        Pseudo-Steady State Hypothesis   111
   5.3 Cooperative Phenomena 118
   5.4 Autocatalysis, Activation and Inhibition   122
   5.5 Multiple Steady States, Mushrooms and Isolas 130
   Exercises   137

6. Biological Oscillators and Switches 140

   6.1 Motivation, History and Background 140
   6.2 Feedback Control Mechanisms   143
   6.3 Oscillations and Switches Involving Two or More Species:
        General Qualitative Results   148
   6.4 Simple Two-Species Oscillators: Parameter Domain
        Determination for Oscillations   156
   6.5 Hodgkin-Huxley Theory of Nerve Membranes:
        FitzHugh-Nagumo Model   161
   6.6 Modelling the Control of Testosterone Secretion   166
   Exercises   175

7. Belousov-Zhabotinskii Reaction 179

   7.1 Belousov Reaction and the Field-Noyes (FN) Model 179
   7.2 Linear Stability Analysis of the FN Model and Existence
        of Limit Cycle Solutions 183
   7.3 Non-local Stability of the FN Model   187
   7.4 Relaxation Oscillators: Approximation for the
        Belousov-Zhabotinskii Reaction 190
   7.5 Analysis of a Relaxation Model for Limit Cycle Oscillations
        in the Belousov-Zhabotinskii Reaction   192
   Exercises   199

8. Perturbed and Coupled Oscillators and Black Holes 200

   8.1 Phase Resetting in Oscillators   200
   8.2 Phase Resetting Curves 204
   8.3 Black Holes 208
   8.4 Black Holes in Real Biological Oscillators 210
   8.5 Coupled Oscillators: Motivation and Model System 215

*8.6   Singular Perturbation Analysis: Preliminary Transformation . 217
*8.7   Singular Perturbation Analysis: Transformed System 220
*8.8   Singular perturbation Analysis: Two-Time Expansion 223
*8.9   Analysis of the Phase Shift Equation and Application
       to Coupled Belousov-Zhabotinskii Reactions 227
Exercises   231

9.   Reaction Diffusion, Chemotaxis and Non-local Mechanisms . . . 232
     9.1   Simple Random Walk Derivation of the Diffusion Equation . . 232
     9.2   Reaction Diffusion Equations 236
     9.3   Models for Insect Dispersal 238
     9.4   Chemotaxis 241
     *9.5 Non-local Effects and Long Range Diffusion 244
     *9.6 Cell Potential and Energy Approach to Diffusion 249
     Exercises   252

10. Oscillator Generated Wave Phenomena and Central Pattern
     Generators 254
     10.1 Kinematic Waves in the Belousov-Zhabotinskii Reaction . . . 254
     10.2 Central Pattern Generator: Experimental Facts in the
           Swimming of Fish   258
     *10.3 Mathematical Model for the Central Pattern Generator . . . 261
     *10.4 Analysis of the Phase-Coupled Model System 268
     Exercises   273

11. Biological Waves: Single Species Models   274
     11.1 Background and the Travelling Wave Form 274
     11.2 Fisher Equation and Propagating Wave Solutions 277
     11.3 Asymptotic Solution and Stability of Wavefront Solutions
           of the Fisher Equation   281
     11.4 Density-Dependent Diffusion Reaction Diffusion Models
           and Some Exact Solutions 286
     11.5 Waves in Models with Multi-Steady State Kinetics:
           The Spread and Control of an Insect Population 297
     11.6 Calcium Waves on Amphibian Eggs: Activation Waves
           on Medaka Eggs 305
     Exercises   309

12. Biological Waves: Multi-species Reaction Diffusion Models 311
     12.1 Intuitive Expectations 311
     12.2 Waves of Pursuit and Evasion in Predator-Prey Systems . . . 315
     12.3 Travelling Fronts in the Belousov-Zhabotinskii Reaction . . . 322
     12.4 Waves in Excitable Media 328

12.5 Travelling Wave Trains in Reaction Diffusion Systems
with Oscillatory Kinetics 336
*12.6 Linear Stability of Wave Train Solutions of λ-ω Systems . . . 340
12.7 Spiral Waves 343
*12.8 Spiral Wave Solutions of λ-ω Reaction Diffusion Systems . . . 350
Exercises 356

***13. Travelling Waves in Reaction Diffusion Systems with
Weak Diffusion: Analytical Techniques and Results** 360

*13.1 Reaction Diffusion System with Limit Cycle Kinetics and
Weak Diffusion: Model and Transformed System 360
*13.2 Singular Perturbation Analysis: The Phase Satisfies
Burgers' Equation 363
*13.3 Travelling Wavetrain Solutions for Reaction Diffusion Systems
with Limit Cycle Kinetics and Weak Diffusion: Comparison
with Experiment 367

**14. Spatial Pattern Formation with Reaction/Population Interaction
Diffusion Mechanisms** 372

14.1 Role of Pattern in Developmental Biology 372
14.2 Reaction Diffusion (Turing) Mechanisms 375
14.3 Linear Stability Analysis and Evolution of Spatial Pattern:
General Conditions for Diffusion-Driven Instability 380
14.4 Detailed Analysis of Pattern Initiation in a Reaction Diffusion
Mechanism 387
14.5 Dispersion Relation, Turing Space, Scale and Geometry Effects
in Pattern Formation in Morphogenetic Models 397
14.6 Mode Selection and the Dispersion Relation 408
14.7 Pattern Generation with Single Species Models:
Spatial Heterogeneity with the Spruce Budworm Model . . . 414
14.8 Spatial Patterns in Scalar Population Interaction-Reaction
Diffusion Equations with Convection: Ecological Control
Strategies 419
*14.9 Nonexistence of Spatial Patterns in Reaction Diffusion
Systems: General and Particular Results 424
Exercises 430

**15. Animal Coat Patterns and Other Practical Applications
of Reaction Diffusion Mechanisms** 435

15.1 Mammalian Coat Patterns – 'How the Leopard Got Its Spots' . 436
15.2 A Pattern Formation Mechanism for Butterfly Wing Patterns . 448
15.3 Modelling Hair Patterns in a Whorl in *Acetabularia* 468

16. **Neural Models of Pattern Formation** 481

    16.1 Spatial Patterning in Neural Firing with a Simple
          Activation-Inhibition Model   481
    16.2 A Mechanism for Stripe Formation in the Visual Cortex . . . 489
    16.3 A Model for the Brain Mechanism Underlying Visual
          Hallucination Patterns   494
    16.4 Neural Activity Model for Shell Patterns   505
    Exercises   523

17. **Mechanical Models for Generating Pattern and Form
    in Development**   525

    17.1 Introduction and Background Biology 525
    17.2 Mechanical Model for Mesenchymal Morphogenesis 528
    17.3 Linear Analysis, Dispersion Relation and Pattern Formation
          Potential 538
    17.4 Simple Mechanical Models Which Generate Spatial Patterns
          with Complex Dispersion Relations   542
    17.5 Periodic Patterns of Feather Germs   554
    17.6 Cartilage Condensations in Limb Morphogenesis 558
    17.7 Mechanochemical Model for the Epidermis   566
    17.8 Travelling Wave Solutions of the Cytogel Model   572
    17.9 Formation of Microvilli 579
    17.10 Other Applications of Mechanochemical Models   586
    Exercises   590

18. **Evolution and Developmental Programmes** 593

    18.1 Evolution and Morphogenesis 593
    18.2 Evolution and Morphogenetic Rules in Cartilage Formation
          in the Vertebrate Limb   599
    18.3 Developmental Constraints, Morphogenetic Rules and
          the Consequences for Evolution 606

19. **Epidemic Models and the Dynamics of Infectious Diseases** 610

    19.1 Simple Epidemic Models and Practical Applications 611
    19.2 Modelling Venereal Diseases   619
    19.3 Multi-group Model for Gonorrhea and Its Control   623
    19.4 AIDS: Modelling the Transmission Dynamics of the Human
          Immunodeficiency Virus (HIV) 624
    19.5 Modelling the Population Dynamics of Acquired Immunity
          to Parasite Infection   630
  *19.6 Age Dependent Epidemic Model and Threshold Criterion   . . 640
    19.7 Simple Drug Use Epidemic Model and Threshold Analysis . . 645
    Exercises   649

20. Geographic Spread of Epidemics 651

20.1 Simple Model for the Spatial Spread of an Epidemic 651

20.2 Spread of the Black Death in Europe 1347-1350 655

20.3 The Spatial Spread of Rabies Among Foxes I: Background and Simple Model 659

20.4 The Spatial Spread of Rabies Among Foxes II: Three Species (SIR) Model 666

20.5 Control Strategy Based on Wave Propagation into a Non-epidemic Region: Estimate of Width of a Rabies Barrier . 681

20.6 Two-Dimensional Epizootic Fronts and Effects of Variable Fox Densities: Quantitative Predictions for a Rabies Outbreak in England 689

Exercises 696

Appendices

1. Phase Plane Analysis 697

2. Routh-Hurwitz Conditions, Jury Conditions, Descarte's Rule of Signs and Exact Solutions of a Cubic 702

3. Hopf Bifurcation Theorem and Limit Cycles 706

4. General Results for the Laplacian Operator in Bounded Domains 720

Bibliography 723

Index 745

1. Continuous Population Models
for Single Species

The increasing study of realistic mathematical models in ecology (basically the study of the relation between species and their environment) is a reflection of their use in helping to understand the dynamic processes involved in such areas as predator-prey and competition interactions, renewable resource management, evolution of pesticide resistant strains, ecological control of pests, multi-species societies, plant-herbivore systems and so on. The continually expanding list of applications is extensive. There are also interesting and useful applications of single species models in the biomedical sciences: in Section 1.5 we discuss two practical examples of these which arise in physiology. Here, and in the following three chapters, we shall consider some deterministic models. The book edited by May (1981) gives an overview of theoretical ecology from a variety of different aspects; experts in diverse fields review their areas. The book by Nisbet and Gurney (1982) is a comprehensive account of mathematical modelling in population dynamics: a good elementary introduction is given in the textbook by Edelstein-Keshet (1988).

1.1 Continuous Growth Models

Single species models are of relevance to laboratory studies in particular but, in the real world, can reflect a telescoping of effects which influence the population dynamics. Let $N(t)$ be the population of the species at time t, then the rate of change

$$\frac{dN}{dt} = \text{births} - \text{deaths} + \text{migration} , \tag{1.1}$$

is a *conservation equation* for the population. The form of the various terms on the right hand side of (1.1) necessitates modelling the situation that we are concerned with. The simplest model has no migration and the birth and death terms are proportional to N. That is

$$\frac{dN}{dt} = bN - dN \quad \Rightarrow \quad N(t) = N_0 \, e^{(b-d)t}$$

where b, d are positive constants and the initial population $N(0) = N_0$. Thus if $b > d$ the population grows exponentially while if $b < d$ it dies out. This approach, due to Malthus in 1798 but actually suggested earlier by Euler, is pretty unrealistic. However if we consider the past and predicted growth estimates for the total world population from the 17th to 21st centuries it is perhaps less unrealistic as seen in the following table.

Date	Mid 17th Century	Early 19th Century	1918-27	1960	1974	1987	1999	2010	2022
Population in billions	0.5	1	2	3	4	5	6	7	8

In the long run of course there must be some adjustment to such exponential growth. Verhulst in 1836 proposed that a self-limiting process should operate when a population becomes too large. He suggested

$$\frac{dN}{dt} = rN(1 - N/K) \,, \tag{1.2}$$

where r and K are positive constants. This is called *logistic growth* in a population. In this model the per capita birth rate is $r(1 - N/K)$, that is, it is dependent on N. The constant K is the *carrying capacity* of the environment, which is usually determined by the available sustaining resources.

There are two *steady states* or *equilibrium states* for (1.2), namely $N = 0$ and $N = K$, that is where $dN/dt = 0$. $N = 0$ is unstable since linearization about it (that is N^2 is neglected compared with N) gives $dN/dt \approx rN$, and so N grows exponentially from any initial value. The other equilibrium $N = K$ is stable: linearization about it (that is $(N - K)^2$ is neglected compared with $|N - K|$) gives $d(N - K)/dt \approx -r(N - K)$ and so $N \to K$ as $t \to \infty$. The carrying capacity K determines the size of the stable steady state population while r is a measure of the rate at which it is reached, that is, it is a measure of the dynamics: we could incorporate it in the time by a transformation from t to rt. Thus $1/r$ is a representative *time scale* of the response of the model to any change in the population.

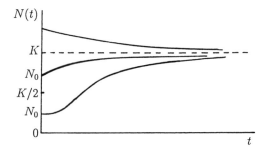

Fig. 1.1. Logistic population growth. Note the qualitative difference for the two cases $N_0 < K/2$ and $K > N_0 > K/2$.

If $N(0) = N_0$ the solution of (1.2) is

$$N(t) = \frac{N_0 K e^{rt}}{[K + N_0(e^{rt} - 1)]} \to K \quad \text{as} \quad t \to \infty , \tag{1.3}$$

and is illustrated in Fig. 1.1. From (1.2), if $N_0 < K$, $N(t)$ simply increases monotonically to K while if $N_0 > K$ it decreases monotonically to K. In the former case there is a qualitative difference depending on whether $N_0 > K/2$ or $N_0 < K/2$: with $N_0 < K/2$ the form has a typical sigmoid character, which is commonly observed.

In the case where $N_0 > K$ this would imply that the per capita birth rate is negative! Of course all it is really saying is that in (1.1) the births plus immigration is less than the deaths plus emigration. The point about (1.2) is that it is more like a metaphor for a class of population models with density dependent regulatory mechanisms – a kind of compensating effect of overcrowding – and must not be taken too literally as the equation governing the population dynamics. It is a particularly convenient form to take when seeking qualitative dynamic behaviour in populations in which $N = 0$ is an unstable steady state and $N(t)$ tends to a finite positive stable steady state. The logistic form will occur in a variety of different contexts throughout the book.

In general if we consider a population to be governed by

$$\frac{dN}{dt} = f(N) , \tag{1.4}$$

where typically $f(N)$ is a *nonlinear* function of N then the equilibrium solutions N^* are solutions of $f(N) = 0$ and are linearly stable to small perturbations if $f'(N^*) < 0$, and unstable if $f'(N^*) > 0$. This is clear from linearizing about N^* by writing

$$n(t) \approx N(t) - N^*, \quad |n(t)| \ll 1$$

and (1.4) becomes

$$\frac{dn}{dt} = f(N^* + n) \approx f(N^*) + n f'(N^*) + \dots ,$$

which to first order in $n(t)$ gives

$$\frac{dn}{dt} \approx n f'(N^*) \quad \Rightarrow \quad n(t) \propto \exp[f'(N^*)t] . \tag{1.5}$$

So n grows or decays according as $f'(N^*) > 0$ or $f'(N^*) < 0$. The time scale of the response of the population to a disturbance is of the order of $1/|f'(N^*)|$: it is the time to change the initial disturbance by a factor e.

There may be several equilibrium, or steady state populations N^* which are solutions of $f(N) = 0$: it depends on the system $f(N)$ models. Graphically plotting $f(N)$ against N immediately gives the equilibria. The gradient $f'(N^*)$

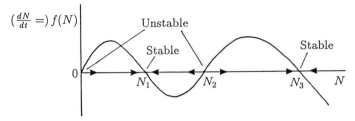

Fig. 1.2. Population dynamics model $dN/dt = f(n)$ with several steady states. The gradient $f'(N)$ at the steady state, that is where $f(N) = 0$, determines the linear stability.

at each steady state then determines its linear stability. Such steady states may, however, be unstable to finite disturbances. Suppose, for example, that $f(N)$ is as illustrated in Fig. 1.2. The gradients $f'(N)$ at $N = 0, N_2$ are positive so these equilibria are unstable while those at $N = N_1, N_3$ are stable to small perturbations: the arrows symbolically indicate stability or instability. If, for example, we now perturb the population from its equilibrium N_1 so that N is in the range $N_2 < N < N_3$ then $N \rightarrow N_3$ rather than returning to N_1. A similar perturbation from N_3 to a value in the range $0 < N < N_2$ would result in $N(t) \rightarrow N_1$. Qualitatively there is a threshold perturbation below which the steady states are always stable, and this threshold depends on the full nonlinear form of $f(N)$. For N_1, for example, the necessary threshold perturbation is $N_2 - N_1$.

1.2 Insect Outbreak Model: Spruce Budworm

A practical model which exhibits two positive linearly stable steady state populations is that for the spruce budworm which can, with ferocious efficiency, defoliate the balsam fir: it is a major problem in Canada. Ludwig et al. (1978) considered the budworm population dynamics to be governed by the equation

$$\frac{dN}{dt} = r_B N \left(1 - \frac{N}{K_B}\right) - p(N) .$$

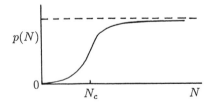

Fig. 1.3. Typical functional form of the predation in the spruce budworm model: note the sigmoid character. The population value N_c is an approximate threshold value. For $N < N_c$ predation is small, while for $N > N_c$ it is "switched on".

Here r_B is the linear birth rate of the budworm and K_B is the carrying capacity which is related to the density of foliage available on the trees. The $p(N)$-term represents predation, generally by birds: the qualitative form of it is important and is illustrated in Fig. 1.3. Predation usually saturates for large enough N. There is an approximate threshold value N_c, below which the predation is small, while above it the predation is close to its saturation value: such a functional form is like a switch with N_c being the critical switch value. For small population densities N, the birds tend to seek food elsewhere and so the predation term $p(N)$ drops more rapidly, as $N \rightarrow 0$, than a linear rate proportional to N. To be specific we take the form for $p(N)$ suggested by Ludwig et al. (1978) namely $BN^2/(A^2 + N^2)$ where A and B are positive constants, and the dynamics of $N(t)$ is then governed by

$$\frac{dN}{dt} = r_B N \left(1 - \frac{N}{K_B} \right) - \frac{BN^2}{A^2 + N^2} \, . \tag{1.6}$$

This equation has four parameters, r_B, K_B, B and A, with A and K_B having the same dimensions as N, r_B has dimension $(\text{time})^{-1}$ and B has the dimensions of $N(\text{time})^{-1}$. A is a measure of the threshold where the predation is 'switched on', that is N_c in Fig. 1.3. If A is small the 'threshold' is small, but the effect is dramatic.

Before analysing the model it is essential, or rather obligatory, to express it in *nondimensional* terms. This has several advantages. For example, the units used in the analysis are then unimportant and the adjectives small and large have a definite relative meaning. It also always reduces the number of relevant parameters to dimensionless groupings which determine the dynamics. A pedagogical article with several practical examples by Segel (1972) discusses the necessity and advantages for nondimensionalisation and scaling in general. Here we introduce nondimensional quantities by

$$u = \frac{N}{A}, \quad r = \frac{Ar_B}{B}, \quad q = \frac{K_B}{A}, \quad \tau = \frac{Bt}{A} \tag{1.7}$$

which on substituting into (1.6) becomes

$$\frac{du}{d\tau} = ru \left(1 - \frac{u}{q} \right) - \frac{u^2}{1 + u^2} = f(u; r, q) \, , \tag{1.8}$$

where f is defined by this equation. Note that it has only two parameters r and q, which are pure numbers, as also is u of course. Now, for example, if $u \ll 1$ it means simply that $N \ll A$. In real terms it means that predation is negligible in this population range. In any model there are usually several different nondimensionalisations possible. The dimensionless groupings to choose depends on the aspects you want to investigate. The reasons for the particular form (1.7) will become clear below.

The steady states are solutions of

$$f(u; r, q) = 0 \quad \Rightarrow \quad ru\left(1 - \frac{u}{q}\right) = \frac{u^2}{1 + u^2} \ . \tag{1.9}$$

Clearly $u = 0$ is one solution with other solutions, if they exist, satisfying

$$r\left(1 - \frac{u}{q}\right) = \frac{u}{1 + u^2} \ . \tag{1.10}$$

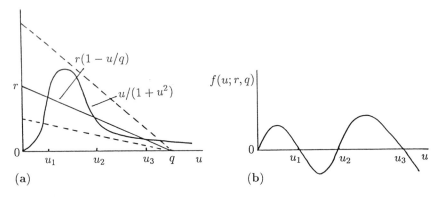

(a)   (b)

Fig. 1.4a,b. Equilibrium states for the spruce budworm population model (1.8). The positive equilibria are given by the intersections of the straight line $r(1 - u/q)$ and $u/(1 + u^2)$. With the solid straight line in (a) there are 3 steady states with $f(u; r, q)$ typically as in (b).

Although we know the analytical solutions of a cubic (Appendix 2), they are often clumsy to use because of their algebraic complexity; this is one of these cases. It is convenient here to determine the existence of solutions of (1.10) graphically as shown in Fig. 1.4 (a). We have plotted the straight line, the left of (1.10), and the function on the right of (1.10); the intersections give the solutions. The actual expressions are not important here. What is important, however, is the existence of one, three, or again, one solution as r increases for a fixed q, as in Fig. 1.4 (a), or as also happens for a fixed r and a varying q. When r is in the appropriate range, which depends on q, there are three equilibria with a typical corresponding $f(u; r, q)$ as shown in Fig. 1.4 (b). The nondimensional groupings which leave the two parameters appearing only in the straight line part of Fig. 1.4 is particularly helpful and was the motivation for the form introduced in (1.7). By inspection $u = 0$, $u = u_2$ are linearly unstable, since $\partial f / \partial u > 0$ at $u = 0, u_2$, while u_1 and u_3 are stable steady states, since at these $\partial f / \partial u < 0$. There is a domain in the r, q parameter space where three roots of (1.10) exist. This is shown in Fig. 1.5: the analytical derivation of the boundary curves is left as an exercise (Exercise 1).

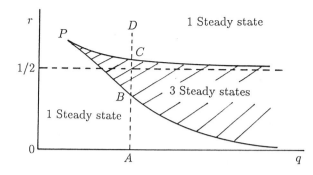

Fig. 1.5. Parameter domain for the number of positive steady states for the budworm model (1.8). The boundary curves are given parametrically (see Exercise 1) by $r(a) = 2a^3/(a^2+1)^2$, $q(a) = 2a^3/(a^2-1)$ for $a \geq \sqrt{3}$, the value giving the cusp point P.

This model exhibits a *hysteresis effect*. Suppose we have a fixed q, say, and r increases from zero along the path $ABCD$ in Fig. 1.5. Then, referring also to Fig. 1.4 (a), we see that if $u_1 = 0$ at $r = 0$ the u_1-equilibrium simply increases monotonically with r until C in Fig. 1.5 is reached. For a larger r this steady state disappears and the equilibrium value jumps to u_3. If we now reduce r again the equilibrium state is the u_3 one and it remains so until r reaches the lower critical value, where there is again only one steady state, at which point there is a jump from the u_3 to the u_1 state. In other words as r increases along $ABCD$ there is a discontinuous jump up at C while as r decreases from D to A there is a discontinuous jump down at B. This is an example of a *cusp catastrophe* which is illustrated schematically in Fig. 1.6 where the letters A, B, C and D correspond to those in Fig. 1.5. Note that Fig. 1.5 is the projection of the surface onto the r, q plane with the shaded region corresponding to the fold.

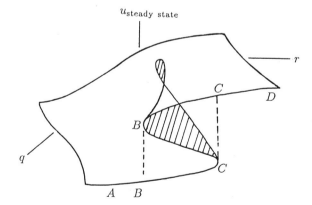

Fig. 1.6. Cusp catastrophe for the equilibria states in the $(u_{\text{steady state}}, r, q)$ parameter space. As r increases from A the path is $ABCCD$ while as r decreases from D the path is $DCBBA$. The projection of this surface onto the r, q plane is given in Fig. 1.5. Three equilibria exist where the fold is.

The parameters from field observation are such that there are three possible steady states for the population. The smaller steady state u_1 is the *refuge* equilibrium while u_3 is the *outbreak* equilibrium. From a pest control point of view, what should be done to try to keep the population at a refuge state rather than allow it to reach an outbreak situation? Here we must relate the real parameters to the dimensionless ones, using (1.7). For example, if the foliage was sprayed to discourage the budworm this would reduce q since K_B is reduced. If the reduction is large enough this could force the dynamics to have only one equilibrium: that is the effective r and q do not lie in the shaded domain of Fig. 1.5. Alternatively we could try and reduce the reproduction rate r_B or increase the threshold number of predators, since both reduce r which would be effective if it is below the critical value for u_3 to exist. Although these give preliminary qualitative ideas for control, it is not easy to determine the optimal strategy, particularly since spatial effects such as budworm dispersal must be taken into effect: we shall discuss this aspect in detail in Chapter 14.

It is appropriate here to mention briefly the time scale with which this model is concerned. An outbreak of budworm during which balsam fir trees are denuded of foliage is about four years. The trees then die and birch trees takeover. Eventually, in the competition for nutrient, the fir trees will drive out the birch trees again. The time scale for fir reforestation is of the order of 50–100 years. A full model would incorporate the tree dynamics as well: see Ludwig et al. (1978). So, the model we have analysed here is only for the short time scale, namely that related to a budworm outbreak.

1.3 Delay Models

One of the deficiencies of single population models like (1.4) is that the birth rate is considered to act instantaneously whereas there may be a time delay to take account of the time to reach maturity, the finite gestation period and so on. We can incorporate such delays by considering delay differential equation models of the form

$$\frac{dN(t)}{dt} = f(N(t), N(t - T)) , \qquad (1.11)$$

where $T > 0$, the delay, is a parameter. One such model, which has been used, is an extension of the logistic growth model (1.2), namely the differential delay equation

$$\frac{dN}{dt} = rN(t) \left[1 - \frac{N(t - T)}{K} \right] , \qquad (1.12)$$

where r, K and T are positive constants. This says that the regulatory effect depends on the population at an earlier time, $t - T$, rather than that at t. This equation is itself a model for a delay effect which should really be an average over past populations and which results in an integro-differential equation. Thus a more accurate model than (1.12) is, for example, the convolution type model

$$\frac{dN}{dt} = rN(t) \left[1 - K^{-1} \int_{-\infty}^{t} w(t - s)N(s) \, ds \right] , \qquad (1.13)$$

where $w(t)$ is a weighting factor which says how much emphasis should be given to the size of the population at earlier times to   determine the present effect

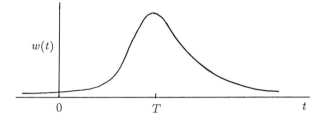

on resource availability. Practically $w(t)$ will tend to zero for large negative and positive t and will probably have a maximum at some representative time T. Typically $w(t)$ is as illustrated in Fig. 1.7. If $w(t)$ is sharper in the sense that the region around T is narrower or larger then in the limit we can think of $w(t)$ as approximating the Dirac function $\delta(T - t)$, where

$$\int_{-\infty}^{\infty} \delta(T - t)f(t)\, dt = f(T) \,.$$

Equation (1.13) in this case then reduces to (1.12)

$$\int_{-\infty}^{t} \delta(t - T - s)N(s)\, ds = N(t - T) \,.$$

The character of the solutions of (1.12), and the type of boundary conditions required are quite different to those of (1.2). Even with the seemingly innocuous equation (1.12) the solutions in general have to be found numerically. Note that to compute the solution for $t > 0$ we require $N(t)$ for *all* $-T \le t \le 0$. We can however get some qualitative impression of the kind of solutions of (1.12) which are possible, by the following heuristic reasoning.

Refer now to Fig. 1.8 and suppose that for some $t = t_1$, $N(t_1) = K$ and that for some time $t < t_1$, $N(t - T) < K$. Then from the governing equation (1.12), since $1 - N(t - T)/K > 0$, $dN(t)/dt > 0$ and so $N(t)$ at t_1 is still increasing. When $t = t_1 + T$, $N(t - T) = N(t_1) = K$ and so $dN/dt = 0$. For $t_1 + T < t < t_2$, $N(t - T) > K$ and so $dN/dt < 0$ and $N(t)$ decreases until $t = t_2 + T$ since then $dN/dt = 0$ again because $N(t_2 + T - T) = N(t_2) = K$. There is therefore the possibility of oscillatory behaviour. For example, with the simple linear delay equation

$$\frac{dN}{dt} = -\frac{\pi}{2T}N(t - T) \quad \Rightarrow \quad N(t) = A \cos\frac{\pi t}{2T} \,,$$

which is periodic in time and which can be easily checked is a solution.

In fact the solutions of (1.12) can exhibit *stable limit cycle* periodic solutions for a large range of values of the product rT of the birth rate r and the delay T. If t_p is the period then $N(t + t_p) = N(t)$ for all t. The point about *stable* limit cycle solutions is that if a perturbation is imposed the solution returns to the original periodic solution as $t \to \infty$, although possibly with a *phase* shift. The periodic behaviour is also independent of any initial data.

From Fig. 1.8 and the heuristic argument above the period of the limit cycle periodic solutions might be expected to be of the order of $4T$. From numerical calculations this is the case for a large range of rT, which is a dimensionless grouping. The reason we take this grouping is because (1.12) in dimensionless form becomes

$$\frac{dN^*}{dt^*} = N^*(t^*)[1 - N^*(t^* - T^*)], \quad \text{where} \quad N^* = \frac{N}{K}, \ t^* = rt, \ T^* = rT \ .$$

What does vary with rT, however, is the amplitude of the oscillation. For example, for $rT = 1.6$, the period $t_p \approx 4.03T$ and $N_{\max}/N_{\min} \approx 2.56$; $rT = 2.1$, $t_p \approx 4.54T$, $N_{\max}/N_{\min} \approx 42.3$; $rT = 2.5$, $t_p \approx 5.36T$, $N_{\max}/N_{\min} \approx 2930$. For large values of rT, however, the period changes considerably.

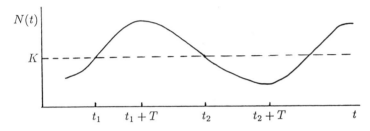

Fig. 1.8. Schematic periodic solution of the delay equation population model (1.12).

This simple delay model has been used for several different practical situations. For example, it has been applied by May (1975) to Nicholson's (1957) careful experimental data for the Australian sheep-blowfly (*Lucila cuprina*), a pest of considerable importance in Australian sheep farming. Over a period of nearly two years Nicholson observed the population of flies which were maintained under carefully regulated temperature and food control. He observed a regular basic periodic oscillation of about 35–40 days. Applying (1.12) to the experimental arrangement, K is set by the food level available. T, the delay, is approximately the time for a larva to mature into an adult. Then the only unknown parameter is r, the intrinsic rate of population increase. Fig. 1.9 illustrates the comparison with the data for $rT = 2.1$ for which the period is about $4.54T$. If we take the observed period as 40 days this gives a delay of about 9 days: the actual delay is closer to 11 days. The model implies that if K is doubled nothing changes from a time periodic point of view since it can be scaled out by writing N/K for N: this lack of change with K is what was observed.

It is encouraging that such a simple model as (1.12) should give such reasonable results. This is some justification for using delay models to study the dynamics of single population populations which exhibit periodic behaviour. It is important however, not to be too easily convinced as to the validity or reasonableness of a model simply because some solutions agree even quantitatively

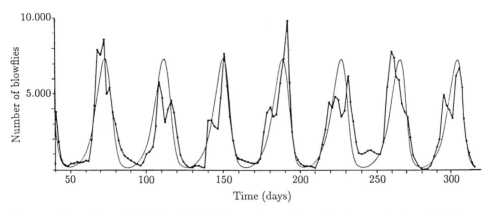

Fig. 1.9. Comparison of Nicholson's (1957) experimental data for the population of the Australian sheep-blowfly and the model solution from equation (1.12) with $rT = 2.1$. (From May 1975)

well with the data: this is a phenomenon, or rather a pitfall, we shall encounter repeatedly later in the book particularly when we discuss models for generating biological pattern and form in Chapters 14–17. From the experimental data reproduced in Fig. 1.9 we see a persistent 'second burst' feature that the solutions of (1.12) do not mimic. Also the difference between the calculated delay of 9 days and the actual 11 days is really too large. Gurney et al. (1980) solved this problem with a more elaborate delay model which agrees even better with the data, including the two bursts of reproductive activity observed: see also the book by Nisbet and Gurney (1982) where this problem is discussed fully as a case study.

Another example of the application of this model (1.12) to extant data is given by May (1981) who considers the lemming population in the Churchill area of Canada. There is approximately a 4 year period where, in this case the gestation time is $T = 0.72$ year. The vole population in the Scottish Highlands, investigated by Stirzaker (1975) using a delay equation model, also undergoes a cycle of just under 4 years, which is again approximately $4T$ where here the gestation time is $T = 0.75$ year. In this model the effect of predation is incorporated into the single equation for the vole population. The articles by Myers and Krebs (1974) and Krebs and Myers (1974) discuss population cycles in rodents in general: they usually have 3 to 4 year cycles.

Not all periodic population behaviour can be treated quite so easily. One such example which is particularly dramatic is the 13 and 17 year cycle exhibited by a species of locusts; that is their emergencies are synchronized to 13 or 17 years. This phenomenon is discussed in detail in Chapter 4, Section 4.3.

It should perhaps be mentioned here that single (non-delay) differential equation models for population growth without delay, that is like $dN/dt = f(N)$, *cannot* exhibit limit cycle behaviour. We can see this immediately as follows. Suppose this equation has a periodic solution with period T, that is $N(t+T) =$

$N(t)$. Multiply the equation by dN/dt and integrate from t to $t + T$ to get

$$\int_t^{t+T} \left(\frac{dN}{dt}\right)^2 dt = \int_t^{t+T} f(N)\frac{dN}{dt}\, dt$$

$$= \int_{N(t)}^{N(t+T)} f(N)\, dN$$

$$= 0\, ,$$

since $N(t + T) = N(t)$. But the left hand integral is positive since $(dN/dt)^2$ cannot be identically zero, so we have a contradiction. So the single scalar equation $dN/dt = f(N)$ cannot have periodic solutions.

1.4 Linear Analysis of Delay Population Models: Periodic Solutions

We saw in the last section how the delay differential equation model (1.12) was capable of generating limit cycle periodic solutions. One indication of their existence is if the steady state is unstable by growing oscillations, although this is certainly not conclusive. We consider here the linearization of (1.12) about the equilibrium states $N = 0$ and $N = K$. Small perturbations from $N = 0$ satisfy $dN/dt \approx rN$, which shows that $N = 0$ is unstable with exponential growth. We thus need only consider perturbations about the steady state $N = K$.

It is again expedient to nondimensionalize the model equation (1.12) by writing

$$N^*(t) = \frac{N(t)}{K}, \quad t^* = rt, \quad T^* = rT\, , \tag{1.14}$$

where the asterisk denotes dimensionless quantities. Then (1.12) becomes, on dropping the asterisks for algebraic convenience, but keeping in mind that we are now dealing with non-dimensional quantities,

$$\frac{dN(t)}{dt} = N(t)[1 - N(t - T)]. \tag{1.15}$$

Linearizing about the steady state, $N = 1$, by writing

$$N(t) = 1 + n(t) \quad \Rightarrow \quad \frac{dn(t)}{dt} \approx -n(t - T)\, . \tag{1.16}$$

We look for solutions for $n(t)$ in the form

$$n(t) = ce^{\lambda t} \quad \Rightarrow \quad \lambda = -e^{-\lambda T}\, , \tag{1.17}$$

from (1.16), where c is a constant and the eigenvalues λ are solutions of the second of (1.17), a transcendental equation in which $T > 0$.

It is not easy to find the analytical solutions of (1.17). However, all we really want to know from a stability point of view is whether there are any solutions with $\text{Re}\,\lambda > 0$ which from the first of (1.17) implies instability since $n(t)$ grows exponentially with time.

Set $\lambda = \mu + i\omega$. There is a real number μ_0 such that all solutions λ of the second of (1.17) satisfy $\text{Re}\,\lambda < \mu_0$. To see this take the modulus to get $|\lambda| = e^{-\mu T}$ and so, if $|\lambda| \to \infty$ then $e^{-\mu T} \to \infty$ which requires $\mu \to -\infty$. Thus there must be a number μ_0 that bounds $\text{Re}\,\lambda$ from above. If we introduce $z = 1/\lambda$ and $w(z) = 1 + ze^{-T/z}$ then $w(z)$ has an essential singularity at $z = 0$. So by Picard's theorem, in the neighbourhood of $z = 0$, $w(z) = 0$ has infinitely many complex roots. Thus there are infinitely many roots λ.

We now take the real and imaginary parts of the transcendental equation in (1.17), namely

$$\mu = -e^{-\mu T} \cos \omega T, \quad \omega = e^{-\mu T} \sin \omega T \, , \tag{1.18}$$

and determine the range of T such that $\mu < 0$. That is, we want to find the conditions such that the upper limit μ_0 on μ is negative. Let us first dispense with the simple case where λ is real, that is $\omega = 0$. From (1.18), $\omega = 0$ satisfies the second equation and the first becomes $\mu = -e^{-\mu T}$. This has no positive roots $\mu > 0$ since $e^{-\mu T} > 0$ for all μT or as can be seen on sketching each side of the equation as a function of μ and noting that they can only intersect with $T > 0$ if $\mu < 0$.

Consider now $\omega \neq 0$. From (1.18) if ω is a solution then so is $-\omega$, so we can consider $\omega > 0$ without any loss of generality. From the first of (1.18), $\mu < 0$ requires $\omega T < \pi/2$ since $-e^{-\mu T} < 0$ for all μT. This, together with the second of (1.18) on multiplying by T, gives

$$Te^{-\mu T} \sin \omega T = \omega T < \frac{\pi}{2}$$
$$\Rightarrow \quad Te^{-\mu T} \sin \omega T < \frac{\pi}{2} \quad \Rightarrow \quad 0 < T < \frac{\pi}{2} \tag{1.19}$$

as the condition on T to satisfy the inequality, since $\sin \omega T < 1$ and $e^{-\mu T} \geq 1$ if $\mu \leq 0$. [This method for determining the critical T is, in effect, equivalent to determining the condition on T such that at the bifurcation, where $\omega \neq 0$, $\mu = 0$, λ in (1.17) is such that $\partial \lambda / \partial \mu > 0$.]

Returning now to dimensional quantities we thus have that the steady state $N(t) = K$ is stable if $0 < rT < \pi/2$ and unstable for $rT > \pi/2$. In the latter case we expect the solution to exhibit stable limit cycle behaviour. The critical value $rT = \pi/2$ is the *bifurcation value*, that is the value of the parameter, rT here, where the character of the solutions of (1.12) changes abruptly, or bifurcates, from a stable steady state to a time varying solution. The effect of delay in models is usually to increase the potential for instability. Here as T is increased beyond the bifurcation value $T_c = \pi/2r$, the steady state becomes unstable.

Near the bifurcation value we can get a first estimate of the period of the bifurcating oscillatory solution as follows. Consider the dimensionless form (1.15) and let

$$T = T_c + \varepsilon = \frac{\pi}{2} + \varepsilon, \quad 0 < \varepsilon \ll 1. \tag{1.20}$$

The solution $\lambda = \mu + i\omega$, of (1.18), with the largest $\text{Re}\,\lambda$ when $T = \pi/2$ is $\mu = 0$, $\omega = 1$. For ε small we expect μ and ω to differ from $\mu = 0$ and $\omega = 1$ also by small quantities so let

$$\mu = \delta, \quad \omega = 1 + \sigma, \quad 0 < \delta \ll 1, \quad |\sigma| \ll 1, \tag{1.21}$$

where δ and σ are to be determined. Substituting these into the second of (1.18) and expanding for small δ, σ and ε gives

$$1 + \sigma = \exp\left[-\delta\left(\frac{\pi}{2} + \varepsilon\right)\right] \sin\left[(1+\sigma)\left(\frac{\pi}{2} + \varepsilon\right)\right] \quad \Rightarrow \quad \sigma \approx -\frac{\pi\delta}{2}$$

to first order, while the first of (1.18) gives

$$\delta = -\exp\left[-\delta\left(\frac{\pi}{2} + \varepsilon\right)\right] \cos\left[(1+\sigma)\left(\frac{\pi}{2} + \varepsilon\right)\right] \quad \Rightarrow \quad \delta \approx \varepsilon + \frac{\pi\sigma}{2}$$

Thus on solving these simultaneously

$$\delta \approx \frac{\varepsilon}{1 + \frac{\pi^2}{4}}, \quad \sigma \approx -\frac{\varepsilon\pi}{2\left(1 + \frac{\pi^2}{4}\right)}, \tag{1.22}$$

and hence, near the bifurcation, the first of (1.17) with (1.16) gives

$$N(t) = 1 + \text{Re}\{c\,\exp[\delta t + i(1 + \sigma)t]\}$$

$$\approx 1 + \text{Re}\left\{c\,\exp\left[\frac{\varepsilon t}{1 + \frac{\pi^2}{4}}\right] \exp\left[it\left\{1 - \frac{\varepsilon\pi}{2(1 + \frac{\pi^2}{4})}\right\}\right]\right\}. \tag{1.23}$$

This shows that the instability is by growing oscillations with period

$$\frac{2\pi}{1 - \frac{\varepsilon\pi}{2(1+\frac{\pi^2}{4})}} \approx 2\pi$$

to $O(1)$ for small ε. In *dimensional* terms this is $2\pi/r$ and, since to $O(1)$, $rT = \pi/2$, the period of oscillation is then $4T$ as we expected from the intuitive arguments above. From the numerical results for limit cycles quoted above the solution with $rT = 1.6$ had period $4.03T$. With $rT = \pi/2 + \varepsilon = 1.6$, this gives $\varepsilon \approx 0.029$ so the dimensional period to $O(\varepsilon)$ is obtained from (1.23) as

$$\frac{2T}{\pi} \frac{2\pi}{1 - \frac{\varepsilon\pi}{2(1+\frac{\pi^2}{4})}} \approx 4.05T,$$

which compares well with the numerical computed value of $4.03T$. When $rT = 2.1$ this gives $\varepsilon \approx 0.53$ and corresponding period $5.26T$ which is to be compared with the computed period of $4.54T$. This ε is too large for the above first order analysis to hold (ε^2 is not negligible compared with ε): a more accurate result would be obtained if the analysis was carried out to second order.

The natural appearance of a 'slow time' εt in $N(t)$ in (1.23) suggests that a full nonlinear solution near the bifurcation value $rT = \pi/2$ is amenable to a *two-time asymptotic procedure* to obtain the (uniformly valid in time) solution. This can in fact be done; see, for example, Murray's (1984) book on asymptotic methods for a description of such techniques and how to use them.

The subject of delay or functional differential equations is now rather large. An introductory mathematical book on the subject is Driver's (1977). The book by MacDonald (1979) is solely concerned with time lags in biological models. Although the qualitative properties of such delay equation models for population growth dynamics and nonlinear analytical solutions near bifurcation can often be determined, in general numerical methods have to be used to get useful quantitative results.

1.5 Delay Models in Physiology: Dynamic Diseases

There are many acute physiological diseases where the initial symptoms are manifested by an alteration or irregularity in a control system which is normally periodic, or by the onset of an oscillation in a hitherto non-oscillatory process. Such physiological diseases have been termed dynamical diseases by Glass and Mackey (1979) who have made a particular study of several important physiological examples. The symposium proceedings of a meeting specifically devoted to temporal disorders in human oscillatory systems edited by Rensing et al. (1987) is particularly apposite to the material and modelling in this section. Here we discuss two specific examples which have been modelled, analysed and related to experimental observations by Mackey and Glass (1977). The review article on dynamic diseases by Mackey and Milton (1988) is of direct relevance to the material discussed here: it also describes some examples drawn from neurophysiology. Although the second model we consider here is concerned with populations of cells, the first does not relate to any population species but rather to the concentration of a gas. It does however fit naturally here since it is a scalar delay differential equation model the analysis for which is directly applicable to the second problem. It is also interesting in its own right.

Cheyne-Stokes respiration. The first example, Cheyne-Stokes respiration, is a human respiratory ailment manifested by an alteration in the regular breathing pattern. Here the amplitude of the breathing pattern, directly related to the breath volume – the ventilation V – regularly waxes and wanes with each period separated by periods of apnea, that is where the volume per breath is exceedingly

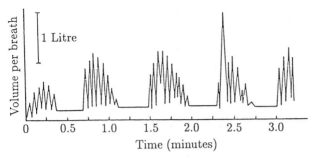

Fig. 1.10. A spirogram of the breathing pattern of a 29 year old man with Cheyne-Stokes respiration. The typical waxing and waning of the volume of breath is interspersed with periods of low ventilation levels; this is apneic breathing. (Redrawn with permission from Mackey and Glass 1977)

low. Fig. 1.10 is typical of spirograms of those suffering from Cheyne-Stokes respiration.

We first need a few physiological facts for our model. The level of arterial carbon dioxide (CO_2), $c(t)$ say, is monitored by receptors which in turn determine the level of ventilation. It is believed that these CO_2-sensitive receptors are situated in the brainstem so there is an inherent time lag, T say, in the overall control system for breathing levels. It is known that the ventilation response curve to CO_2 is sigmoidal in form. We assume the dependence of the ventilation V on c to be adequately described by what is called a Hill function, of the form

$$V = V_{\text{max}} \frac{c^m(t - T)}{a^m + c^m(t - T)} \, , \tag{1.24}$$

where V_{max} is the maximum ventilation possible and the parameter a and the Hill coefficient m, are positive constants which are determined from experimental data. (We discuss the biological relevance of Hill functions and how they arise later in Chapter 5.) We assume the removal of CO_2 from the blood is proportional to the product of the ventilation and the level of CO_2 in the blood.

Let p be the constant production rate of CO_2 in the body. The dynamics of the CO_2 level is then modelled by

$$\frac{dc(t)}{dt} = p - bV c(t) = p - bV_{\text{max}}\, c(t) \frac{c^m(t - T)}{a^m + c^m(t - T)} \, , \tag{1.25}$$

where b is a positive parameter which is also determined from experimental data. The delay time T is the time between the oxygenation of the blood in the lungs and monitoring by the chemoreceptors in the brainstem. This first order differential-delay model exhibits, as we shall see, the qualitative features of both normal and abnormal breathing.

As a first step in analysing (1.25) we introduce the nondimensional quantities

$$x = \frac{c}{a}, \quad t^* = \frac{pt}{a}, \quad T^* = \frac{pT}{a}, \quad \alpha = \frac{abV_{max}}{p}, \quad V^* = \frac{V}{V_{max}} \qquad (1.26)$$

and the model equation becomes

$$x'(t) = 1 - \alpha x(t)\frac{x^m(t-T)}{1 + x^m(t-T)} = 1 - \alpha x V(x(t-T)) . \qquad (1.27)$$

where for algebraic simplicity we have omitted the asterisks on t and T.

As before we get an indication of the dynamic behaviour of solutions by investigating the linear stability of the steady state x_0 given by

$$1 = \alpha \frac{x_0^{m+1}}{1 + x_0^m} = \alpha x_0 V(x_0) = \alpha x_0 V_0 \qquad (1.28)$$

where V_0, defined by the last equation, is the dimensionless steady state ventilation. A simple plot of $1/\alpha x_0$ and $V(x_0)$ as functions of x_0 shows there is a unique positive steady state. If we now consider small perturbations about the steady state x_0 we write $u = x - x_0$ and consider $|u|$ small. Substituting into (1.27) and retaining only linear terms we get, using (1.28),

$$u' = -\alpha V_0 u - \alpha x_0 V_0' u(t-T) . \qquad (1.29)$$

where $V_0' = dV(x_0)/dx_0$ is positive. As in the last Section we look for solutions in the form

$$u(t) \propto e^{\lambda t} \quad \Rightarrow \quad \lambda = -\alpha V_0 - \alpha x_0 V_0' e^{-\lambda T} . \qquad (1.30)$$

If the solution λ with the largest real part is negative, then the steady state is stable. Since here we are concerned with the oscillatory nature of the disease we are interested in parameter ranges where the steady state is unstable, and in particular, unstable by growing oscillations in anticipation of limit cycle behaviour. So, as before we must determine the bifurcation values of the parameters such that $\text{Re}\lambda = 0$.

Set $\lambda = \mu + i\omega$. In the same way as in the last Section it is easy to show that a real number μ_0 exists such that for all solutions λ of (1.30), $\text{Re}\lambda < \mu_0$ and also that no real positive solution exists. For algebraic simplicity let us write the transcendental equation (1.30) as

$$\lambda = -A - Be^{-\lambda T}, \quad A = \alpha V_0 > 0, \quad B = \alpha x_0 V_0' > 0 , \qquad (1.31)$$

Equating real and imaginary parts gives

$$\mu = -A - Be^{-\mu T} \cos \omega T, \quad \omega = Be^{-\mu T} \sin \omega T . \qquad (1.32)$$

Simultaneous solutions of these give μ and ω in terms of A, B and T: we cannot determine them explicitly of course. The bifurcation we are interested in is when

$\mu = 0$ so we consider the parameter ranges which admit such a solution. With $\mu = 0$ the last equations give, with $s = \omega T$,

$$\cot s = -\frac{AT}{s}, \quad \Rightarrow \quad \frac{\pi}{2} < s_1 < \pi \qquad (1.33)$$

for all finite $AT > 0$. We can see that such a solution s_1 exists on sketching $\cot s$ and $-AT/s$ as functions of s. There are of course other solutions s_m of this equation in the ranges $[(2m+1)\pi/2, (m+1)\pi]$ for $m = 1, 2, \ldots$ but we need only consider the smallest positive solution s_1 since that gives the bifurcation for the smallest critical $T > 0$. We now have to determine the parameter ranges so that with $\mu = 0$ and s_1 substituted back into (1.32) a solution exists. That is, what are the restrictions on A, B and T so that

$$0 = -A - B \cos s_1, \quad s_1 = BT \sin s_1$$

are consistent? These imply

$$BT = [(AT)^2 + s_1^2]^{1/2} . \qquad (1.34)$$

If B, A and T, which determine s_1, are such that the last equality cannot hold then no solution with $\mu = 0$ exists.

Since A and B are positive, the solution is stable in the limiting case $T = 0$ since then $\text{Re}\lambda = \mu = -A - B < 0$. Now consider (1.32) and increase T from $T = 0$. From the last equation and (1.33) a solution with $\mu = 0$ cannot exist if

$$BT < [(AT)^2 + s_1^2]^{1/2}$$
$$s_1 \cot s_1 = -AT, \quad \frac{\pi}{2} < s_1 < \pi \qquad (1.35)$$

and, from continuity arguments from $T = 0$ we must have $\mu < 0$. So the bifurcation condition which just gives $\mu = 0$ is (1.34). Or, put in another way, if (1.35) hold the steady state solution of (1.27) is linearly, and in fact globally, stable. In terms of the original dimensionless variables from (1.31) the conditions are thus

$$\alpha x_0 V_0' T < [(\alpha V_0 T)^2 + s_1^2]^{1/2} ,$$
$$s_1 \cot s_1 = -\alpha V_0 T . \qquad (1.36)$$

If we now have A and B fixed, a bifurcation value T_c is given by (1.35) with an equality sign in the inequality.

Actual parameter values for normal humans have been obtained by Mackey and Glass (1977). The concentration of gas in blood is measured in terms of the partial pressure it sustains and so it is measured in mmHg (that is in torr). Relevant to the dimensional system (1.25), they estimated

$$c_0 = 40\,\text{mmHg}, \quad p = 6\,\text{mmHg/min}, \quad V_0 = 7\,\text{litre/min}\,,$$
$$V_0' = 4\,\text{litre/min\,mmHg}, \quad T = 0.25\,\text{min}\,. \tag{1.37}$$

From (1.28), which defines the dimensionless steady state, we have $\alpha V_0 = 1/x_0$. So, with (1.36) in mind, we have, using (1.37) and the nondimensionalisation (1.26),

$$\alpha V_0 T = \frac{T}{x_0} = \frac{pT_{\text{dimensional}}}{c_0} = 0.0375\,.$$

The solution of the second of (1.36) with such a small right hand side is $s_1 \approx \pi/2$ and so $s_1 \gg \alpha V_0 T$ which means that the inequality for stability from the first of (1.36) is approximately, but quite accurately,

$$V_0' < \frac{\pi}{2\alpha x_0 T}\,. \tag{1.38}$$

So, if the gradient of the ventilation at the steady state becomes too large the steady state becomes unstable and limit cycle periodic behaviour ensues. With the values in (1.37) the critical dimensional $V_0' = 7.44\,\text{litre/min\,mmHg}$. The gradient increases with the Hill coefficient m in (1.24). Other parameters can of course also initiate periodic behaviour: all we require is that (1.38) is violated.

In dimensional terms we can determine values for m and a in the expression (1.24) for the ventilation, which result in instability by using (1.38) with (1.26) and V_0 from (1.28). Fig. 1.11 (a), (b) show the dimensional results of numerical simulations of (1.25) with two values for V_0'.

Note that the period of oscillation in both solutions in Fig. 1.11 is about 1 minute, which is $4T$ where $T = 0.25\,\text{min}$ is the estimate for delay given in (1.37). This is as we would now expect from the analysis in the last section. A perturbation analysis in the vicinity of the bifurcation state in a similar way to that given in the last Section shows that the period of the growing unstable solution is approximately $4T$. This is left as an exercise (Exercise 4).

In fact the experimentally observed period is of the order of two to three times the estimated delay time. The model here for carbon dioxide in the blood is a simple one and to make detailed quantitative comparison with the actual process that takes place is not really justified. It does however, clearly show how a delay model can arise in a genuine physiological context and produce oscillatory behaviour such as is observed in Cheyne-Stokes respiration. There are many fascinating and challenging modelling problems associated with breathing: see, for example, the book of articles edited by Benchetrit, Baconnier and Demongeot (1987).

Regulation of hematopoiesis The second example we consider briefly has certain similarities to the last and so we shall not go through the analysis in as much detail. It is concerned with the regulation of hematopoiesis, the formation of

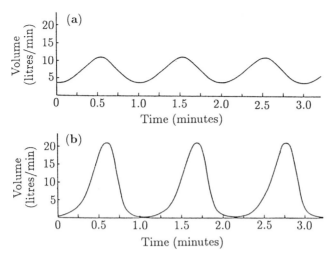

Fig. 1.11a,b. (a) The solution behaviour of the model equation (1.27) presented in dimensional terms for $V_0' = 7.7\,$litre/min mmHg. (b) The solution behaviour for $V_0' = 10.01\,$litre/min mmHg. Note the pronounced apneic regions, that is where the ventilation is very low: this should be compared with the spirogram in Fig. 1.10. (Redrawn from Glass and Mackey 1979)

blood cell elements in the body. For example white and red blood cells, platelets and so on are produced in the bone marrow from where they enter the blood stream. When the level of oxygen in the blood decreases this leads to a release of a substance which in turn causes an increase in the release of the blood elements from the marrow. There is thus a feedback from the blood to the bone marrow. Further details of the process and the model are given in Mackey and Glass (1977) and Glass and Mackey (1979).

Let $c(t)$ be the concentration of cells (the population species) in the circulating blood; the units of c are, say, cells/mm^3. We assume that the cells are lost at a rate proportional to their concentration, that is like gc, where the parameter g has dimensions $(\text{day})^{-1}$. After the reduction in cells in the blood stream there is about a 6 day delay before the marrow releases further cells to replenish the deficiency. We thus assume that the flux λ of cells into the blood stream depends on the cell concentration at an earlier time, namely $c(t - T)$, where T is the delay. Such assumptions suggest a model equation of the form

$$\frac{dc(t)}{dt} = \lambda(c(t - T)) - gc(t) . \qquad (1.39)$$

Mackey and Glass (1977) proposed two possible forms for the function $\lambda(c(t-T))$. The one we consider gives

$$\frac{dc}{dt} = \frac{\lambda a^m c(t - T)}{a^m + c^m(t - T)} - gc , \qquad (1.40)$$

again equal to c_1 when time t increases by the period: Fig. 1.12 (a) is a typical solution $c(t)$ as a function of time in this situation. If we now look at Fig. 1.13 (b) it looks a bit like a double loop trajectory of the kind in (a); you have to go round twice to return to where you started. A typical solution here is like that shown in Fig. 1.14 (a).

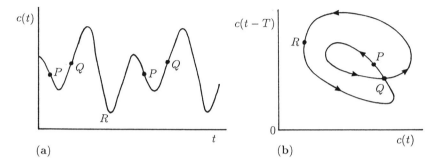

Fig. 1.14a,b. (a) Qualitative solution for $c(t)$ when the parameters in the differential-delay equation (1.25) have a phase plane trajectory as in (b), namely the case in Fig. 1.13 (b).

The solutions $c(t)$ implied by Fig. 1.13 illustrate a common and important feature of many model systems, namely different periodic solution behaviour as a parameter passes through specific bifurcation values; here it is the Hill coefficient m in (1.40).

Referring now to Fig. 1.14 (b) if you start at P the solution first decreases with time and then increases as you move along the trajectory of the first, inner, loop. Now when $c(t)$ reaches Q, instead of going round the same loop past P again it moves onto the outer loop through R. It eventually goes through P again after the second circuit. As before the solution is still periodic of course, but its appearance is like a mixture of two solutions of the type in Fig. 1.12 (a) but with different periods and amplitudes. As m increases, the phase plane trajectories become progressivly more complex suggesting quite complex solution behaviour for $c(t)$. For the case in Fig. 1.13 (e) the solution undergoes very many loops before it *possibly* returns to its starting point. In fact it never does! The solutions in such cases are *not* periodic although they have a quasi-periodic appearance. This is an example of *chaotic* behaviour.

Fig. 1.12 (b) above is a solution of (1.40) which exhibits this chaotic behaviour while Fig. 1.12 (c) shows the dynamic behaviour of the white cell count in the circulating blood of a leukemia patient. Although Figs. 1.12 (b), (c) exhibit similar aperiodic behaviour it is dangerous to presume that this model is therefore the one governing white cell behaviour in leukemia patients. However, what this modelling exercise has demonstrated, among other things, is that delay can play a significant role in physiological pattern disruption. In turn this suggests that a deficiency in bone marrow cell production could account for the erratic behaviour in the white cell count. So although such a model can highlight important questions for a medical physiologist to ask, for it to be of practical use it is essential that close interdisciplinary collaboration is maintained so that realism is retained in making suggestions and drawing conclusions, however plausible they may be.

The numerical simulations of this differential-delay model (1.40), which is clearly illustrative of a whole class, indicates a cascading sequence of bifurcating periodic solutions which become chaotic. The sequence then passes through a coherent periodic stage and again becomes chaotic and so on as a parameter in the model itself passes through successive bifurcating values. As we said, this behaviour arises in a different context in the following chapter on discrete models where it is discussed in some detail: periodic doubling can be shown analytically. The existence of this kind of sequential bifurcating behaviour in such model equations is of considerable potential importance in its biomedical implications.

1.6 Harvesting a Single Natural Population

It is clearly necessary to develop an ecologically acceptable strategy for harvesting any renewable resource be it animals, fish, plants or whatever. We also usually want the maximum *sustainable* yield with the minimum effort. The inclusion of economic factors in population models of renewable resources is increasing and these introduce important constraints: see for example the book by Clark (1976a). The collection of papers edited by Vincent and Skowronski (1981) specifically deals with renewable resource management. The review article by Plant and Mangel (1987) is concerned with insect pest management. The model we describe here is a simple logistic one with the inclusion of a harvesting contribution: it was discussed by Beddington and May (1977). Although it is a particularly simple one it brings out several interesting and important points which more sophisticated models must also take into account. Rotenberg (1987) also considered the logistic model with harvesting, with a view to making the model more quantitative. He also examined the effects of certain stochastic parameters on possible population extinction.

Most species have a growth rate, depending on the population, which more or less maintains a constant population equal to the environment's carrying capacity K. That is the growth and death rates are about equal. Harvesting the species affects the mortality rate and, if it is not excessive, the population adjusts and settles down to a new equilibrium state $N_h < K$. The modelling problem is how to maximize the sustained yield by determining the population growth dynamics so as to fix the harvesting rate which keeps the population at its *maximum growth rate*.

We discuss here a basic model which consists of the logistic population model (1.2) in which the mortality rate is enhanced, as a result of harvesting, by a term linearly proportional to N, namely,

$$\frac{dN}{dt} = rN\left(1 - \frac{N}{K}\right) - EN = f(N).\tag{1.41}$$

Here r, K and E are positive constants and EN is the harvesting yield per unit time with E a measure of the effort expended. K and r are the natural carrying capacity and the linear per capita growth rate respectively. The new non-zero steady state from (1.41) is

$$N_h(E) = K\left(1 - \frac{E}{r}\right) > 0 \quad \text{if} \quad E < r \tag{1.42}$$

which gives a yield

$$Y(E) = EN_h(E) = EK\left(1 - \frac{E}{r}\right).\tag{1.43}$$

Clearly if the harvesting effort is sufficiently large so that it is greater than the linear growth rate when the population is low the species will die out. That is if $E > r$ the only realistic steady state is $N = 0$. If $E < r$ (which was possibly not the case, for example, with whaling in the 1970's) the maximum sustained yield and the new harvesting steady state are, from (1.43) and (1.42),

$$Y_M = Y(E)]_{E=r/2} = \frac{rK}{4}, \quad N_h]_{Y_M} = \frac{K}{2}.\tag{1.44}$$

Does an analysis of the dynamic behavior tell us anything different from the naive, and often used, steady state analysis just given here?

Fig. 1.15 illustrates the growth rate $f(N)$ in (1.41) as a function of N for various efforts E. Linearising (1.41) about $N_h(E)$ gives

$$\frac{d(N - N_h)}{dt} \approx f'(N_h(E))(N - N_h) = (E - r)(N - N_h),\tag{1.45}$$

which shows linear stability if $E < r$: arrows indicate stability or instability in Fig. 1.15.

We can consider the dynamic aspects of the process by determining the time scale of the recovery after harvesting. If $E = 0$ then, from (1.41), the recovery time $T = 0(1/r)$, namely the time scale of the reproductive growth. This is the order of magnitude of the recovery time of N to its carrying capacity K after a small perturbation from K since, for $N(t) - K$ small and $N_h(0) = K$, (1.45) shows

$$\frac{d(N - K)}{dt} \approx -r(N - K) \quad \Rightarrow \quad N(t) - K \propto e^{-rt}.$$

If $E \neq 0$, with $0 < E < r$, then the recovery time in a harvesting situation, from (1.45), is

$$T_R(E) = O\left(\frac{1}{r - E}\right)$$

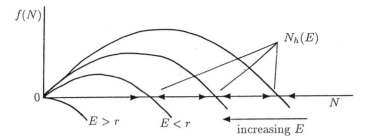

Fig. 1.15. Growth function $f(N)$ for the logistic model with harvesting according to (1.41). Note how the positive steady state decreases with increasing E eventually tending to zero as $E \to r$.

and so

$$\frac{T_R(E)}{T_R(0)} = O\left(\frac{1}{1 - \frac{E}{r}}\right) . \tag{1.46}$$

Thus for a fixed r, a larger E increases the recovery time since $T_R(E)/T_R(0)$ increases with E. When $E = r/2$, the value giving the maximum sustained yield Y_M, $T_R(E) = O(2T_R(0))$.

The usual definition of a recovery time is the time to decrease a perturbation from equilibrium by a factor e. Then, on a linear basis,

$$T_R(0) = \frac{1}{r}, \quad T_R(E) = \frac{1}{r - E} \quad \Rightarrow \quad T_R\left(E = \frac{r}{2}\right) = 2T_R(0) . \tag{1.47}$$

Since it is the yield Y that is recorded, if we solve (1.43) for E in terms of Y we have

$$\frac{T_R(Y)}{T_R(0)} = \frac{2}{1 \pm \left[1 - \frac{Y}{Y_M}\right]^{1/2}} \tag{1.48}$$

which is sketched in Fig. 1.16 (a) where L_+ and L_- denote the positive and negative roots of (1.48). It is clearly advantageous to stay on the L_+ branch and potentially disastrous to get onto the L_- one. Let us now see what determines the branch.

Suppose we start harvesting with a small effort E, then, as is clear from Fig. 1.16 (b), the equilibrium population $N_h(E)$ is close to K and $N_h(E) > K/2$, the equilibrium population for the maximum yield Y_M. The recovery time ratio $T_R(E)/T_R(0)$ from (1.47) is then approximately 1. So increasing E, and hence the yield, we are on branch L_+. As E increases further, $N_h(E)$ decreases towards $K/2$, the value for the maximum sustained yield Y_M and we reach the point A

in Fig. 1.16 (a) when $N_h(E) = K/2$. As E is increased further, $N_h(E) < K/2$ and the recovery time is further increased but with a decreasing yield: we are now on the L_- branch.

We can now see what an optimal harvesting strategy could be, at least with this deterministic point of view. An effort E should be made which keeps the equilibrium population density $N_h(E) > K/2$, but as close as possible to $K/2$, the value for the maximum sustained yield. The closer to $K/2$ however the more delicate the situation becomes since we might inadvertently move onto branch L_- in Fig. 1.16 (a). At this stage, when $N_h(E)$ is close to $K/2$ a stochastic analysis should be carried out as has been done by Beddington and May (1977). Stochastic elements of course reduce the predictability of the catch. In fact they decrease the average yield for a given effort.

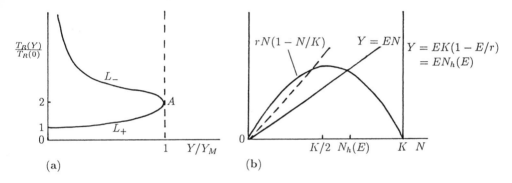

Fig. 1.16a,b. (a) Ratio of the recovery times as a function of the yield for the logistic growth model, with yield proportional to the population: equation (1.41). (b) Graphical method for determining the steady state yield Y for the harvested logistic model (1.41).

As an alternative harvesting resource strategy suppose we harvest with a constant yield Y_0 as our goal. The model equation is then

$$\frac{dN}{dt} = rN\left(1 - \frac{N}{K}\right) - Y_0 = f(N; r, K, Y_0) . \tag{1.49}$$

Fig. 1.17 (a) shows the graphical way of determining the steady states as Y_0 varies. It is trivial to find the equilibria analytically of course, but often the behavioral traits as a parameter varies are more obvious from a figure, such as here. If $0 < Y_0 < rK/4 = Y_M$, the maximum sustainable yield here, there are two positive steady states $N_1(Y_0)$, $N_2(Y_0) > N_1(Y_0)$ which from Fig. 1.17 are respectively unstable and stable. As $Y_0 \to rK/4$, the maximum sustainable yield of the previous model, this model is even more sensitive to fluctuations since if a perturbation from N_2 takes N to a value $N < N_1$ the mechanism then drives N to zero: see Fig. 1.17 (b). Not only that, $N \to 0$ in a *finite* time since for

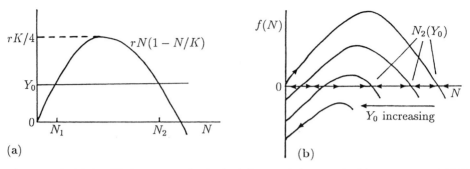

(a)   (b)

Fig. 1.17a,b. (a) Equilibrium states for the logistic growth harvested with a constant yield Y_0: equation (1.49). **(b)** Growth rate $f(N)$ in (1.49) as the yield Y_0 increases.

small enough N, (1.49) becomes $dN/dt \approx -Y_0$ and so for any starting N_0 at t_0, $N(t) \approx N_0 - Y_0(t - t_0)$.

For easy comparison with the constant effort model we evaluate the equivalent recovery time ratio $T_R(Y_0)/T_R(0)$. The recovery time $T_R(Y_0)$ is only relevant to the stable equilibrium $N_2(Y_0)$ which from (1.49) is

$$N_2(Y_0) = \frac{K}{2}\left\{1 + \left[1 - \frac{4Y_0}{Kr}\right]^{1/2}\right\}, \quad Y_0 < rK/4 \ .$$

The linearized form of (1.49) is then

$$\frac{d(N - N_2(Y_0))}{dt} \approx (N - N_2)\left[\frac{\partial f}{\partial N}\right]_{N_2(Y_0)} = -(N - N_2)\, r\left[1 - \frac{4Y_0}{rK}\right]^{1/2} \ .$$

Thus

$$\frac{T_R(Y_0)}{T_R(0)} = \frac{1}{\left(1 - \frac{Y_0}{Y_M}\right)^{1/2}}, \quad Y_M = \frac{rK}{4} \tag{1.50}$$

which shows that $T_R(Y_0)/T_R(0) \to \infty$ as $Y_0 \to Y_M$. This model is thus a much more sensitive one and, as a harvesting strategy, is not very good.

The main conclusion from this modelling exercise is that a constant effort rather than a constant yield harvesting strategy is less potentially catastrophic. It calls into question, even with this simple model, the fishing laws, for example, which regulate catches. A more realistic model, on the lines described here, should take into account the economic costs of harvesting and other factors. This implies a feedback mechanism which can be a stabilizing factor: see Clark (1976a). With the unpredictability of the real world it is probably essential to include feedback. Nevertheless such simple models can pose highly relevant ecological and long term financial questions which have to be considered in any more realistic and more sophisticated model.

*1.7 Population Model with Age Distribution

One of the deficiencies of the above ordinary differential equation models is that they do not take into account any age structure which, in many situations, can influence population size and growth in a major way. It is not always of importance but, when it is, we must know how to incorporate it in a model. So, we consider here a first extension to include age dependence in the birth and death rates. The books by Hoppensteadt (1975), Charlesworth (1980) and Metz and Diekmann (1986) provide a good survey as well as the wide spectrum of applicability of age-structured models.

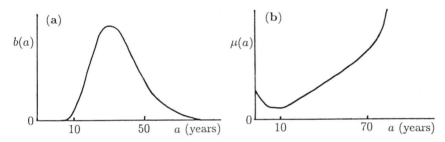

Fig. 1.18a,b. Qualitative birth (a) and death (b) rates for man as functions of age in years.

Let $n(t, a)$ be the population density at time t in the age range a to $a + da$. Let $b(a)$ and $\mu(a)$ be the birth and death rates which are functions of age a: for man, for example, they qualitatively look like the curves in Fig. 1.18. For example, in a small increment of time dt the number of the population of age a that dies is $\mu(a)n(t, a)dt$. The birth rate only contributes to $n(t, 0)$; there can be no births of age $a > 0$. The conservation law for the population now says that

$$dn(t, a) = \frac{\partial n}{\partial t}dt + \frac{\partial n}{\partial a}da = -\mu(a)n(t, a)\, dt \ .$$

The $(\partial n/\partial a)da$ term is the contribution to the change in $n(t, a)$ from individuals getting older. Dividing this equation by dt and noting that $da/dt = 1$ since a is chronological age, $n(t, a)$ satisfies the linear partial differential equation

$$\frac{\partial n}{\partial t} + \frac{\partial n}{\partial a} = -\mu(a)n \ , \tag{1.51}$$

which holds for $t > 0$ and $a > 0$. For example if $\mu = 0$, it reduces to a conservation equation which simply says that the time rate of change of the population at time t and age a, $\partial n/\partial t$, simply changes by the rate at which the population gets older namely $\partial n/\partial a$.

Equation (1.51) is a first order partial differential equation which requires a condition on $n(t, a)$ in t and in a. The initial condition

$$n(0, a) = f(a) , \tag{1.52}$$

says that the population at time $t = 0$ has a given age distribution $f(a)$. The other boundary condition on a comes from the birth rate and is

$$n(t, 0) = \int_0^\infty b(a)n(t, a)\, da , \tag{1.53}$$

where, for mathematical simplicity, we have taken the upper limit of ∞ for the age: $b(a)$ of course will tend to zero for large a, as in Fig. 1.18 (a) for example, and so we could replace ∞ by a_m say where $b(a) = 0$ for $a > a_m$. Note that the birth rate $b(a)$ only appears in the integral equation expression (1.53) and not in the differential equation. Equation (1.51) is often referred to in the ecological literature as the *Von Foerster equation*: the equation arises in a variety of different disciplines and theoretical biology areas, cell proliferation models for example. The main question we wish to answer with the model here is how the birth and death rates $b(a)$ and $\mu(a)$ affect the growth of the population after a long time.

Fig. 1.19. Characteristics for the Von Foerster equation (1.51).

One way to solve (1.51) is by characteristics which are given by

$$\frac{da}{dt} = 1 \quad \text{on which} \quad \frac{dn}{dt} = -\mu n . \tag{1.54}$$

These are the straight lines

$$a = \begin{cases} t + a_0, & a > t \\ t - t_0, & a < t \end{cases} \tag{1.55}$$

as shown in Fig. 1.19. Here a_0, t_0 are respectively the initial age of an individual at time $t = 0$ in the original population and the time of birth of an individual. The second of (1.54), which holds along each characteristic, has a different solution according as $a > t$ and $a < t$, that is one form for the population that was present at $t = 0$, namely $a > t$, and the other for those born after $t = 0$, that is $a < t$. On integrating the second of (1.54), using $da/dt = 1$ and (1.55), the solutions are

$$n(t, a) = n(0, a_0) \exp \left[- \int_{a_0}^{a} \mu(s) \, ds \right], \quad a > t$$

where $n(0, a_0) = n(0, a - t) = f(a - t)$ from (1.52), and so

$$n(t, a) = f(a - t) \exp \left[- \int_{a-t}^{a} \mu(s) \, ds \right], \quad a > t . \tag{1.56}$$

For $a < t$,

$$n(t, a) = n(t_0, 0) \exp \left[- \int_{0}^{a} \mu(s) \, ds \right] ,$$

and so, since $n(t_0, 0) = n(t - a, 0)$

$$n(t, a) = n(t - a, 0) \exp \left[- \int_{0}^{a} \mu(s) \, ds \right], \quad a < t . \tag{1.57}$$

In the last equation $n(t - a, 0)$ is determined by solving the integral equation (1.53), using (1.56) and (1.57) to get

$$n(t, 0) = \int_{0}^{t} b(a) n(t - a, 0) \exp \left[- \int_{0}^{a} \mu(s) \, ds \right] da$$

$$+ \int_{t}^{\infty} b(a) f(a - t) \exp \left[- \int_{a-t}^{a} \mu(s) \, ds \right] da . \tag{1.58}$$

Although this is a linear equation it is not easy to solve: it can be done by iteration however.

We are mainly interested in the long time behaviour of the population and in particular whether or not it will increase or decline. If t is large so that for practical purposes $t > a$ for all a then $f(a - t) = 0$ and all we require in (1.58) is the first integral term on the right hand side. The solution is then approximated by $n(t, a)$ in (1.57), although it does not satisfy the boundary condition on a in (1.53). It is still not trivial to solve so let us return to the original partial differential equation (1.51) and see if other solution forms are possible.

We can look for a *similarity solution* of (1.51) in the form

$$n(t, a) = e^{\gamma t} r(a) . \tag{1.59}$$

That is, the age distribution is simply changed by a factor which either grows or decays with time according to whether $\gamma > 0$ or $\gamma < 0$. Substitution of (1.59) into (1.51) gives

$$\frac{dr}{da} = -[\mu(a) + \gamma] r$$

and so

$$r(a) = r(0) \exp\left[-\gamma a - \int_0^a \mu(s)\,ds\right]. \tag{1.60}$$

With this $r(a)$ in (1.59) the resulting $n(t, a)$ when inserted into the boundary condition (1.53), gives

$$e^{\gamma t} r(0) = \int_0^\infty b(a) e^{\gamma t} r(0) \exp\left[-\gamma a - \int_0^a \mu(s)\,ds\right] da$$

and hence, on cancelling $e^{\gamma t} r(0)$,

$$1 = \int_0^\infty b(a) \exp\left[-\gamma a - \int_0^a \mu(s)\,ds\right] da = \phi(\gamma), \tag{1.61}$$

which defines $\phi(\gamma)$. This equation determines a unique γ, γ_0 say, since $\phi(\gamma)$ is a monotonic decreasing function of γ. The sign of γ is determined by the size of $\phi(0)$: see Fig. 1.20.

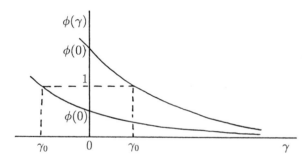

Fig. 1.20. The growth factor γ_0 is determined by the intersection of $\phi(\gamma) = 1$: $\gamma_0 > 0$ if $\phi(0) > 1$, $\gamma_0 < 0$ if $\phi(0) < 1$.

That is γ_0 is determined solely by the birth rate, $b(a)$, and death, $\mu(a)$, rates. The critical threshold S for population growth is thus

$$S = \phi(0) = \int_0^\infty b(a) \exp\left[-\int_0^a \mu(s)\,ds\right] da \tag{1.62}$$

where $S > 1$ implies growth and $S < 1$ implies decay. In (1.62) we can think of $\exp[-\int_0^a \mu(s)\,ds]$ almost like the probability that an individual survives to age a, only the integral over all a is not 1.

The solution (1.59) with (1.60) cannot satisfy the initial condition (1.52). The question arises as to whether it approximates the solution of (1.51)-(1.53), the original problem, after a long time. If t is large so that for all practical purposes $n(t, 0)$ in (1.58) requires only the first integral on the right hand side, then

$$n(t, 0) \sim \int_0^t b(a) n(t - a, 0) \exp\left[-\int_0^a \mu(s)\,ds\right] da, \quad t \to \infty. \tag{1.63}$$

If we now look for a solution of this equation in the similarity form (1.59), substitution of it into (1.63) then gives (1.61) again as the equation for γ. We thus conjecture that the solution (1.59) with $r(a)$ from (1.60) and γ from (1.61) is the solution for large time t of equation (1.51), with initial and boundary conditions (1.52) and (1.53). It is of course undetermined to the extent of a constant $r(0)$ but, since our main question is one of growth or decay, it is not important to know $r(0)$ since it does not affect either. The important parameter is the threshold parameter S in (1.62) from which long time effects of alterations in the birth and death rates can be assessed.

Exercises

1. A model for the spruce budworm population $u(t)$ is governed by

$$\frac{du}{dt} = ru\left(1 - \frac{u}{q}\right) - \frac{u^2}{1+u^2}$$

where r and q are positive dimensionless parameters. The non-zero steady states are thus given by the intersection of the two curves

$$U(u) = r\left(1 - \frac{u}{q}\right), \quad V(u) = \frac{u}{1+u^2}.$$

Show, using the conditions for a double root, that the curve in r, q space which divides it into regions where there are 1 or 3 positive steady states is given parametrically by

$$r = \frac{2a^3}{(1+a^2)^2}, \quad q = \frac{2a^3}{a^2 - 1}.$$

Show that the two curves meet a cusp, that is where $dr/da = dq/da = 0$, at $a = \sqrt{3}$. Sketch the curves in r, q space noting the limiting behaviour of $r(a)$ and $q(a)$ as $a \to 0$ and $a \to 1$.

2. The predation $P(N)$ on a population $N(t)$ is very fast and a model for the prey $N(t)$ satisfies

$$\frac{dN}{dt} = RN\left(1 - \frac{N}{K}\right) - P\left\{1 - \exp\left[-\frac{N^2}{\varepsilon A^2}\right]\right\}, \quad 0 < \varepsilon \ll 1$$

where R, K, P and A are positive constants. By an appropriate nondimensionalization show that the equation is equivalent to

$$\frac{du}{d\tau} = ru\left(1 - \frac{u}{q}\right) - \left(1 - \exp\left[-\frac{u^2}{\varepsilon}\right]\right)$$

where r and q are positive parameters. Demonstrate that there are three possible nonzero steady states if r and q lie in a domain in r, q space given approximately by $rq > 4$. Could this model exhibit hysterysis?

3. A continuous time model for the baleen whale (a slightly more complicated model of which the International Whaling Commission used) is the delay equation

$$\frac{dN}{dt} = -\mu N(t) + \mu N(t - T)[1 + q\{1 - [N(t - T)/K]^z\}] \,.$$

Here $\mu(> 0)$ is a measure of the mortality, $q(> 0)$ is the maximum increase in fecundity the population is capable of, K is the unharvested carrying capacity, T is the time to sexual maturity and $z > 0$ is a measure of the intensity of the density dependent response as the population drops.

Determine the steady state populations. Show that the equation governing small perturbations $n(t)$ about the positive equilibrium is

$$\frac{dn(t)}{dt} \approx -\mu n(t) - \mu(qz - 1)n(t - T) \,,$$

and hence that the stability of the equilibrium is determined by $\text{Re}\,\lambda$ where

$$\lambda = -\mu - \mu(qz - 1)e^{-\lambda T} \,.$$

Deduce that the steady state is stable (by considering the limiting case $\text{Re}\,\lambda = 0$) if

$$\mu T < \mu T_c = \frac{\pi - \cos^{-1}\frac{1}{b}}{(b^2 - 1)^{1/2}}, \quad b = qz - 1 > 1$$

and stable for all T if $b < 1$.

For $T = T_c + \varepsilon$, $0 < \varepsilon \ll 1$ show that to $O(1)$ the period of the growing oscillation is $2\pi/[\mu(b^2 - 1)^{1/2}]$, $b > 1$.

4. The concentration of carbon dioxide in the blood is believed to control breathing levels through a delay feed back mechanism. A simple delay model for the concentration in dimensionless form is

$$\frac{dx(t)}{dt} = 1 - axV(x(t - T)), \quad V(x) = \frac{x^m}{1 + x^m} \,,$$

where a and m are positive constants and T is the delay. For given a and m a critical delay T_c exists such that for $T > T_c$ the steady state solution becomes linearly unstable. For $T = T_c + \varepsilon$, where $0 < \varepsilon \ll 1$, carry out a perturbation analysis and show that the period of the exponentially growing solution is approximately $4T_c$.

5. A model for the concentration $c(t)$ of arterial carbon dioxide, which controls the production of certain blood elements, is given by

$$\frac{dc(t)}{dt} = p - V(c(t-T))c(t) = p - \frac{bV_{\max}c(t)c^m(t-T)}{a^m + c^m(t-T)} \, ,$$

where p, b, a, T and $V_{\max}$ are positive constants. (This model is briefly discussed in Section 1.5.) Nondimensionalise the equation and examine the linear stability of the steady state. Obtain a relation between the parameters such that the steady state is stable and hence establish the existence of a bifurcation value T_c for the delay. Obtain an estimate for the period of the periodic solution which bifurcates off the steady state when $T = T_c + \varepsilon$ for small ε.

6. A similarity solution of the form $n(t, a) = e^{\gamma t}r(a)$ of the age distribution model equation

$$\frac{\partial n}{\partial t} + \frac{\partial n}{\partial a} = -\mu(a)n$$

satisfies the age boundary equation

$$n(t,0) = \int_0^\infty b(a)n(t,a)\, da$$

if

$$\int_0^\infty b(a)\exp\left[-\gamma a - \int_0^a \mu(s)\, ds\right] da = 1 \, .$$

Show that if the birth rate $b(a)$ is essentially zero except over a very narrow range about $a_0 > 0$ the population will die out whatever the mortality rate $\mu(a)$. If there is a high, linear in age, mortality rate, say what you can about the birth rate if the population is not to die out.

2. Discrete Population Models for a Single Species

2.1 Introduction: Simple Models

Differential equation models whether ordinary, delay, partial or stochastic, imply a continuous overlap of generations. Many species have no overlap whatsoever between successive generations and so population growth is in discrete steps. For primitive organisms these can be quite short in which case a continuous (in time) model may be a reasonable approximation. However, depending on the species the step lengths can vary widely. A year is common. In the models we discuss in this chapter and later in Chapter 4 we have scaled the time step to be 1. Models must thus relate the population at time $t + 1$, denoted by N_{t+1}, in terms of the population N_t at time t. This leads us to study difference equations, or discrete models, of the form

$$N_{t+1} = N_t F(N_t) = f(N_t) \qquad (2.1)$$

where $f(N_t)$ is in general a nonlinear function of N_t. The first form is often used to emphasise the existence of a zero steady state. Such equations are usually impossible to solve analytically but again we can extract a considerable amount of information about the population dynamics without an analytical solution. The mathematics of difference equations is now being studied in depth and in diverse fields: it is a fascinating area. From a practical point of view if we know the form of $f(N_t)$ it is a straightforward matter to evaluate N_{t+1} and subsequent generations by simply using (2.1) recursively. Of course, whatever the form of $f(N)$, we shall only be interested in non-negative populations.

The skill in modelling a specific population's growth dynamics lies in determining the appropriate form of $f(N)$ to reflect known observations or facts about the species in question. To do this with any confidence we must understand the major effects on the solutions of changes in the form of $f(N)$ and its parameters, and also what solutions of (2.1) look like for a few specimen examples of practical interest. The mathematical problem is a mapping one, namely that of finding the orbits, or trajectories, of nonlinear maps given a starting value $N_0 > 0$. It should be noted here that there is no simple connection between difference equation models and what might appear to be the continuous differential equation analogue even though a finite difference approximation results in a discrete equation. This will become clear below.

Suppose the function $F(N_t) = r > 0$, that is the population one step later is simply proportional to the current population, then from (2.1),

$$N_{t+1} = rN_t \quad \Rightarrow \quad N_t = r^t N_0 .\tag{2.2}$$

So the population grows or decays geometrically according to whether $r > 1$ or $r < 1$ respectively. This particularly simple model is not very realistic for most populations nor for long times but, even so, it has been used with some justification for the early stages of growth of certain bacteria. A slight modification to bring in crowding effects could be

$$N_{t+1} = rN_S, \quad N_S = N_t^{1-b}, \quad b \text{ constant} ,$$

where N_S is the population that survives to breed. There must be restrictions on b of course, so that $N_S \le N_t$ otherwise those surviving to breed would be more than the population of which they form a part.

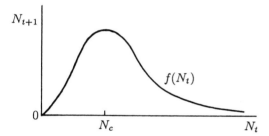

Fig. 2.1. Typical growth form in the model $N_{t+1} = f(N_t)$.

Generally, because of crowding and self-regulation, we expect $f(N_t)$ in (2.1) to have some maximum, at N_m say, as a function of N_t with f decreasing for $N_t > N_m$: Fig. 2.1 illustrates a typical form. A variety of $f(N_t)$ have been used in practical biological situations: see for example the list in May and Oster (1976). One such model, sometimes referred to as the Verhulst process, is

$$N_{t+1} = rN_t \left(1 - \frac{N_t}{K}\right), \quad r > 0, \quad K > 0 ,\tag{2.3}$$

which is a kind of discrete analogue of the continuous logistic growth model. As we shall see, however, the solutions and their dependence on r are very different. An obvious drawback of this specific model is that if $N_t > K$ then $N_{t+1} < 0$.

A more realistic model should be such that for large N_t there should be a reduction in the growth rate but N_{t+1} should remain non-negative; the qualitative form for $f(N_t)$ in Fig. 2.1 is an example. One such frequently used model is

$$N_{t+1} = N_t \exp\left[r\left(1 - \frac{N_t}{K}\right)\right], \quad r > 0, \quad K > 0 \tag{2.4}$$

which we can think of as a modification of (2.2) where there is a mortality factor $\exp(-rN_t/K)$ which is more severe the larger N_t. Here $N_t > 0$ for all t if $N_0 > 0$.

Since t increases by discrete steps there is, in a sense, an inherent *delay* in the population to register change. Thus there is a certain heuristic basis for relating these difference equations to delay differential equations discussed in Chapter 1, which, depending on the length of the delay, could have oscillatory solutions. Since we scaled the time step to be one in the general form (2.1) we should expect the other parameters to be the controlling factors as to whether or not solutions are periodic. With (2.3) and (2.4) the determining parameter is r since K can be scaled out by writing N_t for N_t/K.

2.2 Cobwebbing: A Graphical Procedure of Solution

We can elicit a considerable amount of information about the population growth behaviour by simple graphical means. Consider (2.1) with f as in Fig. 2.1. The steady states are solutions N^* of

$$N^* = f(N^*) = N^*F(N^*) \quad \Rightarrow \quad N^* = 0 \quad \text{or} \quad F(N^*) = 1 . \tag{2.5}$$

Graphically the steady states are intersections of the curve $N_{t+1} = f(N_t)$ and

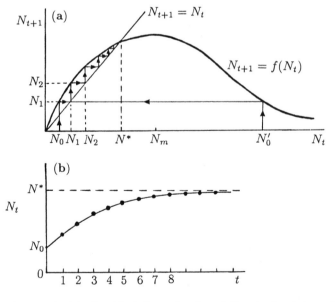

Fig. 2.2a,b. (a) Graphical determination of the steady state and demonstration of how N_t approaches it. (b) Time evolution of the population growth using (a). We use a continuous curve joining up the populations at different time steps for visual clarity: strictly the population changes abruptly at each time step.

the straight line $N_{t+1} = N_t$ as shown in Fig. 2.2 (a) for a case where the maximum of the curve $N_{t+1} = f(N_t)$, at N_m say, has $N_m > N^*$. The dynamic evolution of the solution N_t of (2.1) can be obtained graphically as follows. Suppose we start at N_0 in Fig. 2.2 (a). Then N_1 is given by simply moving along the N_{t+1} axis until we intersect with the curve $N_{t+1} = f(N_t)$, which gives $N_1 = f(N_0)$. The line $N_{t+1} = N_t$ is now used to start again with N_1 in place of N_0. We then get N_2 by proceeding as before and then N_3, N_4 and so on: the arrows show the path sequence. The path is simply a series of reflexions in the line $N_{t+1} = N_t$. We see that $N_t \to N^*$ as $t \to \infty$ and it does so monotonically as illustrated in Fig. 2.2 (b). If we started at $N_0' > N^*$ in Fig. 2.2 (a), again $N_t \to N^*$ and monotonically after the first step. If we start close enough to the steady state N^* the approach to it is monotonic as long as the curve $N_{t+1} = f(N_t)$ crosses $N_{t+1} = N_t$ appropriately: here that means

$$0 < \left[\frac{df(N_t)}{dN_t} \right]_{N_t=N^*} = f'(N^*) < 1 . \tag{2.6}$$

The value $f'(N^*)$, where the prime denotes the derivative with respect to N_t, is an important parameter as we shall see: it is the *eigenvalue* of the system at the steady state N^*. Since any small perturbation about N^* simply decays to zero, N^* is a linearly stable equilibrium state.

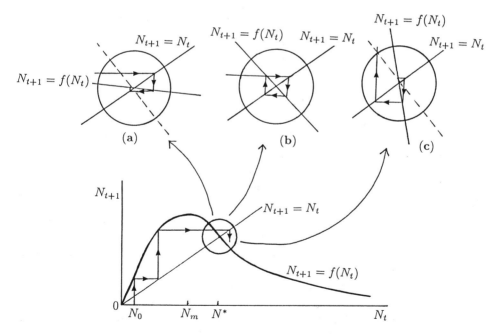

Fig. 2.3a-c. Local behaviour of N_t near a steady state where $f'(N^*) < 0$. The enlargements show the cases where: (a) $-1 < f'(N^*) < 0$, N^* is stable with decreasing oscillations in any small perturbation from the steady state. (b) $f'(N^*) = -1$, N^* is neutrally stable. (c) $f'(N^*) < -1$, N^* is unstable with growing oscillations.

Suppose now $f(N_t)$ is such that the equilibrium $N^* > N_m$ as in Fig. 2.3. The dynamic behaviour of the population depends critically on the geometry of the intersection of the curves at N^* as seen from the inset enlargements in Fig. 2.3 (a), (b), (c): these respectively have $-1 < f'(N^*) < 0$, $f'(N^*) = -1$ and $f'(N^*) < -1$. The solution N_t is oscillatory in the vicinity of N^*. If the oscillations decrease in amplitude and $N_t \to N^*$ then N^* is stable as in Fig. 2.3 (a), while it is unstable if the oscillations grow as in Fig. 2.3 (c). The case Fig. 2.3 (b) exhibits oscillations which are periodic and suggest that periodic solutions to the equation $N_{t+1} = f(N_t)$ are possible. The steady state is strictly unstable if a small perturbation from N^* does not tend to zero. The population's dynamic behaviour for each of the three cases in Fig. 2.3 is illustrated in Fig. 2.4.

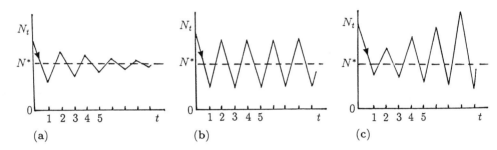

Fig. 2.4a-c. Local behaviour of small perturbations about the equilibrium population N^* with (a), (b) and (c) corresponding to the situations illustrated in Fig. 2.3 (a), (b) and (c) respectively: (a) is the stable case and (c) the unstable case.

The parameter $\lambda = f'(N^*)$, the *eigenvalue* of the equilibrium N^* of $N_{t+1} = f(N_t)$, is crucial in determining the local behaviour about the steady state. The cases in which the behaviour is clear and decisive are when $0 < \lambda < 1$ as in Fig. 2.2 (a) and $-1 < \lambda < 0$ and $\lambda < -1$ as in Fig. 2.3 (a), (c) respectively. The equilibrium is stable if $-1 < \lambda < 1$ and is said to be an *attracting equilibrium*. The critical *bifurcation* values $\lambda = \pm 1$ are where the solution N_t changes its behavioral character. The case $\lambda = 1$ is where the curve $N_{t+1} = f(N_t)$ is tangent to $N_{t+1} = N_t$ at the steady state since $f'(N^*) = 1$ and is called a *tangent bifurcation* for obvious reasons. The case $\lambda = -1$ for reasons that will become clear below is called a *pitchfork bifurcation*.

The reason for the colourful description 'cobwebbing' for this graphical procedure is obvious from Figs. 2.2, 2.3 and 2.5 below. It is an exceedingly useful procedure for suggesting the dynamic behaviour of the population N_t for single equations of the type (2.1). Although we have mainly concentrated on the local behaviour near an equilibrium it also gives the quantitative global behaviour. If the steady state is unstable, it can presage the peculiar behaviour that solutions of such equations can exhibit. As an example suppose $\lambda = f'(N^*) < -1$, that is the local behaviour near the unstable N^* is as in Fig. 2.3 (c). If we now cobweb such a case we have a situation such as shown in Fig. 2.5. The solution trajectory cannot tend to N^*. On the other hand the population must be bounded by $N_{\max}$ in Fig. 2.5 (a) since there is no way we can generate a larger N_t although we would start with one. Thus the solution is globally bounded but does not tend to a steady state. In fact it seems to wander about in a seemingly random way if we look at it as a function of time in Fig. 2.5 (b). Solutions which do this are called *chaotic*.

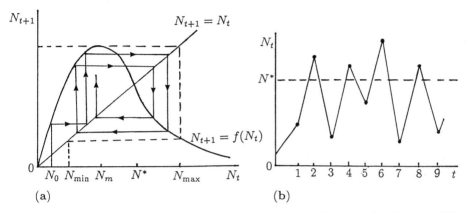

Fig. 2.5a,b. (a) Cobweb for $N_{t+1} = f(N_t)$ where the eigenvalue $\lambda = f'(N^*) < -1$. (b) The corresponding population behaviour as a function of time.

With the different kinds of solutions of models like (2.1), as indicated by the cobweb procedure and the sensitivity hinted at by the special critical values of the eigenvalue λ, we must now investigate such equations analytically. The results suggested by the graphical approach can be very helpful in the analysis.

2.3 Discrete Logistic Model: Chaos

As a concrete example consider the nonlinear logistic model (2.3) which we rescale by writing $u_t = N_t/K$ so that the 'carrying capacity' is 1. The form we study is then

$$u_{t+1} = ru_t(1 - u_t), \quad r > 0 , \tag{2.7}$$

where we assume $0 < u_0 < 1$ and we are interested in solutions $u_t \geq 0$. The steady states and corresponding eigenvalues λ are

$$u^* = 0, \quad \lambda = f'(0) = r ,$$

$$u^* = \frac{r-1}{r}, \quad \lambda = f'(u^*) = 2 - r . \tag{2.8}$$

As r increases from zero but with $0 < r < 1$ the only realistic, that is non-negative, equilibrium is $u^* = 0$ which is stable since $0 < \lambda < 1$. It is also clear from a cobwebbing of (2.7) with $0 < r < 1$ or analytically from equation (2.7) on noting that $u_1 < u_0 < 1$ and $u_{t+1} < u_t$ for all t which implies that $u_t \to 0$ as $t \to \infty$.

The first bifurcation comes when $r = 1$ since $u^* = 0$ becomes unstable since its eigenvalue $\lambda > 1$ for $r > 1$ while the positive steady state $u^* = (r - 1)/r > 0$, for which $-1 < \lambda < 1$ for $1 < r < 3$, is stable for this range of r. The second bifurcation is at $r = 3$ where $\lambda = -1$. Here $f'(u^*) = -1$, and so, locally near u^*, we have the situation in Fig. 2.3 (b) which exhibits a periodic solution.

To see what is happpening when r passes through the bifurcation value $r = 3$ let us first introduce the following notation for the iterative procedure:

$$\begin{cases} u_1 = f(u_0) \\ u_2 = f(f(u_0)) = f^2(u_0) \\ \quad \vdots \\ u_t = f^t(u_0) \end{cases} \tag{2.9}$$

With the example (2.7) the first iteration is simply the equation (2.7) while the second iterate is

$$u_{t+2} = f^2(u_t) = r[ru_t(1 - u_t)][1 - ru_t(1 - u_t)] . \tag{2.10}$$

Fig. 2.6 (a) illustrates the effect on the first iteration as r varies; the eigenvalue $\lambda = f'(u^*)$ decreases as r increases and $\lambda = -1$ when $r = 3$. If we now look at the second iteration (2.10) and ask if it has any equilibria, that is where $u_{t+2} = u_t = u_2^*$. A little algebra shows that u_2^* satisfies

$$u_2^*[ru_2^* - (r - 1)][r^2 u_2^{*2} - r(r + 1)u_2^* + (r + 1)] = 0 \tag{2.11}$$

which has solutions

$$u_2^* = 0 \quad \text{or} \quad u_2^* = \frac{r - 1}{r} > 0 \quad \text{if} \quad r > 1 ,$$
$$u_2^* = \frac{(r + 1) \pm [(r + 1)(r - 3)]^{1/2}}{2r} > 0 \quad \text{if} \quad r > 3 . \tag{2.12}$$

We thus see that there are 2 more real steady states of $u_{t+2} = f^2(u_t)$ with $f(u_t)$ from (2.7) if $r > 3$. This corresponds to the situation in Fig. 2.6 (b) where A, B and C are the positive equilibria u_2^*, with B equal to $(r - 1)/r$, lying between the two new solutions for u_2^* in (2.12) which appear when $r > 3$.

We can think of (2.10) as a first iteration in a model where the iterative time step is 2. The eigenvalues λ of the equilibria can be calculated at the points A, B and C. Clearly $\lambda_B = f'(u_B^*) > 1$ from Fig. 2.6 (b) where u_B^* denotes u_2^* at B and similarly for A and C. For r just greater than 3, $-1 < \lambda_A < 1$ and $-1 < \lambda_C < 1$

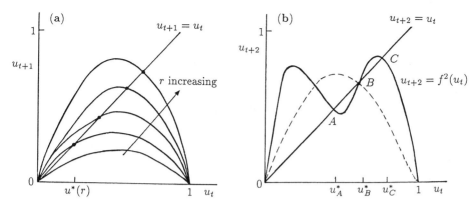

Fig. 2.6a,b. (a) First iteration as a function of r for $u_{t+1} = ru_t(1 - u_t)$: $u^* = (r-1)/r$, $\lambda = f'(u^*) = 2 - r$. (b) Second iteration $u_{t+2} = f^2(u_t)$ as a function of u_t for $r = 3 + \varepsilon$ where $0 < \varepsilon \ll 1$. The dashed line reproduces the first iteration curve of u_{t+1} as a function of u_t; it passes through B, the unstable steady state.

as can be seen visually or, from (2.10), by evaluating $\partial f^2(u_t)/\partial u_t$ at u_A^* and u_C^* given by the last two solutions in (2.12). Thus the steady states, u_A^* and u_C^*, of the second iteration (2.10) are stable. What this means is that there is a stable equilibrium of the second iteration (2.10) and this means that there exists a stable *periodic solution* of period 2 of equation (2.7). In other words if we start at A for example, we come back to it after 2 iterations, that is $u_{A+2}^* = f^2(u_A^*)$ but $u_{A+1}^* = f(u_A^*) \neq u_A^*$. In fact $u_{A+1}^* = u_C^*$ and $u_{C+1}^* = u_A^*$.

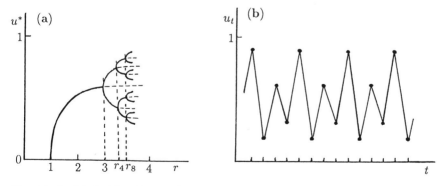

Fig. 2.7a,b. (a) Stable solutions (schematic) for the logistic model (2.7) as r passes through bifurcation values. At each bifurcation the previous state becomes unstable and is represented by the dashed lines. The sequence of stable solutions have periods $2, 2^2, 2^3, \ldots$. (b) An example (schematic) of a 4-cycle periodic solution where $r_4 < r < r_8$ where r_4 and r_8 are the bifurcation values for 4-period and 8-period solutions respectively.

As r continues to increase, the eigenvalues λ at A and C in Fig. 2.6 (b) pass through $\lambda = -1$ and so these 2-period solutions become unstable. At this stage we look at the 4th iterative and we find, as might now be expected, that u_{t+4} as a function of u_t will have four humps as compared with two in Fig. 2.6 (b) and a 4-cycle periodic solution appears. Thus as r passes through a series of bifurcation values the character of the solution u_t passes through a series of bifurcations, here in period doubling of the perodic solutions. The bifurcation situation is illustrated in Fig. 2.7 (a). These bifurcations when $\lambda = -1$ are called *pitchfork bifurcations*, for obvious reasons from the picture they generate in Fig. 2.7 (a). For example if $3 < r < r_4$, where r_4 is the bifurcation value to a 4-period solution, then the periodic solution is between the two u^* in Fig. 2.7 (a) which are the intersections of the vertical line through the r value and the curve of equilibrium states. Fig. 2.7 (b) is an example of a 4-cycle periodic solution, that is $r_4 < r < r_8$ with the actual u_t values again given by the 4 intersections of the curve of equilibrium states with the vertical line through that value of r.

As r increases through successive bifurcations every even p-periodic solution branches into a $2p$-periodic solution and this happens when r is such that the eigenvalue of the p-periodic solution passes through -1. The distance between bifurcations in r-space gets smaller and smaller: this is heuristically plausible since higher order iterates imply more humps (compare with Fig. 2.6 (b)) all of which are fitted into the same interval $(0,1)$. There is thus a hierarchy of solutions of period 2^n for every n, and associated with each, is a parameter interval in which it is stable. There is a limiting value r_c at which instability sets in for all periodic solutions of period 2^n and then for $r > r_c$ locally attracting cycles with periods $k, 2k, 4k, \ldots$ appear but where now k is odd.

This critical parameter value r_c in our model (2.7) is when a 3-period solution is just possible. This happens when the third iterate has 3 steady states which are tangent to the line $u_{t+3} = u_t$: that is the eigenvalue $\lambda = 1$ at these steady states of $u_{t+3} = f^3(u_t)$. This situation is shown schematically in Fig. 2.8. For the model (2.7) the critical $r_c \approx 3.828$.

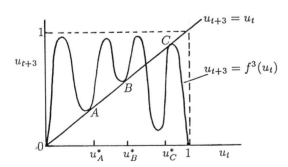

Fig. 2.8. Schematic third iterate $u_{t+3} = f^3(u_t)$ for (2.7) at $r = r_c$, the parameter value where the three steady states A, B and C all have eigenvalue $\lambda = 1$.

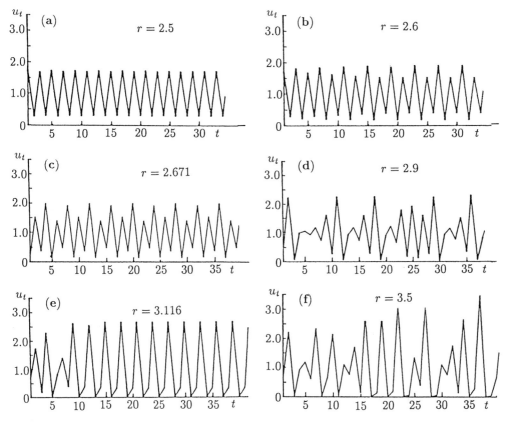

Fig. 2.9a-f. Solutions u_t of the model system $u_{t+1} = u_t \exp[r(1 - u_t)]$ for various r. Here the first bifurcation to periodicity occurs at $r = r_c = 2$. The larger the parameter r the larger the amplitude of the oscillatory solution. (a), (b), and (c) exhibit 2-, 4-, and 8-cycle periodic solutions, (d) and (f) chaotic behaviour and (e) a 3-cycle solution.

Sarkovskii (1964) published an important paper on one-dimensional maps, which has dramatic practical consequences, and is directly related to the situation in Fig. 2.8. He proved, among other things, that if a solution of *odd* (≥ 3) period exists for a value r_c then aperiodic or *chaotic solutions* exist for $r > r_c$. Such solutions simply oscillate in an apparently random manner. The bifurcation here, at r_c, is called a *tangent bifurcation*: the name is suggestive of the situation illustrated in Fig. 2.8. Fig. 2.9 illustrates some solutions for the model equation (2.4) for various r, including chaotic examples in Fig. 2.9 (d), (f). Note the behaviour in Fig. 2.9 (f) for example: there is population explosion, crashback and slow recovery.

Sarkovskii's theorem was further extended by Stefan (1977). Li and Yorke's (1975) result, namely that if a period 3 solution exists then solutions of period n exist for all $n \geq 1$, is a special case of Sarkovskii's theorem.

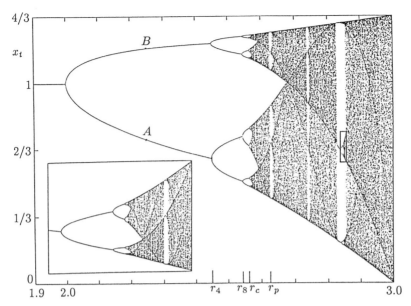

Fig. 2.10. Long time asymptotic iterates for the discrete equation $x_{t+1} = x_t + rx_t(1 - x_t)$ for $1.9 < r < 3$. By a suitable rescaling this can be written in the form (2.7). These are typical of discrete models which exhibit period doubling and eventually chaos and the subsequent path through chaos. Another example is that used in Fig. 2.9: see text for a detailed explanation. The enlargement of the small window (with a greater magnification in the r-direction than in the x_t-direction) shows the fractal nature of the bifurcation sequences. (Reproduced with permission from the book by Peitgen and Richter 1986: some labelling only has been added)

Although we have concentrated here on the logistic model (2.7) this kind of behaviour is typical of difference equation models with the dynamics like (2.1) and schematically illustrated in Fig. 2.1; that is they all exhibit bifurcations to higher periodic solutions eventually leading to chaos.

Fig. 2.9 (d)-(f) illustrate an interesting apsect of the paths to chaos. As r increases from its value giving the aperiodic solution in Fig. 2.9 (d) we again get periodic solutions, as in Fig. 2.9 (e). For larger r aperiodic solutions again appear as in Fig. 2.9 (f). So as r increases beyond where chaos first appears there are windows of parameter values where the solution behaviour is periodic. There are thus parameter windows of periodicity interlaced with windows of aperiodicity. Fig. 2.10 shows a typical figure obtained when the iterative map is run after a long time, the order of several thousand iterations and then run for many more iterations during which the values u_t were plotted.

Refer now to Fig. 2.10 and consider the effect on the solutions of increasing r. For $r_2 < r < r_4$ the solution u_t simply oscillates beteen the two points, A and B for example, which are the intersections of a vertical line through the r-value. For $r_4 < r < r_8$, u_t exhibits a 4-period solution with the values again given by the intersection of the curves with the vertical line through the r-value as shown. For values of $r_c < r < r_p$ the solutions are chaotic. For a small window of r-values greater than r_p the solutions again exhibit regular periodic solutions after which they are again aperiodic. The sequence of aperiodicity – periodicty – aperiodicity is repeated. If we now look at the inset which is an enlargement of the the small rectangle we see the same sequence of bifurcations repeated in a fractal sense. The elegant book by Peitgen and Richter (1986) shows a colourful selection of beautiful and spectacular figures and fractal sequences which can arise from discrete models, particularly with two-dimensional models.

There is now widespread interest and a large amount of research going on in chaotic behaviour such as we have been discussing, much of it prompted by new and potential applications in a variety of different fields. The interest is not restricted to discrete models of course: it was first demonstrated by a system of ordinary differential equations – the Lorenz system (Lorenz 1963: see Sparrow 1982, 1986 for a recent review). This research into chaos has produced many interesting and unexpected results associated with models such as we have been discussing, namely those which exhibit periodic doubling. For example, if $r_2, r_4, \ldots r_{2n}, \ldots$ are the sequence of period doubling bifurcation values, Feigenbaum (1978) proved that

$$\lim_{n \to \infty} \frac{r_{2(n+1)} - r_{2n}}{r_{2(n+2)} - r_{2(n+1)}} = \delta = 4.66920 \ldots .$$

He showed that δ is a universal constant; that is, it is the value for the equivalent ratio for general iterative maps of the form $u_{t+1} = f(u_t)$, where $f(u_t)$ has a maximum similar to that in Fig. 2.1, and which exhibit period doubling.

A useful, practical and quick way to show the existence of chaos has been given by Li et al. (1982). They proved that if, for some u_t and any $f(u_t)$, an *odd* integer n exists such that

$$f^n(u_t; r) < u_t < f(u_t; r)$$

then an *odd* periodic solution exists, which thus implies chaos. For example with

$$u_{t+1} = f(u_t; r) = u_t \exp[r(1 - u_t)]$$

if $r = 3.0$ and $u_0 = 0.1$ a computation of the first few terms shows

$$u_7 = f^5(u_2) < u_2 < f(u_2) = u_3 ,$$

that is $n = 5$ in the above inequality requirement. Hence this $f(u_t; r)$ with $r = 3$ is chaotic.

2.4 Stability, Periodic Solutions and Bifurcations

All relevant ecological models involve at least one parameter, r say. From the above discussion, as this parameter varies the solutions of the general model equation

$$u_{t+1} = f(u_t; r) , \qquad (2.13)$$

will usually undergo bifurcations at specific values of r. Such bifurcations can be to periodic solutions with successively higher periods ultimately generating chaotic solutions for r greater than some finite critical r_c. From the graphical analysis such bifurcations occur when the appropriate eigenvalues λ pass through $\lambda = 1$ or $\lambda = -1$. Here we discuss some analytical results associated with these bifurcations. For algebraic simplicity we shall often omit the r in $f(u_t; r)$ (unless we want to emphasise a point) by writing $f(u_t)$ but the dependence on a parameter will always be understood. The functions f we have in mind are qualitatively similar to that illustrated in Fig. 2.1.

The equilibrium points or fixed points of (2.13) are solutions of

$$u^* = f(u^*; r) \quad \Rightarrow \quad u^*(r) . \qquad (2.14)$$

To investigate the linear stability of u^* we write, in the usual way,

$$u_t = u^* + v_t, \quad |v_t| \ll 1 . \qquad (2.15)$$

Substituting this into (2.13) and expanding for small v_t, using a Taylor expansion, we get

$$u^* + v_{t+1} = f(u^* + v_t)$$
$$= f(u^*) + v_t f'(u^*) + O(v_t^2), \quad |v_t| \ll 1 .$$

Since $u^* = f(u^*)$ the linear (in v_t) equation which determines the linear stability of u^* is then

$$v_{t+1} = v_t f'(u^*) = \lambda v_t, \quad \lambda = f'(u^*) ,$$

where λ is the eigenvalue of the first iterate (2.13) at the fixed point u^*. The solution is

$$v_t = \lambda^t v_0 \rightarrow \begin{cases} 0 \\ \pm\infty \end{cases} \quad \text{as} \quad t \rightarrow \infty \quad \text{if} \quad |\lambda| \begin{cases} < 1 \\ > 1 \end{cases}$$

Thus

$$u^* \text{ is } \begin{cases} \text{stable} \\ \text{unstable} \end{cases} \text{ if } \begin{cases} -1 < f'(u^*) < 1 \\ |f'(u^*)| > 1 \end{cases} \qquad (2.16)$$

If u^* is stable any small perturbation from this equilibrium decays to zero, monotonically if $0 < f'(u^*) < 1$, or with decreasing oscillations if $-1 < f'(u^*) < 0$. On the other hand if u^* is unstable any perturbation grows monotonically if $f'(u^*) > 1$, or by growing oscillations if $f'(u^*) < -1$. This is all as we deduced before by graphical arguments.

As an example the rescaled model (2.4) is

$$u_{t+1} = u_t \exp[r(1 - u_t)], \quad r > 0 . \qquad (2.17)$$

Here the steady states are

$$u^* = 0 \quad \text{or} \quad 1 = \exp[r(1 - u^*)] \quad \Rightarrow \quad u^* = 1 . \tag{2.18}$$

Thus the corresponding eigenvalues are

$$\lambda_{u^*=0} = f'(0) = e^r > 1 \quad \text{for} \quad r > 0 ,$$

so $u^* = 0$ is unstable (monotonically), and

$$\lambda_{u^*=1} = f'(1) = 1 - r . \tag{2.19}$$

Hence $u^* = 1$ is stable for $0 < r < 2$ with oscillatory return to equilibrium if $1 < r < 2$. It is unstable by growing oscillations for $r > 2$. Thus $r = 2$ is the first bifurcation value. On the basis of the above we expect a periodic solution to be the bifurcation from $u^* = 1$ as r passes through the bifurcation value $r = 2$. For $|1 - u_t|$ small (2.17) becomes

$$u_{t+1} \approx u_t[1 + r(1 - u_t)]$$

and so writing this in the form

$$U_{t+1} = (1 + r)U_t[1 - U_t] \quad \text{where} \quad U_t = \frac{r u_t}{1 + r} ,$$

gives the same as the logistic model (2.7) with $r + 1$ in place of r. There we saw that a stable periodic solution with period 2 appeared at the first bifurcation. With example (2.17) the next bifurcation, to a 4-periodic solution, occurs at $r = r_4 \approx 2.45$ and a 6-periodic one at $r = r_6 \approx 2.54$ with aperiodic or chaotic behaviour for $r > r_c \approx 2.57$. The successive bifurcation values of r for period doubling again become progressively closer. The sensitivity of the solutions to small variations in $r > 2$ is quite severe in this model: it is in most of them in fact, at least for the equivalent of r beyond the first few bifurcation values.

After t iterations of u_0, $u_t = f^t(u_0)$, using the notation defined in (2.9). A *trajectory* or *orbit* generated by u_0 is the set of points $\{u_0, u_1, u_2, \ldots\}$ where

$$u_{i+1} = f(u_i) = f^{i+1}(u_0), \quad i = 0, 1, 2, \ldots .$$

We say that a point is periodic of period m or *m-periodic* if

$$f^m(u_0; r) = u_0$$
$$f^i(u_0; r) \neq u_0 \quad \text{for} \quad i = 1, 2, \ldots, m - 1 \tag{2.20}$$

and that u_0, a fixed point of the mapping f^m in (2.20), is a *period-m fixed point* of the mapping f in (2.13). The points $u_0, u_1, \ldots, u_{m-1}$ form an *m-cycle*.

For the stability of a fixed point (solution) we require the eigenvalue: for the equilibrium state u^* it was simply $f'(u^*)$. We now extend this definition to an m-cycle of points $u_0, u_1, \ldots, u_{m-1}$. For convenience, introduce

$$F(u; r) = f^m(u; r), \quad G(u; r) = f^{m-1}(u; r) .$$

Then the eigenvalue λ_m of the m-cycle is defined as

$$\lambda_m = \left. \frac{\partial f^m(u; r)}{\partial u} \right]_{u=u_i} \quad i = 0 \text{ or } 1 \text{ or } 2 \text{ or } \ldots m - 1 , \qquad (2.21)$$

$$= F'(u_i; r)$$

$$= f'(G(u_i; r)) \, G'(u_i; r)$$

$$= f'(u_{i-1}; r) \, G'(u_i; r)$$

$$= f'(u_{i-1}; r) \left[\frac{\partial f^{m-1}(u_i; r)}{\partial u} \right]_{u=u_i}$$

and so

$$\lambda_m = \prod_{i=0}^{m-1} f'(u_i; r) , \qquad (2.22)$$

which shows that the form (2.21) is independent of i.

In summary then a bifurcation occurs at a parameter value r_0 if there is a qualitative change in the dynamics of the solution for $r < r_0$ and $r > r_0$. From the above discussion we now expect it to be from one periodic solution to another with a different period. Also when the sequence of even periods bifurcates to an odd-period solution the Sarkovskii (1964) theorem says that cycles of every integer period exists, which implies chaos. Bifurcations with $\lambda = -1$ are the pitchfork bifurcations while those with $\lambda = 1$ are the tangent bifurcations.

Using one of the several computer packages currently available which carry out algebraic manipulations, it is easy to calculate the eigenvalues λ for each iterate and hence generate the sequence of bifurcation values r using (2.21) or (2.22). There are systematic analytic ways of doing this which are basically extensions of the above; see for example Gumowski and Mira (1980). There are also several approximate methods such as those by May and Oster (1976) and Hoppensteadt and Hyman (1977). Since we are mentioning books here, that by Bergé, Pomeau and Vidal (1984) is a useful introductory text. You get some idea of the widespread interest in chaos from the collection of reprints, put together by Cvitanović (1984), of some of the frequently quoted papers, and the book of survey articels edited by Holden (1986). These show the diverse areas in which chaos has been studied.

2.5 Discrete Delay Models

All of the discrete models we have so far discussed are based on the assumption that each member of the species at time t contributes to the population at time $t+1$: this is implied by the general form (2.1), or (2.13) in a scaled version. This is of course the case with most insects but is not so with many other animals where, for example, there is a substantial maturation time to sexual maturity. Thus the population's dynamic model in such cases, must include a delay effect: it is, in a sense, like incorporating an age structure. If this delay, to maturity say, is T time steps, then we are led to study difference delay models of the form

$$u_{t+1} = f(u_t, u_{t-T}) . \tag{2.23}$$

In the model for baleen whales, which we discuss below, the delay T is of the order of several years.

To illustrate the problems associated with the linear stability analysis of such models and to acquire a knowledge of what to expect from delay equations we consider the following simple model, which, even so, is of practical interest:

$$u_{t+1} = u_t \exp[r(1 - u_{t-1})], \quad r > 0 . \tag{2.24}$$

This is a delay version of (2.17). The equilibrium states are again $u^* = 0$ and $u^* = 1$. The steady state $u^* = 0$ is unstable almost by inspection; a linearization about $u^* = 0$ immediately shows it.

We linearize about $u^* = 1$ by setting, in the usual way,

$$u_t = 1 + v_t, \quad |v_t| \ll 1$$

and (2.24) then gives

$$1 + v_{t+1} = (1 + v_t) \exp[-rv_{t-1}] \approx (1 + v_t)(1 - rv_{t-1})$$

and so

$$v_{t+1} - v_t + rv_{t-1} = 0 . \tag{2.25}$$

We look for solutions of this difference equation in the form

$$v_t = z^t \quad \Rightarrow \quad z^2 - z + r = 0$$

which gives two values for z, z_1 and z_2 where

$$z_1, z_2 = \frac{1}{2}[1 \pm (1 - 4r)^{1/2}], \quad r < \frac{1}{4}, \quad z_1, z_2 = \rho e^{\pm i\theta}, \quad r > \frac{1}{4} \tag{2.26}$$

with

$$\rho = r^{1/2}, \quad \theta = \tan^{-1}(4r - 1)^{1/2}, \quad r > 1/4 .$$

The solution of (2.25), for which the *characteristic equation* is the quadratic in z, is then

$$v_t = A z_1^t + B z_2^t , \qquad (2.27)$$

where A and B are arbitrary constants.

   If $0 < r < 1/4$, z_1 and z_2 are real, $0 < z_1 < 1$, $0 < z_2 < 1$ and so from (2.27), $v_t \to 0$ as $t \to \infty$ and hence $u^* = 1$ is a linearly stable equilibrium state. Furthermore the return to this equilibrium after a small perturbation is monotonic.

   If $1/4 < r$, z_1 and z_2 are complex with $z_2 = \bar{z}_1$, the complex conjugate of z_1. Also $z_1 z_2 = |z_1|^2 = \rho^2 = r$. Thus for $1/4 < r < 1$, $|z_1||z_2| < 1$. In this case the solution is

$$v_t = A z_1^t + B \bar{z}_1^t$$

and, since it is real, we must have $B = \bar{A}$ and so, with (2.26), the real solution

$$v_t = 2|A|\rho^t \cos(t\theta + \gamma), \quad \gamma = \arg A, \quad \theta = \tan^{-1}(4r - 1)^{1/2} . \qquad (2.28)$$

As $r \to 1$, $\theta \to \tan^{-1} \sqrt{3} = \pi/3$.

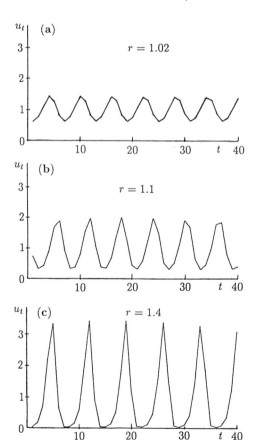

Fig. 2.11a-c. Solutions of the delay difference equation (2.24), $u_{t+1} = u_t \times \exp[r(1 - u_{t-1})]$ for three values of $r > r_c = 1$. (a) $r = 1.02$. This shows the 6-period solution which bifurcates off the steady state at $r = r_c$. (b) $r = 1.1$. Here elements of a 6-cycle still exist but these are lost in (c), where $r = 1.4$.

As r passes through the critical $r_c = 1$, $|z_1| > 1$ and so v_t grows unboundedly with $t \to \infty$ and u^* is then unstable. Since $\theta \approx \pi/3$ for $r \approx 1$ and $v_t \approx 2|A| \cos(t\pi/3 + \gamma)$, which has a period of 6, we expect the solution of (2.24), at least for r just greater than $r_c(= 1)$, to exhibit a 6-cycle periodic solution. Fig. 2.11 illustrates the computed solution for three values of $r > 1$. In Fig. 2.11 (b) there are still elements of a 6-cycle, but they are irregular. In Fig. 2.11 (c) the element of 6-periodicity is lost and the solution becomes more spike-like, often an early indication of chaos.

In the last chapter we saw how delay had a destabilising effect and it increased with increasing delay. It has a similar destabilising effect in discrete models as is clear from comparing the r-values in Fig. 2.9 and Fig. 2.11. In the former the critical $r_c = 2$ and the solution bifurcates to a 2-period solution whereas in the latter delay case the critical $r_c = 1$ and bifurcation is to a 6-period solution. Again, the longer the delay the greater the destabilising effect. This is certainly another reason why the modelling and anlaysis in the following example gave cause for concern. Higher period solutions are often characterised by large population swings and if the crash-back to low population levels from a previous very high one is sufficiently severe extinction is a distinct possibility. Section 2.7 briefly discusses a possible path to extinction.

To conclude this section we briefly describe a practical model used by the International Whaling Commission (IWC) for the baleen whale. The aim of the IWC is to manage the whale population for a sustained yield, prevent extinction and so on. The commercial and cultural pressures on the IWC are considerable. To carry out its charter requirements in a realistic way it must understand the dynamics of whale population growth and its ecology.

A model for the now protected baleen whale which the IWC used is based on the discrete-delay model for the population N_t of sexually mature whales at time t,

$$N_{t+1} = (1 - \mu)N_t + R(N_{t-T}) . \tag{2.29}$$

Here $(1-\mu)N_t$, with $0 < \mu < 1$, is the surviving fraction of whales that contribute to the population a year later and $R(N_{t-T})$ is the number which augments the adult population from births T years earlier. The delay T is the time to sexual maturity and is of the order of 5–10 years. This model assumes that the sex ratio is 1 and the mortality is the same for each sex. The crux of the model is the form of the recruitment term $R(N_{t-T})$ which in the IWC model (see for example, IWC 1979) is

$$R(N) = \frac{1}{2}(1 - \mu)^T N \left\{ P + Q \left[1 - \left(\frac{N}{K} \right)^z \right] \right\} . \tag{2.30}$$

Here K is the unharvested equilibrium density, P is the per capita fecundity of females at $N = K$ with Q the maximum increase in the fecundity possible as the population density falls to low levels, and z is a measure of the severity with which this density is registered. Finally $1 - \mu$ is the probability that a new born whale survives each year and so $(1-\mu)^T$ is the fraction that survive to adulthood after the required T years: the $1/2$ is because half the whales are females and so the fecundity of the females has to be multiplied by $N/2$. This specific model has been studied in detail by Clark (1976b). Further models in whaling, and fisheries management generally, are reviewed by May (1980).

The parameters μ, T and P in (2.29) and (2.30) are not independent. The equilibrium state is

$$N^* = N_{t+1} = N_t = N_{t-T} = K \quad \Rightarrow \quad \mu = \frac{1}{2}(1-\mu)^T P = h \qquad (2.31)$$

which, as well as defining h, relates the fecundity P to the mortality μ and the delay T. Independent measurement of these gives a rough consistency check. If we now rescale the model with $u_t = N_t/K$, (2.29), with (2.30), becomes

$$u_{t+1} = (1-\mu)u_t + hu_{t-T}[1 + q(1 - u_{t-T}^z)] , \qquad (2.32)$$

where h is defined in (2.31) and $q = Q/P$. Linearizing about the steady state $u^* = 1$ by writing $u_t = 1 + v_t$ the equation for the perturbation is

$$v_{t+1} = (1-\mu)v_t + h(1 - qz)v_{t-T} , \qquad (2.33)$$

On setting $v_t \propto s^t$

$$s^{T+1} - (1-\mu)s^T + h(qz - 1) = 0 . \qquad (2.34)$$

which is the characteristic equation. The steady state becomes unstable when $|s| > 1$. Here there are 4 parameters μ, T, h and qz and the analysis centres around a study of the roots of (2.34): see the paper by Clark (1976b). Although they are complicated, we can determine the conditions on the parameters such that $|s| < 1$ by using the Jury conditions (see Appendix 2). The Jury conditions are inequalities that the coefficients of a real polynomial must satisfy for the roots to have modulus less than 1. For polynomials of order greater than about 4 the conditions are prohibitively unwieldy. When $|s| > 1$, as is now to be expected, solutions of (2.29) exhibit bifurcations to periodic solutions with progressively higher periods ultimately leading to chaos: the response parameter z is critical.

2.6 Fishery Management Model

Discrete models have been used in fishery management for some considerable time. They have proven to be very useful in evaluating various harvesting strategies with a view to optimizing the economic yield and to maintaining it. For example, relevant books on management strategies are those by Clark (1976a), Goh (1982) and the series of papers in that edited by Cohen (1987). The following model is applicable in principle to any renewable resource which is harvested;

the detailed analysis applies to any population whose dynamics can be described by a discrete model.

Suppose that the population density is governed by $N_{t+1} = f(N_t)$ in the absence of harvesting. If we let h_t be the harvest taken from the population at time t, which generates the next population at $t+1$, then a model for the population dynamics is

$$N_{t+1} = f(N_t) - h_t . \qquad (2.35)$$

The two questions we shall address are: (i) What is the maximum sustained biological yield? (Compare with Section 1.5 in Chapter 1.) (ii) What is the maximum economic yield?

In equilibrium $N_t = N^* = N_{t+1}$, $h_t = h^*$ where, from (2.35)

$$h^* = f(N^*) - N^* . \qquad (2.36)$$

The maximum sustained steady state yield Y_M is when $N^* = N_M$ where

$$\frac{\partial h^*}{\partial N^*} = 0 \quad \Rightarrow \quad f'(N^*) = 1 \quad \text{and} \quad Y_M = f(N_M) - N_M . \qquad (2.37)$$

The only situation of interest of course is when $Y_M \geq 0$.

A management strategy could be simply to maintain the population so as to get the maximum yield Y_M. Since it is hard to know what the actual fish population is, this can be difficult to accomplish. What is known is the actual yield and how much effort has gone into getting it. So it is better to formulate the optimization problem in terms of yield and effort.

Let us suppose that a unit effort to catch fish results in a harvest cN from a population N. The constant c is the 'catchability' parameter which is independent of the population density N. Then the effort to reduce N by 1 unit is $1/(cN)$ and $f(N)$ by 1 unit is $1/(cf(N))$. Thus the effort E_M to provide for a yield

$$Y_M = f(N_M) - N_M \quad \text{is} \quad E_M = \sum_{N_i = N_M}^{f(N_M)} (cN_i)^{-1} .$$

Now if cN is large compared with 1 unit we can approximate the summation in the last equation by an integral and so

$$E_M \approx c^{-1} \int_{N_M}^{f(N_M)} N^{-1} \, dN = c^{-1} \ln \left\{ \frac{f(N_M)}{N_M} \right\} . \qquad (2.38)$$

The two equations (2.37) and (2.38) give the relation between E_M and Y_M parametrically in N_M.

As an example suppose the unharvested dynamics is governed by $N_{t+1} = f(N_t) = bN_t/(a + N_t)$ with $0 < a < b$ then

$$N_M : \quad 1 = f'(N_M) = \frac{ab}{(a + N_M)^2} \quad \Rightarrow \quad N_M = a^{1/2}(b^{1/2} - a^{1/2}) .$$

Substituting this into (2.37) and (2.38) gives

$$Y_M = \frac{bN_M}{a + N_M} - N_M, \quad E_M = c^{-1} \ln \left\{ \frac{b}{a + N_M} \right\} . \tag{2.39}$$

In this example we can get an explicit relation between Y_M and E_M, on eliminating N_M, as

$$Y_M = [b \exp(-cE_M) - a][\exp(cE_M) - 1] . \tag{2.40}$$

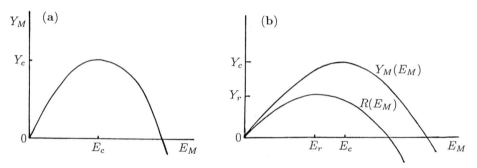

Fig. 2.12a,b. (a) The yield-effort relation (schematic) for the maximum sustained yield with the model dynamics $N_{t+1} = bN_t/(a + N_t)$, $0 < a < b$. (b) The maximum revenue R as a function of the effort E as compared with the $Y_M - E_M$ curve.

Fig. 2.12 (a) illustrates the $Y_M - E_M$ relation. Using this, a crucial aspect of a management strategy is to note that if an increase in effort reduces the yield, then the maximum sustained yield is exceeded, and the effort has to be reduced so that the population can recover. The effort can subsequently be retuned to try and achieve Y_c with E_c in Fig. 2.12 (a), both of which can be calculated from (2.40).

This analysis is for the maximum sustained biological yield. The maximum economic yield must include the price for the harvest and the cost of the effort. As a first model we can incorporate these in the expression for the economic return $R = pY_M - kE_M$ where p is the price per unit yield and k is the cost per unit effort. Using (2.39) for $Y_M(N_M)$ and $E_M(N_M)$ we thus have $R(N_M)$ which we must now maximise. We thus get a curve for the maximum revenue R as a function of the effort E: it is illustrated in Fig. 2.12 (b).

Such 'model' results must not be taken too seriously unless backed up by experimental observation. They can however give some important qualitative pointers. Our analysis here has been based on the fact that the harvested population has a steady state. Fish, in particular, have a high per capita growth rate which, in the detailed models we have analysed, is related to the parameter r. We would expect, therefore, that the fish population would exhibit periodic fluctuations and this is known to be the case. It is possible that the growth rate is sufficiently high that the behaviour may, in some cases, be in the chaotic regime. Since harvesting is, in a sense, an effective lowering of the reproduction rate it is feasible that it could have a stabilizing effect, for example from the chaotic to the periodic or even to a steady state situation.

2.7 Ecological Implications and Caveats

A major reason for modelling the dynamics of a population is to understand the principle controlling features and to be able to predict the likely pattern of development consequent upon a change of environmental parameters. In making the model we may have, to varying degrees, a biological knowledge of the species and observational data with which to compare the results of the analysis of the model. It may be helpful to summarize what we can learn about a population's dynamics from the type of models we have considered and to point out a few of their difficulties and limitations.

When a plausible model for a population's growth dynamics has been arrived at, the global dynamics can be determined. Using graphical methods the changes in the solutions as a major environment parameter varies can also be seen. From Fig. 2.3 for example, we see that if we start with a low population it simply grows for a while, then it can appear to oscillate quasi-regularly and then settle down to a constant state, or exhibit periodic behaviour or just oscillate in a seemingly random way with large populations at one stage and crashing to very low densities in the following time step. Whatever the model, as long as it has a general form such as in Fig. 2.5 the population density is always bounded.

This seemingly random dynamics poses serious problems from a modelling point of view. Are the data obtained which exhibit this kind of behaviour generated by a deterministic model or by a stochastic situation? It is thus a problem to decide which is appropriate and it may not actually be one we can resolve in a specific situation. What modelling can do however is to point to how sensitive the population dynamics can be to changes in environmental parameters, the estimation of which is often difficult and usually important.

The type of dynamics exhibited with $f(N_t)$ such as in Fig. 2.5, shows that the population is always bounded after a long time by some maximum $N_{\max}$ and minimum $N_{\min}$: the first few iterations can lie below $N_{\min}$ if N_0 is sufficiently small. With Fig. 2.5 in mind the maximum $N_{\max}$ is given by the first iteration of the value where $N_{t+1} = f(N_t)$ has a maximum, N_m say. That is

$$\frac{df}{dN_t} = 0 \quad \Rightarrow \quad N_m, \quad N_{\max} = f(N_m) \ .$$

The minimum $N_{\min}$ is then the first iterative of $N_{\max}$ namely

$$N_{\min} = f(N_{\max}) = f(f(N_m)) = f^2(N_m) \ . \tag{2.41}$$

These ultimately limiting population sizes are easy to work out for a given model. For example with

$$N_{t+1} = f(N_t) = N_t \exp\left[r\left(1 - \frac{N_t}{K}\right)\right], \quad f'(N_t) = 0 \quad \Rightarrow \quad N_m = \frac{K}{r}$$

$$N_{\max} = f(N_m) = \frac{K}{r}e^{r-1}, \tag{2.42}$$

$$N_{\min} = f(f(N_m)) = \frac{K}{r}\exp[2r - 1 - e^{r-1}].$$

With a steeply decreasing behaviour of the dynamics curve $N_{t+1} = f(N_t)$ for $N_t > N_m$ the possibility of the dramatic drop in the population to low values close to $N_{\min}$ brings up the question of *extinction* of a species. If the population drops to a value $N_t < 1$ the species is clearly extinct. In fact extinction is almost inevitable if N_t drops to low values. At this stage a stochastic model is required. However an estimate of when the population drops to 1 or less, and hence extinction, can be obtained from the evaluation of $N_{\min}$ for a given model. The condition is, using (2.41),

$$N_{\min} = f^2(N_m) \le 1, \quad \left.\frac{df}{dN}\right]_{N=N_m} = 0. \tag{2.43}$$

With the example in (2.42) this condition is

$$\frac{K}{r}\exp[2r - 1 - e^{r-1}] \le 1.$$

So if $r = 3.5$ say, and if $K < 1600$ approximately, the population will eventually become extinct.

An important phenomenon is indicated by the analysis of this model (2.42); the larger the reproduction parameter r the smaller is $N_{\min}$ and the more likelihood of a population crash which will make the species extinct. Note also that it will usually be the case that the population size immediately before the catastrophic drop is large. With the above example if $r = 3.5$ it is almost 3500, from (2.42). An interesting and potentially practical application of the concept of extinction is that of introducing sterile species of a pest to try and control the numbers: see Exercise 6 below.

An important group of models not specifically discussed up to now but which come into the general class (2.1) are those which exhibit the *Allee effect*. Biological populations which show this effect decrease in size if the population falls below a certain threshold level N_c say. A typical density dependent population model which illustrates this is shown in Fig. 2.13. If we start with a population, N_0 say, such that $f^2(N_0) < N_c$ then $N_t \to 0$. Such models usually arise as a result of predation. The continuous time model for the budworm equation (1.6) in Chapter 1, has such a behaviour. The region $N_t < N_c$ is sometimes called

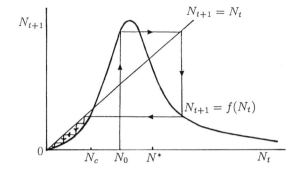

Fig. 2.13. A population model which exhibits the Allee effect whereby if the population $N_t < N_c$ at any time t then $N_t \to 0$, that is extinction.

the *predation pit*. Here $N_t = 0, N_c, N^*$ are all steady states with $N_t = 0$ stable, N_c unstable and N^* stable or unstable depending on $f'(N^*)$ in the usual way. With this type of dynamics extinction is inevitable if $N_t < N_c$, irrespective of how large N_c may be. Models which show an Allee effect display an even richer spectrum of behaviour than those we considered above, namely all of the exotic oscillatory behaviour plus the possibility of extinction if any iterate $f^m(N_t) < N_c$ for some m.

The implications from nonlinear discrete models such as we have considered in this chapter, rely crucially on the biological parameters obtained from an analysis of observational data. Southwood (1976) discusses, among other things, these population parameters and presents hard facts about several species. Hassell et al. (1976) have analysed a large number of species life data and fitted them to the model $N_{t+1} = f(N_t) = rN_t/(1 + aN_t)^b$ with r, a and b positive parameters. With $b > 1$ this $f(N_t)$ has one hump like those in Fig. 2.1. For example the Colorado beetle is well within the stable periodic regime while Nicholson's (1954) blowflies would be well in the chaotic regime.

Finally it should be emphasized here that the richness of solution behaviour is a result of the nonlinearity of these models. It is also interesting that many of the qualitative features can be found by remarkably elementary methods even though they present some sophisticated and challenging mathematical problems.

Exercises

1. All the following discrete time population models are of the form $N_{t+1} = f(N_t)$ and have been taken from the ecological literature and all have been used in modelling real situations. Determine the non-negative steady states, discuss their linear stability and find the first bifurcation values of the parameters, which are all taken to be positive.

$$\text{(i)} \quad N_{t+1} = N_t \left[1 + r \left(1 - \frac{N_t}{K} \right) \right]$$

$$\text{(ii)} \quad N_{t+1} = rN_t^{1-b}, \quad \text{if} \quad N_t > K,$$
$$= rN_t, \quad \text{if} \quad N_t < K,$$

$$\text{(iii)} \quad N_{t+1} = \frac{rN_t}{(1 + aN_t)^b},$$

$$\text{(iv)} \quad N_{t+1} = \frac{rN_t}{1 + \left(\frac{N_t}{K}\right)^b}.$$

2. Construct cobweb maps for:

$$\text{(i)} \quad N_{t+1} = \frac{(1 + r)N_t}{1 + rN_t},$$

$$\text{(ii)} \quad N_{t+1} = \frac{rN_t}{(1 + aN_t)^b}, \quad a > 0, \quad b > 0, \quad r > 0$$

and discuss the global qualitative behaviour of the solutions. Determine, where possible, the maximum and minimum N_t, and the minimum for (ii) when $b \ll 1$.

3. Verify that an exact solution exists for the logistic difference equation

$$u_{t+1} = ru_t(1 - u_t), \quad r > 0$$

in the form $u_t = A\sin^2 \alpha^t$ by determining values for r, A and α. Is the solution (i) periodic? (ii) oscillatory? Describe it! If $r > 4$ discuss possible solution implications.

4. The population dynamics of a species is governed by the discrete model

$$N_{t+1} = f(N_t) = N_t \exp\left[r\left(1 - \frac{N_t}{K}\right)\right],$$

where r and K are positive constants. Determine the steady states and their corresponding eigenvalues. Show that a pitchfork bifurcation occurs at $r = 2$. Briefly describe qualitatively the dynamic bahaviour of the population for $r = 2 + \varepsilon$, where $0 < \varepsilon \ll 1$. In the case $r > 1$ sketch $N_{t+1} = f(N_t)$ and show graphically or otherwise that, for t large, the maximum population is given by $N_M = f(K/r)$ and the minimum possible population by $N_m = f(f(K/r))$. Since a species becomes extinct if $N_t \leq 1$ for any $t > 1$ show that irrespective of the size of $r > 1$ the species could become extinct if the carrying capacity $K < r\exp[1 + e^{r-1} - 2r]$.

5. The population of a certain species subjected to a specific kind of predation is modelled by the difference equation

$$u_{t+1} = a\frac{u_t^2}{b^2 + u_t^2}, \quad a > 0.$$

Determine the equilibria and show that if $a^2 > 4b^2$ it is possible for the population to be driven to extinction if it becomes less that a critical size which you should find.

6. It has been suggested that a means of controlling insect numbers is to introduce and maintain a number of sterile insects in the population. One such model for the resulting population dynamics is

$$N_{t+1} = \frac{RN_t^2}{(R-1)\frac{N_t^2}{M} + N_t + S} \ ,$$

where $R > 1$ and $M > 0$ are constant parameters, and S is the constant sterile insect population.

Determine the steady states and discuss their linear stability, noting whether any type of bifurcation is possible. Find the critical value S_c of the sterile population in terms of R and M so that if $S > S_c$ the insect population is eradicated. Construct a cobweb map and draw a graph of S against the steady-state population density, and hence determine the possible solution behaviour if $0 < S < S_c$.

7. A discrete model for a population N_t consists of

$$N_{t+1} = \frac{rN_t}{1 + bN_t^2} = f(N_t) \ ,$$

where t is the discrete time and r and b are positive parameters. What do r and b represent in this model? Show, with the help of a cobweb, that after a long time the population N_t is bounded by

$$N_{\min} = \frac{2r^2}{(4+r^2)\sqrt{b}} \le N_t \le \frac{r}{2\sqrt{b}} \ .$$

Prove that, for any r, the population will become extinct if $b > 4$.

Determine the steady states and their eigenvalues and hence show that $r = 1$ is a bifurcation value. Show that, for any finite r, oscillatory solutions for N_t are not possible.

Consider a delay version of the model given by

$$N_{t+1} = \frac{rN_t}{1 + bN_{t-1}^2} = f(N_t), \quad r > 1 \ .$$

Investigate the linear stability about the positive steady state N^* by setting $N_t = N^* + n_t$. Show that n_t satisfies

$$n_{t+1} - n_t + 2(r-1)r^{-1}n_{t-1} = 0 \ .$$

Hence show that $r = 2$ is a bifurcation value and that as $r \to 2$ the steady state bifurcates to a periodic solution of period 6.

*8. A basic delay model used by the International Whaling Commission (IWC) for monitoring whale populations is

$$u_{t+1} = su_t + R(u_{t-T}), \quad 0 < s < 1 ,$$

where $T \geq 1$ is an integer.

(i) If u^* is a positive equilibrium show that a sufficient condition for linear stability is $|R'(u^*)| < 1 - s$. [Hint: Use Rouché's theorem on the resulting characteristic polynomial for small perturbations about u^*.]

(ii) If $R(u) = (1 - s)u[1 + q(1 - u)]$, $q > 0$ and the delay $T = 1$ show that the equilibrium state is stable for all $0 < q < 2$. [With this model, T is the time from birth to sexual maturity, s is a survival parameter and $R(u_{t-T})$ the recruitment to the adult population from those born T years ago.]

9. Consider the effect of regularly harvesting the population of a species for which the model equation is

$$u_{t+1} = \frac{bu_t^2}{1 + u_t^2} - Eu_t = f(u_t; E), \quad b > 2, \quad E > 0$$

where E is a measure of the effort expended in obtaining the harvest Eu_t. (This model with $E = 0$ is a special case of that in Exercise 5.) Determine the steady states and hence show that if the effort $E > E_m = (b - 2)/2$ no harvest is obtained. If $E < E_m$ show, by cobwebbing $u_{t+1} = f(u_t; E)$ or otherwise, that the model is realistic only if the population u_t always lies between two positive values which you should determine analytically.

With $E < E_m$ evaluate the eigenvalue of the largest positive steady state. Demonstrate that a tangent bifurcation exists as $E \rightarrow E_m$.

3. Continuous Models
for Interacting Populations

When species interact the population dynamics of each species is affected. In general there is a whole web of interacting species, called a *trophic web*, which makes for structurally complex communities. We consider here systems involving two or more species, concentrating particularly on 2-species systems. There are three main types of interaction. (i) If the growth rate of one population is decreased and the other increased the populations are in a *predator-prey* situation. (ii) If the growth rate of each population is decreased then it is *competition*. (iii) If each population's growth rate is enhanced then it is called *mutualism* or *symbiosis*.

All of the mathematical techniques and analytical methods in this chapter are directly applicable to Chapter 5 on reaction kinetics, where similar equations arise.

3.1 Predator-Prey Models: Lotka-Volterra Systems

Volterra (1926) first proposed a simple model for the predation of one species by another to explain the oscillatory levels of certain fish catches in the Adriatic. If $N(t)$ is the prey population and $P(t)$ that of the predator at time t then Volterra's model is

$$\frac{dN}{dt} = N(a - bP) , \tag{3.1}$$

$$\frac{dP}{dt} = P(cN - d) , \tag{3.2}$$

where a, b, c and d are positive constants.

The assumptions in the model are: (i) The prey in the absence of any predation grows unboundedly in a Malthusian way; this is the aN term in (3.1). (ii) The effect of the predation is to reduce the prey's per capita growth rate by a term proportional to the prey and predator populations; this is the $-bNP$ term. (iii) In the absence of any prey for sustenance the predator's death rate results in exponential decay, that is the $-dP$ term in (3.2). (iv) The prey's contribution to the predators' growth rate is cNP; that is, it is proportional to the available prey as well as to the size of the predator population. The NP terms can be thought of as representing the conversion of energy from one source to another: bNP is taken from the prey and cNP accrues to the predators. We shall see that this model has serious drawbacks. Nevertheless it has been of considerable value in posing highly relevant questions, is a jumping-off place for more realistic models and is the main motivation for studying it here.

The model (3.1) and (3.2) is known as the *Lotka-Volterra model* since the same equations were also derived by Lotka (1920: see also 1925) from a hypothetical chemical reaction which he said could exhibit periodic behaviour in the chemical concentrations. With this motivation the dependent variables represent chemical concentrations: we touch on this again in Chapter 5.

As a first step in analysing the Lotka-Volterra model we non-dimensionalize the system by writing

$$u(\tau) = \frac{cN(t)}{d}, \quad v(\tau) = \frac{bP(t)}{a}, \quad \tau = at, \quad \alpha = d/a, \tag{3.3}$$

and it becomes

$$\frac{du}{d\tau} = u(1 - v), \quad \frac{dv}{d\tau} = \alpha v(u - 1). \tag{3.4}$$

In the u,v phase plane (a brief elementary summary of phase plane methods is given in Appendix 1) these give

$$\frac{dv}{du} = \alpha \frac{v(u - 1)}{u(1 - v)}, \tag{3.5}$$

which has singular points at $u = v = 0$ and $u = v = 1$. We can integrate (3.5) exactly to get the phase trajectories

$$\alpha u + v - \ln u^\alpha v = H, \tag{3.6}$$

where $H > H_{\min}$ is a constant: $H_{\min} = 1 + \alpha$ is the minimum of H over all (u, v) and it occurs at $u = v = 1$. For a given $H > 1 + \alpha$, the trajectories (3.6) in the phase plane are closed as illustrated in Fig. 3.1.

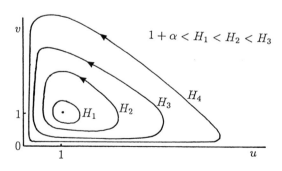

$$1 + \alpha < H_1 < H_2 < H_3$$

Fig. 3.1. Closed phase plane trajectories (schematic), from (3.6) with various H, for the Lotka-Volterra system (3.4). The arrows denote the direction of change with increasing time τ.

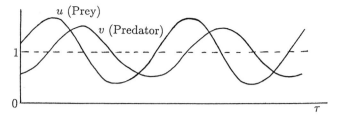

Fig. 3.2. Schematic periodic solutions for the prey $u(\tau)$ and predator $v(\tau)$ for the Lotka-Volterra system (3.4).

A closed trajectory in the u, v plane implies periodic solutions in τ for u and v in (3.4). The initial conditions, $u(0)$ and $v(0)$, determine the constant H in (3.6) and hence the phase trajectory in Fig. 3.1. Typical periodic solutions $u(\tau)$ and $v(\tau)$ are illustrated in Fig. 3.2. From (3.4) we can see immediately that u has a turning point when $v = 1$ and v has one when $u = 1$.

A major inadequacy of the Lotka-Volterra model is clear from Fig. 3.1 – the solutions are not structurally stable. Suppose, for example, $u(0)$ and $v(0)$ are such that u and v for $\tau > 0$ are on the trajectory H_4 which passes close to the u and v axes. Then any small perturbation will move the solution onto another trajectory which does not lie *everywhere* close to the original one H_4. Thus a small perturbation can have a very marked effect, at the very least on the amplitude of the oscillation. This is a problem with any system which has a first integral, like (3.6), which is a closed trajectory in the phase plane. They are called *conservative systems*; here (3.6) is the associated 'conservation law'. They are usually of little use as models for real interacting populations (see one interesting and amusing attempt to do so below). However the method of analysis of the steady states is typical.

Returning to the form (3.4), a linearization about the singular points determines the type of singularity and the stability of the steady states. A similar linear stability analysis has to be carried out on equivalent systems with any number of equations. We first consider the steady state $(u, v) = (0, 0)$. Let x and y be small perturbations about $(0, 0)$. If we keep only linear terms, (3.4) becomes

$$\begin{pmatrix} \dfrac{dx}{d\tau} \\ \dfrac{dy}{d\tau} \end{pmatrix} \approx \begin{pmatrix} 1 & 0 \\ 0 & -\alpha \end{pmatrix} \begin{pmatrix} x \\ y \end{pmatrix} = A \begin{pmatrix} x \\ y \end{pmatrix}. \tag{3.7}$$

The solution is of the form

$$\begin{pmatrix} x(\tau) \\ y(\tau) \end{pmatrix} = B e^{\lambda \tau}$$

where B is an arbitrary constant column vector and the eigenvalues λ are given by the characteristic polynomial of the matrix A and thus are solutions of

$$|A - \lambda I| = \begin{vmatrix} 1 - \lambda & 0 \\ 0 & -\alpha - \lambda \end{vmatrix} = 0 \quad \Rightarrow \quad \lambda_1 = 1, \quad \lambda_2 = -\alpha .$$

Since at least one eigenvalue, $\lambda_1 > 0$, $x(\tau)$ and $y(\tau)$ grow exponentially and so $u = 0 = v$ is linearly unstable. Since $\lambda_1 > 0$ and $\lambda_2 < 0$ this is a *saddle point* singularity (see Appendix 1).

Linearizing about the steady state $u = v = 1$ by setting $u = 1 + x$, $v = 1 + y$ with $|x|$ and $|y|$ small, (3.4) becomes

$$\begin{pmatrix} \dfrac{dx}{d\tau} \\ \dfrac{dy}{d\tau} \end{pmatrix} = A \begin{pmatrix} x \\ y \end{pmatrix}, \quad A = \begin{pmatrix} 0 & -1 \\ \alpha & 0 \end{pmatrix} \tag{3.8}$$

with eigenvalues λ given by

$$\begin{vmatrix} -\lambda & -1 \\ \alpha & -\lambda \end{vmatrix} = 0 \quad \Rightarrow \quad \lambda_1, \lambda_2 = \pm i \sqrt{\alpha} \tag{3.9}$$

Thus $u = v = 1$ is a *centre* singularity since the eigenvalues are purely imaginary. Since $\operatorname{Re} \lambda = 0$ the steady state is *neutrally stable*. The solution of (3.8) is of the form

$$\begin{pmatrix} x(\tau) \\ y(\tau) \end{pmatrix} = \boldsymbol{l} e^{i \sqrt{\alpha} \tau} + \boldsymbol{m} e^{-i \sqrt{\alpha} \tau}$$

where $\boldsymbol{l}$ and $\boldsymbol{m}$ are arbitrary column vectors. So, the solutions in the neighbourhood of the singular point $u = v = 1$ are periodic in τ with period $2\pi/\sqrt{\alpha}$. In dimensional terms from (3.3) this period is $T = 2\pi(a/d)^{1/2}$, that is the period is proportional to the square root of the ratio of the linear growth rate (a) of the prey to the death rate (d) of the predators. Even though we are only dealing with small perturbations about the steady state $u = v = 1$ we see how the period depends on the intrinsic growth and death rates. For example an increase in the growth rate of the prey will increase the period: a decrease in the predator death rate does the same thing. Is this what you would expect intuitively?

In this ecological context the matrix A in the linear equations (3.7) and (3.8) is called the *community matrix*, and its eigenvalues λ determine the stability of the steady states. If $\operatorname{Re} \lambda > 0$ then the steady state is unstable while if both $\operatorname{Re} \lambda < 0$ it is stable. The critical case $\operatorname{Re} \lambda = 0$ is termed *neutral* stability.

There have been many attempts to apply the Lotka-Volterra model to real-world oscillatory phenomena. In view of the system's structural instability, they must essentially all fail to be of quantitative practical use. As we mentioned,

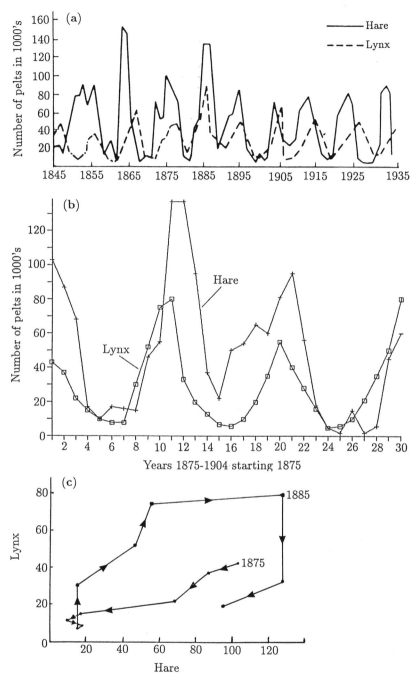

Fig. 3.3a-c. (a) Fluctuations in the number of pelts sold by the Hudson Bay Company (Redrawn from Odum 1953). (b) Detail of the 30 year period starting in 1875 based on the data from Elton and Nichoslon (1942). (c) Phase plane plot of the data represented in (b) (after Gilpin 1973).

however, they can be important as vehicles for suggesting relevant questions that should be asked. One particularly interesting example was the attempt to apply the model to the extensive data on the Canadian lynx-snowshoe hare interaction in the fur catch records of the Hudson Bay Company from about 1845 until the 1930's. We assume that the numbers reflect a fixed proportion of the total population of these animals. Although this assumption is of questionable accuracy, as indicated by what follows, the data nevertheless represents one of the very few long-term records available. Fig. 3.3 reproduces this data. Williamson's (1972) book is a good source of population data which exhibit periodic or quasi-periodic behaviour.

Fig. 3.3 shows reasonable periodic fluctuations and Fig. 3.3 (c) a more or less closed curve in the phase plane as we now expect from a time-periodic behaviour in the variables. Leigh (1968) used the standard Lotka-Volterra model to try and explain the data. Gilpin (1973) did the same with a modified Lotka-Volterra system. Let us examine the results given in Fig. 3.3 a little more carefully. First note that the *direction* of the time arrows in Fig. 3.3 (c) is clockwise in contrast to that in Fig. 3.1. This is reflected in the time curves in Fig. 3.3 (a), (b) where the lynx oscillation, the predators, precedes the hare's. The opposite is the case in the predator-prey situation illustrated in Fig. 3.2. Fig. 3.3 implies that the hares are eating the lynx! This poses a severe interpretation problem! Gilpin (1973) suggested that perhaps the hares could kill the lynx if they carried a disease which they passed on to the lynx. He incorporated an epidemic effect into his model and the numerical results then looked like those in Fig. 3.3 (c): this seemed to provide the explanation for the hare "eating" the lynx. A good try, but no such disease is known. Gilpin (1973) also offered what is perhaps the right explanation, namely that the fur trappers are the "disease". In years of low population densities they probably did something else and only felt it worthwhile to return to the trap lines when the hares were again sufficiently numerous. Since lynx was more profitable to trap than hare they would probably have devoted more time on the lynx than the hare. This would result in the phenomenon illustrated by Fig. 3.3 (b), (c). More recently Schaffer (1984) has suggested that the lynx-hare data could be evidence of a strange attractor (that is, it exhibits chaotic behaviour) in nature. The moral of the story is that it is not enough simply to produce a model which exhibits oscillations but rather to provide a proper explanation of the phenomenon which can stand up to ecological and biological scrutiny.

3.2 Complexity and Stability

To get some indication of the effect of complexity on stability we consider briefly the generalized Lotka-Volterra predator-prey system where there are k prey species and k predators, which prey on all the prey species but with different severity. Then in place of (3.1) and (3.2) we have

$$\frac{dN_i}{dt} = N_i \left[a_i - \sum_{j=1}^{k} b_{ij} P_j \right]$$

$$\frac{dP_i}{dt} = P_i \left[\sum_{j=1}^{k} c_{ij} N_j - d_i \right]$$

$$i = 1, \ldots, k \qquad (3.10)$$

where all of the a_i, b_{ij}, c_{ij} and d_i are positive constants. The trivial steady state is $N_i = P_i = 0$ for all i, and the community matrix is the diagonal matrix

$$A = \left(\begin{array}{ccc|ccc} a_1 & & 0 & & & \\ & \ddots & & & 0 & \\ 0 & & a_k & & & \\ \hline & & & -d_1 & & 0 \\ & 0 & & & \ddots & \\ & & & 0 & & -d_k \end{array} \right)$$

The $2k$ eigenvalues are thus

$$\lambda_i = a_i > 0, \quad \lambda_{k+i} = -d_i < 0, \quad i = 1, \ldots, k \ ,$$

so this steady state is unstable since all $\lambda_i > 0$, $i = 1, \ldots, k$.

The non-trivial steady state is the column vector solution $\boldsymbol{N^*}, \boldsymbol{P^*}$ where

$$\sum_{j=1}^{k} b_{ij} P_j^* = a_i, \quad \sum_{j=1}^{k} c_{ij} N_j^* = d_i, \quad i = 1, \ldots, k$$

or, in vector notation, with $\boldsymbol{N^*}$, $\boldsymbol{P^*}$, $\boldsymbol{a}$ and $\boldsymbol{d}$ column vectors,

$$B\boldsymbol{P^*} = \boldsymbol{a}, \quad C\boldsymbol{N^*} = \boldsymbol{d} \qquad (3.11)$$

where B and C are the $k \times k$ matrices $[b_{ij}]$ and $[c_{ij}]$ respectively.

Equations (3.10) can be written as

$$\frac{d\boldsymbol{N}}{dt} = \boldsymbol{N}^T \cdot [\boldsymbol{a} - B\boldsymbol{P}], \quad \frac{d\boldsymbol{P}}{dt} = \boldsymbol{P}^T \cdot [C\boldsymbol{N} - \boldsymbol{d}]$$

where the superscript T denotes the transpose. So, on linearizing about $(\boldsymbol{N^*}, \boldsymbol{P^*})$ in (3.11) by setting

$$\boldsymbol{N} = \boldsymbol{N^*} + \boldsymbol{u}, \quad \boldsymbol{P} = \boldsymbol{P^*} + \boldsymbol{v} \ ,$$

where $|\boldsymbol{u}|$, $|\boldsymbol{v}|$ are small compared with $|\boldsymbol{N^*}|$ and $|\boldsymbol{P^*}|$, we get

$$\frac{d\boldsymbol{u}}{dt} \approx -\boldsymbol{N}^{*T} \cdot B\boldsymbol{v}, \quad \frac{d\boldsymbol{v}}{dt} \approx \boldsymbol{P}^{*T} \cdot C\boldsymbol{u} \ .$$

Then

$$
\begin{pmatrix} \dfrac{du}{dt} \\[2mm] \dfrac{dv}{dt} \end{pmatrix} \approx A \begin{pmatrix} u \\ v \end{pmatrix}, \quad A = \left(\begin{array}{c|c} 0 & -N^{*T} \cdot B \\ \hline P^{*T} \cdot C & 0 \end{array} \right) \tag{3.12}
$$

where here the community matrix A is a $2k \times 2k$ block matrix with null diagonal blocks. Since the eigenvalues λ_i, $i = 1, \ldots, 2k$ are solutions of $|A - \lambda I| = 0$ the sum of the roots λ_i satisfy

$$
\sum_{i=1}^{2k} \lambda_i = \operatorname{tr} A = 0 , \tag{3.13}
$$

where $\operatorname{tr} A$ is the trace of A. Since the elements of A are real, the eigenvalues, if complex, occur as complex conjugates. Thus from (3.13) there are two cases: all the eigenvalues are purely imaginary or they are not. If all $\operatorname{Re} \lambda_i = 0$ then the steady state (N^*, P^*) is neutrally stable as in the 2-species case. However if there are λ_i such that $\operatorname{Re} \lambda_i \neq 0$ then, since they occur as complex conjugates, (3.13) implies that at least one exists with $\operatorname{Re} \lambda > 0$ and hence (N^*, P^*) is unstable.

We see from this analysis that complexity in the population interaction web introduces the possibility of instability. If a model by chance resulted in only imaginary eigenvalues (and hence perturbations from the steady state are periodic in time) only a small change in one of the parameters in the community matrix would result in at least one eigenvalue with $\operatorname{Re} \lambda \neq 0$ and hence an unstable steady state. This of course only holds for community matrices such as in (3.12). Even so, we get indications of the fairly general and important result that *complexity* usually *results in instability rather than stability.*

3.3 Realistic Predator-Prey Models

The Lotka-Volterra model, unrealistic though it is, does show that simple predator-prey interactions can result in oscillatory behaviour of the populations. Reasoning heuristically this is not unexpected since if a prey population increases, it encourages growth of its predator. More predators however consume more prey the population of which starts to decline. With less food around the predator population declines and when it is low enough, this allows the prey population to increase and the whole cycle starts over again. Depending on the detailed system such oscillations can grow or decay or go into a stable *limit cycle* oscillation or even exhibit chaotic behaviour, although in the latter case there must be at least three interacting species, or the model has to have some delay terms.

A limit cycle solution is a closed trajectory in the predator-prey space which is not a member of a continuous family of closed trajectories such as the solutions of the Lotka-Volterra model illustrated in Fig. 3.1. A stable limit cycle trajectory is such that any small perturbation from the trajectory decays to zero.

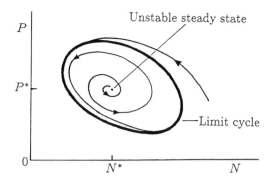

Fig. 3.4. Typical closed predator-prey trajectory which implies a limit cycle periodic oscillation. Any perturbation from the limit cycle tends to zero asymptotically with time.

A schematic example of a limit cycle trajectory in a two species predator(P)-prey(N) interaction is illustrated in Fig. 3.4. Conditions for the existence of such a solution are given in Appendix 1.

One of the unrealistic assumptions in the Lotka-Volterra models, (3.1) and (3.2), and generally (3.10), is that the prey growth is unbounded in the absence of predation. In the form we have written the model (3.1) and (3.2) the bracketed terms on the right are the density dependent per capita growth rates. To be more realistic these growth rates should depend on both the prey and predator densities as in

$$\frac{dN}{dt} = NF(N, P), \quad \frac{dP}{dt} = PG(N, P) , \tag{3.14}$$

where the form of F and G depend on the interaction, the species and so on.

As a reasonable first step we might expect the prey to satisfy a logistic growth, say, in the absence of any predators; that is like (1.2) in Chapter 1, or have some similar growth dynamics which has some maximum carrying capacity. So, for example, a more realistic prey population equation might take the form

$$\frac{dN}{dt} = NF(N, P), \quad F(N, P) = r\left(1 - \frac{N}{K}\right) - PR(N) , \tag{3.15}$$

where $R(N)$ is one of the predation terms discussed below and illustrated in Fig. 3.5 and K is the constant carrying capacity for the prey when $P \equiv 0$.

The predation term, which is the functional response of the predator to change in the prey density, generally shows some saturation effect. Instead of a predator response of bNP, as in the Lotka-Volterra model (3.1), we take $PNR(N)$ where $NR(N)$ saturates for N large. Some examples are

$$R(N) = \frac{A}{N + B}, \quad R(N) = \frac{AN}{N^2 + B^2}, \quad R(N) = \frac{A[1 - e^{-aN}]}{N} , \tag{3.16}$$

where A and B and a are positive constants: these are illustrated in Fig. 3.5 (b)-(d). The second of (3.16), illustrated in Fig. 3.5 (c), is similar to that used in the budworm model in equation (1.6) in Chapter 1. It is also typical of aphid (*Aphidius zbeckistanicus*) predation. The examples in Fig. 3.5 (b), (c)

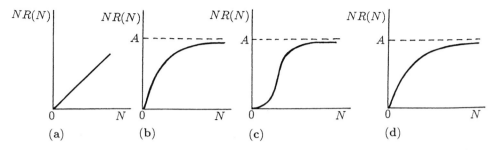

Fig. 3.5a-d. Examples of predator response $NR(N)$ to prey density N. (a) $R(N) = A$, the unsaturated Lotka-Volterra type. (b) $R(N) = A/(N+B)$. (c) $R(N) = AN/(N^2 + B^2)$. (d) $R(N) = A(1 - e^{-aN})/N$.

are approximately linear in N for low densities. The saturation for large N is a reflexion of the limited predator capability, or perseverance, when the prey is abundant.

The predator population equation, the second of (3.14), should also be made more realistic than simply having $G = -d + cN$ as in the Lotka-Volterra model (3.2). Possible forms are

$$G(N, P) = k\left(1 - \frac{hP}{N}\right), \quad G(N, P) = -d + eR(N) \qquad (3.17)$$

where k, h, d and e are positive constants and $R(N)$ is as in (3.16). The first of (3.17) says that the carrying capacity for the predator is directly proportional to the prey density.

The models given by (3.14)-(3.17) are only examples of the many that have been proposed and studied. They are all more realistic than the classical Lotka-Volterra model. Other examples are discussed, for example, in the books by Pielou (1969) and Nisbet and Gurney (1982), to mention but two.

3.4 Analysis of a Predator-Prey Model with Limit Cycle Periodic Behaviour: Parameter Domains of Stability

As an example of how we analyze such realistic 2-species models we consider one of them in detail, namely

$$\begin{aligned}
\frac{dN}{dt} &= N\left[r\left(1 - \frac{N}{K}\right) - \frac{kP}{N+D}\right], \\
\frac{dP}{dt} &= P\left[s\left(1 - \frac{hP}{N}\right)\right],
\end{aligned} \qquad (3.18)$$

where r, K, k, D, s and h are positive constants, 6 in all. It is, as always, extremely useful to write the system in nondimensional form. Although there is no unique way of doing this it is often a good idea to relate the variables to some key relevant parameter. Here, for example, we express N and P as fractions of the predator-free carrying capacity K. Let us write

$$u(\tau) = \frac{N(t)}{K}, \quad v(\tau) = \frac{hP(t)}{K}, \quad \tau = rt,$$

$$a = \frac{k}{hr}, \quad b = \frac{s}{r}, \quad d = \frac{D}{K}$$

(3.19)

and (3.18) become

$$\frac{du}{d\tau} = u(1-u) - \frac{auv}{u+d} = f(u,v),$$

$$\frac{dv}{d\tau} = bv\left(1 - \frac{v}{u}\right) = g(u,v),$$

(3.20)

which have only 3 dimensionless parameters a, b and d. Nondimensionalisation reduces the number of parameters by grouping them in a meaningful way. Dimensionless groupings generally give relative measures of the effect of dimensional parameters. For example b is the ratio of the linear growth rate of the predator to that of the prey and so $b > 1$ and $b < 1$ have definite ecological meanings; with the latter the prey reproduce faster than the predator.

The equilibrium or steady state populations u^*, v^* are solutions of $du/d\tau = 0$, $dv/d\tau = 0$, namely

$$f(u^*, v^*) = 0, \quad g(u^*, v^*) = 0$$

which, from the last equations, are

$$u^*(1 - u^*) - \frac{au^*v^*}{u^* + d} = 0, \quad bv^*\left(1 - \frac{v^*}{u^*}\right) = 0.$$

(3.21)

We are only concerned here with positive solutions, namely the positive solutions of

$$v^* = u^*, \quad u^{*2} + (a + d - 1)u^* - d = 0,$$

of which the only positive one is

$$u^* = \frac{(1 - a - d) + \{(1 - a - d)^2 + 4d\}^{1/2}}{2}, \quad v^* = u^*.$$

(3.22)

We are interested in the stability of the steady states, which are the singular points in the phase plane of (3.20). A linear stability analysis about the steady states is equivalent to the phase plane analysis. For the linear analysis write

$$x(\tau) = u(\tau) - u^*, \quad y(\tau) = v(\tau) - v^*$$

(3.23)

which on substituting into (3.20), linearizing with $|x|$ and $|y|$ small, and using (3.21), gives

$$\begin{pmatrix} \dfrac{dx}{d\tau} \\ \dfrac{dy}{d\tau} \end{pmatrix} = A \begin{pmatrix} x \\ y \end{pmatrix},$$

$$A = \begin{pmatrix} \dfrac{\partial f}{\partial u} & \dfrac{\partial f}{\partial v} \\ \dfrac{\partial g}{\partial u} & \dfrac{\partial g}{\partial v} \end{pmatrix}_{u^*,v^*} = \begin{pmatrix} u^* \left[\dfrac{au^*}{(u^* + d)^2} - 1 \right] & \dfrac{-au^*}{u^* + d} \\ b & -b \end{pmatrix}.$$

(3.24)

A, the community matrix, has eigenvalues λ given by

$$|A - \lambda I| = 0 \quad \Rightarrow \quad \lambda^2 - (\operatorname{tr} A)\lambda + \det A = 0 . \tag{3.25}$$

For stability we require $\operatorname{Re} \lambda < 0$ and so the necessary and sufficient conditions for linear stability are, from the last equation,

$$\operatorname{tr} A < 0 \quad \Rightarrow \quad u^* \left[\dfrac{au^*}{(u^* + d)^2} - 1 \right] < b ,$$

$$\det A > 0 \quad \Rightarrow \quad 1 + \dfrac{a}{u^* + d} - \dfrac{au^*}{(u^* + d)^2} > 0 .$$

(3.26)

Substituting for u^* from (3.22) gives the stability conditions in terms of the parameters a, b and d, and hence in terms of the original parameters r, K, k, D, s and h in (3.18).

In general there is a domain in the a, b, d space such that, if the parameters lie within it, (u^*, v^*) is stable, that is $\operatorname{Re} \lambda < 0$, and if they lie outside it the steady state is unstable. The latter requires at least one of (3.26) to be violated. With (3.22) for u^* and using the first of (3.21) and $v^* = u^*$,

$$\det A = 1 + \dfrac{a}{u^* + d} - \dfrac{au^*}{(u^* + d)^2}$$

$$= 1 + \dfrac{ad}{(u^* + d)^2}$$

$$> 0$$

(3.27)

for all $a > 0$, $d > 0$ and so the second of (3.26) is always satisfied. The instability domain is thus determined solely by the first inequality of (3.26), namely $\operatorname{tr} A < 0$ which, with (3.22) for u^* and again using (3.21), becomes

$$b > \left[a - \{(1 - a - d)^2 + 4d\}^{1/2} \right] \dfrac{\left[1 + a + d - \{(1 - a - d)^2 + 4d\}^{1/2} \right]}{2a} . \tag{3.28}$$

This defines a 3-dimensional surface in (a, b, d) parameter space.

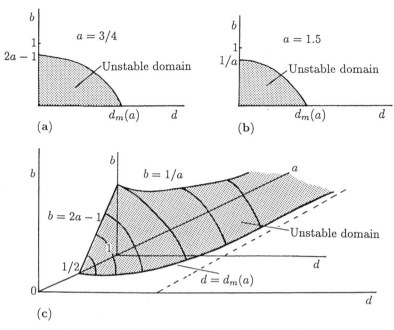

Fig. 3.6a-c. Parameter domains (schematic) of stability of the positive steady state for the predator-prey model (3.20). For $a < 1/2$ and all parameter values $b > 0$, $d > 0$, stability obtains. For a fixed $a > 1/2$ the domain of instability is finite as in (a) and (b). The three-dimensional bifurcation surface between stability and instability is sketched in (c) with $d_m(a) = (a^2+4a)^{1/2}-(1+a)$. When parameter values are in the unstable domain, limit cycle oscillations occur in the populations.

We are only concerned with a, b and d positive. The second square bracket in (3.28) is a monotonic decreasing function of d and always positive. The first square bracket is a monotonic decreasing function of d with a maximum at $d = 0$. Thus, from (3.28),

$$b_{d=0} \begin{cases} > 2a - 1 \\ > 1/a \end{cases} \quad \text{if} \quad \begin{cases} 0 < a \le 1 \\ 1 \le a \end{cases}$$

and so for $0 < a < 1/2$ and all $d > 0$ the stability condition (3.28) is satisfied with any $b > 0$. That is u^*, v^* is linearly stable for all $0 < a < 1/2$, $b > 0$, $d > 0$. On the other hand if $a > 1/2$ there is a domain in the (a, b, d) space with $b > 0$ and $d > 0$ where (3.28) is not satisfied and so the first of (3.26) is violated and hence one of the eigenvalues λ in (3.25) has $\text{Re}\,\lambda > 0$. This in turn implies the steady state u^*, v^* is unstable to small perturbations. The boundary surface is given by (3.28) and it crosses the $b = 0$ plane at $d = d_m(a)$ given by the positive solution of

$$a = \{(1-a-d_m)^2 + 4d_m\}^{1/2} \quad \Rightarrow \quad d_m(a) = d_{b=0} = (a^2+4a)^{1/2} - (1+a) .$$

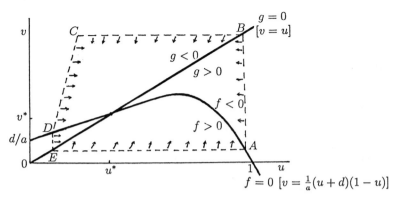

Fig. 3.7. Null clines $f(u, v) = 0$, $g(u, v) = 0$ for the system (3.20): note the signs of f and g on either side of their null clines. $ABCDEA$ is the boundary of the confined set about (u^*, v^*) on which the trajectories all point inwards; that is $\boldsymbol{n} \cdot (du/d\tau, dv/d\tau) < 0$ where $\boldsymbol{n}$ is the unit outward normal on the boundary $ABCDEA$.

Thus $d_m(a)$ is a monotonic increasing function of a bounded above by $d = 1$. Note also that $d < a$ for all $a > 1/2$. Fig. 3.6 illustrates the stability/instability domains in the (a, b, d) space.

When $\operatorname{Re} \lambda < 0$ the steady state is stable and either both λ's are real in (3.25), in which case the singular point u^*, v^* in (3.21) is a stable node in the u, v phase plane of (3.20), or the λ's are complex and the singular point is a stable spiral. When the parameters result in $\operatorname{Re} \lambda > 0$ the singular point is either an unstable node or spiral. In this case we must determine whether or not there is a confined set, or bounding domain, in the (u, v) phase plane so as to use the Poincaré-Bendixson theorem for the existence of a limit cycle oscillation: see Appendix 1. In other words we must find a simple closed boundary curve in the positive quadrant of the (u, v) plane such that on it the phase trajectories always point into the enclosed domain. That is, if $\boldsymbol{n}$ denotes the outward normal to this boundary, we require

$$\boldsymbol{n} \cdot \left(\frac{du}{d\tau}, \frac{dv}{d\tau} \right) < 0$$

for all points on the boundary. If this inequality holds at a point on the boundary it means that the 'velocity' vector $(du/d\tau, dv/d\tau)$ points inwards. Intuitively this means that no solution trajectory can leave the domain if once inside, since, if it did reach the boundary, its 'velocity' points inwards and so the trajectory moves back into the domain.

To find a confined set it is essential and always informative to draw the null-clines of the system, that is the curves in the phase plane where $du/d\tau = 0$ and $dv/d\tau = 0$. From (3.20) these are the curves $f(u, v) = 0$ and $g(u, v) = 0$ and which are illustrated in Fig. 3.7. The sign of the vector components of $(f(u, v), g(u, v))$ indicate the direction of the vector $(du/d\tau, dv/d\tau)$ and hence the direction of the (u, v) trajectory. So if $f > 0$ in a domain, $du/d\tau > 0$

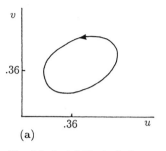

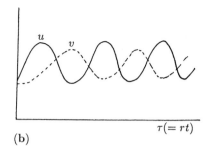

(a)   (b)

Fig. 3.8a,b. (a) Typical phase trajectory limit cycle solution for the predator-prey system (3.20). (b) Corresponding periodic behaviour of the prey(u) and predator(v) populations. Parameter values: $a = 1$, $b = 5$, $d = 0.2$, which give the steady state as $u^* = v^* = 0.36$. Relations (3.19) relate the dimensionless to the dimensional qualities.

and u is thus increasing there. On DE, EA, AB and BC the trajectories clearly point inwards because of the signs of $f(u,v)$ and $g(u,v)$ on them. It can be shown simply but tediously that a line DC exists such that on it $\boldsymbol{n} \cdot (du/d\tau, dv/d\tau) < 0$, that is $\boldsymbol{n} \cdot (f(u,v), g(u,v)) < 0$ where $\boldsymbol{n}$ is the unit vector perpendicular to DC.

We now have a confined set appropriate for the Poincaré-Bendixson theorem to apply when (u^*, v^*) is unstable. Hence the solution trajectory tends to a *limit cycle* when the parameters a, b and d lie in the unstable domain in Fig. 3.6 (c). Basically the Poincaré-Bendixson theorem says that since any trajectory coming out of the unstable steady state (u^*, v^*) cannot cross the confining boundary $ABCDEA$, it must evolve into a closed limit cycle trajectory qualitatively similar to that illustrated in Fig. 3.4. With our model (3.20), Fig. 3.8 (a) illustrates such a closed trajectory with Fig. 3.8 (b) showing the temporal variation of the populations with time. With the specific parameter values used in Fig. 3.8 the steady state is an unstable node in the phase plane, that is both eigenvalues are real and positive. Any perturbation from the limit cycle decays quickly.

This model system, like most which admit limit cycle behaviour, exhibits bifurcation properties as the parameters vary, although not with the complexity shown by discrete models as we see in Chapters 2 and 4 nor, with delay models such as in Chapter 1. We can see this immediately from Fig. 3.6. To be specific consider a fixed $a > 1/2$ so that a finite domain of instability exists, as illustrated in Fig. 3.9, and let us choose a fixed $0 < d < d_m$ corresponding to the line DEF. Suppose b is initially at the value D and is then continuously decreased. On crossing the bifurcation line at E the steady state becomes unstable and a periodic limit cycle solution appears: that is the uniform steady state bifurcates to an oscillatory solution. A similar situation occurs along any parameter variation from the stable to the unstable domains in Fig. 3.6 (c).

The fact that a dimensionless variable passes through a bifurcation value provides useful practical information on equivalent effects of dimensional parameters. For example, from (3.19), $b = s/r$ the ratio of the linear growth rates of the predator and prey. If the steady state is stable then as the predators' growth rate s decreases there is more likelihood of periodic behaviour since b decreases and, if it decreases enough, we move into the instability regime. On the other hand if r decreases, b increases and so probably reduces the possibility of oscillatory behaviour. In this latter case it is not so clear cut since, from (3.19), reducing r also increases a, which from Fig. 3.6 (c) tends to increase the possibility of periodic behaviour. The dimensional bifurcation space is 6-dimensional which is difficult to express graphically: the nondimensionalization reduces it to a simple 3-dimensional space with (3.19) giving clear equivalent effects of different dimensional parameter changes. For example doubling the carrying capacity K is exactly equivalent to halving the predator response parameter D. The dimension*less* parameters are the important bifurcation ones to determine.

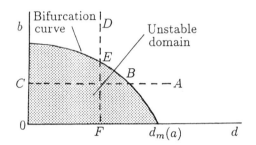

Fig. 3.9. Typical stability bifurcation curve for the predator-prey model (3.20). As the point in parameter space crosses the bifurcation curve the steady state changes stability.

3.5 Competition Models: Principle of Competitive Exclusion

Here two or more species compete for the same limited food source or in some way inhibit each others growth. For example, competition may be for territory which is directly related to food resources. Some interesting phenomena have been found from the study of practical competition models: see for example Hsu et al. (1979). Here we shall discuss a competition model which demonstrates a fairly general principle which is observed to hold in nature, namely that when two species compete for the same limited resources one of the species usually becomes extinct.

Consider the simple 2-species Lotka-Volterra competition model with each species N_1 and N_2 having logistic growth in the absence of the other. Inclusion of logistic growth in the Lotka-Volterra systems makes them much more realistic but to highlight the principle we consider the simpler model which nevertheless reflects many of the properties of more complicated models, particularly as regards stability. We thus consider

$$\frac{dN_1}{dt} = r_1 N_1 \left[1 - \frac{N_1}{K_1} - b_{12} \frac{N_2}{K_1} \right] , \qquad (3.29)$$

$$\frac{dN_2}{dt} = r_2 N_2 \left[1 - \frac{N_2}{K_2} - b_{21}\frac{N_1}{K_2}\right] , \tag{3.30}$$

where r_1, K_1, r_2, K_2, b_{12} and b_{21} are all positive constants and, as before, the r's are the linear birth rates and the K's are the carrying capacities. The b_{12} and b_{21} measure the competitive effect of N_2 on N_1 and N_1 on N_2 respectively: they are generally not equal. Note that the competition model (3.29) and (3.30) is not a conservative system like its predator-prey counterpart.

If we nondimensionalize this model by writing

$$u_1 = \frac{N_1}{K_1}, \quad u_2 = \frac{N_2}{K_2}, \quad \tau = r_1 t, \quad \rho = \frac{r_2}{r_1},$$

$$a_{12} = b_{12}\frac{K_2}{K_1}, \quad a_{21} = b_{21}\frac{K_1}{K_2} \tag{3.31}$$

(3.29) and (3.30) become

$$\frac{du_1}{d\tau} = u_1(1 - u_1 - a_{12}u_2) = f_1(u_1, u_2) ,$$

$$\frac{du_2}{d\tau} = \rho u_2(1 - u_2 - a_{21}u_1) = f_2(u_1, u_2) . \tag{3.32}$$

The steady states, and phase plane singularities, u_1^*, u_2^* are solutions of $f_1(u_1, u_2) = f_2(u_1, u_2) = 0$ which, from (3.32), are

$$u_1^* = 0, u_2^* = 0; \quad u_1^* = 1, u_2^* = 0; \quad u_1^* = 0, u_2^* = 1;$$

$$u_1^* = \frac{1 - a_{12}}{1 - a_{12}a_{21}}, u_2^* = \frac{1 - a_{21}}{1 - a_{12}a_{21}} . \tag{3.33}$$

The last of these is only of relevance if $u_1^* \geq 0$ and $u_2^* \geq 0$ are finite in which case $a_{12}a_{21} \neq 1$. The four possibilities are seen immediately on drawing the null clines $f_1 = 0$ and $f_2 = 0$ in the u_1, u_2 phase plane as shown in Fig. 3.10. The crucial part of the null clines are, from (3.32), the straight lines

$$1 - u_1 - a_{12}u_2 = 0, \quad 1 - u_2 - a_{21}u_1 = 0 .$$

The first of these together with the u_2-axis is $f_1 = 0$ while the second, together with the u_1-axis is $f_2 = 0$.

The stability of the steady states is again determined by the community matrix which, for (3.32), is

$$A = \begin{pmatrix} \dfrac{\partial f_1}{\partial u_1} & \dfrac{\partial f_1}{\partial u_2} \\[2mm] \dfrac{\partial f_2}{\partial u_1} & \dfrac{\partial f_2}{\partial u_2} \end{pmatrix}_{u_1^*, u_2^*}$$

$$= \begin{pmatrix} 1 - 2u_1 - a_{12}u_2 & -a_{12}u_1 \\ -\rho a_{21}u_2 & \rho(1 - 2u_2 - a_{21}u_1) \end{pmatrix}_{u_1^*, u_2^*} \tag{3.34}$$

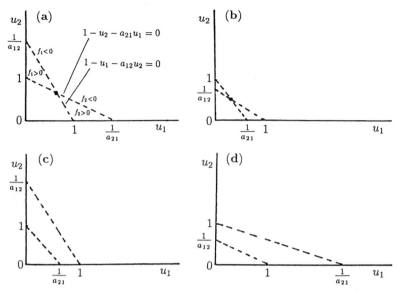

Fig. 3.10a-d. The null clines for the competition model (3.32). $f_1 = 0$ is $u_1 = 0$ and $1 - u_1 - a_{12}u_2 = 0$ with $f_2 = 0$ being $u_2 = 0$ and $1 - u_2 - a_{21}u_1 = 0$. The intersection of the two dashed lines gives the positive steady state if it exists as in (a) and (b): the relative sizes of a_{12} and a_{21} for it to exist are obvious from (a)-(d).

The first steady state in (3.33), that is $(0,0)$, is unstable since the eigenvalues λ of its community matrix, given from (3.34) by

$$|A - \lambda I| = \begin{vmatrix} 1 - \lambda & 0 \\ 0 & \rho - \lambda \end{vmatrix} = 0 \quad \Rightarrow \quad \lambda_1 = 1, \ \lambda_2 = \rho \, ,$$

are positive. For the second of (3.33), namely $(1,0)$, (3.34) gives

$$|A - \lambda I| = \begin{vmatrix} -1 - \lambda & -a_{12} \\ 0 & \rho(1 - a_{21}) - \lambda \end{vmatrix} = 0 \quad \Rightarrow \quad \lambda_1 = -1, \ \lambda_2 = \rho(1 - a_{21})$$

and so

$$u_1^* = 1, \ u_2^* = 0 \quad \text{is} \quad \begin{cases} \text{stable} \\ \text{unstable} \end{cases} \text{if} \begin{cases} a_{21} > 1 \\ a_{21} < 1 \end{cases} \tag{3.35}$$

Similarly, for the third steady state the eigenvalues are $\lambda = -\rho$, $\lambda_2 = (1 - a_{12})$ and so

$$u_1^* = 0, \ u_2^* = 1 \quad \text{is} \quad \begin{cases} \text{stable} \\ \text{unstable} \end{cases} \text{if} \begin{cases} a_{12} > 1 \\ a_{12} < 1 \end{cases} \tag{3.36}$$

Finally for the last steady state in (3.33), when it exists in the positive quadrant, the matrix A from (3.34) is

$$A = (1 - a_{12}a_{21})^{-1} \begin{pmatrix} a_{12} - 1 & a_{12}(a_{12} - 1) \\ \rho a_{21}(a_{21} - 1) & \rho(a_{21} - 1) \end{pmatrix}$$

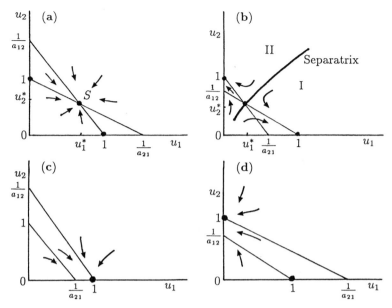

Fig. 3.11a-d. Schematic phase trajectories near the steady states for the dynamic behaviour of competing populations satisfying the model (3.32) for the various cases. (a) $a_{12} < 1$, $a_{21} < 1$. Only the positive steady state S is stable and all trajectories tend to it. (b) $a_{12} > 1$, $a_{21} > 1$. Here (1,0) and (0,1) are stable steady states each of which has a domain of attraction separated by a separatrix which passes through (u_1^*, u_2^*). (c) $a_{12} < 1$, $a_{21} > 1$. Only one stable steady state exists, $u_1^* = 1$, $u_2^* = 0$ with the whole positive quadrant its domain of attraction. (d) $a_{12} > 1$, $a_{21} < 1$. The only stable steady state is $u_1^* = 0$, $u_2^* = 1$ with the positive quadrant as its domain of attraction. Cases (b) to (d) illustrate the competitive exclusion principle whereby 2 species competing for the same limited resource cannot in general coexist.

which has eigenvalues

$$\lambda_1, \lambda_2 = [2(1 - a_{12}a_{21})]^{-1}\Big[[(a_{21} - 1) + \rho(a_{21} - 1)]$$

$$\pm \{[(a_{12}-1) + \rho(a_{21}-1)]^2 - 4\rho(1-a_{12}a_{21})(a_{12}-1)(a_{21}-1)\}^{1/2}\Big]$$

$$(3.37)$$

The sign of λ, or Re λ if complex, and hence the stability of the steady state, depends on the size of ρ, a_{12} and a_{21}. There are several cases we have to consider, all of which have ecological implications which we shall come to below.

Before discussing the various cases note that there is a confined set on the boundary of which the vector of the derivatives, $(du_1/d\tau, du_2/d\tau)$, points along it or inwards: here it is a rectangular box in the (u_1, u_2) plane. From (3.32) this condition holds on the u_1- and u_2-axis. Outer edges of the rectangle are, for example, the lines $u_1 = U_1$ where $1 - U_1 - a_{12}u_2 < 0$ and $u_2 = U_2$ where $1 - U_2 - a_{21}u_1 < 0$. Any $U_1 > 1$, $U_2 > 1$ suffice. So the system is always globally stable.

The various cases are: (i) $a_{12} < 1$, $a_{21} < 1$, (ii) $a_{12} > 1$, $a_{21} > 1$, (iii) $a_{12} < 1$, $a_{21} > 1$, (iv) $a_{12} > 1$, $a_{21} < 1$. All of these are analyzed in a similar way. Fig. 3.10 (a)-(d) and Fig. 3.11 (a)-(d) relate to these cases (i)-(iv) respectively. By way of example, we consider just one of them, namely (ii). The analysis of the other cases is left as an exercise. The results are encapsulated in Fig. 3.11. The arrows indicate the direction of the phase trajectories. The qualitative behaviour of the phase trajectories is given by the signs of $du_1/d\tau$, namely $f_1(u_1,u_2)$, and $du_2/d\tau$ which is $f_2(u_1,u_2)$, on either side of the null clines.

Case $a_{12} > 1$, $a_{21} > 1$. This corresponds to Fig. 3.10 (b). From (3.35) and (3.36), $(1,0)$ and $(0,1)$ are stable. Since $1 - a_{12}a_{21} < 0$, (u_1^*, u_2^*), the fourth steady state in (3.33), lies in the positive quadrant and from (3.37) its eigenvalues are such that $\lambda_2 < 0 < \lambda_1$ and so it is unstable to small perturbations: it is a saddle point. In this case, then, the phase trajectories can tend to either one of the two steady states, as illustrated in Fig. 3.11 (b). Each steady state has a *domain of attraction*. There is a line, a *separatrix*, which divides the positive quadrant into 2 non-overlapping regions I and II as in Fig. 3.11 (b). The separatrix passes through the steady state (u_1^*, u_2^*): it is one of the saddle point trajectories in fact.

Now consider some of the ecological implications of these results. In case (i) where $a_{12} < 1$ and $a_{21} < 1$ there is a stable steady state where both species can exist as in Fig. 3.10 (a). In terms of the original parameters from (3.31) this corresponds to $b_{12}K_2/K_1 < 1$ and $b_{21}K_1/K_2 < 1$. For example if K_1 and K_2 are approximately the same and the *inter*specific competition, as measured by b_{12} and b_{21}, is not too strong, these conditions say that the two species simply adjust to a lower population size than if there was no competition. In other words the competition is not aggresive. On the other hand if the b_{12} and b_{21} are about the same and the K_1 and K_2 are different it is not easy to tell what will happen until we form and compare the *dimensionless* groupings a_{12} and a_{21}.

In case (ii), where $a_{12} > 1$ and $a_{21} > 1$, if the K's are about equal, then the b_{12} and b_{21} are not small. The analysis then says that the competition is such that all three non-trivial steady states can exist, but, from (3.35)-(3.37), only $(1,0)$ and $(0,1)$ are stable, as in Fig. 3.11 (b). It can be a delicate matter which ultimately wins out. It depends crucially on the starting advantage each species has. If the initial conditions lie in domain I then eventually species 2 will die out, $u_2 \to 0$ and $u_1 \to 1$, that is $N_1 \to K_1$ the carrying capacity of the environment for N_1. Thus competition here has eliminated N_2. On the other hand if N_2 has an initial size advantage so that u_1 and u_2 start in region II then $u_1 \to 0$ and $u_2 \to 1$ in which case the N_1-species becomes extinct and $N_2 \to K_2$, its environmental carrying capacity. We expect extinction of one species even if the initial populations are close to the separatrix and in fact if they lie on it, since the ever present random fluctuations will inevitably cause one of u_i, $i = 1, 2$ to tend to zero.

Cases (iii) and (iv) in which the *inter*specific competition of one species is much stronger than the other, or the carrying capacities are sufficiently different so that $a_{12} = b_{12}K_2/K_1 < 1$ and $a_{21} = b_{21}K_1/K_2 > 1$ or alternatively $a_{12} > 1$ and $a_{21} < 1$, are quite   definite in the ultimate result. In case (iii), as in

Fig. 3.11 (c), the stronger dimensionless interspecific competition of the u_1-species dominates and the other species, u_2, dies out. In case (iv) it is the other way round and species u_1 becomes extinct.

Although all cases do not result in species elimination those in (iii) and (iv) always do and in (ii) it is probable due to natural fluctuations in the population levels. This work led to the *principle of competitive exclusion* which was mentioned above. Note that the conditions for this to hold depend on the di*mensionless* parameter groupings a_{12} and a_{21}: the growth rate ratio parameter ρ does not affect the gross stability results, just the dynamics of the system. Since $a_{12} = b_{12}K_2/K_1$, $a_{21} = b_{21}K_1/K_2$ the conditions for competitive exclusion depend critically on the interplay between competition and the carrying capacities as well as the initial conditions in case (ii).

Suppose for example we have 2 species comprised of large animals and small animals, with both competing for the same grass in a fixed area. Suppose also that they are equally competitive with $b_{12} = b_{21}$. With N_1 the large animals and N_2 the small, $K_1 < K_2$ and so $a_{12} = b_{12}K_2/K_1 < b_{21}K_2/K_1 = a_{21}$. As an example if $b_{12} = 1 = b_{21}$, $a_{12} < 1$ and $a_{21} > 1$ then in this case $N_1 \to 0$ and $N_2 \to K_2$: that is the large animals become extinct.

The situation in which $a_{12} = 1 = a_{21}$ is special and, with the usual stochastic variability in nature, is unlikely in the real world to hold exactly. In this case the competitive exclusion of one or other of the species also occurs.

The importance of species competition in nature is obvious. We have discussed only one particularly simple model but again the method of analysis is quite general. A review and introductory article by Pianka (1981) deals with some practical aspects of competition.

3.6 Mutualism or Symbiosis

There are many examples where the interaction of two or more species is to the advantage of all. Mutualism or symbiosis often plays the crucial role in promoting and even maintaining such species; plant and seed dispersers is one example. Even if survival is not at stake the mutual advantage of mutualism or symbiosis can be very important. As a topic of theoretical ecology, even for two species, this area has not been as widely studied as the others even though its importance is comparable to that of predator-prey and competition interactions. This is in part due to the fact that simple models in the Lotka-Volterra vein give silly results. The simplest mutualism model equivalent to the classical Lotka-Volterra predator-prey one is

$$\frac{dN_1}{dt} = r_1 N_1 + a_1 N_1 N_2, \quad \frac{dN_2}{dt} = r_2 N_2 + a_2 N_2 N_1$$

where r_1, r_2, a_1 and a_2 are all positive constants. Since $dN_1/dt > 0$ and

$dN_2/dt > 0$, N_1 and N_2 simply grow unboundedly in, as May (1981) so aptly puts it, 'an orgy of mutual benefaction'.

Realistic models must at least show a mutual benefit to both species, or as many as are involved, and have some positive steady state or limit cycle type oscillation. Some models which do this are described by Whittaker (1975). A practical example is discussed by May (1975).

As a first step in producing a reasonable 2-species model we incorporate limited carrying capacities for both species and consider

$$\frac{dN_1}{dt} = r_1 N_1 \left(1 - \frac{N_1}{K_1} + b_{12}\frac{N_2}{K_1} \right)$$

$$\frac{dN_2}{dt} = r_2 N_2 \left(1 - \frac{N_2}{K_2} + b_{21}\frac{N_1}{K_2} \right)$$

(3.38)

where r_1, r_2, K_1, K_2, b_{12} and b_{21} are all positive constants. If we use the same nondimensionalization as in the competition model (the signs preceding the b's are negative there), namely (3.31), we get

$$\frac{du_1}{d\tau} = u_1(1 - u_1 + a_{12}u_2) = f_1(u_1, u_2)$$

$$\frac{du_2}{d\tau} = \rho u_2(1 - u_2 + a_{21}u_1) = f_2(u_1, u_2)$$

(3.39)

where

$$u_1 = \frac{N_1}{K_1}, \quad u_2 = \frac{N_2}{K_2}, \quad \tau = r_1 t, \quad \rho = \frac{r_2}{r_1},$$

$$a_{12} = b_{12}\frac{K_2}{K_1}, \quad a_{21} = b_{21}\frac{K_1}{K_2}.$$

(3.40)

Analysing the model in the usual way we start with the steady states (u_1^*, u_2^*) which from (3.39) are

$$(0,0), \quad (1,0), \quad (0,1),$$

$$\left(\frac{1 + a_{12}}{\delta}, \frac{1 + a_{21}}{\delta} \right), \quad \text{positive if} \quad \delta = 1 - a_{12}a_{21} > 0.$$

(3.41)

After calculating the community matrix for (3.39) and evaluating the eigenvalues λ for each of (3.41) it is straightforward to show that $(0,0)$, $(1,0)$ and $(0,1)$ are all unstable: $(0,0)$ is an unstable node and $(1,0)$ and $(0,1)$ are saddle point equilibria. If $1 - a_{12}a_{21} < 0$ there are only 3 steady states, the first three in (3.41), and so the populations become unbounded. We see this by drawing the null clines in the phase plane for (3.39), namely $f_1 = 0$, $f_2 = 0$, and noting that the phase trajectories move off to infinity in a domain in which $u_1 \to \infty$ and $u_2 \to \infty$ as in Fig. 3.12 (a).

When $1 - a_{12}a_{21} > 0$ the fourth steady state in (3.41) exists in the positive quadrant. Evaluation of the eigenvalues of the community matrix shows it to

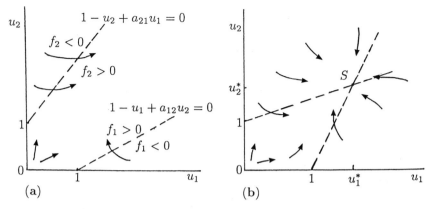

Fig. 3.12a,b. Phase trajectories for the mutualism model for 2 species with limited carrying capacities given by the dimensionless system (3.39). (a) $a_{12}a_{21} > 1$: unbounded growth occurs with $u_1 \to \infty$ and $u_2 \to \infty$ in the domain bounded by the null clines – the dashed lines. (b) $a_{12}a_{21} < 1$: all trajectories tend to a positive steady state S with $u_1^* > 1$, $u_2^* > 1$ which shows the initial benefit that accrues since the carrying capacities for each species is greater than if no interaction was present.

be a stable equilibrium: it is a node singularity in the phase plane. This case is illustrated in Fig. 3.12 (b). Here all the trajectories in the positive quadrant tend to $u_1^* > 1$ and $u_2^* > 1$; that is $N_1 > K_1$ and $N_2 > K_2$ and so each species has increased its steady state population from its maximum value in isolation.

This model has certain drawbacks. One is the sensitivity between unbounded growth and a finite positive steady state. It depends on the inequality $a_{12}a_{21} < 1$, which from (3.40) in dimensional terms is $b_{12}b_{21} < 1$. So if symbiosis of either species is too large this last condition is violated and both populations grow unboundedly.

3.7 General Models and Some General and Cautionary Remarks

All of the models we have discussed in this chapter result in systems of nonlinear differential equations of the form

$$\frac{dN_i}{dt} = N_i F_i(N_1, N_2, \dots, N_n), \quad i = 1, 2, \dots , \tag{3.42}$$

which emphasizes the fact that the vector of populations $\mathbf{N}$ has $\mathbf{N} = 0$ as a steady state. The two species version is sometimes referred to as the Kolmogorov model or as the *Kolmogorov equations*.

Although we have mainly considered 2-species interactions in this chapter, in nature, and in the sea in particular, there are many species or *trophic levels* where energy, in the form of food, flows from one species to another. That is there is a flow from one trophic level to another. The mass of the total number of individuals in a species is often referred to as its *biomass*, here the population times the unit mass. The ultimate source of energy is the sun, and in the sea for example the trophic web runs through plankton, fish, sharks up to whales and finally man, with the miriad of species in between. The species on one trophic level may predate several species below it. In general, models involve interaction between several species.

Multi-species models are of the form

$$\frac{du}{dt} = f(u) \quad \text{or} \quad \frac{du_i}{dt} = f_i(u_1, \ldots, u_n), \quad i = 1, \ldots, n \qquad (3.43)$$

where $u(t)$ is the n-dimensional vector of population densities and $f(u)$ describes the nonlinear interaction between the species. The function $f(u)$ involves parameters which characterize the various growth and interaction features of the system under investigation, with $f_i(u_1, \ldots, u_n)$ specifying the overall rate of growth for the i-th species. The stability of the steady states is determined in exactly the same way as before by linearizing about the steady states u^*, where $f(u^*) = 0$ and examining the eigenvalues λ of the community or stability matrix

$$A = (a_{ij}) = \left(\frac{\partial f_i}{\partial u_j} \right)_{u=u^*} \qquad (3.44)$$

The necessary and sufficient conditions for the eigenvalues λ, solutions of polynomial $|A - \lambda I| = 0$, to have Re $\lambda > 0$ are given by the Routh-Hurwitz conditions which are listed in Appendix 2.

If a steady state is unstable then the solution u may grow unboundedly or evolve into another steady state or into a stable oscillatory pattern like a limit cycle. For 2-species models the theory of such equations is essentially complete: they are phase plane systems and a brief review of their analysis is given in Appendix 1. For three or more interacting species no general theory exists. Some results, at least for solutions near the steady state when it becomes unstable, can often be found using *Hopf bifurcation theory*; see Appendix 3 and, for example, the book by Hassard, Kazarinoff and Wan (1981). At its simplest this theory says that if a parameter of the system, p say, has a critical value p_c such that for $p < p_c$ the eigenvalue with the largest Re $\lambda < 0$, and for $p = p_c$ Re $\lambda = 0$, Im $\lambda \neq 0$ and for $p > p_c$ Re $\lambda > 0$, Im $\lambda \neq 0$ then for $p - p_c > 0$ and small, the solution u will exhibit small amplitude limit cycle behaviour around u^*.

The community matrix A, defined by (3.44), which is so crucial in determining the linear stability of the steady states, has direct biological significance. The elements a_{ij} measure the effect of the j-species on the i-species near equilibrium. For example, if U_i is the perturbation from the steady state u_i^* the equation for U_i is

$$\frac{dU_i}{dt} = \sum_{j=1}^{n} a_{ij} U_j \qquad (3.45)$$

and so $a_{ij}U_j$ is the effect of the species U_j on the growth of U_i. If $a_{ij} > 0$ then U_j directly enhances U_i's growth while if $a_{ij} < 0$ it diminishes it. If $a_{ij} > 0$ and $a_{ji} > 0$ then U_i and U_j enhance each others growth and so they are in a symbiotic interaction. If $a_{ij} < 0$ and $a_{ji} < 0$ then they are in competition. May (1975) gives a survey of some generalized models and, in his discussion on stability versus complexity, gives some results for stability based on properties of the community matrix.

There has been a considerable amount of study of systems where the community matrix has diagonal symmetry or anti-symmetry or has other rather special properties, where general results can be given about the eigenvalues and hence the stability of the steady states. This has had very limited practical value since models of real situations do not have such simple properties. The stochastic element in assessing parameters mitigates against even approximations by such models. However just as the classical Lotka-Volterra system is not relevant to the real world, these special models have often made people ask the right questions. Even so, a pre-occupation with such models or their generalizations must be avoided if the basic aim is to understand the real world.

An important class of models which we have not discussed are interaction models with delay. If the species exhibit different or distributed delays, such models open up a veritable Pandora's box of solution behaviour which to a large extent is quite unexplored.

If we consider three or more species, aperiodic behaviour can arise. Lorenz (1963) first demonstrated this with the model system

$$\frac{du}{dt} = a(v - u), \quad \frac{dv}{dt} = -uw + bu - v, \quad \frac{dw}{dt} = uv - cw$$

where a, b and c are positive parameters. (The equations arose in a fluid flow model.) As the parameters are varied the solutions exhibit period doubling and eventually chaos or aperiodicity. Many authors have considered such systems. For example, Rössler (1976, 1979, 1983) and Sparrow (1982, 1986) have made a particular study of such systems and discovered several other basic examples which show similar properties: see also the book edited by Holden (1986). It would be surprising if certain population interaction models of three or more species did not display similar properties. Competition models of three or more species produce some unexpected and practical results.

Evolutionary development of complex population interactions have generally produced reasonably stable systems. From our study of interaction models up to now we know that a system can be driven unstable if certain parameters are changed appropriately, that is pass through bifurcation values. It should therefore be a matter of considerable scientific study before any system is altered by external manipulation. The use of models to study the effect of artificially interfering in such trophic webs is essential and can be extremely illuminating. Had this been done it is likely that the following catastrophe would have been avoided. Although the use of realistic dynamic models cannot give the complete answer, in the form of predictions, which might result from introducing another species or eradicating one in the chain, they can certainly point to various danger signs that must be seriously considered. By the same arguments it is essential that not too much credence be put on models since the interactions can often be extremely complicated and the modeller might simply not construct a sufficiently good model. To conclude this section we shall describe a major ecological catastrophe which has come from one such attempt to manipulate a complex trophic web in East Africa.

Lake Victoria and the Nile Perch Catastrophe 1960-

In 1960 the Nile perch (*Lates niloticus*) was introduced into Lake Victoria, the largest lake in East Africa. The lake is bordered by Kenya, Tanzania and Uganda and it used to support hundreds of small fishing communities along the shore. It was thought that the introduction of this large carnivorous species, which can weigh up to 100 Kg or more, would provide a high-yielding and valuable source of protein. Its introduction was supported at the time by the United Nations Food and Agriculture Organisation. There were dissenting views from some scientists but these were ignored.

The presence of the large carnivorous perch over the past 25 years has practically wiped out the several hundred smaller cichlid fish in the lake: many of these provided the main basis of the fishing communities economy on the lake's shore. Markets are now flooded with perch. It was estimated that in 1984 the overall productivity of the lake was reduced by about 80% of its pre-1960 level.

Within the lake, the unplanned introduction of such a major, new and unsuitable species was a mistake of horrifying dimensions and has caused an ecological disaster. There are, however, other knock-on effects outside the lake over and above the economic catastrophe which has engulfed the shore communities: these effects should certainly have been anticipated. For example, the large perch are oily and cannot be dried in the sun but have to be preserved by smoking. This has resulted in major felling of valuable trees to provide fuel.

Even more serious is the fact that many of the cichlid species, which have all but disappeared and which used to flourish in the lake, helped to control the level of a particular snail which live in and around the lake. These snails are an essential link in the cycle of the human liver fluke disease called bilharzia or schistisomiasis: it is invariably fatal to humans if not treated in time. Since the best mathematical biology is usually carried out within a truly interdisciplinary environment it is often the case that in trying to make a model certain questions and answers are elicited from the ecologists, which in turn initiate other related questions not directly connected with the model. These knock-on effects would have been important examples.

In spite of the disaster caused by this introduction of an unsuitable species into such a delicate and complex trophic webb there are (in 1987) plans to introduce Nile perch into other large lakes in the region, such as Lake Malawi.

3.8 Threshold Phenomena

With the exception of the Lotka-Volterra predator-prey model, the 2-species models, which we have considered or referred to in this chapter, have either had stable steady states where small perturbations die out, or unstable steady states where perturbations from them grow unboundedly or result in limit cycle periodic solutions. There is an interesting group of models which have a non-zero stable state such that if the perturbation from it is sufficiently large or of the right kind,the population densities undergo large variations before returning to the steady state. Such models are said to exhibit a threshold effect. We study one such group of models here.

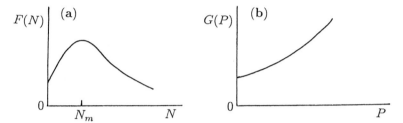

Fig. 3.13a,b. (a) Qualitative form of the prey's per capita growth rate $F(N)$ in (3.46) which exhibits the Allee effect. (b) Predators' per capita mortality rate.

Consider the model predator-prey system

$$\frac{dN}{dt} = N[F(N) - P] = f(N, P) , \tag{3.46}$$

$$\frac{dP}{dt} = P[N - G(P)] = g(N, P) , \tag{3.47}$$

where for convenience all the parameters have been incorporated in the F and G by a suitable rescaling: the $F(N)$ and $G(P)$ are qualitatively as illustrated in Fig. 3.13. The specific form of $F(N)$ demonstrates the *Allee effect* which means that per capita growth rate of the prey initially increases with prey density but reaches a maximum at some N_m and then decreases for larger prey densities.

The steady states N^*, P^* from (3.46) and (3.47) are $N^* = 0 = P^*$ and the non-negative solutions of

$$P^* = F(N^*), \quad N^* = G(P^*) . \tag{3.48}$$

As usual it is again helpful to draw the null clines $f = 0$, $g = 0$ which are sketched in Fig. 3.14. Depending on the various parameters in $F(N)$ and $G(P)$ the steady state can be typically at S or at S'. To be specific we shall consider the case where $N^* > N_m$, that is the steady state is at S in Fig. 3.14.

From (3.46) and (3.47) the community matrix A for the zero steady state $N^* = 0$, $P^* = 0$ is

$$A = \begin{pmatrix} \dfrac{\partial f}{\partial N} & \dfrac{\partial f}{\partial P} \\ \dfrac{\partial g}{\partial N} & \dfrac{\partial g}{\partial P} \end{pmatrix}_{N=0=P} = \begin{pmatrix} F(0) & 0 \\ 0 & -G(0) \end{pmatrix}$$

The eigenvalues are $\lambda = F(0) > 0$ and $\lambda = -G(0) < 0$. So, with the $F(N)$ and $G(N)$ in Fig. 3.13, $(0,0)$ is unstable: it is a saddle point singularity in the (N, P) phase plane.

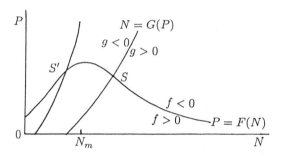

Fig. 3.14. Null clines $N = 0$, $P = 0$, $N = G(P)$, $P = F(N)$ for the predator-prey system (3.46) and (3.47): $f = N[F(N) - P]$, $g = P[N - G(P)]$. S and S' are possible stable steady states.

For the positive steady state (N^*, P^*) the community matrix is, from (3.46)-(3.48),

$$A = \begin{pmatrix} N^* F'(N^*) & -N^* \\ P^* & -P^* G'(P^*) \end{pmatrix}$$

where the prime denotes differentiation and, from Fig. 3.14, $G'(P^*) > 0$ and $F'(N^*) < 0$ when (N^*, P^*) is at S and $G'(P^*) > 0$ and $F'(N^*) > 0$ when at S'. The eigenvalues λ are solutions of

$$|A - \lambda I| = 0 \quad \Rightarrow \quad \lambda^2 - (\text{tr } A)\lambda + \det A = 0 \tag{3.49}$$

where

$$\begin{aligned} \text{tr } A &= N^* F'(N^*) - P^* G'(P^*) \\ \det A &= N^* P^* [1 - F'(N^*) G'(P^*)] \end{aligned} \tag{3.50}$$

When the steady state is at S in Fig. 3.14, $\text{tr } A < 0$ and $\det A > 0$ and so it is stable to small perturbations for all $F(N)$ and $G(P)$ since $\text{Re}\,\lambda < 0$ from (3.49). If the steady state is at S', $\text{tr } A$ and $\det A$ can be positive or negative since now $F'(N^*) > 0$. Thus S' may be stable or unstable depending on the particular $F(N)$ and $G(P)$. If it is unstable then a limit cycle solution results since there is a confined set for the system: refer to Section 3.4 for a worked

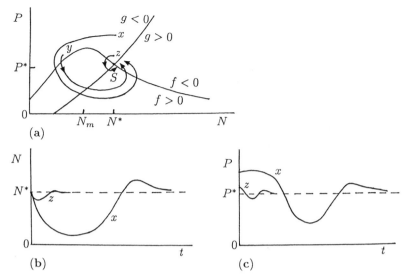

Fig. 3.15a-c. (a) Null clines for the predator-prey threshold model (3.46) and (3.47). The steady state S is always stable. A perturbation to X results in a large excursion in phase space before returning to S. A perturbation to Z is under the threshold and hence returns to S without a large excursion. (b) and (c) Schematic time evolution of the solutions illustrating the effect of a perturbation to X and to Z as in (a).

example of a qualitatively similar problem and Fig. 3.8 which illustrates the solution behaviour.

The case of interest here is when the steady state is at S and is thus always stable. Suppose we perturb the system to the point X in the phase plane as in Fig. 3.15 (a). Since here $f < 0$ and $g < 0$, equations (3.46) and (3.47) imply that $dN/dt < 0$ and $dP/dt < 0$ and so the trajectory starts to move qualitatively as on the trajectory shown in Fig. 3.15 (a): it *eventually* returns to S but only after a large excursion in the phase plane. The path is qualitatively indicated by the signs of f and g and hence of dN/dt and dP/dt. If the perturbation took (N, P) to Y then a similar behaviour occurs. If, however, the perturbation is to Z then the perturbation remains close to S. Fig. 3.15 (b) and (c) illustrate a typical temporal behaviour of N and P.

There is clearly a rough threshold perturbation below which the perturbation always remains close to the steady state and above which it does not, even though the solution ultimately returns to the steady state. The threshold perturbation is more a threshold curve or rather domain and is such that if the perturbation results in the trajectory getting past the maximum N_m in Fig. 3.15 (a) then the trajectories are typically like those from X and Y. If the trajectory crosses $f = 0$ at $N > N_m$ then no large perturbation occurs. The reason that such a threshold property exists is a consequence of the form of the null cline $f = 0$ which has a maximum as shown; in this case this is a consequence of the Allee effect in the dynamics of the model (3.46). With the problems we have discussed earier in this chapter it might appear from looking at the temporal behaviour of the population that we were dealing with an unstable situation. The necessity for a careful drawing of the null clines is clear. The definition of a threshold at this stage is rather imprecise. We shall see later in Chapter 12 that if one of the species is allowed to disperse spatially, for example by diffusion, then threshold travelling waves are possible. These have important biological consequences. In this context the concept of a threshold can be made precise.

This threshold behaviour will arise in an important way later in the book in biochemical contexts which are formally similar since the equations for reaction kinetics are mathematically of the same type as those for the dynamics of interacting populations such as we have discussed here.

A final remark on the problem of modelling interacting populations is that there can be no 'correct' model for a given situation since many models can give qualitatively similar behaviour. Getting the right qualitative characteristics is only the first step and must not be considered justification for a model. This important caveat to all models will be repeated with great regularity throughout the book. What helps to make a model a good one is the plausibility of the growth dynamics based on observation, real facts and whether or not a reasonable assessment of the various parameters is possible and, finally, whether predictions based on the model are borne out by subsequent experiment and observation.

Exercises

1. In the competition model for two species with populations N_1 and N_2

$$\frac{dN_1}{dt} = r_1 N_1 \left(1 - \frac{N_1}{K_1} - b_{12} \frac{N_2}{K_1} \right) ,$$

$$\frac{dN_2}{dt} = r_2 N_2 \left(1 - b_{21} \frac{N_1}{K_2} \right)$$

where only one species, N_1, has limited carrying capacity. Nondimensionalize the system and determine the steady states. Investigate their stability and sketch the phase plane trajectories. Show that irrespective of the size of the parameters the principle of competitive exclusion holds. Briefly describe under what ecological circumstances the species N_2 becomes extinct.

2. Determine the kind of interactive behaviour between two species with populations N_1 and N_2 that is implied by the model

$$\frac{dN_1}{dt} = r_1 N_1 \left[1 - \frac{N_1}{K_1 + b_{12} N_2} \right] ,$$

$$\frac{dN_2}{dt} = r_2 N_2 \left[1 - \frac{N_2}{K_2 + b_{21} N_1} \right] .$$

Draw the nullclines and determine the steady states and their stability. Briefly describe the ecological implications of the results of the analysis.

3. A predator-prey model for herbivore(H)-plankton(P) interaction is

$$\frac{dP}{dt} = rP\left[(K-P) - \frac{BH}{C+P}\right], \quad \frac{dH}{dt} = DH\left[\frac{P}{C+P} - AH\right],$$

where r, K, A, B, C and H are positive constants. Briefly explain the ecological assumptions in the model. Nondimensionalise the system so that it can be written in the form

$$\frac{dp}{d\tau} = p\left[(k-p) - \frac{h}{1+p}\right], \quad \frac{dh}{d\tau} = dh\left[\frac{p}{1+p} - ah\right].$$

Sketch the null clines and note any qualitative changes as the parameter k varies. Hence, or otherwise, demonstrate that a positive steady state (p_0, h_0) exists for all $a > 0$, $k > 0$.

By considering the community matrix determine the signs of the partial derivatives of the right hand sides of the equation system evaluated at (p_0, h_0) for this steady state to be stable. By noting the signs of $dp/d\tau$ and $dh/d\tau$ relative to the null clines in the p,h phase plane, show that (i) for $k < 1$ the positive steady state is stable and (ii) that for $k > 1$, and small enough a, the positive positive steady may be stable or unstable. Hence show that in the a,k parameter plane a necessary condition for a periodic solution to exist is that a,k lie in the domain bounded by $a = 0$ and $a = 4(k-1)/(k+1)^3$. Hence show that if $a < 4/27$ there is a window of values of k where periodic solutions are possible. Under what conditions can the system exhibit a threshold phenomenon?

4. The interaction between two populations with densities N_1 and N_2 is modelled by

$$\frac{dN_1}{dt} = rN_1\left(1 - \frac{N_1}{K}\right) - aN_1N_2(1 - \exp[-bN_1]),$$

$$\frac{dN_2}{dt} = -dN_2 + N_2e(1 - \exp[-bN_1]),$$

where a, b, d, e, r and K are positive constants. What type of interaction exists between N_1 and N_2? What do the various terms imply ecologically?

Nondimensionalize the system by writing

$$u = \frac{N_1}{K}, \quad v = \frac{aN_2}{r}, \quad \tau = rt, \quad \alpha = \frac{e}{r}, \quad \delta = \frac{d}{r}, \quad \beta = bK.$$

Determine the non-negative equilibria and note any parameter restrictions. Discuss the linear stability of the equilibria. Show that a non-zero N_2-

population can exist if $\beta > \beta_c = -\ln(1 - \delta/\alpha)$. Briefly describe the bi-furcation behaviour as β increases with $0 < \delta/\alpha < 1$.

5. The sterile insect release method (SIRM) for pest control releases a number of sterile insects into a population. If a population n of sterile insects is maintained in a population a possible simple model for the population of fertile insects $N(t)$ is

$$\frac{dN}{dt} = \left[\frac{aN}{N+n} - b \right] N - kN(N+n) ,$$

where $a > b > 0$ and $k > 0$ are constant parameters. Briefly discuss the assumptions which lie behind the model.

Determine the critical number of sterile insects n_c which would eradicate the pests and show that this is less than a quarter of the environmental carrying capacity.

Suppose that a single release of sterile insects is made and that the sterile insects have the same death rate as fertile insects. Write down the appropriate model system for $N(t)$ and $n(t)$ and show that it is not possible to eradicate the insect pests with a single release of sterile insects.

If a fraction γ of the insects born are sterile a suggested model is

$$\frac{dN}{dt} = \left[\frac{aN}{N+n} - b \right] N - kN(N+n), \quad \frac{dn}{dt} = \gamma N - bn .$$

Determine the condition on γ for eradication of the pest and briefly discuss the realism of the result.

4. Discrete Growth Models
for Interacting Populations

Here we consider two interacting species, each with non-overlapping generations, which affect each others population dynamics. As in the continuous growth models, there are the same main types of interaction namely predator-prey, competition and mutualism. In a predator-prey situation the growth rate of one is enhanced at the expense of the other whereas in competition the growth rates of both are decreased while in mutualism they are both increased. These topics have been widely studied but nowhere near to the same extent as for continuous models for which, in the case of two species, there is a complete mathematical treatment of the equations. The book by Hassel (1978) deals with predator-prey models. Beddington et al. (1975) present some results on the dynamic complexity of coupled predator-prey systems. The book by Gumowski and Mira (1980) is more mathematical, dealing generally with the mathematics of coupled systems but also including some interesting numerically computed results: see also the introductory article by Lauwerier (1986). The review article by May (1986) is apposite to the material here and that in the previous chapters, the central issue of which is how populations regulate. He also discusses, for example, the problems associated with unpredictable environmental factors superimposed on deterministic models and various practical aspects of resource management. In view of the complexity of solution behaviour with single species discrete models it is not surprising that even more complex behaviour is possible with coupled discrete systems. Even though we expect complex behaviour it is hard not to be overwhelmed by the astonishing solution diversity when we see the baroque patterns that can be generated as has been so beautifully demonstrated by Peitgen and Richter (1986). Their book is devoted in large part to the numerically generated solutions of discrete systems. They show, in striking colour, a wide spectrum of patterns which can arise, for example, with a system of only two coupled equations; the dynamics need not be very complicated. They also show, among other things, how the solutions relate to fractal generation (see, for example, Mandelbrot 1982), Julia sets, Hubbard trees and other exotica. Most of the text is a technical but easily readable discussion of the main topics of current interest in dynamical systems.

Here we shall be mainly concerned with predator-prey models. An important aspect of evolution by natural selection is the favouring of efficient predators and cleverly elusive prey. Within the general class, we shall have in mind primarily insect predator-prey systems, since as well as the availability of a substantial body of experimental data, insects often have life cycles which can be modelled by two-species discrete models.

We consider the interaction for the prey(N) and the predator(P) to be governed by the discrete time(t) system of coupled equations

$$N_{t+1} = rN_t f(N_t, P_t) , \tag{4.1}$$

$$P_{t+1} = N_t g(N_t, P_t) , \tag{4.2}$$

where $r > 0$ is the net linear rate of increase of the prey and f and g are functions which relate the predator-influenced reproductive efficiency of the prey and the searching efficiency of the predator respectively.

4.1 Predator-Prey Models: Detailed Analysis

We first consider a simple model in which predators simply search over a constant area and have unlimited capacity for consuming the prey. This is reflected in the system

$$N_{t+1} = rN_t \exp{[-aP_t]} ,$$
$$P_{t+1} = N_t\{1 - \exp{[-aP_t]}\} . \qquad a > 0 \tag{4.3}$$

Perhaps it should be mentioned here that it is always informative to try and get an intuitive impression of how the interaction affects each species by looking at the qualitative behaviour indicated by the equations. With this system, for example, try and decide what the outcome of the stability analysis will be. In general if the result is not what you anticipated such a preliminary qualitative impression can often help in modifying the model to make it more realistic.

The equilibrium values N^*, P^* of (4.3) are given by

$$N^* = 0, \quad P^* = 0$$

$$\text{or} \quad 1 = r \exp{[-aP^*]}, \quad P^* = N^*(1 - \exp{[-aP^*]})$$

and so positive steady state populations are

$$P^* = \frac{1}{a}\ln r, \quad N^* = \frac{r}{a(r-1)}\ln r, \quad r > 1 . \tag{4.4}$$

The linear stability of the equilibria can be determined in the usual way by writing

$$N_t = N^* + n_t, \quad P_t = P^* + p_t, \quad \left|\frac{n_t}{N^*}\right| \ll 1, \quad \left|\frac{p_t}{P^*}\right| \ll 1 , \tag{4.5}$$

substituting into (4.3) and retaining only linear terms. For the steady state (0,0) the analysis is particularly simple since

$$n_{t+1} = rn_t, \quad p_{t+1} = 0 ,$$

and so it is stable for $r < 1$ since $n_t \to 0$ as $t \to \infty$ and unstable for $r > 1$, that is the range of r when the positive steady state (4.4) exists. For this positive steady state we have the linear system of equations

$$n_{t+1} = n_t - N^* a p_t, \quad p_{t+1} = n_t \left(1 - \frac{1}{r}\right) + \frac{N^* a}{r} p_t , \qquad (4.6)$$

where we have used the relation $1 = r \exp\left[-aP^*\right]$ which defines P^*.

A straightforward way to solve (4.6) is to iterate the first equation and then use the second to get a single equation for n_t. That is

$$n_{t+2} = n_{t+1} - N^* a p_{t+1}$$

$$= n_{t+1} - N^* a \left[n_t \left(1 - \frac{1}{r}\right) + \frac{N^* a}{r} p_t\right]$$

$$= n_{t+1} - N^* a \left[n_t \left(1 - \frac{1}{r}\right) + \frac{1}{r}(n_t - n_{t+1})\right]$$

and so

$$n_{t+2} - \left(1 + \frac{N^* a}{r}\right) n_{t+1} + N^* a n_t = 0 . \qquad (4.7)$$

We now look for solutions in the form

$$n_t = A x^t \quad \Rightarrow \quad x^2 - \left(1 + \frac{N^* a}{r}\right) x + N^* a = 0 .$$

With N^* from (4.4) the characteristic polynomial is thus

$$x^2 - \left\{1 + \frac{1}{r-1} \ln r\right\} x + \frac{r}{r-1} \ln r = 0, \quad r > 1 \qquad (4.8)$$

of which the two solutions x_1 and x_2 are

$$x_1, x_2 = \frac{1}{2} \left\{ \left[1 + \frac{\ln r}{r-1}\right] \pm \left\{ \left[1 + \frac{\ln r}{r-1}\right]^2 - 4 \frac{r \ln r}{r-1}\right\}^{1/2}\right\} . \qquad (4.9)$$

Thus

$$n_t = A_1 x^t + A_2 x^t , \qquad (4.10)$$

where A_1, A_2 are arbitrary constants. With this, or by a similar analysis, we then get p_t as

$$p_t = B_1 x^t + B_2 x^t , \qquad (4.11)$$

where B_1 and B_2 are arbitrary constants.

A more elegant, and easy to generalize, way to find x_1 and x_2 is to write the linear perturbation system (4.6) in matrix form

$$\begin{pmatrix} n_{t+1} \\ p_{t+1} \end{pmatrix} = A \begin{pmatrix} n_t \\ p_t \end{pmatrix}, \qquad A = \begin{pmatrix} 1 & -N^*a \\ 1 - \dfrac{1}{r} & \dfrac{N^*a}{r} \end{pmatrix} \tag{4.12}$$

and look for solutions in the form

$$\begin{pmatrix} n_t \\ p_t \end{pmatrix} = B \begin{pmatrix} 1 \\ 1 \end{pmatrix} x^t$$

where B is an arbitrary constant 2×2 matrix. Substituting this into (4.12) gives

$$B \begin{pmatrix} x^{t+1} \\ x^{t+1} \end{pmatrix} = AB \begin{pmatrix} x^t \\ x^t \end{pmatrix} \quad \Rightarrow \quad xB \begin{pmatrix} x^t \\ x^t \end{pmatrix} = AB \begin{pmatrix} x^t \\ x^t \end{pmatrix}$$

which has a non-trivial solution $B \begin{pmatrix} x^t \\ x^t \end{pmatrix}$ if

$$|A - xI| = 0 \quad \Rightarrow \quad \begin{vmatrix} 1 - x & -N^*a \\ 1 - \dfrac{1}{r} & \dfrac{N^*a}{r} - x \end{vmatrix} = 0$$

which again gives the quadratic characteristics equation (4.8). The solutions x_1 and x_2 are simply the eigenvalues of the matrix A in (4.12). This matrix approach is the discrete equation analogue of the one we used for the continuous interacting population models in Chapter 3. The generalization to higher order discrete model systems is clear.

The stability of the steady state (N^*, P^*) is determined by the magnitude of $|x_1|$ and $|x_2|$. If either of $|x_1| > 1$ or $|x_2| > 1$ then n_t and p_t become unbounded as $t \to \infty$ and hence (N^*, P^*) is unstable since perturbations from it grow with time. A little algebra shows that in (4.9),

$$\left[1 + \frac{\ln r}{r - 1} \right]^2 - \frac{4r \ln r}{r - 1} < 0 \quad \text{for} \quad r > 1$$

and so the roots x_1 and x_2 are complex conjugates. The product of the roots, from (4.8), or (4.9), is

$$x_1 x_2 = |x_1|^2 = \frac{r \ln r}{r - 1} > 1, \quad \text{for all} \quad r > 1, \quad \Rightarrow \quad |x_1| > 1 \,.$$

(An easy way to see that $r \ln r/(r - 1) > 1$ for all $r > 1$ is to consider the graphs of $\ln r$ and $(r - 1)/r$ for $r > 1$ and note that $d(\ln r)/dr > d\left[(r - 1)/r \right]/dr$ for all $r > 1$.) Thus the solutions (n_t, p_t) from (4.10) and (4.11) become unbounded as $t \to \infty$ and so the positive equilibrium (N^*, P^*) in (4.4) is unstable, and by growing oscillations since x_1 and x_2 are complex. Numerical solutions of the system (4.3) indicate that the system is unstable to finite perturbations as well: the solutions grow unboundedly. Thus this simple model is just too simple for any practical applications except possibly under contrived laboratory conditions and then only for a limited time.

Density Dependent Predator-Prey Model

Let us re-examine the underlying assumptions in the simple initial model (4.3). The form of the equations imply that the number of encounters a predator has with a prey increases unboundedly with the prey density: this seems rather unrealistic. It is more likely that there is a limit to the predators' appetite. Another way of looking at this equation as it stands, and which is formally the same, is that if there were no predators $P_t = 0$ and then N_t would grow unboundedly, if $r > 1$, and become extinct if $0 < r < 1$: it is the simple Malthusian model (2.2). It is reasonable to modify the N_t equation (4.3) to incorporate some saturation of the prey population or, in terms of predator encounters, a prey limiting model. We thus take as a more realistic model

$$N_{t+1} = N_t \exp\left[r\left(1 - \frac{N_t}{K}\right) - aP_t\right] ,$$
$$P_{t+1} = N_t\{1 - \exp\left[-aP_t\right]\} .$$

(4.13)

Now with $P_t = 0$ this reduces to the single species model (2.4) in Section 2.1. There is a stable positive equilibrium $N^* = K$ for $0 < r < 2$ and oscillatory and periodic solutions for $r > 2$. We can reasonably expect a similar bifurcation behaviour here although probably not with a first bifurcation at $r = 2$ and certainly not the same values for r with higher bifurcations. This model has been studied in detail by Beddington et al. (1975).

The non-trivial steady states of (4.13) are solutions of

$$1 = \exp\left[r\left(1 - \frac{N^*}{K}\right) - aP^*\right], \quad P^* = N^*(1 - \exp\left[-aP^*\right]) .$$

(4.14)

The first of these gives

$$P^* = \frac{r}{a}\left(1 - \frac{N^*}{K}\right)$$

(4.15)

which on substituting into the second gives N^* as solutions of the transcendental equation

$$\frac{r\left(1 - \frac{N^*}{K}\right)}{aN^*} = 1 - \exp\left[-r\left(1 - \frac{N^*}{K}\right)\right] .$$

(4.16)

Clearly $N^* = K$, $P^* = 0$ is a solution. If we plot the left and right hand sides of (4.16) against N^* as in Fig. 4.1 we see there is another equilibrium

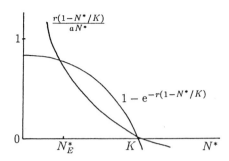

Fig. 4.1. Graphical solution for the positive equilibrium N_E^* of the model system (4.13).

$0 < N_E^* < K$, the other intersection of the curves: it depends on r, a and K. With N_E^* determined, (4.15) then gives P_E^*.

The linear stability of this equilibrium can be treated in exactly the same way as before with the eigenvalues x again being given by the eigenvalues of the matrix of the linearised system. It has to be done numerically. It can be shown that for some $r > 0$ the equilibrium is stable and that it bifurcates for larger r. Beddington et al. (1975) determine the stability boundaries in the $r, N_E^*/K$ parameter space where there is a bifurcation from stability to instability and where the solutions exhibit periodic and ultimately chaotic behaviour. The stability analysis of realistic two-species and models often has to be carried out numerically. For three-species and higher the Jury conditions (see Appendix 2) can be used to determine the conditions which the coefficients must satisfy so that the linear soltuions x satisfy $|x| < 1$. For higher order systems, however, they are of little use except within a numerical scheme.

*4.2 Synchronized Insect Emergence: 13 Year Locusts

A remarkable feature of some insects is the synchronized emergence of large numbers of them. One of these is the 13-year locust or cicada (*Magicicada*) which appear in outbreak proportions every 13 years with essentially none in the time in between. There is also a 17-year cicada with 17-year synchronized emergences. [One of the 17-year locust emergences coincided with the degree giving ceremony at Princeton University in 1970 when the folk singer Bob Dylan got an honorary degree. He subsequently commemorated it in one of his folk songs "Day of the Locusts" (New Morning, Columbia PC30290)]. This phenomenon cannot be explained by the lifespan of the cicadae. We would expect births and deaths every year. A brief critical review of various theories for this long term periodic cicada emergence phenomenon is given by May (1979). Hoppensteadt and Keller (1976) proposed an interesting explanation by showing that synchronized insect emergences are a possible consequence of the interaction of predation and a limited environmental carrying capacity. It is their model which we discuss in detail here. The main interest in their model and analysis from our point of view is to show the remarkable spectrum of possible behaviour the model (or rather the whole class of them) can exhibit rather than suggesting it as the definitive explanation of the specific phenomenon.

Adult cicadae, during the few weeks they live, lay eggs in trees near where they emerge. The eggs fall to the ground and become nymphs on hatching. The nymphs then attach themselves to the underground tree roots for sustenance and they stay there for 13 years. Life spans of the nymphs vary from 3 to 4 to 7 to 13 to 17 years depending on the species. Interestingly it is only the 13 and 17 year cicadae that have synchronized emergences. This extreme periodicity in time is most unlikely to be maintained without some ecological pressure other than simple birth and death processes. The candidates for it are bird predation and the carrying capacity for the cicadae. The model we develop here incorporates a predator satiation and a limited carrying capacity for the cicadae.

Suppose that a species has a life span of k years with nymph production occuring every k-th year after which the parents die. Let n_{t-k} be the number of nymphs which become established under the ground in year $t - k$ and suppose that a fraction μ of them survive each year. Then $\mu^k n_{t-k}$ will survive k years and will emerge as adults to start the cycle over again. When the adults emerge let us suppose that up to p_t are taken by predators. Then if $\mu^k n_{t-k} < p_t$ there are no adults available for reproduction. If $\mu^k n_{t-k} > p_t$ then $\mu^k n_{t-k} - p_t$ are available for mating. In what follows we use the notation

$$x_+ = \begin{cases} x \\ 0 \end{cases} \text{if} \begin{cases} x > 0 \\ x < 0 \end{cases} \tag{4.17}$$

Then, if f, a constant, is a measure of the fertility of the hatched nymphs who established themselves underground the number of nymphs N_t produced in year t is

$$N_t = f(\mu^k n_{t-k} - p_t)_+ . \tag{4.18}$$

At year t let the residual carrying capacity of the underground roots be c_t and assume that all the n_t nymphs are accommodated if the c_t can sustain them, that is if $n_t \leq c_t$. Any excess over c_t die. Thus the number of nymphs n_t which become established in year t is the minimum of N_t and c_t namely

$$n_t = \min (N_t, c_t) . \tag{4.19}$$

We now need c_t. To find it let D be the total environmental carrying capacity in which case,

$$c_t = \left(D - \sum_{i=1}^{k-1} \mu^i n_{t-i} \right)_+ \tag{4.20}$$

since the summation contribution is the number surviving from earlier years. Clearly if enough survive underground to fill the available carrying capacity no extra nymphs can be accomodated, in which case $c_t = 0$. Using (4.18) and (4.20) in (4.19) the equation for the nymphs underground is then

$$n_t = \min \left[f(\mu^k n_{t-k} - p_t)_+, \left(D - \sum_{i=1}^{k-1} \mu^i n_{t-i} \right)_+ \right] . \tag{4.21}$$

We now need an expression for the predation p_t which incorporates all of the predators feeding on the emergent cicada population. Suppose, as is reasonable, that there is a natural decrease in the predators if there are no prey and an increase proportional to the size of the previous years nymph emergence if there is prey. Then

$$p_t = \nu p_{t-1} + a\mu^k n_{t-k-1} , \tag{4.22}$$

where $1 > \nu > 0$ and $a > 0$ are constants. The last two equations represent the model for the interacting populations n_t and p_t in year $t > k$ in terms of the populations $n_0, n_1, \ldots, n_k$ and p_k. The model is a coupled interacting population delay system the solutions of which we would expect to be rich in structure.

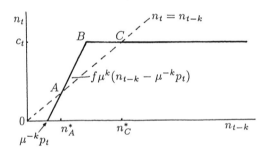

Fig. 4.2. The reproduction curve for n_t as a function of n_{t-k} for equation (4.21). The curve changes each year with the predation amount p_t and the available capacity c_t.

To get a picture of what to expect from such a model consider first, equation (4.21) for n_t. Fig. 4.2 shows the reproduction curve of n_t against n_{t-k} for some typical parameters. The steady states are the intersections of the reference line $n_t = n_{t-k}$ with the reproduction curve. Here there are three equilibria, $n_t = 0, n_A^*, n_C^*$ as shown. At the middle one at A, $n_t = n_{t-k} = n_A^*$, where

$$n_A^* = \frac{f p_A^*}{f\mu^k - 1} , \tag{4.23}$$

is a threshold density: it is given by (4.21) when the first bracketed term is used. The form of the 'curve' is a *depensatory* one which can be compared, for example, with the situation near $N_t = 0$ illustrated in Chapter 2, Fig. 2.12. There, if the population was less than a critical value the species became extinct. Here if c_t and p_t were fixed we can then see why we call n_A^* a threshold to extinction since if n_t is ever less than n_A^* a simple cobweb shows that n_t tends to zero. In fact the whole population becomes extinct unless $f\mu^k > 1$ and $\nu < 1$ both of which we assume hold. Also if there is no emergence in any year t there is no emergence in any year $t + mk$, $m = 1, 2 \ldots$. In general, of course, p_t and c_t vary from year to year and so the reproduction curve Fig. 4.2 also varies.

There are several parameters in the model mechanism (4.21) and (4.22) and we now expect the solutions for n_t and p_t to behave, qualitatively as well as quantitatively, differently for different values of the parameters. There may be extinction, or steady state populations for every year, oscillatory behaviour with short time scales or where the progeny of one specific year approach the carrying capacity while the progeny of all other years are close to extinction. This last possibility is the solution that describes our 13 year cicadae. The analysis of (4.21) and (4.22) even for a linearized theory gets rather complicated but we now know in principle how it can be done. All we shall do here is find conditions such that the solution is in a steady state and hence find a bifurcation situation. Some numerical results are given below for parameter values where the steady state has bifurcated to a periodic solution.

For an unchanging steady state all n_t are equal, to n^* say. From (4.21) using the second expression on the right, the steady state is n_C^*, marked C in Fig. 4.2, and is given by

$$n_C^* = D - \sum_{i=1}^{k-1} \mu^i n_C^* \quad \Rightarrow \quad n_C^* = D \frac{1 - \mu}{1 - \mu^k} , \tag{4.24}$$

which on substituting into (4.22) gives the steady state p_C^* of the predators as

$$p_C^* = \frac{a\mu^k n_C^*}{(1 - \nu)} = \frac{aD\mu^k(1 - \mu)}{(1 - \nu)(1 - \mu^k)}, \quad 0 < \mu < 1, \quad 0 < \nu < 1 . \tag{4.25}$$

This n_C^* from Fig. 4.2 is stable as is clear from a cobweb. Thus we can only have instability if the first term on the right in (4.21) is less than the second. The critical bifurcation state then is when the inequality $n_C^* > n_A^*$ ceases to hold, that is when $n_A^* = n_C^*$ and $p_A^* = p_C^*$ since then it is the second bracketed term on the right of (4.21) that is the minimum. In this case A, B and C in Fig. 4.2 coalesce to a single point lying on $n_t = n_{t-k}$. Thus the critical state is when

$$n_C^* = n_A^* = \frac{fp_A^*}{f\mu^k - 1} = \frac{fp_C^*}{f\mu^k - 1} \tag{4.26}$$

and so, using (4.25), we have as the necessary condition for a stable state of balance between n_t and p_t,

$$n_C^* > \frac{f}{f\mu^k - 1} \frac{a^* \mu^k n_C^*}{1 - \nu}$$

$$\Rightarrow \quad T = \frac{af\mu^k}{(f\mu^k - 1)(1 - \nu)} < 1 . \tag{4.27}$$

A synchronized emergence solution of the kind we want, namely a period-k solution, that is $n_t = n_{t+k}$ in the notation of Section 2.4, is the period-k fixed point

$$n_{mk} = L, \quad n_{mk+j} = 0, \quad j = 1, 2, \ldots, k-1 \tag{4.28}$$

for some $L > 0$ and any integer m. Now with $p_{t+k} = p_t$, from (4.22)

$$p_{t+k+1} = \nu p_{t+k} + a\mu^k n_t$$

and hence

$$p_{t+1} = \nu p_t + a\mu^k n_t = \nu p_t + a\mu^k L .$$

Now increasing the time t by $k-1$ steps, noting (4.28),

$$p_{t+2} = \nu p_{t+1} + 0$$
$$\vdots$$
$$p_{t+k} = \nu p_{t+k-1}$$

we have, since $p_{t+k} = p_t$,

$$p_t = p_{t+k} = \nu^{k-1} p_{t+1} = \nu^{k-1}(\nu p_t + a\mu^k L)$$

and so

$$p_t = \frac{aL\nu^{k-1}\mu^k}{1 - \nu^k} .$$

We thus have

$$p_{mk+j} = \frac{aL\nu^{j-1}\mu^k}{1 - \nu^k}, \quad j = 1, \ldots, k . \tag{4.29}$$

But, it is necessary that $n_{mk} > n_C^*$ if this periodic solution satisfies (4.21). That is, using (4.26),

$$L = n_{mk} > n_C^* = \frac{fp_C^*}{f\mu^k - 1} ,$$

which, on using (4.29), leads to

$$S = \frac{af\nu^{k-1}\mu^k}{(f\mu^k - 1)(1 - \nu^k)} < 1 \tag{4.30}$$

as the necessary condition for a synchronized (period-k) solution in which L nymphs emerge every kth year.

Now since $0 < \nu < 1$, the expressions for T and S in (4.27) and (4.30) show that $0 < S < T$ for $k > 1$. Thus if $T < 1$, that is a stable balanced steady state can exist, it is also the case that $0 < S < T < 1$ and so a synchronized solution exists. If $S > 1$ then necessarily $T > 1$ and so neither a steady state nor a synchronized solution can exist. So finally, if $S < 1 < T$ no steady state exists

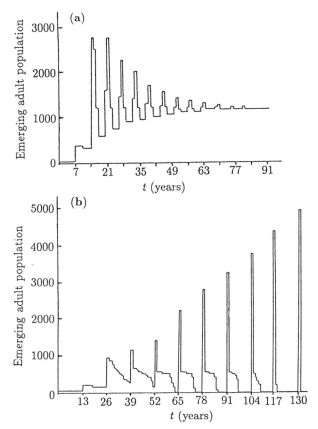

Fig. 4.3a,b. Numerical solution of the mechanism (4.21) and (4.22) for $L = 10,000$, $\nu = \mu = 0.95$, $a = 0.042$ and $f = 10$. The initial population was 100 nymphs. The emergent population $\mu^k n_{t-k}$ in year t is plotted for each t. (a) $k = 7$: the population tends to a uniform equilibrium. (b) $k = 13$: the emergence is synchronized to a 13-year period. (Redrawn from Hoppensteadt and Keller 1976)

but a synchronized solution can, and this is the situation in which an initial state will approach the synchronized solution behaviour.

Fig. 4.3 shows the results of two calculations for the system (4.21) and (4.22) for one set of parameter values and two different k. The initial conditions, k of them, were all chosen to be the same and in each an initial sample of 100 nymphs were taken. Being a coupled delay system this means that $n_t = 100$ for $t = -k, -k+1, \ldots, -1$. With the parameter values used $S < 1 < T$ for $k > 10$ and $S < T < 1$ for $2 < k < 9$. For $k = 7$, the case in Fig. 4.3 (a), the steady equilibrium solution evolved, while for $k = 11, 12 \ldots 17$ a synchronized solution evolved. Fig. 4.3 (b) illustrates a period-13 solution.

This model, with the *specific parameter values* used, is consistent with the observation that cicadae with life spans of 3, 4 and 7 year life spans do not exhibit synchronized emergencies. However before being too convinced by this model and analysis remember that (4.30) is the condition for emergence every k-th year. We could equally well ensure a 15-year emergence or a 9 or whatever. It may be, however, that this model is a reasonable one to start with for the cicadae phenomenon: see May (1979) for a discussion of it and other models. The importance of this model from our point of view here is in showing how stable periodic behaviour can exist if the predation is of some intermediate intensity.

The synchronized emergence behaviour does not rely on the detailed form of the equations (4.21) and (4.22), the schematic form of the former being the piecewise linear reproduction curve in Fig. 4.2. By appropriate choice of parameters a depensatory curve which saturates for large n_t would also suffice: for example $n_t = F(n_{t-k}; n^*, c_t)$ where F is a smooth function satisfying $F(n) < n$ for $0 < n < n^* < c_t$ and $F(n) > n$ for $0 < n^* < n < c_t$. Hoppensteadt and Keller (1976) give as an example $F(n) = [n - n(n^* - n)(c_t - n)]_+$ where the residual carrying capacity c_t and the threshold to extinction n^* are as defined above in the model we have analysed. The analysis would be considerably more complicated and would have to be done numerically.

(Some people have said that 13 and 17 year cycles arise because they are prime numbers! Then why not 11 say – it would then allow resonance with the sun spot cycle of 11 years!)

4.3 Biological Pest Control: General Remarks

The model proposed for the 13-year locust emergence is an example of the remarkable solution behaviour such coupled population models can exhibit. Importantly it demonstrates the effect of delay in the population dynamics. The model is based on a plausible mechanism of the environment and the predation threshold above which the prey tends to extinction. It is a subtle balance between the predation and carrying capacity which effect the synchronized behaviour. If this is indeed a fair representation of the governing mechanism then it is clear how comparatively easy it is to destabilize the process by changing one of the parameters by external control. For example if the predation response to an emergence is increased by, say, human involvement then all the progeny of subsequent years will eventually be eliminated. On the other hand if the predation is reduced, for example by predation of the predators, there could be yearly emergences.

The use of natural predators for pest control is to inhibit any large pest increase by a corresponding increase in the predator population. The aim is to keep both populations at acceptably low levels. The aim is *not* to eradicate the pest, only to control its population. Although many model systems of real predators and real pests are reasonably robust from a stability point of view some can be extremely sensitive. This is why the analysis of realistic models is so important. When the parameters for a model are taken from observations it is fortunate that many result in either steady state equilibria or simple periodic behaviour: chaotic behaviour is much less common. Thus effective parameter manipulation is more predictable in a substantial number of practical situations.

There are many notable successes of biological pest control particularly with long standing crops such as fruit and forest crops and so on, where there is a continuous predator-prey interaction. With the major ecological changes caused by harvesting in perennial crops it has been less successful. The successes have mainly been of the predator-prey variety where the predator is a parasite. This can be extremely important in certain human diseases.

In the models we have analysed we have concentrated particularly on the model building aspects, the study of stability in relation to parameter ranges and the existence of either steady states or periodic behaviour. What we have not discussed is the influence of initial conditions. Although not generally the case, they can be important. One such example is the control of the red spider mite which is a glasshouse tomato plant pest where the initial predator-prey ratio is crucial. We should expect initial data to be important particularly in those cases where the oscillations show outbreak, crashback and slow recovery. The crashback to low levels may bring the species close enough to extinction to actually cause it.

There are several books on biological pest control: see for example DeBach (1974) and Huffaker (1971).

A moderately new and, in effect, virgin territory is the study of coupled systems where the time steps for the predator and prey are not equal. This clearly occurs in the real world. With the wealth of interesting and unexpected behaviour displayed by the models in this chapter and Chapter 2 it would be surprising if different time step models did not produce equally unexpected solution behaviour.

Exercises

1. A general form for models for insect predator(P)-prey(N), or insect parasitism is

$$N_{t+1} = rN_t f(N_t, P_t), \quad P_{t+1} = N_t[1 - f(N_t, P_t)],$$

where f is a nonlinear function which incorporates assumptions about predator searching, and $r > 0$ is the rate of increase of prey population. The scaling is such that $0 < f < 1$. Here f is an increasing function as N_t increases and a decreasing function as P_t increases. Does this model make sense ecologically?

Show that a positive equilibrium state (N^*, P^*) can exist and give any conditions on r required. Show that the linear stability of the steady state is ensured if the roots of

$$x^2 - \left[1 + rN^*\frac{\partial f}{\partial N_t} - N^*\frac{\partial f}{\partial P_t}\right]x - rN^*\frac{\partial f}{\partial P_t} = 0$$

have magnitudes less than 1, where $\partial f / \partial N_t$ and $\partial f / \partial P_t$ are evaluated at (N^*, P^*), and hence determine the conditions for linear stability.

2. A model for the regulation of a host population by a microparasite population u_t which was proposed and studied by May (1985) is, in dimensionless form,

$$1 - I_t = \exp\left[-I_t u_t\right], \quad u_{t+1} = \lambda u_t (1 - I_t),$$

where $\lambda > 0$ and I_t denotes the fraction of the host population which has been infected by the time the epidemic has run its course. The assumption in this specific form is that the parasite epidemic has spread through each generation before the next population change. (This is why the host population equation does *not* involve I_{t+1}.) Determine the steady states and note any restrictions on λ for a positive steady state to exist for both the host and microparasite populations. Investigate the linear stability of the positive steady state. Show that it is *always* unstable and that the instability arises via a pitchfork bifurcation.

[May (1985) studies this model in depth and shows that the positive steady state *and* all periodic solutions are unstable: that is the model only exhibits chaotic behaviour without going through the usual period doubling. He also discusses the epidemiological implications of such a simple, yet interesting and surprising, system.]

5. Reaction Kinetics

5.1 Enzyme Kinetics: Basic Enzyme Reaction

Biochemical reactions are continually taking place in all living organisms and most of them involve proteins called *enzymes*, which act as remarkably efficient catalysts. Enzymes react selectively on definite compounds called *substrates*. For example haemoglobin in red blood cells is an enzyme and oxygen, with which it combines, is a substrate. Enzymes are important in regulating biological processes, for example as activators or inhibitors in a reaction. To understand their role we have to study enzyme kinetics which is mainly the study of rates of reactions, the temporal behaviour of the various reactants and the conditions which influence them. Introductions with a mathematical bent are given in the books by Rubinow (1975), Murray (1977) and the one edited by Segel (1980). A biochemically oriented book, Roberts (1977) for example, goes into the subject in more depth.

The complexity of biological and biochemical processes is such that the development of a simplifying model is often essential in trying to understand the phenomenon under consideration. For such models we should use reaction mechanisms which are plausible biochemically. Frequently the first model to be studied may itself be a model of a more realistic, but still too complicated, biochemical model. Models of models are often first steps since it is a qualitative understanding that we want initially. In this chapter we discuss some model reaction mechanisms, which mirror a large number of real reactions, and some general types of reaction phenomena and their corresponding mathematical realisations: a knowledge of these is essential when constructing models to reflect specific known biochemical properties of a mechanism.

Basic Enzyme Reaction

One of the most basic enzymatic reactions, first proposed by Michaelis and Menten (1913), involves a substrate S reacting with an enzyme E to form a complex SE which in turn is converted into a product P and the enzyme. We represent this schematically by

$$S + E \underset{k_{-1}}{\overset{k_1}{\rightleftharpoons}} SE, \quad SE \overset{k_2}{\to} P + E \,. \tag{5.1}$$

Here k_1, k_{-1} and k_2 are constant parameters associated with the rates of reaction: they are defined below. The double arrow symbol $\rightleftharpoons$ indicates that the reaction is reversible while the single arrow $\rightarrow$ indicates that the reaction can go only one way. The overall mechanism is a conversion of the substrate S, via the enzyme catalyst E, into a product P. In detail it says that one molecule of S combines with one molecule of E to form one of SE, which eventually produces one molecule of P and one molecule of E again.

The *Law of Mass Action* says that the rate of a reaction is proportional to the product of the concentrations of the reactants. We denote the concentrations of the reactants in (5.1) by lower case letters

$$s = [S], \quad e = [E], \quad c = [SE], \quad p = [P], \tag{5.2}$$

where [ ] traditionally denotes concentration. Then the law of mass action applied to (5.1) leads to one equation for each reactant and hence the system of nonlinear reaction equations

$$\frac{ds}{dt} = -k_1 es + k_{-1}c, \quad \frac{de}{dt} = -k_1 es + (k_{-1} + k_2)c$$

$$\frac{dc}{dt} = k_1 es - (k_{-1} + k_2)c, \quad \frac{dp}{dt} = k_2 c \tag{5.3}$$

The k's, called *rate constants*, are constants of proportionality in the application of the Law of Mass Action. For example, the first equation for s is simply the statement that the rate of change of the concentration $[S]$ is made up of a loss rate proportional to $[S][E]$ and a gain rate proportional to $[SE]$.

To complete the mathematical formulation we require initial conditions which we take here as those at the start of the process which converts S to P, so

$$s(0) = s_0, \quad e(0) = e_0, \quad c(0) = 0, \quad p(0) = 0. \tag{5.4}$$

The solutions of (5.3) with (5.4) then give the concentrations, and hence the rates of the reactions, as functions of time. Of course in any reaction kinetics problem we are only concerned with non-negative concentrations.

Equations (5.3) are not all independent. Also the last equation is uncoupled from the first three; it gives the product

$$p(t) = k_2 \int_0^t c(t') \, dt', \tag{5.5}$$

once $c(t)$ has been determined. In the mechanism (5.1) the enzyme E is a catalyst, which only facilitates the reaction, so its total concentration, free plus combined, is a constant. This conservation law for the enzyme also comes immediately from (5.3) on adding the 2nd and 3rd equations, those for the free(e) and combined(c) enzyme concentrations respectively, to get

$$\frac{de}{dt} + \frac{dc}{dt} = 0 \quad \Rightarrow \quad e(t) + c(t) = e_0 \tag{5.6}$$

on using the initial conditions (5.4). With this, the system of ordinary differential equations reduces to only two, for s and c, namely

$$
\begin{aligned}
\frac{ds}{dt} &= -k_1 e_0 s + (k_1 s + k_{-1})c \,, \\
\frac{dc}{dt} &= k_1 e_0 s - (k_1 s + k_{-1} + k_2)c \,,
\end{aligned}
\tag{5.7}
$$

with initial conditions $s(0) = s_0$, $c(0) = 0$.

With the nondimensionalization

$$
\tau = k_1 e_0 t, \quad u(\tau) = \frac{s(t)}{s_0}, \quad v(\tau) = \frac{c(t)}{e_0} \,,
$$

$$
\lambda = \frac{k_2}{k_1 s_0}, \quad K = \frac{k_{-1} + k_2}{k_1 s_0}, \quad \varepsilon = \frac{e_0}{s_0}
\tag{5.8}
$$

the system (5.7) and the initial conditions become

$$
\frac{du}{d\tau} = -u + (u + K - \lambda)v, \quad \varepsilon \frac{dv}{d\tau} = u - (u + K)v
\tag{5.9}
$$

$$
u(0) = 1, \quad v(0) = 0
$$

Note that $K - \lambda > 0$ from (5.8). With the solutions $u(\tau)$, $v(\tau)$ we then immediately get e and p from (5.6) and (5.5) respectively.

From the original reaction (5.1), which converts S into a product P, we clearly have the final steady state $u = 0$ and $v = 0$, that is both the substrate and the substrate-enzyme complex concentrations are zero. We are interested here in the time evolution of the reaction so we need the solutions of the nonlinear system (5.9), which we cannot solve analytically in a simple closed form. However we can see what $u(\tau)$ and $v(\tau)$ look like qualitatively. Near $\tau = 0$, $du/d\tau < 0$ so u decreases from $u = 1$ and since there $dv/d\tau > 0$, v increases from $v = 0$ and continues to do so until $v = u/(u + K)$ where $dv/d\tau = 0$ at which point, from the first of (5.9), u is still decreasing. After v has reached a maximum it then decreases ultimately to zero as does u, which does so monotonically for all τ. The dimensional enzyme concentration $e(t)$ first decreases from e_0 and then increases again to e_0 as $t \to \infty$. Typical solutions are illustrated later in Fig. 5.1. Quite often a qualitative feel for the solution behaviour can be obtained from just looking at the equations: it is always profitable to try.

5.2 Michaelis-Menten Theory: Detailed Analysis and the Pseudo-Steady State Hypothesis

It is widespread in biology that the remarkable catalytic effectiveness of enzymes is reflected in the small concentrations needed in their reactions as compared with the concentrations of the substrates involved. In the Michaelis-Menten model reaction kinetics in dimensionless form (5.9) this means $\varepsilon = e_0/s_0 \ll 1$: typically ε is in the range 10^{-2} to 10^{-7}. We now exploit this fact to obtain a very accurate approximate, or rather asymptotic, solution to (5.9) for $0 < \varepsilon \ll 1$. Before doing this we should note that the specific non-dimensionalisation (5.8) is only one of several we could choose. In a recent review by Segel and Slemrod (1989) they show that a more appropriate small parameter is $e_0/[s_0 + k_1^{-1}(k_{-1} + k_2)]$. They show that this allows a more comprehensive applicability of the pseudo-steady state hypothesis and has recently been exploited by Frenzen and Maini (1989) in a practical example. With this form we can have, for example, $e_0/s_0 = O(1)$ as long as the Michaelis constant $(k_{-1} + k_2)/k_1$ is large compared with e_0. This can occur in many biochemical reactions (see, for example, Frenzen and Maini 1989). Segel and Slemrod (1989) use a similar singular perturbation analysis to that which we now describe.

Suppose we simply look for a regular Taylor expansion solution to u and v in the form

$$u(\tau; \varepsilon) = \sum_{n=0} \varepsilon^n u_n(\tau), \quad v(\tau; \varepsilon) = \sum_{n=0} \varepsilon^n v_n(\tau), \tag{5.10}$$

which, on substituting into (5.9) and equating powers of ε, gives a sequence of differential equations for the $u_n(\tau)$ and $v_n(\tau)$. In other words we assume that $u(\tau; \varepsilon)$ and $v(\tau; \varepsilon)$ are analytic functions of ε as $\varepsilon \to 0$. The $0(1)$ equations are

$$\frac{du_0}{d\tau} = -u_0 + (u_0 + K - \lambda)v_0, \quad 0 = u_0 - (u_0 + K)v_0, \tag{5.11}$$

$$u_0(0) = 1, \quad v_0(0) = 0.$$

We can already see a difficulty with this approach since the second equation is simply algebraic and does not satisfy the initial condition: in fact if $u_0 = 1$, $v_0 = 1/(1 + K) \neq 0$. If we solve (5.11)

$$v_0 = \frac{u_0}{u_0 + K} \quad \Rightarrow \quad \frac{du_0}{d\tau} = -u_0 + (u_0 + K - \lambda)\frac{u_0}{u_0 + K} = -\lambda\frac{u_0}{u_0 + K}$$

and so

$$u_0(\tau) + K \ln u_0(\tau) = A - \lambda\tau.$$

If we require $u_0(0) = 1$ then $A = 1$. Thus we have a solution $u_0(\tau)$, given implicitly, and the corresponding $v_0(\tau)$

$$u_0(\tau) + K \ln u_0(\tau) = 1 - \lambda\tau, \quad v_0(\tau) = \frac{u_0(\tau)}{u_0(\tau) + K}. \tag{5.12}$$

However, this solution is not a uniformly valid approximate solution for all $\tau \geq 0$ since $v_0(0) \neq 0$. This is not surprising since (5.11) involves only one derivative; it was obtained on setting $\varepsilon = 0$ in (5.9). The system of equations (5.11) has only one constant of integration from the u-equation so it is not surprising that we cannot satisfy initial conditions on *both* u_0 and v_0.

The fact that a small parameter $0 < \varepsilon \ll 1$ multiplies a derivative in (5.9) indicates that it is a *singular perturbation* problem. One class of such problems is immediately recognised if, on setting $\varepsilon = 0$, the order of the system of differential equations is reduced: such a reduced system cannot in general satisfy all the initial conditions. Singular perturbation techniques are very important and powerful methods for determining asymptotic solutions of such systems of equations for small ε. Asymptotic solutions are usually remarkably accurate approximations to the exact solutions. A practical and elementary discussion of some of the key techniques is given in Murray's (1984) book on asymptotic analysis. In the following the philosophy and actual technique of the singular perturbation method will be described in detail and the asymptotic solution to (5.9) for $0 < \varepsilon \ll 1$ derived. The main reason for doing this is to indicate when we can neglect the ε-terms in practical situations.

Since the solution (5.12), specifically $v_0(\tau)$, does not satisfy the initial conditions (and inclusion of higher order terms in ε cannot remedy the problem) we must conclude that at least one of the solutions $u(\tau; \varepsilon)$ and $v(\tau; \varepsilon)$ is *not* an analytic function of ε as $\varepsilon \to 0$. By assuming $\varepsilon dv/d\tau$ is $O(\varepsilon)$ to get (5.11) we tacitly assumed $v(\tau; \varepsilon)$ to be analytic: (5.10) also requires analyticity of course. Since the initial condition $v(0) = 0$ could not be satisfied because we neglected $\varepsilon dv/d\tau$ we must therefore retain this term in our analysis, at least near $\tau = 0$. So, a more appropriate time scale *near* $\tau = 0$ is $\sigma = \tau/\varepsilon$ rather than τ; this makes $\varepsilon dv/d\tau = dv/d\sigma$. The effect of the transformation $\sigma = \tau/\varepsilon$ is to magnify the neighbourhood of $\tau = 0$ and lets us look at this region more closely since, for a fixed $0 < \tau \ll 1$, we have $\sigma \gg 1$ as $\varepsilon \to 0$. That is a very small neighbourhood near $\tau = 0$ corresponds to a very large domain in σ. We now use this to analyse (5.9) near $\tau = 0$, after which we shall get the solution away from $\tau = 0$ and finally show how to get a uniformly valid solution for all $\tau \geq 0$.

With the transformations

$$\sigma = \frac{\tau}{\varepsilon}, \quad u(\tau; \varepsilon) = U(\sigma; \varepsilon), \quad v(\tau; \varepsilon) = V(\sigma; \varepsilon) \tag{5.13}$$

the equations in (5.9) become

$$\frac{dU}{d\sigma} = -\varepsilon U + \varepsilon(U + K - \lambda)V, \quad \frac{dV}{d\sigma} = U - (U + K)V,$$
$$U(0) = 1, \quad V(0) = 0. \tag{5.14}$$

If we now set $\varepsilon = 0$ to get the $0(1)$ system in a regular perturbation solution

$$U(\sigma; \varepsilon) = \sum_{n=0} \varepsilon^n U_n(\sigma), \quad V(\sigma; \varepsilon) = \sum_{n=0} \varepsilon^n V_n(\sigma), \tag{5.15}$$

we get

$$\frac{dU_0}{d\sigma} = 0, \quad \frac{dV_0}{d\sigma} = U_0 - (U_0 + K)V_0 \,,$$
$$U_0(0) = 1, \quad V_0(0) = 0 \tag{5.16}$$

which is *not* of lower order than the original system (5.14). The solution of (5.16) is

$$U_0(\sigma) = 1, \quad V_0(\tau) = (1 + K)^{-1}(1 - \exp[-(1 + K)\sigma]) \,. \tag{5.17}$$

The last solution cannot be expected to hold for all $\tau \geq 0$ since if it did it would mean that $dv/d\sigma = \varepsilon dv/d\tau$ is $0(1)$ for all τ. The part of the solution given by (5.17) is the *singular* or *inner* solution for u and v and is valid for $0 \leq \tau \ll 1$, while (5.12) is the *nonsingular* or *outer* solution valid for all τ not in the immediate neighbourhood of $\tau = 0$. If we now let $\varepsilon \to 0$ we have for a fixed $0 < \tau \ll 1$, however small, $\sigma \to \infty$. Thus in the limit of $\varepsilon \to 0$ we expect the solution (5.12) as $\tau \to 0$ to be equal to the solution (5.17) as $\sigma \to \infty$. That is, the singular solution as $\sigma \to \infty$ matches the nonsingular solution as $\tau \to 0$: this is the essence of *matching* in singular perturbation theory. From (5.17) and (5.12) we see in fact that

$$\lim_{\sigma \to \infty} [U_0(\sigma), V_0(\sigma)] = \left[1, \frac{1}{1 + K}\right] = \lim_{\tau \to 0} [u_0(\tau), v_0(\tau)] \,.$$

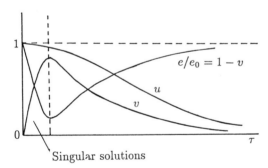

Singular solutions

Fig. 5.1. Schematic behaviour of the solutions of (5.9) for the dimensionless substrate(u), substrate-enzyme complex(v) and free enzyme($e/e_0 = 1 - v$) concentrations as functions of the time τ.

Fig. 5.1 illustrates the solution $u(\tau)$ and $v(\tau)$, together with the dimensionless enzyme concentration e/e_0 given by the dimensionless form of (5.6), namely $e/e_0 = 1 - v(\tau)$. The thin $O(\varepsilon)$ layer near $\tau = 0$ is sometimes called the *boundary layer* and is the τ-domain where there are very rapid changes in the solution. Here, from (5.17),

$$\left.\frac{dV}{d\tau}\right]_{\tau=0} \sim \varepsilon^{-1} \left.\frac{dV_0}{d\sigma}\right]_{\sigma=0} = \varepsilon^{-1} \gg 1 \,.$$

Of course from the original system (5.9) we can see this from the second equation and the boundary conditions.

To proceed in a systematic singular perturbation way, we first look for the outer solution of the full system (5.9) in the form of a regular series expansion (5.10). The sequence of equations is then

$$O(1):\quad \frac{du_0}{d\tau} = -u_0 + (u_0 + K - \lambda)v_0, \quad 0 = u_0 - (u_0 + K)v_0,$$

$$O(\varepsilon):\quad \begin{aligned} \frac{du_1}{d\tau} &= u_1(v_0 - 1) + (u_0 + K - \lambda)v_1, \\[2mm] \frac{dv_0}{d\tau} &= u_1(1 - v_0) - (u_0 + K)v_1, \end{aligned} \qquad (5.18)$$

and so on, which are valid for $\tau > 0$. The solutions involve undetermined constants of integration, one at each order, which have to be determined by matching these solutions as $\tau \to 0$ with the singular solutions as $\sigma \to \infty$.

The sequence of equations for the singular part of the solution, valid for $0 \le \tau \ll 1$, is given on substituting (5.15) into (5.14) and equating powers of ε, namely

$$O(1):\quad \frac{dU_0}{d\sigma} = 0, \quad \frac{dV_0}{d\sigma} = U_0 - (U_0 + K)V_0,$$

$$O(\varepsilon):\quad \begin{aligned} \frac{dU_1}{d\sigma} &= -U_1 + (V_0 + K - \lambda)V_1 \\[2mm] \frac{dV_1}{d\sigma} &= (1 - V_0)V_0 - (V_0 + K)V_0, \end{aligned} \qquad (5.19)$$

and so on. The solutions of these must satisfy the initial conditions at $\sigma = 0$, that is $\tau = 0$,

$$1 = U(0;\varepsilon) = \sum_{n=0} \varepsilon^n U_n(0) \quad \Rightarrow \quad U_0(0) = 1, \quad U_{n \ge 1}(0) = 0,$$

$$0 = V(0;\varepsilon) = \sum_{n=0} \varepsilon^n V_n(0) \quad \Rightarrow \quad V_{n \ge 0}(0) = 0. \qquad (5.20)$$

In this case the singular solutions of (5.19) are determined completely. This is not generally the case in singular perturbation problems (see, for example, Murray 1984). Matching of the inner and outer solutions requires choosing the undetermined constants of integration in the solutions of (5.18) so that to all orders of ε,

$$\lim_{\sigma \to \infty} [U(\sigma;\varepsilon), V(\sigma;\varepsilon)] = \lim_{\tau \to 0} [u(\tau;\varepsilon), v(\tau;\varepsilon)]. \qquad (5.21)$$

Formally from (5.18), but as we had before,

$$u_0(\tau) + K \ln u_0(\tau) = A - \lambda\tau, \quad v_0(\tau) = \frac{u_0(\tau)}{u_0(\tau) + K}$$

where A is the constant of integration we must determine by matching. The solution of the first of (5.19) with (5.20) has, of course, been given before in

(5.17). We get it now by applying the limiting process (5.21) to (5.17) and the last equations

$$\lim_{\sigma \to \infty} V_0(\sigma) = \frac{1}{1+K} = \lim_{\tau \to 0} v_0(\tau)$$

$$\Rightarrow \quad v_0(0) = \frac{1}{1+K} = \frac{u_0(0)}{u_0(0)+K}$$

$$\Rightarrow \quad u_0(0) = 1 \quad \Rightarrow \quad A = 1 .$$

We thus get the uniformly valid asymptotic solution for $0 < \varepsilon \ll 1$ to $O(1)$, derived heuristically before and given by (5.12) for $\tau > 0$ and (5.17) for $0 \leq \tau \ll 1$, although the singular part of the solution is more naturally expressed in terms of $0 \leq \tau/\varepsilon < \infty$.

We can now proceed to calculate $U_1(\sigma)$ and $V_1(\sigma)$ from (5.19) and $u_1(\tau)$ and $v_1(\tau)$ from (5.18) and so on to any order in ε ; the solutions become progressively more complicated even though all the equations are linear. In this way we get a uniformly valid asymptotic solution for $0 < \varepsilon \ll 1$ for all $\tau \geq 0$ of the nonlinear kinetics represented by (5.9). In summary, to $O(1)$ for small ε,

$$u(\tau;\varepsilon) = u_0(\tau) + O(\varepsilon); \quad u_0(\tau) + K \ln u_0(\tau) = 1 - \lambda\tau$$

$$v(\tau;\varepsilon) = V_0(\sigma) + O(\varepsilon); \quad V_0(\sigma) = (1+K)^{-1}\left(1 - \exp\left[-(1+K)\frac{\tau}{\varepsilon}\right]\right), \quad 0 \leq \tau \ll 1$$

$$= v_0(\tau) + O(\varepsilon); \quad v_0(\tau) = \frac{u_0(\tau)}{u_0(\tau)+K}, \quad 0 < \varepsilon \ll \tau .$$

$$(5.22)$$

Since in most biological applications $0 < \varepsilon \ll 1$, we need only evaluate the $O(1)$ terms: the $O(\varepsilon)$ terms' contributions are negligible.

To complete the analysis of the original kinetics problem (5.3) with (5.4), if we write the dimensionless product and free enzyme concentrations as

$$z(\tau) = \frac{p(t)}{s_0}, \quad w(\tau) = \frac{e(t)}{e_0}$$

then, using (5.22) for u and v, (5.5) and (5.6) give

$$z(\tau) = \lambda \int_0^\tau v(\tau')\,d\tau', \quad w(\tau) = 1 - v(\tau) .$$

The rapid change in the substrate-enzyme complex $v(\tau;\varepsilon)$ takes place in dimensionless times $\tau = O(\varepsilon)$ which is very small. The equivalent dimensional time t is also very short, $O(1/k_1 s_0)$ in fact, and for many experimental purposes is not measurable. Thus in many experimental situations the singular solution for $u(\tau)$ and $v(\tau)$ is never observed. The relevant solution is then the $O(1)$ outer solution $u_0(\tau)$, $v_0(\tau)$ in (5.12), obtained from the kinetics system (5.9) on setting $\varepsilon = 0$ and satisfying only the initial condition on $u(\tau)$, the substrate concentration. In other words we say that the reaction for the complex $v(\tau)$ is essentially in a steady state, or mathematically that $\varepsilon dv/d\tau \approx 0$. That is, the v-reaction is so fast it is more or less in equilibrium at all times. This is Michaelis and Menten's *pseudo-steady state hypothesis*.

The form of (5.9) is generally like

$$\frac{du}{d\tau} = f(u,v), \quad \frac{dv}{d\tau} = \varepsilon^{-1}g(u,v), \quad 0 < \varepsilon \ll 1, \qquad (5.23)$$

which immediately shows that $dv/d\tau \gg 1$ if $g(u,v)$ is not approximately equal to zero. So the v-reaction is very fast compared with the u-reaction. The v-reaction reaches a pseudo-steady state very quickly, which means that for times $\tau = O(1)$ it is essentially at equilibrium and the model mechanism is then approximated by

$$\frac{du}{d\tau} = f(u,v), \quad g(u,v) = 0, \quad u(0) = 1 . \qquad (5.24)$$

If we solve the algebraic equation $g(u,v) = 0$ to get $v = h(u)$ then

$$\frac{du}{d\tau} = f(u, h(u)) , \qquad (5.25)$$

which is the rate or *uptake* equation for the substrate concentration. Much modelling of biological processes hinges on qualitative assumptions for the uptake function $f(u, h(u))$.

What is of interest biologically is the *rate of reaction*, or the rate of uptake; that is $du/d\tau$ when $u(\tau)$ has been found. It is usually determined experimentally by measuring the dimensional substrate concentration $s(t)$ at various times, then extrapolating back to $t = 0$, and the magnitude r of the initial rate $[ds/dt]_{t=0}$ calculated. Since the time measurements are almost always for $\tau \gg \varepsilon$, that is $t \gg 1/k_1 s_0$, which is usually of the order of seconds, the equivalent analytical rate is given by the *nonsingular* or *outer solution*. Thus, from the first of (5.22) the $O(1)$ solution for the rate r, with $0 < \varepsilon \ll 1$, r_0 say, is

$$r_0 = \left[\frac{du_0(\tau)}{d\tau} \right]_{\tau=0} = \lambda \frac{u_0(0)}{u_0(0) + K_m} = \frac{\lambda}{1 + K} . \qquad (5.26)$$

In dimensional terms, using (5.8), the $O(1)$ rate of reaction R_0 is

$$R_0 = \frac{k_2 e_0 s_0}{s_0 + K_m} = \frac{Q s_0}{s_0 + K}, \quad K_m = \frac{k_{-1} + k_2}{k_1}, \quad Q = [R_0]_{max} = k_2 e_0 , \qquad (5.27)$$

where Q is the maximum velocity, or rate, of the reaction and K_m is the *Michaelis constant*. This rate, based on the pseudo-steady state hypothesis, is what is usually wanted from a biological point of view. From (5.9), the exact initial rate for the substrate is $[du/d\tau]_{\tau=0} = 1$ while for the complex it is $[dv/d\tau]_{\tau=0} = 1/\varepsilon$.

When the uptake of a substrate, or whatever, is described as a Michaelis-Menten uptake, what is understood is a rate of reaction like (5.27) and which is

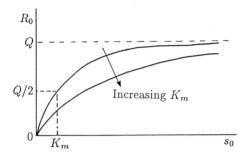

Fig. 5.2. Michaelis-Menten rate of uptake $R_0 = Qs_0/(K_m + s_0)$ as a function of the substrate concentration $s_0 : Q$ is the maximum rate and K_m is the Michaelis constant.

illustrated in Fig. 5.2. The rate of reaction, which in fact varies with time, is the magnitude of ds/dt from the outer solution $du_0/d\tau$ and written in dimensional form. Thus the (Michaelis-Menten) uptake of S is governed by the equation

$$\frac{ds}{dt} = -\frac{Qs}{K_m + s} \, . \tag{5.28}$$

This is simply the dimensional form of (5.25) on carrying out the algebra for $f(u, v)$, $g(u, v)$ in (5.24), with (5.9) defining them. For $s \ll K_m$ the uptake is linear in s, the right hand side of (5.28) is approximately $-Qs/K_m$. The maximum rate $Q = k_2 e_0$, from (5.27), depends on the rate constant k_2 of the product reaction $SE \rightarrow P + E$: this is called the *rate limiting* step in the reaction mechanism (5.1).

Useful and important as the pseudo-state hypothesis is, something is lost by assuming $\varepsilon dv/dt$ is negligible in (5.9) and by applying experimental results to a theory which cannot satisfy all the initial conditions. What can be determined, using experimental results with a Michaelis-Menten theory, is a curve such as in Fig. 5.2, which gives values for the maximum rate Q and the Michaelis constant K_m. This does not determine all three rate constants k_1, k_{-1} and k_2, only k_2 and a relationship between them all. To determine all of them, measurements for $\tau = O(\varepsilon)$ would be required. Usually, however, the rate of uptake from the pseudo-steady state hypothesis, that is a Michaelis-Menten theory, is all that is required.

5.3 Cooperative Phenomena

In the model mechanism (5.1) one enzyme molecule combines with one substrate molecule, that is the enzyme has one binding site. There are many enzymes which have more than one binding site for substrate molecules. For example, haemoglobin(Hb), the oxygen-carrying protein in red blood cells, has 4 binding sites for oxygen(O_2) molecules. A reaction between an enzyme and a substrate is described as *cooperative* if a single enzyme molecule, after binding a substrate molecule at one site can then bind another substrate molecule at another site. Such phenomena are very common.

Another important cooperative behaviour is when an enzyme with several binding sites is such that the binding of one substrate molecule at one site can affect the activity of binding other substrate molecules at another site. This indirect interaction between distinct and specific binding sites is called *allostery*, or an *allosteric effect*, and an enzyme exhibiting it, an allosteric enzyme. If a substrate that binds at one site increases the binding activity at another site then the substrate is an *activator*; if it decreases the activity it is an *inhibitor*. The detailed mathematical analysis for the kinetics of such allosteric reactions is given briefly in the book by Murray (1977) and in more detail in the one by Rubinow (1975). The latter book also gives a graph-theoretic approach to enzyme kinetics.

As an example of a cooperative phenomenon we consider the case where an enzyme has 2 binding sites and calculate an equivalent Michaelis-Menten theory and the substrate uptake function. A model for this consists of an enzyme molecule E which binds a substrate molecule S to form a single bound substrate-enzyme complex C_1. This complex C_1 not only breaks down to form a product P and the enzyme E again, it can also combine with another substrate molecule to form a dual bound substrate-enzyme complex C_2. This C_2 complex breaks down to form the product P and the single bound complex C_1. A reaction mechanism for this model is then

$$
S + E \underset{k_{1-1}}{\overset{k_1}{\rightleftharpoons}} C_1 \overset{k_2}{\rightarrow} E + P ,
$$
$$
S + C_1 \underset{k_{-3}}{\overset{k_3}{\rightleftharpoons}} C_2 \overset{k_4}{\rightarrow} C_1 + P ,
$$

(5.29)

where the k's are the rate constants as indicated.

With lower case letters denoting concentrations, the mass action law applied to (5.29) gives

$$
\frac{ds}{dt} = -k_1 se + (k_{-1} - k_3 s)c_1 + k_{-3} c_2 ,
$$

$$
\frac{dc_1}{dt} = k_1 se - (k_{-1} + k_2 + k_3 s)c_1 + (k_{-3} + k_4)c_2 ,
$$

$$
\frac{dc_2}{dt} = k_3 s c_1 - (k_{-3} + k_4)c_2 ,
$$

(5.30)

$$
\frac{de}{dt} = -k_1 se + (k_{-1} + k_2)c_1 ,
$$

$$
\frac{dp}{dt} = k_2 c_1 + k_4 c_2 .
$$

Appropriate initial conditions are

$$
s(0) = s_0, \quad e(0) = e_0, \quad c_1(0) = c_2(0) = p(0) = 0 .
$$

(5.31)

The conservation of the enzyme is obtained by adding the 2nd, 3rd and 4th equations in (5.30) and using the initial conditions; it is

$$\frac{dc_1}{dt} + \frac{dc_2}{dt} + \frac{de}{dt} = 0 \quad \Rightarrow \quad e + c_1 + c_2 = e_0 . \tag{5.32}$$

The equation for the product $p(t)$ is again uncoupled and given, by integration, once c_1 and c_2 have been found. Thus, using (5.32), the resulting system we have to solve is

$$\frac{ds}{dt} = -k_1 e_0 s + (k_{-1} + k_1 s - k_3 s)c_1 + (k_1 s + k_{-3})c_2 ,$$

$$\frac{dc_1}{dt} = k_1 e_0 s - (k_{-1} + k_2 + k_1 s + k_3 s)c_1 + (k_{-3} + k_4 - k_1 s)c_2 , \tag{5.33}$$

$$\frac{dc_2}{dt} = k_3 s c_1 - (k_{-3} + k_4)c_2 ,$$

with initial conditions (5.31).

As always, we nondimensionalize the system by introducing

$$\tau = k_1 e_0 t, \quad u = \frac{s}{s_0}, \quad v_1 = \frac{c_1}{e_0}, \quad v_2 = \frac{c_2}{e_0} ,$$

$$a_1 = \frac{k_{-1}}{k_1 s_0}, \quad a_2 = \frac{k_2}{k_1 s_0}, \quad a_3 = \frac{k_3}{k_1}, \quad a_4 = \frac{k_{-3}}{k_1 s_0} , \tag{5.34}$$

$$a_5 = \frac{k_4}{k_1 s_0}, \quad \varepsilon = \frac{e_0}{s_0} .$$

and (5.33) becomes

$$\frac{du}{d\tau} = -u + (u - a_3 u + a_1)v_1 + (a_4 + u)v_2 = f(u, v_1, v_2) , \tag{5.35}$$

$$\varepsilon \frac{dv_1}{d\tau} = u - (u + a_3 u + a_1 + a_2)v_1 + (a_4 + a_5 - u)v_2 = g_1(u, v_1, v_2) , \tag{5.36}$$

$$\varepsilon \frac{dv_2}{d\tau} = a_3 u v_1 - (a_4 + a_5)v_2 = g_2(u, v_1, v_2) . \tag{5.37}$$

which, with the initial conditions

$$u(0) = 1, \quad v_1(0) = v_2(0) = 0 , \tag{5.38}$$

represents a well posed mathematical problem.

This problem, just as the Michaelis-Menten one (5.9) analyzed in the last section, is a singular perturbation one for $0 < \varepsilon \ll 1$. The complete inner and outer solution can be found in a comparable way using the method set out in Section 5.3 so we leave it as an exercise. What is of interest here, however, is the form of the uptake function for the substrate concentration u, for times $\tau \gg \varepsilon$;

that is for times in the experimentally measurable regime. So, we only need the outer, or nonsingular, solution which is given to $O(1)$ for $0 < \varepsilon \ll 1$ by (5.35)-(5.38) on setting the ε-terms to zero. This gives

$$\frac{du}{d\tau} = f(u, v_1, v_2), \quad g_1(u, v_1, v_2) = 0, \quad g_2(u, v_1, v_2) = 0 .$$

The last two equations are algebraic which on solving for v_1 and v_2 give

$$v_2 = \frac{a_3 u v_1}{a_4 + a_5}, \quad v_1 = \frac{u}{a_1 + a_2 + u + a_3 u^2 (a_4 + a_5)^{-1}} .$$

Substituting these into $f(u, v_1(u), v_2(u))$ we get the uptake equation, or rate equation, for u as

$$\frac{du}{d\tau} = f(u, v_1(u), v_2(u))$$

$$= -u \frac{a_2 + a_3 a_5 u (a_4 + a_5)^{-1}}{a_1 + a_2 + u + a_3 u^2 (a_4 + a_5)^{-1}} \tag{5.39}$$

$$= -r(u) < 0 .$$

The dimensionless velocity of the reaction is thus $r(u)$. In dimensional terms, using (5.34), the Michaelis-Menten velocity of the reaction for $0 < e_0/s_0 \ll 1$, denoted by $R_0(s_0)$ say, is, from (5.39),

$$R_0(s_0) = \left| \frac{ds}{dt} \right|_{t=0} = e_0 s_0 \frac{k_2 K'_m + k_4 s_0}{K_m K'_m + K'_m s_0 + s_0^2} \tag{5.40}$$

$$K_m = \frac{k_2 + k_{-1}}{k_1}, \quad K'_m = \frac{k_4 + k_{-3}}{k_3}$$

where K_m and K'_m are the Michaelis constants for the mechanism (5.29), equivalent to the Michaelis constant in (5.27).

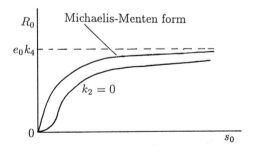

Fig. 5.3. Rate of reaction, or substrate uptake, as a function of substrate concentration s_0 for the cooperative reaction (5.29). Note the inflexion in the cooperative uptake curve when $k_2 = 0$.

The rate of the reaction $R_0(s_0)$ is illustrated in Fig. 5.3. If some of the parameters are zero there is a point of inflexion: for example, if $k_2 = 0$ it is clear from

(5.40) since then for s_0 small, $R_0 \propto s_0^2$. A good example of such a cooperative behaviour is the binding of oxygen by haemoglobin; the experimental measurements give an uptake curve very like the lower curve in Fig. 5.3. Myoglobin(Mb), a protein in abundance in red muscle fibres, on the other hand has only one oxygen binding site and its uptake is of the Michaelis-Menten form also shown in Fig. 5.3 for comparison.

When a cooperative phenomenon in an enzymatic reaction is suspected, a *Hill plot* is often made. The underlying assumption is that the reaction velocity or uptake function is of the form

$$R_0(S_0) = \frac{Qs_0^n}{K_m + s_0^n} \, , \tag{5.41}$$

where $n > 0$ is not usually an integer: this is often called a Hill equation. Solving the last equation for s_0^n we have

$$s_0^n = \frac{R_0 K_m}{Q - R_0} \quad \Rightarrow \quad n \ln s_0 = \ln K_m + \ln \frac{R_0}{Q - R_0} \, .$$

A Hill plot is the graph of $\ln\left[R_0/(Q - R_0)\right]$ against $\ln s_0$ the slope of which gives n, and is a constant if the Hill equation is a valid description for the uptake kinetics. If $n < 1$, $n = 1$ or $n > 1$ we say that there is negative, zero or positive cooperativity respectively. Although the Hill equation may be a reasonable quantitative form to describe a reaction's velocity in a Michaelis-Menten sense the detailed reactions which give rise to it are not too realistic: essentially it is (5.1) but now instead of $E+S$ we require $E+nS$ combining to form the complex in one step. This is somewhat unlikely if n is not an integer although it could be a stoichiometric form. If n is an integer and $n \geq 2$ the reaction is then tri-molecular or higher. Such reactions do not occur except possibly through what is in effect a telescoping together of several reactions because intermediary reactions are very fast.

Even with such drawbacks as regards the implied reaction mechanisms, empirical rate forms like the Hill equation, are extremely useful in modelling. After all, what we want from a model is some understanding of the underlying dynamics and mechanisms governing the phenomena. A very positive first step is to find a biologically reasonable model which qualitatively describes the behaviour. Detailed refinements or amendments come later.

5.4 Autocatalysis, Activation and Inhibition

Many biological systems have feedback controls built into them. These are very important and we must know how to model them. In the next chapter on biological oscillators, we shall describe one area where they are essential. A review of theoretical models and the dynamics of metabolic feedback control systems is given by Tyson and Othmer (1977). Here we describe some of the more important types of feedback control. Basically feedback is when the product of one step in a reaction sequence has an effect on other reaction steps in the sequence. The effect is generally nonlinear and may be to activate or inhibit these reactions. The next chapter gives some specific examples with actual reaction mechanisms. *Autocatalysis* is the process whereby a chemical is involved in its own production. A very simple example is

$$A + X \underset{k_{-1}}{\overset{k_1}{\rightleftharpoons}} 2X \,, \tag{5.42}$$

where a molecule of X combines with one of A to form two molecules of X. If A is maintained at a constant concentration a, the Law of Mass Action applied to this reaction gives the rate of reaction as

$$\frac{dx}{dt} = k_1 ax - k_{-1}x^2 \quad \Rightarrow \quad x(t) \to x_S = \frac{k_1 a}{k_{-1}} \,, \tag{5.43}$$

where $x = [X]$ and x_S is the final nonzero steady state as $t \to \infty$. The zero steady state is unstable by inspection. This autocatalytic reaction exhibits a strong feedback with the 'product' inhibiting the reaction rate. It is obvious that some back reaction ($k_{-1} \neq 0$) is necessary. This is the chemical equivalent of logistic growth discussed in Chapter 1.

Suppose, instead of (5.42), the reaction system is

$$A + X \underset{k_{-1}}{\overset{k_1}{\rightleftharpoons}} 2X, \quad B + X \overset{k_2}{\rightarrow} C \,. \tag{5.44}$$

That is X is used up in the production of C. This mechanism exhibits a simple bifurcation as we shall see. If B, as well as A, are maintained at constant concentrations, a and b, then

$$\frac{dx}{dt} = k_1 ax - k_{-1}x^2 - k_2 bx = (k_1 a - k_2 b)x - k_{-1}x^2 \,. \tag{5.45}$$

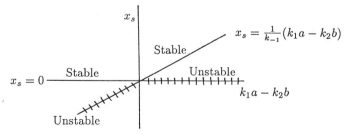

Fig. 5.4. Stability of the steady states x_S of the reaction mechanism (5.44) and (5.45). As the parameter $k_1 a - k_2 b$, the difference between the production and loss rates, changes sign so does the stability, namely from $x_S = 0$ to $x_S \neq 0$.

Here k_1a is the unit production rate of x and k_2b the unit loss rate. From (5.45) we see that if $k_1a > k_2b$ the steady state $x = 0$ is unstable and $x(t) \to x_S = (k_1a - k_2b)/k_{-1} > 0$ as $t \to \infty$, which is stable. On the other hand if $k_1a < k_2b$ then $x = 0$ is stable, which is not surprising since the inequality implies that the loss rate is greater than the production rate. In this case mathematically there is still, of course, another steady state but it is negative and unstable. The simple bifurcation exhibited by this reaction is summarized in Fig. 5.4 where the steady states x_S are given in terms of the parameter $k_1a - k_2b$. The bifurcation is at $k_1a - k_2b = 0$ where the stability changes from one steady state to another.

Anticipating the next chapter on biological oscillators, the classical Lotka (1920) reaction mechanism which he proposed as a hypothetical model oscillator is another example of autocatalysis. It is

$$A + X \overset{k_1}{\to} 2X, \quad X + Y \overset{k_2}{\to} 2Y, \quad Y \overset{k_3}{\to} B , \tag{5.46}$$

where A is maintained at a constant concentration a. The first two reactions are autocatalytic. The Law of Mass Action gives

$$\frac{dx}{dt} = k_1ax - k_2xy, \quad \frac{dy}{dt} = k_2xy - k_3y ,$$

which, with the nondimensional variables

$$u = \frac{k_2x}{k_3}, \quad v = \frac{k_2y}{k_1a}, \quad \tau = k_1at, \quad \alpha = k_3/k_1a ,$$

become

$$\frac{du}{d\tau} = u(1 - v), \quad \frac{dv}{d\tau} = \alpha v(u - 1) .$$

These are the Lotka-Volterra equations (3.4) discussed in detail in Section 3.1 in Chapter 3: the solutions u and v are periodic in time but, as we saw, are structurally unstable.

In almost all biological processes we do not know the detailed biochemical reactions that are taking place. However we often do know the qualitative effect of varying a known reactant or of changing the operating conditions in one way or another. So, in modelling such biological processes it is usually much more productive and illuminating to incorporate such known qualitative behaviour in a model mechanism. It is such model mechanisms which have proved so useful in interpreting and unravelling the basic underlying processes involved, and in making useful predictions in a remarkably wide spectrum of biomedical problems. Since we know how to represent a reaction sequence as a differential equation system we can now construct models which incorporate the various qualitative behaviours directly into the differential equations for the concentrations. It is then the differential equation system which constitutes the model.

Suppose we have a differential equation system, the model for which can be reduced, through asymptotic procedures such as we discussed above, to two key elements which are governed by the dimensionless mechanism

$$\frac{du}{dt} = \frac{a}{b+v} - cu = f(u,v) \,,$$
$$\frac{dv}{dt} = du - ev = g(u,v) \,,$$

(5.47)

where a, b, c, d and e are positive constants. The biological interpretation of this model is that u activates v, through the term du, and both u and v are degraded linearly proportional to their concentrations; these are the $-cu$ and $-ev$ terms. This linear degradation is referred to as *first order kinetics* removal. The term $a/(b+v)$ shows a negative feedback by v on the production of u, since an increase in v decreases the production of u, and hence indirectly a reduction in itself. The larger v, the smaller is the u-production. This is an example of *feedback inhibition*.

We can easily show that there is a stable positive steady state for the mechanism (5.47). The relevant steady state (u_0, v_0) is the positive solution of

$$f(u_0, v_0) = g(u_0, v_0) = 0$$
$$\Rightarrow \quad v_0 = \frac{du_0}{e} \,, \quad u_0^2 + \frac{ebu_0}{d} - \frac{ae}{cd} = 0 \,.$$

The differential equation system (5.47) is exactly the same type that we analysed in detail in Chapter 3. The linear stability then is determined by the eigenvalues λ of the linearized Jacobian or *reaction matrix* or stability matrix (equivalent to the community matrix in Chapter 3), and are given by

$$\begin{vmatrix} \dfrac{\partial f}{\partial u} - \lambda & \dfrac{\partial f}{\partial v} \\ \dfrac{\partial g}{\partial u} & \dfrac{\partial g}{\partial v} - \lambda \end{vmatrix}_{u_0,v_0} \quad \begin{vmatrix} -c - \lambda & -c\dfrac{u_0}{v_0 + b} \\ d & -e - \lambda \end{vmatrix} = 0 \,.$$

Thus

$$\lambda^2 + (c+e)\lambda + \left[ce + \frac{cdu_0}{b+v_0} \right] = 0 \quad \Rightarrow \quad \mathrm{Re}\,\lambda < 0 \,,$$

and so (u_0, v_0) is linearly stable. It is also a globally attracting steady state: it is straightforward to construct a rectangular confined set in the (u, v) space on which the vector $(du/dt, dv/dt)$ points inwards.

Several specific model systems have been proposed as the mechanisms governing certain basic biological phenomena such as oscillatory behaviour, pattern formation in developing embryos, mammalian coat patterns and so on. We study some of these in detail in subsequent chapters. Here we briefly look at two.

The Thomas (1975) mechanism, used later in Section 15.1 in Chapter 15, is based on a specific reaction involving the substrates oxygen and uric acid which react in the presence of the enzyme uricase. The dimensionless form of the empirical rate equations for the oxygen(v) and the uric acid(u) can be written as

$$\frac{du}{dt} = a - u - \rho R(u,v) = f(u,v) \,,$$

$$\frac{dv}{dt} = \alpha(b - v) - \rho R(u,v) = g(u,v) \,, \qquad (5.48)$$

$$R(u,v) = \frac{uv}{1 + u + Ku^2}$$

where a, b, α, ρ and K are positive constants. Basically u and v are supplied at constant rates a and αb, degrade linearly proportional to their concentrations and both are used up in the reaction at a rate $\rho R(u,v)$. The form of $R(u,v)$ exhibits *substrate inhibition*. For a given v, $R(u,v)$ is $O(uv)$ for u small and is thus linear in u, while for u large it is $O(v/Ku)$. So, for u small R increases with u but for u large it decreases with u. This is what is meant by substrate inhibition. The parameter K is a measure of the severity of the inhibition. From Fig. 5.5, giving $R(u,v)$ as a function of u, we see that the uptake rate is like a Michaelis-Menten form for small u, reaches a maximum at $u = 1/\sqrt{K}$ and then decreases with increasing u. The value of the concentration for the maximum $R(u,v)$, and the actual maximum rate, decrease with increasing inhibition, that is as K increases.

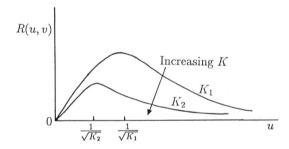

Fig. 5.5. Reaction rate $R(u,v)$ in (5.48) for a fixed v. The reduction in R as u increases for $u > 1/\sqrt{K}$ is a typical example of substrate(u) inhibition: the larger the K the greater the inhibition.

It is always informative to draw the nullclines for the reaction kinetics in the (u,v) phase plane in the same way as for the interacting population models in Chapter 3. Here the nullclines for (5.48) are

$$f(u,v) = 0 \quad \Rightarrow \quad v = (u - a)\frac{1 + u + Ku^2}{\rho u} \,,$$

$$g(u,v) = 0 \quad \Rightarrow \quad v = \alpha b \frac{1 + u + Ku^2}{\rho u + \alpha(1 + u + Ku^2)} \,,$$

which are sketched in Fig. 5.6. Depending on the parameters there can be one or three positive steady states. Although these nullclines are for a specific substrate-

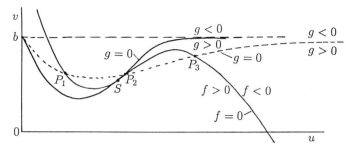

Fig. 5.6. Schematic nullclines for the substrate-inhibition kinetics (5.48). There may be one, S, or three, P_1, P_2, P_3 (dashed $g = 0$ curve) steady states where $f = 0$ and $g = 0$ intersect. Note the signs of f and g on either side of their nullclines.

inhibition mechanism they are fairly typical of general substrate inhibition models, the $f = 0$ nullcline in particular: see also Fig. 5.7.

The question of the stability of the steady states will be discussed in detail and in some generality in the next chapter. At this stage, however, we can get an intuitive indication of the stabiltiy from looking at the nullclines in the (u, v) phase plane. Consider the situation in Fig. 5.6 when there are three steady states at P_1, P_2 and P_3 and, to be specific, look at $P_1(u_1, v_1)$ first. Now let us move along a line, $v = v_1$ say, through P_1 and note the signs of $f(u, v_1)$ as we cross the $f = 0$ nullcline. Let us stay in the neighbourhood of P_1. On the left of the $f = 0$ nullcline, $f > 0$ and on the right $f < 0$. So with $v = v_1$, a constant, $\partial f / \partial u < 0$ at P_1. If we now consider the kinetics equation for u with $v = v_1$, namely $du/dt = f(u, v_1)$, we see that locally $\partial f / \partial u < 0$ at P_1 and so, from our discussion in Section 1.1 in Chapter 1, if this was an uncoupled scalar equation for u it would mean that P_1 is a linearly stable steady state. But of course from (5.48) the u-equation is not uncoupled and maybe the coupling has a destabilising affect.

Let us still consider P_1 and use the same kind of argument to move across the $g = 0$ nullcline along a line, $u = u_1$ say, through P_1. We now see that $\partial g / \partial v < 0$ so locally $dv/dt = g(u_1, v)$ with $\partial g / \partial v < 0$ at P_1 and by the same argument about scalar equations this would reinforce our intuition that P_1 is linearly stable. So intuitively from both these analyses we would expect P_1 to be linearly stable. These kind of arguments will be developed rigorously in the next chapter where we shall see that our intuition is indeed correct. In a similar way we can intuitively deduce that P_3 is also stable. If we apply the above sign arguments to P_2 with $v = v_2$ at P_2 we see, from the Fig. 5.6, that $\partial f(u, v_2)/\partial u > 0$ so we expect P_2 to be unstable. When there is a single steady state at S, the situation needs a careful analysis (see Chapter 6).

Without carrying out any analysis, it is clear that there must be certain parameter ranges where there is a single steady state and where there are three steady states. An informative analysis therefore is to determine the parameter domains for each situation. Although this is simple in principle – you determine the positive steady states from the simulataneous algebraic equations $f(u,v) = g(u,v) = 0$ – it is usually hard algebraically and has to carried out numerically. Such analyses produce some interesting results which we discuss in more detail in Section 5.5.

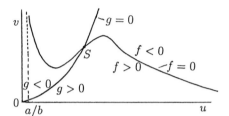

Fig. 5.7. Typical nullclines for the activator-inhibitor model (5.49).

Another model mechanism, algebraically simpler than the Thomas system (5.48), is the hypothetical but biologically plausible reaction scheme

$$\frac{du}{dt} = a - bu + \frac{u^2}{v(1 + Ku^2)} = f(u,v) \,,$$

$$\frac{dv}{dt} = u^2 - v = g(u,v) \,,$$

(5.49)

where a, b and K are constants. This is an *activator(u)-inhibitor(v) system* and is a dimensionless version of the kinetics of a model proposed by Gierer and Meinhardt (1972). It has been used in a variety of modelling situations. Here there is an autocatalytic production of the activator u via the $u^2/[v(1 + Ku^2)]$ term but which saturates to $1/(Kv)$ for u large. The inhibitor v is activated by u according to the second equation, but it inhibits its activator production since $u^2/[v(1 + Ku^2)]$ decreases as v increases. The nullclines $f = 0$ and $g = 0$ from (5.49) are illustrated in Fig. 5.7. Note the qualitative similarity between the nullclines in the Figs. 5.6 and 5.7 particularly in the vicinity of the steady state and for large u: we consider the implications of this later. In the next chapter we introduce other reaction systems while in Chapter 7 we discuss in detail a specific system which is of considerable experimental importance.

For a general system

$$\frac{du}{dt} = f(u,v), \qquad \frac{dv}{dt} = g(u,v)$$

(5.50)

u is an activator of v if $\partial g/\partial u > 0$ while v is an inhibitor of u if $\partial f/\partial v < 0$. Depending on the detailed kinetics a reactant may be an activator, for example, only for a range of concentrations or parameters. There are thus many possibilities of bifurcation phenomena which have biologically important implications as we see later in the book.

With the mathematical parallel between interacting populations and reaction kinetics model systems, we also expect to observe threshold phenomena such as

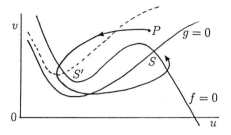

Fig. 5.8. Reaction kinetics nullclines which illustrate a threshold behaviour. With a perturbation to P, the solution embarks on a large excursion in the phase space before returning to the stable steady S. A similar threshold behaviour is possible if the nullclines intersect at the steady state S'.

we discussed in Section 3.7 in Chapter 3. This is indeed the case and the model system (5.48) exhibits a similar threshold behaviour if the parameters are such that the steady state is at S, or at S', as in Fig. 5.8. The analysis in Section 3.7 is directly applicable here.

We can now start to build model reactions to incorporate a variety of reaction kinetics behaviour such as autocatalysis, activation and inhibition and so on, since we know qualitatively what is required. As an example suppose we have cells which react to the local concentration level of a chemical S by activating a gene so that the cells produce a product G. Suppose that the product is autocatalytically produced in a saturable way and that it degrades linearly with its concentration, that is according to first order kinetics. With lower case letters for the concentrations, a rate equation for the product g which incorporates all of these requirements is, for example,

$$\frac{dg}{dt} = k_1 s + \frac{k_2 g^2}{k_3 + g^2} - k_4 g = f(g) , \qquad (5.51)$$

where the k's are positive constants. This model has some useful biological switch properties which we shall consider and use later in Chapter 15 when we discuss models for generating biological spatial patterns.

It is now clear that the study of the reaction kinetics of n reactions results in an n-th order system of first order differential equations of the form

$$\frac{du_i}{dt} = f_i(u_1, \ldots, u_n), \quad i = 1, \ldots, n . \qquad (5.52)$$

This is formally the same type of general system which arose in interacting population models, specifically equations (3.43) in Chapter 3. There we were only concerned with non-negative solutions and so also here, since $u(t)$ is a vector of concentrations. All of the methods for analysing stability of the steady states, that is solutions of $f(u_1, \ldots, u_n) = 0$, are applicable. Thus all of the conditions for limit cycles, threshold phenomena and so on also hold here.

The interaction details between reactants and those for interacting populations are of course quite different both in form and motivation. In biological systems there is generally more complexity as regards the necessary order of the differential equation model. As we have seen, however, this is often compensated by the presence of enzyme catalysts and thus a biological justification for reducing the order considerably. For example a system which results in the dimensionless equations

$$\frac{du_i}{dt} = f_i(u_1, \ldots, u_n), \quad i = 1, 2$$

$$\varepsilon_i \frac{du_i}{dt} = f_i(u_1, \ldots, u_n), \quad i = 3, \ldots, n \tag{5.53}$$

$$0 < \varepsilon_i \ll 1, \quad i = 3, \ldots, n$$

reduces, for almost all practical purposes, to a second order system

$$\frac{du_i}{dt} = f_i(u_1, u_2, u_3(u_1, u_2), \ldots, u_n(u_1, u_2)), \quad i = 1, 2$$

for small enough ε's. Here $f_i(u_1, \ldots, u_n) = 0$ for $i = 3, \ldots, n$ are algebraic equations which are solved to give $u_{n \geq 3}$ as functions of u_1 and u_2. It is this general extension of the pseudo-steady state hypothesis to higher order systems which justifies the extensive study of two-reactant kinetics models. Mathematically the last equation is the $O(1)$ asymptotic system, as $\varepsilon_i \to 0$ for all i, for the nonsingular solution of (5.53). Biologically this is all we generally require since it is the relatively long time behaviour of mechanisms which usually dominates biological development.

5.5 Multiple Steady States, Mushrooms and Isolas

We saw in Fig. 5.6 that it is possible to have multiple positive steady states. The transition from a situation with one steady state to three occurs when some parameter in the model passes through a bifurcation value. Fig. 5.9 illustrates typical scenarios where this occurs. For example referring to Fig. 5.7 and the kinetics in (5.49) the steady state would behave qualitatively like that in Fig. 5.9 (a) with the inhibition parameter K playing the role of p.

Now suppose that as a parameter, k say, varies the u_s versus p curve changes in such a way that for a range of k the qualitative form of the curve is as in Fig. 5.9 (b). For a fixed k and $p_1 < p < p_2$ there are three steady states, one on each branch BC, CD and DE. This is equivalent to the three steady state situation in Fig. 5.6. From the discussion in the last section we expect the steady states lying on the CD branch to be linearly unstable: this will be proved in the next chapter.

The form of the (u_s, p) graph in Fig. 5.9 (b) suggests the possibility of hysteresis (recall Section 1.1) as p varies. Assume, as is the case, that a steady state lying on the branches ABC and DEF is stable. Now suppose we slowly

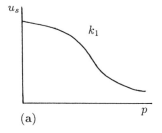

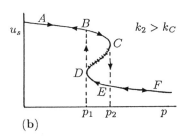

(a)   (b)

Fig. 5.9a,b. (a) Typical variation of the steady state u_s as a function of a parameter p in the kinetics for a fixed value k_1 of another kinetics parameter k. (b) As the parameter k passes through a bifurcation value k_c, multiple steady states are possible when $p_1 < p < p_2$. The steady state that lies on the branch DC is unstable.

increase the parameter p from a value $p < p_1$ to a value $p > p_2$. Until p reaches p_2, u_s simply increases and is given by the appropriate value on the branch ABC. When p passes through p_2, u_s changes abruptly, moving onto the branch EF; with increasing p it is given by the appropriate value on this branch. Now suppose we slowly decrease p. In this situation u_s stays on the lower branch FED until p reaches p_1 since solutions on this branch are stable. Now the abrupt change takes place at p_1 where u_s jumps up onto the upper BA branch. This is a typical hysteresis loop. For increasing p, the path is along $ABCEF$ while the path through decreasing values of p is $FEDBA$.

Mushrooms. Instead of the (u_s, p) variation in Fig. 5.9 (a) another common form simply has u_s increasing with increasing p as in Fig. 5.10 (a): the transition to three steady states is then as illustrated. It is not hard to imagine that even more complicated behaviour is possible with the simple curve in Fig. 5.10 (a) evolving to form the mushroom like shape in Fig. 5.10 (b) with two regions in p-space where there are multi-steady states.

The mushroom like (u_s, p) relationship in Fig. 5.10 (b) has two distinct p-ranges where there are three steady states. Here the steady states lying on the branches CD and GH are unstable. There are two hysteresis loops equivalent to Fig. 5.9 (b), namely $BCED$ and $IHFG$.

Isolas. The situation shown in Fig. 5.10 (c), namely that of a separate breakaway region, is an obvious extension from Fig. 5.10 (b). Such a solution behaviour is called an isola. Now we expect the solutions lying on the branch DCG to be unstable. The physical situation represented by this situation is rather different from that which obtains with a mushroom. First there is no hysteresis in the usual way since u_s simply stays on the branch $ABIJ$ as the parameter p increases from a value $p < p_1$ to a value $p > p_2$: it stays on this branch on the return sweep through the multi-steady state region $p_1 < p < p_2$. *Isolas* are isolated closed curves of solution branches and can only arise as solutions of nonlinear equations.

Referring still to Fig. 5.10 (c), if u_s lies on the branch BI it is only possible

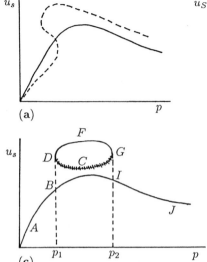

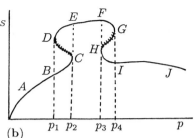

(a)

(b)

(c)

Fig. 5.10a–c. (a) Another typical example of a steady state dependence on a parameter with transition to multiple steady states: compare with Fig. 5.9 (a). **(b)** Typical mushroom dependence of the steady state as a function of a parameter p. **(c)** This shows an example of an isola: it can be a natural evolution from the form in **(b)**.

to move onto the other stable branch DFG if u_s is given a finite perturbation so that u moves into the domain of attraction of the stable steady state on the DFG branch. The various possible scenarios are now clear.

It is possible to predict quite complex solution behaviour by simply manipulating the curves, in effect as we have just done. The appearance of multi-steady states is not difficult to imagine with the right kinetics. Dellwo et al. (1982) present a general theory which describes analytically the structure of a class of isolas, namely those which tend to a point as some parameter tends to a critical value. The question immediately arises as to whether isolas for example can exist in the real world. Isolas have been found in a variety of genuine practical situations including chemical reactions: an early review is given by Uppal et al. (1976) with more recent references in the paper by Gray and Scott (1986).

A simple model kinetics system has been proposed by Gray and Scott (1983, 1986) which exhibits, among other things, multi-steady states with mushrooms and isolas: it involves autocatalysis in a continuously stirred tank reactor (CSTR). It consists of the following hypothetical reactions involving two reactants X and Y with concentrations x and y respectively. The specific mechanism is represented schematically in Fig. 5.11.

The process in the figure involves the trimolecular autocatalytic step $X + 2Y \rightarrow 3Y$ and the specific equation system which describes the process is

$$\frac{dx}{dt} = k_0(x_0 - x) - k_1 x y^2 - k_3 x ,$$

$$\frac{dy}{dt} = k_0(y_0 - y) + k_1 x y^2 + k_3 x - k_2 y ,$$

(5.54)

where the k's are the positive rate constants. An appropriate nondimensionali-

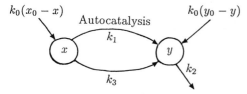

Fig. 5.11. Model autocatalytic mechanism which exhibits multi-steady states with mushrooms and isolas. The system is a continously stirred flow tank reactor ($CSTR$) mechanism with Y being produced autocatalytically and by a simple uncatalysed process. X and Y are fed into the process and Y degrades with first order kinetics. The mechanism is described by the differential equation system (5.54). The lower case letters x and y denote the concentrations of X and Y.

sation is

$$u = \frac{x}{x_0}, \quad v = \frac{y}{x_0}, \quad t^* = tk_1x_0^2, \quad c = \frac{y_0}{x_0},$$

$$a = \frac{k_0}{k_1x_0^2}, \quad b = \frac{k_3}{k_1x_0^2}, \quad d = \frac{k_2}{k_1x_0^2}, \tag{5.55}$$

with which (5.54) become, on omitting the asterisk for notational simplicity, the dimensionless system

$$\frac{du}{dt} = a(1 - u) - uv^2 - bu = f(u, v),$$

$$\frac{dv}{dt} = a(c - v) + uv^2 + bu - dv = g(u, v), \tag{5.56}$$

which now involve four dimensionless parameters a, b, c and d.

Here we are only interested in the steady states u_s and v_s which are solutions of $f(u, v) = g(u, v) = 0$. A little algebra shows that

$$u_s(1 + c - u_s)^2 = a\left(1 + \frac{d}{a}\right)^2(1 - u_s) - bu_s\frac{(a + d)^2}{a^2}, \tag{5.57}$$

which is a cubic, namely

$$u_s^3 - 2(1 + c)u_s^2 + \left[(1 + c)^2 + \frac{(a + d)^2}{a} + b\frac{(a + d)^2}{a^2}\right]u_s - \frac{(a + d)^2}{a} = 0. \tag{5.58}$$

Since there are three changes in sign in the cubic there is thus, using Descarte's rule of signs (see Appendix 2), the possibility of three positive solutions. Certain analytical solutions for these can be found asymptotically for large and small values of the parameters. The full picture however has to be obtained numerically as was done by Gray and Scott (1986). Typical results are illustrated schematically

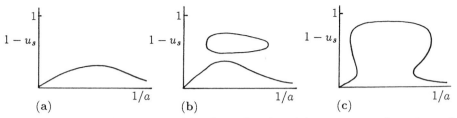

Fig. 5.12a-c. The steady states u_s of (5.56) as a function of the parameter a for various values of b, c and d. For a fixed c, less than a critical value, and an increasing d from $d = 0$ the progression of steady state behaviours is from the mushroom situation (c), through the isola region (b), to the single steady state situation (a).

in Fig. 5.12. A good review of this reaction and its complex behaviour together with analytical and numerical results is given by Gray (1988).

It is, of course, always possible to construct more and more complex solution behaviours mathematically and to postulate hypothetical reactions which exhibit them. So, the key question at this stage is to ask whether there are any real reaction processes which exhibit these interesting phenomena, such as mushrooms and isolas. The inorganic iodate-arsenous acid reaction under appropriate conditions has been shown experimentally to have the required kinetics. This has been convincingly demonstrated by Ganapathisubramanian and Showalter (1984) whose model and experimental results are described below. Although this is not an enzymatic or biological reaction it nevertheless shows that real reaction mechanisms, which have mushroom and isola solution behaviour, exist. With the richness and complexity of biological processes it would be unbelievable if such reaction systems did not exist within the biomedical sciences. So, it is with this belief in mind that we describe here the elements of this inorganic reaction and present the relevant experimental results.

Iodate-Arsenous Acid Reaction: Bistability, Mushrooms, Isolas

The iodate-arsenous acid reaction in a continuous flow stirred tank reactor (CSTR) can be described by two composite reactions, namely

$$IO_3^- + 5I^- + 6H^+ \quad \rightarrow \quad 3I_2 + 3H_2O \,, \tag{5.59}$$

$$I_2 + H_3AsO_3 + H_2O \quad \rightarrow \quad 2I^- + H_3AsO_4 + 2H^+ \,. \tag{5.60}$$

The net reaction, given by the (5.59) $+ 3 \times$ (5.60), is

$$IO_3^- + 3H_3AsO_3 \quad \rightarrow \quad I^- + 3H_3AsO_4 \,. \tag{5.61}$$

The rate of the reaction (5.59) is slow compared with (5.60) and so it is the rate limiting step in the overall process (5.60). If we denote this rate for (5.59) by R,

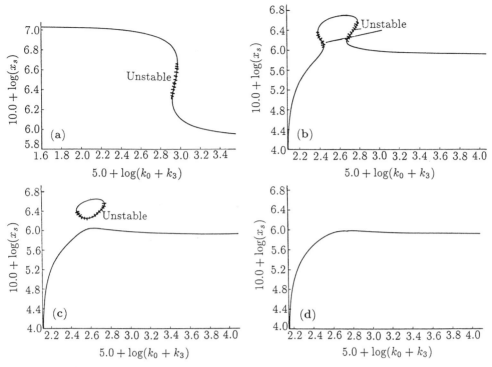

Fig. 5.13a-d. Computed steady state iodide concentration X_s from (5.66) as a function of $k_0 + k_3$. The continuous lines represent stable solution branches and the dashed lines unstable branches. Parameter values: $k_1 = 4.5 \times 10^3 M^{-3} s^{-1}$, $k_2 = 4.5 \times 10^8 M^{-4} s^{-1}$, $Y_0 = 1.01 \times 10^{-3} M$, $X_0 = 8.40 \times 10^{-5} M$, $[H^+] = 7.59 \times 10^{-3} M$; (a) $k_3 = 0$, (b) $k_3 = 1.20 \times 10^{-3} s^{-1}$. (c) $k_3 = 1.30 \times 10^{-3} s^{-1}$, (d) $k_3 = 1.42 \times 10^{-3} s^{-1}$. Compare (a)-(d) respectively with the schematic forms in Fig. 5.9 (b), Fig. 5.10 (b), (c), (a). (Redrawn from Ganapathisubramanian and Showalter 1984)

an empirical form has been determined experimentally as

$$R = -\frac{d[IO_3^-]}{dt} = (k_1 + k_2[I^-])[I^-][H^+]^2[IO_3^-] . \tag{5.62}$$

A simple model reaction mechanism which quantitatively describes the iodate-arsenous acid reaction in a continuous flow stirred tank reactor consists of rate equations for the iodide, I^-, and iodate, IO_3^-, in (5.61), with appropriate flow terms and decay terms, given by

$$\frac{d[I^-]}{dt} = R + k_0[I^-]_0 - (k_0 + k_3)[I^-] , \tag{5.63}$$

$$\frac{d[IO_3^-]}{dt} = -R + k_0[IO_3^-]_0 - (k_0 + k_3)[IO_3^-]) , \tag{5.64}$$

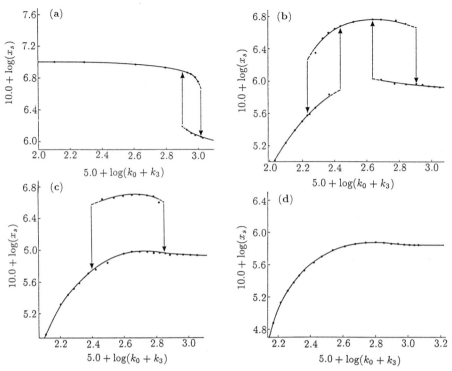

Fig. 5.14a-d. Experimentally determined steady state iodide concentrations for the iodate-arsenous acid reaction as a function of $k_0 + k_3$ for different values of k_3. Parameter values: $X_0 = 1.01 \times 10^{-3} M$, $Y_0 = 8.40 \times 10^{-5} M$ with the flow of $[H_3AsO_3]_0 = 4.99 \times 10^{-3} M$; (a) $k_3 = 0$, (b) $k_3 = 1.17 \times 10^{-3} s^{-1}$, (c) $k_3 = 9.71 \times 10^{-4} s^{-1}$, (d) $k_3 = 1.37 \times 10^{-3} s^{-1}$. Compare with Fig. 5.13 (a), (b), (c), (d) respectively. (Redrawn from Ganapathisubramanian and Showalter 1984)

where k_0 and k_3 are positive constants, $[I^-]_0$ and $[IO_3^-]_0$ are the concentrations in the inflow and R is given by (5.62).

If we now write

$$X = [I^-], \quad Y = [IO_3^-], \quad X_0 = [I^-]_0 ,$$
$$Y_0 = [IO_3^-]_0, \quad k_1^* = k_1[H^+]^2, \quad k_2^* = k_2[H^+]^2 ,$$

(5.65)

the steady states X_s and Y_s are given by the solutions of

$$0 = R + k_0 X_0 - (k_0 + k_3)X, \quad 0 = -R + k_0 Y_0 - (k_0 + k_3)Y ,$$
$$R = (k_1^* + k_2^* X)XY .$$

These give the cubic polynomial for X_s

$$k_2^*(k_0 + k_3)X_s^3 + [k_1^*(k_0 + k_3) - k_2^* k_0(X_0 + Y_0)]X_s^2$$
$$+ [(k_0 + k_3)^2 - k_1^* k_0(X_0 + Y_0)]X_s - k_0(k_0 + k_3)X_0 = 0 .$$

(5.66)

Values for k_1 and k_2 have been determined experimentally and X_0 and Y_0 and $[H^+]$ can be imposed, and so, from (5.65), k_1^* and k_2^* can be determined. Fig. 5.13 shows the positive steady state iodide concentration X_s claculated numerically from the cubic equation (5.66) as a function of $k_0 + k_3$ for different values of k_3.

When the above iodate-arsenous acid reaction model is compared with the full reaction system good quantitative results are obtained. Fig. 5.13 shows that mushroom and isola multi-steady state behaviour is possible. The final step in demonstrating the existence of this type of behaviour is experimental confirmation. This has also been done by Ganapathisubramanian and Showalter (1984) whose results are reproduced in Fig. 5.14. Note the comparison between these experimental results and those obtained with the model mechanism for this iodate-arsenous acid reaction. The results in Fig. 5.14 clearly show the various hysteresis behaviours suggested by Figs. 5.9 and 5.10.

Exercises

1. An allosteric enzyme E reacts with a substrate S to produce a product P according to the mechanism

$$S + E \underset{k_{-1}}{\overset{k_1}{\rightleftharpoons}} C_1 \overset{k_2}{\rightarrow} E + P$$

$$S + C_1 \underset{k_{-3}}{\overset{k_3}{\rightleftharpoons}} C_2 \overset{k_4}{\rightarrow} C_1 + P$$

where the k's are rate constants and C_1 and C_2 enzyme-substrate complexes. With lower case letters denoting concentrations, and initial conditions $s(0) = s_0$, $e(0) = e_0$, $c_1(0) = c_2(0) = p(0) = 0$ write down the differential equation model based on the Law of Mass Action. If

$$\varepsilon = \frac{e_0}{s_0} \ll 1, \quad \tau = k_1 e_0 t, \quad u = \frac{s}{s_0}, \quad v_i = \frac{c_i}{e_0}$$

show that the nondimensional reaction mechanism reduces to

$$\frac{du}{d\tau} = f(u, v_1, v_2), \quad \frac{dv_i}{d\tau} = g_i(u, v_1, v_2), \quad i = 1, 2 .$$

Determine f, g_1 and g_2 and hence show that for $\tau \gg \varepsilon$ the uptake of u is governed by

$$\frac{du}{d\tau} = -r(u) = -u \frac{A + Bu}{C + u + Du^2} ,$$

where A, B, C and D are positive parameters.

When $k_2 = 0$ sketch the uptake rate $r(u)$ as a function of u and compare it with the Michaelis-Menten uptake.

2. Two dimensionless activator-inhibitor mechanisms have reaction kinetics described by

$$(i) \quad \frac{du}{dt} = a - bu + \frac{u^2}{v}, \quad \frac{dv}{dt} = u^2 - v ,$$

$$(ii) \quad \frac{du}{dt} = a - u + u^2 v, \quad \frac{dv}{dt} = b - u^2 v ,$$

where a and b are positive constants. Which is activator and which the inhibitor in each of (i) and (ii)? What phenomena are indicated by the nonlinear terms? Sketch the nullclines. Is it possible to have positive multi-steady states with these kinetics? What can you say if substrate inhibition is included in (i), that is u^2/v is replaced by $u^2/[v(1 + Ku^2)]$?

3. A gene product with concentration g is produced by a chemical S, is auto-catalyzed and degrades linearly according to the kinetics equation

$$\frac{dg}{dt} = s + k_1 \frac{g^2}{1 + g^2} - k_2 g = f(g; s)$$

where k_1 and k_2 are positive constants and $s = [S]$ is a given concentration. First show that if $s = 0$ there are two positive steady states if $k_1 > 2k_2$ and determine their stability. Sketch the reaction rate dg/dt as a function of g for $s = 0$ (that is $f(g; 0)$). By considering $f(g; s)$ for $s > 0$ show that a critical value s_c exists such that the steady state switches to a higher value for all $s > s_c$. Thus demonstrate that, if $g(0) = 0$ and s increases from $s = 0$ to a sufficiently large value and then decreases to zero again, a biochemical switch has been achieved from $g = 0$ to $g = g_2 > 0$, which you should find.

4. Consider the reaction system whereby two reactants X and Y degrade linearly and X activates Y and Y activates X according to

$$\frac{dx}{dt} = k_1 \frac{y^2}{K + y^2} - k_2 x ,$$

$$\frac{dy}{dt} = h_1 \frac{x^2}{H + x^2} - h_2 y ,$$

where $x = [X]$, $y = [Y]$ and k_1, k_2, h_1, h_2, K and H are positive constants. Nondimensionalise the system to reduce the relevant number of parameters. Show (i) graphically and (ii) analytically that there can be two or zero positive steady states. [Hint for (ii): use Descarte's Rule of Signs (see Appendix 2)]

5. If the reaction kinetics $\mathbf{f}(\mathbf{u})$ in a general mechanism

$$\frac{d\mathbf{u}}{dt} = \mathbf{f}(\mathbf{u})$$

is a gradient system, that is

$$\mathbf{f}(\mathbf{u}) = \nabla_{\mathbf{u}} F(\mathbf{u}) \ ,$$

which is guaranteed if curl $\mathbf{f}(\mathbf{u}) = 0$, show that the solution $\mathbf{u}$ cannot exhibit limit cycle behaviour. [Hint: Use an energy method; that is, first multiply the system by $d\mathbf{u}/dt$]

6. Biological Oscillators and Switches

6.1 Motivation, History and Background

Although living biological systems are immensely complex they are at the same time highly ordered and compactly put together in a remarkably efficient way. Such systems concisely store the information and means of generating the mechanisms required for repetitive cellular reproduction, organisation, control and so on. To see how efficient they can be you need only compare the information storage efficiency per weight of the most advanced computer chip with, say, the ribonucleic acid molecule (mRNA) or a host of others: we are talking here of factors of the order of billions. This chapter, and the next two, will be mainly concerned with oscillatory processes. In the biomedical sciences these are common, appear in widely varying contexts and can have periods from a few seconds to hours and even days and weeks. We shall consider some in detail in this chapter but mention here a few others from the large number of areas of current research involving biological oscillators.

The periodic pacemaker in the heart is, of course, an important example, which will be touched on in Chapter 8. The approximately 24 hour periodic emergence of fruit flies from their pupae might appear to be governed by the external daily rhythm, but this is not the case: see the book by Winfree (1980) for a full exposition. We also briefly discuss this phenomenon in Chapter 8. There is the now classical work of Hodgkin and Huxley (1952) on nerve action potentials, which are the electrical impulses which propagate along a nerve fibre. This is now a highly developed mathematical biology area (see, for example, the review article by Rinzel 1981). Under certain circumstances such nerve fibres exhibit regular periodic firing. The propagation of impulses in neurons normally relies on a threshold stimulus being applied, and is an important practical example of an excitable medium. We discuss the major model for the regular periodic firing behaviour and threshold behaviour in Section 6.5 below and its application to the wave phenomena in Chapter 12.

Breathing is a prime example of another physiological oscillator, here the period is of the order of a second. There are many others, such as certain neural activity in the brain, where the cycles have very small periods. A different kind of oscillator is that observed in the glycolytic pathway. Glycolysis is the process that breaks down glucose to provide the energy for cellular metabolism: oscillations with periods of several minutes are observed in the concentrations of certain chemicals in the process: see, for example the review of this phenomenon and its modelling in Goldbeter's (1980) chapter in the book edited by Segel (1980). Blood testosterone levels in man are often observed to oscillate with periods of the order of 2–3 hours. In Section 6.6 we shall discuss the modelling of this physiological process.

At certain stages in the life cycle of the cellular slime mold, *Dictyostelium discoideum*, the cells emit the chemical cyclic-AMP periodically, with a period of a few minutes. This important topic has been extensively studied theoretically and experimentally: see, for example, the relevant chapter on the periodic aspects in Segel (1984) and the recent models proposed by Martiel and Goldbeter (1987) and Monk and Othmer (1989). Wave phenomena associated with this slime mold are rich in structure as we shall see in Chapter 12. The process of regular cell division in *Dictyostelium*, where the period is measured in hours, indicates a governing biological oscillator of some kind (see for example, the compilation of articles edited by Edmunds 1984).

All of the above examples are different to the biological clocks associated with circadian or daily rhythms, which are associated with external periodicities, in that they are more reasonably described as autonomous oscillators. Limit cycle oscillators, of the kind we consider here, must of course be open systems from thermodynamic arguments, but they are *not* periodic by virtue of some external periodic forcing function.

Since the subject of biological oscillators is now so large, it is not feasible to give a comprehensive coverage of the field here. Instead we shall concentrate on a few general results and some useful simple models which highlight different concepts; we analyse these in detail. We shall also discuss some of the major areas and mechanisms of practical importance and current interest. A knowledge of these is essential in extending the mathematical modelling ideas to other situations. We have already seen periodic behaviour in population models such as discussed in Chapters 1–4, and, from Chapter 5, that it is possible in enzyme kinetics reactions. Other well known examples, not yet mentioned, are the more or less periodic outbreaks of a large number of common diseases: we shall briefly touch on these in Chapter 19 and give references there.

The history of oscillating reactions really dates from Lotka (1910) who suggested a theoretical reaction which exhibits damped oscillations. Later Lotka (1920a,b) proposed the reaction mechanism which now carries the Lotka-Volterra label and which we discussed in its ecological context in Chapter 3 and briefly in its chemical context in the last chapter. Experimentally oscillations were found by Bray (1921) in the hydrogen peroxide-iodate ion reaction where temporal oscillations were observed in the concentrations of iodine and rate of oxygen evolution. He specifically refered to Lotka's early paper. This interesting and important work was dismissed and widely disbelieved since, among other criticisms, it was mistakenly thought that it violated the second law of thermodynamics. It doesn't of course since the oscillations eventually die out, but they only do so slowly.

The next discovery of an oscillating reaction was made by Belousov (1951, 1959), the study of which was continued by Zhabotinskii (1964) and is now known as the Belousov-Zhabotinskii reaction. This important reaction is the subject matter of Chapter 7. There are now many reactions which are known to admit periodic behaviour: the book of articles edited by Field and Burger (1985) describes some of the more recent research in the area, in particular that associated with the Belousov-Zhabotinskii reaction. The book by Winfree (1980), among other things, also discusses it together with a variety of other oscillatory phenomena.

In the rest of this section we comment generally about differential equation systems for oscillators and in the following section we describe some special control mechanisms, models which have proved particularly useful for demonstrating typical and unusual behaviour of oscillators. They are reasonable starting points for modelling real and specific biological phenomena associated with periodic behaviour. Some of the remarks are extensions or generalizations of what we did for two species systems in Chapters 3 and 5.

The models for oscillators which we are concerned with here, with the exception of that in Section 6.6, all give rise to systems of ordinary differential equations (of the type (3.43) studied in Chapter 3) for the concentration vector $\mathbf{u}(t)$ namely

$$\frac{d\mathbf{u}}{dt} = \mathbf{f}(\mathbf{u}) \,, \tag{6.1}$$

where $\mathbf{f}$ describes the nonlinear reaction kinetics, or underlying biological oscillator mechanism. The mathematical literature on nonlinear oscillations is large and daunting, but much of it is not of relevance to real biological modelling. A good practical review, with a list of some relevant references from a mathematical biology point of view, is given by Howard (1979). Mostly we are interested here in periodic solutions of (6.1) such that

$$\mathbf{u}(t + T) = \mathbf{u}(t) \tag{6.2}$$

where $T > 0$ is the period. In the phase space of concentrations this solution trajectory is a simple closed orbit, γ say. If $\mathbf{u}_0(t)$ is a limit cycle solution then it is asymptotically stable (globally) if any perturbation from $\mathbf{u}_0$, or γ, eventually tends to zero as $t \to \infty$.

It is always the case with realistic, qualitative as well as quantitative, biological models that the differential equations involve parameters, generically denoted by λ, say. The behaviour of the solutions $\mathbf{u}(t; \lambda)$ varies with the values or ranges of the parameters as we saw, for example, in Chapter 3. Generally steady state solutions of (6.1), that is solutions of $\mathbf{f}(\mathbf{u}) = 0$, are stable to small perturbations if λ is in a certain range, and become unstable when λ passes through a critical value λ_c, a *bifurcation point*. When the model involves only two dependent variables the analysis of (6.1) can be carried out completely in the phase plane (see Appendix 1) as we saw in Chapters 3 and 5. For higher order systems the theory is certainly not complete and each case usually has to be studied individually.

A major exception is provided by the *Hopf bifurcation theorem*, a detailed proof of which is given in Appendix 3 together with some simple illustrative examples (see also Hassard et al. 1981). Hopf bifurcation results strictly hold only near the bifurcation values. A basic, useful and easily applied result of the Hopf theorem is the following.

Let us suppose that $\mathbf{u} = 0$ is a steady state of (6.1) and that a linearization about it gives a simple complex conjugate pair of eigenvalues $\alpha(\lambda) = \operatorname{Re}\alpha \pm i\operatorname{Im}\alpha$. Now suppose this pair of eigenvalues has the largest real part of all the eigenvalues and is such that in a small neighbourhood of a bifurcation value λ_c, (i) $\operatorname{Re}\alpha < 0$ if $\lambda < \lambda_c$, (ii) $\operatorname{Re}\alpha = 0$ and $\operatorname{Im}\alpha \neq 0$ if $\lambda = \lambda_c$ and (iii) $\operatorname{Re}\alpha > 0$ if $\lambda > \lambda_c$. Then, in a small neighbourhood of λ_c, $\lambda > \lambda_c$ the steady state is unstable by growing oscillations and, at least, a small amplitude *limit cycle periodic solution exists* about $\mathbf{u} = 0$. Furthermore the period of this limit cycle solution is given by $2\pi/T_0$ where $T_0 = \operatorname{Im}[\alpha\lambda_c)]$. The value λ_c is a *Hopf bifurcation* value. The theorem says nothing about the stability of such limit cycle solutions: see Hassard et al. (1981), for example, for a full description of Hopf bifurcation and some applications.

6.2 Feedback Control Mechanisms

It is well documented that in a large number of cell cultures some of the enzymes involved show periodic increases in their activity during division, and these reflect periodic changes in the rate of enzyme synthesis: Tyson (1979) lists several specific cases where this happens. Regulatory mechanisms require some kind of feedback control. In a classic paper, mainly on regulatory mechanisms in cellular physiology, Monod and Jacob (1961) proposed several models which were capable of self-regulation and control and which are known to exist in bacteria. One of these models suggests that certain metabolites repress the enzymes which are essential for their own synthesis. This is done by inhibiting the transcription of the molecule DNA to messenger RNA (mRNA), which is the template which makes the enzyme. Goodwin (1965) proposed a simple model for this process which is schematically shown in Fig. 6.1 in the form analysed in detail by Hastings et al. (1977).

A slight generalization of the Goodwin (1965) model which reflects the process in Fig. 6.1 is

$$\frac{dM}{dt} = \frac{V}{K + P^m} - aM \,,$$

$$\frac{dE}{dt} = bM - cE \,, \tag{6.3}$$

$$\frac{dP}{dt} = dE - eP \,,$$

where M, E and P represent respectively the concentrations of the mRNA, the enzyme and the product of the reaction of the enzyme and a substrate, assumed

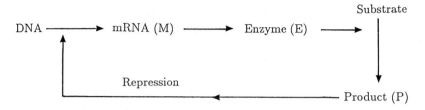

Fig. 6.1. Schematic control system for the production of an enzyme (E) according to the model system (6.3). Here the enzyme combines with the substrate to produce a product (P) which represses the transcription of DNA to mRNA (M), the template for making the enzyme.

to be available at a constant level. All of V, K, m (the Hill coefficient) and a, b, c, d and e are constant positive parameters. Since DNA is externally supplied in this process we do not need an equation for its concentration. With the experience gained from Chapter 5 we interpret this model (6.3) as follows. The creation of M is inhibited by the product P and is degraded according to first order kinetics, while E and P are created and degraded by first order kinetics. Clearly more sophisticated kinetics could reasonably be used with the methods described in Chapter 5. By considering the stability of the steady state, Griffith (1968) showed that oscillations are not possible unless the Hill coefficient m in the first of (6.3) is sufficiently large (see Exercise 4), roughly greater than 8 – an unnaturally high value. For m in this range the system does exhibit limit cycle oscillations.

A more biologically relevant modification is to replace the P-equation in (6.3) by

$$\frac{dP}{dt} = dE - \frac{eP}{k+P} \, .$$

That is, degradation of the product saturates for large P according to Michaelis-Menten kinetics. With this in place of the linear form, limit cycle oscillations can occur for low values of the Hill coefficient m – even as low as $m = 2$.

The concept of a sequence of linked reactions is a useful one and various modifications have been suggested. In one, which has been widely used and studied, the number of reactions has been increased generally to n and the feedback function made more general and hence widely applicable. In a suitable nondimensional form the system is

$$\frac{du_1}{dt} = f(u_n) - k_1 u_1 \, ,$$
$$\frac{du_r}{dt} = u_{r-1} - k_r u_r, \quad r = 2, 3, \ldots, n \tag{6.4}$$

where the $k_r > 0$ and $f(u)$, which is always positive, is the nonlinear feedback function. If $f(u)$ is an increasing function of u, $f'(u) > 0$, (6.4) represents a *positive feedback* loop, while if $f(u)$ is a monotonic decreasing function of u, $f'(u) < 0$, the system represents a *negative feedback* loop or *feedback inhibition*. Positive feedback loops are not common metabolic control mechanisms, whereas negative ones are: see Tyson and Othmer (1978). Yagil and Yagil (1971) have suggested forms for $f(u)$ for several biochemical situations.

Steady state solutions of (6.4) are given by

$$f(u_n) = k_1 k_2 \ldots k_n u_n \, ,$$
$$u_{n-1} = k_n u_n, \ldots, \quad u_1 = k_2 k_3 \ldots k_n u_n \tag{6.5}$$

the first of which is most easily solved graphically by plotting $f(u)$ and noting the intersections with the straight line $k_1 k_2 \ldots k_n u$. With positive feedback functions $f(u)$, multiple steady states are possible whereas with feedback inhibition there is always a unique steady state (see Exercise 3).

Although with higher dimensional equation systems there is no equivalent of the Poincaré-Bendixson theorem for the two-dimensional phase plane (see Appendix 1), realistic systems must have some enclosing domain with boundary B, that is a confined set, such that

$$\mathbf{n} \cdot \frac{d\mathbf{u}}{dt} < 0 \quad \text{for} \quad \mathbf{u} \text{ on } B \, , \tag{6.6}$$

where $\mathbf{n}$ is the outward unit normal to B.

In the case of the more important negative feedback systems of the type (6.4), the determination of such a domain is quite simple. As we noted, we are, of course, only interested in non-negative values for $\mathbf{u}$. Consider first the two-species case of (6.4), namely

$$\frac{du_1}{dt} = f(u_2) - k_1 u_1, \quad \frac{du_2}{dt} = u_1 - k_2 u_2 \, ,$$

where $f(u_2) > 0$ and $f'(u_2) < 0$. Consider first the rectangular domain bounded by $u_1 = 0$, $u_2 = 0$, $u_1 = U_1$ and $u_2 = U_2$, where U_1 and U_2 are to be determined. On the boundaries

$$u_1 = 0, \quad \mathbf{n} \cdot \frac{d\mathbf{u}}{dt} = -\frac{du_1}{dt} = -f(u_2) < 0 \quad \text{for all} \quad u_2 \geq 0 \, ,$$

$$u_2 = 0, \quad \mathbf{n} \cdot \frac{d\mathbf{u}}{dt} = -\frac{du_2}{dt} = -u_1 < 0 \quad \text{for} \quad u_1 > 0 \, ,$$

$$u_1 = U_1, \quad \mathbf{n} \cdot \frac{d\mathbf{u}}{dt} = f(u_2) - k_1 U_1 < 0$$

$$\text{if } U_1 > \frac{f(u_2)}{k_1} \quad \text{for all} \quad 0 \leq u_2 \leq U_2 \quad \Rightarrow \quad U_1 > \frac{f(0)}{k_1} \tag{6.7}$$

$$u_2 = U_2, \quad \mathbf{n} \cdot \frac{d\mathbf{u}}{dt} = u_1 - k_2 U_2 < 0$$

$$\text{if } U_2 > \frac{u_1}{k_2} \quad \text{for all} \quad 0 < u_1 \leq U_1 \, .$$

If we now choose U_1 and U_2 to satisfy the inequalities

$$U_1 > \frac{f(0)}{k_1}, \quad U_2 > \frac{U_1}{k_2} \tag{6.8}$$

then (6.7) shows that there is a confined set B on which (6.6) is satisfied. We can always find such U_1 and U_2 when $f(u)$ is a monotonic decreasing function of u. Note that the positive steady state, given by the unique solution of

$$u_1 = k_2 u_2, \quad f(u_2) = k_1 k_2 u_2$$

always lies inside the domain B defined by (6.7) and (6.8), and, since $f'(u) < 0$, it is always linearly stable, since the eigenvalues of the stability (or community) matrix are both negative. Thus the two-species model cannot admit limit cycle oscillations.

It is now clear how to generalize the method to determine a domain boundary B on which (6.6) is satisfied for an n-species negative feedback loop. The appropiate confined set is given by the box bounded by the planes $u_r = 0$, $r = 1, 2, \ldots, n$ and $u_r = U_r$, $r = 1, 2, \ldots, n$ where any U_r, $r = 1, \ldots, n$ satisfying

$$U_1 > \frac{f(0)}{k_1}, \quad U_2 > \frac{U_1}{k_2}, \ldots, \quad U_n > \frac{U_1}{k_1 k_2 \ldots k_n} \tag{6.9}$$

will suffice. As in the two-species case the steady state always lies inside such a boundary B.

Whether or not such systems with $n \geq 3$ admit periodic solutions is more difficult to determine than in the two species case (see Exercise 4). As the order of the system goes up the possibility of periodic solutions increases. If we consider the Goodwin oscillator (6.3) or, in its dimensionless form (6.4) for u_1, u_2, u_3 with $f(u_3) = 1/(1 + u_3)$, it can be shown that the steady state is always stable (Exercise 4). If we have $f(u_3) = 1/(1 + u_3^m)$ then (Exercise 4), using the Routh-Hurwitz conditions on the cubic for the eigenvalues of the stability matrix, the steady state is only unstable if $m > 8$, which as we have mentioned is an unrealistically high value for the implied cooperativity. As the number of reactions, n, goes up Tyson and Othmer (1978) have shown that the steady state goes unstable if the cooperativity m and the length of the feedback loop n are such that $m > m_0(n) = \sec^n(\pi/n)$. When $n = 3$ this gives $m = 8$ as above: some values for higher n are $n = 4$, $m = 4$; $n = 10$, $m = 1.65$ and $n \to \infty$, $m \to 1$.

By linearizing (6.4) about the steady state (6.5), conditions on the function and parameters can be found such that limit cycle periodic solutions exist: MacDonald (1977), for example, uses bifurcation theory while Rapp (1976) has developed a numerical search procedure for the full nonlinear system and gives quantitative estimates for the period of oscillation.

We can get some analytical approximations for the period of the solutions, when they exist, using a method suggested by Tyson (1979). First we use a result pointed out by Hunding (1974), namely that most of the kinetics parameters

$k_1, k_2, \ldots, k_n$ must be approximately equal or oscillatory solutions will not be possible for low values of m. To see this, first note that each k_r is associated with the inverse of the dimensionless half-life time of u_r. Suppose for example, that one of the constants, say k_s, is much larger than all the others, and choose a time t_1 such that $t_1 \gg 1/k_s$ and $t_1 \ll 1/k_r$ for all $r \neq s$. As the system evolves over a time interval $0 \le t \le t_1$, since $k_r t_1 \ll 1$ for all $r \neq s$, from (6.4) u_{s-1} does not change much in this time interval. So, the solution of the ordinary differential equation for $u_S(t)$ from (6.4) with u_{s-1} constant, is

$$u_s(t) \approx u_s(0) \exp\left[-k_s t\right] + \frac{u_{s-1}}{k_s}\{1 - \exp\left[-k_s t\right]\}, \quad 0 \le t \le t_1 .$$

But, since $k_s t_1 \gg 1$, the last equation gives $u_s(t) \approx u_{s-1}/k_s$ which means that the sth species is essentially at its pseudo-steady state (since $du_s/dt = u_{s-1} - k_s u_s \approx 0$) over the time interval that all the other species change appreciably. This says that the sth species is effectively not involved in the feedback loop process and so the order of the loop is reduced by one to $n-1$.

Now let K be the smallest of all the kinetics parameters and denote the half-life of u_K by H; this is the longest half-life of all the species. Using the above result, the effective length of the feedback loop is equal to the number N of species whose half-lives are all roughly the same as H or, what is the same thing, have rate constants $k \approx K$. All the other reactions take place on a faster time scale and so are not involved in the reaction scheme.

Suppose now we have a periodic solution and consider one complete oscillation in which each of the species undergoes an increase, then a decrease, to complete the cycle. Start off with u_1 which first increases, then u_2, then u_3 and so on to u_N. Then u_1 decreases, then u_2 and so on until u_N decreases. There is thus a total of $2N$ steps involved in the oscillation with each increase and decrease taking approximately the same characteristic time $1/K$. So, the approximate period T of the oscillation is $T \approx 2N/K$. A more quantitative result for the period has been given by Rapp (1976) who showed that the frequency Ω is given by

$$\Omega = K \tan\left(\frac{\pi}{N}\right) \quad \Rightarrow \quad T = \frac{2\pi}{\Omega} ,$$

which reduces to $T \approx 2N/K$ for large N.

The dynamic behaviour of the above feedback control circuits, and generalizations of them, in biochemical pathways have been treated in depth by Tyson and Othmer (1978), and from a more mathematical point of view by Hastings et al. (1977). The latter prove useful results for the existence of periodic solutions for systems with more general reactions than the first order kinetics feedback loops we have just considered.

It is encouraging from a practical point of view that it is very often the case that if (i) a steady state becomes unstable by growing oscillations at some bifurcation value of a parameter, and (ii) there is a confined set enclosing the steady state, then a limit cycle oscillation solution exists. Of course in any specific example it has to be demonstrated, and if possible proved that this is indeed the case. But, as this can often be difficult to do, it is better to try predicting from experience and heuristic reasoning and then simulate the system numerically rather than wait for a mathematical proof which may not be forthcoming. An unstable steady state with its own confined set (6.6), although necessary, are not sufficient conditions for an oscillatory solution of (6.1) to exist. One particularly useful aspect of the rigorous mathematical treatment of Hastings et al. (1977) is that it gives some general results which can be used on more realistic feedback circuits which better mimic real biochemical feedback control mechanisms.

Tyson (1983) proposed a negative feedback model similar to the above to explain periodic enzyme synthesis. He gives an explanation as to why the period of synthesis is close to the cell cycle time when cells undergo division.

6.3 Oscillators and Switches Involving Two or More Species: General Qualitative Results

We have already seen in Chapter 3 that two species models of interacting populations can exhibit limit cycle periodic oscillations. Here we shall derive some general results as regards the qualitative character of the reaction kinetics which may exhibit such periodic solutions.

Let the two species u and v satisfy reaction kinetics given by

$$\frac{du}{dt} = f(u, v), \quad \frac{dv}{dt} = g(u, v), \tag{6.10}$$

where, of course, f and g are nonlinear. Steady state solutions (u_0, v_0) of (6.10) are given by

$$f(u_0, v_0) = g(u_0, v_0) = 0, \tag{6.11}$$

of which only the positive solutions are of interest. Linearizing about (u_0, v_0) we have, in the usual way (see Chapter 3),

$$\begin{pmatrix} \dfrac{d(u - u_0)}{dt} \\ \dfrac{d(v - v_0)}{dt} \end{pmatrix} = A \begin{pmatrix} u - u_0 \\ v - v_0 \end{pmatrix}, \quad A = \begin{pmatrix} f_u & f_v \\ g_u & g_v \end{pmatrix}_{u_0, v_0}. \tag{6.12}$$

The linear stability of (u_0, v_0) is determined by the eigenvalues λ of the stability matrix A, given by

$$|A - \lambda I| = 0 \quad \Rightarrow \quad \lambda^2 - (\operatorname{tr} A)\lambda + |A| = 0.$$

$$\Rightarrow \quad \lambda = \frac{1}{2} \left\{ \operatorname{tr} A + [(\operatorname{tr} A)^2 - 4|A|]^{1/2} \right\}. \tag{6.13}$$

Necessary and sufficient conditions for stability are

$$\operatorname{tr} A = f_u + g_v < 0, \quad |A| = f_u g_v - f_v g_u > 0 \tag{6.14}$$

where here, and in what follows unless stated otherwise, the derivatives are evaluated at the steady state (u_0, v_0).

Near the steady state $S(u_0, v_0)$ in the (u, v) phase plane the null clines $f(u, v) = 0$ and $g(u, v) = 0$ locally can intersect in different ways, for example as illustrated in Fig. 6.2. Note that Fig. 6.2 (b) is effectively equivalent to Fig. 6.2 (a); it is simply Fig. 6.2 (a) rotated. Fig. 6.2 (c) is qualitatively different from the others.

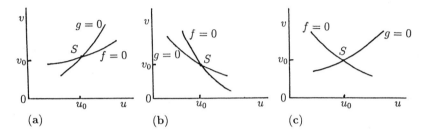

Fig. 6.2a-c. Local behaviour of the reaction null clines $f = 0$, $g = 0$ at a steady state $S(u_0, v_0)$.

Let us assume that the kinetics $f(u, v)$ and $g(u, v)$ are such that (6.10) has a confined set in the positive quadrant. Then, by the Poincaré-Bendixson theorem, limit cycle solutions exist if (u_0, v_0) is an unstable spiral or node, but not if it is a saddle point (see Appendix 1). For an unstable node or spiral to occur, we require

$$\text{tr } A > 0, \quad |A| > 0, \quad (\text{tr } A)^2 \begin{Bmatrix} > \\ < \end{Bmatrix} 4|A| \quad \Rightarrow \quad \text{unstable} \begin{cases} \text{node} \\ \text{spiral} \end{cases} \qquad (6.15)$$

Consider now Fig. 6.2 (a). At the steady state (u_0, v_0) on each of $f = 0$ and $g = 0$ the gradient $dv/du > 0$ with $dv/du]_{g=0} > dv/du]_{f=0}$, so

$$\frac{dv}{du}\Bigg]_{g=0} = -\frac{g_u}{g_v} > \frac{dv}{du}\Bigg]_{f=0} = -\frac{f_u}{f_v} > 0$$

$$\Rightarrow \quad |A| = f_u g_v - f_v g_u > 0 ,$$

providing f_v and g_v have the same sign. Since $dv/du > 0$ it also means that at S, f_u and f_v have different signs, as do g_u and g_v. Now from (6.13), tr $A > 0$ requires at least that f_u and g_v are of opposite sign or are both positive. So, the matrix A (the stability matrix or community matrix in interaction population terms) in (6.12) has terms with the following possible signs for the elements:

$$A = \begin{pmatrix} + & - \\ + & - \end{pmatrix} \quad \text{or} \quad \begin{pmatrix} - & + \\ - & + \end{pmatrix} \qquad (6.16)$$

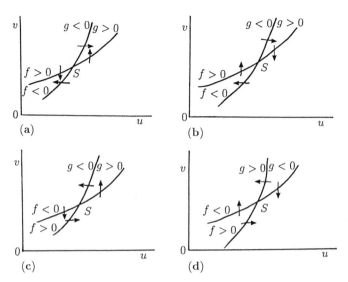

Fig. 6.3a-d. The various possible signs of the kinetics functions $f(u, v)$ and $g(u, v)$ on either side of their null clines for the case illustrated in Fig. 6.2 (a). The arrows indicate, qualitatively, directions of typical trajectories in the neighbourhood of the steady state S.

with each of which it is possible to have $\mathrm{tr}\,A > 0$. We have already shown that $|A| > 0$. To proceed further we need to know individually the signs of f_u, f_v, g_u and g_v at the steady state. With Fig. 6.2 (a) there are 4 possibilities as illustrated in Fig. 6.3. These imply that the elements in the matrix A in (6.12) have the following signs:

$$A = \begin{pmatrix} - & + \\ + & - \end{pmatrix} \quad \text{or} \quad \begin{pmatrix} - & + \\ - & + \end{pmatrix} \quad \text{or} \quad \begin{pmatrix} + & - \\ + & - \end{pmatrix} \quad \text{or} \quad \begin{pmatrix} + & - \\ - & + \end{pmatrix} .$$
$$\text{(a)} \qquad\qquad \text{(b)} \qquad\qquad \text{(c)} \qquad\qquad \text{(d)} \qquad (6.17)$$

For example, to get the sign of f_u at S in Fig. 6.3 (a) we simply note that as we move along a line parallel to the u-axis through S, f decreases since $f > 0$ on the lower u-side and $f < 0$ on the higher u-side. If we now compare these forms with those in (6.16) we see that the only possible forms in (6.17) are (b) and (c). With (d), $|A| < 0$ which makes S a saddle point (which is unstable of course) and so there can be no limit cycle solution enclosing S (see Appendix 1).

   For any given kinetics functions it is easy to determine from the null clines the qualititative behaviour in the neighbourhood of a steady state, and hence the signs in the matrix A in (6.12). If the nullclines look locally like those in Fig. 6.2 (b) and (c) similar results can easily be obtained for the allowable type of kinetics which can admit periodic solutions for (6.10).

   Let us now consider two typical examples which illustrate the qualitative approach we have just described. Let us suppose a parameter λ of the kinetics is such that the null clines for (6.10) look like those in Fig. 6.4 for different ranges of

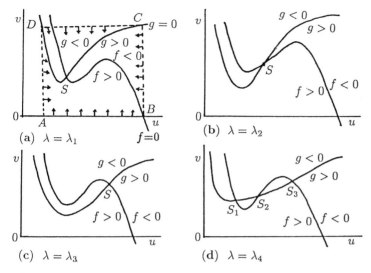

Fig. 6.4a-d. Qualitative form of the null clines for a specimen kinetics in (6.10) as a parameter λ varies: $\lambda_1 \neq \lambda_2 \neq \lambda_3 \neq \lambda_4$. With the signs of f and g as indicated, there is a confined set for (6.10): it is, for example, the rectangular box $ABCDA$ as indicated in (a).

the parameter λ. (This is in fact the nullcline situation for the real biological oscillator (Thomas 1975) briefly discussed in Chapter 5, Section 5.5.) To be specific we choose specific signs for f and g on either side of the nullclines as indicated (these are in accord with the practical Thomas 1975 kinetics situation). Note that there is a confined set on the boundary of which the vector $(du/dt, dv/dt)$ points into the set: one such set is specifically indicated by $ABCDA$ in Fig. 6.4 (a).

Let us now consider each case in Fig. 6.4 in turn. Fig. 6.4 (a) is equivalent to that in Fig. 6.2 (c). Here, in the neighbourhood of S,

$$\left.\frac{dv}{du}\right]_{f=0} = -\frac{f_u}{f_v} < 0, \quad f_u < 0, \quad f_v < 0,$$

$$\left.\frac{dv}{du}\right]_{g=0} = -\frac{g_u}{g_v} > 0, \quad g_u > 0, \quad g_v < 0.$$

So, the stability matrix A in (6.12) has the signs

$$A = \begin{pmatrix} - & - \\ + & - \end{pmatrix} \quad \Rightarrow \quad \operatorname{tr} A < 0, \quad |A| > 0$$

which does not correspond to any of the forms in (6.16): from (6.13), $\operatorname{Re}\lambda < 0$ and so the steady state in Fig. 6.4 (a) is always stable and periodic solutions are not possible for (6.10) in this situation. This case, however, is exactly the same as that in Fig. 6.4 (c) and so the same conclusion also holds for it. By a similar analysis we get for Fig. 6.4 (b)

$$A = \begin{pmatrix} + & - \\ + & - \end{pmatrix}$$

which is the same as (c) in (6.17), and is one of the possible forms for (6.10) to admit periodic solutions.

If we now consider the multi-steady state situation in Fig. 6.4 (d), we have already dealt with S_1 and S_3, which are the same as in Fig. 6.4 (a), (c) – they are always *linearly* stable. For the steady state S_2 we have

$$f_u > 0, \quad f_v < 0, \quad g_u > 0, \quad g_v < 0$$

$$0 < \frac{dv}{du}\Big]_{g=0} < \frac{dv}{du}\Big]_{f=0} \quad \Rightarrow \quad 0 < -\frac{g_u}{g_v} < -\frac{f_u}{f_v}$$

$$\Rightarrow \quad |A| = f_u g_v - f_v g_u < 0 ,$$

which, from (6.15), shows that the steady state is a saddle point, and although it means S_2 is unstable it is the type of singularity which does not admit periodic solutions for (6.10) according to the Poincaré-Bendixson theorem (Appendix 1).

This last case, Fig. 6.4 (d), is of considerable general importance. Recall the threshold phenomenon described in the last chapter in Section 5.5. There we saw that in a situation similar to that in Fig. 6.4 (a), (c) that, although the steady state is linearly stable, if a perturbation is sufficiently large the values of u and v can undergo large perturbations before returning to the steady state (refer to Fig. 5.8). This phenomenon is illustrated in Fig. 6.5 (a), (b).

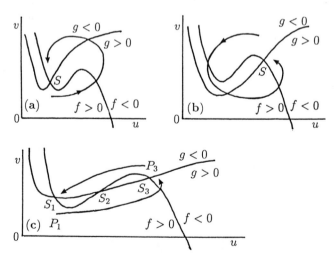

Fig. 6.5a-c. Threshold phenomena for various kinetics for (6.10). In (c) a suitable perturbation from one linearly stable steady state can effect a permanent change to the other stable steady state.

Now consider Fig. 6.4 (d). S_1 and S_3 are respectively equivalent to the S in Fig. 6.5 (a), (b). We now see, in Fig. 6.5 (c), that if we perturb (u, v) from say S_1 to P_1, the solution trajectory will be qualitatively as shown. Now, instead of returning to S_1 the solution moves to S_3, the second stable steady state. In this way a *switch* has been effected from S_1 to S_3. In a similar way a switch can be effected from S_3 to S_1 by, for example, a perturbation from the steady state S_3 to P_3. It is possible that a parameter in the kinetics function g, say, can be varied in such a way that the null cline is translated vertically as the parameter is, for example, increased. In this case it is possible for the system to exhibit *hysteresis* such as we discussed in detail in Chapter 1, Section 1.2 and Chapter 5, Section 5.5. If the reaction kinetics give rise to mushrooms and isolas, even more baroque dynamic, threshold and limit cycle behaviour is possible. Biological switches, not only those exhibiting hysteresis and more exotic behaviour, are of considerable importance in biology. We discuss one important example below in Section 6.5. We also see a specific example of its practical importance in the wave phenomenon observed in certain eggs after fertilization, a process and mechanism for which is discussed in detail in Chapter 11, Section 11.6 and Chapter 17, Section 17.8.

It is clear from the above that the qualitative behaviour of the solutions can often be deduced from a gross geometric study of the null clines and the global phase plane behaviour of trajectories. We can carry this approach much further, as has been done, for example by Rinzel (1985), to predict even more complex solution behaviour of such differential equation systems. Here I only want to give a flavour of what can be found.

Let us suppose we have a general system governed by

$$\frac{d\mathbf{u}}{dt} = \mathbf{f}(\mathbf{u}, \alpha), \quad \frac{d\alpha}{dt} = \varepsilon g(\mathbf{u}, \alpha) , \tag{6.18}$$

where $0 < \varepsilon \ll 1$, $\mathbf{u}$ is a vector of concentrations and α is a parameter which is itself governed by an equation but in a slowly varying way. The *fast* subsystem of (6.18) is the $O(1)$ system, as $\varepsilon \to 0$, in which α is simply a constant parameter, since $d\alpha/dt \approx 0$. The *slow* dynamics governs the change in α with time. We analyse some specific systems like this in the following chapter, when we discuss relaxation oscillators.

Suppose a uniform steady state $\mathbf{u}_0$ depends on α as indicated schematically in Fig. 6.6 (a). That is there is a region $\alpha_1 < \alpha < \alpha_2$ where three possible steady states $\mathbf{u}_0$ exist: recall also the discussion in Section 5.5 in the last chapter. To be more specific let us suppose that α varies periodically in such a way that in each cycle it sweeps back and forth through the window which gives three solutions for $\mathbf{u}_0$, the one on the dashed line in Fig. 6.6 (a) being unstable as to be expected. At the start suppose $\alpha = \alpha_1$ and $\mathbf{u}_0$ is at A in Fig. 6.6 (a). Now as α increases, $\mathbf{u}_0$ slowly varies until α passes through α_2. At α_2, $\mathbf{u}_0$ jumps discontinuously from B to C, after which it again varies slowly with α. On the return α-trip, $\mathbf{u}_0$ remains on the lower branch of the S-curve until it reaches D, where it jumps up to A again. The limit cycle behaviour of this system is

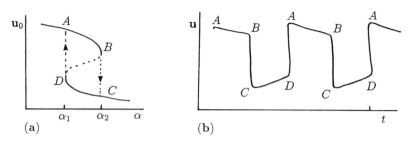

Fig. 6.6a,b. (a) Schematic steady state u_0 dependence on the parameter α : steady states on the dashed line are unstable. (b) Typical limit cycle behaviour of u if α slowly varies in a periodic way. The oscillation is described as a *relaxation oscillator*: that is there are slowly varying sections of the solution interspersed with rapidly varying regions.

illustrated schematically in Fig. 6.6 (b). The rapidly varying region is where u drops from B to C and increases from D to A. This is a typical *relaxation oscillator* behaviour: see Chapter 7, Section 7.4 below.

The fast dynamics subsystem in (6.18) may, of course, have as its steady state a periodic solution, say, u_{per}. Now the parameter α affects an oscillatory solution. A relevant bifurcation diagram is then one which shows, for example, a transition from one oscillation to another. Fig. 6.7 (a) illustrates such a possibility. The branch AB represents, say, a small amplitude stable limit cycle oscillation around u_0 for a given α. Solutions on the branch BC are unstable. Now as α increases there is a slow variation in the solution until it passes through α_2 at B, after which the periodic solution undergoes a bifurcation to a larger amplitude oscillation with bounds for u on the curves EF and HI. The transition from one solution type to another is fast, as in the relaxation oscillator situation in Fig. 6.6. Now let α decrease. The bifurcation to the AB branch now occurs at D, where $\alpha = \alpha_1$. So, as α varies periodically such that it includes a window with $\alpha < \alpha_1$ and $\alpha > \alpha_2$, the solution behaviour will be qualitatively like that shown in Fig. 6.7 (b).

Fig. 6.7 (c) shows another possible example. The line AB represents a non-oscillatory solution which bifurcates for $\alpha = \alpha_0$ to a periodic solution at B. These branches terminate at D and C, where $\alpha = \alpha_2$. The branch EF is again a uniform stable steady state. Suppose we now consider α to vary periodically between $\alpha > \alpha_2$ and $\alpha_0 < \alpha < \alpha_1$. To be specific let us start at F in Fig. 6.7 (c). As α decreases we move along the branch FE; that is the uniform steady state $\underline{u}_{ss}$ varies slowly. At E, where $\alpha = \alpha_1$, the uniform steady state bifurcates to a periodic solution on the branches BD and BC. Now as α increases the periodic solution remains on these branches until α reaches α_2 again after which the solution jumps down again to the homogeneous steady state branch EF. A typical time behaviour for the solution is illustrated in Fig. 6.7 (d). Both this behaviour and that in Fig. 6.7 (b) are described as 'periodic bursting'.

The complexity of solution behaviour of such systems (6.18) can be spectacular. The specific behaviour just described in Figs. 6.6 and 6.7 has been found in

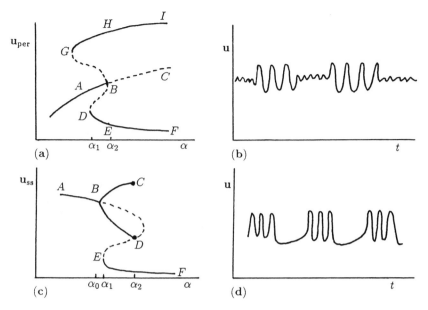

Fig. 6.7a-d. (a) Schematic bifurcation for periodic solutions of the fast dynamics subsystem of (6.18) as α varies periodically. The dashed lines are unstable branches. (b) Typical periodic behaviour as α slowly varies in a periodic way back and forth through the (α_1,α_2) window, for the bifurcation picture in (a). (c) Another example of a periodic solution bifurcation diagram for the subsystem of (6.18) as α varies. (d) Qualitative periodic solution behaviour as α varies periodically through α_1 and α_2 in (c). These are examples of periodic 'bursting'.

models for real biological systems, an example of the former is in the following chapter, while qualitatively similar curves to those in Fig. 6.7 have been found by Rinzel (1985). The system studied by Rinzel (1985) is specifically related to the model described below in Section 6.5 on neural periodic behaviour. The model system given by (5.56) in Section 5.5 in the last chapter, and the iodate-arsenous model reaction scheme (5.63)–(5.64), exhibit comparable solution behaviour but with the potential for even more complex dynamic phenomena. Decroly and Goldbeter (1987) consider a 3-variable system associated with cyclic-AMP emission by cells of the slime mold *Dictyostelium discoideum* as a vehicle to demonstrate the transition from simple to complex oscillatory behaviour. As well as obtaining increasingly complex patterns of bursting they show period doubling leading to chaos.

Fig. 6.7 shows some of the complex effects which appear when oscillators interact or when reaction schemes have fast and slow subschemes. This is mathematically a very interesting and challenging field and one of current widespread research. We consider in some detail some important aspects of oscillator interaction in Chapter 8. In Chapter 10 we discuss another important and quite different aspect of interacting oscillations.

6.4 Simple Two-Species Oscillators: Parameter Domain Determination for Oscillations

If we restrict our reaction system to only two species it was shown by Hanusse (1972) and independently by Tyson and Light (1973) that limit cycle solutions can only exist if there are tri-molecular reactions. These would be biochemically unrealistic if they were the only reactions involved, but as we have shown in Chapter 5 such two reactant models can arise naturally from a higher order system if typical enzyme reactions, for example, are part of the mechanism being considered. So, it is reasonable to consider tri-molecular two species models and not just for algebraic and mathematical convenience in demonstrating principles and techniques. Schnackenberg (1979) considered the class of two-species 'simplest', but chemically plausible, tri-molecular reactions which will admit periodic solutions. The simplest such reaction mechanism is

$$ X \underset{k_{-1}}{\overset{k_1}{\rightleftharpoons}} A, \quad B \overset{k_2}{\rightarrow} Y, \quad 2X + Y \overset{k_3}{\rightarrow} 3X \qquad (6.19) $$

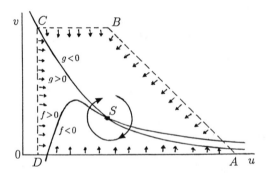

Fig. 6.8. Typical null clines $f = 0$ and $g = 0$ for the 'simplest' oscillator (6.20) for $a > 0$ and $b > 0$. The quadrilateral $ABCDA$ is a boundary of a confined set enclosing the steady state S.

which, using the Law of Mass Action, results in the nondimensional equations for u and v, the dimensionless concentrations of X and Y, given by

$$ \frac{du}{dt} = a - u + u^2 v = f(u, v), \quad \frac{dv}{dt} = b - u^2 v = g(u, v) \, , \qquad (6.20) $$

where a and b are positive constants. Typical null clines are illustrated in Fig. 6.8. In the vicinity of the steady state S these are equivalent to the situation in Fig. 6.2 (b). With (6.20) it is easy to construct a confined set on the boundary of which the vector $(du/dt, dv/dt)$ points inwards or along it: the quadrilateral in Fig. 6.8 is one example. Hence, because of the Poincaré-Bendixson theorem, the existence of a periodic solution is assured if, for (6.20), the stability matrix A for the steady state satisfies (6.15).

Determination of Parameter Space for Oscillations

For any model involving parameters it is always useful to know the ranges of parameter values where oscillatory solutions are possible and where they are not. For all but the simplest kinetics this has to be done numerically, but the principles involved are the same for them all. Here we shall carry out the detailed analysis for the simple model reaction (6.20) to illustrate the general principles: the model involves only 2 parameters a and b and we can calculate the (a, b) parameter space analytically. The requisite space is the range of the parameters a and b which make the steady state an unstable node or spiral: that is the parameter range where, from (6.15), tr $A > 0$ and $|A| > 0$. Later we shall develop a more powerful and general parametric method which can be applied to less simple kinetics.

The steady state (u_0, v_0) for (6.20) is given by

$$f(u_0, v_0) = a - u_0 + u_0^2 v_0 = 0, \quad g(u_0, v_0) = b - u_0^2 v_0 = 0 \ ,$$

$$\Rightarrow \quad u_0 = b + a, \quad v_0 = \frac{b}{(a+b)^2}, \quad \text{with} \quad b > 0, \ a + b > 0 \ . \tag{6.21}$$

Substituting these into the stability matrix A in (6.12), we get

$$\text{tr} \, A = f_u + g_v = (-1 + 2u_0 v_0) + (-u_0^2) = \frac{b-a}{a+b} - (a+b)^2 \ , \tag{6.22}$$

$$|A| = f_u g_v - f_v g_u = (a+b)^2 > 0 \quad \text{for all} \quad a, b \ .$$

The domain in (a, b) space where (u_0, v_0) is an unstable node or spiral is, from (6.15), where tr $A > 0$ and so the domain boundary is

$$\text{tr} \, A = 0 \quad \Rightarrow \quad b - a = (a+b)^3 \ . \tag{6.23}$$

Even with this very simple model, determination of the boundary involves the solution of a cubic, not admittedly a major problem but a slightly tedious one. Care has to be taken since the solution, say, of b in terms of a, involves three branches. Fig. 6.9 gives the parameter domain where oscillations are possible for $b > 0$.

There is another more powerful way (Murray 1982) of determining the boundary, namely parametrically, which is much easier and which also avoids the multiple branch problem. Furthermore, it is a method which has wider applicability, can be used with more complicated systems and provides the numerical procedure for the determination of the parameter domain for systems where it is not feasible to do it analytically. We shall again use the simple model system (6.20) to illustrate the method: see the exercises for other examples.

Let us consider the steady state u_0 as a parameter and determine b and a in terms of u_0. From (6.21),

$$v_0 = \frac{u_0 - a}{u_0^2}, \quad b = u_0^2 v_0 = u_0 - a \ . \tag{6.24}$$

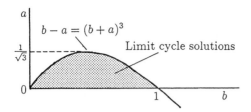

Fig. 6.9. Parameter space where limit cycle periodic solutions of (6.20) exist for $a > 0$ and $b > 0$. The boundary curve is given by (6.23), although it was in fact calculated using the more easily applied parameteric form (6.27) below.

and

$$
A = \begin{pmatrix} f_u & f_v \\ g_u & g_v \end{pmatrix} = \begin{pmatrix} -1 + 2u_0v_0 & u_0^2 \\ -2u_0v_0 & -u_0^2 \end{pmatrix} = \begin{pmatrix} 1 - \dfrac{2a}{u_0} & u_0^2 \\ -2 - \dfrac{2a}{u_0} & -u_0^2 \end{pmatrix}
$$

Since $|A| = u_0^2 > 0$ the required necessary condition for oscillations from (6.15) is tr $A > 0$, that is

$$
f_u + g_v > 0 \quad \Rightarrow \quad 1 - \frac{2a}{u_0} - u_0^2 > 0 \quad \Rightarrow \quad a < \frac{u_0(1 - u_0^2)}{2} . \tag{6.25}
$$

We also have from (6.24)

$$
b = u_0 - a > \frac{u_0(1 + u_0^2)}{2} . \tag{6.26}
$$

The last two inequalities define, parametrically in u_0, the boundary curve where tr $A = 0$. Since the parameter u_0 is the steady state, the only parameter range of interest is $u_0 \geq 0$. Thus, one of the boundary curves in (a, b) space, which defines the domain where the necessary condition for oscillations are satisfied (in this example it is only tr $A > 0$), is defined by

$$
a = \frac{u_0(1 - u_0^2)}{2}, \quad b = \frac{u_0(1 + u_0^2)}{2}, \quad \text{for all} \quad u_0 > 0 . \tag{6.27}
$$

Sufficient conditions for an oscillatory solution are given by (6.15) together with the existence of a confined set. Since a confined set has been obtained for this mechanism (see Fig. 6.8) the conditions (6.25) and (6.26) are sufficient. Fig. 6.9 was calculated using (6.27) and shows the space given by (6.23). The mechanism (6.20) will exhibit a limit cycle oscillation for *any* parameter values which lie in the shaded region; for all other values in the positive quadrant the steady state is stable.

This pedagogically very useful model (6.20) is a particularly simple one for which to determine the parameter space for periodic solutions. This is because the requirement $|A| > 0$ was automatically satisfied for all values of the parameter and the necessary and sufficient condition for existence boiled down to finding the domain where $\operatorname{tr} A$ was positive. Generally, once a confined set has been found (which in itself can often put constraints on the parameters), the parameter space for periodic solutions is determined by the *two* boundary curves in parameter space defined by $\operatorname{tr} A = 0$, $|A| = 0$.

Although we envisage the biochemical mechanism (6.20) to have $a > 0$ the mathematical problem need not have such a restriction as long as u_0 and v_0 are non-negative.

To show how the parameteric procedure works in general, let us allow a to be positive or negative. Now the necessary and sufficient conditions are satisfied if $\operatorname{tr} A > 0$, namely (6.25) with (6.26), and $|A| > 0$. Since $|A| = u_0^2 > 0$, the condition $|A| > 0$ is automatically satisfied. With the requirement $u_0 \geq 0$ this gives the curve in (a, b) space as

$$b + a > 0 \quad \Rightarrow \quad a > -u_0, \quad b > u_0 , \qquad (6.28)$$

as a particularly simple parametetric representation. Thus the two sets of inequalities are bounded in parameter space by the curves

$$\left.\begin{array}{cc} a = \dfrac{u_0(1 - u_0^2)}{2}, & b = \dfrac{u_0(1 + u_0^2)}{2}, \\[2mm] a = -u_0, & b = u_0. \end{array}\right\} \quad \text{for all} \quad u_0 \geq 0 \qquad (6.29)$$

Fig. 6.10 gives the general parameter space defined by (6.29). The inequality (6.28) is satisfied by values (a, b) which lie above the the straight line given by (6.29) while the inequality (6.26) is satisfied for values lying below the curve given by (6.29). Together they define a closed domain.

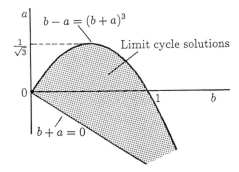

Fig. 6.10. Parameter space in which solutions (u, v) of (6.20) are periodic limit cycles. Note that $a < 0$ is possible, although it is not of biochemical interest.

λ-ω Systems

These are particularly simple systems of equations which have exact limit cycle solutions, and which have been widely used in prototype studies of reaction diffusion systems. The equations can be written in the form

$$\frac{du}{dt} = \lambda(r)u - \omega(r)v, \quad \frac{dv}{dt} = \omega(r)u + \lambda(r)v ,$$

$$r = (u^2 + v^2)^{1/2} , \tag{6.30}$$

where λ is a positive function of r for $0 \leq r \leq r_0$ and negative for $r > r_0$, and so $\lambda(r_0) = 0$, and $\omega(r)$ is a positive function of r. It does not seem possible to derive such equations from any sequence of reasonable biochemical reactions. However their advantage primarily lies in the fact that explicit analytic results can be derived when they are used as the kinetics in the study of wave phenomena in reaction diffusion models. Such analytical solutions can often provide indications of what to look for in more realistic systems. So although their use is in an area to be discussed later, it is appropriate to introduce them here simply as examples of nontrivial mathematical oscillators.

If we express the variables (u, v) in the complex form $c = u + iv$, equations (6.30) become the complex equation

$$\frac{dc}{dt} = [\lambda(|c|) + i\omega(|c|)]c, \quad c = u + iv . \tag{6.31}$$

From this, or by multiplying the first of (6.30) by u and adding it to v times the second, we see that a limit cycle solution is the circle in the (u, v) plane or complex c-plane since

$$\frac{d|c|}{dt} = \lambda(|c|)|c| \quad \Rightarrow \quad |c| = r_0 , \tag{6.32}$$

because $\lambda(|c|)$ is positive if $0 \leq |c| \leq r_0$ and negative if $|c| \geq r_0$.

An alternative way to write (6.31) in the complex plane is to set

$$c = re^{i\theta} \quad \Rightarrow \quad \frac{dr}{dt} = r\lambda(r), \quad \frac{d\theta}{dt} = \omega(r) \tag{6.33}$$

for which the limit cycle solution is

$$r = r_0, \quad \theta(t) = \omega(r_0)t + \theta_0 , \tag{6.34}$$

where θ_0 is a constant.

6.5 Hodgkin-Huxley Theory of Nerve Membranes: FitzHugh-Nagumo Model

Neural communication is clearly a very important field. We make no attempt here to give other than an introduction to it and discuss one of the key mathematical models which has been studied extensively. Rinzel (1981) gives a short review of models in neurobiology: see also the references there.

Electric signalling or firing by individual nerve cells or neurons is particularly common. The seminal and now classical work by Hodgkin and Huxley (1952) on this aspect of nerve membranes was on the nerve axon of the giant squid. (They were awarded a Nobel prize for their work.) Basically the axon is a long cylindrical tube which extends from each neuron and electrical signals propagate along its outer membrane, about 50–70 Ångströms thick. The electrical pulses arise because the membrane is preferentially permeable to various chemical ions with the permeabilities affected by the currents and potentials present. The key elements in the system are potassium (K^+) ions and sodium (Na^+) ions. In the rest state there is a transmembrane potential difference of about –70 millivolts(mV) due to the higher concentration of K^+ ions within the axon as compared with the surrounding medium. The deviation in the potential across the membrane, measured from the rest state, is a primary observable in experiments. The membrane permeability properties change when subjected to a stimulating electrical current I: they also depend on the potential. Such a current can be generated, for example, by a local depolarization relative to the rest state.

In this section we shall be concerned with the *space-clamped* dynamics of the system; that is we consider the spatially homogeneous dynamics of the membrane. With a real axon the space-clamped state can be obtained experimentally by having a wire down the middle of the axon maintained at a fixed potential difference to the outside. Later, in Chapter 12, we shall discuss the important spatial propagation of action potential impulses along the nerve axon: we shall refer back to the model we discuss here. We derive here the Hodgkin-Huxley (1952) model and the reduced analytically tractable FitzHugh-Nagumo mathematical model (FitzHugh 1961, Nagumo, Arimoto and Yoshizawa 1962) which captures the key phenomena. The analysis of the various mathematical models has indicated phenomena which have motivated considerable experimental work. The theory of neuron firing and propagation of nerve action potentials is one of the major successes of real mathematical biology.

Basic Mathematical Model

Let us take the positive direction for the membrane current, denoted by I, to be outwards from the axon. The current $I(t)$ is made up of the current due to the individual ions which pass through the membrane and the contribution from the time variation in the transmembrane potential, that is the membrane capacitance contribution. Thus we have

$$I(t) = C\frac{dV}{dt} + I_i , \tag{6.35}$$

where C is the capacitance and I_i is the current contribution from the ion movement across the membrane. Based on experimental observation Hodgkin and Huxley (1952) took

$$\begin{aligned} I_i &= I_{Na} + I_K + I_L , \\ &= g_{Na}m^3h(V - V_{Na}) + g_K n^4(V - V_K) + g_L(V - V_L) , \end{aligned} \tag{6.36}$$

where V is the potential and I_{Na}, I_K and I_L are respectively the sodium, potassium and 'leakage' currents: I_L is the contribution from all the other ions which contribute to the current. The g's are constant conductances with, for example, $g_{Na}m^3h$ the sodium conductance, and V_{Na}, V_K and V_L are constant equilibrium potentials. The m, n and h are variables, bounded by 0 and 1, which are determined by the differential equations

$$\begin{aligned} \frac{dm}{dt} &= \alpha_m(V)(1 - m) - \beta_m(V)m , \\ \frac{dn}{dt} &= \alpha_n(V)(1 - n) - \beta_n(V)n , \\ \frac{dh}{dt} &= \alpha_h(V)(1 - h) - \beta_h(V)h , \end{aligned} \tag{6.37}$$

where the α and β are given functions of V (again empirically determined by fitting the results to the data): see for example FitzHugh (1969). α_n and α_m are qualitatively like $(1 + \tanh V)/2$ while $\alpha_h(V)$ is qualitatively like $(1 - \tanh V)/2$, which is a 'turn-off' switch if V is moderately large.

If an applied current $I_a(t)$ is imposed the governing equation using (6.35) becomes

$$C\frac{dV}{dt} = -g_{Na}m^3h(V - V_{Na}) - g_K n^4(V - V_K) - g_L(V - V_L) + I_a . \tag{6.38}$$

The system (6.38) with (6.37) constitute the 4-variable model which was solved numerically by Hodgkin and Huxley (1952).

If $I_a = 0$, the rest state of the model (6.37) and (6.38) is linearly stable but is excitable in the sense discussed in Chapter 5. That is, if the perturbation from the steady state is sufficiently large there is a large excursion of the variables in their phase space before returning to the steady state. If $I_a \neq 0$ there is a range of values where regular repetitive firing occurs; that is the mechanism displays limit cycle characteristics. Both types of phenomena have been observed experimentally. Because of the complexity of the equation system various simpler mathematical models, which capture the key features of the full system, have been proposed, the best known and particularly useful one of which is the FitzHugh-Nagumo model (FitzHugh 1961, Nagumo et al. 1962), which we now derive.

The time scales for m, n and h in (6.37) are not all of the same order. The time scale for m is much faster than the others, so it is reasonable to assume it is sufficiently fast that it relaxes immediately to its value determined by setting $dm/dt = 0$ in (6.37). If we also set $h = h_0$, a constant, the system still retains many of the features experimentally observed. The resulting 2-variable model in V and n can then be qualitatively approximated by the dimensionless system

$$\frac{dv}{dt} = f(v) - w + I_a, \quad \frac{dw}{dt} = bv - \gamma w \,,$$

$$f(v) = v(a - v)(v - 1) \,,$$

(6.39)

where $0 < a < 1$ and b and γ are positive constants. Here v is like the membrane potential V, and w plays the role of all three variables m, n and h in (6.37).

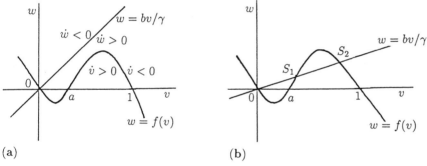

(a)   (b)

Fig. 6.11a,b. Phase plane for the model system (6.39) with $I_a = 0$. As the parameters vary there can be **(a)** one stable, but excitable steady state or, **(b)** three possible steady states, one unstable, namely S_1, and two stable but excitable, namely $(0,0)$ and S_2.

With $I_a = 0$, or just a constant, the system (6.39) is simply a two-variable phase plane system, the null clines for which are illustrated in Fig. 6.11. Note how the phase portrait varies with different values of the parameters a, b and γ. There can, for example, be 1 or 3 steady states as shown in Fig. 6.11 (a), (b) respectively. The situation corresponds to that illustrated in Fig. 6.5, except that here it is possible for v to be negative – it is an electric potential. The excitability characteristic, a key feature in the Hodgkin-Huxley system, is now quite evident. That is a perturbation, for example from 0 to a point on the v-axis with $v > a$, undergoes a large phase trajectory excursion before returning to 0, Fig. 6.12 shows a specific example.

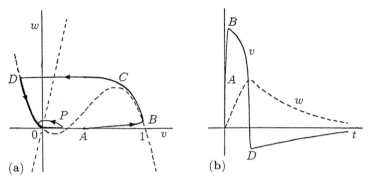

Fig. 6.12a,b. (a) The phase portrait for (6.39) with $I_a = 0$, $a = 0.25$, $b = \gamma = 2 \times 10^{-3}$ which exhibits the threshold behaviour. With a perturbation from the steady state $v = w = 0$ to a point, P say, where $w = 0$, $v < a$, the trajectory simply returns to the origin with v and w remaining small. A perturbation to A initiates a large excursion along $ABCD$ and then back to $(0,0)$, effectively along the null cline since b and γ are small. (b) The time variation of v and w corresponding to the excitable trajectory $ABCD0$ in (a). (Redrawn from Rinzel 1981)

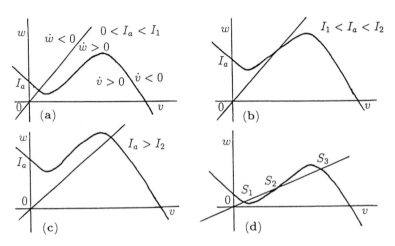

Fig. 6.13a-d. Null clines for the FitzHugh-Nagumo model (6.39) with different applied currents I_a. Cases (a), where $I_a < I_1$, and (c), where $I_a > I_2$, have linearly stable, but excitable, steady states, while in (b), where $I_1 < I_a < I_2$, the steady state can be unstable and limit cycle periodic solutions are possible. With the configuration (d), the steady states S_1, S_3 are stable with S_2 unstable. Here a perturbation from either S_1 or S_3 can effect a switch to the other.

Periodic Neuron Firing

With $I_a = 0$ the possible phase portraits, as illustrated in Fig. 6.11, shows there can be no periodic solutions (see Section 6.3). Suppose now that there is an applied current I_a. The corresponding null clines for (6.39) are illustrated in Fig. 6.13 (a)-(c) for several $I_a > 0$. The effect on the null clines is simply to move the v null cline, with $I_a = 0$, up the w-axis. With parameters values such that the null clines are as in Fig. 6.13 (a) we can see that by varying only I_a there is a window of applied currents (I_1, I_2) where the steady state can be unstable and limit cycle oscillations possible; that is a null cline situation like that in Fig. 6.13 (b). The algebra to determine the various parameter ranges for a, b, γ and I_a for each of these various possibilities to hold is straightforward. It is just an exercise in elementary analytical geometry, and is left as an exercise (Exercise 7). With the situation exhibited in Fig. 6.13 (d) limit cycle solutions are not possible. On the other hand this form can exhibit switch properties.

The FitzHugh-Nagumo model (6.39) is a *model* of the Hodgkin-Huxley *model*. So, a further simplification of the mechanism (6.39) is not unreasonable if it simplifies the analysis or makes the various solution possibilities simpler to see. Of course such a simplification must retain the major elements of the original, so care must be exercised. From Fig. 6.11 we can reasonably approximate the v null cline by a piece-wise linear approximation as in Fig. 6.14, which in Fig. 6.14 (a) has zeros at $v = 0$, a, 1. The positions of the minimum and maximum, (v_1, w_1) and (v_2, w_2) are obtained from (6.39) as

$$v_2, v_1 = \frac{(a+1) \pm \{(a+1)^2 - 3a\}^{1/2}}{3},$$
(6.40)

$$w_i = -v_i(a - v_i)(1 - v_i) + I_a, \quad i = 1, 2.$$

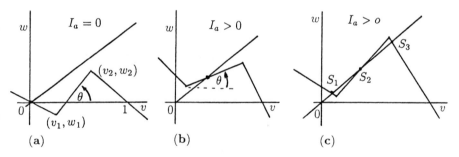

Fig. 6.14a-c. (a) Phase plane null clines for a piece-wise linear approximation to the v null cline in the FitzHugh-Nagumo model (6.39) with $I_a = 0$, where (v_1, w_1) and (v_2, w_2) are given by (6.40). (b) The geometric conditions for possible periodic solutions, which requires $I_a > 0$, are shown in terms of the angle $\theta = \tan^{-1}[(w_2 - w_1)/(v_2 - v_1)]$. (c) Geometric conditions for multiple roots and threshold switch possibilities from one steady state S_1 to S_3 and vice-versa.

The line from (v_1, w_1) to (v_2, w_2) passes through $v = a$ if $a = 1/2$. The acute angle θ the null cline makes with the v-axis in Fig. 6.14 is given by

$$\theta = \tan^{-1}\left[\frac{w_2 - w_1}{v_2 - v_1}\right].$$
(6.41)

We can now write down simply a necessary condition for limit cycle oscillations for the piece-wise model, that is conditions for the null clines to be as in Fig. 6.14 (b). The gradient of the v null cline at the steady state must be less than the gradient, b/γ, of the w null cline, that is

$$\tan \theta = \frac{w_2 - w_1}{v_2 - v_1} < \frac{b}{\gamma} . \tag{6.42}$$

Sufficient conditions for a limit cycle solution to exist are obtained by applying the results of Section 6.3 and demonstrating that a confined set exists. Analytical expressions for the limits on the applied current I_a for limit cycles can also be found (Exercise 7).

A major property of this model for the space-clamped axon membrane is that it can generate regular beating of a limit cycle nature when the applied current I_a is in an appropriate range $I_1 < I_a < I_2$. The bifurcation to a limit cycle solution when I_a increases past I_1 is essentially a Hopf bifurcation and so the period of the limit cycle is given by an application of the Hopf bifurcation theorem (see Appendix 3). This model with periodic beating solutions will be referred to again in Chapter 8 when we consider the effect of perturbations on the oscillations. All of the solution behaviour found with the model (6.39) have also been found in the full Hodgkin-Huxley model, numerically of course. The various solution properties have also been demonstrated experimentally.

Some neuron cells fire with periodic bursts of oscillatory activity like that illustrated in Fig. 6.7 (b), (d). We would expect such behaviour if we considered coupled neuronal cells which independently undergo continuous firing. By modifying the above model to incorporate other ions, such as a calcium(Ca^{++}) current, periodic bursting is obtained: see Plant (1978, 1981). There are now several neural phenomena where periodic bursts of firing are observed experimentally. With the knowledge we now have of the qualitative nature of the terms and solution behaviour in the above models and some of their possible modifications, we can now build these into other models to reflect various observations which indicate similar phenomena. The field of neural signalling, both temporal and spatial, is a fascinating and important one which will be an area of active research for many years.

6.6 Modelling the Control of Testosterone Secretion

The hormone testosterone, although present in very small quantities in the blood, is an extremely important hormone; any regular imbalance can cause dramatic changes. In man the blood levels of testosterone can fluctuate periodically with periods of the order of 2–3 hours. In this section we shall discuss the physiology of testosterone production and construct and analyse a model, rather different from those we have so far discussed in this chapter, to try and explain the periodic levels of testosterone observed. Although the phenomenon is interesting in its own right, another reason for discussing it is to demonstrate the procedure used to analyse this type of model.

Before describing the important physiological elements in the process of testosterone production there are some very interesting effects and ideas associated with this important hormone. Men have a testosterone level of between 10–35 nanomoles per litre of blood with women having between 0.7–2.7 nanomoles per litre. Reduced levels of testosterone, or rather the level of a sex hormone binding globulin (SHBG), directly related to free testosterone, are often accompanied by personality changes – the individual tends to become less forceful and commanding. On the other hand increased levels of testosterone induce the converse. In women the testosterone level tends to increase after menopause (perhaps explaining why at that stage more women often become increasingly interested in leadership roles). Although the actual differences in testosterone levels are minute the effects can be major.

In men the high level of tesosterone primarily comes from the testes, which produces about 90%, with the rest from other parts of the endocrine system, which is why women also produce it. The drug goserelin, which was introduced to treat cancer of the prostate, can achieve chemical castration within a few weeks after the start of treatment. The patient's testosterone level is reduced to what would be achieved by removal of the testes. The body does not seem to adjust to the drug and so effective castration continues only as long as the treatment is maintained. How the drug works in blocking the production of testosterone will be pointed out below when we discuss the physiological production process. Enthusiasm for sex, or sex drive, depends on many factors and not only the level of testosterone, which certainly plays a very significant role. If we consider the problem of an excessive sex drive, one man recently sentenced to 14 years for rape asked to be treated with goserelin. Drug induced castration thus opens up a very controversial area of treatment of sex offenders.

The full physiological process is not yet fully understood although there is general agreement on certain key elements. The following shows how a first model had to be modified to incorporate key physiological facts and points the way to more recent and complex models. We first derive a model for testosterone (T) production in the male suggested by Smith (1980): it is based on accepted basic experimental facts. We shall then discuss a modification which results in a delay model which incorporates more realistic physiology associated with the spatial separation of the various control regions. A more complicated delay model which is consistent with a wider range of experiments has been presented by Cartwright and Husain (1986): it represents the current thinking as regards modelling this particular physiological process. We shall discuss it very briefly later.

Let us now consider the basic physiology. The secretion of testosterone from the gonads is stimulated by a pituitary hormone called the luteinizing hormone (LH). The secretion of LH from the pituitary gland is stimulated by the luteinizing hormone releasing hormone (LHRH). This LHRH is normally secreted by the hypothalamus (part of the third ventricle in the brain) and carried to the pituitary gland by the blood. Testosterone is believed to have a feedback effect on the secretion of LH and LHRH. Based on these, Smith (1980) proposed a simple negative feedback compartment model, such as we discussed in Section 6.2, involving the three hormones T, LH and LHRH and is represented schematically in Fig. 6.15.

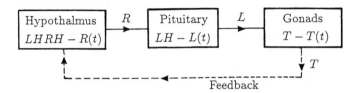

Fig. 6.15. Compartment model for the control of testosterone production in the male. The hypothalmus secretes luteinizing hormone release hormone (LHRH), denoted by $R(t)$, which controls the release of luteinizing hormone (LH), denoted by $L(t)$, by the pituitary (P) which controls the production of testosterone, $T(t)$, by the gonads. The dashed line denotes the feedback control to the hypothalmus from the testes.

Denote the concentrations of the LHRH, LH and T respectively by $R(t)$, $L(t)$ and $T(t)$. At the simplest modelling level (Smith 1980) considered each of the hormones to be cleared from the bloodstream according to first order kinetics with LH and T produced by their precursors according to first order kinetics. There is a nonlinear negative feedback by $T(t)$ on $R(t)$. The governing system reflecting this scheme is essentially the model feedback system (6.3), which here is written as

$$\frac{dR}{dt} = f(T) - b_1 R \,,$$

$$\frac{dL}{dt} = g_1 R - b_2 L \,,$$

$$\frac{dT}{dt} = g_2 L - b_3 T \,,$$

$$(6.43)$$

where b_1, b_2, b_3, g_1 and g_2 are positive parameters and the negative feedback function $f(T)$ is a positive monotonic decreasing function of T. At this stage we do not need a specific form for $f(T)$ although we might reasonably take it to be typically of the form $A/(K + T^m)$ as in the prototype feedback model (6.3).

As we mentioned, the blood level of testosterone in men oscillates in time. Experiments in which the natural state is disturbed have also been carried out: see the brief surveys given by Smith (1980) and Cartwright and Husain (1986). Our interest here will be mainly related to the observed periodic fluctuations.

From the analysis in Section 6.2, or simply by inspection, we know that a positive steady state R_0, L_0 and T_0 exists for the model (6.43). With the specific form (6.3) for $f(T)$ oscillations exist for a Hill coefficient $m \geq 8$ (Exercise 4), which we noted before is an unrealistically high figure. We can modify the specific form of $f(T)$ so that periodic solutions exist but this is essentially the same as choosing the form in (6.3) with $m \geq 8$. We assume therefore that the feedback function $f(T)$ is such that the steady state is always stable. Thus we must modify the model to include more of the physiology. If we consider the actual process that is taking place there must be a delay between production of the hormone at one level and its effect on the production of the hormone it stimulates simply because of their spatial separation and the fact that the hormones are transported by circulating blood. Accordingly W.R. Smith and J.D. Murray suggested a simple delay model based on the modification to the system (6.3) with $m = 1$, similar to the delay control model suggested by Murray (1977), in which the production of testosterone is delayed. Although it is reasonable to consider a delay in each hormone's production they incorporated them all in the T-equation so as to be able to investigate the system analytically and hence get an intuitive feel for the effect of delay on the system. They took, in place of (6.43),

$$\frac{dR}{dt} = f(T) - b_1 R ,$$

$$\frac{dL}{dt} = g_1 R - b_2 L ,$$

$$\frac{dT}{dt} = g_2 L(t - \tau) - b_3 T ,$$

(6.44)

where τ is a delay associated with the blood circulation time in the body. The steady state is again (R_0, L_0, T_0) determined by

$$L_0 = \frac{b_3 T_0}{g_2} , \quad R_0 = \frac{b_3 b_2 T_0}{g_1 g_2} , \quad f(T_0) - \frac{b_1 b_2 b_3 T_0}{g_1 g_2} = 0 ,$$

(6.45)

which always exists if $f(0) > 0$ and $f(T)$ is a monotonic decreasing function. If we now investigate the stability of the steady state by writing

$$x = R - R_0, \quad y = L - L_0, \quad z = T - T_0$$

(6.46)

the linearised system from (6.44) is

$$\frac{dx}{dt} = f'(T_0)z - b_1 x ,$$

$$\frac{dy}{dt} = g_1 x - b_2 y ,$$

$$\frac{dz}{dt} = g_2 y(t - \tau) - b_3 z .$$

(6.47)

Now look for solutions in the form

$$\begin{pmatrix} x \\ y \\ z \end{pmatrix} = \mathbf{A} \exp [\lambda t]$$

(6.48)

which on substitution into (6.47) gives

$$\lambda^3 + a\lambda^2 + b\lambda + c + de^{-\lambda\tau} = 0 ,$$

$$a = b_1 + b_2 + b_3, \quad b = b_1 b_2 + b_2 b_3 + b_3 b_1 , \tag{6.49}$$

$$c = b_1 b_2 b_3, \quad d = -f'(T_0) g_1 g_2 > 0 .$$

We now want to determine the conditions for the steady state to be linearly unstable; that is we require the conditions on a, b, c, d and τ such that there are solutions of (6.49) with $\mathrm{Re}\,\lambda > 0$.

We know that with $\tau = 0$ the steady state (R_0, L_0, T_0) is stable, that is if $\tau = 0$ in (6.49), $\mathrm{Re}\,\lambda < 0$. Using the Routh-Hurwitz conditions on the cubic in λ given by (6.49) with $\tau = 0$ this means that a, b, c and d necessarily satisfy

$$a > 0, \quad c + d > 0, \quad ab - c - d > 0 . \tag{6.50}$$

We know that delay can be destabilising so we now try to determine the critical delay $\tau_c > 0$, in terms of a, b, c and d so that a solution with $\mathrm{Re}\,\lambda > 0$ exists for $\tau > \tau_c$. We determine the conditions by considering (6.49) as a complex variable mapping problem.

From the analysis on transcendental equations in Chapter 1 we know that all solutions of (6.49) have $\mathrm{Re}\,\lambda$ bounded above. The critical τ_c is that value of τ such that $\mathrm{Re}\,\lambda = 0$, that is the bifurcation between stability and instability in the solutions (6.48). Consider the transformation from the λ-plane to the w-plane defined by

$$w = \lambda^3 + a\lambda^2 + b\lambda + c + de^{-\lambda\tau}, \quad \tau > 0 . \tag{6.51}$$

Setting $\lambda = \mu + i\nu$ this gives

$$\begin{aligned} w = &[\mu^3 - 3\mu\nu^2 + a(\mu^2 - \nu^2) + b\mu + c + de^{-\mu\tau}\cos\nu\tau] \\ &+ i[-\nu^3 + 3\mu^2\nu + 2a\mu\nu + b\nu - de^{-\mu\tau}\sin\nu\tau] . \end{aligned} \tag{6.52}$$

We wish to find the conditions on a, b, c, d and τ such that $w = 0$ has solutions $\mu > 0$; the bifurcation state is $\mu = 0$.

Consider the contour in the λ-plane consisting of the imaginary axis and a semi-circle of infinite radius as shown schematically in Fig. 6.16 (a). Without any delay $\tau = 0$ and we know that in this case $\mathrm{Re}\,\lambda < 0$ so w as a function of λ in (6.51) does not pass through the origin in the w-plane: that is the map in the w-plane of AGHA in Fig. 6.16 (a) does not pass through $w = 0$. Consider first $\tau = 0$ and the mapping (6.52). AG in Fig. 6.16 (a), and on which $\mu = 0$, is mapped by

$$w = [(c - a\nu^2) + d] + i[b\nu - \nu^3] \tag{6.53}$$

onto $A'G'$ in Fig. 6.16 (b) with $D'(= c + d + i0)$ in the w-plane corresponding to $D(= 0 + i0)$ in the λ-plane. The domain V is mapped into V'; the hatches in

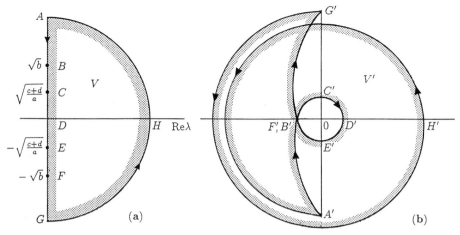

Fig. 6.16a,b. (a) λ-plane. (b) w-plane.

Fig. 6.16 point into the respective domains. The points A, B, C, D, E, F and G in Fig. 6.16 (a) are mapped onto their primed equivalents as

$$
\begin{array}{llll}
A & (\infty e^{i\pi/2}) & A' & (\infty e^{3i\pi/2}) \\[2mm]
B & (\sqrt{b}\,e^{i\pi/2}) & B' & ((ab - c - d)e^{i\pi}) \\[2mm]
C & \left(\left[\dfrac{c+d}{a}\right]^{1/2} e^{i\pi/2}\right) & C' & \left(\left[\dfrac{c+d}{a}\right]^{1/2}\left(b - \dfrac{c+d}{a}\right)e^{i\pi/2}\right) \\[3mm]
D & (0) & D' & (c + d) \\[2mm]
E & \left(\left[\dfrac{c+d}{a}\right]^{1/2} e^{-i\pi/2}\right) & E' & \left(\left[\dfrac{c+d}{a}\right]^{1/2}\left(b - \dfrac{c+d}{a}\right)e^{-i\pi/2}\right) \\[3mm]
F & (\sqrt{b}\,e^{-i\pi/2}) & F' & ((ab - c - d)e^{-i\pi}) \\[2mm]
G & (\infty e^{-i\pi/2}) & G' & (\infty e^{-3i\pi/2})
\end{array}
\tag{6.54}
$$

As the semi-circle GHA is traversed λ moves from $\infty e^{-i\pi/2}$ to $\infty e^{i\pi/2}$ and so $w(\sim \lambda^3)$ moves from $\infty e^{-3i\pi/2}$, namely G', to ∞, namely H', and then to $\infty e^{3i\pi/2}$, that is A', all as shown in Fig. 6.16 (b).

Now consider the mapping in the form (6.52) with $\tau > 0$. The line AG in Fig. 6.16 (a) has $\mu = 0$ as before so it now maps onto

$$
w = [(c - a\nu^2) + d\cos\nu\tau] + i[b\nu - \nu^3 - d\sin\nu\tau].
\tag{6.55}
$$

The effect of the trigonometric terms is simply to add oscillations to the line $A'B'C'D'E'F'G'$ as shown schematically in Fig. 6.17 (a) which is now the map of $ABCDEFGHA$ under (6.55). Now let τ increase. As soon as the transformed curve passes through the origin in the w-plane this gives the critical τ_c. For $\tau > \tau_c$ the mapping is schematically as shown in Fig. 6.17 (b) and the origin in the w-plane is now enclosed in V', that is the transformation of V under (6.55).

If we now traverse the boundary of V' and compute the change in $\arg w$ we immediately get the number of roots of $w(\lambda)$ defined by (6.51). It helps to refer to both Fig. 6.17 (a) and (b). Let us start at A' where $\arg w = 3\pi/2$. On reaching B', $\arg w = \pi$ and so on, giving

Point	A'	B'	C'	D'	E'	F'	G'	A'(via H')
$\arg w$	$\dfrac{3\pi}{2}$	π	$\dfrac{3\pi}{2}$	2π	$\dfrac{5\pi}{2}$	3π	$\dfrac{5\pi}{2}$	$\dfrac{5\pi}{2}+3\pi\left(=\dfrac{11\pi}{2}\right)$

Thus the change in $\arg w$ is 4π which implies two roots with $\operatorname{Re}\lambda > 0$ in the domain V; the roots are complex conjugates.

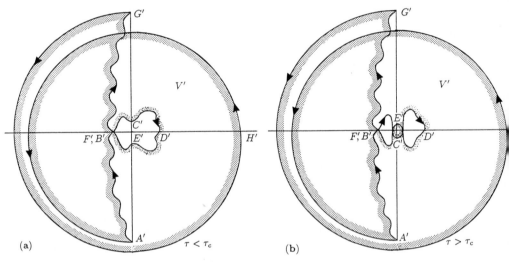

Fig. 6.17a,b. Map in the w-plane of the contour in Fig. 6.16 (a) in the λ-plane under the transformation (6.55): (a) $\tau < \tau_c$; (b) $\tau > \tau_c$. In this case, the map encloses the origin.

Let us now obtain expressions for the critical τ_c such that the curve $A'B'C'D'E'F'G'H'A'$ just passes through the origin in the w-plane. The bifurcation value we are interested in for $\operatorname{Re}\lambda$ is of course $\mu = 0$ so we require the value of τ such that $w = 0$ in (6.55), namely

$$c - a\nu^2 + d\cos\nu\tau = 0, \quad b\nu - \nu^3 - d\sin\nu\tau = 0 , \qquad (6.56)$$

from which we get

$$\cot\nu\tau = \frac{a\nu^2 - c}{\nu(b - \nu^2)} \quad \Rightarrow \quad \nu = \nu(\tau) . \qquad (6.57)$$

If we plot each side of (6.57) as a function of ν, as shown schematically in Fig. 6.18 where for illustration we have taken $\sqrt{b} < \pi/\tau$, we see that there is always a solution, given by the intersection of the curves, such that

$$0 < \nu(\tau) < \frac{\pi}{\tau}, \quad 0 < \nu(\tau) < \sqrt{b}. \tag{6.58}$$

Further, the solution $\nu(\tau)$ as a function of τ satisfies

$$\nu(\tau_1) < \nu(\tau_2) \quad \text{if} \quad \tau_1 > \tau_2 .$$

If $\sqrt{b} > \pi/\tau$ then (6.58) still holds but another solution exists with $\pi/\tau < \nu(\tau) < \sqrt{b}$.

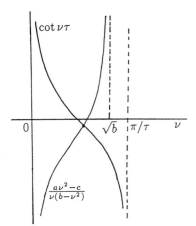

Fig. 6.18. Schematic graphical solution ν of (6.57) in the case where $\sqrt{b} < \pi/\tau$.

Consider now the solution $\nu(\tau)$ which satisfies (6.58). A solution of the simultaneous equations (6.56) must satisfy (6.57) and one of (6.56), which to be specific we take here to be the second. With $\nu(\tau)$ the solution of (6.57) satisfying (6.58), the second of (6.56) gives

$$d = \frac{[b - \nu^2(\tau)]\nu(\tau)}{\sin(\nu(\tau)\tau)}. \tag{6.59}$$

If τ is such that this equation cannot be satisfied with the corresponding $\nu(\tau)$ then no solution can exist with $\mathrm{Re}\,\lambda > 0$. We can not get an analytical solution of (6.57) for $\nu(\tau)$ but an indication of how the solution behaves is easily obtained for τ large and small. For the case in (6.58), we have from (6.57)

$$\nu(\tau) = \sqrt{b} - (ab - c)\frac{\tau}{2\sqrt{b}} + O(\tau^2) \quad \text{for} \quad 0 < \tau \ll 1 \tag{6.60}$$

and the second of (6.56) becomes

$$b\nu(\tau) - \nu^3(\tau) - d\sin[\nu(\tau)\tau] \approx (ab - c)\tau\sqrt{b} - d\tau\sqrt{b} + \ldots$$
$$= \tau(ab - c - d)\sqrt{b} + O(\tau^2)$$
$$> 0$$

from conditions (6.50). Thus for small enough τ no solution exists with $\mathrm{Re}\,\lambda > 0$. Now let τ increase until a solution to (6.59) can be satisfied: this determines the critical τ_c as the solution (6.58) of (6.57) which also satifies (6.59). The procedure is to obtain $\nu(\tau)$ from (6.57) and then the value τ_c which satisfies (6.59).

Now suppose $\tau \gg 1$. From (6.57) we get

$$\nu(\tau) \sim \frac{\pi}{\tau} - \frac{\pi b}{c\tau^2} + O\left(\frac{1}{\tau^3}\right) \quad \text{for} \quad \tau \gg 1,$$

and now the second of (6.56) gives

$$b\nu - \nu^3 - d\sin[\nu(\tau)\tau] \sim \frac{b\pi}{\tau} - \frac{bd\pi}{c\tau} + O\left(\frac{1}{\tau^2}\right)$$
$$< 0 \quad \text{if} \quad d > c.$$

Thus there is a range of $\tau > \tau_c > 0$ such that a solution λ exists with $\mathrm{Re}\,\lambda > 0$ if $d > c$.

An approximate range for $\nu(\tau)$ can be found by noting that with $\nu < \sqrt{b} < \pi/\tau$, the specific case we are considering and which is sketched in Fig. 6.18,

$$g(\nu) = c - a\nu^2 + d\cos\nu\tau$$

is monotonic decreasing in $0 \le \nu\tau < \pi$. Further

$$g\left(\left[\frac{c+d}{a}\right]^{1/2}\right) = -d + d\cos\left(\tau\left[\frac{c+d}{a}\right]^{1/2}\right) < 0, \quad g(0) = c + d > 0$$

which implies that

$$0 < \nu < \left[\frac{c+d}{a}\right]^{1/2} < \sqrt{b} < \frac{\pi}{\tau}.$$

We have thus demonstrated that there is a critical delay τ_c such that the steady state (R_0, L_0, T_0) is linearly unstable by growing oscillations. Since the model system (6.44) has a confined set we might thus expect limit cycle periodic solutions to be generated: this is indeed what happens when the parameters are chosen so that the steady state is linarly unstable.

The model proposed by Cartwright and Husain (1986) is based on more recent experimental results and includes further delays in the production of each of R, L and T. They also incorporate feedback by LH as well as T. Their model is necessarily more complex. Analytical results such as we have derived above, even if possible, would necessarily be much more complicated. Numerical simulations of their model system with reasonable parameter values show stable periodic solutions in all of R, L and T. They also carry out mathematical "experiments" which mimic certain laboratory experiments, with encouraging results.

Returning to the effect of the drug goserelin, mentioned above, it effects chemical castration by blocking the production of the hormone LH produced by the pituitary. That is, in the model system (6.44), $g_1 = 0$. In this case the governing equation for $L(t)$, that is the concentration of LH, is uncoupled from the other equations and $L \to 0$ with time, which in turn implies from the T-equation in (6.44) that $T \to 0$ with time, which is the equivalent of castration. This castration procedure could be used to replace the widely used surgical methods currently used by veterinarians on farm and domestic animals. Such a vaccine has been developed by Carelli et al. (1982).

Exercises

1. The 'Brusselator' reaction mechanism proposed by Prigogene and Lefever (1968) is

$$A \xrightarrow{k_1} X, \quad B + X \xrightarrow{k_2} Y + D, \quad 2X + Y \xrightarrow{k_3} 3X, \quad X \xrightarrow{k_4} E$$

where the k's are the rate constants, and the reactant concentrations of A and B are kept constant. Write down the governing differential equation system for the concentrations of X and Y and nondimensionalize the equations so that they become

$$\frac{du}{d\tau} = 1 - (b+1)u + au^2v, \quad \frac{dv}{d\tau} = bu - au^2v ,$$

where u and v correspond to X and Y, $\tau = k_4 t$, $a = k_3(k_1 A)^2/k_4^3$ and $b = k_2 B/k_4$. Determine the positive steady state and show that there is a bifurcation value $b = b_c = 1 + a$ at which the steady state becomes unstable in a Hopf bifurcation way. Hence show that in the vicinity of $b = b_c$ there is a limit cycle periodic solution with period $2\pi/\sqrt{a}$.

2. In the reaction mechanism $du/dt = \mathbf{f}(\mathbf{u})$ the kinetics $\mathbf{f}$ is curl free, that is $\operatorname{curl}_{\mathbf{u}}\mathbf{f}(\mathbf{u}) = 0$. This implies that $\mathbf{f}$ can be written in terms of a gradient of a potential $F(\mathbf{u})$, that is

$$\operatorname{curl}_{\mathbf{u}}\mathbf{f}(\mathbf{u}) = 0 \quad \Rightarrow \quad \mathbf{f}(\mathbf{u}) = \nabla_{\mathbf{u}}F(\mathbf{u}) .$$

The model system becomes

$$\frac{d\mathbf{u}}{dt} = \mathbf{f}(\mathbf{u}) = \nabla_{\mathbf{u}}F(\mathbf{u}) ,$$

which is called a gradient system. By supposing that $\mathbf{u}(t)$ is a periodic solution with period T show, by considering

$$\int_t^{t+T} \left(\frac{d\mathbf{u}}{ds}\right)^2 ds \ ,$$

that a gradient system cannot have periodic solutions.

3. In the feedback control system governed by

$$\frac{du_1}{dt} = f(u_n) - k_1 u_1 \ ,$$

$$\frac{du_r}{dt} = u_{r-1} - k_r u_r, \quad r = 2, 3, \ldots n$$

the feedback function is given by

$$\text{(i)} \quad f(u) = \frac{a + u^m}{1 + u^m} \qquad \text{(ii)} \quad f(u) = \frac{1}{1 + u^m}$$

where a and m are positive constants. Determine which of these represents a positive feedback control and which a negative feedback control. Determine the steady states and hence show that with positive feedback multi-steady states are possible while if $f(u)$ represents negative feedback there is only a unique steady state.

4. Consider the negative feedback mechanism

$$\frac{du_1}{dt} = \frac{1}{1 + u_3^m} - k_1 u_1 \ ,$$

$$\frac{du_i}{dt} = u_{i-1} - k_i u_i, \quad i = 2, 3 \ .$$

(i) For the case $m = 1$, show that a confined set B is given by a rectangular box whose sides are bounded by the $u_i = 0$, $i = 1,2,3$ axes and U_i, $i = 1,2,3$ where

$$\frac{1}{1 + U_3} < k_1 U_1 < k_1 k_2 U_2 < k_1 k_2 k_3 U_3 \ ,$$

and hence determine U_i, $i = 1,2,3$. For the case $m \neq 1$ show that U_3 is given by the appropriate solution of the equation

$$U_3^{m+1} + U_3 - \frac{1}{k_1 k_2 k_3} = 0 \ .$$

(ii) Prove, using the Routh-Hurwitz conditions (see Appendix 2), that the model with $m = 1$ cannot have limit cycle periodic solutions.

*(iii) Prove that limit cycle solutions are possible if $m > 8$. [At one stage in the analysis for (iii) you will need to use the general inequality

$$\frac{k_1 + k_2 + k_3}{3} \geq \left[\frac{k_1 + k_2 + k_3}{3}\right]^{1/2} \geq (k_1 k_2 k_3)^{1/3} \ .]$$

5. Sketch the null clines for the system (Exercise 1):

$$\frac{du}{d\tau} = 1 - (b+1)u + au^2 v = f(u,v), \qquad \frac{dv}{d\tau} = bu - au^2 v = g(u,v) \ .$$

Note the signs of f and g in the (u, v) phase plane and find a confined set (not a trivial exercise) enclosing the steady state. Determine the (a, b) parameter domain where the system has periodic solutions.

6. Consider the dimensionless activator(u)-inhibitor(v) system represented by

$$\frac{du}{dt} = a - bu + \frac{u^2}{v} = f(u,v), \qquad \frac{dv}{dt} = u^2 - v = g(u,v) \ ,$$

where $a, b (> 0)$ are parameters. Sketch the null clines, append the signs of f and g, and examine the signs in the stability matrix for the steady state. Is there a confined set? Show that the (a, b) parameter space in which u and v may exhibit periodic behaviour is bounded by the curve

$$b = \frac{2}{1-a} - 1 \ .$$

and hence sketch the domain in which the systems could have periodic solutions.

Consider the modified system in which there is inhibition by u. In this case u^2/v is replaced by $u^2/[v(1 + Ku^2)]$ in the u equation, where $K(> 0)$ is the inhibition parameter. Sketch the null clines. Show that the boundary curve in (a, b) space for the domain in which periodic solutions may exist is given parametrically by

$$b = \frac{2}{u_0(1 + Ku_0^2)} - 1 \ ,$$

$$a = \frac{2}{(1 + Ku_0^2)} - u_0 - \frac{1}{(1 + Ku_0^2)^2} \ ,$$

$$b \geq 0$$

and sketch the domain. Indicate how the domain for periodic solutions changes as the inhibition parameter K varies.

7. The 2-variable FitzHugh-Nagumo model for space-clamped nerve axon firing with an external applied current I_a is

$$\frac{dv}{dt} = v(a - v)(v - 1) - w + I_a, \qquad \frac{dw}{dt} = bv - \gamma w \ ,$$

where $0 < a < 1$ and b, γ and I_a are positive constants. Here v is directly related to the transmembrane potential and w is the variable which represents the effects of the various chemical ion-generated potentials.

Determine the local maximum and minimum for the v null cline in terms of a and I_a and hence give the corresponding piece-wise linear approximate form.

Show that there is a confined set for the model system. Using the piece-wise linear model, determine the conditions on the paramaters such that the positive steady state is stable but excitable. Find the conditions on the parameters, and the relevant window (I_1, I_2) of applied currents, for the positive steady state to be linearly unstable and hence for a limit cycle solution to exist.

*   For a fixed set of parameters a, b and γ find the period of the small amplitude limit cycle when I_a is just greater than the bifurcation value I_1. [Use the Hopf bifurcation result from Appendix 3.]

8. Consider a simplified model for the control of testosterone secretion given by

$$\frac{dR}{dt} = f(T) - b_1 R \ ,$$

$$\frac{dT}{dt} = b_2 R(t - \tau) - b_3 T \ ,$$

where R denotes the luteinizing hormone releasing hormone($LHRH$), T denotes the hormone testosterone and $f(T)$ is a positive monotonic decreasing function of T. The delay τ is associated with the blood circulation time in the body and b_1, b_2 and b_3 are positive constants. When $\tau = 0$ show that the steady state is stable. Using the method in Section 6.6 investigate the possibility of periodic solution behaviour when $\tau > 0$.

7. Belousov-Zhabotinskii Reaction

7.1 Belousov Reaction and the Field-Noyes (FN) Model

This important oscillating reaction was discovered by Belousov (1951) and is described in an unpublished paper (which was contemptuously rejected by a journal editor), a translation of which has now appeared in the book edited by Field and Burger (1985). Eventually Belousov (1959) published a brief note in the obscure proceedings of a Russian medical meeting. He found oscillations in the ratio of concentrations of the catalyst in the oxidation of citric acid by bromate. The study of this reaction was continued by Zhabotinskii (1964) and is now known as the Belousov-Zhabotinskii reaction or simply the BZ reaction. When the details of this important reaction and some of its dramatic oscillatory and wave-like properties finally reached the West in the 1970's it provoked widespread interest and research. Belousov's seminal work was finally, but posthumously, recognised in 1980 by his being awarded the Lenin Prize. Winfree (1984) gives a brief interesting description of the history of the Belousov-Zhabotinskii reaction.

Although it is a chemical rather than a biochemical oscillator the BZ reaction is now considered *the* prototype oscillator. The detailed reactions involved are more or less understood, as are many, but not all, of the complex spatial phenomena it can exhibit: we shall describe some of the wave properties later in Chapter 12. The book of articles edited by Field and Burger (1985) describes some of the current research on the BZ reaction. It also has several articles on chemical oscillators and wave phenomena. The book by Winfree (1980), among other things, also discusses some of the reaction's properties, both temporal and spatial. In this chapter we consider the reaction in detail not only because of its seminal importance in the field, but also because it illustrates techniques of analysis which have wide applicability. Almost all the phenomena theoretically exhibited by reaction and reaction-diffusion mechanisms have been found in this real and practical reaction – but many of these only *after* the mathematics predicted them.

The BZ reaction is probably the most widely studied oscillating reaction both theoretically and experimentally. Here we shall briefly describe the key steps in the reaction and develop the Field-Noyes (Field and Noyes 1974) model system which quantitatively mimics the actual chemical reactions (Field, Körös and Noyes 1972). The models for the BZ reaction are prototypes to study since the theoretical developments can be tested against experiments. The experience gained from this is directly transferable to biochemical oscillators. The literature on the subject is now large, but a succinct review of the detailed reaction and its properties is given by Tyson (1985).

Although there are now several variants of the original Belousov (1951) reaction, the basic mechanism consists of the oxidation of malonic acid, in an acid medium, by bromate ions, BrO_3^-, and catalyzed by cerium, which has two states Ce^{3+} and Ce^{4+}. Sustained periodic oscillations are observed in the cerium ions. With other metal ion catalysts and appropriate dyes, for example iron Fe^{2+} and Fe^{3+} and phenanthroline, the regular periodic colour change is visually dramatic, oscillating between a reddish-orange to blue. It is not only the catalyst ion concentrations which vary with time, of course, other reactants also vary. Fig. 7.1 illustrates the temporal variations in the bromide ion concentration $[Br^-]$ and the cerium ion concentration ratio $[Ce^{4+}]/[Ce^{3+}]$ measured by Field, Körös and Noyes (1972), who studied the mechanism in depth: see Tyson (1985) for more recent references and technical details.

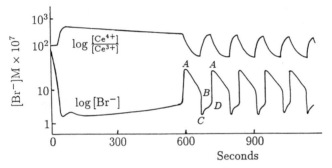

Fig. 7.1. Experimentally measured periodic limit cycle type of temporal variation in the concentrations in the ratio of the cerium metal ion concentration $[Ce^{4+}]/[Ce^{3+}]$ and the bromide ion concentration $[Br^-]$ in the Belousov-Zhabotinskii reaction. (Redrawn from Field, Körös and Noyes 1972)

Basically the reaction can be separated into two parts, say I and II, and the concentration $[Br^-]$ determines which is dominant at any time. When $[Br^-]$ is high, near A in Fig. 7.1, I is dominant and during this stage Br^- is consumed, that is we move along AB, and the cerium ion is mainly in the Ce^{3+} state. As $[Br^-]$ decreases further it passes through a critical value, B, and then drops quickly to a low level, that is C in Fig. 7.1. At this stage process II takes over from I. During II the Ce^{3+} changes to Ce^{4+}. However, in the II-process Ce^{4+} reacts to produce Br^- again while it reverts to the Ce^{3+} state. Now $[Br^-]$ increases, that is along CDA, and, when its value is sufficiently high, process I again becomes dominant. The whole sequence is continually repeated and hence produces the observed oscillations. The rapid variation along BC and DA is typical of a *relaxation* oscillator, which is just an oscillator in which parts of the limit cycle are traversed quickly. This behaviour suggests a particular asymptotic technique which often allows us to get analytical results for the period in terms of the parameters: we discuss this later in Section 7.4.

Although there are many reactions involved they can be rationally reduced to five key reactions, with known values for the rate constants, which capture the basic elements of the mechanism. These 5 reactions can then be represented by a 3-chemical system in which the overall rate constants can be assigned with reasonable confidence. This model is known as the Field-Noyes or FN model and is the model proposed by Field and Noyes (1974) based on the Field-Körös-Noyes (1972) mechanism. We give this simpler model system and derive the 3-species model. A complete derivation from the chemistry together with estimates for the various rate constants are given by Tyson (1985).

The key chemical elements in the 5-reaction FN model are

$$X = HBrO_2, \quad Y = Br^-, \quad Z = Ce^{4+} ,$$
$$A = BrO_3^-, \quad P = HOBr , \tag{7.1}$$

and the model reactions can be approximated by the sequence

$$A + Y \xrightarrow{k_1} X + P, \quad X + Y \xrightarrow{k_2} 2P ,$$
$$A + X \xrightarrow{k_3} 2X + 2Z, \quad 2X \xrightarrow{k_4} A + P, \quad Z \xrightarrow{k_5} fY \tag{7.2}$$

where the rate constants $k_1, \ldots, k_5$ are known and f is a stoichiometric factor, usually taken to be 0.5. The first two reactions are roughly equivalent to the process I, described above, while the last three relate approximately to process II. It is reasonable to take the concentration $[A]$ of the bromate ion to be constant: the concentration $[P]$ is not of interest here. So, using the Law of Mass Action, we get the following third order system of kinetics equations for the concentrations, denoted by lower case letters:

$$\frac{dx}{dt} = k_1 ay - k_2 xy + k_3 ax - k_4 x^2 ,$$
$$\frac{dy}{dt} = -k_1 ay - k_2 xy + f k_5 z , \tag{7.3}$$
$$\frac{dz}{dt} = 2k_3 ax - k_5 z .$$

This oscillator system is sometimes referred to as the 'Oregonator' since it exhibits limit cycle oscillations and the research by Field et al. (1972) was done at the University of Oregon.

Oscillatory behaviour of (7.3) depends critically on the parameters involved. For example if $k_5 = 0$, the bromide ion (Br^-) concentration y decays to zero according to the second equation, so no oscillations can occur. On the other hand if $f = 0.5$ and k_5 is very large, the last reaction in (7.2) is very fast and the third and fifth reactions in (7.2) effectively collapse into the single reaction

$$A + X \overset{k_3}{\rightarrow} 2X + Y .$$

The system then reduces to a 2-species mechanism, which is bimolecular, and so cannot oscillate (Hanusse 1972, Tyson and Light 1973). There is clearly a domain in the (f, k_5) plane where periodic behaviour is not possible.

The only sensible way to analyse the system (7.3) is in a dimensionless form. There are usually several ways to nondimensionalise the equations: see for example Murray (1977) and Tyson (1982, 1985), both of whom give a fuller description of the chemistry and justification for the model. Here we give only the more recent one suggested by Tyson (1985) which incorporates current thinking about the values of the rate constants in the above model (7.3). Often different nondimensionalizations highlight different features of the oscillator. Following Tyson (1985), introduce

$$x^* = \frac{x}{x_0}, \quad y^* = \frac{y}{y_0}, \quad z^* = \frac{z}{z_0}, \quad t^* = \frac{t}{t_0}$$

$$x_0 = \frac{k_3 a}{k_4} \approx 1.2 \times 10^{-7} M, \quad y_0 = \frac{k_3 a}{k_2} \approx 6 \times 10^{-7} M ,$$

$$z_0 = \frac{2(k_3 a)^2}{k_4 k_5} \approx 5 \times 10^{-3} M, \quad t_0 = \frac{1}{k_5} \approx 50s , \tag{7.4}$$

$$\varepsilon = \frac{k_5}{k_3 a} \approx 5 \times 10^{-5}, \quad \delta = \frac{k_4 k_5}{k_2 k_3 a} \approx 2 \times 10^{-4} ,$$

$$q = \frac{k_1 k_4}{k_2 k_3} \approx 8 \times 10^{-4}, \quad (f \approx 0.5)$$

which we substitute into (7.3). Field and Noyes (1974) suggested the value for f, based on experiment. As said before, since the model telescopes a number of reactions the parameters cannot be given unequivocally; the values are current 'best' estimates. For our purposes here we need only the fact that ε, δ and q are small. With (7.4) we get the following dimensionless system, where for algebraic convenience we have omitted the asterisks:

$$\varepsilon \frac{dx}{dt} = qy - xy + x(1 - x) ,$$

$$\delta \frac{dy}{dt} = -qy - xy + 2fz , \tag{7.5}$$

$$\frac{dz}{dt} = x - z .$$

In vector form with $\mathbf{r} = (x, y, z)^T$ we can write this as

$$\frac{d\mathbf{r}}{dt} = \mathbf{F}(\mathbf{r}; \varepsilon, \delta, q, f) = \begin{pmatrix} \varepsilon^{-1}(qy - xy + x - x^2) \\ \delta^{-1}(-qy - xy + 2fz) \\ (x - z) \end{pmatrix} \tag{7.6}$$

7.2 Linear Stability Analysis of the FN Model and Existence of Limit Cycle Solutions

Even though the system (7.5) is third order, the linear stability analysis procedure is standard and described in detail in Chapter 3, namely first find the positive steady state or states, determine the eigenvalues of the linear stability matrix and look for a confined set, which is a finite closed surface S enclosing the steady state such that any solution at time t_0 which lies inside S always remains there for all $t > t_0$. This was done by Murray (1974) for the original but only slightly different FN equation system: it is his type of analysis we apply here to (7.6).

The non-negative steady states (x_s, y_s, z_s) of (7.5) are given by setting the left hand sides to zero and solving the resulting system of algebraic equations, to get

$$(0, 0, 0) \quad \text{or} \quad z_s = x_s, \quad y_s = \frac{2fx_s}{q + x_s},$$

$$2x_s = (1 - 2f - q) + [(1 - 2f - q)^2 + 4q(1 + 2f)]^{1/2} . \tag{7.7}$$

The other non-zero steady state is negative.

Linearizing about $(0, 0, 0)$ we obtain the stability matrix A with eigenvalues λ given by

$$|A - \lambda I| = \begin{vmatrix} \varepsilon^{-1} - \lambda & \dfrac{q}{\varepsilon} & 0 \\ 0 & -q\delta^{-1} - \lambda & 2f\delta^{-1} \\ 1 & 0 & -1 - \lambda \end{vmatrix} = 0$$

$$\Rightarrow \quad \lambda^3 + \lambda^2(1 + q\delta^{-1} - \varepsilon^{-1}) - \lambda[\varepsilon^{-1}(1 + q\delta^{-1}) - q\delta^{-1}] - \frac{q(1 + 2f)}{\varepsilon\delta} = 0 .$$

If we simply sketch the left hand side of this cubic as a function of λ for $\lambda \geq 0$ we see that there is at least one positive root. Alternatively note that the product of the 3 roots is $q(1 + 2f)/\varepsilon\delta > 0$, which implies the same thing. Thus the steady state $(0, 0, 0)$ is always linearly unstable.

If we now linearize (7.5) about the positive steady state (x_s, y_s, z_s) in (7.7) the eigenvalues λ of its stability matrix are given, after a little algebra, by

$$|A - \lambda I| = \begin{vmatrix} \dfrac{1 - 2x_s - y_s}{\varepsilon} - \lambda & \dfrac{q - x_s}{\varepsilon} & 0 \\[2ex] \dfrac{-y_s}{\delta} & -\dfrac{x_s + q}{\delta} - \lambda & \dfrac{2f}{\delta} \\[2ex] 1 & 0 & -1 - \lambda \end{vmatrix} = 0 \tag{7.8}$$

$$\Rightarrow \quad \lambda^3 + A\lambda^2 + B\lambda + C = 0 ,$$

where, on using the quadratic for x_s, the simultaneous equations for x_s, y_s from (7.6) and some tedious but elementary algebra, we get

$$A = 1 + \frac{q + x_s}{\delta} + \frac{E}{\varepsilon} ,$$

$$E = 2x_s + y_s - 1 = \frac{x_s^2 + q(x_s + 2f)}{q + x_s} > 0 ,$$

$$B = \frac{q + x_s}{\delta} + \frac{E}{\varepsilon} + \frac{(q + x_s)E + y_s(q - x_s)}{\varepsilon\delta} , \tag{7.9}$$

$$C = \frac{(q + x_s)E - 2f(q - x_s) + y_s(q - x_s)}{\varepsilon\delta}$$

$$= \frac{x_s^2 + q(2f + 1)}{\varepsilon\delta} > 0 .$$

Note that $A > 0$, since $E > 0$, and that $C > 0$, on using the expression for x_s from (7.7): B can be positive or negative. It follows from Descarte's rule of signs (see Appendix 2) that at least one eigenvalue λ in (7.8) is real and negative. The remaining necessary and sufficient condition for all of the solutions λ to have negative real parts is, from the Routh-Hurwitz conditions (see Appendix 2), $AB - C > 0$. Substituting for A, B and C from (7.9) gives a quadratic in $1/\delta$ for the left hand side and hence the condition for stability of the positive steady state in (7.7). This is given by

$$AB - C = \phi(\delta, f, \varepsilon) = \frac{N\delta^2 + M\delta + L}{\delta^2} > 0 ,$$

$$L = (q + x_s)\left\{ (q + x_s) + \frac{x_s(1 - q - 4f) + 2q(1 + 3f)}{\varepsilon} \right\} , \tag{7.10}$$

$$N = [x_s^2 + q(x_s + 2f)]\frac{1 + \dfrac{E}{\varepsilon}}{\varepsilon(q + x_s)} > 0$$

with M also determined as a function of x_s, f, q and ε: we do not require it in the subsequent analysis and so do not give it here. With x_s from (7.7), L, M and N are functions of f, q and ε. Thus for the steady state to be linearly unstable, δ, f and ε must lie in a domain in (δ, f, ε) space where $\phi(\delta, f, \varepsilon) < 0$. The boundary or bifurcation surface in (δ, f, ε) space is given by $\phi(\delta, f, \varepsilon) = 0$.

We can get an indication of the eigenvalue behaviour asymptotically for large positive and negative B. If $B \gg 1$ the asymptotic solutions of (7.8) are given by

$$\lambda \sim -\frac{C}{B}, \quad -\frac{A}{2} \pm i\sqrt{B} , \tag{7.11}$$

while if $B < 0$ and $B \gg 1$

$$\lambda \sim \frac{C}{|B|}, \quad \pm\sqrt{|B|} . \tag{7.12}$$

So, for large positive B condition (7.10) is satisfied and from (7.11), Re $\lambda < 0$ and the steady state is linearly stable, while if B is large and negative, it is unstable.

When the parameters are such that $B = C/A$, the bifurcation situation, we can solve for the roots λ in (7.8), namely

$$\lambda = -A, \quad \pm i\sqrt{B}, \quad \text{when} \quad B = \frac{C}{A} . \tag{7.13}$$

If $B = (C/A) - \omega$, $0 < \omega \ll 1$, it can be seen by looking for asymptotic solutions to (7.8) in the form $\lambda = \pm i(C/A)^{1/2} + O(\omega)$ that the $O(\omega)$ term has a positive real part. Thus, near the bifurcation surface in the unstable region, the steady state is unstable by growing oscillations. The conditions of the Hopf bifurcation theorem (see Appendix 3) are satisfied and so in the vicinity of the surface $\phi(\delta, f, \varepsilon) = 0$ the system exhibits a small amplitude limit cycle solution with period

$$T = \frac{2\pi}{\left(\frac{C}{A}\right)^{1/2}} . \tag{7.14}$$

With the parameter values obtained from experiment, and given in (7.4), the amplitudes of the oscillations in fact are not small, so the last expression is of pedagogical rather than practical use. However, the bifurcation surface $\phi(\delta, f, \varepsilon) = 0$ given by (7.10) is of practical use and this we must discuss further.

The surface $\phi(\delta, f, \varepsilon) = 0$, namely $N\delta^2 + M\delta + L = 0$ in (7.10) where $N > 0$, is quadratic in δ. So, for the steady state to be unstable, that is δ, f and ε make $\phi < 0$, δ must be such that

$$0 < \delta < -\frac{M}{2N} + \frac{[M^2 - 4LN]^{1/2}}{2N} . \tag{7.15}$$

But there is a non-zero range of positive δ only if the right-hand side of this inequality is positive, which requires, with L from (7.10),

$$L = (q + x_s)\left\{(q + x_s) + \frac{x_s(1 - q - 4f) + 2q(1 + 3f)}{\varepsilon}\right\} < 0 .$$

With x_s from (7.7) this gives an algebraic equation relating f, q and ε. From (7.4) ε is small, so to first order the last inequality gives the square-bracketed expression equal to zero, which reduces to

$$(1-4f-q)\{(1-2f-q)+[(1-2f-q)^2+4q(1+2f)]^{1/2}\}+4q(1+3f) < 0 , \quad (7.16)$$

which (with an equals sign in place of $<$) defines the critical f, f_c, for a given q. In fact there are 2 critical f, namely $_1f_c$ and $_2f_c$ and f must lie between them

$$_1f_c < f < {_2f_c} .$$

With $q = 8 \times 10^{-4}$ we can determine accurate values for $_1f_c$ and $_2f_c$ by exploiting the fact that $0 < q \ll 1$. The critical f_c are given by (7.16) on replacing the inequality sign with an equals sign. Suppose first that $(1 - 2f - q) > 0$, that is $2f < 1$ to $O(1)$. Then on letting $q \to 0$,

$$(1 - 4f)(1 - 2f) \approx 0 \quad \Rightarrow \quad {_1f_c} \approx \frac{1}{4} .$$

With $(1 - 2f - q) < 0$, that is $2f > 1$ to $O(1)$, the limiting situation $q \to 0$ has to be done carefully: this gives $_2f_c$. To $O(q)$, (7.16), again with an equals sign, now becomes

$$(1 - 4f)\left\{-(2f + q - 1) + (2f + q - 1)\frac{1 + 2q(1 + 2f)}{(2f + q - 1)^2}\right\} + 4q(1 + 3f) \approx 0 ,$$

which reduces to

$$4f^2 - 4f - 1 \approx 0 \quad \Rightarrow \quad {_2f_c} \approx \frac{1 + \sqrt{2}}{2} .$$

So, for small q, the range of f for which the positive steady state in (7.7) is linearly unstable is

$$\frac{1}{4} \approx {_1f_c} < f < {_2f_c} \approx \frac{1 + \sqrt{2}}{2} . \qquad (7.17)$$

Finally the stability bifurcation curve of δ against f for each ε is given by (7.15), namely

$$\delta = -\frac{M}{2N} + \frac{[M^2 - 4LN]^{1/2}}{2N} . \qquad (7.18)$$

with $_1f_c < f < {_2f_c}$, where the critical f_c are obtained from (7.16), and L, M and N are defined by (7.10), with (7.9), in terms of f, q and ε.

7.3 Nonlocal Stability of the FN Model

We showed in the last section that for each ε, if δ and f lie in the appropriate
domain the positive steady state is linearly unstable, and indeed by growing
oscillations if (δ,f) are close to the bifurcation curve. Wherever (δ,f) lies in
the unstable domain, we must now consider global stability. Even though we do
not have the equivalent of the Poincaré-Bendixson theorem here since we are
dealing with a third order system, the existence of a periodic solution with finite
amplitudes requires the system to have a confined set, S say. That is with $\mathbf{n}$ the
unit outward normal to S, we must have

$$\mathbf{n} \cdot \frac{d\mathbf{r}}{dt} < 0, \quad \mathbf{r} \text{ on } S, \tag{7.19}$$

where $d\mathbf{r}/dt$ is given by (7.6).

Although the existence of a confined set and a single unstable steady state is
not sufficient to prove the existence of a periodic limit cycle, they give sufficient
encouragement to pursue the analysis further. With 3 equations it is possible to
have chaotic solutions, such as we found with discrete models in Chapter 2. (The
classical Lorenz (1963) equation system which exhibits chaos involves 3 ordinary
differential equations.) Hastings and Murray (1975) have given a rigorous proof,
together with a procedure which determines the general trajectory path over a
cycle, showing that the FN model system (7.5) possesses at least one limit cycle
periodic solution. The procedure they developed has wider applications to a fairly
broad class of feedback control systems as has been demonstrated by Hastings,
Tyson and Webster (1977).

Let us look for the simplest surface S, namely a rectangular box defined by
the faces

$$x = x_1, \; x = x_2; \quad y = y_1, \; y = y_2; \quad z = z_1, \; z = z_2$$

enclosing the steady state (x_s, y_s, z_s) in (7.7). Let us first determine the planes
$x = x_1$ and x_2 where $0 < x_1 < x_s < x_2$. Let $\mathbf{i}, \mathbf{j}$ and $\mathbf{k}$ be the unit normals in
the positive x, y and z directions. On $x = x_1$, $\mathbf{n} = -\mathbf{i}$ and (7.19) requires

$$-\mathbf{i} \cdot \frac{d\mathbf{r}}{dt}\bigg]_{x=x_1} = -\frac{dx}{dt}\bigg]_{x=x_1} < 0 \quad \Rightarrow \quad qy - xy + x - x^2\big]_{x=x_1} > 0 \,.$$

Since, from (7.4), $0 < q \ll 1$, and assuming $x_1 = O(q)$, the last inequality
requires x_1 to satisfy

$$y(q - x_1) + x_1 - x_1^2 \approx y(q - x_1) + x_1 > 0 \quad \text{for all} \quad y_1 \le y \le y_2 \,.$$

So at the least a natural boundary for $x < x_s$ is $x_1 = q$, which we choose as a
first approximation. Then on $x = x_1 = q$

$$-\mathbf{i} \cdot \frac{d\mathbf{r}}{dt}\bigg]_{x=x_1=q} = -\frac{q(1-q)}{\varepsilon} < 0 \quad \text{if} \quad q < 1 \,.$$

On $x = x_2$, $\mathbf{n} = \mathbf{i}$ and (7.19) now requires

$$\mathbf{i} \cdot \left. \frac{d\mathbf{r}}{dt} \right]_{x=x_2} = \left. \frac{dx}{dt} \right]_{x=x_2} < 0 \quad \Rightarrow \quad \left[y(q-x) + x - x^2 \right]_{x=x_2} < 0 \, .$$

If we choose $x_2 = 1$, we get

$$\mathbf{i} \cdot \left. \frac{d\mathbf{r}}{dt} \right]_{x=x_2} = \varepsilon^{-1} y(q-1) < 0 \quad \text{if} \quad q < 1, \quad \text{for all} \quad y > 0 \, .$$

With x_s as given by (7.7) a little algebra shows that

$$q = x_1 < x_s < 1 \quad \text{if} \quad q < 1 \, .$$

With typical values for the parameters these conditions are satisfied.

Consider now the planes $z = z_1$ and $z = z_2$, where $z_1 < z_s < z_2$. On $z = z_1$, $\mathbf{n} = -\mathbf{k}$ and (7.19) requires

$$-\mathbf{k} \cdot \left. \frac{d\mathbf{r}}{dt} \right]_{z=z_1} = -\left. \frac{dz}{dt} \right]_{z=z_1} = -\left. (x-z) \right]_{z=z_1} < 0$$

and since, on the boundary S, $x \geq x_1$ we have a natural lower boundary for z of $z = z_1 = q$. Strictly z_1 should be just less than q since $x_1 = q$. Now on $z = z_2$, $\mathbf{n} = \mathbf{k}$ and we require

$$\mathbf{k} \cdot \left. \frac{d\mathbf{r}}{dt} \right]_{z=z_2} < 0 \quad \Rightarrow \quad \left. (x-z) \right]_{z=z_2} < 0 \, .$$

Since $x \leq 1$, an upper boundary for z is $z = z_2 = 1$: again we should have z_2 just greater than 1.

Finally let us consider the planes $y = y_1$ and $y = y_2$ where $y_1 < y_s < y_2$. On $y = y_1$, $\mathbf{n} = -\mathbf{j}$ and (7.19) requires

$$-\mathbf{j} \cdot \left. \frac{d\mathbf{r}}{dt} \right]_{y=y_1} = \left[y(q+x) - 2fz \right]_{y=y_1} < 0$$

and so we must have

$$y_1 < \frac{2fz}{q+x} \quad \text{for all} \quad q \leq x \leq 1 \quad \text{and} \quad q \leq z \leq 1 \, .$$

So,

$$y_1 < \frac{2fq}{q+1} = \frac{2f z_{\text{minimum}}}{1 + x_{\text{maximum}}} \, .$$

Thus, an appropriate lower boundary for y is

$$y_1 = \frac{2fq}{q+1} \, .$$

When $y = y_2$, $\mathbf{n} = \mathbf{j}$ and we need

$$\left. \mathbf{j} \cdot \frac{d\mathbf{r}}{dt} \right]_{y=y_2} < 0 \quad \Rightarrow \quad 2fz - y(q+x)]_{y=y_2} < 0$$

which implies

$$y_2 > \frac{2fz}{q+x} \quad \text{for all} \quad q \leq x \leq 1 \quad \text{and} \quad q \leq z \leq 1$$

and so we can take

$$y_2 = \frac{2fz_{\text{maximum}}}{q + x_{\text{minimum}}} = \frac{f}{q} .$$

Again using (7.4), for typical values of q and f, $y_1 < y_s < y_2$.

Finally we have (7.19) satisfied on the surface S of the rectangular box given by

$$x = q, \; x = 1; \quad y = \frac{2fq}{q+1}, \; y = \frac{f}{q}; \quad z = q, \; z = 1 \qquad (7.20)$$

within which the steady state (7.7) lies if f and q satisfy certain inequalities, which are satisfied by the parameter values in the Belousov-Zhabotinskii reaction.

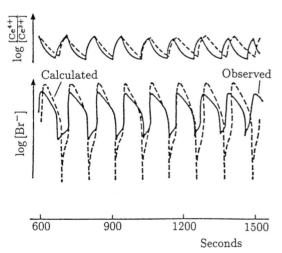

Fig. 7.2. Comparison of the observed oscillations in the BZ reaction with the numerically computed solution of the limit cyle solution of the model FN system (7.5) with $f = 0.3$, $\delta = 1/3$, $q = 5 \times 10^{-3}$, $\varepsilon = 0.01$. (Redrawn from Tyson 1977)

This bounding surface could be refined to give more accurate bounds on any solutions of (7.5). Since the ultimate limit cycle solutions have to be found numerically, or asymptotically as in the following section, all that is needed is a demonstration that such a confined set exists.

Fig. 7.2 shows the numerical solution of the system (7.5) as compared with the observed oscillations.

7.4 Relaxation Oscillators: Approximation for the Belousov-Zhabotinskii Reaction

If we look again at Fig. 7.1 we see that certain parts of the cycle are covered very quickly. This is particularly evident in the trace of the Br^- ion where it suddenly rises along DA and drops as quickly along BC. As we mentioned above when parts of a limit cycle are traversed quickly in comparison with other parts it is often referred to as a relaxation oscillator. What this means from a modelling point of view is that a small parameter must be present in the differential equation system in a crucial place to cause this rapid variation in the solution.

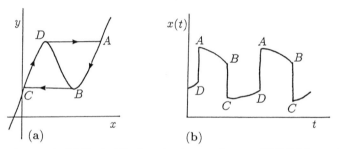

Fig. 7.3a,b. (a) Typical limit cycle phase trajectory $ABCDA$ for a relaxation oscillator governed by (7.21). The two horizontal parts of the trajectory are traversed very quickly. (b) The solution $x(t)$ corresponding to the limit cycle trajectory in (a).

To be specific, and to show how we can exploit such behaviour, consider first the simple relaxation oscillator

$$\varepsilon \frac{dx}{dt} = y - f(x), \quad \frac{dy}{dt} = -x, \quad 0 < \varepsilon \ll 1 , \tag{7.21}$$

where $f(x)$ is a continuous function such that, say $f(x) \to \pm\infty$ as $x \to \pm\infty$. The classic example, where $f(x) = \frac{1}{3}x^3 - x$, is known as the *Van der Pol oscillator*. System (7.21) is a typical singular perturbation problem (see, for example, the book by Murray 1984) since ε multiplies one of the derivatives. Fig. 7.3 (a) illustrates a typical limit cycle phase plane trajectory for (7.21), with Fig. 7.3 (b) the corresponding solution $x(t)$.

From the first of (7.21), except where $y \approx f(x)$, x changes rapidly by $O(1/\varepsilon)$.

So, referring now to Fig. 7.3 (a), along DA and BC, $x(t)$ changes quickly. Along these parts of the trajectory the appropriate independent variable is $\tau = t/\varepsilon$ rather than t. With this transformation the second of (7.21) becomes, as $\varepsilon \to 0$,

$$\frac{dy}{d\tau} = -\varepsilon x \quad \Rightarrow \quad y \approx \text{constant} ,$$

as it is on DA and BC in Fig. 7.3 (a). From (7.21), along the nullcline $y = f(x)$ between AB and CD the second equation becomes

$$f'(x)\frac{dx}{dt} \approx -x , \tag{7.22}$$

which can be integrated to give x implicitly as a function of t. If $f(x)$ is the Van der Pol cubic above, or can be reasonably approximated by a piecewise linear function, then we can integrate this equation exactly. We can then estimate (which we do in detail below) the period T of the oscillation since the major contribution comes from the time it takes to traverse the branches AB and CD: the time to move across DA and BC is small, $o(1)$. It can be shown that if T is the limit cycle period calculated in this manner from (7.22), then the asymptotic limit cycle period of (7.21) is $T + O(\varepsilon^{2/3})$. For our purposes all we shall need is the $O(1)$ approximation T.

By way of example, suppose $f(x) = x^3/3 - x$, that is (7.21) is the simple Van der Pol oscillator, and let us calculate the $O(1)$ period T. The nullcline $y = f(x)$ here and the limit cycle relaxation trajectory are very similar in shape to those illustrated in Fig. 7.3 except that the origin has been moved to a point half way down DB in Fig. 7.3 (a) and to a point in Fig. 7.3 (b) such that the solution is symmetrical about the $x = 0$ axis. From the above analysis, on integrating from the equivalent of A to B in Fig. 7.3 (a) and from the equivalent of C to D, as $\varepsilon \to 0$ the period T to $O(1)$ is given by (7.22), with $f(x) = x^3/3 - x$. A little algebra gives A as $(2, 2/3)$ and B as $(1, -2/3)$. Because of the symmetry, and to be specific if we take $t = 0$ at A, the $O(1)$ period T is given by

$$\int_2^1 \left(x - \frac{1}{x}\right) dx = -\int_0^{T/2} dt \quad \Rightarrow \quad T = 3 - 2\ln 2 . \tag{7.23}$$

Note that even if we cannot integrate $f'(x)/x$ simply, to get the period, we can still determine the maxima and minima of the limit cycle variables simply from the algebra of the null clines $y = f(x)$. These correspond to A and C for $x(t)$ and D or A and B or C for $y(t)$.

On comparing the bromide ion concentration [Br$^-$] as a function of time in Fig. 7.1 and the limit cycle time dependent solution sketched in Fig. 7.3 (b) it is reasonable to look for a relaxation oscillator type of approximation for the BZ oscillator. This has been done by Tyson (1976, 1977) whose analysis we effectively follow in the next section.

Let us again consider the FN mechanism in the dimensionless form (7.5), namely

$$\varepsilon \frac{dx}{dt} = qy - xy + x(1-x) \, ,$$

$$\delta \frac{dy}{dt} = -qy - xy + 2fz \, , \tag{7.24}$$

$$\frac{dz}{dt} = x - z \, .$$

with the dimensionless parameters given by (7.4). Note that $\varepsilon \ll \delta$, in which case we can reduce the system (7.24) by setting $\varepsilon dx/dt \approx 0$. This gives

$$0 = qy - xy + x(1-x) \quad \Rightarrow \quad x = x(y) = \frac{(1-y) + [(1-y)^2 + 4qy]^{1/2}}{2} \, . \tag{7.25}$$

With this (7.24) reduces to the 2nd order differential equation system in y and z:

$$\delta \frac{dy}{dt} = 2fz - y[x(y) + q] \, ,$$

$$\frac{dz}{dt} = x(y) - z \, , \tag{7.26}$$

which of course can now be analyzed completely in the (y, z) phase plane. We can, in the usual way, determine the steady state, analyze the linear stability, show there is a confined set and hence determine the conditions on the parameters for a limit cycle solution to exist.

7.5 Analysis of a Relaxation Model for Limit Cycle Oscillations in the Belousov-Zhabotinskii Reaction

Here we exploit the relaxation oscillator aspects of (7.26) and hence determine the approximate period of the limit cycle, the maxima and minima of the dependent variables and then compare the results with the experimental observations of the oscillating reaction. To do this we first give approximations for $x(y)$ using the fact that $0 < q \ll 1$ from (7.4), and then sketch the nullclines.

From (7.25), with $q \ll 1$, the z-nullcline from (7.26) is

$$z = x(y) \approx \begin{cases} 1 - y, \\ \dfrac{qy}{y-1} \end{cases} \quad \text{for} \quad \begin{cases} q \ll 1 - y \le 1 \\ q \ll y - 1 \end{cases} \tag{7.27}$$

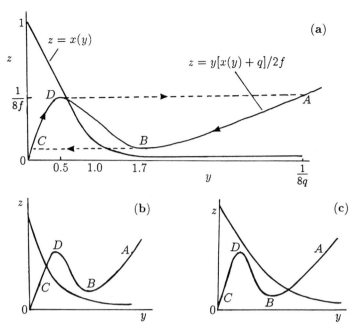

Fig. 7.4a-c. Schematic nullclines for the reduced BZ model system (7.26) using the asymptotic forms for $0 < q \ll 1$ from (7.27) and (7.28). The points B and C correspond to $z_{\min}$ given by (7.30) and the points D and A to $z_{\max}$ in (7.29). The asymptotic expressions for A, B, C and D are gathered together in (7.31) below. Note how the position of the steady state changes with f: (a) $1/4 < f < (1 + \sqrt{2})/2$; (b) $f < 1/4$; (c) $(1 + \sqrt{2})/2 < f$.

The y-nullcline, from (7.26), is

$$z = \frac{y[x(y) + q]}{2f} \approx \begin{cases} \dfrac{y(1 - y)}{2f} \\[2mm] \dfrac{y\left[\frac{qy}{y-1} + q\right]}{2f} \\[2mm] \dfrac{qy}{f} \end{cases} \quad \text{for} \quad \begin{cases} q \ll 1 - y \ll 1 \\ q \ll y - 1 \\ y \gg 1 \end{cases} \qquad (7.28)$$

The z-nullcline is a monotonically decreasing function of y: it is sketched in Fig. 7.4. The y-nullcline, also shown in Fig. 7.4 for various ranges of f, has a local maximum (z_D and z_A in Fig. 7.4 (a))

$$z_{\max} = \frac{1}{8f} \quad \text{at} \quad y_{\max} = \frac{1}{2}, \qquad (7.29)$$

obtained from the first of (7.28). From the second of (7.28), z has a local minimum

$(z_B$ and z_C in Fig. 7.4 (a)) at

$$\frac{dz}{dy} = \left\{ \frac{q}{2f(y-1)^2} \right\} \{2y^2 - 4y + 1\} = 0$$

$$\Rightarrow \quad y_{min} = \frac{2+\sqrt{2}}{2}, \tag{7.30}$$

$$\Rightarrow \quad z_{min} = \frac{q(1+\sqrt{2})^2}{2f} = \frac{q(3+2\sqrt{2})}{2f}.$$

The values of z and y at the relevant points A, B, C and D are obtained from (7.27) and (7.28), with $z_D(= z_A)$ and y_D given by (7.29) and $z_B(= z_C)$ and y_B from (7.30). For y_C, we have from the first of (7.28), with $q \ll 1$ and $z_C = z_{min}$ from (7.30),

$$z_C = \frac{y_C(1 - y_C)}{2f} \quad \Rightarrow \quad y_C = \frac{1 - [1 - 8fz_C]^{1/2}}{2}$$

$$\approx \frac{1 - [1 - 4q(3 + 2\sqrt{2})]^{1/2}}{2}$$

$$\approx q(3 + 2\sqrt{2}).$$

For y_A, we have, from the second of (7.27) for y large, $z \sim qy/f$, and $z_A = z_{max}$ from (7.29),

$$\frac{1}{8f} = z_A \approx \frac{qy_A}{f} \quad \Rightarrow \quad y_A \approx \frac{1}{8q} \gg 1.$$

Gathering together these results, and those from (7.29) and (7.30), we have, for the points $ABCD$ in Fig. 7.4 (a),

$$y_A \approx \frac{1}{8q}, \ z_A = \frac{1}{8f}; \quad y_B \approx \frac{2+\sqrt{2}}{2}, \ z_B \approx \frac{q(3+2\sqrt{2})}{2f};$$

$$y_C \approx q(3+2\sqrt{2}), \ z_C \approx \frac{q(3+2\sqrt{2})}{2f}; \quad y_D \approx \frac{1}{2}, \ z_D \approx \frac{1}{8f}. \tag{7.31}$$

Fig. 7.4 (a)-(c) illustrate the various nullcline possibilities for the the reduced BZ model (7.26). From (7.27), (7.28) and (7.31) we can get the f-ranges where each holds. Fig. 7.4 (b) is when the local maximum at z_D lies to the right of the steady state. This requires that on the z-nullcline, given by (7.27),

$$z]_{y=y_D} < z_D \quad \Rightarrow \quad x(y_D) \approx 1 - y_D < z_D \quad \Rightarrow \quad f < \frac{1}{4}, \tag{7.32}$$

on using (7.31). Fig. 7.4 (a) holds when z on the z-nullcline is such that

$$z]_{y=y_D} > z_D \quad \text{and} \quad z]_{y=y_B} < z_B ,$$

which gives

$$x(y_D) \approx 1 - y_D > \frac{1}{8f} \quad \text{and} \quad x(y_B) \approx \frac{q y_B}{y_B - 1} < \frac{q(3 + 2\sqrt{2})}{2f} \; ,$$

which reduces to

$$\frac{1}{4} < f < \frac{1 + \sqrt{2}}{2} \; . \tag{7.33}$$

Finally Fig. 7.4 (c) holds when, on the z-nullcline,

$$z]_{y=y_B} > z_B \quad \Rightarrow \quad f > \frac{1 + \sqrt{2}}{2} \; . \tag{7.34}$$

We know from Chapter 6 that Fig. 7.4 (a) is a case in which a limit cycle oscillation is possible. So with f in the range (7.33) and the steady state unstable, the reduced BZ model system (7.26) will exhibit limit cycle solutions. Now comparing Fig. 7.4 (a) with the relaxation limit cycle oscillator in Fig. 7.3 (a) gives the $O(1)$ period of the oscillatory solution of (7.26) for $0 < \delta \ll 1$ as

$$T \approx \int_{AB} dt + \int_{CD} dt = \left(\int_{z_A}^{z_B} + \int_{z_C}^{z_B} \right) \left(\frac{dz}{dt} \right)^{-1} dz \tag{7.35}$$

with dz/dt given by (7.26). So,

$$T_{AB} = \int_{z_A}^{z_B} [x(y) - z]^{-1} \, dz \; . \tag{7.36}$$

To get an exact evaluation to $O(1)$ for $q \ll 1$, it is convenient to change the variable to y, with z as a function of y given by the second of (7.28), since AB is part of the y-nullcline. It is a tedious integration. All we want here is a reasonable approximation to the period. So, using the expressions for $q \ll 1$ in (7.27), we have, along most of AB,

$$x(y) \approx \frac{q y}{y - 1} \sim q \; ,$$

and so, using (7.31),

$$T_{AB} = \int_{z_A}^{z_B} (q - z)^{-1} \, dz = \ln \left[\frac{z_A - q}{z_B - q} \right]$$

$$\sim \ln \left\{ \frac{\left[\frac{1}{8f} \right]}{\left[\frac{q(3 + 2\sqrt{2})}{2f} - q \right]} \right\} \tag{7.37}$$

$$\sim -\ln[4(3 - 2f + 2\sqrt{2})q], \quad q \ll 1$$

This, in fact, is an upper bound for T_{AB} since on AB, $x(y) \approx qy/(y-1)$, which asymptotes to q only for $y \gg 1$. On AB, $x(y)$ goes from

$$x(y_A) = \frac{q \left[\frac{1}{8q}\right]}{\left[\frac{1}{8q} - 1\right]} \sim q + O(q^2), \quad q \ll 1,$$

to

$$x(y_B) = \frac{q \left[\frac{2+\sqrt{2}}{2}\right]}{\left[\frac{2+\sqrt{2}}{2} - 1\right]} = q(1 + \sqrt{2}).$$

Returning to the integral in (7.36) we have T_{AB} bounded above by (7.37) and below by the expression there with $q(1 + \sqrt{2})$ replacing q. That is

$$- \ln \left[4(3 - 2f + 2\sqrt{2})q\right] < T_{AB} < - \ln \left[4(3 - 2f + 2\sqrt{2})(1 + \sqrt{2})q\right]. \quad (7.38)$$

Let us now evaluate the T_{CD} contribution to the period in (7.35), namely

$$T_{CD} = \int_{z_C}^{z_D} [x(y) - z]^{-1} \, dz.$$

It is also convenient here to change to y as the integration variable. Between C and D in Fig. 7.4 (a) and on CD

$$x(y) \approx 1 - y, \quad z \approx \frac{y(1-y)}{2f} \quad \Rightarrow \quad dz = \left[\frac{1-2y}{2f}\right] dy$$

so the last integral gives, after some tedious algebra,

$$T_{CD} \approx \int_{y_C}^{y_D} \frac{\left[\frac{1-2y}{2f}\right]}{\left[(1-y) - \frac{y(1-y)}{2f}\right]} \, dy \qquad (7.39)$$

$$= - \left[\frac{4f - 1}{2f - 1}\right] \ln \left[2^{1/(4f-1)} \frac{4f - 1}{4f}\right].$$

The period to $O(1)$ for $q \ll 1$ is then given as a function of f and q by $T_{AB} + T_{CD}$, using (7.38) and (7.39). The dimensional period is given by multiplying T by t_0 from (7.4).

To complete the analysis of the relaxation oscillator we have to integrate the approximate equations on the branches AB and CD. We have effectively already done this when we evaluated T_{AB} and T_{CD}. Let us take $t = t_A = 0$ and hence $z(0) = z_A$ for algebraic convenience. Then on AB, with $x(y) \sim q$ for y large from (7.27), (7.26) gives

$$\frac{dz}{dt} \approx q - z \quad \Rightarrow \quad z(t) = z_A e^{-t} + q(1 - e^{-t}), \qquad (7.40)$$

and so there is an exponential decay in time from z_A to $z_B = O(q)$ for times of $O(1)$.

On CD, $x(y) \approx 1 - y$ and $z \approx y(1-y)/2f$ so the second of (7.26) on changing to the variable y, gives

$$\left[\frac{1 - 2y}{2f} \right] \frac{dy}{dt} \approx (1 - y) - \frac{y(1-y)}{2f}$$

which integrates to give y implicitly as a function of t from

$$\ln \left[\frac{1 - y}{(2f - y)^{4f-1}} \right] = K + (2f - 1)t , \qquad (7.41)$$

where K is an integration constant. Since the time to traverse the horizontal part of the trajectory, namely BC, is negligible, we can determine K by taking $y(t_C) = y_C$, where $t = t_C \approx t_B$, with t_B given by (7.40) on setting $z = z_B$, that is

$$t_B = \ln \left[\frac{z_A(1-q)}{z_B - q} \right] . \qquad (7.42)$$

From (7.41) and the last equation we thus have

$$K = \ln \left[\frac{1 - y_C}{(2f - y_C)^{4f-1}} \right] - (2f - 1)t_B . \qquad (7.43)$$

Now with $y(t)$ given implicitly by (7.41), we can obtain z from $z \sim y(1-y)/(2f)$. Finally the time to traverse DA is also negligible. We thus have analytical expressions for the dependent variables in the relaxation oscillator as functions of time for $0 < q \ll 1$.

Let us now return to the actual BZ reaction and recall from (7.1) that z and y are the dimensionless variables associated with the catalyst form Ce^{4+} and bromide ion Br^- respectively. Referring to Fig. 7.4 (a) and starting at A, say, the limit cycle oscillation trajectory goes then through BCD to A again. This trajectory corresponds with the experimentally obtained cycle illustrated in Fig. 7.1 with corresponding letters. The [Br^-] decreases exponentially to B, then there is a rapid drop to the value at C. The value at B is the threshold Br^- concentration described in Section 7.1 where process II takes over from process I. There is then an increase in times $O(1)$ from C to D, given by (7.42) with (7.43), followed by a rapid increase in [Br^-] to its value at A again. Thus the relaxation oscillator approximation (7.26) mimics the real experimental oscillator of Fig. 7.1. To go further we must compare actual measurable quantities.

The experimentally suggested value for f from Field and Noyes (1974) is $f = 0.5$. Taking the limit $f \to 0.5$ in (7.39), by setting $f = 0.5 + \omega$ and then letting $\omega \to 0$, $T_{CD} = 2\ln(2) - 1$. Because of the smallness of q the major part of the period comes from T_{AB}. Now take the values in (7.4), which also gives y_0 used in the nondimensionalisation of the bromide ion concentration y, and substitute them into the limiting values for $y_0y(= [\text{Br}^-])$ from (7.31), and in the expressions for the T_{AB} bounds and T_{CD} from (7.38) and (7.39); Table 7.1 lists the values obtained. The table also gives the experimentally observed values of Field, Körös and Noyes (1972). Considering the complexity of the reaction and the number of approximations used in reducing the mechanism to manageable proportions and finally to the relaxation oscillator model, the results are very good.

Table 7.1. Comparison of values obtained from the relaxation oscillator approximation (7.26) for the BZ reaction with observed values from Field, Körös and Noyes (1972)

	Calculated values	Experimental values
Period	183–228 s	110 s
$[\text{Br}^-]_B = [\text{Br}^-]_{\text{crit}}$	$1.7 \times 10^{-5} [\text{BrO}_3^-]$	$2 \times 10^{-5} [\text{BrO}_3^-]$
$[\text{Br}^-]_C = [\text{Br}^-]_{\text{jump up}}$	$0.3 [\text{Br}^-]_{\text{crit}}$	$0.3 [\text{Br}^-]_{\text{crit}}$
$[\text{Br}^-]_A = [\text{Br}^-]_{\text{max}}$	$1.6 \times 10^{-3} [\text{BrO}_3^-] = 90 [\text{Br}^-]_{\text{crit}}$	$3 [\text{Br}^-]_{\text{crit}}$

When the parameter f is in the ranges which give nullclines typically as in Fig. 7.4 (b) and Fig. 7.4 (c), oscillations are not possible for the mechanism (7.26) as we saw in Chapter 6, Section 6.3. However the mechanism in these cases can exhibit threshold behaviour: compare with Fig. 6.5 (a) and Fig. 6.5 (b). Further, if reversibility is allowed in the basic FN model mechanism (7.2) it is possible to have three positive steady states of which two are stable, as in Fig. 6.4 (d). This corresponds to the biological switch behaviour also discussed in Chapter 6, Section 6.3 and schematically illustrated in Fig. 6.5 (c). Numerical studies of the reversible model also indicate bursting behaviour and chaos. Various plausible models for the Belousov-Zhabotinskii reaction have predicted a variety of unexpected phenomena which should be experimentally exhibited by the real reaction, and several have now been confirmed: see, for example, Tyson (1985) and the book edited by Field and Burger (1985). A recent paper by Barkley et al. (1987) demonstrates the existence of quite complex behaviour including periodic bursting, hysterisis and periodic-chaotic sequences.

Here we have considered only homogeneous or well-stirred systems. When we investigate, in later chapters, coupled biological oscillators and unstirred systems, where diffusion effects must be included, a new and surprising range of phenomena appear. Once again the Belousov-Zhabotinskii reaction is the key reaction used in experiments to verify the theoretical results.

Exercises

1. Another scaling (Murray 1977) of the FN model results in the third order system

$$\varepsilon\frac{dx}{dt} = y - xy + x(1 - qx), \quad \frac{dy}{dt} = -y - xy + 2fz, \quad \frac{dz}{dt} = \delta(x - z).$$

where ε and q are small. Determine the steady states, discuss their linear stability and show that a confined set for the positive steady state is

$$1 < x < \frac{1}{q}, \quad \frac{2fq}{1+q} < y < \frac{f}{q}, \quad 1 < z < \frac{1}{q}.$$

2. With the system in Exercise 1. derive the relevant reduced second order system on the basis that $0 < \varepsilon \ll 1$. Sketch the nullclines in the phase plane, exploiting the fact that $0 < q \ll 1$, and hence determine the necessary conditions on f for a limit cycle solution to exist.

*3. A relaxation oscillator is given by

$$\varepsilon\frac{dx}{dt} = f(x) - y, \quad \frac{dy}{dt} = x, \quad f(x) = \frac{x^3}{3} - x$$

where $0 < \varepsilon \ll 1$. Sketch the limit cycle trajectory in the y, x phase plane, noting the direction of motion. Determine the period T to $O(1)$ as $\varepsilon \to 0$.

Approximate the function $f(x)$ by a piecewise linear function, and sketch the corresponding phase plane limit cycle. Integrate the equations using the piecewise linear approximation. Hence sketch the solution x as a function of t. Also evaluate the $O(1)$ period and compare the result with that in the first part of the question.

*4. A possible relaxation oscillator model for the FN mechanism is governed by the dimensionless system

$$\varepsilon\frac{dx}{dt} = \frac{2fz(q - x)}{x + q} + x(1 - x), \quad \frac{dz}{dt} = x - z,$$

where $0 < \varepsilon \ll 1$, $0 < q \ll 1$ and $f = O(1)$. Using the results in Exercise 2, when a limit cycle solution is possible, sketch the relaxation oscillator trajectory, determine the maximum and minimum values for x and z, and obtain expressions for the $O(1)$ estimate for the limit cycle period.

8. Perturbed and Coupled Oscillators and Black Holes

8.1 Phase Resetting in Oscillators

With the plethora of known biological oscillators, and their generally accepted importance, it is natural to ask what effects external perturbations can have on the subsequent oscillations. In his pioneering work on circadian rhythms in the 1960's, A.T. Winfree asked this basic and deceptively simple question in a biological context in connection with his experimental work on the periodic emergence of the fruit fly, *Drosophila melonogaster*, from their pupae. Since then a series of spectacular discoveries of hitherto unknown properties of perturbed oscillators, spatially coupled oscillators, oscillators coupled to diffusion processes and so on (see, for example, Chapters 10 and 12 below), have been made as a result of this simple yet profound question. Winfree has developed a new conceptual geometric theory of biological time, which poses many challenging and interesting mathematical problems. Winfree's (1980) seminal book, which has a full bibliography, discusses the area in detail. He also gives numerous important examples of biological situations where a knowledge of such effects are crucial to understanding certain phenomena which are observed.

The periodic pacemaker in the heart is an important oscillator and one which is being widely studied, and in particular the effects of imposed perturbations. For example, Jalife and Antzelevitch (1979), whose results we shall discuss in Section 8.4 below, deal with pacemaker activity in cardiac tissue; Krinsky (1978) discusses cardiac wave arrhythmias; Winfree (1983) discusses, among other things, the topological aspects of sudden cardiac death. There is also interesting work on the sophisticated neural control of synchrony of breathing to stride in runners and horses (see, for example, Hoppensteadt 1985 and the references there).

We saw in Section 6.5 in the last chapter that under certain conditions nerve cells can exhibit regular periodic firing. In view of the crucial importance of neuronal signalling it is clearly of considerable interest to study the effect of external stimuli on such oscillations. The work of Best (1979) is of particular relevance to this and the following three sections. He subjected one of the accepted models for the propagation of nerve action potentials, namely the FitzHugh-Nagumo model (see Section 6.5, equations (6.39)), to periodic impulses and demonstrated some of the important phenomena we shall discuss in this chapter.

As we shall see in Chapter 12, the spatial propagation of impulses in neurons normally relies on a threshold stimulus being applied, and is an important practical example of an excitable medium. The heart pacemaker problem and some kinds of cardiac failure are probably related to wave phenomena associated with perturbed oscillators: see for example the general scientific article by Winfree (1983). This area has been studied over a period of some years, specifically in relation to heart failure and is one of the motivations for the material described in this section. The results and conclusions in this section, however, are quite general and will, in effect, be model independent, even though we use specific models for pedagogical reasons.

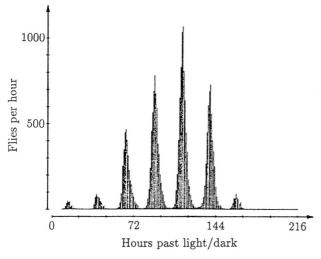

Fig. 8.1. Fruit fly pupae were placed in a completely dark environment. Fly emergence (eclosion) takes place approximately every 24 hours over a period of about 7 hours until all the pupae have matured. Here we are concerned with the periodic peak timing, not the number emerging at each peak. (Redrawn from Winfree 1980 with permission)

By way of introduction we briefly describe some of the experimental observations made on the approximately 24 hour rhythmic emergence of fruit flies from their pupae. During the pupal stage of the flies' development, a metamorphosis takes place which culminates in the emergence of an adult fruit fly. If metamorphosing pupae are simply left alone in a typical diurnal cycle of light and dark the flies emerge in quanta over a period of about 6–8 hours roughly every 24 hours. If such pupae are now placed in complete darkness the flies continue to emerge in almost exactly the same way: Fig. 8.1 illustrates the aggregated results of numerous experiments.

If the pupae, in the dark environment, are now subjected to a brief pulse of light, the timing, or *phase*, of the periodic emergence of the flies is shifted. In other words there is a *phase shift* in the underlying biological clock. The phase shift depends both on the timing T of the light pulse and its duration or rather the number D in ergs/cm^2 transmitted by the light. We are interested in the emergence time T_E after the pulse of light: T_E depends on T and D. If the dose $D = 0$ is given at T then clearly the phase shift is zero; $T_E = 24 - T$ hours. Winfree (1975) gives the results of numerous experiments in which T and D are varied and T_E recorded. The important point to note at this stage about these experiments is that there is a critical dose D^* which, if administered at a specific time T^*, results in no further periodic emergences but rather a continuous emergence. In other words the periodic behaviour has been destroyed. What is also surprising from the data is just how small the dose was which caused this: see Winfree (1975). Basically these experimental results suggest that there is a critical phase and stimulus which destroy the basic underlying periodic behaviour or biological clock. This has important implications for oscillators in general.

This section and the following three are principally concerned with biological oscillators, the effect of stimulus and timing on the periodic behaviour and the experimental evidence and implications. With the fruit fly experiments there is a singularity (or singularities) in the stimulus-timing-response space of the oscillator at which point the oscillator simply quits or does unpredictable things. Away from this singularity the subsequent behaviour is more or less predictable. Later, in Section 8.4, we shall describe other stimulus experiments, namely on cardiac tissue, which exhibit similar phase singularity behaviour.

Prior to doing the analysis, which is very easy for the illustrative example we consider, it is helpful to consider the simple pendulum to demonstrate the phenomena of *phase resetting* and *stimulus-timing-phase singularity* that we have just described. Suppose a pendulum is swinging with period ω, and suppose we measure zero phase or time $t = 0$ from the time the pendulum bob is at S, its highest point, at the right say. Then every time $t = n\omega$ for all integers n, the bob is again at S. If, during the regular oscillation, we give an impulse to the bob, we can clearly upset the regular periodic swinging. After such an impulse or stimulus, eventually the pendulum again exhibits simple harmonic motion, but now the bob does not arrive at S every $t = n\omega$ but at some other time $t = t_s + n\omega$, where t_s is some constant. In other words the phase has been *reset*. If we now give a stimulus to the bob when it is exactly at the bottom of its swing we can, if the stimulus is just right, stop the pendulum altogether. That is, if we give a stimulus of the right size at the right phase or time we can stop the oscillation completely: this is the singular point in the stimulus-phase-response space we referred to above in the fruit fly experiments.

Suppose that an oscillator is described by some vector state variable **u** which satisfies the differential equation system

$$\frac{d\mathbf{u}}{dt} = \mathbf{f}(\mathbf{u}, \boldsymbol{\lambda}) , \tag{8.1}$$

where **f** is the nonlinear rate function and $\boldsymbol{\lambda}$ denotes the parameters of the oscillator. For visual clarity and algebraic simplicity, suppose (8.1) describes a

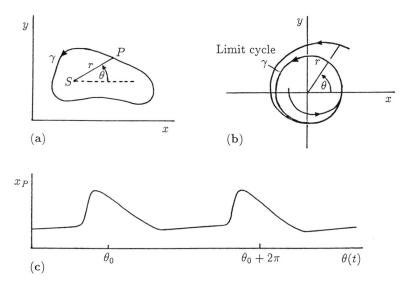

Fig. 8.2a-c. (a) A typical limit cycle solution trajectory γ in the phase plane. (b) Typical solutions to the system (8.2) with conditions (8.3). With any initial conditions, the solution evolves to the limit cycle given by $r = 1$, $d\theta/dt = 1$. (c) Typical time periodic behaviour of the point P in (a): note that the velocity of the point P is in general not constant as is the case in (b).

limit cycle oscillator involving only 2 species, x and y. Then typically the limit cycle trajectory is a simple plane closed curve, γ say, in the two-dimensional species plane as in Fig. 8.2 (a). By a suitable change of variable we can transform this limit cycle into one in which the closed trajectory is a circle and the state of the oscillator is essentially described by an angle θ, the 'phase', with its origin at some arbitrary point on the circle. The limit cycle is traversed with speed $v = d\theta/dt$. In one complete traversal of the orbit, θ increases by 2π.

A simple example of such a limit cycle system is

$$\frac{dr}{dt} = R(r), \quad \frac{d\theta}{dt} = \Phi(r) , \tag{8.2}$$

where

$$R(r) \begin{cases} > 0 \\ < 0 \end{cases} \text{ for } \begin{cases} 0 < r < r_0 \\ r > r_0 \end{cases}, \quad R(r_0) = 0, \quad \Phi(r_0) = 1 . \tag{8.3}$$

These conditions imply that (8.2) has a unique attracting limit cycle $r = r_0$, $d\theta/dt = 1$. (A particularly simple case, mentioned before in Chapter 3, has $R(r) = r(1 - r)$, $\Phi(r) = 1$, for which the solution can be given trivially.) If we normalize the circle with respect to r_0 we can then take the limit cycle to be $r = 1$.

Fig. 8.2 (b) illustrates a typical phase plane limit cycle solution. Fig. 8.2 (c) shows, for example, how the point $x_P = x_S + \cos\theta(t)$, where x_S is the steady state in Fig. 8.2 (a), might vary as a function of t, with equivalent values of θ marked at two points.

The fact that limit cycle solutions can be visualized as motion around a circle has been developed in an intuitive way by Winfree (1980) under the general topic of ring dynamics. The topological aspects are interesting and produce some unexpected results and new concepts. Here we consider only the basic elements of the subject but they are sufficient to demonstrate certain important concepts.

With the modelling of physiological oscillators in mind we envisage some event, a heart beat for example, to occur at some specific value of the phase, which we can normalize to $\theta = 0$. The pacemaker goes through a repeating cycle during which it fires at this specific phase (that is, time), then is refractory for part of the cycle, after which it again fires, and so on. With the ring or circle concept for an oscillator, we can think of the pacemaker as a point moving round a ring at a constant velocity with firing occurring every time the point passes through the position on the circle with phase $\theta = 0$. Although from a time point of view, t increases linearly, at specific times (multiples of the period) the pacemaker fires. To appreciate the basic concept of phase resetting of an oscillator by a stimulus we shall take, as an illustrative example, the simplest non-trivial limit cycle oscillator system

$$\frac{dr}{dt} = r(1 - r), \quad \frac{d\theta}{dt} = 1 , \tag{8.4}$$

for which the phase $\theta(t) = \theta_0 + t$, modulo 2π: see Fig. 8.2 (b). With it we shall discuss the two basic types of phase resetting, namely Type 1 and Type 0.

8.2 Phase Resetting Curves

Type 1 Phase Resetting Curves

Suppose we first perturb only the phase so that the governing equation becomes

$$r = 1, \quad \frac{d\theta}{dt} = 1 + v(\theta, I) , \tag{8.5}$$

where $v(\theta, I)$ represents the imposed velocity change, that is the stimulus, on the angular velocity $d\theta/dt$. I is a parameter which represents the magnitude of the impulse imposed on the oscillator. Again for pedagogical reasons let us take a simple, but nontrivial, v which depends on θ and I, and which was used by Winfree (1980), namely

$$\frac{d\theta}{dt} = 1 + I \cos 2\theta , \tag{8.6}$$

where I may be positive or negative. If the stimulus I is imposed at $t = 0$ and maintained for a time T then integrating (8.6) gives the new phase ϕ in terms of the old phase θ when the stimulus was started. From (8.6)

$$\int_\theta^\phi (1 + I \cos 2s)^{-1} \, ds = \int_0^T dt = T \,, \tag{8.7}$$

which integrates to give

$$|I| < 1 : \quad \tan \phi = A \tan \left[TB + \tan^{-1}(A^{-1} \tan \theta) \right] \,,$$

$$I = 1 : \quad \tan \phi = 2T + \tan \theta$$

$$I = -1 : \quad \tan \phi = \frac{\tan \theta}{1 - 2T \tan \theta}$$

$$|I| > 1 : \quad \tan \phi = A \frac{|K| + 1}{|K| - 1} \quad \text{if} \quad |\tan \phi| > A \tag{8.8}$$

$$\tan \phi = A \frac{|K| - 1}{|K| + 1} \quad \text{if} \quad |\tan \phi| < A$$

$$K = \left[\frac{A + \tan \theta}{A - \tan \theta} \right] \exp (2TB) \,,$$

where

$$A = \left[\frac{|1 + I|}{|1 - I|} \right]^{1/2} \,, \quad B = \left[|1 - I^2| \right]^{1/2} \,. \tag{8.9}$$

These give, explicitly, the new phase ϕ as a function of the old phase θ and the strength I and duration T of the stimulus. So, applying a stimulus causes a *phase shift* in the oscillator; in other words it *resets* the phase. For $t > T$ the oscillator simply reverts to $d\theta/dt = 1$ but now there is a phase shift. This means that the oscillator will fire at different times but at the same value of the phase that it did before: the subsequent period, of course, is also the same as it was before the stimulus. That is the periodic 'wave' form such as in Fig. 8.2 (c) will simply be moved along a bit.

We are interested in the *phase resetting curve* of ϕ as a function of θ for various stimulus magnitudes, which depend on I and its duration T. An important point with stimuli like that in (8.5) is that $d\phi/d\theta > 0$ for all I, T and θ. This says that a later new phase ϕ results if the impulse is applied at a later old phase. This is seen immediately with the v in (8.6) by differentiating (8.8) with respect to θ and noting that it gives an expression for $d\phi/d\theta$ which is strictly positive. If we now plot the new phase ϕ against the old phase θ when the impulse was applied, we obtain the phase resetting curve, which is typically as shown in Fig. 8.3. Note that whatever the stimulus I, the values of the new phase ϕ cover the complete phase cycle, here 0 to 2π. In other words any new phase $0 < \phi \le 2\pi$ can be obtained by a suitable choice of an old phase $0 < \theta \le 2\pi$ and the stimulus I. For a given I and T the new phase ϕ is uniquely determined by the old phase θ. This is known as a *Type 1* phase resetting curve and it is characterized by the

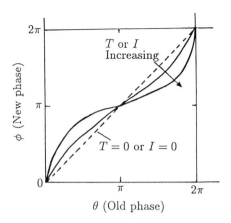

Fig. 8.3. Typical Type 1 phase resetting curves, giving the new phase in terms of the old phase for the phase velocity stimulus given by (8.6) for $I \geq 0$ and $T \geq 0$. This case shows a phase advance for an impulse in $0 < \theta < \pi$ and a delay for $\pi < \theta < 2\pi$.

fact that $d\phi/d\theta > 0$ for all $0 < \theta \leq 2\pi$: the average gradient over a cycle is 1 – hence the name. Although in Fig. 8.3 there is an advance for I in $0 < \theta < \pi$, another oscillator could well display a delay. The main point is that $d\phi/d\theta > 0$ in Type 1 resetting.

If I is sufficiently strong, $|I| > 1$ in fact, the phase velocity $d\theta/dt = 1+v(\theta, I)$ can become negative for some phases: in the case of (8.6) this is for θ satisfying $1+I\cos 2\pi\theta < 0$. This means that during the time of stimulation there is a phase attractor and a phase repellor, where $d\theta/dt = 0$ and where $d[d\theta/dt]/d\theta$ is negative and positive respectively (recall the stability analysis of single population models in Chapter 1). The stimulus is not sustained for all time, so the oscillator resumes its periodic cycle after the stimulus is removed – but of course with a different phase as determined by (8.8).

Type 0 Phase Resetting Curves

Consider the same limit cycle (8.4) but now let us subject it to a stimulus I which moves the solution off the limit cycle $r = 1$. To be specific let us take I as an impulse parallel to the y-axis as shown in Fig. 8.4. The analysis goes through with any perturbation but the algebra is more complicated and simply tends to obscure the main point. Let us decide on the notation that $I > 0$ is the situation illustrated in Fig. 8.4, that is with $0 < \theta < \pi/2$ the new phase ϕ is less than the old phase θ and the new position in general has $r = \rho \neq 1$. We now want the new phase ϕ in terms of the old phase θ and the stimulus I. From the figure

$$\rho \cos \phi = \cos \theta, \quad \rho \sin \phi + I = \sin \theta \qquad (8.10)$$

which, on eliminating ρ gives $\phi = \phi(\theta, I)$ implicitly; this is a three-dimensional surface in (ϕ, θ, I) space. As we shall see, it is the projection of this surface onto the (I, θ) plane which is of particular interest. Before considering this, however, let us construct phase resetting curves equivalent to those in Fig. 8.3, namely the new phase ϕ as a function of the old phase θ for various stimuli I: these are the projections of the surface $\phi = \phi(\theta, I)$ onto the (ϕ, θ) plane for various I.

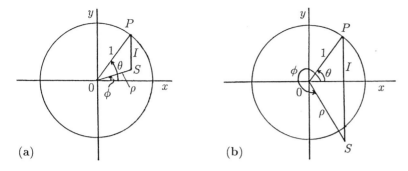

Fig. 8.4a,b. The impulse I takes the point $P(r = 1$, old phase $= \theta)$ instantaneously to $S(r = \rho$, new phase $= \phi)$. (a) $0 < I < 1$. (b) $I > 1$.

From (8.10),

$$\tan \phi = \tan \theta - \frac{I}{\cos \theta} , \qquad (8.11)$$

which gives ϕ in terms of θ for a given I. Let us suppose first that $0 < I < 1$. Then it is clear qualitatively from Fig. 8.4 that for $0 < \theta < \pi/2$ and $3\pi/2 < \theta < 2\pi$, $\phi < \theta$, while for $\pi/2 < \theta < 3\pi/2$, $\phi > \theta$. Thus the qualitative phase resetting curve ϕ against θ is as shown in Fig. 8.5 (a): it crosses the zero stimulus diagonal at $\theta = \pi/2, 3\pi/2$. The quantitative details are not important here. From (8.11), differentiating with respect to θ gives

$$\begin{aligned}
(1 + \tan^2 \phi)\frac{d\phi}{d\theta} &= 1 + \tan^2 \theta - \frac{I \sin \theta}{\cos^2 \theta} \\
&= \frac{1 - I \sin \theta}{\cos^2 \theta}
\end{aligned} \qquad (8.12)$$

$$\begin{cases} > 0 & \text{for all } 0 \le \theta \le 2\pi, \text{ if } I < 1. \\ < 0 & \text{for } \theta \text{ such that } \sin \theta > \dfrac{1}{I}. \end{cases}$$

So, on the phase resetting curves, if $0 < I < 1$, $d\phi/d\theta > 0$ for all θ as illustrated in Fig. 8.5 (a). Comparing these with the curves in Fig. 8.3 they are all topologically equivalent, so Fig. 8.5 (a) is a Type 1 phase resetting curve. The same remarks hold if $-1 < I < 0$.

Let us now consider $I > 1$. From (8.12) there is a range of θ where $d\phi/d\theta < 0$. Refer now to Fig. 8.4 (b) and let P move round the circle. We see that S never moves into the upper half plane. That is, as θ varies over the complete period of 2π, at the very least ϕ never takes on any phase in the range $(0, \pi)$: in fact the exact range can easily be calculated from (8.11) or (8.12). The phase resetting curve in this case is qualitatively as shown in Fig. 8.5 (b). This curve is *not* topologically equivalent to those in Fig. 8.5 (a). All phase resetting curves with $I > 1$ are topologically different from Type 1 resetting curves. Phase resetting curves like those in Fig. 8.5 (b), namely curves in which as the old phase θ takes

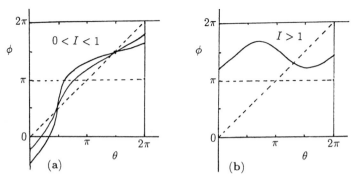

Fig. 8.5a,b. (a) Phase resetting curves from (8.11) for $0 < I < 1$. From (8.12) note that $d\phi/d\theta > 0$ for all such I. (b) Phase resetting curves from (8.11) for $I > 1$. Here for a range of θ, $d\phi/d\theta < 0$ and not all new phases ϕ can be obtained.

on all phase values in $(0, 2\pi)$ the new phase ϕ only takes on a *subset* of the full cycle range, are called *Type 0* resetting curves. Note that on such curves the gradient $d\phi/d\theta < 0$ for some range of θ: the average gradient on this curve is 0, which accounts for the name for this type of resetting curve. Note also that Type 0 resetting curves cannot be obtained from only a *phase* stimulus. The same type of resetting curves, namely Type 0, are obtained for stimuli $I < -1$.

8.3 Black Holes

From the analysis in the last section we see that as the stimulus I is increased from 0 there is a distinct bifurcation in phase resetting type as I passes through $I = 1$. That is there is a *singularity* in phase resetting for $I = 1$. To see clearly what is going on physically we must consider the projection of the $\phi = \phi(\theta, I)$ surface, given by (8.11), onto the (I, θ) plane for various ϕ. That is, we construct curves

$$I = \sin\theta - \cos\theta \tan\phi \tag{8.13}$$

for various ϕ in the range $0 \leq \phi \leq 2\pi$. Although this is an exercise in elementary curve drawing, using simple calculus, it has to be done with considerable care. The results are schematically shown in Fig. 8.6. Let us first consider the old phase range $0 \leq \theta \leq \pi$ and suppose, for the moment, $\phi \neq \pi/2, 3\pi/2$. Irrespective of the value of ϕ all curves pass through the point $I = 1$, $\theta = \pi/2$, since there, $\cos\theta\tan\phi = 0$ and $I = \sin\pi/2 = 1$ for all ϕ. All the curves with $\pi/2 > \phi > 0$, $2\pi > \phi > 3\pi/2$ intersect the $\theta = 0$ axis at $I = -\tan\phi$. For $\pi/2 < \phi < 3\pi/2$ a little calculus on (8.13) gives the curves shown. The special values $\phi = \pi/2, 3\pi/2$ give the vertical singularity line through $\theta = \pi/2$, as can be seen by taking the singular limit $\phi \to \pi/2$, or by simply observing the behaviour of the constant ϕ phase curves as ϕ approaches $\pi/2$. Having dealt with the θ-range $(0, \pi)$ the $(\pi, 2\pi)$ range is treated similarly and the overall picture obtained is shown in

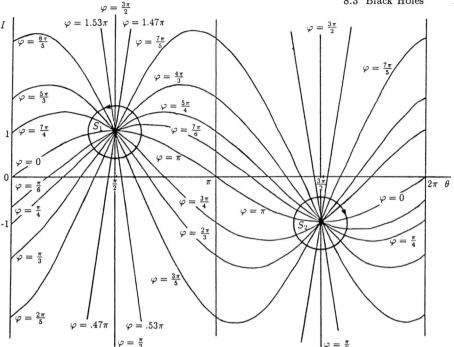

Fig. 8.6. Projections of the new phase(ϕ)-old phase(θ)-stimulus(I) surface, given by (8.13), onto the (I, θ) plane for various ϕ in the period cycle range $0 \leq \phi \leq 2\pi$. Note that S_1 and S_2 are singularities into each of which goes a complete selection of phases ϕ, $0 \leq \phi \leq 2\pi$; in one case they are traversed anti-clockwise, that is for S_1, and in the other, namely S_2, clockwise.

Fig. 8.6. The important thing to note is that there are two singular points S_1 and S_2 into each of which goes a constant phase curve of every phase in $(0, 2\pi)$, in the one case curves of increasing ϕ are arranged clockwise and in the other anti-clockwise.

Let us now consider the implications of this important Fig. 8.6. Suppose we have such an oscillator and we give it a stimulus I at a given phase θ. As long as $|I| < 1$ we can simply read off the new phase given I and the old phase θ, and what is more, the result is unique. For all $|I| > 1$, given the old phase θ, once again the new phase is determined uniquely. In this situation however we can get the same new phase ϕ for a given I for *two* different old phases θ. In the former we have, referring to Fig. 8.5, a Type 1 phase resetting while in the latter it is a Type 0 phase resetting.

Now suppose we take the particular stimulus $I = 1$ and impose it on the oscillator at phase $\theta = \pi/2$, the resulting point in Fig. 8.6 is the singular point S_1, which has no one specific phase ϕ associated with it, but rather the whole range $0 \leq \phi \leq 2\pi$. In other words the effect of this particular stimulus at this specific phase gives an *indeterminate* result. These singular points S_1 and S_2 are *black holes* in the stimulus-phase space, and are points where the outcome of a stimulus is unknown. If I is not exactly equal to 1, but close to it, the result is clearly a delicate matter, since all phases ϕ pass through the singularity. From a practical point of view the result of such a stimulus on a biological oscillator is unpredictable. Mathematically, however, if the exact stimulus $I = 1$ is imposed at exactly $\theta = \pi/2$ there is no resultant new phase ϕ. This is what happens in the simple pendulum situation when exactly the right impulse is given when the pendulum is just passing through the vertical position. In practice to stop a real pendulum dead is clearly quite difficult, and even if we could get quite close to the mathematically calculated conditions the resulting phase outcome would be far from obvious.

It is clear that the above concepts, due to Winfree (1970; see also 1980), are applicable to any endogenous oscillator, and so the results and implications are quite general. A key feature then of biological oscillators which can exhibit Type 1 and Type 0 phase resetting is that there are impulses and phases in their old phase – stimulus space which correspond to black holes. Perhaps the most important application of this is that there is thus, for such oscillators, a stimulus, which, if applied at specific phase, will annihilate the oscillation completely. The continuity argument for the existence of black holes is that if, as the stimulus is continuously increased, a transition from Type 1 to Type 0 resetting occurs at a specific value, then a black hole exists at the transition values of phase and stimulus.

Let us now consider some of the experimental evidence of black holes and annihilation in real oscillators.

8.4 Black Holes in Real Biological Oscillators

There are now several well documented experimental cases of Type 0 phase resetting and of annihilation of the basic oscillation by appropriate stimuli at the right phase – all as predicted above. Other than the cases we shall discuss in this section, there is, for example, the Type 0 phase response curve measured in *Hydra attenuata* by Taddei-Ferretti and Cordella (1976); the work of Pinsker (1977) on the bursting neurons of *Aplysia* perturbed by synaptic input – again a Type 0 case; and the work of Guttman et al. (1980) which displays annihilation in the squid axon membrane neuron oscillator. Before describing in detail an experimental case, we give Best's (1979) direct verification of the existence of black holes in the Hodgkin-Huxley model discussed in Chapter 6, Section 6.5, which models the oscillations in the space-clamped membrane of the squid giant axon.

The Hodgkin and Huxley (1952) model for the space-clamped neuronal firing of the squid axon given by the equation system (6.37) and (6.38) exhibits limit cycle oscillations. We showed in Section 6.5 that the FitzHugh-Nagumo model of this model had limit cycle periodic behaviour. Best (1979) numerically investigated the full Hodgkin-Huxley model (6.37) and (6.38) with an applied current in the range where limit cycle oscillations occurred. He then perturbed the os-

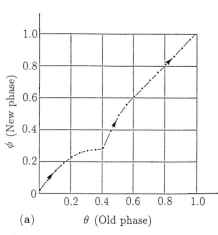

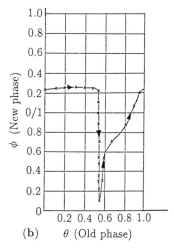

(a) θ (Old phase)   (b) θ (Old phase)

Fig. 8.7a,b. (a) Type 1 phase resetting curve obtained for the Hodgkin and Huxley (1952) model when the endogenous oscillator was subjected to voltage perturbations of 2 mV. The period of the cycle has been normalized to 1. The average slope across the graph is 1. **(b)** Type 0 phase resetting curve with voltage perturbations of 60 mV: here the average gradient is zero. (Redrawn from Best 1979)

cillator by subjecting it to voltage changes, these are the stimuli, with a view to experimental implimentation of his results. He found, as anticipated, Type 1 and Type 0 phase resetting curves: Fig. 8.7 shows one example of each.

The existence of a black hole, or null space was indicated in Best's (1979) simulations by a transition from a Type 1 to Type 0 resetting curve as he increased the voltage stimulus. This is as we might expect from Fig. 8.7, where Fig. 8.7 (a) and Fig. 8.7 (b) are topologically different and hence are separated by some bifurcation state. Because of the approximations inherent in any numerical simulation it is not possible to determine a single singular point as in Fig. 8.6. Instead there is a region around the singularity, the black holes or null space, where, after a suitable perturbation in an appropriate range of old phase, the new phase is indeterminate. Fig. 8.8 illustrates the results found by Best (1979). Except for the shaded regions there is a unique reset phase ϕ, for a given old phase θ, and stimulus I: note, however, that there is a (I, θ) subspace where it is possible to have the same ϕ for two θ's and a single I. Note also that the new phase values vary through a complete cycle in a clockwise way round hole 1 and anti-clockwise round hole 2 as indicated in the figure. A key feature to remember about stimulus-old phase contour maps like Fig. 8.8 is the convergence of contour lines to a black hole, one for positive stimuli and one for negative stimuli.

Another crucial property of black holes is that if the endogenous oscillator is subjected to a critical stimulus at the appropriate phase the oscillation simply disappears. Best (1979) demonstrated this with the Hodgkin-Huxley model system; the result is shown in Fig. 8.9. Note the annihilation of the endogenous oscillation. Guttman et al. (1979) showed experimentally that repetitive firing

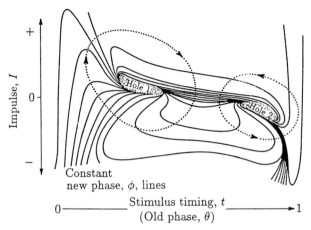

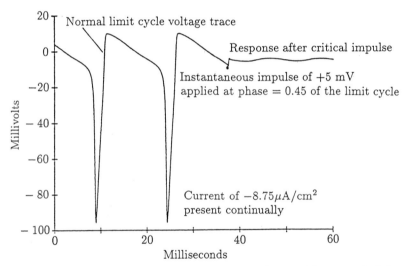

Fig. 8.8. Black holes or null space (after Winfree 1982) found by Best (1979) for the Hodgkin-Huxley (1952) model (equations (6.37) and (6.38)) for repetitive firing of the space-clamped giant axon of the squid. A voltage stimulus and phase which gives a point in the shaded black hole regions produces unpredictable phase resetting values. A complete path round either of the dashed curves gives a full range of phases.

Fig. 8.9. Voltage oscillations in the Hodgkin-Huxley model system (6.37) and (6.39) and the response when subjected to a critical stimulus (here 5 mV) at 0.45 through the phase, normalized to 1 (after Best 1979): the same applied current was used as in Fig. 8.7 and Fig. 8.8.

in space-clamped axons immersed in a weak calcium solution was stopped by a stimulus of the right size applied at a specific time in the cycle.

Jalife and Antzelevitch (1979) carried out similar work on the regular periodic beating of cardiac pacemaker cells, which is, of course, directly related to the cardiac pacemaker. They used tissue from the hearts of dogs, cats and calves and subjected the basic oscillation to electrical stimuli. They obtained from their

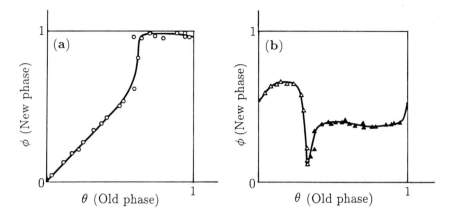

Fig. 8.10a,b. Phase resetting curves, normalized and in the notation used above, obtained by Jalife and Antzelevitch (1979) by applying brief current stimuli to pacemaker cells which spontaneously fire periodically. (a) Type 1 resetting, obtained when the stimulus duration was sufficiently short, here 10 msec. (b) Type 0 resetting with a stimulus time of 50 msec. (Photographs courtesy of J. Jalife and reproduced with permission)

experiments phase resetting curves which exhibited Type 1 and Type 0 resetting curves: Fig. 8.10 shows some of their results.

From the resetting curves in Fig. 8.10 we would expect there to be a transition value or values for stimulus duration, which destroys the oscillation, namely in the null space or black hole of the endogenous cardiac oscillator. This is indeed what was found, as shown in Fig. 8.11 (a). Fig. 8.11 (b) shows the resetting curve with the stimulus close to the transition value, intermediate between the values in Fig. 8.10 (a) and Fig. 8.10 (b). Fig. 8.11 (c) shows stimulus destruction of the regular oscillation in heart fibres from a dog.

A.T. Winfree for some years has been investigating the possible causes of sudden cardiac death and their connections with pacemaker oscillator topology, both temporal and spatial. Although the contraction of the heart involves a pacemaker, departures from the norm frequently involve the appearance of circulating contraction waves rather than the interruption of the firing mechanism. In the case of fibrillation, when the arrhythmias make the heart look a bit like a handful of squirming worms, it may be that a thorough understanding of the appearance of singular points or black holes could help to shed some light on this problem. The *Scientific American* article by Winfree (1983) is specifically concerned with the topology of sudden cardiac failure. Later in Chapter 12 we shall discuss spiral rotating waves which have a direct bearing on such heart problems. The mathematical problems associated with coupled and spatially distributed oscillators which are subjected to spatially heterogeneous applied stimuli are clearly challenging and fascinating, and of considerable biological importance.

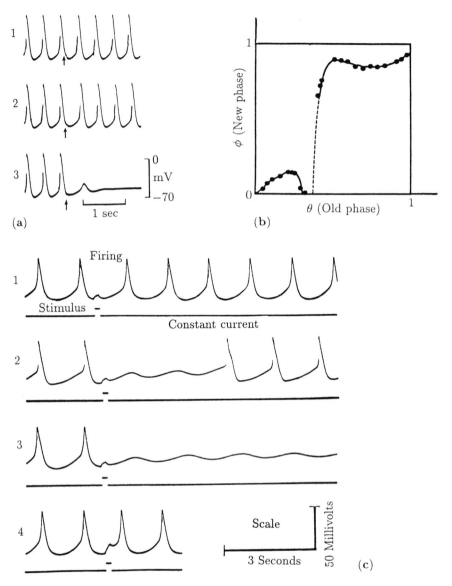

Fig. 8.11a-c. Experimental results obtained by Jalife and Antzelevitch (1979): (a) The micro-electrode traces of the transmembrane potentials of oscillating cardiac tissue (taken from a kitten) when subjected to a depolarizing current stimulus of 50 msec duration at successively later times in the cycle. When the time of stimulus was applied at 130 msec through the cycle, as in trace 3, the oscillation was completely suppressed. (b) Resetting curve for an interme-diate stimulus duration of 30 msec, that is between those in Fig. 8.10 (a) and Fig. 8.10 (b): here the periodic activity of the pacemaker can be destroyed as shown in the third trace in (a). (c) Oscillation annihilation by a stimulus in heart tissue of the dog. The small extra current stimulus lasts for 200 msec and is applied at progressively later stages in the cycle in 1,2,3,4. In 1 the next firing is slightly delayed while in 4 it is advanced. (Photographs courtesy of J. Jalife and reproduced with permission)

8.5 Coupled Oscillators: Motivation and Model System

The appearance of biological oscillators and periodic processes in ecology, epidemiology, developmental biology and so on is an accepted fact. It is inevitable that in a large number of situations oscillators are coupled in some way to obtain the required output. We have just seen how important it is to have some understanding of the effects of perturbations on oscillators. So, here we consider some of the effects of oscillator coupling and describe one of the key analytical techniques used to study such problems.

Coupled limit cycle oscillators have been widely studied mathematically for many years and the analytical problems are far from trivial. Not surprisingly the range of phenomena which they can corporately exhibit is very much larger than any single oscillator is capable of (see, for example, Winfree 1980). The subject is currently one of increasing research effort, not only in biology but also under the general heading of dynamical systems. Many of the processes which have been observed are still only partially understood. In the rest of this chapter we shall mainly be concerned with synchronization processes and when they break down. These synchronization phenomena may be phase locking, frequency coordination and so on, and they all arise from the interactive coupling of limit cycle oscillators. Here we shall restrict our study to the coupling of two oscillators and consider only weak coupling: we shall essentially follow the analysis of Neu (1979). Later in Chapter 10 we shall consider an important phenomenon associated with a chain of coupled oscillators when we model the neural arrangement in certain swimming vertebrates.

Before considering the mathematical problem it is relevant to describe briefly one of the experimental motivations for the specific model system we study. Marek and Stuchl (1975) investigated the effect of coupling two Belousov-Zhabotinskii reaction systems with different parameters, and hence different periodic oscillations. They did this by having each reaction in a separate stirred tank reactor and coupled them via an exchange of material between them through a common perforated wall. They observed that if the autonomous oscillators had almost the same frequency then the phase difference tended to a constant value as time went on: this is known as *phase locking*. However, if the difference in the autonomous frequencies was too large then phase locking did not persist but instead the coupled system had long intervals of slow variation in the phase difference separated by rapid fluctuations over very short intervals. The analysis we now give will explain these phenomena.

In our analytical study of coupled limit cycle oscillators, it is not necessary to know in detail the specific system they model. However, in view of the above experiments, we have the Belousov reaction system in mind. Suppose that the limit cycle oscillators are identical and that each, on its own, is governed by the equations

$$\frac{dx_i}{dt} = F(x_i, y_i), \quad \frac{dy_i}{dt} = G(x_i, y_i), \quad i = 1, 2 \qquad (8.14)$$

where the nonlinear functions F and G represent the dynamics of the oscillator. (They could be, for example, the functions on the right hand side of (7.27), one of the two-reactant models for the Belousov reaction discussed in Sections 7.4 and 7.5, or the interactive dynamics in a predator-prey model such as the one given by (3.18) in Chapter 3, Section 3.3.) We assume that the solutions of (8.14) exhibit a stable limit cycle behaviour with period T given by

$$x_i = X(t + \psi_i), \quad y_i = Y(t + \psi_i), \quad i = 1, 2 \tag{8.15}$$

where here the ψ_i are arbitrary constants. So

$$X(t + \psi_i + T) = X(t + \psi_i), \quad Y(t + \psi_i + T) = Y(t + \psi_i), \quad i = 1, 2 .$$

So as to formulate the weak coupling in a convenient (as we shall see) yet still general way we consider the nondimensional model system

$$\frac{dx_1}{dt} = F(x_1, y_1) + \varepsilon\{k(x_2 - x_1) + \lambda f(x_1, y_1)\} ,$$

$$\frac{dy_1}{dt} = G(x_1, y_1) + \varepsilon\{k(y_2 - y_1) + \lambda g(x_1, y_1)\} ,$$

$$\frac{dx_2}{dt} = F(x_2, y_2) + \varepsilon k(x_1 - x_2) , \tag{8.16}$$

$$\frac{dy_2}{dt} = G(x_2, y_2) + \varepsilon k(y_1 - y_2) ,$$

where $0 < \varepsilon \ll 1$ and $k > 0$ is a coupling constant. When $\varepsilon = 0$ the oscillators are uncoupled and these equations reduce to (8.14). The generality in the form (8.16) comes from the λ-terms. If $\varepsilon \neq 0$ and $\lambda = 0$ the two oscillators are identical with uncoupled solutions like (8.15). If $\varepsilon \neq 0$ and $\lambda \neq 0$, two different oscillators are coupled, with the $\varepsilon\lambda$-terms in the first two equations of (8.16) simply part of the isolated limit cycle oscillator given by these two equations. The specific coupling we have chosen, represented by the k-terms, is proportional to the differences $x_1 - x_2$ and $y_1 - y_2$. In the case of Marek and Stuchl's (1975) experiments this reflects the fact that there is a mass transfer. In the case of interacting populations it can be thought of as a mass transfer of species, a kind of diffusion flux approximation. In fact, when considering inter-habitat influence on the population dynamics, it is often incorporated in this way: it takes gross spatial effects into account without diffusion terms as such, which of course would make the models partial differential equation systems: these we consider later.

*8.6 Singular Perturbation Analysis: Preliminary Transformation

Equations (8.16) in general are hard to analyse. Even numerically it is not easy to see how the solution behaviour depends on the various parameters, particularly in the non-identical autonomous oscillator case. Since in many situations of interest the coupling is weak, and as we anticipate this to be the case in many biological applications, we shall exploit the fact that $0 < \varepsilon \ll 1$ and use singular perturbation theory (see for example, Murray 1984 for a simple discussion of the basic techniques).

Each oscillator has its own limit cycle solution which can be represented by a closed trajectory γ in the x-y phase plane. We can introduce a new coordinate system using this curve as the basis of the local coordinate system. We can characterize the periodic limit cycle by a *phase* θ which goes from 0 to T as we make a complete circuit round γ and any perturbation from it by the perpendicular distance A measured from γ; on γ, $A = 0$. It turns out to be particularly convenient algebraically to use this characterization in our coupled oscillator analysis. So in place of (8.15) as our autonomous limit cycle solutions we have

$$x_i = X(\theta_i), \quad y_i = Y(\theta_i), \quad i = 1, 2 \tag{8.17}$$

where $X(\theta_i)$ and $Y(\theta_i)$ are T-periodic functions of θ_i. Note that θ_i and t are related by $d\theta_i/dt = 1$.

The idea of representing the solution of a phase plane system, which admits a periodic limit cycle solution, in terms of the phase and a perturbation perpendicular to the limit cycle can be illustrated by the following example, which, although admittedly contrived, is still instructive.

Consider the differential equation system

$$\frac{dx_1}{dt} = x_1(1 - r) - \omega y_1, \quad \frac{dy_1}{dt} = y_1(1 - r) + \omega x_1, \quad r = (x_1^2 + y_1^2)^{1/2}, \tag{8.18}$$

where ω is a positive constant. A phase plane analysis (see Appendix 1) shows that $(0, 0)$ is the only singular point and it is an unstable spiral, spiralling anti-clockwise. A confined set can be found (just take r large and note that on this large circle the vector of the trajectories $(dx_1/dt, dy_1/dt)$ points inwards), so by the Poincaré-Bendixson theorem a limit cycle periodic solution exists and is represented by a closed orbit γ, in the (x_1, y_1) plane. If we now change to polar coordinates (r, θ) with

$$x_1 = r \cos \theta, \quad y_1 = r \sin \theta \tag{8.19}$$

the system (8.18) becomes

$$\frac{dr}{dt} = r(1 - r), \quad \frac{d\theta}{dt} = \omega . \tag{8.20}$$

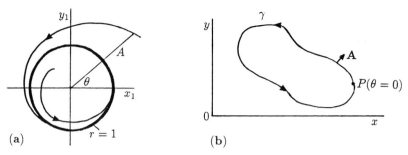

(a)   (b)

Fig. 8.12a,b. (a) The phase plane solution of the differential equation system (8.20). The asymptotically stable limit cycle is $r = 1$ with the phase $\theta = \omega t$, on taking $\theta = 0$ at $t = 0$. (b) Schematic example illustrating the local limit cycle coordinates. The point P has phase $\theta = 0$ and the phase increases by 2π on returning to P after moving round γ once.

The limit cycle, the trajectory γ, is then seen to be $r = 1$. The solution is illustrated in Fig. 8.12 (a). The limit cycle is asymptotically stable since from (8.20) any perturbation from $r = 1$ will die out with r simply winding back onto $r = 1$, in an anti-clockwise way because $d\theta/dt > 0$. In this example if the perturbation from the limit cycle is to a point $r < 1$ then, from (8.20), r increases while if the perturbation is to a point $r > 1$, r decreases as it tends to the orbit $r = 1$. In this case $r = 1$ is the equivalent of the orbit γ and A, the perpendicular distance from it is simply $r - 1$. The differential equation system in terms of $A(= r - 1)$ and θ is, from (8.20),

$$\frac{dA}{dt} = -A(1 + A), \quad \frac{d\theta}{dt} = \omega . \tag{8.21}$$

We can, of course, integrate (8.20) exactly to get

$$r(t) = \frac{r_0 e^t}{(1 - r_0) + r_0 e^t}, \quad \theta(t) = \omega t + \theta_0 , \tag{8.22}$$

where $r(0) = r_0$, $\theta(0) = \theta_0$ and from (8.19)

$$x_1(t) = r(t) \cos \theta(t), \quad y_1(t) = r(t) \sin \theta(t) . \tag{8.23}$$

As $t \to \infty$, $r(t) \to 1$ (so $A(t) \to 0$) and $x_1 \to \cos \theta$, $y_1 \to \sin \theta$, which are the equivalent of the $X(\theta)$ and $Y(\theta)$ in (8.17): they are 2π-periodic functions of θ. Here the rate of traversing γ (that is $r = 1$) is $d\theta/dt = \omega$ from (8.21). Fig. 8.12 (b) schematically illustrates the general situation. There $\theta = 0$ is taken to be at some point P and the phase increases by 2π as the orbit γ is traversed once in an anti-clockwise sense.

If we now consider our two oscillators, each with its autonomous closed limit cycle orbit γ_i, $i = 1,2$, the effect of coupling will be to alter the orbits and phase of each. We can characterize the effect in local coordinate terms by a phase θ_i which parametrizes points on γ_i and a perturbation A_i perpendicular to the original limit cycle orbit. Recall that for the coupled oscillator system (8.16) we are interested in weak coupling and so $0 < \varepsilon \ll 1$. With $\varepsilon = 0$ each oscillator has its limit cycle solution which in terms of the phase we can write as in (8.17), namely

$$x_i = X(\theta_i), \quad y_i = Y(\theta_i), \quad i = 1, 2 . \tag{8.24}$$

$$\theta_i = t + \psi_i \quad \Rightarrow \quad \frac{d\theta_i}{dt} = 1 . \tag{8.25}$$

We expect that the effect of the $O(\varepsilon)$ coupling is to cause the orbits γ_i, given by (8.17), to be displaced by $O(\varepsilon)$. We can thus see that an appropriate change of variables is from (x_i, y_i) for $i = 1,2$ to the local variables A_i and the new phase θ_i for $i = 1,2$, where A_i is the distance perpendicular to the orbit γ_i.

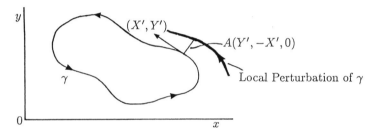

Fig. 8.13. Schematic visualization of the effect of coupling on the limit cycle orbit γ. The perpendicular displacement **A** of the velocity vector is given by the appropriate vector product, namely $A(X', Y', 0) \times (0, 0, 1)$, that is $A(Y', -X', 0)$

To motivate the specific variable transformation we shall use, refer now to Fig. 8.13. In the absence of coupling, the trajectory γ is traversed with velocity $(dx/dt, dy/dt)$ parallel to γ. In terms of the phase θ, which increases monotonically as the orbit is traversed, the velocity from (8.24) and (8.25) is equal to $(X'(\theta), Y'(\theta))$ where primes denote differentiation with respect to θ. This velocity vector is perturbed, due to the coupling, and the orbit γ will be displaced. This displacement can be described at each point by the perpendicular distance it is displaced, denoted by the vector **A** in the figure. Since **A** is the vector product of the velocity $(X'(\theta), Y'(\theta), 0)$ and the unit vector perpendicular to the (x, y) plane, that is $(0, 0, 1)$, this gives

$$\mathbf{A} = A(X'(\theta), Y'(\theta), 0) \times (0, 0, 1) = (AY'(\theta), -AX'(\theta), 0) . \tag{8.26}$$

Now consider the system (8.16) with $0 < \varepsilon \ll 1$ with our assumption that the autonomous orbits are perturbed $O(\varepsilon)$. An appropriate change of variable from (x_i, y_i) to (A_i, θ_i) is then, using (8.24) and (8.26),

$$x_i = X(\theta_i) + \varepsilon A_i Y'(\theta_i), \quad y_i = Y(\theta_i) - \varepsilon A_i X'(\theta_i), \quad i = 1, 2 . \tag{8.27}$$

Here we have used εA in place of A to emphasize the fact that since ε is small in our analysis, so is the orbit perturbation. Later in Chapter 12 we shall again use this transformation.

*8.7 Singular Perturbation Analysis: Transformed System

Let us now use the change of variable (8.27) in the coupled system (8.16) with $0 < \varepsilon \ll 1$. That is we use (8.27) in the right hand sides and expand in a Taylor series in ε: the algebra is complicated and tedious, but the concise, interesting and important end result is worth it, not just for the results we shall exhibit in this chapter but also for two other dramatic phenomena we shall discuss later in Chapters 10 and 13. We shall carry out enough of the algebra to show how to get the equations (8.16) in terms of the variables θ_i and A_i: however, use of separate pen and paper is recommended for those who want to follow the details of the algebra. (Those readers who wish to skip this algebra can proceed to equations (8.35) although later reference will be made to some of the definitions included here.)

In the following, the argument of the various functions, mainly X and Y, is θ_1 unless otherwise stated or included for emphasis. The first of (8.16), using (8.27), becomes

$$
\begin{aligned}
\frac{dx_1}{dt} &= X'\frac{d\theta_1}{dt} + \varepsilon Y'\frac{dA_1}{dt} + \varepsilon A_1 Y''\frac{d\theta_1}{dt} \\
&= F(X,Y) + \varepsilon A_1[Y'F_X(X,Y) - X'F_Y(X,Y)] + \varepsilon k[X(\theta_2) - X(\theta_1)] \\
&\quad + \varepsilon \lambda f(X,Y) + \varepsilon^2 k[A_2 Y'(\theta_2) - A_1 Y'(\theta_1)] + \\
&\quad \varepsilon^2 \lambda A_1[Y' f_X(X,Y) - X' f_Y(X,Y)] + O(\varepsilon^3) ,
\end{aligned}
\tag{8.28}
$$

while the second becomes

$$
\begin{aligned}
\frac{dy_1}{dt} &= Y'\frac{d\theta_1}{dt} - \varepsilon X'\frac{dA_1}{dt} - \varepsilon A_1 X''\frac{d\theta_1}{dt} \\
&= G(X,Y) + \varepsilon A_1[Y'G_X(X,Y) - X'G_Y(X,Y)] + \varepsilon k[Y(\theta_2) - Y(\theta_1)] \\
&\quad + \varepsilon \lambda g(X,Y) + \varepsilon^2 k[A_1 X'(\theta_1) - A_2 X'(\theta_2)] + \\
&\quad \varepsilon^2 \lambda A_1[Y' g_X(X,Y) - X' g_Y(X,Y)] + O(\varepsilon^3) .
\end{aligned}
\tag{8.29}
$$

When $\varepsilon = 0$ we have from (8.14) and (8.24)

$$
X'(\theta_1) = F(X,Y), \quad Y'(\theta_1) = G(X,Y) .
\tag{8.30}
$$

Now multiply (8.28) by $X'(\theta_1)$ and add to it $Y'(\theta_1)$ times (8.29) to get

$$(X'^2 + Y'^2)\frac{d\theta_1}{dt} + \varepsilon A_1(X'Y'' - Y'X'')\frac{d\theta_1}{dt}$$
$$= [X'F(X,Y) + Y'G(X,Y)] + \varepsilon A_1\{X'Y'[F_X(X,Y) - G_Y(X,Y)]$$
$$- X'^2 F_Y(X,Y) + Y'^2 G_X(X,Y)\} + \varepsilon k\{X'[X(\theta_2) - X(\theta_1)]$$
$$+ Y'[Y(\theta_2) - Y(\theta_1)]\} + \varepsilon^2 k A_2[Y'(\theta_2)X'(\theta_1)$$
$$- X'(\theta_2)Y'(\theta_1)] + \varepsilon\lambda[X'f(X,Y) + Y'g(X,Y)]$$
$$+ \varepsilon^2\lambda A_1\{X'Y'[f_X(X,Y) - g_Y(X,Y)] - X'^2 f_Y(X,Y)$$
$$+ Y'^2 g_X(X,Y)\} + O(\varepsilon^3) .$$

From (8.30), $X'F(X,Y) = X'^2$ and $Y'G(X,Y) = Y'^2$ so the last equation becomes

$$R^2(1 + \varepsilon\Gamma A_1)\frac{d\theta_1}{dt} = R^2 + \varepsilon[R^2\Omega A_1 + R^2 kr + R^2 kV + R^2\gamma\lambda] + O(\varepsilon^2) ,$$

where $R^2 = X'^2 + Y'^2 \neq 0$ and

$$R^2\Gamma = X'Y'' - Y'X'' ,$$
$$R^2\gamma = X'f(X,Y) + Y'g(X,Y), \quad R^2 r = -XX' - YY' ,$$
$$R^2\Omega = X'Y'[F_X(X,Y) - G_Y(X,Y)] - X'^2 F_Y(X,Y) + Y'^2 G_X(X,Y) ,$$
$$R^2 V = X'(\theta_1)X(\theta_2) + Y'(\theta_1)Y(\theta_2) .$$

$$(8.31)$$

If we now divide both sides by $R^2(1 + \varepsilon\Gamma A_1)$ and expand the right hand side as a series for $0 < \varepsilon \ll 1$, we get

$$\frac{d\theta_1}{dt} = 1 + \varepsilon[\{\Omega(\theta_1) - \Gamma(\theta_1)\}A_1 + \lambda\gamma(\theta_1) + kr(\theta_1) + kV(\theta_1, \theta_2)] + O(\varepsilon^2) . \quad (8.32)$$

In a similar way we get the equation for A_1 by multiplying (8.28) by $Y'(\theta_1)$ and subtracting from it, $X'(\theta_1)$ times (8.29). Using (8.30) and (8.32) for $d\theta_1/dt$, remembering that $\varepsilon \ll 1$, we get

$$\frac{dA_1}{dt} = \Phi(\theta_1)A_1 + kU(\theta_1, \theta_2) + \lambda\phi(\theta_1) + \varepsilon\Psi(\mathbf{A}, \boldsymbol{\theta}) + O(\varepsilon^2) \quad (8.33)$$

where

$$R^2 U(\theta_1, \theta_2) = X(\theta_2)Y'(\theta_1) - Y(\theta_2)X'(\theta_1) \quad (8.34)$$

and Φ, ϕ and Ψ are all determined: $\mathbf{A}$ and $\boldsymbol{\theta}$ are the vectors (A_1, A_2) and (θ_1, θ_2). The only functions whose exact form we shall require are $U(\theta_1, \theta_2)$ and $V(\theta_1, \theta_2)$, given by (8.34) and (8.31) respectively.

If we now do the same with the 3rd and 4th equations of (8.16) we find that the effect of the transformation to the (A_i, θ_i) dependent variables is to replace the model coupled oscillator system (8.16) by

$$\frac{dA_1}{dt} = \Phi(\theta_1)A_1 + kU(\theta_1, \theta_2) + \lambda\phi(\theta_1) + \varepsilon\Psi_1(\mathbf{A}, \boldsymbol{\theta}) + O(\varepsilon^2) ,$$

$$\frac{d\theta_1}{dt} = 1 + \varepsilon[\{\Omega(\theta_1) - \Gamma(\theta_1)\}A_1 + \lambda\gamma(\theta_1) + kr(\theta_1) + kV(\theta_1, \theta_2)] + O(\varepsilon^2)$$

$$(8.35)$$

$$\frac{dA_2}{dt} = \Phi(\theta_2)A_2 + kU(\theta_2, \theta_1) + \varepsilon\Psi_2(\mathbf{A}, \boldsymbol{\theta}) + O(\varepsilon^2)$$

$$(8.36)$$

$$\frac{d\theta_2}{dt} = 1 + \varepsilon[\{\Omega(\theta_2) - \Gamma(\theta_2)\}A_2 + kr(\theta_2) + kV(\theta_2, \theta_1)] + O(\varepsilon^2) .$$

The functions V and U, given by (8.31) and (8.34), will be referred to later. The exact forms of the functions Φ, ϕ, Γ, γ, Ω, Ψ_1, Ψ_2 and r are not essential for the following analysis, but what is important is that all of them are T-periodic in θ_1 and θ_2 and that Φ satisfies the relation

$$\int_0^T \Phi(\sigma) \, d\sigma < 0 . \tag{8.37}$$

This last relation comes from the fact that the original limit cycle solutions of the uncoupled oscillators is stable: we digress briefly to prove this.

Limit Cycle Stability Condition for the Uncoupled Oscillators

The oscillators are uncoupled when k and λ are zero. We want to keep $\varepsilon \neq 0$ since we are going to study the perturbed limit cycle oscillator using the transformation (8.27). In terms of the variables A and θ the governing system from (8.35) and (8.36), with $k = \lambda = 0$ is then

$$\frac{dA}{dt} = \Phi(\theta)A + O(\varepsilon), \quad \frac{d\theta}{dt} = 1 + O(\varepsilon) , \tag{8.38}$$

where Φ is a T-periodic function of θ. For times $O(1)$ the second equation gives $\theta \approx t$ and the first becomes

$$\frac{dA}{dt} = \Phi(t)A + O(\varepsilon) ,$$

which on integrating from t to $t + T$ gives

$$\frac{A(t+T)}{A(t)} = [1 + O(\varepsilon)] \exp\left[\int_t^{t+T} \Phi(\sigma) \, d\sigma\right]$$

$$= [1 + O(\varepsilon)] \exp\left[\int_0^T \Phi(\sigma) \, d\sigma\right] , \tag{8.39}$$

where the limits of integration have been changed because Φ is a T-periodic function. The unperturbed limit cycle is $A \equiv 0$, $\theta = t + \psi$. So, the limit cycle is stable if all the solutions of (8.38) have $A(t) \to 0$ as $t \to \infty$. From (8.39) we see that if

$$\int_0^T \Phi(\sigma)\, d\sigma < 0$$

then $A(t + T) < A(t)$ for all t and so $A(t) \to 0$ as $t \to \infty$.

*8.8 Singular Perturbation Analysis: Two-Time Expansion

The $O(\varepsilon)$ terms in the equations (8.35) and (8.36) will have an effect after a long time, $O(1/\varepsilon)$ in fact. This suggests looking for an asymptotic solution as $\varepsilon \to 0$ for the A_i and θ_i in (8.35) and (8.36) in the following form:

$$A_i \sim {}^0 A_i + \varepsilon {}^1 A_i, \quad \theta_i \sim {}^0 \theta_i + \varepsilon {}^1 \theta_i , \tag{8.40}$$

where the A's and θ's are functions of the time t, the fast time, and $\tau = \varepsilon t$, the long or slow time. In other words only after times $\tau = O(1)$, that is $t = O(1/\varepsilon)$, do the ε-effects show. (See, for example, Murray (1984) for an elementary exposition to this two-time expansion procedure.) Now all time derivatives

$$\frac{d}{dt} = \frac{\partial}{\partial t} + \left(\frac{d\tau}{dt}\right)\frac{\partial}{\partial \tau} = \frac{\partial}{\partial t} + \varepsilon \frac{\partial}{\partial \tau} , \tag{8.41}$$

and the system (8.36) becomes a partial differential equation system.

The algebra in the rest of this section is also rather involved. The end result, namely equation (8.59) is required in the following Section *8.9, where a very important result for coupled oscillators is derived.

If we now substitute (8.40) with (8.41) into (8.35) and (8.36) and equate powers of ε we get the following hierarchy of equations:

$$
\begin{aligned}
O(1): \quad & \frac{\partial {}^0 A_1}{\partial t} - \Phi({}^0\theta_1)\, {}^0 A_1 = kU({}^0\theta_1, {}^0\theta_2) + \lambda\Phi({}^0\theta_1) , \\[1mm]
& \frac{\partial {}^0\theta_1}{\partial t} = 1 , \\[1mm]
& \frac{\partial {}^0 A_2}{\partial t} - \Phi({}^0\theta_2){}^0 A_2 = kU({}^0\theta_2, {}^0\theta_1) , \\[1mm]
& \frac{\partial {}^0\theta_2}{\partial t} = 1 .
\end{aligned}
\tag{8.42}
$$

$$O(\varepsilon): \quad \frac{\partial {}^1A_1}{\partial t} - \Phi({}^0\theta_1){}^1A_1 = \{\Phi'({}^0\theta_1){}^0A_1 +$$

$$k\frac{\partial U}{\partial \theta_1}({}^0\theta_1, {}^0\theta_2) + \lambda\phi'({}^0\theta_1)\}{}^1\theta_1 +$$

$$k\frac{\partial U}{\partial \theta_2}({}^0\theta_1, {}^0\theta_2){}^1\theta_2 + \Psi_1({}^0\underline{A}, {}^0\underline{\theta}) - \frac{\partial {}^0A_1}{\partial \tau},$$

$$\frac{\partial {}^1\theta_1}{\partial t} = [\Omega({}^0\theta_1) - \Gamma({}^0\theta_1)]\,{}^0A_1 + \lambda\gamma({}^0\theta_1) + kr({}^0\theta_1) +$$

$$kV({}^0\theta_1, {}^0\theta_2) - \frac{\partial {}^0\theta_1}{\partial \tau},$$

$$\frac{\partial {}^1A_2}{\partial t} - \Phi({}^0\theta_2){}^1A_2 = \{\Phi'({}^0\theta_2){}^0A_2 + k\frac{\partial U}{\partial \theta_2}({}^0\theta_2, {}^0\theta_1)\}\,{}^1\theta_2 +$$

$$k\frac{\partial U}{\partial \theta_1}({}^0\theta_2, {}^0\theta_1)\,{}^1\theta_1 + \Psi_2({}^0\underline{A}, {}^0\underline{\theta}) - \frac{\partial {}^0A_2}{\partial \tau},$$

$$\frac{\partial {}^1\theta_2}{\partial t} = [\Omega({}^0\theta_2) - \Gamma({}^0\theta_2)]\,{}^0A_2 + kr({}^0\theta_2) + kV({}^0\theta_2, {}^0\theta_1) - \frac{\partial {}^0\theta_2}{\partial \tau}.$$

$$(8.43)$$

If we integrate the 2nd and 4th of (8.42) we get

$$ {}^0\theta_i = t + \psi_i(\tau), \quad i = 1, 2 \tag{8.44}$$

where, at this stage the $\psi_i(\tau)$ are arbitrary functions of τ. If we now substitute these into the 1st and 3rd equations of (8.42) we get

$$\frac{\partial {}^0A_1}{\partial t} - \Phi(t + \psi_1)\,{}^0A_1 = kU(t + \psi_1, t + \psi_2) + \lambda\phi(t + \psi_1),$$

$$\frac{\partial {}^0A_2}{\partial t} - \Phi(t + \psi_2)\,{}^0A_2 = kU(t + \psi_2, t + \psi_1). \tag{8.45}$$

To get the required solutions let us digress again, and consider the less cluttered equations:

$$\frac{dx}{ds} - \Phi(s)x = \phi(s),$$

$$\frac{dy}{ds} - \Phi(s)y = U(s, s + \chi), \tag{8.46}$$

where $\chi \equiv \psi_2 - \psi_1$. Remember that the functions Φ, ϕ and U are all T-periodic. The complementary or homogeneous solution of each is

$$\exp[v(s)], \quad v(s) = \int_0^s \Phi(\sigma)\,d\sigma.$$

From (8.37) we have $v(s) < 0$, which is necessary for the stability of the uncoupled limit cycle oscillators. So, the complementary solutions will decay, eventually, to zero. We can now show that each of (8.46) has a unique T-periodic solution. To do this consider, say, the first of (8.46). The exact solution is

$$x(s) = x(0) \exp \left[\int_0^s \Phi(\sigma) \, d\sigma \right] + \int_0^s \exp \left[\int_\alpha^s \Phi(\sigma) \, d\sigma \right] \phi(\alpha) \, d\alpha \, . \qquad (8.47)$$

The equation for x is unchanged if we replace s by $s+T$. Since the above solution for $x(s)$ is periodic

$$x(0) = x(T) = x(0) \exp \left[\int_0^T \Phi(\sigma) \, d\sigma \right] + \int_0^T \exp \left[\int_\alpha^T \Phi(\sigma) \, d\sigma \right] \phi(\alpha) \, d\alpha \, ,$$

which is an equation for the inital value $x(0)$, substitution of which into (8.47) gives the unique T-periodic solution of the first of (8.46). Similarly the second of (8.46) has a unique T-periodic solution. Denote these periodic solutions by

$$x = p(s), \quad y = \rho(s, \chi) \, . \qquad (8.48)$$

In terms of these solutions (8.48) the general solutions of (8.45) are

$$^0A_1 = k\rho(t + \psi_1, \chi) + \lambda p(t + \psi_1) + h_1(\tau) \exp \left[v(t + \psi_1) \right] , \qquad (8.49)$$
$$^0A_2 = k\rho(t + \psi_2, -\chi) + h_2(\tau) \exp \left[v(t + \psi_2) \right] ,$$

where the h_1 and h_2 are arbitrary functions of τ. If we now substitute these into the θ-equations in (8.43) we get

$$\frac{\partial^1 \theta_1}{\partial t} = \{ k\rho(t + \psi_1, \chi) + \lambda p(t + \psi_1) \} \{ \Omega(t + \psi_1) - \Gamma(t + \psi_1) \} + \lambda \gamma(t + \psi_1)$$
$$+ kr(t + \psi_1) + kV(t + \psi_1, t + \psi_2) - \frac{d\psi_1}{d\tau}$$
$$+ h_1(\tau) \exp \left[v(t + \psi_1) \right] \{ \Omega(t + \psi_1) - \Gamma(t + \psi_1) \} \, ,$$

$$\frac{\partial^1 \theta_2}{\partial t} = k\rho(t + \psi_2, -\chi) \{ \Omega(t + \psi_2) - \Gamma(t + \psi_2) \}$$
$$+ kr(t + \psi_2) + kV(t + \psi_2, t + \psi_1)$$
$$- \frac{d\psi_2}{d\tau} + h_2(\tau) \exp \left[v(t + \psi_2) \right] \{ \Omega(t + \psi_2) - \Gamma(t + \psi_2) \} \, .$$

$$(8.50)$$

If $f(t)$ is a T-periodic function we can write

$$f(t) = \mu + \omega(t), \quad \mu = \frac{1}{T} \int_0^T f(s) \, ds \qquad (8.51)$$

where $\omega(t)$ is T-periodic and has zero mean. If we now use this fact in (8.50) we get

$$
\frac{\partial\,{}^1\theta_1}{\partial t} = \mu_1(\chi) + \omega_1(t,\tau)\,,
$$

$$
\frac{\partial\,{}^1\theta_2}{\partial t} = \mu_2(\chi) + \omega_2(t,\tau)\,,
$$

(8.52)

where

$$
\mu_1 = H(\chi) + \lambda\beta - \frac{d\psi_1}{d\tau}\,,
$$

$$
\mu_2 = H(-\chi) - \frac{d\psi_2}{d\tau}\,,
$$

(8.53)

with

$$
\beta = \frac{1}{T}\int_0^T \{p(s)[\Omega(s) - \Gamma(s)] + \gamma(s)\}\,ds\,,
$$

$$
H(\chi) = \frac{1}{T}\int_0^T \{k\rho(s,\chi)[\Omega(s) - \Gamma(s)] + kr(s) + kV(s,s+\chi)\}\,ds\,.
$$

(8.54)

The functions $\omega_1(t,\tau)$ and $\omega_2(t,\tau)$ are made up of exponentially decaying terms and periodic terms of zero mean, so on integrating (8.52) we get

$$
{}^1\theta_1 = \mu_1(\chi)t + W_1(t,\tau)\,,
$$

$$
{}^1\theta_2 = \mu_2(\chi)t + W_2(t,\tau)\,,
$$

(8.55)

where W_1 and W_2 are bounded functions. If we now substitute these solutions into the third of (8.43) and use (8.44) and (8.47), the equation for ${}^1A_2$ becomes

$$
\frac{\partial\,{}^1A_2}{\partial t} - \Phi(t+\psi_2)\,{}^1A_2 = \{S_1(t,\tau)\mu_1 + S_2(t,\tau)\mu_2\}t + B(t,\tau)\,,
$$

$$
S_1 = k\frac{\partial U}{\partial\theta_1}(t+\psi_2, t+\psi_1)\,,
$$

$$
S_2 = k\frac{\partial U}{\partial\theta_2}(t+\psi_2, t+\psi_1) + k\rho(t+\psi_2, -\chi)\Phi'(t+\psi_2)\,,
$$

(8.56)

where $B(t,\tau)$ is another function which consists of an exponentially decaying term and a T-periodic part.

In the usual asymptotic singular perturbation way we now require the $O(\varepsilon)$ part of the amplitude, that is ${}^1A_2$, and its time derivative $\partial\,{}^1A_2/\partial t$, to be bounded for all time – this ensures that the series solution (8.40) is uniformly valid for all time. From (8.56) this requires that

$$
\{S_1(t,\tau)\mu_1 + S_2(t,\tau)\mu_2\}t
$$

must be bounded for all time. However, since S_1 and S_2 are T-periodic functions (and so cannot tend to zero as $t \to \infty$) the only way we can get boundedness is if

$$S_1(t,\tau)\mu_1 + S_2(t,\tau)\mu_2 \equiv 0 \qquad (8.57)$$

In general S_1 and S_2 are two different periodic functions, so the only way to ensure (8.57) is if μ_1 and μ_2 are both zero. That is, from (8.53), we require

$$\mu_1 = H(\chi) + \lambda\beta - \frac{d\psi_1}{d\tau} = 0, \quad \mu_2 = H(-\chi) - \frac{d\psi_2}{d\tau} = 0$$

and so

$$\frac{d\psi_1}{d\tau} = H(\chi) + \lambda\beta, \quad \frac{d\psi_2}{d\tau} = H(-\chi) . \qquad (8.58)$$

Recalling that $\chi = \psi_2 - \psi_1$, if we subtract the last two equations we get the following single equation for χ:

$$\frac{d\chi}{d\tau} = P(\chi) - \lambda\beta , \qquad (8.59)$$

$$\text{where} \quad P(\chi) = H(-\chi) - H(\chi), \quad \chi = \psi_2 - \psi_1 .$$

The dependent variable χ is the phase shift due to the coupling: the ordinary differential equation (8.59) governs the time evolution of this phase shift. The derivation of this equation is the main purpose of the singular perturbation analysis in the the last two Sections 8.7 and 8.8. It is also the equation which we shall use to advantage not only here but also in Chapter 13.

*8.9 Analysis of the Phase Shift Equation and Application to Coupled Belousov-Zhabotinskii Reactions

The functions $H(\chi)$ and $H(-\chi)$ are T-periodic functions from their definition (8.54). So, $P(\chi)$ in the phase shift equation (8.59) is also a T-periodic function. If $\chi = 0$, $P(0) = H(0) - H(0) = 0$. From the form of $H(\chi)$ in (8.54) its derivative at $\chi = 0$ is

$$H'(0) = \frac{1}{T} \int_0^T k \left\{ \frac{\partial\rho}{\partial\chi}(s,0)[\Omega(s) - \Gamma(s)] + \frac{\partial V}{\partial\theta_2}(s,s) \right\} ds . \qquad (8.60)$$

The function $\rho(s,\chi)$ is the T-periodic solution of the second of (8.46), namely

$$\frac{\partial\rho}{\partial s} - \Phi(s)\rho = U(s, s + \chi) ,$$

which on differentiating with respect to χ and setting $\chi = 0$ gives

$$\frac{\partial}{\partial s} \left\{ \left[\frac{\partial\rho}{\partial\chi} \right]_{(s,0)} \right\} - \Phi(s) \left[\frac{\partial\rho}{\partial\chi} \right]_{(s,0)} = \left[\frac{\partial U}{\partial\theta_2} \right]_{(s,s)} .$$

From the definition of U in (8.34) we see that $[\partial U/\partial \theta_2]_{(s,s)} = 0$ and so the only periodic solution of the last equation is $[\partial \rho/\partial \chi]_{(s,0)} = 0$. Using (8.31) which gives the definition of V, we get $[\partial V/\partial \theta_2]_{(s,s)} = 1$. With these values (8.60) gives

$$H'(0) = \frac{1}{T} \int_0^T k \, ds = k \ .$$

So $P'(0) = -H'(0) - H'(0) = -2k$. Fig. 8.14 (a) illustrates a typical $P(\chi)$.

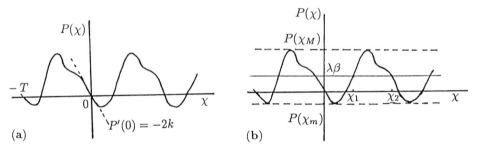

(a)

(b)

Fig. 8.14a,b. (a) Schematic form of the T-periodic function $P(\chi)$ in the phase shift equation (8.59). (b) Determination of steady state solutions χ_0 from the intersection of $P(\chi) = \lambda \beta$.

Let us now consider the time evolution equation (8.59) for the phase shift $\chi = \psi_2 - \psi_1$. *Phase locking* is when the phase difference is constant for all time, that is, when $\chi = \chi_0$ is constant. Phase locked solutions are given by (8.59) on setting $d\chi/d\tau = 0$, namely solutions χ_0 of

$$P(\chi) - \lambda \beta = 0 \ . \tag{8.61}$$

The linear stability of χ_0 is given by linearizing (8.59) about it, which gives

$$\frac{d(\chi - \chi_0)}{d\tau} \approx P'(\chi_0)(\chi - \chi_0) \ ,$$

and so

$$\chi_0 \quad \text{is} \quad \begin{cases} \text{stable} \\ \text{unstable} \end{cases} \quad \text{if} \quad P'(\chi_0) \begin{cases} < 0 \\ > 0 \end{cases} \tag{8.62}$$

For example, if the coupled oscillators are identical, $\lambda = 0$ from (8.16), and from Fig. 8.14 (a) $\chi = 0$ is a solution of $P(\chi) = 0$ and its derivative $P'(0) < 0$. So $\chi = 0$ is stable. This means that coupling *synchronizes* identical oscillators.

Suppose now that the coupled oscillators are not identical, that is $\lambda \neq 0$ and $\varepsilon \neq 0$ in (8.16). In this case steady states χ_0 of (8.61) will depend on whether the horizontal line $\lambda \beta$ intersects the curve of $P(\chi)$ in Fig. 8.14 (a). So there are at least two steady state solutions in $0 \leq \chi_0 \leq T$ if

$$\min P(\chi) = P(\chi_m) < \lambda \beta < P(\chi_M) = \max P(\chi) \ . \tag{8.63}$$

Referring to Fig. 8.14 (b), we see that two typical solutions, χ_1 and χ_2, are respectively unstable and stable from (8.62), since by inspection $P'(\chi_1) > 0$ and $P'(\chi_2) < 0$. In this situation the coupled oscillator system will evolve to stable limit cycle oscillations with a *constant phase shift* χ_2 between the two oscillators after a long time, by which we mean τ large, which in turn means εt large. Whether or not there are more than two steady state solutions for the phase shift χ depends on the form of $P(\chi)$: the example in Fig. 8.14 is illustrative of the simplest situation.

As long as $\lambda\beta$ lies between the maximum and minimum of $P(\chi)$, that is (8.63) is satisfied, the steady state solutions χ_0, of (8.61) depend continuously on $\lambda\beta$. This is clear from Fig. 8.14 (b) if the line $P = \lambda\beta$ is moved continuously between the upper and lower bounds $P(\chi_m)$ and $P(\chi_M)$. For example, as $\lambda\beta$ increases the stable steady state phase shift between the oscillators decreases.

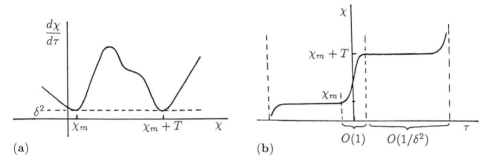

(a)   (b)

Fig. 8.15a,b. (a) Situation for the χ-equation (8.65) where there is no steady state solution: here $0 < \delta^2 \ll 1$ (b) Schematic solution, during which there are long periods of very little change in χ interspersed with rapid variations.

Suppose now that $\lambda\beta$ is such that the two solutions coalesce, either at $P(\chi_m)$ or $P(\chi_M)$, and we ask what happens when $\lambda\beta$ is such that these solutions no longer exist. To examine this situation let us be specific and consider the case where $\lambda\beta$ is slightly less than the critical value, $(\lambda\beta)_c = P(\chi_m)$. With $\lambda\beta$ just less than $P(\chi_m)$ there is no intersection of $P(\chi)$ and the line $\lambda\beta$. Let

$$\lambda\beta = (\lambda\beta)_c - \delta^2 = P(\chi_m) - \delta^2, \quad 0 < \delta^2 \ll 1 . \tag{8.64}$$

Now write the phase difference equation (8.59) in the form

$$\begin{aligned}
\frac{d\chi}{d\tau} &= \{P(\chi) - P(\chi_m)\} + \{P(\chi_m) - \lambda\beta\} \\
&= \{P(\chi) - P(\chi_m)\} + \delta^2 .
\end{aligned} \tag{8.65}$$

Fig. 8.15 (a) illustrates $d\chi/d\tau$ as a function of χ for $\delta^2 > 0$. This is the same curve as in Fig. 8.14 (a) simply moved up a distance $-P(\chi_m) + \delta^2$, which in our situation is sufficient to make (8.65) have no steady state solutions for χ: that is the curve of $d\chi/d\tau$ as a function of χ does not cross the $d\chi/d\tau = 0$ axis.

The solution problem posed by (8.65) is a *singular perturbation* one when $0 < \delta^2 \ll 1$ and can be dealt with by standard singular perturbation techniques (see Murray 1984). However it is not necessary to carry out the asymptotic analysis to see what is going on with the solutions. First note that when $\delta = 0$ (that is $\lambda\beta = P(\chi_m)$) there are solutions $\chi = \chi_m + nT$ for $n = 0, 1, 2, \ldots$ since $P(\chi)$ is a T-periodic function. To be specific, let us start with $\chi \approx \chi_m$ in Fig. 8.15 (a). From (8.65) and from the graph in Fig. 8.15, $d\chi/d\tau = O(\delta^2) > 0$ which implies that $\chi \approx \chi_m + \delta^2\tau$ and so for times $\tau = O(1)$, χ does not vary much from χ_m. However, for all $\tau > 0$, $d\chi/d\tau > 0$ and so χ slowly increases. For τ sufficiently large so that χ starts to diverge significantly from χ_m, $P(\chi) - P(\chi_m)$ is no longer approximately zero, in which case $d\chi/d\tau$ is $O(1)$ and χ changes measurably in times $\tau = O(1)$ and $\chi \to \chi_m + T$. When χ is close to $\chi_m + T$, once again $P(\chi) - P(\chi_m) \approx 0$ and $d\chi/d\tau = O(\delta^2)$ again. The qualitative picture is now clear. The solution stays in the vicinity of the solutions $\chi = \chi_m + nT$ for a long time $\tau = O(1/\delta^2)$, then changes in times $\tau = O(1)$ to the next solution (with $\delta = 0$), where again it stays for a long time. This process repeats itself in a quasi-periodic but quite different way to the T-periodic limit cycle behaviour we got before. The solution is illustrated in Fig. 8.15 (b). The rapidly changing regions are the singular regions while the roughly constant regions are the nonsingular parts of the solutions. This behaviour is known as *rhythm splitting*. So, as $\lambda\beta \to (\lambda\beta)_c = P(\chi_m)$ the solution for χ *bifurcates* from phase synchronization to rhythm splitting.

Now that we know the qualitative behaviour of $\chi = \psi_2 - \psi_1$ we can determine ψ_1 and ψ_2 from (8.58), namely

$$\frac{d\psi_1}{d\tau} = H(\chi) + \lambda\beta, \quad \frac{d\psi_2}{d\tau} = H(-\chi). \tag{8.66}$$

The solutions x_i and y_i are then given by

$$x_i = X(t + \psi_i(\tau)), \quad y_i = Y(t + \psi_i(\tau)), \quad i = 1, 2. \tag{8.67}$$

The frequency of the oscillations is given to $O(\varepsilon)$ by

$$\frac{d\theta_i}{dt} = 1 + \varepsilon \frac{d\psi_i}{d\tau} = \begin{cases} 1 + \varepsilon[H(\chi) + \lambda\beta], & i = 1 \\ 1 + \varepsilon H(-\chi), & i = 2 \end{cases} \tag{8.68}$$

Fig. 8.16 (a) illustrates the rhythm splitting solution. The small bumps are the rapid variations in χ in times $O(1)$ while the long flat regions correspond to the slowly varying solutions where the phase difference $\psi_2 - \psi_1(= \chi)$ is approximately constant. Fig. 8.16 (b) illustrates a typical solution x_i as a function of t corresponding to this rhythm splitting. Note the sudden change in frequency which occurs when the phase difference χ goes through its region of rapid change from one solution χ_m to the next $\chi_m + T$.

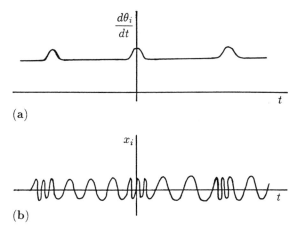

$\frac{d\theta_i}{dt}$

t

(a)

x_i

t

(b)

Fig. 8.16a,b. (a) The frequency $d\theta/dt$ of one of the oscillators when $\lambda\beta$ is slightly less than the minimum $P(\chi_m)$ and rhythm splitting occurs. (b) The time variation in x_i when phase locking just bifurcates to rhythm splitting.

This important bifurcation phenomenon exhibited by coupled oscillators, where there is an abrupt change from phase locking to rhythm splitting is exactly what was demonstrated experimentally by Marek and Stuchl (1975), as described in Section 8.5. That is they first observed *phase locking*. Then when the parameters were changed so that the autonomous limit cycle frequencies were sufficiently different they observed that the phase difference was slowly varying for long periods of time but punctuated by short periods of rapid fluctuations. This is the *rhythm splitting* phenomenon we have discussed above.

Exercises

1. Consider the limit cycle oscillator

$$\frac{dr}{dt} = r(1-r), \quad \frac{d\theta}{dt} = 1$$

to be perturbed by an impulse I parallel to the x-axis (refer to Fig. 8.4). Determine the resulting phase resetting curves and discuss the possible existence of black holes.

*2. Discuss the bifurcation situation for coupled oscillators, using equation (8.63) for the phase difference χ, when the interaction parameters $\lambda\beta$ is just greater than $P(\chi_M)$, the maximum $P(\chi)$ in Fig. 8.14 (a). Sketch the equivalent functions to those illustrated in Fig. 8.15 and Fig. 8.16 and discuss the practical implications.

9. Reaction Diffusion, Chemotaxis and Non-local Mechanisms

9.1 Simple Random Walk and Derivation of the Diffusion Equation

In an assemblage of particles, for example cells, bacteria, chemicals, animals and so on, each particle usually moves around in a random way. The particles spread out as a result of this irregular individual motion of each particle. When this microscopic irregular movement results in some macroscopic or gross regular motion of the group we think of it as a *diffusion* process. Of course there may be interaction between particles for example, or the environment may give some bias in which case the gross movement is not simple diffusion. To get the macroscopic behaviour from a knowledge of the individual microscopic behaviour is much too hard so we derive a continuum model equation for the global behaviour in terms of a particle density or concentration. It is instructive to start with a random process which we shall study probabilistically in an elementary way, and then derive a deterministic model.

For simplicity we consider initially only one-dimensional motion and the simplest random walk process. The generalisation to higher dimensions is then intuitively obvious from the one-dimensional equation.

Suppose a particle moves randomly back and forward along a line in fixed steps Δx that are taken in a fixed time Δt. If the motion is unbiased then it is equally probable that the particle takes a step to the right or left. After time $N\Delta t$ the particle can be anywhere from $-N\Delta x$ to $N\Delta x$ if we take the starting point of the particle as the origin. The spatial distribution is clearly not going to be uniform if we release a group of particles about $x = 0$ since the probability of a particle reaching $x = N\Delta x$ after N steps is very small compared with that for x nearer $x = 0$.

We want the probability $p(m, n)$ that a particle reaches a point m space steps to the right (that is to $x = m\Delta x$) after n time steps (that is after a time $n\Delta t$). Let us suppose that to reach $m\Delta x$ it has moved a steps to the right and b to the left. Then

$$m = a - b, \quad a + b = n \quad \Rightarrow \quad a = \frac{n + m}{2}, \quad b = n - a \,.$$

The number of possible paths that a particle can reach this point $x = m\Delta x$ is

$$\frac{n!}{a!b!} = \frac{n!}{a!(n-a)!} \equiv C_a^n$$

where C_a^n is the binomial coefficient defined, for example, by

$$(x+y)^n = \sum_{a=0}^{n} C_a^n x^{n-a} y^a .$$

The total number of possible n-step paths is 2^n and so the probability $p(m,n)$ (the favorable possibilities/total possibilities) is

$$p(m,n) = \frac{1}{2^n} \frac{n!}{a!(n-a)!}, \quad a = \frac{n+m}{2} . \tag{9.1}$$

$n + m$ is even.

Note that

$$\sum_{m=-n}^{n} p(m,n) = 1 ,$$

as it must since the sum of all probabilities must equal 1. It is clear mathematically since

$$\sum_{m=-n}^{n} p(m,n) = \sum_{a=0}^{n} C_a^n \left(\frac{1}{2}\right)^{n-a} \left(\frac{1}{2}\right)^a = \left(\frac{1}{2} + \frac{1}{2}\right)^n = 1 .$$

$p(m,n)$ is the *binomial distribution*.

If we now let n be large so that $n \pm m$ are also large we have asymptotically

$$n! \sim (2\pi n)^{1/2} n^n e^{-n}, \quad n \to \infty , \tag{9.2}$$

which is Stirling's formula. (This is derived by noting that

$$n! = \Gamma(n+1) = \int_0^\infty e^{-t} t^n \, dt ,$$

where Γ is the gamma function, and using Laplace's method for the asymptotic approximation for such integrals for n large (see for example Murray's (1984) *Asymptotic Analysis*). Using (9.2) in (9.1) we get, after a little algebra, the *normal* or *Gaussian probability distribution*

$$p(m,n) \sim \left[\frac{2}{\pi n}\right]^{1/2} \exp\left[\frac{-m^2}{2n}\right], \quad m \gg 1, \quad n \gg 1 . \tag{9.3}$$

m and n need not be very large for (9.3) to be an accurate approximation to (9.1). For example with $n = 8$ and $m = 6$, (9.3) is within 5% of the exact value from (9.1): with $n = 10$ and $m = 4$ it is accurate to within 1%. In fact for all practical purposes we can use (9.3) for $n > 6$. Asymptotic approximations can often be remarkably accurate over a wider range than might be imagined.

Now set

$$m\Delta x = x, \quad n\Delta t = t ,$$

where x and t are the continuous space and time variables. If we anticipate letting $m \to \infty$, $n \to \infty$, $\Delta x \to 0$, $\Delta t \to 0$ so that x and t are finite, then it is not appropriate to have $p(m, n)$ as the quantity of interest since this probability must tend to zero: the number of points on the line tends to ∞ as $\Delta x \to 0$. The relevant dependent variable is more appropriately $u = p/2\Delta x$: $u2\Delta x$ is the probability of finding a particle in the interval $(x, x + \Delta x)$ at time t. From (9.3) with $m = x/\Delta x$, $n = t/\Delta t$

$$\frac{p\left(\frac{x}{\Delta x}, \frac{t}{\Delta t}\right)}{2\Delta x} \sim \left\{\frac{\Delta t}{2\pi t(\Delta x)^2}\right\}^{1/2} \exp\left\{-\frac{x^2}{2t}\frac{\Delta t}{(\Delta x)^2}\right\} .$$

If we assume

$$\lim_{\substack{\Delta x \to 0 \\ \Delta t \to 0}} \frac{(\Delta x)^2}{2\Delta t} \to D \neq 0$$

the last equation gives

$$u(x, t) = \lim_{\substack{\Delta x \to 0 \\ \Delta t \to 0}} \frac{p\left(\frac{x}{\Delta x}, \frac{t}{\Delta t}\right)}{2\Delta x} = \left(\frac{1}{4\pi Dt}\right)^{1/2} \exp\left[-\frac{x^2}{4Dt}\right] . \tag{9.4}$$

D is the *diffusion coefficient* or *diffusivity* of the particles: note that it has dimensions $(\text{length})^2/(\text{time})$. It is a measure of how efficiently the particles disperse from a high to a low density. For example in blood, haemoglobin molecules have a diffusion coefficient of the order of $10^{-7} \text{cm}^2 \text{sec}^{-1}$ while that for oxygen in blood is of the order of $10^{-5} \text{cm}^2 \text{sec}^{-1}$.

Let us now relate this result to the classical approach to diffusion, namely *Fickian diffusion*. This says that the flux, J, of material, which can be cells, amount of chemical, number of animals and so on, is proportional to the gradient of the concentration of the material. That is, in one dimension

$$J \propto -\frac{\partial c}{\partial x} \quad \Rightarrow \quad J = -D\frac{\partial c}{\partial x}, \tag{9.5}$$

where $c(x, t)$ is the concentration of the species and D is its diffusivity. The minus sign simply indicates that diffusion transports matter from a high to a low concentration.

We now write a general conservation equation which says that the rate of change of the amount of material in a region is equal to the rate of flow across the boundary plus any that is created within the boundary. If the region is $x_0 < x < x_1$ and no material is created,

$$\frac{\partial}{\partial t} \int_{x_0}^{x_1} c(x,t)\, dx = J(x_0,t) - J(x_1,t) \, . \tag{9.6}$$

If we take $x_1 = x_0 + \Delta x$, take the limit as $\Delta x \to 0$ and use (9.5) we get the *classical diffusion equation* in one dimension, namely

$$\frac{\partial c}{\partial t} = -\frac{\partial J}{\partial x} = \frac{\partial \left(D \frac{\partial c}{\partial x} \right)}{\partial x} \, , \tag{9.7}$$

which, if D is constant, becomes

$$\frac{\partial c}{\partial t} = D \frac{\partial^2 c}{\partial x^2} \, . \tag{9.8}$$

If we release an amount Q of particles per unit area at $x = 0$ at $t = 0$, that is

$$c(x,0) = Q\delta(x) \, , \tag{9.9}$$

where $\delta(x)$ is the Dirac delta function, then the solution of (9.8) is (see, for example, Crank's 1975 book)

$$c(x,t) = \frac{Q}{2(\pi D t)^{1/2}} e^{-x^2/(4Dt)}, \quad t > 0 \tag{9.10}$$

which, with $Q = 1$, is the same result as (9.4), obtained from a random walk approach when the step and time sizes are small compared with x and t. Fig. 9.1 illustrates the concentration $c(x,t)$ from (9.10) as a function of x for various times.

This way of relating the diffusion equation to the random walk approach essentially uses circumstantial evidence. We now derive it by extending the random walk approach and start with $p(x,t)$, from (9.4), as the probability that a particle released at $x = 0$ at $t = 0$ reaches x in time t. At time $t - \Delta t$ the particle was at $x - \Delta x$ or $x + \Delta x$. Thus if α and β are the probabilities that a particle will move to the right or left

$$p(x,t) = \alpha p(x - \Delta x, t - \Delta t) + \beta p(x + \Delta x, t - \Delta t), \quad \alpha + \beta = 1 \, . \tag{9.11}$$

If there is no bias in the random walk, that is it is isotropic, $\alpha = 1/2 = \beta$. Expanding the right hand side of (9.11) in a Taylor series we get

$$\frac{\partial p}{\partial t} = \left[\frac{(\Delta x)^2}{2\Delta t} \right] \frac{\partial^2 p}{\partial x^2} + \left(\frac{\Delta t}{2} \right) \frac{\partial^2 p}{\partial t^2} + \dots$$

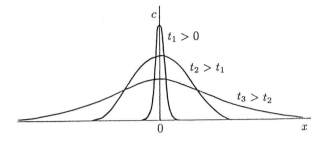

Fig. 9.1. Schematic particle concentration distribution arising from Q particles released at $x = 0$ at $t = 0$ and diffusing according to the diffusion equation (9.8).

If we now let $\Delta x \to 0$ and $\Delta t \to 0$ such that, as before

$$\lim_{\substack{\Delta x \to 0 \\ \Delta t \to 0}} \frac{(\Delta x)^2}{2\Delta t} = D$$

we get

$$\frac{\partial p}{\partial t} = D \frac{\partial^2 p}{\partial x^2} .$$

If the total number of released particles is Q, then the concentration of particles $c(x, t) = Qp(x, t)$ and the last equation becomes (9.8).

The random walk derivation is not completely satisfactory since it relies on Δx and Δt tending to zero in a rather specific way so that D exists. A better and more sophisticated way is to derive it from the Fokker-Planck equations using a probability density function with a Markov process, that is a process at time t depending only on the state at time $t - \Delta t$; in other words a one-generation-time dependency: see, for example, Skellam (1973) or the excellent book by Okubo (1980). The latter gives some justification to the limiting process used above. The review article by Okubo (1986) also discusses the derivation of various diffusion equations.

9.2 Reaction Diffusion Equations

Consider now diffusion in three space dimensions. Let S be an arbitrary surface enclosing a volume V. The general conservation equation says that the rate of change of the amount of material in V is equal to the rate of flow of material across S into V plus the material created in V. Thus

$$\frac{\partial}{\partial t} \int_V c(\mathbf{x}, t) \, dv = - \int_S \mathbf{J} \cdot \mathbf{ds} + \int_V f \, dv \tag{9.12}$$

where $\mathbf{J}$ is the flux of material and f, which represents the source of material, may be a function of c, $\mathbf{x}$ and t. Applying the divergence theorem to the surface integral and assuming $c(\mathbf{x}, t)$ is continuous, the last equation becomes

$$\int_V \left[\frac{\partial c}{\partial t} + \nabla \cdot \mathbf{J} - f(c, \mathbf{x}, t) \right] dv = 0 . \tag{9.13}$$

Since the volume V is arbitrary the integrand must be zero and so the *conservation equation* for c is

$$\frac{\partial c}{\partial t} + \nabla \cdot \mathbf{J} = f(c, \mathbf{x}, t) . \tag{9.14}$$

This equation holds for a general flux transport $\mathbf{J}$, whether by diffusion or some other process.

If classical diffusion is the process then the generalization of (9.12), for example, is

$$\mathbf{J} = -D\nabla c \tag{9.15}$$

and (9.14) becomes

$$\frac{\partial c}{\partial t} = f + \nabla \cdot (D\nabla c) , \tag{9.16}$$

where D may be a function of $\mathbf{x}$ and c and f a function of c, $\mathbf{x}$ and t.

For example, the source term f in an ecological context could represent the birth-death process and c the population density, n. With logistic population growth $f = rn(1 - n/K)$ where r is the linear reproduction rate and K the carrying capacity of the environment. The resulting equation with D constant is

$$\frac{\partial n}{\partial t} = rn \left(1 - \frac{n}{K} \right) + D\nabla^2 n , \tag{9.17}$$

now known as the *Fisher equation* after Fisher (1937) who proposed the one-dimensional version as a model for the spread of an advantageous gene in a population. This is an equation we shall study in detail later in Chapter 11.

If we further generalize (9.16) to the situation in which there are, for example, several interacting species or chemicals we then have a vector $u_i(\mathbf{x}, t)$ $i = 1, \ldots m$ of densities or concentrations each diffusing with its own diffusion coefficient D_i and interacting according to the vector source term $\mathbf{f}$. Then (9.16) becomes

$$\frac{\partial \mathbf{u}}{\partial t} = \mathbf{f} + \nabla \cdot (D\nabla \mathbf{u}) , \tag{9.18}$$

where now D is a matrix of the diffusivities which, if there is no cross diffusion among the species, is simply a diagonal matrix. In (9.18) $\nabla \mathbf{u}$ is a tensor so $\nabla \cdot D\nabla \mathbf{u}$ is a vector. Cross diffusion does not arise often in genuinely practical models: one example where it does will be described below in Chapter 12, Section 12.2. Cross diffusion systems can pose very interesting mathematical problems particularly regarding their well-posedness. Equation (9.18) is referred to as a *reaction diffusion* or an *interacting population diffusion* system. Such a mechanism was proposed as a model for the chemical basis of morphogenesis by Turing (1952) in one of the most important papers in theoretical biology this century. Such systems have been widely studied since about 1970. We shall be concerned with reaction diffusion systems when D is diagonal and constant and $\mathbf{f}$ is a function only of $\mathbf{u}$. Further generalisation can include, in the case of population models for example, integral terms in $\mathbf{f}$ which reflect the population history. The generalisations seem endless. For most practical models of real world situations it is premature, to say the least, to spend too much time on sophisticated generalisations[1] before the simpler versions have been shown to be inadequate when compared with experiments.

9.3 Models for Insect Dispersal

Diffusion models form a reasonable basis for studying insect and animal dispersal: this and other aspects of population models are discussed in detail, for example, by Okubo (1980, 1986) and Shigesada (1980). Dispersal of interacting species is discussed by Shigesada, Kawasaki and Teramoto (1979) and of competing species by Shigesada and Roughgarden (1982). Kareiva (1983) has shown that many species appear to disperse according to a reaction diffusion model with a constant diffusion coefficient. He gives actual values for the diffusion coefficients which he measured from experiments on a variety of insect species.

One extension of the classical diffusion model which is of particular relevance to insect dispersal is when there is an increase in diffusion due to population pressure. One such model has the diffusion coefficient, or rather the flux $\mathbf{J}$, depending on the population density n such that D increases with n, that is

$$\mathbf{J} = -D(n)\nabla n, \quad \frac{dD}{dn} > 0 . \tag{9.19}$$

A typical form for $D(n)$ is $D_0(n/n_0)^m$, where $m > 0$ and D_0 and n_0 are positive constants. The dispersal equation for n without any growth term is then

$$\frac{\partial n}{\partial t} = D_0 \nabla \cdot \left[\left(\frac{n}{n_0} \right)^m \nabla n \right] .$$

In one dimension

$$\frac{\partial n}{\partial t} = D_0 \frac{\partial}{\partial x} \left[\left(\frac{n}{n_0} \right)^m \frac{\partial n}{\partial x} \right] , \tag{9.20}$$

which has an exact analytical solution of the form

$$n(x,t) = n_0 [\lambda(t)]^{-1} \left[1 - \left\{ \frac{x}{r_0 \lambda(t)} \right\}^2 \right]^{1/m} , \quad |x| \le r_0 \lambda(t) \tag{9.21}$$

$$= 0, \quad |x| > r_0 \lambda(t) ,$$

---

[1]   As de Toqueville has remarked, there is no point in generalizing since God knows all the special cases.

where

$$\lambda(t) = \left(\frac{t}{t_0}\right)^{1/(2+m)} , r_0 = \frac{Q\Gamma\left(\frac{1}{m} + \frac{3}{2}\right)}{\left\{\pi^{1/2} n_0 \Gamma\left(\frac{1}{m} + 1\right)\right\}}$$

$$t_0 = \frac{r_0^2 m}{2D_0(m+2)} ,$$

(9.22)

where Γ is the gamma function and Q is the initial number of insects released at the origin. It is straightforward to check that (9.21) is a solution of (9.20) for all r_0. The evaluation of r_0 comes from requiring the integral of n over all x to be Q. (In another context (9.20) is known as the *porous media equation*.) The population is identically zero for $x > r_0\lambda(t)$. This solution is fundamentally different from that when $m = 0$, namely (9.10). The difference is due to the fact that $D(0) = 0$. The solution represents a kind of wave with the front at $x = x_f = r_0\lambda(t)$. The derivative of n is discontinuous here. The wave 'front', which we define here as the point where $n = 0$, propagates with a speed $dx_f/dt = r_0 d\lambda/dt$, which, from (9.22), decreases with time for all m. The solution for n is illustrated schematically in Fig. 9.2. The dispersal patterns for grasshoppers exhibit a similar behaviour to this model (Aikman and Hewitt 1972). Without any source term the population n, from (9.21), tends to zero as $t \to \infty$. Shigesada (1980) proposed such a model for animal dispersal in which she took the linear diffusion dependence $D(n) \propto n$.

The equivalent plane radially symmetric problem with Q insects released at $r = 0$ at $t = 0$ satisfies the equation

$$\frac{\partial n}{\partial t} = \left(\frac{D_0}{r}\right) \frac{\partial}{\partial r}\left[r\left(\frac{n}{n_0}\right)^m \frac{\partial n}{\partial r}\right]$$

(9.23)

with solution

$$n(r,t) = \frac{n_0}{\lambda^2(t)}\left[1 - \left\{\frac{r}{r_0\lambda(t)}\right\}^2\right]^{1/m} , \quad r \le r_0\lambda(t)$$

$$= 0, \quad r > r_0\lambda(t)$$

$$\lambda(t) = \left(\frac{t}{t_0}\right)^{1/2(m+1)} , \quad t_0 = \frac{r_0^2 m}{4D_0(m+1)}$$

$$r_0^2 = \frac{Q}{\pi n_0}\left(1 + \frac{1}{m}\right) .$$

(9.24)

As $m \to 0$, that is $D(n) \to D_0$, the solutions (9.21) and (9.24) tend to the usual constant diffusion solutions – (9.10) for example in the case of (9.21). To show this involves some algebra and use of the exponential definition $\exp[s] = \lim_{m\to 0}(1 + ms)^{1/m}$.

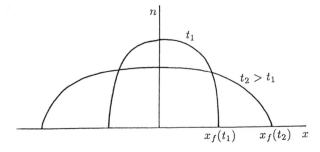

Fig. 9.2. Schematic solution, from (9.21), of equation (9.20) as a function of x at different times t. Note the discontinuous derivative at the wave front $x_f(t) = r_0\lambda(t)$.

Insects at low population densities frequently tend to aggregate. One model (in one dimension) which reflects this has the flux

$$J = Un - D(n)\frac{\partial n}{\partial x}$$

where U is a transport velocity. For example if the centre of attraction is the origin and the velocity of attraction is constant, Shigesada et al. (1979) took $U = -U_0 \, \text{sgn}\,(x)$ and the resulting dispersal equation becomes

$$\frac{\partial n}{\partial t} = U_0\frac{\partial}{\partial x}[\text{sgn}\,(x)n] + D_0\frac{\partial}{\partial x}\left[\left(\frac{n}{n_0}\right)^m\frac{\partial n}{\partial x}\right], \tag{9.25}$$

which is not easy to solve. We can however get some idea of the solution behaviour for parts of the domain.

Suppose Q is again the initial flux of insects released at $x = 0$. We expect that gradients in n near $x = 0$ for $t \approx 0$ are large and so, in this region, the convection term is small compared with the diffusion term, in which case the solution is approximately given by (9.21). On the other hand after a long time we expect the population to reach some steady, spatially inhomogeneous state where convection and diffusion effects balance. Then the solution is approximated by (9.25) with $\partial n/\partial t = 0$. Integrating this equation twice using the conditions $n \to 0$, $\partial n/\partial x \to 0$ as $|x| \to \infty$ we get the steady state spatial distribution

$$\lim_{t\to\infty} n(x,t) \to n(x) = n_0\left(1 - \frac{mU_0|x|}{D_0}\right)^{1/m}, \quad |x| \leq \frac{D_0}{mU_0}$$

$$= 0, \quad |x| > \frac{D_0}{mU_0}. \tag{9.26}$$

The derivation of this is left as an exercise. The solution (9.26) shows that the dispersal is *finite* in x. The form obtained when $m = 1/2$ is similar to the population distribution observed by Okubo and Chiang (1974) for a special type of mosquito swarm (see Okubo 1980, Fig. 9.6). Fig. 9.3 schematically illustrates the steady state insect population.

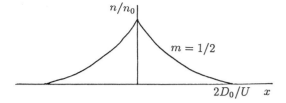

Fig. 9.3. Schematic form of the steady state insect population distribution from (9.26), for insects which tend to aggregate at low densities according to (9.25).

Insect dispersal is a very important subject which is still not understood. The above model is a fairly simple one but even so it gives some pointers as to possible insect dispersal behaviour. If there is a population growth/death term we simply include it on the right hand side of (9.25) (Exercises 3 and 4). In these the insect population dies out as expected, since there is no birth only death, but what is interesting is that the insects move only a finite distance from the origin.

The use of diffusion models for animal and insect dispersal is increasing and has been applied to a variety of practical situations: Okubo (1980) gives other examples. The definitive review by Okubo (1986) discusses various models and specifically addresses animal grouping, insect swarms, flocking and so on.

9.4 Chemotaxis

A large number of insects and animals rely on an acute sense of smell for conveying information between members of the species. Chemicals which are involved in this process are called *pheromones*. For example, the female silk moth *Bombyx mori* exudes a pheromone, called bombykol, as a sex attractant for the male, which has a remarkably efficient antenna filter to measure the bombykol concentration, and it moves in the direction of increasing concentration. The modelling problem here is a fascinating and formidable one (Murray 1977). The acute sense of smell of many deep sea fish is particularly important for communication and predation. Other than for territorial demarcation the simplest important exploitation of pheromone release is the directed movement it can generate in the population. Here we model this chemically directed movement, that is *chemotaxis*, which, unlike diffusion, directs the motion *up* a concentration gradient.

It is not only in animal and insect ecology that chemotaxis is important. It can be equally crucial in biological processes where there are numerous examples. For example when a bacterial infection invades the body it may be attacked by movement of cells towards the source as a result of chemotaxis. Convincing evidence suggests that leukocyte cells in the blood move towards a region of bacterial inflammation, to counter it, by moving up a chemical gradient caused by the infection (see, for example, Lauffenburger and Keller 1979, Tranquillo and Lauffenburger 1986, 1988, Alt and Lauffenburger 1987).

A widely studied chemotactic phenomenon is that exhibited by the slime mold *Dictyostelium discoideum* where single-cell amoebae move towards regions of relatively high concentrations of a chemical called cyclic-AMP which is produced by the amoebae themselves. Interesting wave-like movement and spatial patterning are observed experimentally: see Chapter 12 below. A discussion of the phenomenon and some of the mathematical models which have been proposed together with some analysis are given, for example, in the book by Segel (1984). The kinetics involved have been modelled by several authors. As more was found out about the biological system the models changed. Recently new, more complex and more biologically realistic models have been proposed by Martiel and Goldbeter (1987) and Monk and Othmer (1989). Both of these new models exhibit oscillatory behaviour.

Let us suppose that the presence of a gradient in an attractant, $a(\mathbf{x}, t)$, gives rise to a movement, of the cells say, up the gradient. The flux of cells will increase with the number of cells, $n(\mathbf{x}, t)$, present. Thus we may reasonably take as the chemotactic flux

$$\mathbf{J} = n\chi(a)\nabla a , \qquad (9.27)$$

where $\chi(a)$ is a function of the attractant concentration. In the general conservation equation for $n(\mathbf{x}, t)$, namely

$$\frac{\partial n}{\partial t} + \nabla \cdot \mathbf{J} = f(n) ,$$

where $f(n)$ represents the growth term for the cells, the flux

$$\mathbf{J} = \mathbf{J}_{\text{diffusion}} + \mathbf{J}_{\text{chemotaxis}}$$

where the diffusion contribution is from (9.15) with the chemotaxis flux from (9.27). Thus the *reaction* (or *population*) *diffusion-chemotaxis equation* is

$$\frac{\partial n}{\partial t} = f(n) - \nabla \cdot n\chi(a)\nabla a + \nabla \cdot D\nabla n . \qquad (9.28)$$

where D is the diffusion coefficient of the cells.

Since the attractant $a(\mathbf{x}, t)$ is a chemical it also diffuses and is produced, by the amoebae for example, so we need a further equation for $a(\mathbf{x}, t)$. Typically

$$\frac{\partial a}{\partial t} = g(a, n) + \nabla \cdot D_a \nabla a , \qquad (9.29)$$

where D_a is the diffusion coefficient of a and $g(a, n)$ is the kinetics/source term, which may depend on n and a. Normally we would expect $D_a > D$. If several species or cell types all respond to the attractant the governing equation for the species vector is an obvious generalization of (9.28) to a vector form with $\chi(a)$ probably different for each species.

In the slime mold model of Keller and Segel (1971), $g(a,n) = hn - ka$ where h,k are positive constants. Here hn represents the spontaneous production of the attractant and is proportional to the number of amoebae n, while $-ka$ represents decay of attractant activity: that is there is an exponential decay if the attractant is not produced by the cells.

One simple version of the model has $f(n) = 0$: that is the amoebae production rate is negligible. This is the case during pattern formation phase in the mold's life cycle. The chemotactic term $\chi(a)$ is taken to be a positive constant χ_0. The form of this term in any case is speculative. With constant diffusion coefficients, together with the above linear form for $g(a,n)$, the model in one space dimension becomes the nonlinear system

$$\frac{\partial n}{\partial t} = D\frac{\partial^2 n}{\partial x^2} - \chi_0 \frac{\partial}{\partial x}\left(n\frac{\partial a}{\partial x}\right) ,$$

$$\frac{\partial a}{\partial t} = hn - ka + D_a \frac{\partial^2 a}{\partial x^2} ,$$

$$(9.30)$$

which we study later in Chapter 12. There we consider n to be a bacterial population and a the food which it consumes.

Other forms have been proposed for the chemotactic factor $\chi(a)$. For example

$$\chi(a) = \frac{\chi_0}{a}, \quad \chi(a) = \frac{\chi_0 K}{(K+a)^2}, \quad \chi_0 > 0, \quad K > 0 \qquad (9.31)$$

which are known respectively as the log law and receptor law. In these, as a decreases the chemotactic effect increases.

There are various ways to define a practical measureable *chemotaxis index* I which reflects the strength of the chemoattractant. To be specific consider the planar movement of a cell, say, towards a source of chemoattractant at position x_s. Suppose the cell starts at x_A and the source is distance D_1 away. In the absence of chemotaxis the cell's movement is purely random and the mean distance, D_2 say, that the cell moves in a given time T in the direction of x_s is zero. In the presence of chemotaxis the random movement is modified so that there is a general tendency for the cell to move towards the chemoattractant source and over the same time T, $D_2 > 0$. We can define the index $I = D_2/D_1$: the larger I the stronger the chemotaxis. Tranquillo and Lauffenburger (1988) have analysed the detailed chemosensory movement of leukocyte cells with a view to determining its chemotaxis parameters.

The movement of certain cells can be influenced by the presence of applied electric fields and the cells tend to move in a direction parallel to the applied field. This is called *galvanotaxis*. The strength of galvanotaxis can be defined in a similar way to chemotaxis. If V is an electric potential the galvanotaxis flux $\mathbf{J}$ of cells can reasonably be taken as proportional to $nG(V)\nabla V$ where G may be a function of the applied voltage V.

Before leaving this topic, note the difference in sign in (9.28) and (9.30) in the diffusion and chemotaxis terms. Each has a Laplacian contribution. Whereas diffusion is generally a stabilizing force, chemotaxis is generally *destabilizing*, like a kind of negative diffusion. At this stage therefore it is reasonable to suppose that the balance between stabilizing and destabilizating forces in the model system (9.30) could result in some steady state spatial patterns in n and a, or in some unsteady wave-like spatially heterogeneous structure. That is, nonuniform spatial patterns in the cell density appear: see Chapters 12 and 13 below.

*9.5 Non-local Effects and Long Range Diffusion

The classical approach to diffusion, which we have used above is strictly only applicable to dilute systems, that is where the concentrations c, or densities n, are small. Its applicability in practice is much wider than this of course, and use of the Fickian form (9.15) for the diffusional flux, namely $\mathbf{J} = -D\nabla c$, or $\mathbf{J} = -D(n)\nabla c$ from (9.19) in which the diffusion is dependent on n, is usually sufficient for most practical modelling purposes. What these forms in effect imply, is that diffusion is a *local* or *short range* effect. We can see this if we consider the Laplacian operator $\nabla^2 n$ in the simple diffusion equation $\partial n/\partial t = D\nabla^2 n$. The Laplacian averages the neighbouring densities and formally (see, for example, Hopf 1948, Morse and Feshbach 1953)

$$\nabla^2 n \propto \frac{\langle n(\mathbf{x},t)\rangle - n(\mathbf{x},t)}{R^2}, \quad \text{as} \quad R \to 0 \tag{9.32}$$

where $\langle n \rangle$ is the average density in a sphere of radius R about $\mathbf{x}$, that is

$$n_{av} = \langle n(\mathbf{x},t)\rangle \equiv \left[\frac{3}{4\pi R^3}\right] \int_V n(\mathbf{x}+\mathbf{r},t)\, d\mathbf{r} \tag{9.33}$$

where V is the sphere of radius R. This interpretation of the Laplacian was first suggested by James Clerk Maxwell in 1871 (see, J.C. Maxwell, Scientific Papers. Dover, New York 1952, p.264).

Because the radius $R \to 0$ we can expand $n(\mathbf{x}+\mathbf{r},t)$ in a Taylor series about $\mathbf{x}$ for small $\mathbf{r}$, namely

$$n(\mathbf{x}+\mathbf{r},t) = n(\mathbf{x},t) + (\mathbf{r}\cdot\nabla)n + \frac{1}{2}(\mathbf{r}\cdot\nabla)^2 n + \ldots$$

and substitute this into the integral in (9.33) for n_{av} to get

$$n_{av} = \left[\frac{3}{4\pi R^3}\right]\int_V \left[n(\mathbf{x},t) + (\mathbf{r}\cdot\nabla)n + \frac{1}{2}(\mathbf{r}\cdot\nabla)^2 n + \ldots\right] d\mathbf{r} .$$

Because of the symmetry the second integral is zero. If we neglect all terms $O(r^3)$ and higher in the integrand, integration gives

$$
n_{av} = \left(\frac{3}{4\pi R^3}\right) \left[n(\mathbf{x},t) \int_V d\mathbf{r} + \nabla^2 n(\mathbf{x},t) \int_V \frac{r^2}{2} d\mathbf{r} \right]
$$

$$
= n(\mathbf{x},t) + \frac{3}{10} R^2 \nabla^2 n(\mathbf{x},t) . \tag{9.34}
$$

If we now substitute this into the expression (9.32) we see that the proportionality factor is $10/3$.

In many biological areas, such as embryological development, the densities of cells involved are not small and a local or short range diffusive flux proportional to the gradient is not sufficiently accurate. When we discuss the mechanical approach to generating biological pattern in Chapter 17 we shall see how important and intuitively necessary it is to include long range effects.

Instead of simply taking $\mathbf{J} \propto \nabla n$ we now consider

$$
\mathbf{J} = \underset{\mathbf{r} \in N(\mathbf{x})}{G} [\nabla n(\mathbf{x}+\mathbf{r},t)] \tag{9.35}
$$

where $N(\mathbf{x})$ is some neighbourhood of the point $\mathbf{x}$ over which effects are noticed at $\mathbf{x}$, and G is some functional of the gradient. From symmetry arguments and assumptions of isotropy in the medium we are modelling, be it concentration or density, it can be shown that the first correction to the simple linear ∇n for the flux $\mathbf{J}$ is a $\nabla(\nabla^2 n)$ term. The resulting form for the flux in (9.35) is then

$$
\mathbf{J} = -D_1 \nabla n + \nabla D_2 (\nabla^2 n) , \tag{9.36}
$$

where $D_1 > 0$ and D_2 are constants. D_2 is a measure of the long range effects and in general is smaller in magnitude than D_1. This approach is due to Othmer (1969), who goes into the formulation, derivation and form of the general functional G in detail. We give different motivations for the long range D_2-term below and in Section 9.6.

If we now take the flux $\mathbf{J}$ as given by (9.36) and use it in the conservation equation ((9.14) with $f \equiv 0$) we get

$$
\frac{\partial n}{\partial t} = -\nabla \cdot \mathbf{J} = \nabla \cdot D_1 \nabla n - \nabla \cdot \nabla(D_2 \nabla^2 n) . \tag{9.37}
$$

In this form, using (9.32), we can see that whereas the first term represents an average of nearest neighbours, the second – the *biharmonic term* – is a contribution from the *average of nearest averages*.

The biharmonic term is stabilizing if $D_2 > 0$, or de-stabilizing if $D_2 < 0$. We can see this if we look for solutions of (9.37) in the form

$$
n(\mathbf{x},t) \propto \exp[\sigma t + i\mathbf{k} \cdot \mathbf{x}], \quad k = |\mathbf{k}| \tag{9.38}
$$

which represents a wave-like solution with wave vector $\mathbf{k}$ (and so has a wavelength $2\pi/k$). Since (9.37) is a linear equation we can use this last solution to obtain the solution to the general initial value problem using an appropriate Fourier series or integral technique. Substitution of (9.38) into equation (9.37) gives what is called the *dispersion relation* for σ in terms of the wave number k as

$$\sigma = -D_1 k^2 - D_2 k^4 . \tag{9.39}$$

The growth or decay of the solution is determined by $\exp[\sigma t]$ in (9.38). Dispersion relations are very important in many different contexts. We discuss some of these in detail in later chapters: see Section 14.5 in particular. With σ as a function of k, the solution (9.38) shows the time behaviour of each wave, that is for each k. In fact on substituting (9.39) into the solution (9.38) we see that

$$n(\mathbf{x}, t) \propto \exp\left[-(D_1 k^2 + D_2 k^4)t + i\mathbf{k} \cdot \mathbf{x}\right]$$

so, for large enough wavenumbers k, $k^2 > D_1/|D_2|$ in fact, we always have

$$n(x, t) \rightarrow \begin{cases} 0 \\ \infty \end{cases} \quad \text{as} \quad t \rightarrow \infty \quad \text{if} \quad D_2 \begin{cases} > 0 \\ < 0 \end{cases} \tag{9.40}$$

In classical Fickian diffusion $D_2 \equiv 0$ and $n \rightarrow 0$ as $t \rightarrow \infty$ for all k. From (9.40) we see that if $D_2 > 0$ the biharmonic contribution (that is the long range diffusion effect) to the diffusion process is stabilizing, while it is destabilizing if $D_2 < 0$.

Another important concept and approach to modelling long range effects uses an integral equation formulation. (This approach provides a useful unifying concept we shall come back to later when we consider a specific class of models for the generation of steady state spatial patterns: see Chapter 16.) Here the rate of change of n at position x at time t depends on the influence of neighbouring n at all other positions x'. A model, in one space dimension for example, is represented mathematically by

$$\frac{\partial n}{\partial t} = f(n) + \int_{-\infty}^{\infty} w(x - x')n(x', t) \, dx' , \tag{9.41}$$

where $w(x - x')$ is the *kernel function* which quantifies the effect the neighbouring $n(x', t)$ has on $n(x, t)$. The form here assumes that the influence depends only on the distance from x to x'. The function $f(n)$ is the usual source or kinetics term – the same as we included in the reaction diffusion mechanisms (9.17) and (9.18). We assume reasonably that the influence of neighbours tends to zero for $|x - x'|$ large and that this influence is spatially symmetric; that is

$$w \rightarrow 0 \quad \text{as} \quad |x - x'| \rightarrow \infty, \quad w(x - x') = w(x' - x) . \tag{9.42}$$

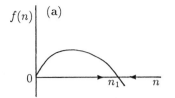
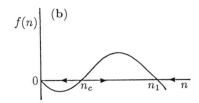

Fig. 9.4a,b. (a) Typical firing rate function with a single non-zero steady state: $n \to n_1$ in the absence of spatial effects. (b) A typical multi-steady state firing function. If $n < n_c$, a critical firing rate, then $n \to 0$, that is extinction. If $n > n_c$ then $n \to n_1$.

Such a model (9.41) directly incorporates long range effects through the kernel: if w tends to zero quickly, for example like $\exp[-(x - x')^2/s]$ where $0 < s \ll 1$, then the long range effects are weak, whereas if $s \gg 1$ they are strong.

To determine the spatio-temporal properties of the solutions of (9.41) the kernel w has to be specified. This involves modelling the specific biological phenomenon under consideration. Suppose we have neural cells which are cells which can fire spontaneously; here n represents the cells' firing rate. Then $f(n)$ represents the autonomous spatially independent firing rate, and, in the absence of any neighbouring cells' influence, the firing rate simply evolves to a stable steady state, determined by the zeros of $f(n)$. The mathematics is exactly the same as we discussed in Chapter 1. For example if $f(n)$ is as in Fig. 9.4 (a) the rate evolves to the single steady state firing rate n_1. If $f(n)$ is as in Fig. 9.4 (b) then there is a threshold firing rate above which n goes to a non-zero steady state and below which it goes to extinction.

If we now incorporate spatial effects we must include the influence of neighbouring cells, that is we must prescribe the kernel function w. Suppose we assume that the cells are subjected to both excitatory and inhibitory inputs from neighbouring cells, with the strongest excitatory signals coming from the cells themselves. That is if a cell is in a high firing state n tends to increase; it is like autocatalysis. A kernel which incorporates such behaviour is illustrated in Fig. 9.5.

We can relate this integral equation approach to the long range diffusion approximation which gave (9.37). Let

$$y = x - x' \quad \Rightarrow \quad \int_{-\infty}^{\infty} w(x - x')n(x', t)\, dx' = \int_{-\infty}^{\infty} w(y)n(x - y, t)\, dy \ .$$

If we now expand $n(x - y)$ about x as a Taylor series, as we did for the integral in (9.33),

$$\int_{-\infty}^{\infty} w(x - x')n(x', t)\, dx' = \int_{-\infty}^{\infty} w(y)\left[n(x, t) - y\frac{\partial n(x, t)}{\partial x} + \frac{y^2}{2}\frac{\partial^2 n(x, t)}{\partial x^2}\right.$$
$$\left. -\frac{y^3}{3!}\frac{\partial^3 n(x, t)}{\partial x^3} + \frac{y^4}{4!}\frac{\partial^4 n(x, t)}{\partial x^4} - \ldots\right] dy \ . \quad (9.43)$$

Because of the assumed symmetry of the kernel $w(y)$,

$$\int_{-\infty}^{\infty} y^{2m+1} w(y)\, dy = 0, \quad m = 0, 1, 2, \ldots \tag{9.44}$$

If we now define the *moments* w_m of the kernel $w(y)$ by

$$w_{2m} = \frac{1}{(2m)!} \int_{-\infty}^{\infty} y^{2m} w(y)\, dy, \quad m = 0, 1, 2, \ldots \tag{9.45}$$

equation (9.41) becomes

$$\frac{\partial n}{\partial t} = f(n) + w_0 n + w_2 \frac{\partial^2 n}{\partial x^2} + w_4 \frac{\partial^4 n}{\partial x^4} + \ldots \tag{9.46}$$

Higher moments of typical kernels get progressively smaller: this is intutively clear from (9.45). If we truncate the series in (9.46) at the 4th moment we get an approximate model equation with a biharmonic $\partial^4 n / \partial x^4$ contribution comparable to that in (9.37).

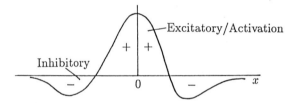

Fig. 9.5. Typical excitatory-inhibitory kernel w for spatial influence of neighbours in the model (9.41).

The solution behaviour of (9.41) depends crucially on the signs of the kernel moments and hence on the detailed form of the kernel, one typical form of which is shown qualitatively in Fig. 9.5. For example, if $w_2 < 0$ the 'short range diffusion' term is destabilizing, and if $w_4 < 0$ the 'long range diffusion' is stabilizing (cf.(9.39)).

This integral equation approach is in many ways a much more satisfactory way to incorporate long range effects since it reflects, in a more descriptive way, what is going on biologically. As we have said above we shall discuss such models in depth in Chapter 16.

*9.6 Cell Potential and Energy Approach to Diffusion

We now discuss an alternative approach to motivate the higher order, long range diffusion terms. To be specific, in the following we shall have cell population densities in mind and for pedagogical reasons give the derivation of the classical (Fickian) diffusion before considering the more general case. The treatment here follows that given by Cohen and Murray (1981).

In general phenomenological terms if there is a gradient in a potential μ it can drive a flux $\mathbf{J}$ which, classically, is proportional to $\nabla\mu$. We can, still in classical terms, think of the potential as the work done in changing the state by a small amount or, in other words, the variational derivative of an energy. Let $n(\mathbf{x},t)$ be the cell density. We associate with a spatial distribution of cells, an energy density $e(n)$, that is an internal energy per unit volume of an evolving spatial pattern so that the total energy $E[n]$ in a volume V is given by

$$E[n] = \int_V e(n)\, \mathbf{dx} \; . \tag{9.47}$$

The change in energy δE, that is the work done in changing states by an amount δn, is the variational derivative $\delta E/\delta n$, which defines a potential $\mu(n)$. So

$$\mu(n) = \frac{\delta E}{\delta n} = e'(n) \; . \tag{9.48}$$

The gradient of the potential μ produces a flux $\mathbf{J}$; that is the flux $\mathbf{J}$ is proportional to $\nabla\mu$ and so

$$\mathbf{J} = -D\nabla\mu(n) \; , \tag{9.49}$$

where D is the proportionality parameter, which in this derivation may depend on $\mathbf{x}$, t and n. The continuity equation for n becomes

$$\frac{\partial n}{\partial t} = -\nabla \cdot \mathbf{J} = \nabla \cdot [D\nabla\mu(n)] = \nabla \cdot [De''(n)\nabla n] \tag{9.50}$$

on using (9.48) for $\mu(n)$, and so

$$\frac{\partial n}{\partial t} = \nabla \cdot [D^*(n)\nabla n] \tag{9.51}$$

where

$$D^*(n) = De''(n) \; . \tag{9.52}$$

In the simple classical diffusion situation with constant diffusion, the internal energy density is the usual quadratic with $e(n) = n^2/2$. With this, $\mu(n) = n$ and (9.51) becomes the usual diffusion equation $\partial n/\partial t = D\nabla^2 n$, with $D^* = D$, the constant diffusion coefficient. If D is a function of $\mathbf{x}$, t and n, the derivation is the same and the resulting conservation equation for n is then

$$\frac{\partial n}{\partial t} = \nabla \cdot [D^*(\mathbf{x}, t, n) \nabla n] . \qquad (9.53)$$

Here n can be a vector of cell species.

This derivation assumes that the energy density $e(n)$ depends only on the density n. If the cells are sensitive to the environment other than in their immediate neighbourhood, it is reasonable to suppose that the energy required to maintain a spatial heterogeneity depends on neighbouring gradients in the cell density. It is the spatial heterogeneity which ultimately is of importance in biological pattern formation.

We take a more realistic energy functional, which is chosen so as to be invariant under reflexions $(x_i \to -x_i)$ and rotations $(x_i \to x_j)$, as

$$E[n] = \int_V [e(n) + k_1 \nabla^2 n + k_2 (\nabla n)^2 + \ldots] \, \mathbf{dx} . \qquad (9.54)$$

where the k's may be functions of n. Using Green's theorem

$$\int_V k_1 \nabla^2 n \, \mathbf{dx} + \int_V \nabla k_1 \cdot \nabla n \, \mathbf{dx} = \int_S k_1 \frac{\partial n}{\partial N} \, \mathbf{ds}$$

where $\mathbf{N}$ is the outward pointing normal to the surface S which encloses V and where we let k_1 depend on n so that $\nabla k_1 = k_1'(n) \nabla n$. From the last equation

$$\int_V k_1 \nabla^2 n \, \mathbf{dx} = -\int_V k_1'(n)(\nabla n)^2 \, \mathbf{dx} + \int_S k_1 \frac{\partial n}{\partial N} \, \mathbf{ds} . \qquad (9.55)$$

We are not concerned with effects at the external boundary, so we can choose the bounding surface S such that $\partial n/\partial N = 0$ on S; that is zero flux at the boundary. So (9.54) for the energy functional in a spatially hetereogeneous situation becomes

$$E[n] = \int_V \left[e(n) + \frac{k}{2}(\nabla n)^2 + \ldots \right] \, \mathbf{dx} ,$$
$$\frac{k}{2} = -k_1'(n) + k_2 . \qquad (9.56)$$

Here $e(n)$ is the energy density in a spatially homogeneous situation with the other terms representing the energy density (or 'gradient' density) which depends on the neighbouring spatial density variations.

We now carry out exactly the same steps that we took in going from (9.48) to (9.53). The potential μ is obtained from the energy functional (9.56) as

$$\mu = \mu(n, \nabla n) = \frac{\delta E[n]}{\delta n} = -k \nabla^2 n + e'(n) , \qquad (9.57)$$

using the calculus of variations to evaluate $\delta E[n]/\delta n$ and where we have taken k to be a constant. The flux $\mathbf{J}$ is now given by

$$\mathbf{J} = -D^*\nabla\mu(n, \nabla n) .$$

The generalized diffusion equation is then

$$\frac{\partial n}{\partial t} = -\nabla \cdot \mathbf{J} = \nabla \cdot (D^*\nabla\mu)$$

$$= D^*\nabla^2[-\nabla^2 n + e'(n)]$$

$$= -kD^*\nabla^4 n + D^*\nabla \cdot [e''(n)\nabla n] . \tag{9.58}$$

Here we have taken D^*, as well as k, to be constant.

A basic assumption about $e(n)$ is that it can involve only even powers of n since the energy density cannot depend on the sign of n. The Landau-Ginzburg free energy form (see, for example, Cahn and Hilliard 1958, 1959, Cahn 1959, Huberman 1976) has

$$e(n) = \frac{an^2}{2} + \frac{bn^4}{4} ,$$

which on substituting into (9.58) gives

$$\frac{\partial n}{\partial t} = -D^*k\nabla^4 n + D^*a\nabla^2 n + D^*b\nabla^2 n^3 .$$

If we now write

$$D_1 = D^*a, \quad D_2 = D^*k, \quad D_3 = D^*b ,$$

the generalized diffusion equation (9.58) becomes

$$\frac{\partial n}{\partial t} = D_1\nabla^2 n - D_2\nabla^4 n + D_3\nabla^2 n^3 . \tag{9.59}$$

Note the appearance of the extra nonlinear term involving D_3. If the energy $e(n)$ only involves the usual quadratic in n^2, $b = 0$ and (9.59) is exactly the same as (9.37) in Section 9.4. If we now include a reaction or dynamics term $f(n)$ in (9.59) we get the generalized reaction diffusion equation equivalent to (9.14). With the one space dimensional scalar version of (9.59) and a logistic growth form for $f(n)$, Cohen and Murray (1981) have shown that the equation can exhibit steady state spatially inhomogeneous solutions. Lara Ochoa (1984) has analysed their model in a two-dimensional setting and shown that it reflects certain morphogenetic aspects of multicellular systems formed by motile cells.

Exercises

1. Let $p(x, t)$ be the probability that an organism initially at $x = 0$ is at x after a time t. In a random walk there is a slight bias to the right, that is the probabilities of moving to the right and left, α and β, are such that $\alpha - \beta = \varepsilon > 0$ where $0 < \varepsilon \ll 1$. Show that the diffusion equation for the concentration $c(x, t) = Qp(x, t)$ where Q particles are released at the origin at $t = 0$ is

$$\frac{\partial c}{\partial t} + V \frac{\partial c}{\partial x} = D \frac{\partial^2 c}{\partial x^2},$$

    where V and D are constants which you should define.

2. In a one-dimensional domain suppose insects are attracted to the origin $x = 0$ and are convected there by a constant velocity V. If the population pressure is approximated by a density dependent diffusion coefficient $D(n) = D_0(n/n_0)^m$ where n is the population density and D_0, n_0 and m are positive constants, show that the model equation for dispersal, in the absence of any population growth, is

$$\frac{\partial n}{\partial t} = -\frac{\partial J}{\partial x} = \frac{\partial}{\partial x}[V \operatorname{sgn}(x)n] + D_0 \frac{\partial}{\partial x}\left[\left(\frac{n}{n_0}\right)^m \frac{\partial n}{\partial x}\right].$$

    Show that if $n \to 0$, $\partial n/\partial x \to 0$ as $|x| \to \infty$ a steady state, spatially inhomogeneous population density exists and can be represented by

$$n(x) = n_0 \left(1 - \frac{mV|x|}{D_0}\right)^{1/m}, \quad \text{if} \quad |x| \leq \frac{D_0}{mV}$$

$$= 0, \quad \text{if} \quad |x| > \frac{D_0}{mV}$$

3. The larvae of the parasitic worm (*Trichostrongylus retortaeformis*) hatch from eggs in sheep and rabbit excreta. The larvae disperse randomly on the grass and are consequently eaten by sheep and rabbits. In the intestines the cycle starts again. Consider the one-dimensional problem in which the larvae disperse with constant diffusion and have a mortality proportional to the population. Show that n satisfies

$$\frac{\partial n}{\partial t} = D \frac{\partial^2 n}{\partial x^2} - \mu n, \quad D > 0, \quad \mu > 0$$

    where n is the larvae population. Find the population distribution at any x and t arising from N_0 larvae being released at $x = 0$ at $t = 0$. Show that as $t \to \infty$ the population dies out.

If the larvae lay eggs at a rate proportional to the population of the larvae, that is

$$\frac{\partial E}{\partial t} = \lambda n, \quad \lambda > 0$$

where $E(x,t)$ is the egg population density, show that in the limit as $t \to \infty$ a nonzero spatial distribution of eggs persists. [The result for $E(x,t)$ is an integral from which the asymptotic approximation can be found using Laplace's method (see for example, Murray's (1984) *Asymptotic Analysis*): the result gives $E(x,t) \sim O(\exp[-(\mu/D)^{1/2}|x|])$ as $t \to \infty$].

4. Consider the density dependent diffusion model for insect dispersal which includes a linear death process which results in the following equation for the population $n(x,t)$:

$$\frac{\partial n}{\partial t} = D_0 \frac{\partial}{\partial x}\left[\left(\frac{n}{n_0}\right)^m \frac{\partial n}{\partial x}\right] - \mu n, \quad D_0 > 0, \quad \mu > 0.$$

If Q insects are released at $x = 0$ at $t = 0$, that is $n(x,0) = Q\delta(x)$, show, using appropriate transformations in n and t, that the equation can be reduced to an equivalent equation with $\mu = 0$. Hence show that the population wave front reaches a finite distance $x_{\max}$ from $x = 0$ as $t \to \infty$, where

$$x_{\max} = \frac{r_0}{(\mu m \tau_0)^{1/(m+2)}},$$

where

$$r_0 = \frac{Q\Gamma\left(\frac{1}{m} + \frac{3}{2}\right)}{\pi^{1/2} n_0 \Gamma\left(\frac{1}{m} + 1\right)}, \quad \tau_0 = \frac{r_0^2 m}{2D_0(m+2)}.$$

10. Oscillator Generated Wave Phenomena and Central Pattern Generators

In Chapter 9 we saw how diffusion, chemotaxis and convection mechanisms could generate spatial patterns: in later chapters we discuss mechanisms of biologiocal pattern formation extensively. In Chapters 11, 12 and 20 we shall see how diffusion effects, for example, can also generate travelling waves, which have been used to model the spread of pest outbreaks, travelling waves of chemical concentration, colonization of space by a population, spatial spread of epidemics and so on. The existence of such travelling waves is usually a consequence of the coupling of various effects such as diffusion or chemotaxis or convection. There are, however, other wave phenomena of a quite different kind, called *kinematic waves*, which exhibit wave-like spatial patterns, which depend on the coupling of biological oscillators whose properties relating to phase or period vary spatially. The two phenomena described in this chapter are striking, and the models we shall discuss are based on the experiments or biological phenomena which so dramatically exhibit them. The first involves the Belousov-Zhabotinskii reaction and the second, which is specifically associated with the swimming of, for example, lamprey and dogfish, illustrates the very important concept of a *Central Pattern Generator*. The results we derive here apply to spatially distributed oscillators in general.

10.1 Kinematic Waves in the Belousov-Zhabotinskii Reaction

When the reactants in the oscillating Belousov reaction involve an iron catalyst (with Fe^{2+} going to Fe^{3+} and vice versa) the oscillations are dramatically illustrated with an appropriate dye which reflects the state of the catalyst: the colour change is from red (or rather a reddish orange) to blue. When the reactants are left unstirred in a vertical cylindrical tube horizontal bands of blue and red form. These bands usually start to appear at the bottom of the cylinder and move slowly upwards with successive bands moving progressively more slowly. Eventually the cylinder is filled by these bands but with a non-uniform density, the closer to the bottom the denser the wave packing. Diffusion plays a negligible role in the formation and propagation of these bands, unlike the waves we shall discuss later. Beck and Váradi (1972) provided a kinematical explanation for these spatial patterns of bands. The analysis explaining them, which we give here, is that of Kopell and Howard (1973). Although the analysis was originally given for the bands observed in the Belousov reaction, and the experimental results shown in Fig. 10.1 (b) are also for this reaction, the phenomenon and analysis applies equally to any biological oscillator under similar circumstances. The important point to note is that spatial patterns can be obtained without diffusion, convection or chemotaxis playing any role.

Consider each position in the vertical cylinder to be an independent oscillator with period T, which may be a function of position. If these independent oscillators are out of phase or have different frequencies then spatial patterns will appear simply as a consequence of the spatial variation in the phase or frequency. (A simple but illustrative physical demonstration of the phenomenon is given by a row of simple pendula all hanging from the same horizontal rod but with a very slight gradient in their lengths. The slight gradient in their lengths gives a slight gradient in their periods. If they are all set swinging at the same time, then after a very short time it looks as if there is a wave propagating along the line of pendula, the wave length of which gets smaller and smaller with time.)

Returning to the Belousov oscillator, the cause of a gradient in phase or frequency can be due to a concentration gradient in one of the chemicals, or a temperature variation. The experiments (see Fig. 10.1 (b)) carried out by Kopell and Howard (1973) used the former while Thoenes (1973) used the latter. The vertical chemical concentration gradient was in sulphuric acid. This resulted in a monotonic gradient in the period of oscillation and horizontal bands appeared quite quickly, moved slowly upwards and after a few minutes filled the cylinder. It is clear that if a barrier, impermeable to any of the chemicals, were put in the cylinder it would neither affect the pseudo-wave propagation nor the density of bands: spatial transport processes are simply not involved in the generation of this spatial pattern. These 'waves' are indeed only pseudo-waves since nothing is actually being transported.

Let z be the spatial coordinate measured vertically from the bottom of the cylinder, taken to be $z = 0$, and the cylinder height to be normalized so that the top is $z = 1$. Because of the initial concentration gradient there is a gradient in the oscillator period. At position z let the period of oscillation be $T(z)$ defined for all $0 \le z \le 1$. We characterize the state of the oscillator by a 2π-periodic function of its phase denoted by $\phi(z, t)$. In the Belousov reaction, for example, the front of the wave, defined as the point where ϕ has a specific value, can be distinguished by the sharp blue front. Let the initial distribution of the phases be $\phi_0(z)$. We can then represent the phase $\phi(z, t)$ by

$$\phi(z, t) = \psi(z, t) + \phi_0(z), \quad \psi(z, 0) = 0 , \qquad (10.1)$$

where $\psi(z, t)$ is a function which increases by 2π if the time t increases by the periodic time $T(z)$: that is

$$\phi(z, t + T(z)) = 2\pi + \psi(z, t) + \phi_0(z) .$$

Let us now take some reference phase point say $\phi = 0$. Define $t^*(z)$ as the time at position z at which the phase is zero: that is it satisfies

$$0 = \phi(z, t^*(z)) = \psi(z, t^*(z)) + \phi_0(z) . \tag{10.2}$$

Then, for any integer n and time $t = t^*(z) + nT(z)$ we have

$$\phi(z, t^* + nT) = \psi(z, t^* + nT) + \phi_0(z)$$
$$= 2n\pi + \psi(z, t^*) + \phi_0(z)$$
$$= 2n\pi , \tag{10.3}$$

using (10.1) and (10.2). So, in the (z, t) plane the point (z, t) which corresponds to the phase $2n\pi$ moves on the curve given by

$$t = t^*(z) + nT(z) . \tag{10.4}$$

We can continue the analysis with complete generality but it is just as instructive and easier to see what is going on if we are more specific and choose $T(z)$ to be, say, a smooth monotonic increasing function of z in $0 \le z \le 1$. Also for simplicity, let us take the initial distribution of phases to be a constant, which we can take to be zero, that is $\phi_0(z) = 0$. This means that at $t = 0$ all the oscillators are in phase. From the definition of $t^*(z)$ in (10.2) this means that $t^*(z) = 0$.

If we now define

$$t_n(z) = nT(z) \tag{10.5}$$

then (10.3) gives

$$\phi(z, t_n(z)) = 2n\pi . \tag{10.6}$$

This means that $t_n(z)$ is the time at which the n-th wavefront passes the point z in the cylinder. The velocity $v_n(z)$ of this n-th wavefront is given by the rate of change of the position of the front. That is, using the last two equations,

$$v_n(z) = \left[\frac{dz}{dt} \right]_{\phi = 2n\pi} = \left[\frac{dt_n(z)}{dz} \right]^{-1} = \frac{1}{nT'(z)} . \tag{10.7}$$

With $T(z)$ a monotonic increasing function of z, $T'(z) > 0$ and so the n-th wavefront, the leading edge say, starts at $z = 0$ at time $t = nT(0)$ and, from (10.7), propagates up the cylinder at $1/n$ times the velocity of the first wave. This n-th wave reaches $z = 1$, the top of the cylinder, at time $t = nT(1)$. Since the $T(z)$ we have taken here is a monotonic increasing function of z there will be more and more waves in $0 \le z \le 1$ as time goes on, since, with the velocity decreasing proportionally to $1/n$ from (10.7), more waves enter at $z = 0$ than leave $z = 1$ in the same time interval.

Let us consider a specific example and take $\phi_0(z) = 0$ and $T(z) = 1 + z$. So the phase $2n\pi$ moves, in the (z, t) plane, on the lines $t = n(1 + z)$, and the phase $\phi(z,t) = 2\pi t/T(z) = 2\pi t/(1+z)$. From (10.7) the velocity of the n-th wave is $v_n = 1/n$. The space-time picture of the wavefronts, given by $\phi = 2n\pi$, is illustrated in Fig. 10.1 (a). From the figure we see that at time $t = 1$ the wave $\phi = 2\pi$ enters the cylinder at $z = 0$ and moves up with a velocity $v_1 = 1$. At $t = 2$ the wave with phase 4π enters the cylinder at $z = 0$: it moves with velocity $v_2 = 1/2$. At $t = 3$ the wave with $\phi = 6\pi$ moves with velocity $v_3 = 1/3$ and so on. The first wave takes a time $t = 1$ to traverse the cylinder, the second takes a time $t = 2$ and so on. It is clear that as time goes on more and more waves are in $0 \le z \le 1$. This is clear from Fig. 10.1 (a) and the experimental counterpart given in Fig. 10.1 (b): for example at $t = 3.5$ there are two waves while at $t = 7.5$ there are 4. From the figure, it is also clear that as time increases the waves are progressively more tightly packed nearer the bottom, $z = 0$.

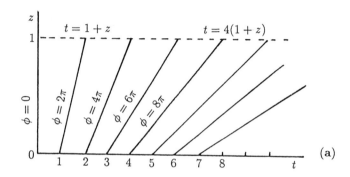

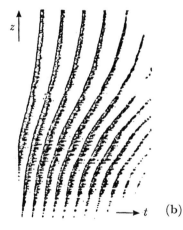

Fig. 10.1a,b. (a) Wavefronts for the period distribution $T(z) = 1 + z$ and initial phase distribution $\phi_0(z) = 0$. Note how many waves there are for different t's: for $t = 1.5$ there is 1 wave in the cylinder, while for $t = 7.5$ there are 4 waves. (b) The experimental space-time situation equivalent to the theoretical results in (a). The experiments were carried out for a Belousov reaction with an initial sulphuric acid gradient in the cylinder. The total time in the figure is approximately 7 minutes and the vertical height about 20 cm: the figure is a sketch from a negative. (After Kopell and Howard 1973)

Suppose the initial phases $\phi_0(z) \ne 0$ then $t^*(z) \ne 0$ and from (10.4), $T(z) = [t - t^*(z)]/n$. So, asymptotically for large time t, $T(z) \sim t/n$, and the above analysis for the illustrative example in which $\phi_0(z) = 0$ still applies asymptotically in time. Since it is unlikely that the experimental arrangement which gave rise to Fig. 10.1 (b) had a strictly linear $T(z)$, such as used in the analysis for Fig. 10.1 (a), the experimental results illustrate this asymptotic result quite dramatically.

The biological implications of the above analysis are of considerable importance. Since biological oscillators and biological clocks are common, a time varying spatial pattern may be a consequence of a spatial variation in the oscillator parameters and not, as might be supposed, a consequence of some reaction-diffusion situation or other such pattern formation mechanism such as we shall be considering in detail in later chapters.

In this section the wave pattern is a continuously changing one. In the following sections we investigate the possibility of a more coherent wave pattern generator of considerable importance.

10.2 Central Pattern Generator: Experimental Facts in the Swimming of Fish

A fish propels itself through water by a sequence of travelling waves which progress down the fish's body from head to tail and its speed is a function of the wave frequency. It is the network of neurons arrayed down the back that controls the muscle movements which generate the actual waves and coordinate them to produce the right effect. It is a widely held lay belief that in mammals the generator, or rather the controlling nerve centre for the rhythmic control of these waves, is the brain. However, in many animals swimming occurs *after* the spinal cord has been severed from the brain – the technical term is spinal transection. In the case of the dogfish for example, the phenomenon has been known since the end of the 19th century. The swimming movement observed in such situations shows the proper intersegmental muscle coordination.

The basis for the required rhythmic behaviour and its intersegmental coordination is a central network of neurons in the spinal cord. It is known that there are neural networks which can generate temporal sequences of signals, which here produce the required cyclic patterns of muscle activity. Such networks are called *central pattern generators* and by definition require no external input control for them to produce the required rhythmic output. It is obvious how important it is to understand such neural control of locomotion. However to do so requires modelling realism and, at the very least, detailed information from experiments. The recent book on neural control of rhythmic movements, edited by Cohen et al. (1988), is specifically about the subject matter of this section and the theory and modelling chapters by Kopell (1988) and Rand et al. (1988) are particularly relevant.

In the case of higher vertebrates there are possibly millions of neurons involved. So it is clear that experimentation and its associated modelling should at least start with as small a spinal cord as possible but one which still exhibits this post-transection activity. Such neural activity, which produces essentially normal swimming, is called "fictive swimming" or "fictive locomotion". This description

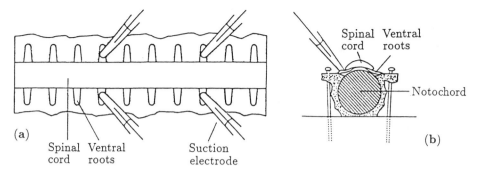

Fig. 10.2a,b. (a) Schematic diagram of the exposed lamprey spinal cord and the experimental arrangement. The preparation was pinned through membrane tissue as shown in the schematic cross-section in (b). There are ventral roots (*VR*) at each segment and the electrical activity of these was measured by the electrodes. (From Cohen and Wallén 1980)

also includes the situation where, even when the muscles which produce the actual locomotion are removed, the neural output from the spinal cord is the same as that of an intact swimming fish.

Grillner (1974), Grillner and Kashin (1976) and Grillner and Wallén (1982) present good experimental data on the dogfish. Kopell (1988) uses this work as a case study for the theory described in detail in her paper. The lamprey, which is rather a primitive vertebrate, was the animal used in a series of interesting and illuminating experiments by Cohen and Wallén (1977) (see also Cohen and Harris-Warrick 1984 and references given there). They studied a specific species of the lamprey which varies from about 13–30 cm in length and has a spinal cord about 0.3 mm thick and 1.5 mm wide. Its advantages (or disadvantages from the lamprey's point of view) are that it has relatively few cells, but still has the necessary basic vertebrate organisation, and it exhibits fictive swimming behaviour. In the experiments of Cohen and Wallén (1980) and Cohen and Harris-Warrick (1984), dissected lengths of spinal cord from about 25–50 segments were used: the lamprey has about 100. The animal was decapitated and the spinal cord exposed but with most of the musculature intact. Motoneuron activity was monitored with electrodes placed on two opposing ventral roots of a single segment: Fig. 10.2 shows schematically the experimental set-up of the spinal cord.

In these experiments the cord was placed in a saline solution and the fictive swimming, that is the periodic rhythmic activity, was induced chemically (by L-DOPA or by the amino acid D-glutamate): the fictive swimming can go on for hours. The ventral root (*VR*) recordings from the electrodes showed alternating bursts of impulses between the left and right *VR* of a single segment. That is the periodic bursts on either side of a segment are 180° out of phase. Fig. 10.3 illustrates the *VR* activity obtained from the left and right sides of the ventral roots from two different segment levels in the cord. An important point to note, and which we shall use in the model, is that the left and right *VR* of a segment

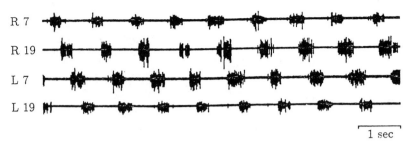

R 7

R 19

L 7

L 19

1 sec

Fig. 10.3. Bursting activity recorded from the left (L) and right (R) sides of the ventral roots (VR) at segment 7 and 19 as measured from the head-end of the specimen, which consists of 27 segments. (From Cohen and Wallén 1980) The time between bursts is approximately 1 sec. Note the approximately constant phase lag as you go from segment 7 to 19.

are like individual oscillators which are phase locked 180° out of phase. This *intra*segmental coupling is very stable.

The results in Fig. 10.3 are from an isolated piece of spinal cord consisting of 27 segments and with the numbering starting from the head side, the recordings were taken at segments 7 and 19. The period of the bursts of activity is about 1 per second. Another point to note is the nearly constant phase lag between the two segments: the lag between the right VR of segment 7 and segment 19 is to a first approximation the same as between the left VR of segment 7 and 19. In Section 10.4 we incorporate this type of behaviour in one of the specific cases in the model developed below. A piece of spinal cord of only about 10 segments can produce a stable neural fictive swimming output.

This periodic activity of the isolated lamprey spinal cord, which is directly related to the undulatory wave-like movements of the swimming fish, schematically shown in Fig. 10.4, is the phenomenon controlled by the central pattern generator, which we now wish to model. Various models can be suggested for the generation of these patterns: see for example Kopell (1988). In the following Sections 10.3–10.4 we shall describe in detail a model proposed by Cohen et al. (1982): it has been used with considerable success to explain certain experimental results associated with selective surgical lesions in the spinal cord (Cohen and Harris-Warrick 1984).

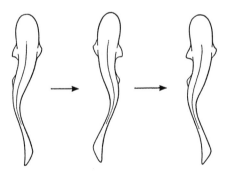

Fig. 10.4. Typical swimming pattern illustrating a propagating wave generated by a ventral root output such as illustrated in Fig. 10.3.

*10.3 Mathematical Model for the Central Pattern Generator

The basic characteristics of the phenomenon, as exemplified by the experiments and indicated in particular by Fig. 10.3, are that the left and right *VR* of a segment are phase-locked oscillators and that there is approximately a constant phase lag from the head to tail of the spinal cord. The key assumptions in the model are: (i) Each segment in the back has associated with it a pair of neuronal oscillators each of which exhibits, in isolation, a stable limit cycle periodic oscillation. The amplitude of the oscillation depends only on internal parameters, and is not usually affected by external factors such as drugs or electrical stimulations. (ii) Each of the oscillators is coupled to its nearest neighbour but with the possibility of long range coupling: there is experimental evidence for the latter in Buchanan and Cohen (1982).

We saw in Chapter 6 on biological oscillators that many biochemical reaction systems can exhibit stable limit cycle periodic oscillations. It is such a biological oscillator, or one coupled to some neuronal electrical property, which we envisage to be the driving force in each of the oscillators associated with the spinal segments. It is not necessary to know the actual details of the biological oscillator for our model here – we do not know what it is in fact. As a preliminary to studying the intersegmental linking of the oscillators we first consider a single oscillator to set up the mathematical treatment and notation and introduce the analytical procedure we shall use.

Single Oscillator and Oscillator Pair

Denote the vector of limit cycle variables of relevance by the vector

$$\mathbf{x}(t) = (x_1(t), x_2(t), \ldots, x_n(t)) . \tag{10.8}$$

Again we do not know what quantity it is that oscillates, only that something does which gives rise to the periodic *VR* neuronal activity which is observed experimentally. For example $\mathbf{x}(t)$ could include the level of the neurotransmitter substance and the periodically varying electric potential. We denote the vector differential equation governing the limit cycles by

$$\frac{d\mathbf{x}}{dt} = \mathbf{f}(\mathbf{x}) , \tag{10.9}$$

where t denotes time. Just as we do not need to know the specific biological oscillator involved, we do not require the detailed functional form of the function $\mathbf{f}(\mathbf{x})$. Let us consider the limit cycle to be the closed orbit γ in the phase space. Using this curve as the local coordinate system we can think of the periodic limit cycle as having a *phase* θ which goes from 0 to 2π as we make a complete circuit round the closed orbit γ. Assume that at some point P on the closed curve the bursting, which is observed experimentally by the electrodes, occurs at the phase

$\theta = 0$. Starting at P, the phase increases from 0 and reaches 2π when we get back to P, where bursting again occurs. Let us further assume that the coordinate system for the limit cycle is chosen so that the speed of the solution round γ, as measured now by $d\theta/dt$, is constant.

The above idea of representing a limit cycle in terms of the phase can be illustrated by the following simple pedagogical example touched on in Section *8.6. Consider the differential equation system given by

$$\frac{dx_1}{dt} = x_1(1 - \rho) - \omega x_2, \quad \frac{dx_2}{dt} = x_2(1 - \rho) + \omega x_1 ,$$

$$\rho = (x_1^2 + x_2^2)^{1/2} \tag{10.10}$$

where ω is a positive constant. Although the solution of this system can be obtained trivially, as we see below, a formal phase plane analysis (see Appendix 1) of (10.10) shows that $(0,0)$ is the only singular point and it is an unstable spiral, spiralling anti-clockwise. A confined set can be found (just take ρ large), so by the Poincaré-Bendixson theorem a limit cycle periodic solution exists and is represented by a closed orbit in the (x_1, x_2) plane. If we now change to polar coordinates ρ and θ in the phase plane with

$$x_1 = \rho \cos \theta, \quad x_2 = \rho \sin \theta \tag{10.11}$$

the system (10.10) becomes

$$\frac{d\rho}{dt} = \rho(1 - \rho), \quad \frac{d\theta}{dt} = \omega , \tag{10.12}$$

and the limit cycle is then seen to be $\rho = 1$. The solution is illustrated in Fig. 10.5. The limit cycle is asymptotically stable since any perturbation from $\rho = 1$ will die out by ρ simply winding back onto $\rho = 1$ and in an anti-clockwise way because $d\theta/dt > 0$. If the perturbation from the circle is to a point $\rho < 1$ then, from (10.12), ρ increases, while if the perturbation is to a point $\rho > 1$, ρ decreases as it tends to the orbit $\rho = 1$. In this case $\rho = 1$ is the equivalent of the orbit γ and θ is the phase, which runs from $\theta = 0$ to 2π as a circuit is completed.

Returning to the model system (10.9), we take θ to be one of the n variables: its value is always modulo 2π, that is any value $\theta = \theta + 2m\pi$ for all integers m. Let us now consider the remaining $n - 1$ variables, denoted by $\mathbf{r}$, to be perturbations perpendicular to the limit cycle orbit γ, where we have taken local coordinates such that the actual orbit γ in an undisturbed state is $\mathbf{r} = 0$. (With the above example in Fig. 10.5, this would be equivalent to changing the variables from (ρ, θ) to (r, θ) where $r = 1 - \rho$: the limit cycle orbit γ, $\rho = 1$, becomes $r = 0$. Now $r \neq 0$ represents a perturbation from γ.) Fig. 10.6 is an example of this system in the case $n = 3$.

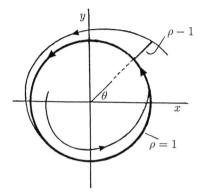

Fig. 10.5. The phase plane solution of the differential equation system (10.12). The asymptotically stable limit cycle is $\rho = 1$ and the phase $\theta = \omega t$.

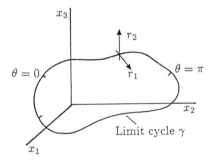

Fig. 10.6. An example of a phase space for the system (10.9) for $n = 3$. Here $\mathbf{r}$ has two components r_1 and r_2 with the phase θ being the third component.

With the coordinate transformation and parametrization above, the system (10.9) can be written as

$$\frac{d\mathbf{r}}{dt} = \mathbf{f}_1(\mathbf{r}, \theta) \, , \tag{10.13}$$

$$\frac{d\theta}{dt} = \omega + f_2(\mathbf{r}, \theta) \, . \tag{10.14}$$

where $\mathbf{f}_1(0, \theta) = 0 = f_2(0, \theta)$ and the period of the oscillator is $T = 2\pi/\omega$. The functions $\mathbf{f}_1$ and f_2 are periodic in θ with period 2π by virtue of the coordinate system we have set up for the limit cycle behaviour of (10.9). If there is no external excitation the limit cycle is simply

$$\mathbf{r} = 0, \quad \frac{d\theta}{dt} = \omega \quad \Rightarrow \quad \theta(t) = \theta(0) + \omega t \, . \tag{10.15}$$

At each segment there are two coupled oscillators such as we have just described. They are linked in such a way that if there are no intersegmental influences the outputs from the right and left oscillator, denoted by $\mathbf{x}_R(t)$ and $\mathbf{x}_L(t)$ respectively, simply oscillate 180° out of phase. Each oscillator is of the form (10.9), that is

$$\frac{d\mathbf{x}_R}{dt} = \mathbf{f}_R(\mathbf{x}_R, \mathbf{x}_L), \quad \frac{d\mathbf{x}_L}{dt} = \mathbf{f}_L(\mathbf{x}_L, \mathbf{x}_R) . \tag{10.16}$$

Associated with each of the oscillators is a phase θ_R and θ_L and a vector $\mathbf{r}_R$ and $\mathbf{r}_L$ which is the deviation from each oscillator in isolation caused by the intrasegmental coupling. That is, if there was no coupling the equations in (10.16) would be uncoupled, $d\theta_R/dt$ and $d\theta_L/dt$ would each be equal to ω and $\mathbf{r}_R$, $\mathbf{r}_L$ both equal to zero.

From the experimental observations described in Section 10.2 the pair of segmental oscillators are $180°$ out of phase so we assume in the model that

$$\theta_L(t) = \theta_R(t) + \pi . \tag{10.17}$$

Later we shall include some weak intersegmental coupling so this relationship will only be a first order approximation. With such a phase relationship the outputs $\mathbf{x}_R$ and $\mathbf{x}_L$ are also $180°$ out of phase which implies

$$\mathbf{x}_L(t) = \mathbf{x}_R(t + T/2) , \tag{10.18}$$

where T is the common period, which gives a relationship between $\mathbf{x}_R$ and $\mathbf{x}_L$. This means that the pair of equations (10.16) can be reduced to a single equation for either the left or right oscillator. The point is that with this assumed intrasegmental coupling we end up again with the reduced system (10.13) and (10.14).

With the single oscillator, as the phase θ increases, the output levels of the variables $\mathbf{x}(t)$ vary periodically; typically as illustrated schematically in Fig. 10.7. Let us suppose that bursting starts when some output $x(t)$ reaches a threshold value and remains on as long as $x(t)$ is above this threshold value. We can take the value of the phase where this threshold for $x(t)$ occurs to be $\theta = 0$: we can set the origin of the phase where we like. Because of the periodic rise and fall of the variable $x(t)$ its value will eventually pass through the critical threshold value again and the bursting will be shut off. (We shall see in Chapters 11 and 12 that such threshold phenomena occur in other wave situations.)

Experimentally bursting is observed for only about 0.4 of the period, which in our scaling is $2\pi \times 0.4$. If we let θ_0 be the phase at which bursting ceases, the bursting based on this then occurs as illustrated in Fig. 10.7 for $\theta + 2m\pi$, $0 < \theta < \theta_0$ for all $m = 0, 1, 2, \ldots$

Sequence of Coupled Oscillators

We now need to look at the effect on each of the segmental oscillators if they are coupled. So, we now consider the series of segmental oscillators (at each stage there is a pair but as we showed above the analysis requires a study of only one) which consists of N equations of the form (10.9), a typical one of which we write

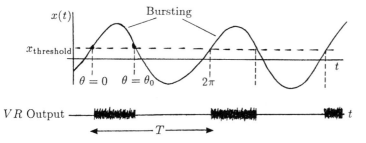

Fig. 10.7. Schematic periodic output $x(t)$ from a single oscillator. When the output level is above a threshold it triggers the VR bursting.

as

$$\frac{dx_j}{dt} = f_j(x_j) + g_j(x_1, \ldots, x_N, c), \quad j = 1, \ldots, N \tag{10.19}$$

where g represents the coupling effect of the other $N-1$ oscillators and c is a vector of coupling parameters. If $c = 0$ the oscillators are uncoupled, that is $g_j(x_1, \ldots, x_N, 0) = 0$ and the N oscillators then simply have their own periodic limit cycle γ_j.

The general mathematical problem (10.19) is essentially intractable without some simplifying assumptions or specializations. The problem of determining the interaction properties of coupled oscillators has been widely studied for many years. Hard quantitative results are not trivial to get and most studies have been on a limited number of oscillators, like two, or when the coupling between the oscillators is weak. The mathematics used covers a wide spectrum, with singular perturbation techniques being among the most powerful and helpful. We discussed some properties of coupled oscillator systems in Chapter 8, Sections 8.5–8.9. With the above system, some useful and experimentally pertinent results can be obtained if we consider the coupling effects to be weak; that is if we assume $|c| \ll 1$ and $|g| \ll |f|$, and then use perturbation methods. The implications, or rather assumptions, of this are that the limit cycles $\gamma_j, j = 1, 2, \ldots, N$ of the isolated oscillators will only be slightly perturbed by the coupling effects (recall Section 8.5). So, it is still appropriate to use the oscillator equation form (10.13) and (10.14) involving the phase θ and deviations r_j from the limit cycle γ_j, but now we have to include an extra small coupling term in the equation. (In fact we shall make even further simplifying assumptions based on what has been observed experimentally, but it is instructive to proceed a little further with the current line since it is the basis for a rigorous justification of the assumptions we shall make later.) The set of N equations we have to study is then

$$\frac{dr_j}{dt} = f_{j1}(r_j, \theta_j) + g_{j1}(r_1, \ldots, r_N, \theta_1, \ldots, \theta_N, c),$$

$$\frac{d\theta_j}{dt} = \omega_j + f_{j2}(r_j, \theta_j) + g_{j2}(r_1, \ldots, r_N, \theta_1, \ldots, \theta_N, c), \quad j = 1, 2, \ldots, N \tag{10.20}$$

Experimentally it has been observed that the individual oscillators when uncoupled, by severing and thus isolating them from their neighbours, have different frequencies ω_j and hence different periods $T_j = 2\pi/\omega_j$. A crucially important point to keep in mind is that when the segmental oscillators are coupled they still perform limit cycle oscillations. So, even when coupled we can still characterize them in terms of their phase θ_j. Since fictive swimming is a reflexion of phase coupling we need only consider a *phase coupling model* for the system (10.19). So, instead of studying the system (10.19) perturbed about $\mathbf{r}_j = 0$ we can consider a system of phase coupled equations of the form

$$\frac{d\theta_j}{dt} = \omega_j + h_j(\theta_1, \ldots, \theta_N, \mathbf{c}), \quad j = 1, \ldots, N \tag{10.21}$$

where h_j includes the (weak) coupling effect of all the other oscillators. Equations (10.21) do not involve the amplitudes of the oscillators. The problem of weak coupling in a population of oscillators has been studied in some depth, for example, by Neu (1979, 1980), Rand and Holmes (1980), Ermentrout (1981) and in the book by Guckenheimer and Holmes (1983). Carrying out a perturbation of (10.19) about $\mathbf{r}_j = 0$ eventually results in a phase-coupled system of equations (10.21) (see Chapter 8). So, there is a mathematical, as well as biological, justification for considering the simpler model (10.21).

Since we assume small perturbations from the individual limit cycle oscillators to come from the coupling, it is reasonable to consider a linear coupling model where the effect of the j-th oscillator on the i-th one is simply proportional to $\mathbf{x}_j$. In this situation the coupled oscillator system is of the form (10.9) perturbed by linear terms, namely

$$\frac{d\mathbf{x}_i}{dt} = \mathbf{f}_i(\mathbf{x}_i) + \sum_{\substack{j=1 \\ j \neq i}}^{N} \mathbf{A}_{ij}\mathbf{x}_j , \tag{10.22}$$

where $\mathbf{A}_{ij}$ are matrices of the coupling coefficients.

Equations (10.22) include both the phases and amplitudes. We argued above that we need only consider a phase coupled model so even (10.22) is more complicated than we need consider. It was shown in Chapter 8, Section *8.8 (see also Neu 1979a,b and 1980 and Rand and Holmes 1980) that, in the case of weak coupling, (10.22) leads to a *phase* coupled system of differential equations of the form

$$\frac{d\theta_i}{dt} = \omega_i + \sum_{j=1}^{N} a_{ij} h(\theta_j - \theta_i), \quad i = 1, \ldots, N \tag{10.23}$$

where h is a periodic function of its argument. We can argue heuristically however to justify the model system (10.23). Since we have suggested that we need only consider a phase model, a linear coupling would reasonably involve a coupling term which was a function of the phase differences of the oscillator and all the others in the system. The periodic nature of the function h is suggested by the fact that we would also reasonably expect the phase difference between any two oscillators to be periodic. There is, as well, experimental evidence to support such a conjecture from Buchanan and Cohen (1982), who found that the slowly varying intrasegmental potentials of the motoneurons is quasi-sinusoidal. As the specific model to study therefore we shall take the simple periodic function $h(\phi) = \sin \phi$ in (10.23). For a discussion of a more general h in the case of coupled oscillators, see Kopell (1988).

The phase coupled model we shall analyse in detail is the set of N phase equations

$$\frac{d\theta_i}{dt} = \omega_i + \sum_{\substack{j=1 \\ j \neq i}}^{N} a_{ij} \sin(\theta_j - \theta_i), \quad i = 1, \ldots, N \qquad (10.24)$$

The a_{ij}'s are a measure of the effect of the j-th oscillator on the i-th one with the effect being *excitatory*, meaning θ_j tends to pull θ_i towards its value if a_{ij} is *positive*, while it is *inhibitory* if a_{ij} is *negative*. In the inhibitory case θ_j tends to increase the difference between it and θ_i. With the specific interaction function $\sin \phi$ we choose for h in (10.23) the maximimum excitatory effect of the j-th oscillator on the i-th one is when they are $\pi/2$ out of phase, the maximum inhibitory effect is when they are $-\pi/2$ out of phase with no effect when they are in phase. Fig. 10.8 schematically illustrates the model and the sinusoidal coupling effect.

We wish to retain in our model the periodic character of the oscillators when they are coupled. More particularly we want $d\theta_i/dt$ always to be a positive and monotonic increasing function of t, so that movement on the limit cycle γ_i is always in one direction, or in other words so that the output has a regular up and down character as in Fig. 10.8 (a). We assume therefore that the magnitudes of ω_i and a_{ij} are such that the phase $\theta_i(t)$ behaves in this way. We now wish to analyse the model system (10.24) and see whether or not it is a reasonable model when the solutions are compared with experimental results.

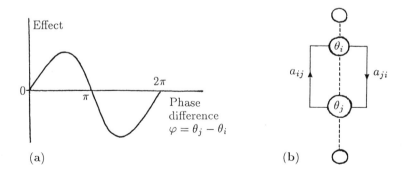

(a)   (b)

Fig. 10.8a,b. (a) Measure of the effect on the i-th oscillator of the j-th oscillator for a sinusoidal form for the interaction function $h(\phi)$ of the phase difference. (b) Schematic representation of the coupling of the i-th and j-th oscillators according to (10.24).

*10.4 Analysis of the Phase-Coupled Model System

We have to decide what kind of coupling we wish to include in the model system
(10.24). Here we shall analyse the simplest, namely when each oscillator is only
coupled to its nearest neighbours. In this case (10.24) becomes the coupled system
of N equations

$$\frac{d\theta_1}{dt} = \omega_1 + a_{12} \sin(\theta_2 - \theta_1)$$

$$\frac{d\theta_2}{dt} = \omega_2 + a_{21} \sin(\theta_1 - \theta_2) + a_{23} \sin(\theta_3 - \theta_2)$$

$$\vdots$$

$$\frac{d\theta_j}{dt} = \omega_j + a_{j,j-1} \sin(\theta_{j-1} - \theta_j) + a_{j,j+1} \sin(\theta_{j+1} - \theta_j) \qquad (10.26)$$

$$\vdots$$

$$\frac{d\theta_N}{dt} = \omega_N + a_{N,N-1} \sin(\theta_{N-1} - \theta_N) .$$

The form of the right hand sides suggests that we introduce

$$\phi_j = \theta_j - \theta_{j+1}, \quad \Omega_j = \omega_j - \omega_{j+1} \qquad (10.26)$$

and rewrite the system (10.25) in terms of the ϕ's, the phase differences, and the
Ω's, the frequency differences, by subtracting the θ-equations pairwise to get the
$N - 1$ equations

$$\frac{d\phi_1}{dt} = \Omega_1 - (a_{12} + a_{21}) \sin \phi_1 + a_{23} \sin \phi_2$$

$$\frac{d\phi_2}{dt} = \Omega_2 + a_{21} \sin \phi_1 - (a_{23} + a_{32}) \sin \phi_2 + a_{34} \sin \phi_3$$

$$\vdots$$

$$\frac{d\phi_j}{dt} = \Omega_j + a_{j,j-1} \sin \phi_{j-1} \qquad (10.27)$$
$$- (a_{j,j+1} + a_{j+1,j}) \sin \phi_j + a_{j+1,j+2} \sin \phi_{j+1}$$

$$\vdots$$

$$\frac{d\phi_{N-1}}{dt} = \Omega_{N-1} + a_{N-1,N-2} \sin \phi_{N-2}$$
$$- (a_{N-1,N} + a_{N,N-1}) \sin \phi_{N-1} .$$

Since we are looking for some regular periodic pattern which we associate with
fictive swimming, we can make some assumptions about the coupling coefficients
a_{ij}. (We are also trying to get the simplest reasonable model to mimic the

experimental phenomenon.) Let us assume that all the upward (in number that is – in our model this is in the head to tail direction) coupling coefficients $a_{j,j+1} = a_u$ and all the downwards coefficients $a_{j,j-1} = a_d$. The system (10.27) in vector form is then

$$\frac{d\phi}{dt} = \Omega + \mathbf{BS} , \qquad (10.28)$$

where the vectors

$$\phi = \begin{pmatrix} \phi_1 \\ \vdots \\ \phi_{N-1} \end{pmatrix}, \quad \mathbf{S} = \begin{pmatrix} \sin \phi_1 \\ \vdots \\ \sin \phi_{N-1} \end{pmatrix}, \quad \Omega = \begin{pmatrix} \Omega_1 \\ \vdots \\ \Omega_{N-1} \end{pmatrix} \qquad (10.29)$$

and $\mathbf{B}$ is the $(N-1) \times (N-1)$ matrix

$$\mathbf{B} = \begin{pmatrix} -(a_d + a_u) & a_u & \cdot & \cdot \\ a_d & -(a_d + a_u) & a_u & \cdot \\ \cdot & \cdot & \cdot & \cdot \\ \cdot & \cdot & \cdot & \cdot \\ \cdot & \cdot & a_d & -(a_d + a_u) \end{pmatrix} \qquad (10.30)$$

For the application of the model to the fictive swimming of the lamprey, we are interested in *phase-locked* solutions of (10.28). That is the coupling must be such that all the oscillators have the same period. This is the same as saying that the phase differences ϕ_j between the oscillators is always constant for all $j = 1, 2, \ldots, N-1$. This in turn means that $d\phi_j/dt = 0$ and so we are looking for equilibrium solutions of (10.28), that is the solutions of

$$0 = \Omega + \mathbf{BS} \quad \Rightarrow \quad \mathbf{S} = -\mathbf{B}^{-1}\Omega . \qquad (10.31)$$

Since $\mathbf{S}$ involves only $\sin \phi_j$, $j = 1, 2, \ldots, N-1$, solutions exist only if all the elements of $\mathbf{B}^{-1}\Omega$ lie between ± 1.

2-Oscillator System

Here we have only a single equation for the phase difference $\phi = \theta_1 - \theta_2$, namely

$$\frac{d\phi}{dt} = \Omega - (a_d + a_u) \sin \phi . \qquad (10.32)$$

This has phase-locked solutions, where $d\phi/dt = 0$, if and only if

$$|\Omega| \leq |a_d + a_u| . \qquad (10.33)$$

If we denote the solutions by ϕ_S then if

$$\begin{aligned} |\Omega| &= |a_d + a_u| \quad \text{then} \quad \phi_S = \pi/2 \quad \text{or} \quad 3\pi/2, \\ |\Omega| &< |a_d + a_u| \quad \text{then} \quad |\phi_S| < \pi/2 \quad \text{or} \quad \pi/2 < |\phi_S| \leq \pi . \end{aligned} \qquad (10.34)$$

We can determine the stability of these steady state solutions by linearizing about them in the usual way (as in Chapter 1, Section 1.1). If we denote perturbations about ϕ_S by ψ, the linear stability equation from (10.32) is

$$\frac{d\psi}{dt} \approx -[(a_d + a_u)\cos\phi_S]\psi .$$

The first two possible solutions in (10.34) are neutrally stable, because $\cos\phi_S = 0$, while one of the second set is stable and the other unstable. So, as the coupling strength $|a_d + a_u|$ *decreases* relative to the frequency difference $|\Omega|$, $|\Omega|/|a_d + a_u|$ increases and the stable steady state, as well as the unstable one, tends to the neutrally stable one as $|a_d + a_u| \to |\Omega|$ after which no solution exists. Thus as the coupling strength $|a_d + a_u|$ becomes weaker relative to the *detuning* $|\Omega| = \omega_1 - \omega_2$ a bifurcation takes place where the stable phase-locked solution ceases to exist. The stable steady state when it exists, is the phase-locked solution we are interested in. From the discussion above in Section 10.2, a model with two oscillators has two pairs of ventral roots and, with the threshold bursting implied by Fig. 10.7 and phase difference at each VR pair, the bursting output is as illustrated in Fig. 10.9 (a).

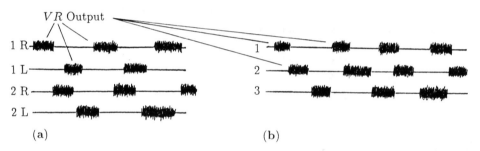

Fig. 10.9.a,b (a) Ventral root output from a phase-locked 2-oscillator model. (b) Ventral root output from a phase-locked solution of a 3-oscillator system: note that the phase difference between the 1st and 2nd root is not necessarily the same as between the 2nd and 3rd.

3-Oscillator System

The steady states for this system are, from (10.31),

$$\begin{pmatrix} \sin\phi_1 \\ \sin\phi_2 \end{pmatrix} = \frac{1}{a_d^2 + a_d a_u + a_u^2} \begin{pmatrix} a_d + a_u & a_u \\ a_d & a_d + a_u \end{pmatrix} \begin{pmatrix} \Omega_1 \\ \Omega_2 \end{pmatrix} \tag{10.35}$$

For algebraic simplicity let us take the coupling coefficients $a_d = a_u = a$, in which case the last equation gives

$$\sin\phi_1 = (2\Omega_1 + \Omega_2)/3a, \quad \sin\phi_2 = (\Omega_1 + 2\Omega_2)/3a . \tag{10.36}$$

Thus a phase locked solution (ϕ_1, ϕ_2) exists if and only if

$$\max\{|2\Omega_1 + \Omega_2|/3a, |\Omega_1 + 2\Omega_2|/3a\} < 1 . \tag{10.37}$$

In this 3-oscillator case the phase differences ϕ_1 and ϕ_2 are unequal in general, and a typical VR output from it is schematically illustrated in Fig. 10.9 (b).

   As the number of oscillators increases the algebraic complexity quickly gets out of hand but it is clear how to set up the algebraic problem to be solved; that is how to get the set of conditions that must hold between the coupling coefficients and the detuning parameters Ω for a phase-locked solution to exist.

Constant Phase Lag System of N-Oscillators

To keep the generality in a multi-oscillator model such as we did with the last two systems is perhaps unnecessarily cumbersome. What we wish to show is that such coupled oscillators can give a stable phase-locked system, such as required by a central pattern generator to produce the required VR output necessary for fictive swimming. So, here we consider a system where there is a constant phase lag between neighbouring segments, that is we assume that the phase difference $\phi_j = \theta_j - \theta_{j+1} = \delta$, a positive constant. This situation is of particular relevance to the experimental facts related to Fig. 10.3. Such a line of oscillators is characteristic of a uniform travelling wave. Although the analysis here can be done with any periodic h in (10.23) (see Exercise 2), we shall continue to use the example $h(\phi) = \sin \phi$ for consistency. If we set $\Delta = \sin \delta > 0$ the system of equations for the steady states becomes, from (10.31) with (10.29) and (10.30),

$$\Omega_1 + [-(a_d + a_u) + a_u]\Delta = 0 \quad \Rightarrow \quad \Omega_1 = a_d \Delta$$

$$\vdots$$

$$\Omega_j + [a_d - (a_d + a_u) + a_u]\Delta = 0 \quad \Rightarrow \quad \Omega_j = 0, \ j = 2, \ldots, N-2 \tag{10.38}$$

$$\vdots$$

$$\Omega_{N-1} + [a_d - (a_d + a_u)]\Delta = 0 \quad \Rightarrow \quad \Omega_{N-1} = a_u \Delta$$

In terms of the original frequencies, since $\Omega_j = \omega_j - \omega_{j+1}$, this gives

$$\omega_j = \omega \quad \text{for all} \quad j = 2, \ldots, N-1$$
$$\omega_1 = \omega + a_d \sin \delta > \omega \tag{10.39}$$
$$\omega_N = \omega - a_u \sin \delta < \omega$$

What this solution means is that all the oscillators except the first, the rostral or head oscillator, and the last, the caudal or tail oscillator, have the natural frequency of each segmental oscillator in isolation. The head oscillator is tuned up, that is to a higher frequency, while the tail one is tuned down. This assumes that the coupling coefficients a_d and a_u in (10.39) are positive, that is the coupling is excitatory. The resulting wave which results from this is one which travels from head to tail. Another solution to this constant phase lag problem is when the head oscillator is tuned down and the tail one tuned up. This results in the wave propagating from tail to head: that is the lamprey swims backwards, which in fact it can do.

We finally have to consider the stability of this constant phase lag solution. We do this in the usual way by linearizing about $\phi_j = \delta$ by writing

$$\phi_j = \delta + \psi_j, \quad j = 1, \ldots, N \tag{10.40}$$

where $\|\psi\| \ll 1$. Substituting this into the time dependent equation (10.28) and linearizing gives

$$\frac{d\psi}{dt} = \mathbf{B}\psi \cos \delta \tag{10.41}$$

where the matrix $\mathbf{B}$ is given by (10.30), depends only on the coupling coefficients a_d and a_u and is a tri-diagonal $(N-1) \times (N-1)$ matrix. If we now look for solutions of (10.41) in the form

$$\psi(t) = e^{\lambda t}\psi_0 \tag{10.42}$$

the eigenvalues λ are solutions of

$$|\mathbf{B} \cos \delta - \lambda \mathbf{I}| = 0 . \tag{10.43}$$

where $\mathbf{I}$ is the unit $(N-1) \times (N-1)$ matrix. From experimental observation phase lags between segments is quite small and $0 < \cos \delta < 1$. Under these circumstances it can be shown, for example by using the Routh-Hurwitz conditions on the polynomial in λ, that $\mathrm{Re}\,\lambda < 0$. So from (10.42) $\psi(t) \to 0$ as $t \to \infty$ which means that the phase-locked constant phase difference solution above is stable to linear perturbations.

There are of course two solutions for the phase difference equation $\sin \phi_j = \sin \delta$, which gives $2^{N-1}-2$ other phase-locked solutions. However, it can be shown, from a study of the eigenvalue matrix, that all of them are linearly unstable, that is there exists at least one eigenvalue λ in (10.42) with $\mathrm{Re}\,\lambda > 0$. We conclude therefore, that the solution (10.39) is the relevant one when there is a constant phase difference between neighbouring segmental oscillators, and hence is the one which gives rise to a stable wave which propagates down (or up) the spinal chord.

Perhaps it should be pointed out here that, as far as the lamprey is concerned, isolated parts of the cord can "swim" forward and backward, so we have to postulate that something automatically tunes the two end segments. Another problem with this simple model is that it does not seem able to account for the experimental fact that the phase lag appears to be constant even with changes in the swimming speed. This is a serious point since the phase lag determines the wave length and hence the shape of the swimming fish. There are however other possible ways of coupling which can be considered (see, for example, Cohen et al. 1982).

One purpose of these Sections 10.2–10.4 has been to show how such a relatively simple model can be the pattern generator for the wave propagation in experimentally observed fictive swimming. A major point to note is that various simple intersegmental coupling of oscillators can generate stable travelling waves. Particularly striking is the fact that even the simple model we analysed is sufficient to generate the required coordination of phase coupling for both forward and backward swimming – only the head and tail oscillators had to be retuned. Cohen et al. (1982) discuss in more detail the comparison with the experimental observations on lamprey. Although there are still problems, the results are encouraging. All of this does not imply that such a model mechanism is *the* central pattern generator, only that it is a possible candidate.

Exercises

1. Consider a 4-oscillator system in which the coupling coefficients $a_d = a_u = a$ and each oscillator frequency differs from its predecessor by a small amount ε: that is $\omega_j = \omega_{j-1} - \varepsilon$. First look for steady state phase locked solutions for $\phi_j = \theta_j - \theta_{j-1}$, $j = 1, 2, 3$ from (10.29). Show that solutions exist for ϕ_j, $j = 1, 2, 3$ only if $\varepsilon \le a/2$. Generalize the result to N oscillators to show that solutions exist only if $\varepsilon \le 8a/N^2$.

2. Consider an N-oscillator system in which there is only nearest neighbour coupling between which there is a constant phase lag δ. Start with equation (10.23) with a general interaction function $h(\delta)$ and derive the equivalent of (10.38) for the steady state frequency differences. Hence determine the frequency of the first and last oscillator in terms of $h(\delta)$.

11. Biological Waves: Single Species Models

11.1 Background and the Travelling Wave Form

There is a vast number of phenomena in biology where a key element or precursor to a developmental process seems to be the appearance of a travelling wave of chemical concentration or mechanical deformation. Looking at almost any film of a developing embryo it is hard not to be struck by the number of wave-like events that appear after fertilisation. There are, for example, both chemical and mechanical waves which propagate on the surface of many vertebrate eggs. In the case of the egg of the fish *Medaka* a calcium (Ca^{++}) wave sweeps over the surface; it emanates from the point of sperm entry: we briefly discuss this problem in Section 11.6 below. Chemical concentration waves such as those found with the Belousov-Zhabotinskii reaction are visually dramatic examples (see Fig. 12.1 in the following chapter). From the analysis on insect dispersal in Section 9.3 in Chapter 9 we can also expect wave phenomena in that area, and in interacting population models where spatial effects are important. Another example, related to interacting populations, is the progressing wave of an epidemic, of which the rabies epizootic currently spreading across Europe is a dramatic and disturbing example: we study a model for this in some detail in Chapter 20. The movement of microorganisms moving into a food source, chemotactically directed, is another. The slime mold *Dictyostelium discoideum* is a particularly widely studied example of chemotaxis: we discuss this phenomenon later (see the photograph in Fig. 12.16 which shows associated waves).

The book by Winfree (1980) is replete with wave phenomena in biology. The introductory text on mathematical models in molecular and cellular biology edited by Segel (1980) also deals with some aspects of wave motion. Although not application oriented, the books on reaction diffusion equations by Fife (1979), Smoller (1983) and Britton (1986) are all relevant. Zeeman (1977) considers wave phenomena in development and other biological areas from a catastrophe theory standpoint.

The point to be emphasized is the widespread existence of wave phenomena in biology. This clearly necessitates a study of travelling waves in depth and of the modelling and analysis involved. This chapter and the next deals with various aspects of wave behaviour where *diffusion* plays a crucial role. The waves studied here are quite different to those discussed in Chapter 10. The mathematical literature on them is now vast, so the number of topics and the depth of the discussions have to be limited. I shall try, however, to cover not only some of the major areas which are now accepted as part of the basic theory in the field but also some of the more important current areas of exciting research which will clearly become part of the accepted theory of biological wave phenomena.

In developing living systems there is almost continual interchange of information at both the inter- and intra-cellular level. Such communication is necessary for the sequential development and generation of the required pattern and form. Propagating wave forms of varying biochemical concentrations are one means of transmitting such biochemical information. In the developing embryo, diffusion coefficients of biological chemicals can be very small; values of the order of 10^{-9}–10^{-11} cm^2 sec^{-1} are fairly common. Such small diffusion coefficients imply that to cover macroscopic distances of the order of several millimetres requires a very long time if diffusion is the principle process involved. Estimation of diffusion coefficients for insect dispersal in interacting populations is now being studied with care and sophistication, for example by Kareiva (1983): not surprisingly the values are larger and species dependent.

With a standard diffusion equation in one space dimension, which from Section 9.1 is typically of the form

$$\frac{\partial u}{\partial t} = D\frac{\partial^2 u}{\partial x^2} \,, \tag{11.1}$$

for a chemical of concentration u, the time to convey information in the form of a changed concentration over a distance L is $O(L^2/D)$. You get this order estimate from the equation using dimensional arguments, similarity solutions or more obviously from the classical solution given by equation (9.10) in Chapter 9. So, if L is of the order of 1mm, typical times with the above diffusion coefficients are $O(10^7$–10^9 sec), which is excessively long for most processes in the early stages of embryonic development. Simple diffusion therefore is unlikely to be the main vehicle for transmitting information over significant distances. A possible exception is the generation of butterfly wing patterns, which takes place during the pupal stage and involves several days (Murray 1981a).

In contrast to simple diffusion we shall see that when reaction kinetics and diffusion are coupled, travelling waves of chemical concentration exist, can effect a biochemical change, very much faster than straight diffusional processes governed by equations like (11.1): see the end of Section 11.4 below. This coupling gives rise to reaction diffusion equations which (cf. Section 9.1, Equation (9.16)) in a simple one-dimensional scalar case can look like

$$\frac{\partial u}{\partial t} = f(u) + D\frac{\partial^2 u}{\partial x^2} \,, \tag{11.2}$$

where u is the concentration, $f(u)$ represents the kinetics and D is the diffusion coefficient, here taken to be constant.

We must first decide what we mean by a travelling wave. We saw in Chapter 9 that the solutions (9.21) and (9.24) described a kind of wave, where the shape and speed of propagation of the front continually changed. Customarily a travelling wave is taken to be a wave which travels *without change of shape*, and this will be our understanding here. So, if a solution $u(x, t)$ represents a travelling wave, the *shape* of the solution will be the same for all time and the speed of propagation of this shape is a constant, which we denote by c. If we look at this wave in a travelling frame moving at speed c it will appear stationary. A mathematical way of saying this is that if the solution

$$u(x, t) = u(x - ct) = u(z), \quad z = x - ct \tag{11.3}$$

then $u(x, t)$ is a travelling wave, and it moves at constant speed c in the positive x-direction. Clearly if $x - ct$ is constant, so is u. If $x - ct$ is constant this means the coordinate system moves with speed c. A wave which moves in the negative x-direction is of the form $u(x + ct)$. The wave speed c generally has to be determined. The dependent variable z is sometimes called the *wave variable*. When we look for travelling wave solutions of an equation or system of equations in x and t in the form (11.3), we have $\partial u / \partial t = -c \, du/dz$ and $\partial u / \partial x = du/dz$. So *partial* differential equations in x and t become *ordinary* differential equations in z. To be physically realistic $u(z)$ has to be bounded for all z and non-negative with the quantities with which we are concerned, such as chemicals and populations.

It is part of the classical theory of linear parabolic equations, such as (11.1), that there are no physically realistic travelling wave solutions. Suppose we look for solutions in the form (11.3), then (11.1) becomes

$$D\frac{d^2 u}{dz^2} + c\frac{du}{dz} = 0 \quad \Rightarrow \quad u(z) = A + Be^{-cz/D} .$$

where A and B are integration constants. Since u has to be bounded for all z, B must be zero since the exponential becomes unbounded as $z \to -\infty$. $u(z) = A$, a constant, is not a wave solution. In marked contrast the parabolic reaction diffusion equation (11.2) can exhibit travelling wave solutions, depending on the form of the reaction/interaction term $f(u)$. This solution behaviour was a major factor in starting the whole mathematical field of reaction diffusion theory.

Although most realistic models of biological interest involve more than one dimension and more than one species, whether concentration or population, there are several multi-species systems which reasonably reduce to a one-dimensional single species mechanism which captures the key features. This chapter therefore is not simply a pedagogical mathematical exposition of some common techniques and basic theory. We shall discuss two very practical problems, one in ecology and the other in developmental biology: both deal with important phenomena.

11.2 Fisher Equation and Propagating Wave Solutions

The classic and simplest case of the nonlinear reaction diffusion equation (11.2) is

$$\frac{\partial u}{\partial t} = ku(1-u) + D\frac{\partial^2 u}{\partial x^2}, \tag{11.4}$$

where k and D are positive parameters. It was suggested by Fisher (1937) as a deterministic version of a stochastic model for the spatial spread of a favoured gene in a population. It is also the natural extension of the logistic growth population model discussed in Chapter 1 when the population disperses via linear diffusion. This equation and its travelling wave solutions have been widely studied, as has the more general form with an appropriate class of functions $f(u)$ replacing $ku(1-u)$. The seminal and now classical paper is that by Kolmogoroff, Petrovsky and Piscounoff (1937). The books by Fife (1979) and Britton (1986) mentioned above give a full discussion of this equation and an extensive bibliography. We discuss this model equation in the following section in some detail, not because in itself it has such wide applicability but because it is the prototype equation which admits travelling wavefront solutions. It is also a convenient equation from which to develop many of the standard techniques for analysing single species models with diffusive dispersal.

Although (11.4) is now referred to as the Fisher equation, the discovery, investigation and analysis of travelling waves in chemical reactions was first reported by Luther (1906). This recently re-discovered paper has been translated by Arnold et al. (1988). Luther's paper was first presented at a conference: the discussion at the end of his presentation (and it is included in the Arnold et al. 1988 translation) is very interesting. There, Luther states that the wave speed is a simple consequence of the differential equations. Showalter and Tyson (1988) put Luther's (1906) remarkable discovery and analysis of chemical waves in a modern context. Luther obtained the wave speed in terms of parameters associated with the reactions he was studying. The analytical form is the same as that found by Fisher (1937) for (11.4).

Let us now consider (11.4). It is convenient at the outset to rescale (11.4) by writing

$$t^* = kt, \quad x^* = x\left(\frac{k}{D}\right)^{1/2} \tag{11.5}$$

and, omitting the asterisks for notational simplicity, (11.4) becomes

$$\frac{\partial u}{\partial t} = u(1-u) + \frac{\partial^2 u}{\partial x^2}. \tag{11.6}$$

In the spatially homogeneous situation the steady states are $u = 0$ and $u = 1$, which are respectively unstable and stable. This suggests that we should look for travelling wavefront solutions to (11.6) for which $0 \le u \le 1$: negative u has no physical meaning with what we have in mind for such models.

If a travelling wave solution exists it can be written in the form (11.3), say

$$u(x,t) = U(z), \quad z = x - ct ,\tag{11.7}$$

where c is the wave speed. We use $U(z)$ rather than $u(z)$ to avoid any nomenclature confusion. Since (11.6) is invariant if $x \to -x$, c may be negative or positive. To be specific we shall assume $c \geq 0$. Substituting this travelling wave form into (11.6), $U(z)$ satisfies

$$U'' + cU' + U(1 - U) = 0 ,\tag{11.8}$$

where primes denote differentiation with respect to z. A typical wave*front* solution is where U at one end, say as $z \to -\infty$, is at one steady state and as $z \to \infty$ it is at the other. So here we have an eigenvalue problem to determine the value, or values, of c such that a non-negative solution U of (11.8) exists which satisfies

$$\lim_{z \to \infty} U(z) = 0, \quad \lim_{z \to -\infty} U(z) = 1 .\tag{11.9}$$

At this stage we do not address the problem of how such a travelling wave solution might evolve from the partial differential equation (11.6) with given initial conditions $u(x,0)$; we come back to this point later.

We study (11.8) for U in the (U, V) phase plane where

$$U' = V, \quad V' = -cV - U(1 - U) ,\tag{11.10}$$

which gives the phase plane trajectories as solutions of

$$\frac{dV}{dU} = \frac{-cV - U(1 - U)}{V} .\tag{11.11}$$

This has two singular points for (U, V), namely $(0,0)$ and $(1,0)$: these are the steady states of course. A linear stability analysis (see Appendix 1) shows that the eigenvalues λ for the singular points are

$$(0,0): \quad \lambda_\pm = \frac{1}{2}[-c \pm (c^2 - 4)^{1/2}] \quad \Rightarrow \quad \begin{cases} \text{stable node} & \text{if } c^2 \geq 4 \\ \text{stable spiral} & \text{if } c^2 < 4 \end{cases}$$

$$(1,0): \quad \lambda_\pm = \frac{1}{2}[-c \pm (c^2 + 4)^{1/2}] \quad \Rightarrow \quad \text{saddle point}\tag{11.12}$$

Fig. 11.1 (a) illustrates the phase plane trajectories.

If $c \geq c_{\min} = 2$ we see from (11.12) that the origin is a stable node, the case when $c = c_{\min}$ giving a degenerate node. If $c^2 < 4$ it is a stable spiral; that is, in the vicinity of the origin U oscillates. By continuity arguments, or simply by heuristic reasoning from the phase plane sketch of the trajectories in Fig. 11.1 (a), there is a trajectory from $(1,0)$ to $(0,0)$ lying entirely in the quadrant $U \geq 0$,

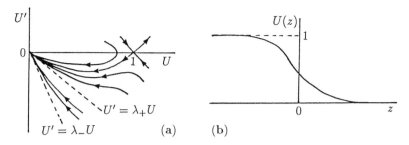

Fig. 11.1a,b. (a) Phase plane trajectories for equation (11.8) for the travelling wavefront solution: here $c^2 > 4$. (b) Travelling wavefront solution for the Fisher equation (11.6): the wave velocity $c \geq 2$.

$U' \leq 0$ with $0 \leq U \leq 1$ for *all* wave speeds $c \geq c_{min} = 2$. In terms of the original dimensional equation (11.4), the range of wave speeds satisfy

$$c \geq c_{min} = 2(kD)^{1/2} \ . \tag{11.13}$$

Fig. 11.1 (b) is a sketch of a typical travelling wave solution. There are travelling wave solutions for $c < 2$ but they are physically unrealistic since $U < 0$, for some z, because in this case U spirals around the origin. In these, $U \to 0$ at the leading edge with decreasing oscillations about $U = 0$.

A key question at this stage is what kind of initial conditions $u(x,0)$ for the original Fisher equation (11.6) will evolve to a travelling wave solution and, if such a solution exists, what is its wave speed c. This problem and its generalisations have been widely studied analytically: see the references in the books cited above in Section 11.1. Kolmogoroff et al. (1937) proved that if $u(x,0)$ has compact support,

$$u(x,0) = u_0(x) \geq 0, \quad u_0(x) = \begin{cases} 1 & \text{if } x \leq x_1 \\ 0 & \text{if } x \geq x_2 \end{cases} \tag{11.14}$$

where $x_1 < x_2$ and $u_0(x)$ is continuous in $x_1 < x < x_2$, then the solution $u(x,t)$ of (11.6) evolves to a travelling wavefront solution $U(z)$ with $z = x - 2t$. That is it evolves to the wave solution with *minimum* speed $c_{min} = 2$. For initial data other than (11.14) the solution depends critically on the behaviour of $u(x,0)$ as $x \to \pm\infty$.

The dependence of the wave speed c on the initial conditions at infinity can be seen easily from the following simple analysis suggested by Mollison (1977). Consider first the leading edge of the evolving wave where, since u is small, we can neglect u^2 in comparison with u. Equation (11.6) is linearized to

$$\frac{\partial u}{\partial t} = u + \frac{\partial^2 u}{\partial x^2} \ . \tag{11.15}$$

Consider now

$$u(x,0) \sim Ae^{-ax} \quad \text{as} \quad x \to \infty \tag{11.16}$$

where $a > 0$ and $A > 0$ is arbitrary, and look for travelling wave solutions of (11.15) in the form

$$u(x,t) = Ae^{-a(x-ct)} . \qquad (11.17)$$

We think of (11.17) as the leading edge form of the wavefront solution of the nonlinear equation. Substitution of the last expression into the linear equation (11.15) gives the *dispersion relation*, that is a relationship between c and a,

$$ca = 1 + a^2 \quad \Rightarrow \quad c = a + \frac{1}{a} . \qquad (11.18)$$

If we now plot this dispersion relation for c as a function of a, we see that $c_{min} = 2$ the value at $a = 1$. For all other values of $a(> 0)$ the wave speed $c > 2$.

Now consider $\min[e^{-ax}, e^{-x}]$ for x large and positive (since we are only dealing with the range where $u^2 \ll u$). If

$$a < 1 \quad \Rightarrow \quad e^{-ax} > e^{-x} ,$$

and so the velocity of propagation with asymptotic initial condition behaviour like (11.16) will depend on the *leading edge* of the wave, and the wave speed c is given by (11.18). On the other hand, if $a > 1$ then e^{-ax} is bounded above by e^{-x} and the front with wave speed $c = 2$. We are thus saying that if the initial conditions satisfy (11.16), then the asymptotic wave speed of the travelling wave solution of (11.6) is

$$c = a + \frac{1}{a}, \quad 0 < a \le 1, \quad c = 2, \quad a \ge 1 . \qquad (11.19)$$

The first of these has been proved by McKean (1975), the second by Larson (1978) and both verified numerically by Manoranjan and Mitchell (1983).

The Fisher equation is invariant under a change of sign of x, as mentioned before, so there is a wave solution of the form $u(x,t) = U(x + ct)$, $c > 0$ where now $U(-\infty) = 0$, $U(\infty) = 1$. So if we start with (11.6) for $-\infty < x < \infty$ and an initial condition $u(x,0)$ which is zero outside a finite domain such as illustrated in Fig. 11.2 the solution $u(x,t)$ will evolve into two travelling wavefronts, one moving left and the other to the right, both with speed $c = 2$. Note that if $u(x,0) < 1$ the $u(1 - u)$ term causes the solution to grow until $u = 1$. Clearly $u(x,t) \to 1$ as $t \to \infty$ for all x.

The axisymmetric form of Fisher's equation, namely

$$\frac{\partial u}{\partial t} = \frac{\partial^2 u}{\partial r^2} + \frac{1}{r}\frac{\partial u}{\partial r} + u(1 - u) \qquad (11.20)$$

does not possess travelling wavefront solutions in which a wave spreads out with constant speed, because of the $1/r$ term: the equation does not become an ordinary differential equation in the variable $z = r - ct$. Intuitively we can see

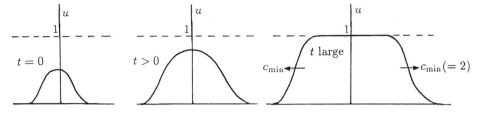

Fig. 11.2. Schematic time development of a wavefront solution of Fisher's equation on the infinite line.

what happens given $u(r, 0)$ qualitatively like the u in the first figure of Fig. 11.2. The u will grow because of the $u(1 - u)$ term since $u < 1$. At the same time diffusion will cause a wave-like dispersal outwards. On the 'wave' $\partial u/\partial r < 0$ so it effectively reduces the value of the right hand side in (11.20). This is equivalent to reducing the diffusion by an apparent convection or alternatively to reducing the source term $u(1 - u)$. The effect is to reduce the velocity of the outgoing wave. For large r the $(1/r)\partial u/\partial r$ term becomes negligible so the solution will tend asymptotically to a travelling wavefront solution with speed $c = 2$ as in the one-dimensional case. So, we can think of the axisymmetric wave-like solutions as having a 'wave speed' $c(r)$, a function of r, where, for r bounded away from $r = 0$, it increases monotonically with $c(r) \sim 2$ for r large.

The Fisher equation (11.4) has been the basis for a variety of models for spatial spread. Aoki (1987), for example, discusses gene-culture waves of advance. Ammerman and Cavali-Sforza (1971, 1984), in an interesting direct application of the model, applied it to the spread of early farming in Europe.

11.3 Asymptotic Solution and Stability of Wavefront Solutions of the Fisher Equation

Travelling wavefront solutions $U(z)$ for Fisher's equation (11.6) satisfy (11.8), namely

$$U'' + cU' + U(1 - U) = 0 , \tag{11.21}$$

and monotonic solutions exist, with $U(-\infty) = 1$ and $U(\infty) = 0$, for all wave speeds $c \geq 2$. The phase plane trajectories are solutions of (11.11), that is

$$\frac{dV}{dU} = \frac{-cV - U(1 - U)}{V} . \tag{11.22}$$

No analytical solutions of these equations for general c have been found although there is an exact solution for a particular $c(> 2)$, as we show below in Section 11.4. There is, however, a small parameter in the equations, namely $\varepsilon = 1/c^2 \leq 0.25$, which suggests we look for asymptotic solutions for $0 < \varepsilon \ll 1$ (see, for example the book by Murray (1984) for a simple description of these asymptotic techniques). Canosa (1973) obtained such asymptotic solutions to (11.21) and (11.22).

Since the wave solutions are invariant to any shift in the origin of the coordinate system (the equation is unchanged if $z \to z + \text{constant}$) let us take $z = 0$ to be the point where $U = 1/2$. We now use a standard singular perturbation technique. The procedure is to introduce a change of variable in the vicinity of the front, which here is at $z = 0$, in such a way that we can find the solution as a Taylor expansion in the small parameter ε. We can do this with the transformation

$$U(z) = g(\xi), \quad \xi = \frac{z}{c} = \varepsilon^{1/2} z . \tag{11.23}$$

The actual transformation in many cases is found by trial and error until the resulting transformed equation gives a consistent perturbation solution satisfying the boundary conditions. With (11.23), (11.21), together with the boundary conditions on U, becomes

$$\varepsilon \frac{d^2 g}{d\xi^2} + \frac{dg}{d\xi} + g(1 - g) = 0$$

$$g(-\infty) = 1, \quad g(\infty) = 0, \quad 0 < \varepsilon \le \frac{1}{c_{\min}^2} = 0.25 , \tag{11.24}$$

and we further require $g(0) = 1/2$.

The equation for g as it stands looks like the standard singular perturbation problem since ε multiplies the highest derivative: that is setting $\varepsilon = 0$ reduces the order of the equation and usually causes difficulties with the boundary conditions. With this equation, and in fact frequently with such singular perturbation analysis of shockwaves and wavefronts, the reduced equation gives a uniformly valid first order approximation: the reason for this is the form of the nonlinear term $g(1 - g)$ which is zero at both boundaries.

Now look for solutions of (11.24) as a regular perturbation series in ε, that is let

$$g(\xi; \varepsilon) = g_0(\xi) + \varepsilon g_1(\xi) + \dots . \tag{11.25}$$

The boundary conditions at $\pm\infty$ and the choice of $U(0) = 1/2$, which requires $g(0; \varepsilon) = 1/2$ for all ε, gives from (11.25) the conditions on the $g_i(\xi)$ for $i = 0, 1, 2, \dots$ as

$$g_0(-\infty) = 1, \quad g_0(\infty) = 0, \quad g_0(0) = \frac{1}{2} ,$$

$$g_i(\pm\infty) = 0, \quad g_i(0) = 0 \quad \text{for} \quad i = 1, 2, \dots \tag{11.26}$$

On substituting (11.25) into (11.24) and equating powers of ε we get

$$O(1): \quad \frac{dg_0}{d\xi} = -g_0(1 - g_0) \quad \Rightarrow \quad g_0(\xi) = (1 + e^{\xi})^{-1} ,$$

$$O(\varepsilon): \quad \frac{dg_1}{d\xi} + (1 - 2g_0)g_1 = -\frac{d^2 g_0}{d\xi^2} , \tag{11.27}$$

and so on, for higher orders in ε. The constant of integration in the g_0-equation was chosen so that $g_0(0) = 1/2$ as required by (11.26). Using the first of (11.27), the g_1-equation becomes

$$\frac{dg_1}{d\xi} - \left(\frac{g_0''}{g_0'}\right)g_1 = -g_0'' \, ,$$

which on integration and using the conditions (11.26) gives

$$g_1 = -g_0' \ln[4|g_0'|] = e^\xi(1 + e^\xi)^{-2} \ln\left[\frac{4e^\xi}{(1+e^\xi)^2}\right] . \tag{11.28}$$

In terms of the original variables U and z from (11.23) the uniformly valid asymptotic solution for all z is given by (11.25)-(11.28) as

$$U(z;\varepsilon) = (1 + e^{z/c})^{-1} + c^{-2}e^{z/c}(1 + e^{z/c})^{-2} \ln[4e^{z/c}(1 + e^{z/c})^{-2}]$$
$$+ O(c^{-4}), \quad c \geq c_{min} = 2 . \tag{11.29}$$

This asymptotic solution is least accurate for $c = 2$. However, when this solution is compared with the computed wavefront solution of Fisher's equation (11.6), the one with speed $c = 2$, the *first* term alone, that is the $O(1)$ term $(1+e^{z/c})^{-1}$, is everywhere within a few percent of it. It is an encouraging fact that asymptotic solutions with 'small' parameters, even of the order of that used here, frequently give remarkably accurate solutions.

Let us now use the asymptotic solution (11.29) to investigate the relationship between the steepness or slope of the wavefront solution and its speed of propagation. Since the gradient of the wavefront is everywhere negative a measure of the steepness, s say, of the wave is the magnitude of the maximum of the gradient $U'(z)$, that is the point where $U'' = 0$, namely the point of inflexion of the wavefront solution. From (11.23) and (11.25) that is where

$$g_0''(\xi) + \varepsilon g_1''(\xi) + O(\varepsilon^2) = 0 \, ,$$

which, from (11.27) and (11.28), gives $\xi = 0$, that is $z = 0$. The gradient at $z = 0$, using (11.29), gives

$$-U'(0) = s = \frac{1}{4c} + O\left(\frac{1}{c^5}\right) , \tag{11.30}$$

which, of course, only holds for $c \geq 2$. This result implies that the *faster* the wave moves, that is the larger the c, the *less steep* is the wavefront. Although the width of the wave is strictly from $-\infty$ to ∞, a practical measure of the width, L say, is the inverse of the steepness, that is $L = 1/s = 4c$ from (11.30). Fig. 11.3 illustrates this effect.

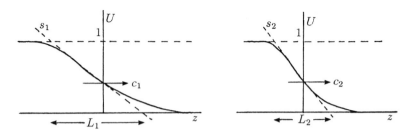

Fig. 11.3. Steepness $s(= |U'(0)|)$ and a practical measure of the width $L(= 1/s)$ of wavefront solutions of the Fisher equation (11.6) for two wave speeds c_2 and $c_1 > c_2 \geq 2$. The flatter the wave the faster it moves.

The results in this section can be generalized to single species population models where logistic growth is replaced by an appropriate $f(u)$, so that (11.6) becomes

$$\frac{\partial u}{\partial t} = f(u) + \frac{\partial^2 u}{\partial x^2} ,\qquad (11.31)$$

where $f(u)$ has only two zeros, say u_1 and $u_2 > u_1$. If $f'(u_1) > 0$ and $f'(u_2) < 0$ then by a similar analysis to the above, wavefront solutions evolve with u going monotonically from u_1 to u_2 with wave speeds

$$c \geq c_{\min} = 2[f'(u_1)]^{1/2} .\qquad (11.32)$$

These results are as expected, with (11.32) obtained by linearising $f(u)$ about the leading edge where $u \approx u_1$ and comparing the resulting equation with (11.15).

Stability of Travelling Wave Solutions

The stability of solutions of biological models is an important aspect and is often another reliability test of model mechanisms. The travelling wavefront solutions of Fisher's equation present an interesting and important stability case study.

We saw above that the speed of propagation of the wavefront solutions (see (11.19) with (11.16)) depends sensitively on the explicit behaviour of the initial conditions $u(x, 0)$ as $|x| \to \infty$. This implies that the wavefront solutions are unstable to perturbations in the far field. On the other hand if $u(x, 0)$ has compact support, that is the kind of initial conditions (11.14) used by Kolmogoroff et al. (1937), then the ultimate wave does not depend on the detailed form of $u(x, 0)$. Unless the numerical analysis is carefully performed, with a priori knowledge of the wave speed expected, the evolving wave has speed $c = 2$. Random effects introduced by the numerical scheme are restricted to the *finite* domain. Any practical model deals, of course, with a finite domain. So it is of importance to consider the stability of the wave solutions to perturbations which are zero outside a finite domain, which includes the wavefront. We shall show, following Canosa (1973), that the solutions are stable to such finite perturbations, if they are perturbations in the moving frame of the wave.

Let $u(x,t) = u(z,t)$, where $z = x - ct$; that is we take z and t as the independent variables in place of x and t. Equation (11.6) becomes for $u(z,t)$

$$u_t = u(1 - u) + cu_z + u_{zz} \tag{11.33}$$

where subscripts now denote partial derivatives. We are concerned with $c \geq c_{min} = 2$ and we denote the wavefront solution $U(z)$, namely the solution of (11.21), by $u_c(z)$: it satisfies the right hand side of (11.33) set equal to zero. Now consider a small perturbation on $u_c(z)$ of the form

$$u(z,t) = u_c(z) + \omega v(z,t), \quad 0 < \omega \ll 1 . \tag{11.34}$$

Substituting this into (11.33) and keeping only the first order terms in ω we get the equation governing $v(z,t)$ as

$$v_t = [1 - 2u_c(z)]v + cv_z + v_{zz} . \tag{11.35}$$

The solution $u_c(z)$ is stable to perturbations $v(z,t)$ if

$$\lim_{t \to \infty} v(z,t) = 0 \quad \text{or} \quad \lim_{t \to \infty} v(z,t) = \frac{du_c(z)}{dz} .$$

The fact that $u_c(z)$ is stable if the second of these holds is because $v(z,t)$ then represents a small translation of the wave along the x-axis since

$$u_c(z + \delta z) \approx u_c(z) + \delta z \frac{du_c(z)}{dz} .$$

Now look for solutions to the linear equation (11.35) by setting

$$v(z,t) = g(z)e^{-\lambda t} , \tag{11.36}$$

which on substituting into (11.35) gives, on cancelling the exponentials,

$$g'' + cg' + [\lambda + 1 - 2u_c(z)]g = 0 . \tag{11.37}$$

Note that if $\lambda = 0$, $g(z) = du_c(z)/dz$ is a solution of this equation, which as we showed, implies that the travelling wave solution is invariant under translation along the z-axis.

Now use the fact that $v(z,t)$ is non-zero only in a finite domain, which from (11.36) means that boundary conditions $g(\pm L) = 0$ for some L are appropriate for g in (11.37). If we introduce $h(z)$ by

$$g(z) = h(z)e^{-cz/2} ,$$

the eigenvalue problem, to determine the possible λ, becomes

$$h'' + \left[\lambda - \left\{2u_c(z) + \frac{c^2}{4} - 1\right\}\right]h = 0, \quad h(\pm L) = 0 \tag{11.38}$$

in which

$$2u_c(z) + \frac{c^2}{4} - 1 \geq 2u_c(z) > 0$$

since $c \geq 2$ and $u_c(z) > 0$ in the finite domain $-L \leq z \leq L$. Standard theory (for example, Titchmarsh 1946, Chapter 1) now gives the result that all eigenvalues λ of (11.38) are real and positive. So, from (11.36), $v(z,t)$ tends to zero as $t \to \infty$. Thus the travelling wave solutions $u_c(z)$ are stable to all small finite domain perturbations of the type $v(z,t)$ in (11.34). In fact such perturbations are not completely general since they are perturbations in the moving frame. The general problem has been treated by several authors, for example Hoppensteadt (1975) and Larson (1978): the analysis is somewhat more complex. The fact that Fisher waves are stable to finite domain perturbations makes it clear why typical numerical simulations of Fisher's equation result in stable wavefront solutions with speed $c = 2$.

11.4 Density-Dependent Diffusion Reaction Diffusion Models and Some Exact Solutions

We saw in Section 9.2 in Chapter 9 that in certain insect dispersal models the diffusion coefficient D depended on the population u. There we did not include any growth dynamics. If we wish to consider longer time scales then we should include such growth terms in the model. A natural extension to incorporate density-dependent diffusion is thus, in the one-dimensional situation, to consider equations of the form

$$\frac{\partial u}{\partial t} = f(u) + \frac{\partial}{\partial x}\left[D(u)\frac{\partial u}{\partial x}\right] , \tag{11.39}$$

where typically $D(u) = D_0 u^m$, with D_0 and m positive constants. Here we shall consider functions $f(u)$ which have two zeros, one at $u = 0$ and the other at $u = 1$. Equations in which $f \equiv 0$ have been studied much more widely than those with non-zero f: see, for example Chapter 9. To be even more specific we shall consider $f(u) = ku^p(1 - u^q)$, where p and q are positive constants. By a suitable rescaling of t and x we can absorb the parameters k and D_0 and the equations we shall thus consider in this section are then of the general form

$$\frac{\partial u}{\partial t} = u^p(1 - u^q) + \frac{\partial}{\partial x}\left[u^m \frac{\partial u}{\partial x}\right] . \tag{11.40}$$

where p, q and m are positive parameters. If we write out the diffusion term in full we get

$$\frac{\partial u}{\partial t} = u^p(1 - u^q) + mu^{m-1}\left(\frac{\partial u}{\partial x}\right)^2 + u^m \frac{\partial^2 u}{\partial x^2}$$

which shows that the nonlinear diffusion can be thought of as contributing an equivalent *convection* with 'velocity' $-mu^{m-1}\partial u/\partial x$.

It might be argued that the forms in (11.40) are rather special. However with the considerable latitude to choose p, q and m such forms can qualitatively mimic more complicated forms for which only numerical solutions are possible. The usefulness of analytical solutions, of course, is the ease with which we can see how solutions depend analytically on the parameters. In this way we can then infer the qualitative behaviour of the solutions of more complicated but more realistic model equations. There are, however, often hidden serious pitfalls, one of which is important and which we shall point out below.

To relate the exact solutions to the above results for the Fisher equation we consider first $m = 0$ and $p = 1$ and (11.40) becomes

$$\frac{\partial u}{\partial t} = u(1 - u^q) + \frac{\partial^2 u}{\partial x^2}, \quad q > 0 . \tag{11.41}$$

Since $u = 0$ and $u = 1$ are the uniform steady states, we look for travelling wave solutions in the form

$$u(x,t) = U(z), \quad z = x - ct, \quad U(-\infty) = 1, \quad U(\infty) = 0 \tag{11.42}$$

where $c > 0$ is the wave speed we must determine. The ordinary differential equation for $u(z)$ is

$$L(U) = U'' + cU' + U(1 - U^q) = 0 , \tag{11.43}$$

which defines the operator L. This equation can of course be studied in the (U', U) phase plane. With the form of the first term in the asymptotic wavefront solution to Fisher's equation given by (11.29) let us optimistically look for solutions of (11.43) in the form

$$U(z) = \frac{1}{(1 + ae^{bz})^s} , \tag{11.44}$$

where a, b and s are positive constants which have to be found. This form automatically satisfies the boundary conditions at $z = \pm\infty$ in (11.42). Because of the translational invariance of the equation we can say at this stage that a is arbitrary: it can be incorporated into the exponential as a translation $b^{-1} \ln a$ in z. It is, however, useful to leave it in as a way of keeping track of the algebraic manipulation. Another reason for keeping it in is that if b and s can be found so that (11.44) is an exact solution of (11.43) then they cannot depend on a.

Substitution of (11.44) into (11.43) gives, after some trivial but tedious algebra,

$$L(U) = (1 + ae^{bz})^{-s-2} \left\{ [s(s+1)b^2 - sb(b+c) + 1]a^2 e^{2bz} \right.$$
$$\left. + [2 - sb(b+c)]ae^{bz} + 1 - [1 + ae^{bz}]^{2-sq} \right\}. \tag{11.45}$$

For $L(U) = 0$ for all z, the coefficients of e^0, e^{bz} and e^{2bz} within the curly brackets must all be identically zero. This implies that

$$2 - sq = 0, \ 1 \ \text{or} \ 2 \quad \Rightarrow \quad s = \frac{2}{q}, \frac{1}{q} \quad \text{or} \quad sq = 0 .$$

Clearly $sq = 0$ is not possible since s and q are positive constants. Consider the other two possibilities.

With $s = 1/q$ the coefficients of the exponentials from (11.45) give

$$\begin{array}{ll} e^{bz}: & 2 - sb(b+c) - 1 = 0 \ \Rightarrow \ sb(b+c) = 1 \\ e^{2bz}: & s(s+1)b^2 - sb(b+c) + 1 = 0 \end{array} \left. \right\} \Rightarrow s(s+1)b^2 = 0 \ \Rightarrow \ b = 0$$

since $s > 0$. This case is therefore also not a possibility since necessarily $b > 0$.

Finally if $s = 2/q$ the coefficients of e^{bz} and e^{2bz} are

$$e^{bz}: \quad sb(b+c) = 2; \qquad e^{2bz}: \quad s(s+1)b^2 = 1$$

which together give b and c as

$$s = \frac{2}{q}, \quad b = [s(s+1)]^{-1/2}, \quad c = \frac{2}{sb} - b$$

which then determine s, b and a *unique* wave speed c in terms of q as

$$s = \frac{2}{q}, \quad b = q[2(q+2)]^{-1/2}, \quad c = (q+4)[2(q+2)]^{-1/2} . \tag{11.46}$$

From these we see that the wave speed c increases with $q(> 0)$. A measure of the steepness, S, given by the magnitude of the gradient at the point of inflexion, is easily found from (11.44). The point of inflexion, z_i, is given by $z_i = -b^{-1} \ln{(as)}$ and hence the gradient at z_i gives the steepness

$$S = \frac{b}{\left(1 + \frac{1}{s}\right)^{s+1}} = \frac{\frac{1}{2}q}{\left(1 + \frac{q}{2}\right)^{3/2 + 2/q}} .$$

So, with increasing q the wave speed c increases and the steepness decreases, as was the case with the Fisher wavefront solutions.

When $q = 1$, equation (11.41) becomes the Fisher equation (11.6) and from (11.46)

$$s = 2, \quad b = \frac{1}{\sqrt{6}}, \quad c = \frac{5}{\sqrt{6}} .$$

We then get *an* exact analytical travelling wave solution from (11.44). The arbitrary constant a can be chosen so that $z = 0$ corresponds to $U = 1/2$, in which case $a = \sqrt{2} - 1$ and the solution is

$$U(z) = \frac{1}{\left[1 + (\sqrt{2} - 1)e^{z/\sqrt{6}}\right]^2} . \tag{11.47}$$

This solution has a wave speed $c = 5/\sqrt{6}$ and on comparison with the asymptotic solution (11.29) to $O(1)$ it is much steeper.

This example highlights one of the serious problems with such exact solutions which we alluded to above, namely they often do not determine all possible solutions and indeed, may not even give the most relevant one, as is the case here. This is not because the wave speed is not 2, in fact $c = 5/\sqrt{6} \approx 2.04$, but rather that the quantitative wave form is so different. To analyse this general form (11.43) properly, a careful phase plane analysis has to be carried out.

Another class of exact solutions can be found for (11.40) with $m = 0, p = q+1$ with $q > 0$, which gives the equation as

$$\frac{\partial u}{\partial t} = u^{q+1}(1 - u^q) + \frac{\partial^2 u}{\partial x^2} . \tag{11.48}$$

Substituting $U(z)$ from (11.44) into the travelling wave form of the last equation and proceeding exactly as before we find a travelling wavefront solution exists, with a unique wave speed, given by

$$U(z) = \frac{1}{(1 + ae^{bz})^s}, \quad s = \frac{1}{q}, \quad b = \frac{q}{(q+1)^{1/2}}, \quad c = (q+1)^{-1/2} . \tag{11.49}$$

A more interesting and useful exact solution has been found for the case $p = q = 1, m = 1$ with which (11.40) becomes

$$\frac{\partial u}{\partial t} = u(1 - u) + \frac{\partial}{\partial x}\left[u\frac{\partial u}{\partial x}\right] , \tag{11.50}$$

a nontrivial example of density dependent diffusion with logistic population growth. Physically this model implies that the population disperses to regions of lower density more rapidly as the population gets more crowded. The solution, derived below, was found independently by Aronson (1980) and Newman (1980). Newman (1983) studies more general forms and carries the work further.

Let us look for the usual travelling wave solutions of (11.50) with $u(x,t) = U(z)$, $z = x - ct$, and so we consider

$$(UU')' + cU' + U(1 - U) = 0 \, ,$$

for which the phase plane system is

$$U' = V, \quad UV' = -cV - V^2 - U(1 - U) \, . \tag{11.51}$$

We are interested in wavefront solutions for which $U(-\infty) = 1$ and $U(\infty) = 0$: we anticipate $U' < 0$. There is a singularity at $U = 0$ in the second equation. We remove this singularity by defining a new variable ζ as

$$U\frac{d}{dz} = \frac{d}{d\zeta} \quad \Rightarrow \quad \frac{dU}{d\zeta} = UV, \quad \frac{dV}{d\zeta} = -cV - V^2 - U(1 - U) \, , \tag{11.52}$$

which is not singular. The critical points in the (U,V) phase plane are

$$(U,V) = (0,0), \quad (1,0), \quad (0,-c) \, .$$

A linear analysis about $(1,0)$ and $(0,-c)$ shows them to be saddle points while $(0,0)$ is like a stable nonlinear node – nonlinear because of the UV in the U-equation in (11.52). Fig. 11.4 illustrates the phase trajectories for (11.52) for various c. From Section 9.2 we can expect the possibility of a wave with a discontinuous tangent at a specific point z_c, the one where $U \equiv 0$ for $z \geq z_c$. This corresponds to a phase trajectory which goes from $(1,0)$ to a point on the $U = 0$ axis at some finite non-zero negative V. Referring now to Fig. 11.4 (a), if $0 < c < c_{min}$ there is no trajectory possible from $(1,0)$ to $U = 0$ except unrealistically for infinite V. As c increases there is a bifurcation value c_{min} for which there is a unique trajectory from $(1,0)$ to $(0,-c_{min})$ as shown in Fig. 11.4 (b). This means that at the wavefront z_c, where $U = 0$, there is a discontinuity in the derivative from $V = U' = -c_{min}$ to $U' = 0$ and $U = 0$ for all $z > z_c$: see Fig. 11.4 (d). As c increases beyond c_{min} a trajectory always exists from $(1,0)$ to $(0,0)$ but now the wave solution has $U \to 0$ and $U' \to \infty$ as $z \to \infty$: this type of wave is also illustrated in Fig. 11.4 (d).

As regards the exact solution, the trajectory connecting $(1,0)$ to $(0,-c)$ in Fig. 11.4 (b) is in fact a straight line $V = -c_{min}(1 - U)$ if c_{min} is appropriately chosen. In other words this is a solution of the phase plane equation which, from (11.51), is

$$\frac{dV}{dU} = \frac{-cV - V^2 - U(1 - U)}{UV} \, .$$

Substitution of $V = -c_{min}(1 - U)$ in this equation, with $c = c_{min}$, shows that $c_{min} = 1/\sqrt{2}$. If we now return to the first of the phase equations in (11.51), namely $U' = V$ and use the phase trajectory solution $V = -(1 - U)/\sqrt{2}$ we get

$$U' = -\frac{1 - U}{\sqrt{2}} \, ,$$

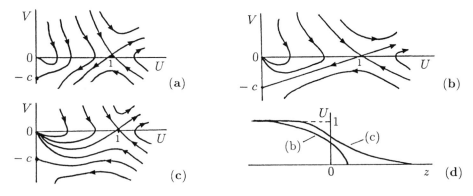

Fig. 11.4a-d. Qualitative phase plane trajectories for the travelling wave equations (11.52) for various c (after Aronson 1980). In (a) no trajectory is possible from $(1,0)$ to $U = 0$ at a finite V. In (b) and (c) travelling wave solutions from $U = 1$ to $U = 0$ are possible but with different characteristics: the travelling wave solutions in (d) illustrate these differences. Note that the solution corresponding to (b) has a discontinuous derivative at the leading edge.

which, on using $U(-\infty) = 1$, gives

$$U(z) = 1 - \exp\left[\frac{z - z_c}{\sqrt{2}}\right] \qquad z < z_c$$
$$\qquad = 0 \qquad\qquad\qquad z > z_c \qquad\qquad (11.53)$$

where z_c is the front of the wave: it can be arbitrarily chosen in the same way as the a in the solutions (11.44). This is the solution sketched in Fig. 11.4 (d).

This analysis, showing the existence of the travelling waves, can be extended to more general cases in which the diffusion coefficient is u^m, for $m \neq 1$, or even more general $D(u)$ in (11.40) if it satisfies certain criteria.

It is perhaps appropriate to state briefly here the travelling wave results we have derived for the Fisher equation and its generalizations to a general $f(u)$ normalized such that $f(0) = 0 = f(1)$, $f'(0) > 0$ and $f'(1) < 0$. In dimensionless terms we have shown that there is a travelling wavefront solution with $0 < u < 1$ which can evolve, with appropriate initial conditions, from (11.31). Importantly these solutions have speeds $c \geq c_{min} = 2[f'(0)]^{1/2}$ with the usual computed form having speed c_{min}. For the Fisher equation (11.4) this *dimensional* wave speed, c^* say, using the nondimensionalization (11.5), is $c^* = 2[kD]^{1/2}$: here k is a measure of the linear growth rate or of the linear kinetics. If we consider typical biological values for D of 10^{-9}–10^{-11} cm^2 sec^{-1} and k is $O(1$ sec$^{-1})$ say, the speed of propagation is then $O(2 \times 10^{-4.5}$–$10^{-5.5}$ cm. sec$^{-1})$. With this, the time it takes to cover a distance of the order of $1\,$mm is $O(5 \times 10^{2.5}$–$10^{3.5}$ sec.) which is *very* much shorter than the pure diffusional time of $O(10^7$–10^9 sec.). It is the combination of reaction and diffusion which greatly enhances the efficiency of information transferral via travelling waves of concentration changes. This reaction diffusion interaction, as we shall see in later chapters, totally changes our concept of the role of diffusion in a large number of important biological situations.

Before leaving this Section let us go back to something we mentioned earlier in the Section when we noted that nonlinear diffusion could be thought of as equivalent to a nonlinear convection effect: the equation following (11.40) demonstrates this. If the convection arises as a natural extension of a conservation law we get, instead, equations like

$$\frac{\partial u}{\partial t} + \frac{\partial h(u)}{\partial x} = f(u) + \frac{\partial^2 u}{\partial x^2} , \tag{11.54}$$

where $h(u)$ is a given function of u. Here the left-hand side is in standard 'conservation' form, that is, it is in the form of a divergence, namely $(\partial/\partial t, \partial/\partial x) \cdot (u, h(u))$: the convective 'velocity' is $h'(u)$. Such equations arise in a variety of contexts, for example in ion-exchange columns and chromotagraphy: see Goldstein and Murray (1959). They have also been studied by Murray (1968, 1970a, 1970b, 1973), where other practical applications of such equations are given, together with analytical techniques for solving them.

The effect of nonlinear convection in reaction diffusion equations can have a dramatic effect on the solutions. This is to be expected since we have another major transport process, namely convection, which depends nonlinearly on u. This process may or may not enhance the diffusional transport. If the diffusion process is negligible compared with the convection effects the solutions can exhibit shock-like solutions (see Murray 1968, 1970a, 1970b, 1973).

Although the analysis is harder than for the Fisher equation, we can determine conditions for the existence of wavefront solutions. For example consider the simple, but non-trivial, case where $h'(u) = ku$ with k a positive or negative constant and $f(u)$ logistic. Equation (11.54) is then

$$\frac{\partial u}{\partial t} + ku\frac{\partial u}{\partial x} = u(1 - u) + \frac{\partial^2 u}{\partial x^2} . \tag{11.55}$$

With $k = 0$ this reduces to the Fisher equation (11.6) the wavefront solutions of which we studied in detail in Section 11.2.

Suppose $k \neq 0$ and we look for travelling wave solutions to (11.55) in the form (11.7), namely

$$u(x, t) = U(z), \quad z = x - ct , \tag{11.56}$$

where, as usual, the wavespeed c has to be found. Substituting into (11.55) gives

$$U'' + (c - kU)U' + U(1 - U) = 0 \tag{11.57}$$

for which appropriate boundary conditions are given by (11.9), namely

$$\lim_{z \to \infty} U(z) = 0, \quad \lim_{z \to -\infty} U(z) = 1 . \tag{11.58}$$

Equations (11.57) and (11.58) define the eigenvalue problem for $c(k)$.

From (11.57), with $V = U'$, the phase plane trajectories are solutions of

$$\frac{dV}{dU} = \frac{-(c - kU)V - U(1 - U)}{V} .$$

(11.59)

Singular points of the last equation are (0,0) and (1,0). We require conditions on $c = c(k)$ such that a monotonic solution exists in which $0 \le U \le 1$ and $U'(z) \le 0$: that is we require a phase trajectory lying in the quadrant $U \ge 0$, $V \le 0$ which joins the singular points. A standard linear phase plane analysis about the singular points shows that $c \ge 2$, which guarantees that $(0,0)$ is a stable node and $(1,0)$ a saddle point. The specific equation (11.55) and the travelling wave form (11.59) have been studied analytically and numerically by the author and R.J. Gibbs (see Murray 1977). It can be shown that a travelling wave solution exists for all $c \ge c(k)$ where

$$c(k) = \left\{ \begin{array}{l} 2 \\ \dfrac{k}{2} + \dfrac{2}{k} \end{array} \right. \quad \text{if} \quad \left\{ \begin{array}{l} 2 > k > -\infty \\ 2 \le k < \infty \end{array} \right.$$

(11.60)

We thus see that here $c = 2$ is a lower bound for only a limited range of k, a more accurate bound being given by the last equation. We present the main elements of the analysis below.

The expression $c = c(k)$ in the last equation gives the wave speed in terms of a key parameter in the model. It is another example of a *dispersion relation*, here associated with wave phenomena. The general concept of dispersion relations are of considerable importance and is a subject we shall be very much involved with later in the book, particularly in Chapters 14–17.

Brief Derivation of the Wave Speed Dispersion Relation (11.60)

Linearizing (11.59) about (0,0) gives

$$\frac{dV}{dU} = \frac{-cV - U}{V}$$

with eigenvalues

$$e_\pm = \frac{-c \pm (c^2 - 4)^{1/2}}{2} .$$

(11.61)

Since we require $U \ge 0$ these must be real and so we must have $c \ge 2$. Thus $0 > e_+ > e_-$ and so $(0,0)$ is a stable node and, for large z

$$\begin{pmatrix} V \\ U \end{pmatrix} \rightarrow a \begin{pmatrix} e_+ \\ 1 \end{pmatrix} \exp[e_+ z] + b \begin{pmatrix} e_- \\ 1 \end{pmatrix} \exp[e_- z]$$

where a and b are constants. This implies that

$$\frac{dV}{dU} \rightarrow \left\{ \begin{array}{l} e_+ \\ e_- \end{array} \right. \quad \text{as} \quad z \rightarrow \infty \quad \text{if} \quad \left\{ \begin{array}{l} a \ne 0 \\ a = 0 \end{array} \right.$$

(11.62)

An exact solution of (11.59) is

$$V = -\frac{k}{2}U(1 - U) \quad \text{if} \quad c = \frac{k}{2} + \frac{2}{k} \ . \tag{11.63}$$

With this expression for c,

$$(c^2 - 4)^{1/2} = \begin{cases} \dfrac{k}{2} - \dfrac{2}{k} \\[2mm] \dfrac{2}{k} - \dfrac{k}{2} \end{cases} \quad \text{if} \quad \begin{cases} k \geq 2 \\ k < 2 \end{cases}$$

and so from (11.61)

$$e_+ = \begin{cases} -\dfrac{2}{k} \\[2mm] -\dfrac{k}{2} \end{cases} \text{if} \begin{cases} k \geq 2 \\ k < 2 \end{cases}, \quad e_- = \begin{cases} -\dfrac{k}{2} \\[2mm] -\dfrac{2}{k} \end{cases} \text{if} \begin{cases} k \geq 2 \\ k < 2 \end{cases}$$

But, from (11.63)

$$\left. \frac{dV}{dU} \right]_{U=0} = -\frac{k}{2} = \begin{cases} e_- \\ e_+ \end{cases} \quad \text{for} \quad \begin{cases} k \geq 2 \\ k < 2 \end{cases}$$

So, from (11.62), for $k \geq 2$ we see that $V(U)$ satisfies $dV/dU \to e_-$ as $z \to \infty$. This gives the second result in (11.60), namely that the wave speed

$$c = \frac{k}{2} + \frac{2}{k} \quad \text{for} \quad k \geq 2 \ . \tag{11.64}$$

Now consider $k < 2$ and $z \to -\infty$. Linearizing about $(1,0)$ gives the eigenvalues $E_\pm$ as

$$E_\pm = \frac{-(c - k) \pm \{(c - k)^2 + 4\}^{1/2}}{2} \tag{11.65}$$

so $E_+ > 0 > E_-$ and $(1,0)$ is a saddle point. As $z \to -\infty$, $U \to 1 - O(\exp[E_+ z])$ from which we see that

$$\frac{dV}{dU} \to E_+(c, k) \quad \text{as} \quad z \to -\infty.$$

With $c \geq 2$ we see from (11.65) that

$$\frac{dE_+(k)}{dk} = [(c - k)^2 + 4]^{1/2} E_+ > 0 \tag{11.66}$$

and so, for U sufficiently close to $U = 1$, dV/dU increases with increasing k. Thus, for U close enough to $U = 1$, the phase plane trajectory $V(U, c, k)$ satisfies

$$V(U, c = 2, k) < V(U, c = 2, k = 2) \quad \text{for} \quad k < 2 \ . \tag{11.67}$$

Now let us suppose that a number d exists, where $0 < d < 1$, such that

$$V(d, c = 2, k = 2) = V(d, c = 2, k),$$
$$V(U, c = 2, k = 2) < V(U, c = 2, k) \quad \text{for} \quad d < U < 1.$$

This implies that

$$\frac{dV(d, c = 2, k = 2)}{dU} \leq \frac{dV(d, c = 2, k)}{dU}. \tag{11.68}$$

But, from (11.59),

$$\frac{dV(d, c = 2, k)}{dU} = -2 + kd - \frac{d(1 - d)}{V(d, c = 2, k)}$$

which, with (11.68), implies

$$-2 + 2d - \frac{d(1 - d)}{V(d, c = 2, k = 2)} \leq -2 + kd - \frac{d(1 - d)}{V(d, c = 2, k)}$$

which, together with the first of (11.67), in turn implies

$$2d \leq kd \quad \Rightarrow \quad 2 \leq k.$$

But this contradicts $k < 2$, so supposition (11.67) is not possible and so implies that the wave speed $c \geq 2$ for all $k < 2$. This together with (11.64) is the result in (11.60).

We have only given the essentials here; to prove the result more rigorously we have to examine the possible trajectories more carefully to show that everything is consistent, such as the trajectories not cutting the U-axis for $U \in (0, 1)$: this can all be done. The result (11.60) is related to the analysis in Section 11.2, where we showed how the wave speed could depend on either the wave front or the wave tail.

When $k \neq 0$ we can cast (11.55) in a different form which highlights the nonlinear convective contribution as opposed to the diffusion contribution to the wave solutions. Suppose $k > 0$ and set

$$\varepsilon = \frac{1}{k^2}, \quad y = \frac{x}{k} = \varepsilon^{1/2} x \quad (k > 0)$$
$$\Rightarrow \quad u_t + u u_y = u(1 - u) + \varepsilon u_{yy}. \tag{11.69}$$

If $k < 0$ we take

$$\varepsilon = \frac{1}{k^2}, \quad y = \frac{x}{k} = \varepsilon^{1/2} x \quad (k < 0)$$
$$\Rightarrow \quad u_t - u u_y = u(1 - u) + \varepsilon u_{yy}. \tag{11.70}$$

We now consider travelling wave solutions as $\varepsilon \to 0$.

With $u(x,t)$ a solution of (11.55), $u(ky,t)$ is a solution of (11.69). So with $U(x-ct)$ a solution of (11.59) satisfying $U(-\infty) = 1$, $U(\infty) = 0$, $U(ky-ct)$ is a solution of (11.69) and the wave speed $\lambda = c/k = c\varepsilon^{1/2}$. So, using the wave speed estimates from (11.60), equation (11.69) has travelling wave solutions for all

$$\lambda \geq \lambda(\varepsilon) = \frac{c(k)}{k} = c(\varepsilon^{-1/2})\varepsilon^{1/2}$$

and so

$$\lambda(\varepsilon) = \begin{cases} 2\varepsilon^{1/2} \\ \dfrac{1}{2} + 2\varepsilon \end{cases} \quad \text{if} \quad \begin{cases} \varepsilon > \dfrac{1}{4} \\ \dfrac{1}{4} \geq \varepsilon > 0 \end{cases}$$

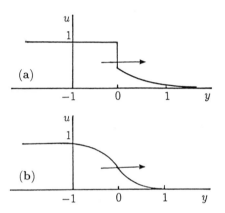

Fig. 11.5a,b. Travelling wave solutions computed from (11.69). Each has wave speed $\lambda = 0.74$ but with different ε; (a) $\varepsilon = 0$, (b) $\varepsilon = 0.12$. The origin is where $u = 0.5$.

Now let $\varepsilon \to 0$ in (11.69) to get

$$u_t + uu_y = u(1-u) .$$

Solutions of this equation can be discontinuous (these are the weak solutions discussed in detail by Murray 1970a). For ε small the wave steepens into a shock-like solution. On the other hand for (11.70) with the same boundary conditions discontinuous solutions do not occur (see Murray 1970a). Fig. 11.5 gives numerically computed travelling wave solutions for (11.69) for a given wave speed and two different values for ε: note the discontinuous solution in Fig. 11.5 (a). Fig. 11.6 shows computed wave solutions for (11.70) for small ε. Note that here the wave steepens but does not display discontinuities like that in Fig. 11.5 (a).

To conclude this section we should note the results of Satsuma (1987) on exact solutions of scalar density dependent reaction diffusion equations. The method he develops is novel and is potentially of wider applicability. The work

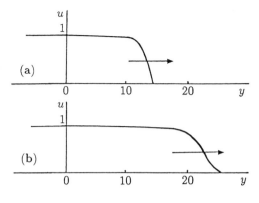

Fig. 11.6a,b. Travelling wave solutions, computed from (11.70), with minimum speed $c = k/2 + 2/k$, $\varepsilon = 1/k^2$, for two different values of ε: (a) $\varepsilon = 10^{-4}$, wave speed $c \approx 2.2$; (b) $\varepsilon = 10^{-1}$, wave speed $c \approx 50$. The origin is where $u = 1 - 10^{-6}$.

on the existence and stability of monotone wave solutions of such equations by Hosono (1986) is also of particular relevance to the material in this section.

An important point about the material in this discussion of nonlinear convection reaction diffusion equations is that it shows how much more varied the solutions of such equations can be.

11.5 Waves in Models with Multi-Steady State Kinetics: The Spread and Control of an Insect Population

The kinetics such as the uptake function in an enzyme reaction system discussed in Chapter 5 or the gross population growth-interaction function $f(u)$ such as we introduced in Chapter 1, can often have more than two steady states. That is $f(u)$ in (11.31) can have three or more positive zeros. The wave phenomena associated with such $f(u)$ is quite different to that in the previous sections. A practical example is the growth function for the behaviour of the spruce budworm, the spatially uniform situation of which was discussed in detail in Chapter 1, Section 1.2. The specific dimensionless $f(u)$ in that model is

$$ f(u) = ru \left(1 - \frac{u}{q} \right) - \frac{u^2}{1 + u^2} , \tag{11.71} $$

where r and q are dimensionless parameters involving real field parameters (see equation (1.17)). For a range of the positive parameters r and q, $f(u)$ is as in Fig. 1.4, which is reproduced in Fig. 11.7 (a) for convenience. Recall the dependence of the number and size of the steady states on r and q; a typical curve is shown again in Fig. 11.7 (b) for convenience. In the absence of diffusion, that is the spatially uniform situation, there can be three positive steady states, two linearly stable ones u_1 and u_3, and one unstable one, u_2. The steady state $u = 0$ is also unstable.

We saw in Section 1.2 that the lower steady state u_1 corresponds to a *refuge* for the budworm while u_3 corresponds to an *outbreak*. The key questions we

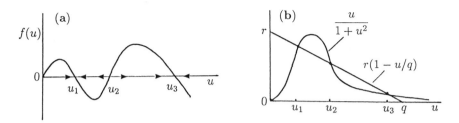

Fig. 11.7a,b. (a) Growth-interaction kinetics for the spruce budworm population u: u_1 corresponds to a refuge population and u_3 corresponds to an infestation outbreak. (b) Schematic dependence of the steady states in (a) on the parameters r and q in (11.71).

consider here are (i) how does an infestation or outbreak propagate when we include spatial dispersal of the budworm, and (ii) can we use the results of the analysis to say anything about a control strategy to prevent an outbreak from spreading. To address both of these questions, we consider the budworm to disperse by linear diffusion and investigate the travelling wave possibilities. Although the practical problem is clearly two-dimensional we discuss here the one-dimensional case since, even with that, we can still offer reasonable answers to the questions, and at the very least pose those that the two-dimensional model must address. In fact there are intrinsically no new conceptual difficulties with the two space dimensional model. The model we consider then is, from (11.31),

$$\frac{\partial u}{\partial t} = f(u) + \frac{\partial^2 u}{\partial x^2} , \qquad (11.72)$$

with $f(u)$ typically as in Fig. 11.7 (a).

Let us look for travelling wave solutions in the usual way. Set

$$u(x,t) = U(z), \quad z = x - ct \quad \Rightarrow \quad U'' + cU' + f(U) = 0 , \qquad (11.73)$$

the phase plane system for which is

$$U' = V, \quad V' = -cV - f(U) \quad \Rightarrow \quad \frac{dV}{dU} = -\frac{cV + f(U)}{V} , \qquad (11.74)$$

which has four singular points

$$(0,0), \quad (u_1,0), \quad (u_2,0), \quad (u_3,0) . \qquad (11.75)$$

We want to solve the eigenvalue problem for c, such that travelling waves, of the kind we seek, exist. As a first step we determine the type of singularities given by (11.75).

Linearizing (11.74) about the singular points $U = 0$ and $U = u_i$, $i = 1, 2, 3$ we get

$$\frac{dV}{d(U - u_i)} = -\frac{cV + f'(u_i)(U - u_i)}{V}, \quad i = 1, 2, 3 \quad \text{and} \quad u_i = 0 \qquad (11.76)$$

which, using standard linear phase plane analysis, gives the following singular point classification:

$$(0,0): \quad f'(0) > 0 \quad \Rightarrow \quad \text{stable} \begin{cases} \text{spiral} \\ \text{node} \end{cases} \text{if} \quad c^2 \begin{cases} < \\ > \end{cases} 4f'(0), \quad c > 0$$

$$(u_2, 0): \quad f'(u_2) > 0 \quad \Rightarrow \quad \text{stable} \begin{cases} \text{spiral} \\ \text{node} \end{cases} \text{if} \quad c^2 \begin{cases} < \\ > \end{cases} 4f'(u_2), \quad c > 0$$

$$(u_i, 0): \quad f'(u_i) < 0 \quad \Rightarrow \quad \text{saddle point} \quad \text{for all } c, \quad i = 1, 3 \, .$$

$$(11.77)$$

If $c < 0$ then $(0,0)$ and $(u_2, 0)$ become unstable – the type of singularity is the same. There are clearly several possible phase plane trajectories depending on the size of $f'(u_i)$ where u_i has $i = 1, 2, 3$ plus $u_i = 0$. Rather than give a complete catalogue of all the possibilities we shall analyse just two to show how the others can be studied.

The existence of the various travelling wave possibilities for various ranges of c can become quite an involved book-keeping process. This particular type of equation has been rigorously studied by Fife and McLeod (1977). The approach we shall use here is intuitive and does not actually prove the existence of the waves we are interested in, but it certainly gives a very strong indication that they exist. The procedure then is in line with the philosophy adopted throughout this book.

Let us suppose that $c^2 > 4 \max [f'(0), f'(u_2)]$ in which case $(0,0)$ and $(u_2, 0)$ are stable nodes. A possible phase portrait is illustrated in Fig. 11.8 (a), which gives possible singular point connections. If we divide the phase plane into the domains shown, for example d_1 includes the node at the origin and the saddle point at $(u_1, 0)$, and if we compare this with Fig. 11.1 (b) they are similar. So, it is reasonable to suppose that a similar wave solution can exist, namely one from $U(-\infty) = u_1$ to $U(\infty) = 0$ and that it exists for all wave speeds $c \geq 2[f'(0)]^{1/2}$. This situation is sketched in Fig. 11.8 (b). In a similar way other domains admit the other travelling wave solutions shown in Fig. 11.8 (b).

As c varies other possible singular point connections appear. In particular let us focus on the points $(u_1, 0)$ and $(u_3, 0)$, both of which are saddle points. The eigenvalues λ_1, λ_2 are found from (11.76) as

$$\lambda_1, \lambda_2 = \frac{-c \pm \{c^2 - 4f'(u_i)\}^{1/2}}{2}, \quad i = 1, 3 \qquad (11.78)$$

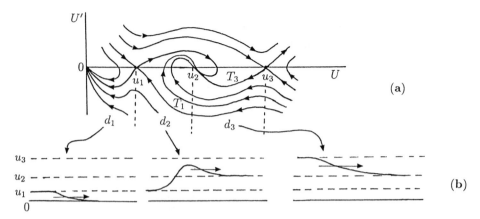

Fig. 11.8a,b. (a) Possible phase plane portrait when $c > 0$ is in an appropriate range relative to $f'(u)$ evaluated at the singular points. **(b)** Possible wavefront solutions if we restrict the domains in the phase portrait as indicated by d_1, d_2 and d_3.

where $f'(u_i) < 0$. The corresponding eigenvectors e_{i1} and e_{i2} are

$$ \mathbf{e}_{i1} = \begin{pmatrix} 1 \\ \lambda_{i1} \end{pmatrix}, \quad \mathbf{e}_{i2} = \begin{pmatrix} 1 \\ \lambda_{i2} \end{pmatrix}, \quad i = 1, 3 \tag{11.79} $$

which vary as c varies. A little algebra shows that as c increases the eigenvectors tend to move towards the U-axis. As c varies the phase trajectory picture varies; in particular the trajectories marked T_1 and T_3 in Fig. 11.8 (a) change. By continuity arguments it is clearly possible, if $f'(u_1)$ and $f'(u_3)$ are in an approriate range, that as c varies there is a *unique* value for c, c^* say, such that the T_1 trajectory joins up with the T_3 trajectory. In this way we then have a phase path connecting the two singular points $(u_1, 0)$ and $(u_3, 0)$ as illustrated in Fig. 11.9 (a), with the corresponding wave solution sketched in Fig. 11.9 (b): this wave moves with a unique speed c^* which depends on the nonlinear interaction term $f(u)$. The solution $U(z)$ in this case has

$$ U(-\infty) = u_3, \quad U(\infty) = u_1 . $$

It is this situation we now consider with the budworm problem in mind.

Suppose we start with $u = u_1$ for all x, that is the budworm population is in a stable refuge state. Now suppose there is a local increase of population to u_3 in some finite domain, that is there is a local outbreak of the pest. To investigate the possibility of the outbreak spreading it is easier to ask the algebraically simpler problem, does the travelling wavefront solution in Fig. 11.9 (b) exist which joins a region where $u = u_1$ to one where $u = u_3$, and if so, what is its speed and direction of propagation? From the above discussion we expect that such a wave to exist. If $c > 0$ the wave moves into the u_1-region and the outbreak spreads; if $c < 0$ it not only does not spread, it is reduced.

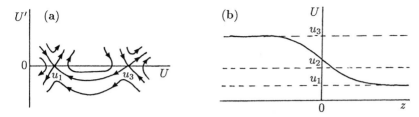

Fig. 11.9a,b. (a) Schematic phase plane portrait for a wave connecting the steady states u_3 and u_1. (b) Typical wave front solution from u_3 to u_1. The unique speed of the wave and its direction of propagation are determined by $f(u)$ in (11.72).

The *sign* of c can easily be found by multiplying the U-equation in (11.73) by U' and integrating from $-\infty$ to ∞. This gives

$$\int_{-\infty}^{\infty} [U'U'' + cU'^2 + U'f(U)] \, dz = 0 \, .$$

Since $U'(\pm\infty) = 0$, $U(-\infty) = u_3$ and $U(\infty) = u_1$, this integrates to give

$$c \int_{-\infty}^{\infty} [U']^2 \, dz = -\int_{-\infty}^{\infty} f(U)U' \, dz = -\int_{u_3}^{u_1} f(U) \, dU$$

and so, since the multiple of c is always positive,

$$c \begin{array}{c} > \\ = \\ < \end{array} 0 \quad \text{if} \quad \int_{u_1}^{u_3} f(u) \, du \begin{array}{c} > \\ = \\ < \end{array} 0 \, . \tag{11.80}$$

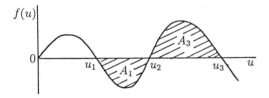

Fig. 11.10. If $A_1 > A_3$ the wave velocity c is negative and the outbreak, where $u = u_3$, is reduced. If $A_1 < A_3$ the outbreak spreads into the refuge region where $u = u_1$.

So, the sign of c is determined solely by the integral of the interaction function $f(u)$. From Fig. 11.10, the sign of the integral is thus given simply by comparing the areas A_1 and A_3. If $A_3 > A_1$ the wave has $c > 0$ and the outbreak spreads into the refuge area. In this case we say that u_3 is dominant; that is as $t \to \infty$, $u \to u_3$ everywhere. On the other hand if $A_3 < A_1$, $c < 0$ and u_1 is dominant and $u \to u_1$ as $t \to \infty$; that is, the outbreak is eliminated.

From the point of view of infestation control, if an insect outbreak occurs and is spreading, we want to know how to alter the local conditions so that the infestation or outbreak wave is either contained or reversed. From the above, we must thus locally change the budworm growth dynamics so that effectively the new areas A_1 and A_3 in Fig. 11.10 satisfy $A_1 > A_3$. We can achieve this if the zeros u_2 and u_3 of $f(u)$, that is the two largest steady states, are closer together. From Fig. 11.7 (b) we see that this can be effected by reducing the dimensionless parameter q in (11.71). The nondimensionalisation used in the budworm model (see Section 1.2 in Chapter 1) relates q to the basic budworm carrying capacity K_B of the environment. So a practical reduction in q could be made by, for example, spraying a strip to reduce the carrying capacity of the tree foliage. In this way an infestation "break" would be created, that is one in which u_1 is dominant, and hence the wave speed c in the above analysis is no longer positive. A practical question, of course, is how wide such a "break" has got to be to stop the outbreak getting through. This problem needs careful modelling consideration since there is a long leading edge, because of the parabolic character of the equations, albeit with $0 < u \ll 1$. A closely related concept will be discussed in detail later in Chapter 20 when the problem of containing the spread of rabies is discussed. The methodology described there is directly applicable to the "break" problem here for containing the spread of the budworm infestation.

Exact Solution for the Wave Speed for an Excitable Kinetics Model: The Calcium-Stimulated-Calcium-Release Mechanism

In Chapter 5 we briefly described possible kinetics, namely equation (5.51), which models a biochemical switch. With such a mechanism, a sufficiently large perturbation from one steady state can move the system to another steady state. An important example which arises experimentally is known as the calcium-stimulated-calcium-release mechanism. This is a process whereby calcium, Ca^{++}, if perturbed above a given threshold concentration, causes the further release, or dumping, of the sequestered calcium; that is the system moves to another steady state. This happens, for example, from calcium sites on the membrane enclosing certain fertilized amphibian eggs (the next section deals with one such real example). As well as releasing calcium, such a membrane also resequesters it. If we denote the concentration of Ca^{++} by u, we can model the kinetics by the rate law

$$\frac{du}{dt} = A(u) - r(u) + L , \qquad (11.81)$$

where L represents a small leakage, $A(u)$ is the autocatalytic release of calcium and $r(u)$ its resequestration. We assume that calcium resequestration is governed by first order kinetics, and the autocatalytic calcium production saturates for high Ca^{++}. With these assumptions, we arrive at the reaction kinetics model equation with typical forms which have been used for $A(u)$ and $r(u)$ (for example, Odell et al. 1981, Murray and Oster 1984a, Cheer et al. 1985, Lane et al. 1987). The specific form of the last equation, effectively the same as (5.51), becomes

$$\frac{du}{dt} = L + \frac{k_1 u^2}{k_2 + u^2} - k_3 u = f(u) , \qquad (11.82)$$

where the k's and L are positive parameters. If the k's are in a certain relation to each other (see Exercise 3 at the end of Chapter 5) this $f(u)$ can have have three positive steady states for L sufficiently small. The form of $f(u)$ in this excitable kinetics situation is illustrated in Fig. 11.11 (a). Although there are two kinds of excitable processes exhibited by this mechanism, they are closely related. We briefly consider each in turn.

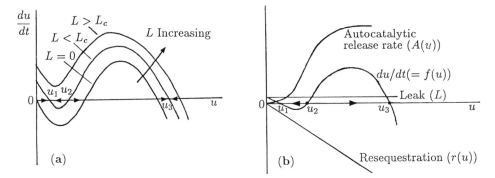

Fig. 11.11a,b. (a) Excitable kinetics example. For $0 < L < L_c$ there are 3 positive steady states u_i, $i = 1, 2, 3$ of (11.82) with two of these coalescing when $L = L_c$. Suppose initially $u = u_1$, with $L < L_c$. If we now increase L beyond the threshold, only the largest steady state exists. So, as L is again reduced to its original value $u \to u_3$, where it remains. A switch from u_1 to u_3 has been effected. (b) The schematic form of each of the terms in the kinetics in (11.81) and (11.82). When added together they give the gross kinetics form in (a).

If $L = 0$ there are three steady states, two stable and one unstable. If L is increased from zero there are first three positive steady states $u_i(L)$, $i = 1, 2, 3$ with u_1 and u_3 linearly stable and u_2 unstable. As L increases above a certain threshold value L_c, u_1 and u_2 first coalesce and then disappear. So if initially $u = u_1$, a pulse of L sufficiently large can result in the steady state shifting to u_3, the larger of the two stable steady states, where it will remain. Although qualitatively it is clear that this happens, the quantitative analysis of such a switch is not simple and has been treated by Kath and Murray (1985) in connection with a model mechanism for generating butterfly wing patterns, a topic we consider in Chapter 15.

The second type of excitability has L fixed and the kinetics $f(u)$ as in the curve marked $du/dt(= f(u))$ in Fig. 11.11 (b). The directions of the arrows there indicate how u will change if a perturbation with a given concentration is introduced. For all $0 < u < u_2$, $u \to u_1$, while for all $u > u_2$, $u \to u_3$. The concentration u_2 is thus a threshold *concentration*. Whereas in the above threshold situation L was the bifurcation parameter, here it is in the imposed perturbation as it relates to u_2.

The complexity of this calcium-stimulated calcium-release process in reality is such that the model kinetics in (11.81) and its quantitative form in (11.82) can only be a plausible caricature. It is reasonable, therefore, to make a further simplifying caricature of it, as long as it preserves the qualitative dynamic behaviour for u and the requisite number of zeros: that is, $f(u)$ is like the curve in Fig. 11.11 (b). We do this by replacing $f(u)$ with a cubic with three positive zeros, namely

$$f(u) = A(u - u_1)(u_2 - u)(u - u_3) ,$$

where A is a positive constant and $u_1 < u_2 < u_3$. This is qualitatively like the curve in Fig. 11.11 (a) where $0 < L < L_c$.

Let us now consider the reaction diffusion equation with such reaction kinetics, namely

$$\frac{\partial u}{\partial t} = A(u - u_1)(u_2 - u)(u - u_3) + D\frac{\partial^2 u}{\partial x^2} , \tag{11.83}$$

where we have not renormalized the equation so as to highlight the role of A and the diffusion D. This equation is very similar to (11.72), the one we have just studied in depth for wavefront solutions. We can assume then that (11.83) has wavefront solutions of the form

$$u(x,t) = U(z), \quad z = x - ct, \quad U(-\infty) = u_3, \quad U(\infty) = u_1 , \tag{11.84}$$

which on substituting into (11.83) gives

$$L(U) = DU'' + cU' + A(U - u_1)(u_2 - U)(U - u_3) = 0 . \tag{11.85}$$

With the experience gained from the exact solutions above and the form of the asymptotic solution obtained for the Fisher equation waves, we might optimistically expect the wavefront solution of (11.85) to have an exponential behaviour. Rather than start with some explicit form of the solution, let us rather start with a differential equation which might reasonably determine it, but which is simpler than (11.85). The procedure, then, is to suppose U satisfies a simpler equation (with exponential solutions of the kind we now expect) but which can be made to satisfy (11.85) for various values of the parameters. It is in effect seeking solutions of a differential equation with a simpler differential equation that we can solve.

Let us try making U satisfy

$$U' = a(U - u_1)(U - u_3) , \tag{11.86}$$

the solutions of which tend exponentially to u_1 and u_3 as $z \to \infty$, which is the appropriate kind of behaviour we want. Substituting this equation into (11.85) we get

$$\begin{aligned} L(U) &= (U - u_1)(U - u_3)\{Da^2(2U - u_1 - u_3) + ca - A(U - u_2)\} \\ &= (U - u_1)(U - u_3)\{(2Da^2 - A)U - [Da^2(u_1 + u_3) - ca - Au_2]\} . \end{aligned}$$

and so for $L(U)$ to be zero we must have

$$2Da^2 - A = 0, \quad Da^2(u_1 + u_3) - ca - Au_2 = 0 ,$$

which determine a and the unique wave speed c as

$$a = \left(\frac{A}{2D}\right)^{1/2} , \quad c = \left(\frac{AD}{2}\right)^{1/2}(u_1 - 2u_2 + u_3) . \tag{11.87}$$

So, by using the differential equation (11.86) we have shown that its solutions can satisfy the full equation if a and c are as given by (11.87). The actual solution U is then obtained by solving (11.86); it is

$$U(z) = \frac{u_3 + Ku_1 \exp\left[a(u_3 - u_1)z\right]}{1 + K \exp\left[a(u_3 - u_1)z\right]} , \tag{11.88}$$

where K is an arbitrary constant which simply lets us set the origin in the z-plane in the now usual way. This solution has

$$U(-\infty) = u_3 \quad \text{and} \quad U(\infty) = u_1 .$$

The sign of c, from (11.87), is determined by the relative sizes of the u_i, $i = 1, 2, 3$: if u_2 is greater than the average of u_1 and u_3, $c < 0$ and positive otherwise. This, of course, is the same result we would get if we used the integral result from (11.80) with the cubic for $f(U)$ from (11.83).

Equation (11.83) and certain extensions of it have been studied by McKean (1970). It arose there in the context of a simple model for the propagation of a nerve action potential, a topic we shall touch on in the following chapter. Equation (11.83) is sometimes referred to as the reduced *Nagumo equation*, which is related to the FitzHugh-Nagumo model for nerve action potentials discussed in Section 6.5.

11.6 Calcium Waves on Amphibian Eggs: Activation Waves on *Medaka* Eggs

The cortex of an amphibian egg is a kind of membrane shell enclosing the egg. Just after fertilisation, and before the first cleavage of the egg, several chemical waves of calcium, Ca^{++}, sweep over the cortex. The top of the egg, near where the waves start, is the *animal pole*, and is effectively determined by the sperm entry point, while the bottom is the *vegetal pole*. The wave emanates from the sperm entry point. Each wave is a precursor of some major event in development and each is followed by a mechanical event. Such waves of Ca^{++} are called *activation waves*. Fig. 11.12 (a) illustrates the progression of such a calcium wave over the egg of the teleost fish *Medaka*. The figure was obtained from the

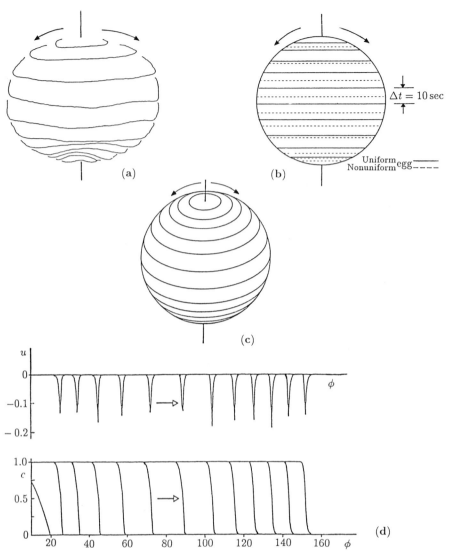

Fig. 11.12a–d. (a) Wavefront propagation of the Ca^{++} wave which passes over the surface of the egg, from the sperm entry point near the animal pole to bottom (the vegetal pole), of the fish *Medaka* prior to cleavage. The wavefronts are 10 sec apart. Note how the wave slows down in the lower hemisphere – the fronts are closer together. (After Cheer et al. 1985 from the experimental data of Gilkey et al. 1978) (b) Computed Ca^{++} wavefront solutions from the reaction diffusion model with uniform surface properties compared with the computed solutions with non-uniform properties. (After Cheer et al. 1987) (c) Computed Ca^{++} wavefront solutions from Lane et al. (1987): here the wave accelerates in the upper hemisphere and slows down in the lower hemisphere because of the variation in a parameter in the calcium kinetics. The lines represent wave fronts at equal time intervals. (d) The Ca^{++} wave and mechanical deformation wave which accompanies it. (From Lane et al. 1987) Here $u(\phi)$, where ϕ is the polar angle measured from the sperm entry point (SEP), is the dimensionless mechanical deformation of the egg surface from its rest state $u = 0$. The spike-like waves are surface contraction waves. (These mechanical waves will be discussed in detail Chapter 17.)

experimental data of Gilkey et al. (1978). The model we shall describe in this section is a simplified mechanism for the chemical wave, and comes from the papers on cortical waves in vertebrate eggs by Cheer et al. (1987) and Lane et al. (1987). They model both the mechanical and mechano-chemical waves observed in amphibian eggs but with different model assumptions. Lane et al. (1987) also present some analytical results based on a piece-wise linear approach and these compare well with the numerical simulations of the full nonlinear system: this will be discussed in Chapter 17. The mechano-chemical process is described in detail in the papers and the model constructed on the basis of the biological facts. The results of their analysis are compared with experimental observations on the egg of the fish *Medaka* and other vertebrate eggs. Cheer et al. (1987) conclude with relevant statements about what must be occurring in the biological process and on the nature of the actual cortex. The paper by Lane et al. (1987) highlights the key elements in the process and displays the analytical dependence of the various phenomena on the model parameters. The mechanical surface waves which accompany the calcium waves are shown in Fig. 11.12 (d). We consider this problem again in Chapter 17 where we consider mechanochemical models.

Here we construct a simple model for the Ca^{++} based on the fact that the calcium kinetics is excitable: we use the calcium-stimulated-calcium-release mechanism described in the last section. We assume that the Ca^{++} diffuses on the cortex (surface) of the egg. We thus have a reaction diffusion model where both the reaction and diffusion take place on a spherical surface. Since the Ca^{++} wavefront is actually a ring propagating over the surface its mathematical description will involve only one independent variable θ, the polar angle measured from the top of the sphere, so $0 \leq \theta \leq \pi$. The kinetics involve the release of calcium from sites on the surface via the calcium-stimulated-calcium-release mechanism. The small leakage here is due to a small amount of Ca^{++} diffusing into the interior of the egg. So, there is a threshold value for the calcium which triggers a dumping of the calcium from the surface sites. The phenomenological model which captures the excitable kinetics and some of the known facts about the process is given by (11.82). We again take the simpler cubic kinetics caricature used in (11.83) and thus arrive at the model reaction diffusion system

$$\frac{\partial u}{\partial t} = f(u) + D \left(\frac{1}{R}\right)^2 \left[\frac{\partial^2 u}{\partial \theta^2} + \cot \theta \frac{\partial u}{\partial \theta}\right] ,$$

$$f(u) = A(u - u_1)(u_2 - u)(u - u_3) ,$$

(11.89)

where A is a positive parameter and R is the radius of the egg: R is simply a parameter in this model.

Refer now to the middle curve in Fig. 11.11 (a), that is like the $f(u)$-curve in Fig. 11.11 (b). Suppose the calcium concentration on the surface of the egg is uniformly at the lower steady state u_1. If it is subjected to a perturbation larger than the threshold value u_2, u will tend towards the higher steady state u_3. If the perturbation is to a value less than u_2, u will return to u_1. There is thus a *firing threshold*, above which $u \to u_3$.

Consider now the possible wave solutions of (11.89). If the $\cot \theta$ term were not in this equation we know that it would have wavefront solutions of the type equivalent to (11.84), that is of the form

$$u(\theta, t) = U(z), \quad z = R\theta - ct, \quad U(-\infty) = u_3, \quad U(\infty) = u_1 . \tag{11.90}$$

Of course with our spherical egg problem, if time t starts at $t = 0$, z here cannot tend to $-\infty$. Not only that, $\cot \theta$ is of course in the equation. However, to get some feel for what happens to waves like those found in the last Section when the mechanism operates on the surface of a sphere, we can intuitively argue in the following way.

At each *fixed* θ let us suppose there is a wavefront solution of the form

$$u(\theta, t) = U(z), \quad z = R\theta - ct . \tag{11.91}$$

Substituting this into (11.89) we get

$$DU'' + \left[c + \frac{D}{R} \cot \theta \right] U' + A(U - u_1)(u_2 - U)(U - u_3) = 0 . \tag{11.92}$$

Since we are considering θ fixed here, this equation is exactly the same as (11.85) with $[c + (D/R) \cot \theta]$ in place of the c there. We can therefore plausibly argue that a quantitative expression for the wave speed c on the egg surface is given by (11.87) with $[c + (D/R) \cot \theta]$ in place of c. So we expect wavefront-like solutions of (11.89) to propagate over the surface of the egg with speeds

$$c = \left(\frac{AD}{2} \right)^{1/2} (u_1 - 2u_2 + u_3) - \frac{D}{R} \cot \theta . \tag{11.93}$$

What (11.93) implies is that as the wave moves over the surface of the egg from the animal pole, where $\theta = 0$, to the vegetal pole, where $\theta = \pi$, the wave speed varies. Since $\cot \theta > 0$ for $0 < \theta < \pi/2$ the wave moves slower in the upper hemisphere, while for $\pi/2 < \theta < \pi$, $\cot \theta < 0$, which means that the wave speeds are higher in the lower hemisphere. We can get this qualitative result from the reaction diffusion equation (11.89) by similar arguments to those used in Section 11.2 for axisymmetric wave-like solutions of Fisher's equation. Compare the diffusion terms in (11.89) with that in the one-dimensional version of the model in (11.83), for which the wave speed is given by (11.87) or (11.93) without the $\cot \theta$ term. If we think of a wave moving into a $u = u_1$ domain from the higher u_3-domain then $\partial u/\partial \theta < 0$. In the animal hemisphere $\cot \theta > 0$ so the term $\cot \theta \, \partial u/\partial \theta < 0$ implies an effective reduction in the diffusional process, which is a critical factor in propagating the wave. So, the wave is slowed down in the upper hemisphere of the egg. By the same token, $\cot \theta \, \partial u/\partial \theta > 0$ in the lower hemisphere, and so the wave speeds up there. This is intuitively clear if we think of the upper hemisphere as where the wave front has to continually expand its perimeter with the converse in the lower hemisphere.

The wave speed given by (11.89) implies that, for surface waves on spheres, it is probably not possible to have travelling wave solutions, with $c > 0$, for all θ: it clearly depends on the parameters which would have to be delicately spatially dependent.

In line with good mathematical biology practice let us now go back to the real biology. What we have shown is that a simplified model for the calcium-stimulated-calcium-release mechanism gives travelling calcium wavefront-like solutions over the surface of the egg. Comparing the various times involved with the experiments, estimates for the relevant parameters can be determined. There is however a serious qualitative difference between the front behaviour in the real egg and the model egg. In the former the wave slows down in the vegetal hemisphere whereas in the model it speeds up. One important prediction or conclusion we can draw from this (Cheer et al. 1987) is that the non-uniformity in the cortex properties are such that they overcome the natural speeding up tendencies for propagating waves on the surface. If we look at the wave speed given by (11.93) it means that AD and the u_i, $i = 1, 2, 3$ must vary with θ. This formula for the speed will also hold if the parameters are slowly varying over the surface of the sphere. So, it is analytically possible to determine qualitative behaviour in the model properties to effect the correct wave propagation properties on the egg, and hence deduce possible parameter variations in the egg cortex properties. Fig. 11.12 (b) illustrates some numerical results given by Cheer et al. (1987) using the above model with non-uniform parameter properties. The reader is referred to that paper for a detailed discussion of the biolgy, the full model and the biological conclusions drawn from the analysis. In Chapter 17 we introduce and discuss in detail the new mechanochemical approach to biological pattern formation of which this section and the papers by Cheer et al. (1987) and Lane et al. (1987) are examples.

Exercises

1. Consider the dimensionless reaction diffusion equation

$$u_t = u^2(1 - u) + u_{xx}.$$

Obtain the ordinary differential equation for the travelling wave solution where with $u(x, t) = U(z)$, $z = x - ct$ where c is the wave speed. Assume that a non-negative monotone solution for $U(z)$ exists with $U(-\infty) = 1$, $U(\infty) = 0$ for a wave speed such that $0 < 1/c = \varepsilon^{1/2}$ where ε is sufficiently small to justify seeking asymptotic solutions for $0 < \varepsilon \ll 1$. With $\xi = \varepsilon^{1/2}z$, $U(z) = g(\xi)$ show that the $O(1)$ asymptotic solution such that $g(0) = 1/2$ is given explicitly by

$$\xi = -2 + \frac{1}{g(\xi)} + \ln\left[\frac{1 - g(\xi)}{g(\xi)}\right], \quad \xi = \frac{x - ct}{c}.$$

Derive the (V, U) phase plane equation for travelling wave solutions where $V = U'$, where the prime denotes differentiation with respect to z. By setting $\phi = V/\varepsilon^{1/2}$ in the equation obtain the asymptotic solution, up to $O(\varepsilon)$, for ϕ as a function of U as a Taylor series in ε. Hence show that the slope of the wave where $U = 1/2$ is given to $O(\varepsilon)$ by

$$-\frac{1}{8c} - \frac{1}{2^5 c^3}.$$

2. Show that an exact travelling wave solution exists for the scalar reaction diffusion equation

$$\frac{\partial u}{\partial t} = u^{q+1}(1 - u^q) + \frac{\partial^2 u}{\partial x^2}.$$

where $q > 0$, by looking for solutions in the form

$$u(x, t) = U(z) = \frac{1}{(1 + de^{bz})^s}, \quad z = x - ct$$

where c is the wave speed and b and s are positive constants. Determine the unique values for c, b and s in terms of q. Choose a value for d such that the magnitude of the wave's gradient is at its maximum at $z = 0$.

12. Biological Waves: Multi-species Reaction Diffusion Models

12.1 Intuitive Expectations

In the last chapter we saw that if we allowed spatial dispersal in the single re-actant or species, travelling wave front solutions were possible. Such solutions effected a smooth transition between two steady states of the space indepen-dent system. For example, in the case of the Fisher equation (11.6), wavefront solutions joined the steady state $u = 0$ to the one at $u = 1$ as shown in the evo-lution to a propagating wave in Fig. 11.2. In Section 11.5, where we considered a model for the spatial spread of the spruce budworm, we saw how such travelling wave solutions could be found to join any two steady states of the spatially inde-pendent dynamics. In this and the next three chapters, we shall be considering systems where several species or reactants are involved, concentrating on reac-tion diffusion mechanisms, of the type derived in Section 9.2 (Equation (9.18)) namely

$$\frac{\partial \mathbf{u}}{\partial t} = \mathbf{f}(\mathbf{u}) + D\nabla^2\mathbf{u} \,, \tag{12.1}$$

where $\mathbf{u}$ is the vector of reactants, $\mathbf{f}$ the nonlinear reaction kinetics and D the matrix of diffusivities, taken here to be constant.

Before analysing such systems let us try to get some intuitive idea of what kind of solutions we might expect to find. As we shall see, a very rich spectrum of solutions it turns out to be. Because of the analytical difficulties and algebraic complexities that can be involved in the study of nonlinear systems of reaction diffusion equations an intuitive approach can often be the key to getting started and to what might be expected. In keeping with the philosophy in this book such intuition is a crucial element in the modelling and analytical processes. We should add the usual cautionary caveat, that it is mainly stable travelling wave solutions that are of principal interest, but not always. The study of the stability of such solutions is not usually very easy, and in many cases has not yet been done.

Consider first a single reactant model in one space dimension x, with multiple steady states, such as we discussed in Section 11.5, where there are 3 steady states u_i, $i = 1, 2, 3$ of which u_1 and u_3 are stable in the spatially homogeneous situation. Suppose that initially u is at one steady state, $u = u_1$ say, for all x. Now suppose we suddenly change u to u_3 in $x < 0$. With u_3 dominant the situation. Suppose that initially u is at one steady state, $u = u_1$ say, for all x. Now suppose we suddenly change u to u_3 in $x < 0$. With u_3 dominant the effect of diffusion is to initiate a travelling wavefront, which propagates into the $u = u_1$ region and so eventually $u = u_3$ everywhere. As we saw, the inclusion of diffusion effects in this situation resulted in a smooth travelling wavefront solution for the reaction diffusion equation. In the case of a multi-reactant system, where $\mathbf{f}$ has several steady states, we should reasonably expect similar travelling wave solutions, which join steady states. Although mathematically a spectrum of solutions may exist we are, of course, only interested here in non-negative solutions. Such multi-species wavefront solutions are usually more difficult to determine analytically but the essential concepts involved are more or less the same, although there are some interesting differences. One of these can arise with interacting predator-prey models with spatial dispersal by diffusion. Here the travelling front is like a wave of pursuit by the predator and of evasion by the prey: we discuss one such case in the following Section 12.2. In Section 12.3 we consider a model for travelling wavefronts in the Belousov-Zhabotinskii reaction and compare the analytical results with experiment.

In the case of a single reactant or population we saw in Chapter 1 that limit cycle periodic solutions are not possible, unless there are delay effects, which we shall not be considering here. With multi-reactant kinetics or interacting species however, as we saw in Chapter 3, we can have stable periodic limit cycle solutions which bifurcate from a stable steady state as a parameter, γ say, increases through a critical γ_c. Let us now suppose we have such reaction kinetics in our reaction diffusion system (12.1) and that initially $\gamma > \gamma_c$ for all x: that is, the system is oscillating. If we now locally perturb the oscillation for a short time in a small spatial domain, say $0 < |x| \leq \varepsilon \ll 1$, then the oscillation there will be at a different phase from the surrounding medium. We then have a kind of localized 'pacemaker' and the effect of diffusion is to try and smooth out the differences between this pacemaker and the surrounding medium. As we noted above, a sudden change in u can initiate a propagating wave. So, in this case as u regularly changes in the small circular domain relative to the outside domain, it is like regularly initiating a travelling wave from the pacemaker. In our reaction diffusion situation we would thus expect a travelling *wave train* of concentration differences moving through the medium. We discuss such wave train solutions in Sections 12.5 and 12.6.

It is possible to have chaotic oscillations when 3 or more equations are involved, as we noted in Chapter 3, and indeed with only a single *delay* equation in Chapter 1. There is thus the possibility of quite complicated wave phenomena if we introduce, say, a small chaotic oscillating region in an otherwise regular oscillation. These more complicated wave solutions can occur with only one space dimension. In two or three space dimensions the solution behaviour can become quite baroque. Interestingly, chaotic behaviour can occur without a chaotic pacemaker: see Fig. 12.21 in Section 12.8 below.

Suppose we now consider two space dimensions. If we have a small circular domain, which is oscillating at a different frequency from the surrounding medium, we should expect a travelling wave train of concentric circles propagating out from the pacemaker centre: they are often referred to as *target patterns* for obvious reasons. Such waves were originally found experimentally by Zaikin and Zhabotinskii (1970) in the Belousov-Zhabotinskii reaction: Fig. 12.1 (a) is an example. Tyson and Fife (1980) discuss target patterns in the Field-Noyes model for the Belousov-Zhabotinskii reaction, which we considered in detail in Chapter 7.

We can think of an oscillator as a pacemaker which continuously moves round a circular ring. If we carry this analogy over to reaction diffusion systems, as the 'pacemaker' moves round a small core ring it continuously creates a wave, which propagates out into the surrounding domain, from each point on the circle. This would produce, not target patterns, but spiral waves with the 'core' the limit cycle pacemaker. Once again these have been found in the Belousov-Zhabotinskii reaction: see Fig. 12.1(b) and, for example, Winfree (1974), Müller, Plesser and Hess (1985) and Agladze and Krinsky (1982). See also the dramatic experimental examples in Figs. 12.14–12.18 below in Section 12.7 on spiral waves. Kuramoto and Koga (1981) and Agladze and Krinsky (1982), for example, demonstrate the onset of chaotic wave patterns: see Fig. 12.21 below. If we consider such waves in three space dimensions the topological structure is remarkable, each part of the basic 'two-dimensional' spiral is itself a spiral: see, for example, Winfree (1974), Welsh, Gomatam and Burgess (1983) for photographs of actual three dimensional waves, and Winfree and Strogatz (1984) for a discussion of the topological aspects.

Such target patterns and spiral waves are common in biology. Spiral waves, in particular, are of considerable practical importance in a variety of medical situations, particularly in cardiology and neurobiology. We shall touch on some of these aspects below. A particularly good biological example is provided by the slime mold *Dictyostelium discoideum* (Newell 1983) and illustrated in Fig. 12.1 (c): see also Fig. 12.16 below.

Suppose we now consider the reaction diffusion situation in which the reaction kinetics has a single stable steady state but which, if perturbed enough, can exhibit a threshold behaviour, such as we discussed in Section 3.8 and also in Section 6.5; the latter is the FHN model for the propagation of Hodgkin-Huxley nerve action potentials. Suppose initially the spatial domain is everywhere at the stable steady state and we perturb a small region so that the perturbation locally initiates a threshold behaviour. Although eventually the perturbation will disappear it will undergo a large excursion in phase space before doing so. So for a time the situation will appear to be like that described above in which there are two quite different states which, because of the diffusion, try to initiate a travelling wavefront. The effect of a threshold capability is thus to provide a basis for a travelling pulse wave. We discuss these threshold waves in Section 12.4.

It is clear that the variety of spatial wave phenomena in multi-species reaction diffusion mechanisms is very much richer than in single species models. If we allow chaotic pacemakers, delay kinetics, and so on, the spectrum of phenomena is extremely wide. Many of them have still to be studied, and will certainly generate

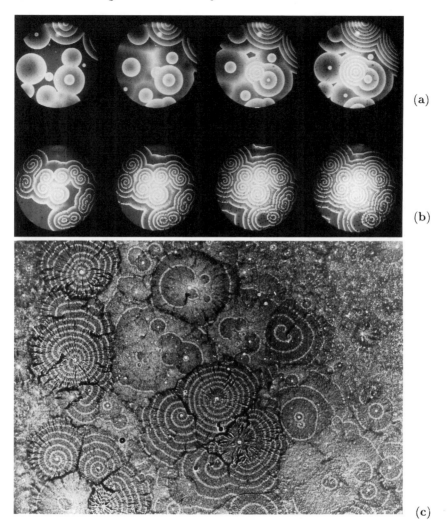

(a)

(b)

(c)

Fig. 12.1a-c. (a) Target patterns (circular waves) generated by pacemaker nuclei in the Belousov-Zhabotinskii reaction. The photographs are about 1 minute apart. (b) Spiral waves, initiated by gently stirring the reagent. The spirals rotate with a period of about 2 minutes. (Reproduced with permission of A.T. Winfree) (c) In the slime mold *Dictyostelium*, the cells (amoebae) at a certain stage in their group development, emit a periodic signal of the chemical cyclic AMP, which is a chemoattractant for the cells. Certain pacemaker cells initiate target-like and spiral waves. The light and dark bands arise from the different optical properties between moving and stationary amoebae. The cells look bright when moving and dark when stationary. (Courtesy of P.C. Newell from Newell 1983)

dramatic and new spatio-temporal phenomena of relevance in the biomedical sciences. Here we have mentioned only a few which we shall now study in more detail. Later in Chapter 20 we shall see another application of some of these studies to the spatial spread of epidemics.

12.2 Waves of Pursuit and Evasion in Predator-Prey Systems

If predators and their prey are spatially distributed it is obvious that there will be temporal spatial variations in the populations as the predators move to catch the prey and the prey move to evade the predators. Travelling bands have been observed, for example in oceanic plankton, a small marine organism (Wyatt 1973). They are also fairly common, for example in the movement of primitive organisms invading a source of nutrient. In this section we consider, mainly for illustration of the analytical technique, a simple predator-prey system with diffusion and show how travelling wavefront solutions occur. The specific model we study is a modified Lotka-Volterra system (see Section 3.1) with logistic growth of the prey and with both predator and prey dispersing by diffusion. Dunbar (1983, 1984) has discussed this model in detail. The model mechanism we consider is

$$\frac{\partial U}{\partial t} = AU\left(1 - \frac{U}{K}\right) - BUV + D_1\nabla^2 U \ ,$$

$$\frac{\partial V}{\partial t} = CUV - DV + D_2\nabla^2 V \ , \tag{12.2}$$

where U is the prey, V is the predator, A, B, C, D and K, the prey carrying capacity, are positive constants and D_1 and D_2 are the diffusion coefficients. We nondimensionalize the system by setting

$$u = \frac{U}{K}, \quad v = \frac{BV}{A}, \quad t^* = At, \quad x^* = x\left(\frac{A}{D_2}\right)^{1/2} \ ,$$

$$D = \frac{D_1}{D_2}, \quad a = \frac{CK}{A}, \quad b = \frac{D}{CK} \ .$$

We consider here only the one-dimensional problem, so the Equations (12.2) become, on dropping the asterisks for notational simplicity,

$$\frac{\partial u}{\partial t} = u(1 - u - v) + D\frac{\partial^2 u}{\partial x^2} \ ,$$

$$\frac{\partial v}{\partial t} = av(u - b) + \frac{\partial^2 v}{\partial x^2} \ , \tag{12.3}$$

and, of course, we are only interested in non-negative solutions.

The analysis of the spatially independent system is a direct application of the procedure in Chapter 3; it is simply a phase plane analysis. There are three steady states (i) (0,0), (ii) (1,0), that is no predator and the prey at its carrying capacity, and (iii) $(b, 1 - b)$, that is coexistence of both species if $b < 1$, which henceforth we assume to be the case. It is left as a revision exercise to show that both (0,0) and (1,0) are unstable and $(b, 1 - b)$ is a stable node if $4a \leq b/(1 - b)$, and a stable spiral if $4a > b/(1 - b)$. In fact in the positive (u, v) quadrant it is a globally stable steady state since (12.3), with $\partial/\partial x \equiv 0$, has a Lyapunov function given by

$$L(u,v) = a \left[u - b - b \ln\left(\frac{u}{b}\right) \right] + \left[v - 1 + b - (1-b) \ln\left(\frac{v}{1-b}\right) \right] .$$

That is $L(b, 1-b) = 0$, $L(u,v)$ is positive for all other (u,v) in the positive quadrant and $dL/dt < 0$ (see, for example, Jordan and Smith 1977 for a readable exposition of Lyapunov functions and their use). Recall, from Section 3.1 that in the simplest Lotka-Volterra system, namely (12.2) without the prey saturation term, the nonzero coexistence steady state was only neutrally stable and so was of no use practically. The modified system (12.2) is considerably more realistic.

Let us now look for constant shape travelling wavefront solutions of (12.3) by setting

$$u(x,t) = U(z), \quad v(x,t) = V(z), \quad z = x + ct , \tag{12.4}$$

in the usual way (see Chapter 11) where c is the positive wave speed which has to be determined. If solutions of the type (12.4) exist they represent travelling waves moving to the left in the z-plane. Substitution of these forms into (12.3) gives the ordinary differential equation system

$$\begin{aligned} cU' &= U(1 - U - V) + DU'' , \\ cV' &= aV(U - b) + V'' , \end{aligned} \tag{12.5}$$

where the prime denotes differentiation with respect to z.

The analysis of (12.5) involves the study of a four-dimensional phase space. Here we shall consider a simpler case, namely that in which the diffusion, D_1, of the prey is very much smaller than that of the predator, namely D_2, and so to a first approximation we take $D = D_1/D_2 = 0$. This would be the equivalent of thinking of a plankton-herbivore system in which only the herbivores were capable of moving. We might reasonably expect the qualitative behaviour of the solutions of the system with $D \neq 0$ to be more or less similar to those with $D = 0$ and this is indeed the case (Dunbar 1984). With $D = 0$ in (12.5) we write the system as a set of first order ordinary equations, namely

$$U' = \frac{U(1 - U - V)}{c}, \quad V' = W, \quad W' = cW - aV(U - b) . \tag{12.6}$$

In the (U, V, W) phase space there are two unstable steady states, $(0,0,0)$ and $(1,0,0)$, and one stable one, $(b, 1-b, 0)$: we are, as noted above, only interested in the case $b < 1$. From the experience gained from the analysis of Fisher equation, discussed in detail in Section 11.2, there is thus the possibility of a travelling wave solution from $(1,0,0)$ to $(b, 1-b, 0)$ and from $(0,0,0)$ to $(b, 1-b, 0)$. So we should look for solutions $(U(z), V(z))$ of (12.6) with the boundary conditions

$$U(-\infty) = 1, \quad V(-\infty) = 0, \quad U(\infty) = b, \quad V(\infty) = 1 - b \tag{12.7}$$

and

$$U(-\infty) = 0, \quad V(-\infty) = 0, \quad U(\infty) = b, \quad V(\infty) = 1 - b . \tag{12.8}$$

We shall consider here only the boundary value problem (12.6) with (12.7). First linearize the system about the singular point (1,0,0), that is the steady state $u = 1, v = 0$, and determine the eigenvalues λ in the usual way as described in detail in Chapter 3. They are given by the roots of

$$\begin{vmatrix} -\lambda - \dfrac{1}{c} & -\dfrac{1}{c} & 0 \\ 0 & -\lambda & 1 \\ 0 & -a(1-b) & c - \lambda \end{vmatrix} = 0$$

namely

$$\lambda_1 = -\frac{1}{c}, \quad \lambda_2, \lambda_3 = \frac{c \pm [c^2 - 4a(1-b)]^{1/2}}{2} . \tag{12.9}$$

Thus there is an unstable manifold defined by the eigenvectors associated with the eigenvalues λ_2 and λ_3 which are positive for all $c > 0$. Further, (1,0,0) is unstable in an oscillatory manner if $c^2 < 4a(1-b)$. So, the only possibility for a travelling wavefront solution to exist with non-negative U and V is if

$$c \geq [4a(1-b)]^{1/2}, \quad b < 1 . \tag{12.10}$$

With c satisfying this condition a realistic solution, with a lower bound on the wave speed, may exist which tends to $u = 1$ and $v = 0$ as $z \to -\infty$. This is reminiscent of the travelling wavefront solutions described in the last chapter.

The solutions here, however, can be qualitatively different from those in the last chapter, as we see by considering the approach of (U, V) to the steady state $(b, 1-b)$. Linearizing (12.6) about the singular point $(b, 1-b, 0)$ the eigenvalues λ are given by

$$\begin{vmatrix} -\lambda - \dfrac{b}{c} & -\dfrac{b}{c} & 0 \\ 0 & -\lambda & 1 \\ -a(1-b) & 0 & c - \lambda \end{vmatrix} = 0$$

and so are the roots of the characteristic polynomial

$$p(\lambda) \equiv \lambda^3 - \lambda^2 \left(c - \frac{b}{c} \right) - \lambda b - \frac{ab(1-b)}{c} = 0 . \tag{12.11}$$

To see how the solutions of this polynomial behave as the parameters vary we consider the plot of $p(\lambda)$ for real λ and see where it crosses $p(\lambda) = 0$. Differentiating $p(\lambda)$, the local maximum and minimum are at

$$\lambda_M, \lambda_m = \frac{\left(c - \frac{b}{c} \right) \pm \left[\left(c - \frac{b}{c} \right)^2 + 3b \right]^{1/2}}{3}$$

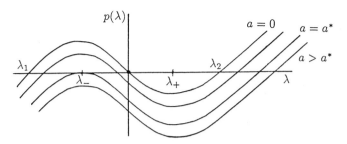

Fig. 12.2. The characteristic polynomial $p(\lambda)$ from (12.11) as a function of λ as a varies. There is a critical value a^* such that for $a > a^*$ there is only one real positive root and two complex ones with negative real parts.

and are independent of a. For $a = 0$ the roots of (12.11) are

$$\lambda = 0, \quad \lambda_1, \lambda_2 = \frac{\left(c - \frac{b}{c}\right) \pm \left[\left(c - \frac{b}{c}\right)^2 + 4b\right]^{1/2}}{2},$$

as illustrated in Fig. 12.2. We can now see how the roots vary with a. From (12.11), as a increases from zero the effect is simply to subtract $ab(1 - b)/c$ everywhere from the $p(\lambda; a = 0)$ curve. Since the local extrema are independent of a, we then have the situation illustrated in the figure. For $0 < a < a^*$ there are 2 negative roots and one positive one. For $a = a^*$ the negative roots are equal while for $a > a^*$ the negative roots become complex with negative real parts. This latter result is certainly the case for a just greater than a^* by continuity arguments. The determination of a^* can be carried out analytically. The same conclusions can be derived using the Routh-Hurwitz conditions (see Appendix 2) but here with these it is intuitively less clear.

The existence of a critical a^* means that, for $a > a^*$, the wavefront solutions (U, V) of (12.6) with boundary conditions (12.7) approach the steady state $(b, 1 - b)$ in an *oscillatory* manner while for $a < a^*$ they are monotonic. Fig. 12.3 illustrates the two types of behaviour.

The full predator-prey system (12.3), in which both the predator and prey diffuse, also gives rise to travelling wavefront solutions which can display oscillatory behaviour (Dunbar 1983, 1984). The proof of existence of these waves involves a careful analysis of the phase plane system to show that there is a trajectory, lying in the positive quadrant, which joins the relevant singular points. These waves are sometimes described as 'waves of pursuit and evasion' even though there is little evidence of prey evasion in the solutions in Fig. 12.3, since other than quietly reproducing, the prey simply wait to be consumed.

Convective Predator-Prey Pursuit and Evasion Models

A totally different kind of 'pursuit and evasion' predator-prey system is one in which the prey try to evade the predators and the predators try to catch the

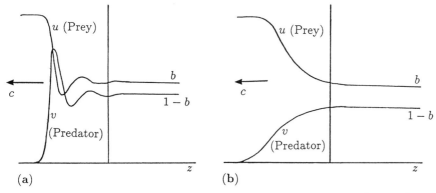

Fig. 12.3a,b. Typical examples of the two types of waves of pursuit given by wavefront solutions of the predator(v)-prey(u) system (12.3) with negligible dispersal of the prey. The waves move to the left with speed c. (a) Oscillatory approach to the steady state $(b, 1 - b)$, when $a > a^*$. (b) Monotonic approach of (u, v) to $(b, 1 - b)$ when $a \leq a^*$.

prey only if they interact. This results in a basically different kind of spatial interaction. Here we briefly describe one possible model, in its one-dimensional form, by way of illustration. Let us suppose that the prey (u) and predator (v) can move with speeds c_1 and c_2 respectively, that diffusion plays a negligible role in the dispersal of the populations and that each population obeys its own dynamics with its own steady state or states. Refer now to Fig. 12.4 and consider first Fig. 12.4 (a). Here the populations do not interact and, since there is no diffusive spatial dispersal, the population at any given spatial position simply grows or decays until the whole region is at that population's steady state. The dynamic situation is then as in Fig. 12.4 (a) with both populations simply moving at their undisturbed speeds c_1 and c_2 and without spatial dispersion, so the width of the bands remain fixed as u and v tend to their steady states. Now suppose that when the predators overtake the prey, the prey try to evade the predators by moving away from them with an extra burst of speed proportional to the predator gradient. In other words, if the overlap is as in Fig. 12.4 (b), the prey try to move away from the increasing number of predators. By the same token the predators try to move further into the prey and so move in the direction of increasing prey. We can model this situation by writing the conservation equations (see Chapter 9) to include convective effects as

$$u_t - [(c_1 + h_1 v_x)u]_x = f(u, v) , \tag{12.12}$$

$$v_t - [(c_2 - h_2 u_x)v]_x = g(v, u) , \tag{12.13}$$

where f and g represent the population dynamics and h_1 and h_2 are the positive parameters associated with the retreat and pursuit of the prey and predator. These are conservation laws for u and v so the terms on the left hand sides of the equations must be in divergence form. We now motivate the various terms in the equations.

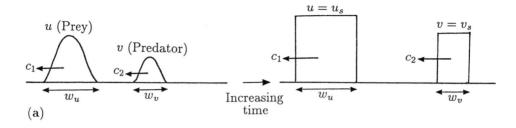

(a)

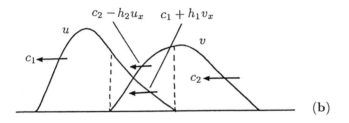

(b)

Fig. 12.4a,b. (a) The prey and predator populations are spatially separate and each satisfies its own dynamics: they do not interact and simply move at their own undisturbed speed c_1 and c_2. Each population grows until it is at the steady state (u_s, v_s) determined by its individual dynamics. Note that there is no dispersion so the spatial width of the 'waves' w_u and w_v remain fixed. (b) When the two populations overlap the prey put on an extra burst of speed $h_1 v_x$, $h_1 > 0$, to try and get away from the predators while the predators put on an extra spurt of speed, namely $-h_2 u_x$, $h_2 > 0$, to pursue them: the motivation for these terms is dicussed in the text.

The interaction terms f and g are whatever predator-prey situation we are considering. Typically $f(u, 0)$ represents the prey dynamics where the population simply grows or decays to a nonzero steady state. The effect of the predators is to reduce the size of the prey's steady state, so $f(u, 0)$ is larger than $f(u, v > 0)$. By the same token the steady state generated by $g(v, u \neq 0)$ is larger than that produced by $g(v, 0)$.

To see what is going on physically with the convective terms, suppose, in (12.12), $h_1 = 0$. Then

$$u_t - c_1 u_x = f(u, v) \,,$$

which simply represents the prey dynamics in a travelling frame moving with speed c_1. We see this if we use $z = x + c_1 t$ and t as the independent variables in which case the equation simply becomes $u_t = f(u, v)$. If $c_2 = c_1$, the predator equation, with $h_2 = 0$, becomes $v_t = g(v, u)$. Thus we have travelling waves of changing populations until they have reached their steady states as in Fig. 12.4 (a), after which they become travelling (top hat) waves of constant shape.

Consider now the more complex case where h_1 and h_2 are positive and $c_1 \neq c_2$. Referring to the overlap region in Fig. 12.4 (b) the effect in (12.12) of the $h_1 v_x$ term, positive because $v_x > 0$, is to increase locally the speed of the wave of the prey to the left. The effect of $-h_2 u_x$, positive because $u_x < 0$, is to increase the local convection of the predator. The intricate nature of interaction depends on the form of the solutions, specifically u_x and v_x, the relative size of the parameters c_1, c_2, h_1 and h_2 and the interaction dynamics. Because the equations are nonlinear through the convection terms (as well as the dynamics) the possibility exists of shock solutions in which u and v undergo discontinuous jumps: see, for example, Murray (1968, 1970, 1973) and, for a reaction diffusion example, Section 11.4 in the last chapter.

Before leaving this topic it is interesting to write the model system (12.12), (12.13) in a different form. Carrying out the differentiation of the left hand sides the equation system becomes

$$u_t - [(c_1 + h_1 v_x)]u_x = f(u, v) + h_1 u v_{xx} ,$$
$$v_t - [(c_2 - h_2 u_x)]v_x = g(v, u) - h_2 v u_{xx} .$$
$$(12.14)$$

In this form we see that the h_1 and h_2 terms on the right hand sides represent *cross diffusion*, one positive and the other negative. Cross diffusion, which, of course, is only of relevance in multi-species models was defined in Section 9.2: it occurs when the diffusion matrix is not strictly diagonal. It is a diffusion-type term in the equation for one species which involves another species. For example, in the u-equation $h_1 u v_{xx}$ is like a diffusion term in v, with 'diffusion' coefficient $h_1 u$. Typically a cross diffusion would be a term $\partial(D v_x)/\partial x$ in the u-equation. The above is an example where cross diffusion arises in a practical modelling problem – it is not common.

The mathematical analysis of systems like (12.12)–(12.14) is a challenging one which is largely undeveloped. Some analytical work has been done, for example, by Hasimoto (1974), Yoshikawa and Yamaguti (1974), who investigated the situation in which $h_1 = h_2 = 0$, and Murray and Cohen (1983), who studied the system with h_1 and h_2 nonzero. Hasimoto (1974) obtained analytical solutions to the system (12.12) and (12.13) where $h_1 = h_2 = 0$ and with the special forms $f(u, v) = l_1 uv$, $g(u, v) = l_2 uv$, where l_1 and l_2 are constants. He showed how blow-up can occur in certain circumstances. Interesting new solution behaviour is likely for general systems of the type (12.12)–(12.14).

Two dimensional problems involving convective pursuit and evasion are of ecological significance and are particularly challenging: they have not been investigated. For example, it would be very interesting to try and model a predator-prey situation in which species territory is involved. With the wolf-moose predator-prey situation in Canada, for example, it should be possible to build into a model the effect of wolf territory boundaries to see if the territorial 'no man's land' provides a partial safe haven for the prey. The reasoning for this speculation is that there is less tendency for the wolves to stray into the neighbouring territory. There seems to be some evidence that moose do travel along wolf territory boundaries.

A related class of wave phenomena occurs when convection is coupled with kinetics, such as occurs in biochemical ion exchange in fixed columns. The case of a single reacion kinetics equation coupled to the convection process, has been investigated in detail by Goldstein and Murray (1959). Interesting shock wave solutions evolve from smooth initial data. The mathematical techniques developed there are of direct relevance to the above problems. When several ion exchanges are occurring at the same time in this convective situation we then have chromatography, a powerful analytical technique in biochemistry.

12.3 Travelling Fronts in the Belousov-Zhabotinskii Reaction

The waves in Fig. 12.1 (a) are travelling bands of chemical concentrations in the Belousov-Zhabotinskii reaction; they are generated by a pacemaker. In this section we shall derive and analyse a model for the propagating *front* of such a wave. Far from the centre the wave is essentially plane, so we consider here the one dimensional problem. We follow in part the analysis of Murray (1976). The reason for investigating this specific problem is the assumption that the speed of the wave front depends primarily on the concentrations of the key chemicals, bromous acid (HBrO$_2$) and the bromide ion (Br$^-$) denoted respectively by x and y. Refer to Chapter 7, specifically Section 7.1, for the details of the model reaction kinetics. This section can, however, be read independently by starting with the reaction scheme in (12.15) below. We assume that these reactants diffuse with diffusion coefficient D. We believe that the wave front is dominated by process I of the reaction, namely the sequence of reactions which (i) reduces the bromide concentration to a small value, (ii) increases the bromous acid to its maximum concentration and in which (iii) the cerium ion catalyst is in the Ce^{3+} state. Since the concentration of Ce^{4+} was denoted by z in Section 7.1 the last assumption implies that $z = 0$. The simplified reaction sequence, from (7.2) without the cerium reaction and with $z = 0$, is then

$$A + Y \overset{k_1}{\to} X + P, \quad X + Y \overset{k_2}{\to} 2P, \quad A + X \overset{k_3}{\to} 2X, \quad 2X \overset{k_4}{\to} P + A, \quad (12.15)$$

where X and Y denote the bromous acid and bromide ion respectively and the k's are rate constants. P (the compound HOBr) does not appear in our analysis and the concentration $A(\text{BrO}_3^-)$ is constant.

Applying the Law of Mass Action (see Chapter 5) to this scheme, using lower case letters for concentrations, and including diffusion of X and Y, we get

$$\frac{\partial x}{\partial t} = k_1 ay - k_2 xy + k_3 ax - k_4 x^2 + D\frac{\partial^2 x}{\partial s^2},$$

$$\frac{\partial y}{\partial t} = -k_1 ay - k_2 xy + D\frac{\partial^2 y}{\partial s^2},$$

$$(12.16)$$

where s is the space variable. An appropriate nondimensionlization here is

$$u = \frac{k_4 x}{k_3 a}, \quad v = \frac{k_2 y}{k_3 a r}, \quad s^* = \left(\frac{k_3 a}{D}\right)^{1/2} s ,$$

$$t^* = k_3 a t, \quad L = \frac{k_1 k_4}{k_2 k_3}, \quad M = \frac{k_1}{k_3}, \quad b = \frac{k_2}{k_4} ,$$

(12.17)

where r is a parameter which reflects the fact that the bromide ion concentration far ahead of the wavefront can be varied experimentally. With these, (12.16) becomes, on omitting the asterisks for notational simplicity,

$$\frac{\partial u}{\partial t} = Lrv + u(1 - u - rv) + \frac{\partial^2 u}{\partial s^2}$$

$$\frac{\partial v}{\partial t} = -Mv - buv + \frac{\partial^2 v}{\partial s^2} .$$

(12.18)

Using the estimated values for the various rate constants and parameters from Chapter 7, equations (7.4), we find

$$L \approx M = O(10^{-4}), \quad b = O(1) .$$

The parameter r can be varied from about 5–50.

With the nondimensionalization (12.17) the realistic steady states are

$$u = v = 0; \quad u = 1, \quad v = 0 ,$$

(12.19)

so we expect u and v to be $O(1)$-bounded. So, to a first approximation, since $L \ll 1$ and $M \ll 1$ in (12.18), we may neglect these terms and thus arrive at a model for the leading edge of travelling waves in the Belousov-Zhabotinskii reaction, namely

$$\frac{\partial u}{\partial t} = u(1 - u - rv) + \frac{\partial^2 u}{\partial s^2}$$

$$\frac{\partial v}{\partial t} = -buv + \frac{\partial^2 v}{\partial s^2} ,$$

(12.20)

where r and b are positive parameters of $O(1)$. Note that this model approximation introduces a new steady state $(0, P)$, where $P > 0$ can take any value. The reason for this is that this is only a model for the *front*, not the whole wave pulse on either side of which $v \to 0$.

Let us now look for travelling wavefront solutions of (12.20) where the wave moves from a region of high bromous acid concentration to one of low bromous acid concentration as it reduces the level of the bromide ion. With (12.19) we therefore look for waves with boundary conditions

$$u(-\infty, t) = 0, \quad v(-\infty, t) = 1, \quad u(\infty, t) = 1, \quad v(\infty, t) = 0$$

(12.21)

and the wave moves to the left.

Before looking for travelling wave solutions we should note that there are some special cases which reduce the problem to a Fisher equation, discussed in detail in Section 11.2 of the last chapter. Setting

$$v = \frac{1-b}{r}(1-u), \quad b \neq 1, \quad r \neq 0 \tag{12.22}$$

the system (12.20) reduces to

$$\frac{\partial u}{\partial t} = bu(1-u) + \frac{\partial^2 u}{\partial s^2},$$

the Fisher equation (11.4), which has travelling monotonic wavefront solutions going from $u = 0$ to $u = 1$ which travel at speeds $c \geq 2\sqrt{b}$. Since we are only concerned here with non-negative u and v, we must have $b < 1$ in (12.22). If we take the initial condition

$$u(s, 0) \sim O(\exp[-\beta s]) \quad \text{as} \quad s \to \infty$$

we saw in Section 11.2 that the asymptotic speed of the resulting travelling wavefront is

$$c = \begin{cases} \beta + \dfrac{b}{\beta}, & 0 < \beta \leq \sqrt{b} \\ 2\sqrt{b}, & \beta > \sqrt{b}. \end{cases} \tag{12.23}$$

The wavefront solutions given by the Fisher wave with v as in (12.22) are not, however, of practical relevance unless $1 - b = r$ since we require u and v to satisfy the boundary conditions (12.21), where $v = 0$ when $u = 1$ and $v = 1$ when $u = 0$. The appropriate Fisher solution with suitable initial conditions, namely

$$u(s, 0) = \begin{cases} 0 \\ h(s) \\ 1 \end{cases} \quad \text{for} \quad \begin{cases} s < s_1 \\ s_1 < s < s_2 \\ s_2 < s \end{cases} \tag{12.24}$$

where $h(s)$ is a positive monotonic continuous function with $h(s_1) = 0$ and $h(s_2) = 1$, then has wave speed $c = 2\sqrt{b} = 2\sqrt{1-r}$ from (12.23) and $v = 1 - u$. Necessarily $0 < r \leq 1$.

We can further exploit the results for the Fisher equation by using the maximum principle for parabolic equations. Let $u_f(s, t)$ denote the unique Fisher solution of

$$\frac{\partial u_f}{\partial t} = u_f(1 - u_f) + \frac{\partial^2 u_f}{\partial s^2} \tag{12.25}$$

$$u_f(-\infty, t) = 0, \quad u_f(\infty, t) = 1$$

with initial conditions (12.24). The asymptotic travelling wavefront solution has speed $c = 2$. Now write

$$w(s, t) = u(s, t) - u_f(s, t)$$

and let $u(s,t)$ have the same initial conditions as u_f, namely (12.24). Subtracting equation (12.25) from the equation for u given by (12.20) and using the definition of w in the last equation we get

$$w_{ss} - w_t + [1 - (u + u_f)]w = ruv \ .$$

We are restricting our solutions to $0 \leq u \leq 1$ and, since $0 \leq u_f \leq 1$, we have $[1-(u+u_f)] \leq 1$ and so we cannot use the usual maximum principle immediately. If we set $W = w \exp[-Kt]$, where $K > 0$ is a finite constant, the last equation becomes

$$W_{ss} - W_t + [1 - (u + u_f) - K]W = ruve^{-Kt} \geq 0 \ .$$

Choosing $K > 1$ we then have $[1 - (u + u_f) - K] < 0$ and the maximum principle can now be used on the W-equation. It says that W, and hence w, has its maximum at $t = 0$ or at $s = \infty$. But $w_{max} = (u - u_f)_{max} = 0$ at $t = 0$ and at $s = \pm\infty$ so we have the result

$$u(s,t) \leq u_f(s,t) \quad \text{for all} \quad s, \quad t > 0 \ .$$

This says that the solution for u of (12.20) is at all points less than or equal to the Fisher solution u_f which evolves from initial conditions (12.24). So, if the solutions of (12.20) have travelling wave solutions with boundary conditions (12.21) and equivalent initial conditions to (12.24), then their wave speeds c must be bounded by the Fisher speed and so we have the upper bound $c(r, b) \leq 2$ for all values of the parameters r and b. Intuitively we would expect any such travelling wave solution of (12.20) to have speed $c \leq 2$ since with $uv \geq 0$ the term $-ruv$ in the first of (12.20) is like a sink term in addition to the Fisher kinetics $u(1 - u)$. This inhibits the growth of u at any point as compared with the Fisher wave solution so we would expect u and its speed to be bounded above by the Fisher solution.

Various limiting values for the wave speed c, as a function of r and b, can be derived from the equation system (12.20). Care, however, has to be taken in their derivation because of nonuniform limiting situations: these will be pointed out at the appropriate places.

If $b = 0$ the equation for v from (12.20) becomes the basic diffusion equation $v_t = v_{ss}$ which cannot have wave solutions. This means that neither can the first of (12.20) for u since a wave solution requires u and v to have the same speed of propagation. This suggests that the limit $c(b \to 0, r) = 0$ for $r > 0$. If $b \to \infty$, (12.20) says that $v = 0$ (we exclude the trivial solution $u = 0$) in which case $c(b \to \infty, r) = 2$ for all $r \geq 0$. Now if $r = 0$, the u and v equations are uncoupled with the u-equation being the basic Fisher equation (12.25) which, with initial conditions (12.24), has wavefront solutions with $c = 2$: this means that the relevant v-solution also has speed 2. This gives the limiting case $c(b, r \to 0) = 2$ for $b > 0$. If $r \to \infty$ then $u = 0$ or $v = 0$, either of which implies that there is no wave solution, so $c(b, r \to \infty) = 0$. Note the nonuniform limiting situation with this case: the limit $r \to \infty$ with $v \neq 0$ is not the same as the situation with $v = 0$

and then letting $r \to \infty$. In the latter u is governed by the Fisher equation and r is irrelevant. As we said above, however, we are here concerned with travelling waves in which neither u nor v are identically zero. With that in mind we then have in summary

$$
\begin{aligned}
c(0,r) = 0, \quad r > 0; \qquad c(\infty, r) = 2, \quad r \geq 0 \\
c(b,0) = 2, \quad b > 0; \qquad c(b,\infty) = 0, \quad b \geq 0 .
\end{aligned}
\tag{12.26}
$$

The first of these does not give the whole story for small b as we see below.

The travelling wavefront problem for the system (12.20), on using the travelling wave transformation

$$
u(s,t) = f(z), \quad v(s,t) = g(z), \quad z = s + ct ,
$$

and the boundary conditions (12.21), becomes

$$
\begin{aligned}
f'' - cf' + f(1 - f - rg) = 0, \quad g'' - cg' - bfg = 0 \\
f(\infty) = g(-\infty) = 1, \quad f(-\infty) = g(\infty) = 0 .
\end{aligned}
\tag{12.27}
$$

Using various bounds and estimation techniques for monotonic solutions of (12.27) with $f \geq 0$ and $g \geq 0$, Murray (1976) obtained the general bounds on c in terms of the parameters r and b given by

$$
\left[\left(r^2 + \frac{2b}{3} \right)^{1/2} - r \right] [2(b + 2r)]^{-1/2} \leq c \leq 2 .
\tag{12.28}
$$

The system (12.20), with initial and boundary conditions (12.21) and (12.24), were solved numerically (Murray 1976) and some of the results are shown in Fig. 12.5. Note, in Fig. 12.5 (b), the region bounded by $b = 0$, $c^2 = 4b$ and $c = 2$ within which non negative solutions do not exist. The limit curve $c^2 = 4b$ is obtained using the special case solution in which $v = 1 - u$ and $b = 1 - r$, $r < 1$. A fuller numerical study of the model system (12.20) has been carried out by Manoranjan and Mitchell (1983).

Let us now return to the experimental situation. From Fig. 12.5 (a), if we keep b fixed, the effect of increasing r (which with the nondimensionalisation (12.17) is the equivalent of increasing the upstream bromide ion (Br$^-$) concentration) is to flatten the v-curve. That is the wavefront becomes less sharp. On the other hand for a fixed r and increasing b the front becomes sharper. Although it is imprecise, we can get some estimate of the actual width of the wave front from the width, ω say, of the computed wavefront solution. In dimensional terms this is ωD where, from (12.17),

$$
\omega D = \left(\frac{D}{k_3 a} \right)^{1/2} \omega \approx 4.5 \times 10^{-4} \omega \ \text{cm} ,
$$

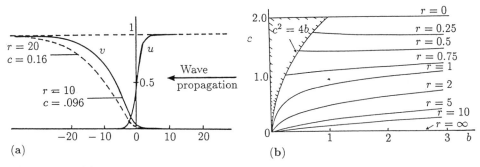

Fig. 12.5a,b. (a) Typical computed wavefront solution of the Belousov-Zhabotinskii model system (12.20) for u and v for $b = 1.25$ and two values of the upstream bromide parameter r. The u-curves for both values of r are effectively indistinguishable (b) Wave speed c of wavefront solutions as a function of b for various r. (From Murray 1976)

where we have taken $D \approx 2 \times 10^{-5}\, \text{cm}^2\text{s}^{-1}$, a typical value for reasonably small molecules such as we are concerned with here, and $k_3 a \approx 10^2\, \text{s}^{-1}$ obtained from the parameter values in (7.4) in Chapter 7. From Fig. 12.5 (a), ω is around 10 which then gives ω_D of the order of 10^{-3} cm. This is of the order found experimentally: the front is very thin.

Another practical prediction, from Fig. 12.5 (b), is that for b larger than about 2, we see that for a fixed r the wave speed is fairly independent of b. Computations for values of b up to about 50 confirm this observation. This has also been observed experimentally.

From the nondimensionalization (12.17) the dimensional wave speed, c_D say, is given by

$$c_D = (k_3 a D)^{1/2} c(r, b) \, ,$$

where r is a measure of the upstream bromide concentration and $b = k_2/k_4$. From the parameter estimates given in Chapter 7, equations (7.4), we get $b \approx 1$. Assigning r is not very easy and values of 5–50 are reasonable experimentally. With r of $O(10)$ and b about 1 we get the dimensionless wave speed from Fig. 12.5 (b) to be $O(10^{-1})$: the precise value for c can be calculated from the model system. With the values for D and $k_3 a$ above we thus get c_D to be $O(4.5 \times 10^{-3}\, \text{cm s}^{-1})$ or $O(2.7 \times 10^{-1}\, \text{cm min}^{-1})$, which is in the experimental range observed. In view of the reasonable quantitative comparison with experiment and the results derived here from a model which mimics the propagation of the wave front, we suggest that the speed of propagation of Belousov-Zhabotinskii wavefronts is mainly determined by the leading edge and not the trailing edge.

Finally in relation to the speed of propagation of a reaction diffusion wavefront compared to simple diffusion we get the time for a wavefront to move 1 cm as $O(10/2.7\, \text{min})$, that is about 4 minutes as compared to the diffusional time which is $O(1\, \text{cm}^2/D)$, namely $O(5 \times 10^4\, \text{s})$ or about 850 minutes. So, as a means of transmitting information via a change in chemical concentration we can safely say that reaction diffusion waves are orders of magnitude faster than pure diffusion, if the distances involved are other than very small. Later we shall be discussing in detail the problem of pattern formation in embryological contexts where distances of interest are of the order of cell diameters, so diffusion is again a relevant mechanism for conveying information. However, as we shall see it is not the only one in embryological contexts.

The model system (12.20) has been studied by several authors. Gibbs (1980), for example, proves the existence and monotonicity of the travelling waves. An interesting formulation of such travelling wave phenomena as Stefan problems, together with a singular perturbation analysis of the Stefan problem associated with the Murray model (12.20) in which the parameters r and b are both large, is given by Ortoleva and Schmidt (1985: see also other references there).

12.4 Waves in Excitable Media

One of the most widely studied systems with excitable behaviour is neural communication by nerve cells via electrical signalling. We discussed the important Hodgkin-Huxley model in Chapter 6, Section 6.5, and derived a mathematical caricature, the Fitzhugh-Nagumo equations (FHN). Here we first consider, simply by way of example, the spatio-temporal FHN model and demonstrate the existence of travelling pulses which only propagate if a certain threshold perturbation is exceeded. By a pulse here, we mean a wave which represents an excursion from a steady state and back to it – like a solitary wave on water: see for example Fig. 12.7 below. We shall consider kinetics models of the form

$$u_t = f(u, v), \quad v_t = g(u, v)$$

and the approach we shall discuss is quite general and applies to a wide class of qualitative models of excitable media whose null clines are qualitatively similar to those in Fig. 12.6 (a). This section can be read without reference to the actual physiological situation if the equation system (12.29) below is simply considered as a specific model example for an excitable medium.

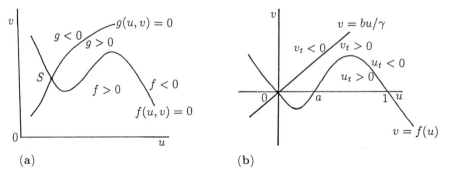

Fig. 12.6a,b. (a) Typical null clines for excitable kinetics. The kinetics here have only one steady state, S, which is globally stable but excitable. (b) Null clines for the excitable Fitzhugh-Nagumo system (12.29): the origin is the single steady state.

As we saw in Section 6.5, without any spatial variation, that is the space clamped situation, the FHN equations exhibited a threshold behaviour in time as illustrated in Fig. 6.12 (refer also to Section 3.8). The FHN system without any applied current ($I_a = 0$), but where we allow spatial 'diffusion' in the transmembrane potential and with a slight change in notation for consistency in this chapter, is

$$\frac{\partial u}{\partial t} = f(u) - v + D\frac{\partial^2 u}{\partial x^2}, \quad \frac{\partial v}{\partial t} = bu - \gamma v \,, \tag{12.29}$$

$$f(u) = u(a - u)(u - 1) \,.$$

Here u is directly related to the membrane potential (V in Section 6.5) and v plays the role of several variables associated with terms in the contribution to the membrane current from sodium, potassium and other ions. The 'diffusion' coefficient D is associated with the axial current in the axon and, referring to the conservation of current equation (6.38) in Section 6.5, the spatial variation in the potential V gives a contribution $(d/4r_i)V_{xx}$ on the right hand side, where r_i is the resistivity and d is the axon diameter. The parameters $0 < a < 1$, b and γ are all positive. The null clines of the 'kinetics' in the (u, v) plane are shown in Fig. 12.6 (b).

We want to demonstrate in this section how travelling wave solutions arise for reaction diffusion systems with excitable kinetics. There are several important physiological applications in addition to that for the propagation of nerve action potentials modelled with the FHN model. One such application, which is currently being studied in depth, is the waves which arise in muscle tissue, particularly heart muscle: in their two and three-dimensional context these excitable waves are intimately related to the problem of atrial flutter and fibrillation (see, for example, Winfree 1983). Another example is the reverberating cortical depression waves in the brain cortex (Shibata and Bureš (1974)). Two and three-dimensional excitable waves can also arise in the Belousov-Zhabotinskii reaction. We shall come back to these applications below.

The system (12.29) has been studied in some detail and the following are only a very small sample from the long list of references. The review by Rinzel (1981) specifically discusses models in neurobiology. Rinzel and Keller (1973) considered the piecewise linear caricature of (12.29) where $\gamma = 0$ and obtained analytical results for travelling pulses and periodic wavetrains. The caricature form when $f(u)$ is replaced by the piecewise linear approximation $f(u) = H(u - a)^\dagger - u$ has been studied by McKean (1970) and Feroe (1982) looked at the stability of multiple pulse solutions of this caricature. Ikeda et al. (1986) considered the Hodgkin-Huxley system and demonstrated the instability of certain slow wave solutions. The situation when b and γ in (12.29) are such that $v = bu/\gamma$ intersects the u-nullcline to give three steady states has been studied by Rinzel and Terman (1982). General discussions and reviews of waves in excitable media have been

---

[†] H denotes the Heaviside function

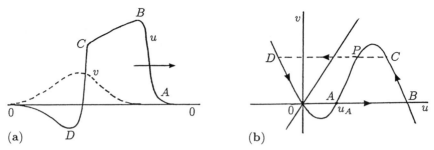

Fig. 12.7a,b. (a) Typical travelling pulse, or solitary wave, solution for the excitable system (12.29). (b) Corresponding phase trajectory in the (u, v) plane. Note the threshold characteristic of the null clines. A perturbation from the origin to a value $u < u_A$ will simply return to the origin with u always less than u_A. A perturbation to $u > u_A$ initiates a large excursion qualitatively like $ABCD0$. The position of C is obtained from the analysis as explained in the text.

given, for example, by Keener (1980), Zykov (1988), Tyson and Keener (1988) and of periodic bursting phenomena in excitable membranes by Carpenter (1979).

Travelling wave solutions of (12.29), in which u and v are functions only of the travelling coordinate variable $z = x - ct$, satisfy the travelling coordinate form of (12.29), namely

$$Du'' + cu' + f(u) - v = 0, \quad cv' + bu - \gamma v = 0, \quad z = x - ct \qquad (12.30)$$

where the prime denotes differentiation with respect to z and the wave speed c is to be determined. The boundary conditions corresponding to a solitary pulse are

$$u \to 0, \quad u' \to 0, \quad v \to 0 \quad \text{as} \quad |z| \to \infty \qquad (12.31)$$

and the pulse is typically as illustrated in Fig. 12.7 (a). The corresponding phase trajectory in the (u, v) plane is as in Fig. 12.7(b).

Initial conditions play a crucial role in the existence of travelling pulses. Intuitively we can see why as follows. Suppose we have a spatial domain with (u, v) initially at the zero rest state and we perturb it by a local rise in u over a small domain, keeping $v = 0$, as in Fig. 12.8 (a). If the perturbation has a maximum u less than the threshold u_A in Fig. 12.7 (b) (and Fig. 12.8 (c)) then the kinetics cause u to return to the origin and the spatial perturbation simply dies out. On the other hand if the perturbation is larger than the threshold u_A then the kinetics initiate a large excursion in both u and v as shown by $0BCD0$ in Fig. 12.8 (c). When a wave is initiated the trailing edge is represented in the phase plane by CD. Whereas it is intuitively clear that the leading edge should be at $0B$ the positioning of CD is not so obvious. We now consider this important aspect of travelling pulses.

It is analytically easier to see what is going on if we consider (12.29) with b and γ small, so we write

$$b = \varepsilon L, \quad \gamma = \varepsilon M, \quad 0 < \varepsilon \ll 1$$

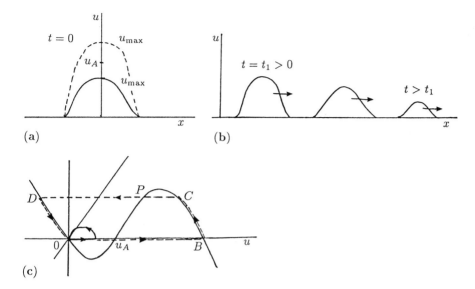

Fig. 12.8a-c. (a) The perturbation given by the solid line has $u_{max} < u_A$ where u_A is the threshold value in Fig. 12.7 (b). The solution is then simply a decaying transient such as illustrated in (b). With the dashed line as initial conditions the maximum value of u is larger than the threshold u_A and this initiates a travelling pulse such as in Fig. 12.7 (a). (c) Typical phase trajectories for u and v depending on whether the inital u is greater than or less than u_A. The positioning of the trailing edge part of the trajectory, CD, is discussed in the text.

and (12.29) becomes

$$u_t = Du_{xx} + f(u) - v, \quad v_t = \varepsilon(Lu - Mv) . \tag{12.32}$$

Now refer back to Fig. 12.7 (a) and consider the leading front $0AB$. In the limiting situation $\varepsilon \to 0$ the last equation says that $v \approx$ constant and from Fig. 12.7 (a) and Fig. 12.7 (b), this constant is zero. The u-equation in (12.32) then becomes

$$u_t = Du_{xx} + f(u), \quad f(u) = u(a - u)(u - 1) \tag{12.33}$$

where $f(u)$ is sketched as a function of u in Fig. 12.7 (b). It has three steady states $u = 0$, $u = a$ and $u = 1$. In the absence of diffusion (12.33) implies that $u = 0$ and $u = 1$ are linearly stable and $u = a$ is unstable. We can thus have a travelling wave solution which joins $u = 0$ to $u = 1$ as we showed in the last chapter. Equation (12.33) is a specific example of the one studied in Section 11.5, specifically equation (11.83), which has an exact analytical solution (11.88) with a unique wave speed given by (11.87). For the wave solution here we thus get

$$u = u(z), \quad z = x - ct; \quad c = \left(\frac{D}{2}\right)^{1/2}(1 - 2a) , \tag{12.34}$$

and so the wave speed is positive only if $a < 1/2$. Recalling the analysis in Chapter 11 this is the same condition we get from the sign determination given by (11.80), which, for (12.33), is

$$c \gtreqless 0 \quad \text{if} \quad \int_0^1 f(u)\, du \gtreqless 0 .$$

Referring now to Fig. 12.7 (b), the area bounded by $0A$ and the curve $v = f(u)$ is less than the area enclosed by AB and the curve $v = f(u)$ so $c > 0$: carrying out the integration gives $c > 0$ for all $a < 1/2$.

We arrived at the equation system for the wave pulse front by neglecting the ε-terms in (12.32). This gave us the contribution to the pulse corresponding to $0AB$ in Fig. 12.7 (b). Along BC, v changes. From (12.32) a change in v will take a long time, $O(1/\varepsilon)$ in fact, since $v_t = O(\varepsilon)$. To get this part of the solution we would have to carry out a singular perturbation analysis (see, for example, Keener 1980), the upshot of which gives a slow transition period where u does not change much but v does. This is the part of the pulse designated BC in Fig. 12.7 (a) and Fig. 12.7 (b).

The crucial question immediately arises as to where the next fast transition takes place; in other words where C is on the phase trajectory. Remember we are investigating the existence of pulse solutions which travel without change of shape. For this to be so the wave speed of the trailing edge, namely the speed of a wavefront solution that goes from C to D via P in Fig. 12.7 (a) (and Fig. 12.8 (c)), has to be the same as that for the leading edge $0AB$. On this part of the trajectory $v \approx v_C$ and the equation for the trailing edge wave front from (12.32) is then given by

$$u_t = D u_{xx} + f(u) - v_C . \tag{12.35}$$

A travelling wavefront solution of this equation has to have

$$u = u(z), \quad z = x - ct; \quad u(-\infty) = u_D, \quad u(\infty) = u_C .$$

The analytical solution and its unique wave speed are again given in terms of u_C, u_P and u_D by the analysis in Section 11.5. It gives the wave speed as

$$c = \left(\frac{D}{2}\right)^{1/2} (u_C - 2u_P + u_D) . \tag{12.36}$$

From the expression for $f(u)$ in (12.33), the roots u_C, u_D and u_P of $f(u) = v_C$ are determined in terms of v_C. The wave speed c in (12.36) is then $c(v_C)$, a function of v_C. We now determine the value of v_C by requiring this $c(v_C)$ to be equal to the previously calculated wave speed for the pulse front, namely $c = (D/2)^{1/2}(1 - 2a)$ from (12.34). In principle it is possible to determine this since the expression for v_C in this way is the solution of a polynomial.

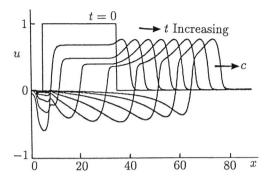

Fig. 12.9. Development of a travelling wave pulse solution from square initial data for the excitable system (12.32) with $f(u) = H(u - a) - u$ with $a = 0.25$, $D = 1$, $\varepsilon = 0.1$, $L = 1$, $M = 0.5$. The wave is moving to the right. (Redrawn from Rinzel and Terman 1982 with permission of J. Rinzel)

To complete the analytical determination of the wave pulse we now have to consider the part of the solution and the phase trajectory $D0$ in Fig. 12.7 and Fig. 12.8 (c). As for the part BC, during this stage v again changes by $O(1)$ in a time $O(1/\varepsilon)$. This is referred to as the *refractory phase* of the phenomenon. Fig. 12.9 shows a computed example for the system (12.29) where the cubic $f(u)$ is approximated by the piecewise linear expression $f(u) = H(u - a) - u - v$.

Threshold waves are also obtained for more general excitable media models. To highlight the analytical concepts let us consider the two species system in which one of the reactions is fast. To facilitate the analysis we consider the reaction diffusion system

$$\varepsilon u_t = \varepsilon^2 D_1 u_{xx} + f(u, v), \quad v_t = \varepsilon^2 D_2 v_{xx} + g(u, v), \tag{12.37}$$

where $0 < \varepsilon \ll 1$ and the kinetics f and g have null clines like those in Fig. 12.10 (a), and we exploit the fact that ε is small. The key qualitative shape for $f(u, v) = 0$ is a cubic. This form is typical of many reactions where activation and inhibition are involved (cf. Section 5.5).

The system (12.37) is excitable in the absence of diffusion. In Section 3.8 the description of a threshold mechanism was rather vague. We talked there of a system where the reactants underwent a large excursion in the phase plane if the perturbation was of the appropriate kind and of sufficient size. A better, and much more precise, definition is that a mechanism is excitable if a stimulus of sufficient size can initiate a travelling pulse which will propagate through the medium.

For $0 < \varepsilon \ll 1$, the $O(1)$ form of (12.37) is $f(u, v) = 0$ which we assume can be solved to give u as a multivalued function of v. From Fig. 12.10 (a) we see that for all given $v_m < v < v_M$ there are three solutions for u of $f(u, v) = 0$: they are the intersections of the line $v = $ constant with the null cline $f(u, v) = 0$. In an analogous way to the above discussion of the FHN system we can thus have a

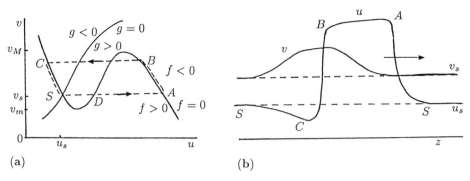

Fig. 12.10a,b. (a) Schematic null clines $f(u, v) = 0$, $g(u, v) = 0$ for the excitable system (12.37). The travelling pulse solution corresponds to the dashed trajectory. (b) Typical pulse solutions for u and v.

wavefront type solution joining S to A and a trailing wavefront from B to C with slow transitions in between. The time for u to change from its value at S to that at A is fast. This is what $f(u, v) = 0$ means since (referring to Fig. 12.10 (a)) if there is a perturbation to a value u to the right of D, u goes to the value at A instantaneously since u moves so that $f(u, v)$ is again zero. It takes, in fact, a time $O(\varepsilon)$. On the other hand it takes a relatively long time, $O(1)$, to traverse the AB and CS parts of the curve, while BC is covered again in $O(\varepsilon)$.

The analytical investigation of the pulse solution is quite involved and the detailed analysis of this general case has been given by Keener (1980). Here we shall consider by way of illustration how to go about carrying out the analysis for the leading front: that is we consider the transition from S to A in Fig. 12.10 (a) and Fig. 12.10 (b). The transition takes place quickly, in a time $O(\varepsilon)$, and spatially it is a sharp front of thickness $O(\varepsilon)$; see Fig. 12.10 (b). These scales are indicated by a singular perturbation appraisal of equations (12.37). This suggests that we introduce new independent variables by the transformations

$$\tau = \frac{t}{\varepsilon}, \quad \xi = \frac{x - x_T}{\varepsilon\sqrt{D_1}}, \tag{12.38}$$

where x_T is the position of the transition front, which we do not need at this stage or level of analysis. All the introduction of x_T does is to make the leading edge of the wave pulse at the origin $\xi = 0$ in the ξ-plane. Substitution into (12.37) and letting $\varepsilon \to 0$, keeping τ and ξ fixed in the usual singular perturbation way (see Murray 1984) gives the $O(1)$ system as

$$u_\tau = u_{\xi\xi} + f(u, v), \quad v_\tau = 0 . \tag{12.39}$$

So, considering the line SA, the second equation is simply $v = v_S$ and we then have to solve

$$u_\tau = u_{\xi\xi} + f(u, v_S) . \tag{12.40}$$

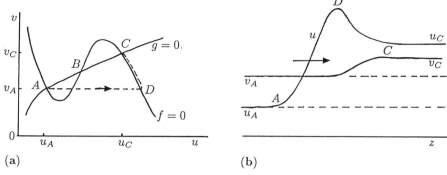

Fig. 12.11a,b. (a) Nullclines $f(u,v) = 0$, $g(u,v) = 0$ where there are 3 steady states: with these kinetics A and C are linearly stable while B is unstable. (b) Typical travelling wave effecting a transition from A to C if the initial perturbation from A is sufficiently large and ε is small in (12.37).

This is just a scalar equation for u in which f has three steady states, namely u_S, u_D and u_A and is qualitatively the same as those studied in the last chapter. It is essentially the same as equation (11.72) which was discussed in detail in Section 11.5. In the absence of diffusion the steady states at S, D and A are respectively stable, unstable and stable. We have already shown in Section 11.5 how a travelling monotonic wave solution with a unique wave speed exists which can join u_S and u_A as in Fig. 12.10 (b).

The complete solution requires determining the wave speed and the other parts of the pulse, namely AB, BC and CS, making sure that they all join up consistently. It is an interesting singular perturbation analysis. This was carried out by Keener (1980), who also presents numerical solutions as well as an analysis of threshold waves in two space dimensions.

Another type of threshold wave of practical interest occurs when the null clines $f(u,v) = 0$, $g(u,v) = 0$ intersect such as in Fig. 12.11 (a). That is there are 3 steady states. With the scaling as in (12.37) we now have the sharp front from A to D and the slower DC transition essentially the same as above. We can get the sign of the wave speed in the same way as described in Section 11.5. Now there is a tail to the wave since C is a linearly stable steady state. Fig. 12.11 (b) is a typical example of such threshold front waves. There is also the possibility of a transition wave obtained by perturbing the uniform steady state at C thus effecting a transition to A. Rinzel and Terman (1982) have studied such waves in the FHN context.

Threshold waves exist for quite a wide spectrum of real world systems – any in fact which can exhibit threshold kinetics. For example, Britton and Murray (1979) studied them in a class of substrate inhibition oscillators (see also the book by Britton 1987). The waves also exist, for a certain parameter domain, in the model chemotaxis mechanism proposed for the slime mold *Dictyostelium* (cf. Section 9.4): see, for example, Keller and Segel (1971) and Keller and Odell (1975).

12.5 Travelling Wave Trains in Reaction Diffusion Systems with Oscillatory Kinetics

Wave train solutions for general reaction diffusion systems with limit cycle kinetics have been studied in depth; the mathematical papers by Kopell and Howard (1973) and Howard and Kopell (1977) are seminal. Several review articles in the book edited by Field and Burger (1985) are apposite to this section. Further references will be given at appropriate places below.

The general evolution system we shall be concerned with is (12.1), which, for our purposes we restrict to one spatial dimension and for algebraic simplicity we incorporate the diffusion coefficient in a new scaled space variable $x \to x/D^{1/2}$. The equation system is then

$$\frac{\partial \mathbf{u}}{\partial t} = \mathbf{f}(\mathbf{u}) + \frac{\partial^2 \mathbf{u}}{\partial x^2} \ . \tag{12.41}$$

We shall assume that the spatially homogeneous system

$$\frac{d\mathbf{u}}{dt} = \mathbf{f}(\mathbf{u}; \gamma) \ , \tag{12.42}$$

where γ is a bifurcation parameter, has a stable steady state for $\gamma < \gamma_c$ and, via a Hopf bifurcation (see Appendix 3), evolves to a stable limit cycle solution for $\gamma > \gamma_c$: that is for $\gamma = \gamma_c + \varepsilon$, where $0 < \varepsilon \ll 1$, a small amplitude limit cycle solution exists and is stable.

Travelling plane wavetrain solutions are of the form

$$\mathbf{u}(x, t) = \mathbf{U}(z), \quad z = \sigma t - kx \ , \tag{12.43}$$

where $\mathbf{U}$ is a 2π-periodic function of z, the 'phase'. Here $\sigma > 0$ is the frequency and k the wavenumber; the wavelength $w = 2\pi/k$. The wave travels with speed $c = \sigma/k$. This form is only a slight variant of the general travelling wave form used in the last chapter and above and can be reduced to that form by rescaling the time. Substituting (12.43) into (12.41) gives the following system of ordinary differential equations for $\mathbf{U}$:

$$k^2 \mathbf{U}'' - \sigma \mathbf{U}' + \mathbf{f}(\mathbf{U}) = 0 \ , \tag{12.44}$$

where prime denotes differentiation with respect to z. We want to find σ and k so that the last equation has a 2π-periodic solution for $\mathbf{U}$.

Rather than consider the general situation (see Kopell and Howard 1973 and the comments below) it is instructive and algebraically simpler to discuss, by way of demonstration, the analysis of the λ-ω model system described in Section 6.4 and given by equations (6.30). Later we shall relate it to general reaction diffusion systems which can arise from real biological situations. This two reactant model mechanism, for (u, v) say, is

$$\frac{\partial}{\partial t}\begin{pmatrix} u \\ v \end{pmatrix} = \begin{pmatrix} \lambda(r) & -\omega(r) \\ \omega(r) & \lambda(r) \end{pmatrix}\begin{pmatrix} u \\ v \end{pmatrix} + \frac{\partial^2}{\partial x^2}\begin{pmatrix} u \\ v \end{pmatrix} \quad \text{where} \quad r^2 = u^2 + v^2 \,. \quad (12.45)$$

Here $\omega(r)$ and $\lambda(r)$ are real functions of r. If r_0 is an isolated zero of $\lambda(r)$ for some $r_0 > 0$ and $\lambda'(r_0) < 0$ and $\omega(r_0) \neq 0$, then the spatially homogeneous system, that is with $\partial^2/\partial x^2 = 0$, has a limit cycle solution (see Section 6.4 and (12.48) below).

It is convenient to change variables from (u, v) to polar variables (r, θ), where θ is the phase, defined by

$$u = r\cos\theta, \quad v = r\sin\theta \quad (12.46)$$

with which (12.45) becomes

$$r_t = r\lambda(r) + r_{xx} - r\theta_x^2 \,,$$
$$\theta_t = \omega(r) + r^{-2}(r^2\theta_x)_x \,. \quad (12.47)$$

If $r_0 > 0$ exists and $\lambda'(r_0) < 0$ the asymptotically stable limit cycle solution of the kinetics is given immediately by

$$r = r_0, \quad \theta = \theta_0 + \omega(r_0)t \,, \quad (12.48)$$

where θ_0 is some arbitrary phase. Substituting into (12.46) gives the limit cycle solutions u and v as

$$u = r_0\cos[\omega(r_0)t + \theta_0], \quad v = r_0\sin[\omega(r_0)t + \theta_0] \,, \quad (12.49)$$

which have frequency $\omega(r_0)$ and amplitude r_0.

Suppose we look for travelling plane wave solutions of the type (12.43) in the polar form

$$r = \alpha, \quad \theta = \sigma t - kx \,. \quad (12.50)$$

Substituting into (12.47) we get the necessary and sufficient conditions for these to be travelling wave solutions as

$$\sigma = \omega(\alpha), \quad k^2 = \lambda(\alpha) \,. \quad (12.51)$$

So, with α the convenient parameter, there is a one parameter family of travelling wave train solutions of (12.45) given by

$$u = \alpha\cos[\omega(\alpha)t - x\lambda^{1/2}(\alpha)], \quad v = \alpha\sin[\omega(\alpha)t - x\lambda^{1/2}(\alpha)] \quad (12.52)$$

The wave speed is given by

$$c = \frac{\sigma}{k} = \frac{\omega(\alpha)}{\lambda^{1/2}(\alpha)} . \tag{12.53}$$

If $r = \alpha \to r_0$, that is there is a limit cycle solution of the λ-ω dynamics, the wave number of the plane waves tends to zero. This suggests that we should look for travelling plane wave train solutions near the limit cycle. Kopell and Howard (1973) show how to do this in general. Here we shall consider a specific simple, but nontrivial, example where $\lambda(r)$ and $\omega(r)$ are such that the kinetics satisfy the Hopf requirements of (12.42), and on which we can carry out the analysis simply, to derive travelling wave train solutions: we mainly follow the analysis of Ermentrout (1981).
Suppose

$$\omega(r) \equiv 1, \quad \lambda(r) = \gamma - r^2 , \tag{12.54}$$

The dynamics in (12.45) then has $u = v = 0$ as a steady state which is stable for $\gamma < 0$ and unstable for $\gamma > 0$. $\gamma = 0$ is the bifurcation value γ_c such that at $\gamma = 0$ the eigenvalues of the linearization about $u = v = 0$ are $\pm i$. This is a standard Hopf bifurcation requirement (Appendix 3) so we expect small amplitude limit cycle solutions for small positive γ; that is $\gamma = \gamma_c + \varepsilon$ with $0 < \varepsilon \ll 1$. With the above general solutions (12.52), since $\lambda = 0$ when $r = \sqrt{\gamma}$, these limit cycle solutions are given by

$$u_\gamma(t) = \sqrt{\gamma} \cos t, \quad v_\gamma(t) = \sqrt{\gamma} \sin t, \quad \gamma > 0 \tag{12.55}$$

and in polar variables by

$$r_0 = \sqrt{\gamma}, \quad \theta = t + \theta_0 , \tag{12.56}$$

where θ_0 is some arbitrary phase which we can take to be zero.
Now consider the reaction diffusion system (12.47) with λ and ω from (12.54). On substituting travelling plane wave solutions of the form

$$r = r_0, \quad \theta = \sigma t - kx$$

we find, as expected from (12.51),

$$\sigma = 1, \quad k^2 = \gamma - r_0^2, \quad 0 < r_0 < \sqrt{\gamma} ,$$

which result in the small amplitude travelling wave train solutions

$$\begin{aligned}
u &= r_0 \cos \left(t - x[\gamma - r_0^2]^{1/2} \right) , \\
v &= r_0 \sin \left(t - x[\gamma - r_0^2]^{1/2} \right) .
\end{aligned} \tag{12.57}$$

Fig. 12.12 illustrates these solutions, which have amplitude $r_0 < \sqrt{\gamma}$ and wavelength $L = 2\pi/(\gamma - r_0^2)^{1/2}$.

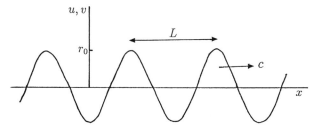

Fig. 12.12. Small amplitude travelling wave solution for the λ-ω system (12.45) with λ and ω given by (12.54). The wave speed $c = \sigma/k = 1/[\gamma - r_0^2]^{1/2}$ and the wavelength $L = 2\pi/[\gamma - r_0^2]^{1/2}$ depend on the amplitude r_0; $0 < r_0 < \sqrt{\gamma}$.

Such travelling wave trains are only of relevance, for example to the target patterns in Fig. 12.1, if they are stable. We consider their linear stability in the following section. It is a rare example where we can carry out the analysis fairly easily.

The effect of diffusion on reaction kinetics which exhibit periodic behaviour is to generate travelling periodic wave train solutions. The specific nonlinearity in the above λ-ω example, namely $\lambda(r) = \gamma - r^2$, is typical of a Hopf bifurcation problem. It seems likely that reaction diffusion mechanisms where the reaction kinetics alone exhibit periodic limit cycle behaviour via a Hopf bifurcation will also generate periodic wave train solutions. To show this it suffices to demonstrate that general reaction diffusion systems with this property are similar to λ-ω systems in the vicinity of the Hopf bifurcation.

Consider the two species system

$$u_t = F(u, v; \gamma) + D\nabla^2 u, \quad v_t = G(u, v; \gamma) + D\nabla^2 v , \tag{12.58}$$

where F and G are the reaction kinetics. For algebraic simplicity suppose (12.58) has a steady state at $u = v = 0$ and the diffusionless ($D = 0$) system exhibits a Hopf bifurcation to a limit cycle at the bifurcation value γ_c. Now consider u and v as the perturbations about the zero steady state and write

$$T = \begin{pmatrix} u \\ v \end{pmatrix}, \quad M = \begin{pmatrix} F_u & F_v \\ G_u & G_v \end{pmatrix}_{u=v=0}, \quad P = \begin{pmatrix} D & 0 \\ 0 & D \end{pmatrix}.$$

The terms in M are functions of the bifurcation parameter γ. The linearized form of (12.58) is then

$$T_t = MT + P\nabla^2 T \tag{12.59}$$

and the full system (12.58) can be written as

$$T_t = MT + P\nabla^2 T + H, \quad H = \begin{pmatrix} f(u, v, \gamma) \\ g(u, v, \gamma) \end{pmatrix} , \tag{12.60}$$

where f and g are the nonlinear contributions in u and v to F and G in the vicinity of $u = v = 0$.

Since the kinetics undergo a Hopf bifurcation at $\gamma = \gamma_c$, the eigenvalues, σ say, of the matrix M are such that $\operatorname{Re}\sigma(\gamma) < 0$ for $\gamma < \gamma_c$, $\operatorname{Re}\sigma(\gamma_c) = 0$, $\operatorname{Im}\sigma(\gamma_c) \neq 0$ and $\operatorname{Re}\sigma(\gamma) > 0$ for $\gamma > \gamma_c$. So, at $\gamma = \gamma_c$

$$\operatorname{Tr} M = 0, \quad \det M > 0 \quad \Rightarrow \quad \sigma(\gamma_c) = \pm i(\det M)^{1/2} . \tag{12.61}$$

Introduce the constant matrix N and the non-constant matrix W by

$$T = NW \quad \Rightarrow \quad W_t = N^{-1}MNW + N^{-1}PN\nabla^2 W + N^{-1}H \tag{12.62}$$

from (12.60). Now choose N such that

$$N^{-1}MN = \begin{pmatrix} 0 & -k \\ k & 0 \end{pmatrix} \quad \text{at} \quad \gamma = \gamma_c \quad \Rightarrow \quad k^2 = \det M]_{\gamma=\gamma_C} .$$

In the transformed system (12.62) we now have the coefficients in the linearized matrix

$$N^{-1}MN = \begin{pmatrix} \alpha(\gamma) & -\beta(\gamma) \\ \beta(\gamma) & \delta(\gamma) \end{pmatrix} \tag{12.63}$$

where

$$\alpha(\gamma_c) = 0 = \delta(\gamma_c), \quad \beta(\gamma_c) \neq 0 . \tag{12.64}$$

That is, the general system (12.58) with a Hopf bifurcation at the steady state can be transformed to a form in which near the bifurcation γ_c, it has a λ-ω form (cf. (12.45)). This result is of some importance since analysis valid for λ-ω systems can be carried over in many situations to real reaction diffusion systems. This result is not restricted to equal diffusion coefficients for u and v as was shown by Duffy, Britton and Murray (1980) who discuss the implications for spiral waves.

*12.6 Linear Stability of Wave Train Solutions of λ-ω Systems

The stability of travelling waves can often be quite difficult to demonstrate analytically: the paper by Feroe (1982) on the stability of excitable FHN waves amply illustrates this. However, some stability results can be obtained, without long and complicated analysis, in the case of the wave train solutions of the λ-ω system derived in the last section.

Because of the simplicity of the plane wave solutions in their polar form (12.50), we consider their linear stability in that form, and set

$$r = \alpha + \rho(x,t), \quad \theta = \sigma t - kx + \phi(x,t) , \tag{12.65}$$

where $|\rho| \ll 1$, $|\phi| \ll 1$. Linearizing the polar form of the λ-ω system, namely (12.47), we get

$$
\begin{aligned}
\rho_t &= \alpha[\lambda'(\alpha)\rho + 2k\phi_x] + \rho_{xx}, \\
\phi_t &= \omega'(\alpha)\rho - 2k\rho_x/\alpha + \phi_{xx}.
\end{aligned}
\tag{12.66}
$$

We want to find conditions on the parameters k and σ such that solutions ρ and ϕ tend to zero as $t \to \infty$. The coefficients in (12.66) are constants so we look for solutions in the usual Fourier form by setting

$$
\begin{pmatrix} \rho \\ \phi \end{pmatrix} = \begin{pmatrix} \rho_0 \\ \phi_0 \end{pmatrix} = \exp[st + iqx],
\tag{12.67}
$$

where q is the perturbation wave number and ρ_0 and ϕ_0 are constants. We require the conditions such that $\operatorname{Re} s < 0$ which implies stability. Substituting this form into (12.66) gives

$$
\begin{pmatrix} s + q^2 - \alpha\lambda' & -2ik\alpha q \\ -\omega' + 2ik\dfrac{q}{\alpha} & s + q^2 \end{pmatrix} \begin{pmatrix} \rho_0 \\ \phi_0 \end{pmatrix} = 0.
\tag{12.68}
$$

So that (12.67) is a nontrivial solution we require ρ_0 and ϕ_0 to be nonzero, so, from (12.68) the determinant of the 2×2 matrix must be zero. This gives a quadratic for s, the solutions of which are

$$
s_1, s_2 = -q^2 + \frac{\alpha\lambda'}{2} \pm \left[\left(\frac{\alpha\lambda'}{2}\right)^2 + 4k^2q^2 + 2ik\alpha\omega'q \right]^{1/2}.
\tag{12.69}
$$

If either s_1 or s_2 has a positive real part for *any* q, the plane wave solutions (12.50) are linearly *unstable*. Remember that s and q refer to the perturbation (12.67) from the plane wave solutions (12.50).

If the perturbation wave number $q = 0$, then $s_1 = 0$ and $s_2 = \alpha\lambda'$. The former, which corresponds to neutral stability, is equivalent to a constant phase shift in the original plane wave solution, while the latter implies stability when $\lambda' < 0$, as is the case here. When $q > 0$ the root s with maximum real part comes from s_2. So, conditions which make $s_2 < 0$ are necessary and sufficient for linear stability.

From (12.69), a little complex variable algebra gives

$$
\operatorname{Re} s_2(q) = -q^2 + \frac{\alpha\lambda'}{2} + \frac{1}{\sqrt{2}} \left[\left(\frac{\alpha\lambda'}{2}\right)^2 + 4k^2q^2 + \left[\left(\left(\frac{\alpha\lambda'}{2}\right)^2 + 4k^2q^2 \right)^2 + 4(kq\alpha\omega')^2 \right]^{1/2} \right]^{1/2}.
\tag{12.70}
$$

from which we see that

$$\text{Re}\, s_2(0) = \frac{\alpha\lambda'}{2} + \frac{|\alpha\lambda'|}{2} = 0$$

and

$$\left[\frac{d\text{Re}\, s_2}{dq^2}\right]_{q=0} = -1 + \frac{4k^2\left(1 + \frac{\omega'^2}{\lambda'^2}\right)}{\alpha|\lambda'|}.$$

So for small enough q^2, $\text{Re}\, s_2(q) < 0$ if and only if the last derivative $[d\text{Re}\, s_2/dq^2]_{q=0} < 0$. Since $\lambda' < 0$ this gives the condition

$$4k^2\left\{1 + \left[\frac{\omega'(\alpha)}{\lambda'(\alpha)}\right]^2\right\} + \alpha\lambda'(\alpha) \le 0. \tag{12.71}$$

We shall now show that this is the necessary and sufficient condition for the plane wave solutions (12.50)–(12.52), with amplitude α and wave number k, to be linearly stable.

Now from (12.70), remembering that $\lambda' < 0$, a little algebra shows that $\text{Re}\, s_2(q) \le 0$ for all q if

$$\left\{\left[\left(\frac{\alpha\lambda'}{2}\right)^2 + 4k^2q^2\right]^2 + 4(kq\alpha\omega')^2\right\}^{1/2} \le 2q^4 - 4q^2\left(k^2 + \frac{\alpha\lambda'}{2}\right) + \left(\frac{\alpha\lambda'}{2}\right)^2. \tag{12.72}$$

Now from (12.71), since $[\omega'/\lambda']^2 > 0$, we also have that

$$\alpha\lambda' + 4k^2 < 0 \quad \Rightarrow \quad \frac{\alpha\lambda'}{2} + k^2 < 0$$

so the right hand side of (12.72) is always positive. Thus, squaring both sides of (12.72) and carrying out some more elementary algebra, we get the condition $\text{Re}\, s_2(q) \le 0$ for all q if and only if

$$k^2 \le \left\{1 - \frac{(\alpha\omega')^2}{\alpha^2\omega'^2 + 4\left(q^2 - \frac{\alpha\lambda'}{2}\right)^2}\right\}\frac{q^2 - \alpha\lambda'}{4}. \tag{12.73}$$

Since $\lambda' < 0$ the right hand side is an increasing function of q for all q so $\text{Re}\, s_2(q) \le 0$ for all q if and only if the last inequality holds for $q = 0$; this gives

$$k^2 \le -\left[1 - \frac{\omega'^2}{\omega'^2 + \lambda'^2}\right]\frac{\alpha\lambda'}{4}.$$

But this condition is exactly the same as (12.71), which is therefore the necessary and sufficient condition that the wave solutions, (12.50) with (12.51), with amplitude α and wave number k, are linearly stable.

Let us apply this stability condition to the small amplitude plane wave solutions (12.57) which we derived above. Here, with (12.54), $\omega' = 0$, $\lambda'(\alpha) = -2\alpha$ and since $\alpha < \sqrt{\gamma}$, $k^2 = \gamma - \alpha^2$ we thus get

$$\sqrt{\gamma} \geq \alpha \geq \sqrt{\frac{2\gamma}{3}} \tag{12.74}$$

as the conditions for the existence of small amplitude stable periodic wave trains which bifurcate from the steady state for any small γ. Fig. 12.13 illustrates the γ-domain where these solutions (12.57) exist and are stable.

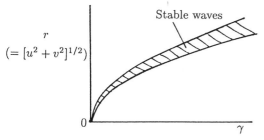

Fig. 12.13. Bifurcation diagram for the linearly stable periodic plane wave solutions (12.57) for the λ-ω system with $\omega \equiv 1$ and $\lambda = \gamma - r^2$. Typical wavetrain solutions are illustrated in Fig. 12.12.

Remember that, since $\sqrt{\gamma}$ is small, these are stable periodic wave solutions of small amplitude, less than $\sqrt{\gamma}$, with wavelength $2\pi/(\gamma - r_0^2)^{1/2}$, that is $O(1/\sqrt{\gamma})$. This means that they are of very long wavelength.

12.7 Spiral Waves

Rotating spiral waves occur naturally in a wide variety of biological, physiological and chemical contexts. Probably the most widely studied are those which arise in the Belousov-Zhabotinskii reaction. Relatively, it is a considerably simpler system than those which arise in physiology. They have been demonstrated experimentally by, for example, Winfree (1974), Krinsky et al. (1986) and by Müller et al. (1985, 1986, 1987). The latter's novel experimental technique, using light absorption, highlights actual concentration levels quantitatively. Fig. 12.14 as well as Fig. 12.17 and Fig. 12.18 below, show some experimentally observed spiral waves in the Belousov-Zhabotinskii reaction: refer also to Fig. 12.1 (b). Although the spirals in these figures are symmetric, this is by no means the only pattern form: see, for example, Winfree (1974) in particular, and Müller et al. (1986), who exhibit dramatic examples of complex spiral patterns. Although in a different context, see Fig. 12.16 below for other examples of non-symmetric, as well as symmetric, spirals.

Considerable effort has gone into the mathematical study of such spiral waves and in particular the diffusion version of the Field-Noyes model system, discussed in detail in Chapter 7, Section 7.1. Keener and Tyson (1986) present a thorough analysis of spiral waves in excitable reaction diffusion systems with general excitable kinetics (see the references there to earlier work in the area). They apply their technique to the Field-Noyes model with diffusion and the results are in good agreement with experiment.

Fig. 12.14. Spiral waves in a thin (1 mm) layer of an excitable Belousov-Zhabotinskii reaction. The section shown is 9 mm square. (Courtesy of T. Plesser from Müller, Plesser and Hess 1986)

There are many other important occurrences of spiral waves. Brain tissue can exhibit electrochemical waves of 'spreading depression' which spread through the cortex of the brain. These waves are characterized by a depolarisation of the neuronal membrane and decreased neural activity. Shibata and Bureš (1972, 1974) have studied this phenomenon experimentally and demonstrated the existence of spiral waves which rotate about a lesion in the brain tissue from the cortex of a rat. Fig. 12.15 (a) schematically shows the wave behaviour they observed.

When death results from a disruption of the coordinated contractions of heart muscle fibres, the cause is often due to fibrillation. In a fibrillating heart, small regions undergo contractions essentially independent of each other. The heart looks, as noted before, like a handful of squirming worms – it is a quivering mass of tissue. If this disruption lasts for more than a few minutes death usually results. Krinsky (1978) and Krinsky et al. (1986), for example, discuss spiral waves in mathematical models of cardiac arrhythmias. Winfree (1983) considers the possible application to sudden cardiac death. He suggests that the precursor to fibrillation is the appearance of rotating waves of electrical impulses. Fig. 12.15 (b) illustrates such waves induced in rabbit heart tissue by Allessie et al. (1977). These authors (Allessie et al. 1973, 1976; Smeets et al. 1986, which gives other references) have carried out an extensive experimental programme on rotating wave propagation in heart muscle.

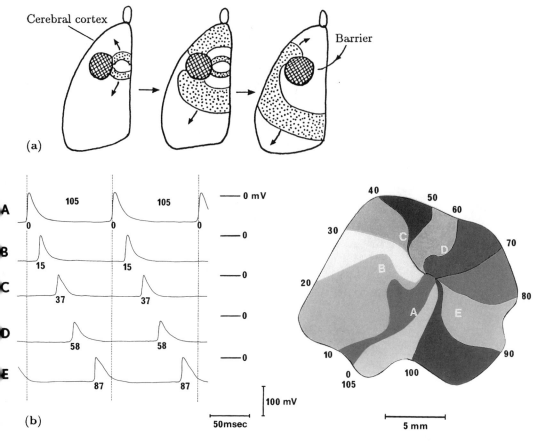

Fig. 12.15a,b. (a) Evolution of spiralling reverberating waves of cortical spreading depression about a lesion (a thermal coagulation barrier) in the right hemisphere of a rat cerebral cortex. The waves were initiated chemically. The shaded regions have different potential from the rest of the tissue. (After Shibata and Bureš 1974) (b) Rotating spiral waves experimentally induced in rabbit heart (left atria) muscle: the numbers represent milliseconds. Each region was traversed in 10 msec with a complete rotation in 105 msec. On the left the transmembrane potentials are shown with the lettering corresponding to the points in the heart muscle on the right. The right also shows the isochronic lines, that is lines where the potential is the same during passage of the wave. (Reproduced from Allessie Bonke and Schopman 1977, courtesy of M.A. Allessie and the American Heart Association, Inc.)

The spirals that arise in signalling patterns of the slime mold *Dictyostelium discoideum* are equally dramatic as seen in Fig. 12.16. A model for these, based on a recently proposed kinetics scheme, has been proposed by Tyson et al. (1988a,b).

It is important that although the similarity between Fig. 12.1 (b) and Fig. 12.16 is striking, one must *not* be tempted to assume that the model for the Belousov-Zhabotinskii reaction is then an approriate model for the slime mold patterns – the mechanisms are quite different. Although producing the

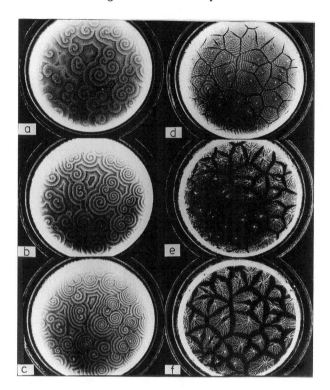

Fig. 12.16. Spiral signalling patterns in the slime mold *Dictyostelium discoideum* which show the increasi chemoattractant (cyclic AMP) signalling. The photographs ar taken about 10 minutes apart, and each shows about 5×10^7 amoebae. The Petri dish is 50 mm in diameter. The amoebae move periodically and the light and dark bands which show up unde dark-field illumination arise from the differences in optica properties between moving and stationary amoebae. The cells are bright when moving and dark when stationary. The patterns eventually lead to the formation of bacterial territories. (Courtesy of P.C. Newell from Newell 1983)

right kind of patterns is an important and essential aspect of successful modelling, understanding the basic mechanism is the ultimate objective.

The possible existence of large scale spirals in interacting population situations does not seem to have been considered with a view to practical applications, but, given the reaction diffusion character of the models, they certainly exist in theory.

General discussions of spiral waves have been given, for example, by Keener (1986), who presents a geometric theory, and by Zykov (1988) in his book on wave processes in excitable media.

From a mathematical point of view, what do we mean by a spiral wave? In the case of the Belousov-Zhabotinskii reaction, for example, it is a rotating, time periodic, spatial structure of reactant concentrations: see Figs. 12.15 and 12.18. At a fixed time a snapshot shows a typical spiral pattern. A movie of the process shows the whole spiral pattern moving like a rotating clock spring.

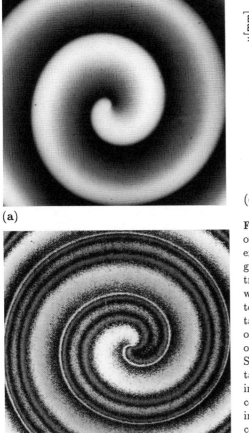

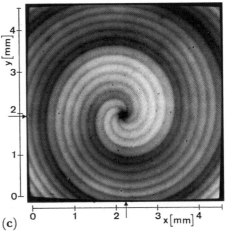

(a)

(b)

(c)

Fig. 12.17a-c. (a) Snapshot (4.5 mm square) of a spiral wave in a thin (1 mm) layer of an excitable Belousov-Zhabotinskii reagent. The grey scale image is a measure of the level of transmitted light intensity (7 intensity levels were measured), which in turn corresponds to isoconcentration lines of one of the reactants. (b) The grey scale highlights the geometric details of the isoconcentration lines of one of the reactants in the reaction. (c) Superposition of snapshots (4.5 mm square) taken at 3 second intervals, including the one in (a). The series covers approximately one complete revolution of the spiral. Here 6 light intensity levels were measured. Note the small core region. (From Müller, Plesser and Hess 1985 courtesy of T. Plesser and the American Association for the Advancement of Science: Copyright 1985 AAAS)

Fig. 12.17 shows such a snapshot and a superposition of them taken at fixed time intervals. The sharp wave fronts are contours of constant concentration, that is, isoconcentration lines.

Consider now a spiral wave rotating around its centre. If you stand at a fixed position in the medium it seems locally as though a periodic wave train is passing you by since every time the spiral turns a wave front moves past you.

As we saw in Chapter 8, the state or concentration of a reactant can be described by a function of its phase, ϕ. It is clearly appropriate to use polar coordinates r and θ when discussing spiral waves. A simple rotating spiral is described by a periodic function of the phase ϕ with

$$\phi = \Omega t \pm m\theta + \psi(r) , \qquad (12.75)$$

where Ω is the frequency, m is the number of arms on the spiral and $\psi(r)$ is a function which describes the type of spiral. The $\pm$ in the $m\theta$ term determines the sense of rotation. Fig. 12.18 shows examples of 1-armed and 3-armed spirals including an experimental example of the latter. Suppose, for example, we set $\phi = 0$ and look at the steady state situation, we get a simple geometric description of a spiral from (12.75): a 1-armed spiral, for example, is given by $\theta = \psi(r)$. Specific $\psi(r)$ are

$$\theta = ar, \quad \theta = a\ln r \qquad (12.76)$$

with $a > 0$: these are respectively Archimedian and logarithmic spirals. For a spiral about a central core the corresponding forms are

$$\theta = a(r - r_0), \quad \theta = a\ln(r - r_0). \qquad (12.77)$$

Fig. 12.18 (a) is a typical Archimedian spiral with Fig. 12.18 (b) an example with m=3.

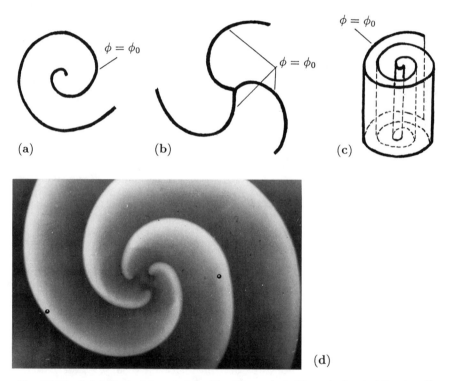

Fig. 12.18a-d. (a) Typical 1-armed Archimedian spiral. The actual spiral line is a line of constant phase ϕ; that is a line of constant concentration. (b) Typical 3-armed spiral. (c) Three dimensional spiral. These have a scroll-like quality and have been demonstrated experimentally by Welsh, Gomatam and Burgess (1983) with the Belousov-Zhabotinskii reaction. (d) Experimentally demonstrated 3-armed spiral in the Belousov-Zhabotinskii reaction. (From Agladze and Krinsky 1982 courtesy of V. Krinsky)

A mathematical description of a spiral configuration in a reactant, u say, could then be expressed by

$$u(r, \theta, t) = F(\phi) \,, \tag{12.78}$$

where $F(\phi)$ is a 2π-periodic function of the phase ϕ given by (12.75). If t is fixed we get a snapshot of a spiral, the form of which puts certain constraints on $\psi(r)$ in (12.75): ar and $a \ln r$ in (12.76) are but two simple cases. A mixed type, for example, has $\psi(r) = ar + b \ln r$ with a and b constants. In (12.78) with ϕ as in (12.75), if we fix r and t and circle around the centre we have m-fold symmetry where m is the number of arms; an example with $m = 3$ is shown in Fig. 12.18 (b) and in Fig. 12.18 (d), one obtained experimentally by Agladze and Krinsky (1982). If we fix r and θ, that is we stay at a fixed point, we see a succession of wavefronts as we described above. If a wavefront passes at $t = t_0$ with say $\phi = \phi_0$, the next wave passes by at time $t = t_0 + 2\pi/\Omega$ which is when $\phi = \phi_0 + 2\pi$.

If we look at a snapshot of a spiral and move out from the centre along a ray we see intuitively that there is a wavelength associated with the spiral; it varies however as we move out from the centre. If one wavefront is at r_1 and the next, moving out, is at r_2, we can define the wavelength λ by

$$\lambda = r_2 - r_1, \quad \theta(r_2) = \theta(r_1) + 2\pi \,.$$

From (12.75), with t fixed, we have, along the curve $\phi = \text{constant}$,

$$\phi_\theta + \phi_r \left[\frac{dr}{d\theta} \right]_{\phi=\text{constant}} = 0$$

and so, if, to be specific, we take $-m$ in (12.75),

$$\left[\frac{dr}{d\theta} \right]_{\phi=\text{constant}} = -\frac{\phi_\theta}{\phi_r} = \frac{m}{\psi'(r)} \,.$$

The wavelength $\lambda(r)$ is now given by

$$\lambda(r) = \int_{\theta(r)}^{\theta(r)+2\pi} \left[\frac{dr}{d\theta} \right] d\theta = \int_{\theta(r)}^{\theta(r)+2\pi} \left[\frac{m}{\psi'(r(\theta))} \right] d\theta \,.$$

where r, as a function of θ, is given by (12.75) with $t = \text{constant}$ and $\phi = \text{constant}$ which we can take to be zero. For an Archimedian spiral $r = \theta/a$, so $\psi' = a$ and the wavelength is $\lambda = m/a$.

The pitch of the spiral is defined by

$$\left[\frac{dr}{d\theta} \right]_{\phi=\text{constant}} = \frac{m}{\psi'(r)}$$

which, for an Archimedian spiral where $\psi'(r) = a$, gives a constant pitch m/a, while for a logarithmic spiral gives the pitch as mr/a since $\psi'(r) = a/r$. For large r, the pitch of the latter is large, that is loosely wound, while for small r the pitch is small, that is the spiral is tightly wound.

Before discussing the analytic solutions of a specific reaction diffusion system we should note some numerical studies on the birth of spiral waves carried out by Krinsky et al. (1986) and Tsujikawa et al. (1988). The latter considered the Fitzhugh-Nagumo excitable mechanism (12.29) and investigated numerically the propagation of a wave of excitation of finite spatial extent. Fig. 12.19 shows a time sequence of the travelling wave of excitation and shows the evolution of spiral waves: similar evolution figures were obtained by Krinsky et al. (1986) who discuss the evolution of spirals waves in some detail. The evolution patterns in this figure are similar to developing spirals observed experimentally in the Belousov-Zhabotinskii reaction.

Fig. 12.19. Evolution of spiral waves for the two-dimensional Fitzhugh-Nagumo model mechanism (12.29), namely

$$u_t = u(a - u)(1 - u) - v + D\nabla^2 u, \quad v_t = bu - \gamma v$$

for excitable nerve action potentials. Parameter values: $D = 2 \times 10^{-6}$, $a = 0.25$, $b = 10^{-3}$, $\gamma = 3 \times 10^{-3}$. The dark regions are where $u \geq a$, that is u is in the excited state. (From Tsujikawa et al. 1989 courtesy of M. Mimura)

*12.8 Spiral Wave Solutions of λ-ω Reaction Diffusion Systems

Numerous authors have investigated spiral wave solutions of general reaction diffusion models, such as Cohen, Neu and Rosales (1978), Duffy, Britton and Murray (1980), Kopell and Howard (1981), Mikhailov and Krinsky (1983). The papers by Keener and Tyson (1986), dealing with the Belousov-Zhabotinskii reaction, and Tyson et al. (1989) with the slime mold *Dictyostelium*, are specific examples. The analysis is usually quite involved with much use being made of asymptotic methods. The λ-ω system, which exhibits wavetrain solutions as we saw above in Section 12.5, has been used as a model system because of the relative algebraic simplicity of the analysis. Spiral wave solutions of λ-ω systems have been investigated, for example, by Greenberg (1981), Hagan (1982), Kuramoto and Koga (1981) and Koga (1982). The list of references is fairly extensive; other relevant references are given in these papers. In this section we develop some solutions for the λ-ω system, keeping in mind the direct relevance to real reaction diffusion mechanisms as we discussed in Section 12.5.

The λ-ω reaction diffusion mechanism for two reactants is

$$\frac{\partial}{\partial t} \begin{pmatrix} u \\ v \end{pmatrix} = \begin{pmatrix} \lambda(A) & -\omega(A) \\ \omega(A) & \lambda(A) \end{pmatrix} \begin{pmatrix} u \\ v \end{pmatrix} + D\nabla^2 \begin{pmatrix} u \\ v \end{pmatrix}$$

$$A^2 = u^2 + v^2 .$$

(12.79)

where $\omega(A)$ and $\lambda(A)$ are real functions of A. (The change of notation from (12.45) is so that we can use r as the usual polar coordinate.) We assume the kinetics sustain limit cycle oscillations; this puts the usual constraints on λ and ω, namely if A_0 is an isolated zero of $\lambda(A)$ for some $A_0 > 0$ and $\lambda'(A_0) < 0$ and $\omega(A_0) \neq 0$, then the spatially homogeneous system, that is with $D = 0$, has a stable limit cycle solution $u^2 + v^2 = A_0$ with cycle frequency $\omega(A_0)$ (see Section 6.4).

Setting $w = u + iv$, (12.79) becomes the single complex equation

$$w_t = (\lambda + i\omega)w + D\nabla^2 w .$$

(12.80)

The form of this equation suggests setting

$$w = A \exp[i\phi] ,$$

(12.81)

where A is the amplitude of w and ϕ its phase. Substituting this into (12.80) and equating real and imaginary parts gives the following equation system for A and ϕ:

$$A_t = A\lambda(A) - DA|\nabla\phi|^2 + D\nabla^2 A,$$

$$\phi_t = \omega(A) + \frac{2D(\nabla A \cdot \nabla \phi)}{A} + D\nabla^2 \phi ,$$

(12.82)

which is the polar form of the λ-ω system. Polar coordinates r and θ are the appropriate ones to use for spiral waves. Motivated by (12.75) and the discussion in the last section, we look for solutions of the form

$$A = A(r), \quad \phi = \Omega t + m\theta + \psi(r) ,$$

(12.83)

where Ω is the unknown frequency and m the number of spiral arms. Substituting these into (12.82) gives the ordinary differential equations for A and ψ:

$$DA'' + Dr^{-1}A' + A\left[\lambda(A) - D\psi'^2 - \frac{Dm^2}{r^2}\right] = 0 ,$$

$$D\psi'' + D\left(\frac{1}{r} + \frac{2A'}{A}\right)\psi' = \Omega - \omega(A) .$$

(12.84)

On multiplying the second equation by rA^2, integration gives

$$\psi'(r) = \frac{1}{DrA^2(r)} \int_0^r sA^2(s)[\Omega - \omega(A(s))]\,ds . \tag{12.85}$$

The form of (12.84) and (12.85) are convenient for analysis and are the equations which have been the basis for most of the papers on spiral waves of λ-ω systems, using asymptotic methods, fixed point theorems, phase space analysis and so on.

Before analysing (12.84) we have to decide on suitable boundary conditions. We want the solutions to be regular at the origin and bounded as $r \to \infty$. The former, together with the form of the equations for A and ψ', thus requires

$$A(0) = 0, \quad \psi'(0) = 0 . \tag{12.86}$$

If $A \to A_\infty$ as $r \to \infty$ we have, from (12.85),

$$\psi'(r) \sim \frac{1}{DrA_\infty^2} \int_0^r sA_\infty^2[\Omega - \omega(A_\infty)]\,ds$$

$$= \frac{[\Omega - \omega(A_\infty)]r}{2D} \quad \text{as} \quad r \to \infty$$

and so ψ' is bounded only if $\Omega = \omega(A_\infty)$. The first of (12.84) determines $\psi'(\infty)$ as $[\lambda(A_\infty)/D]^{1/2}$. We thus have the dispersion relation

$$\psi'(\infty) = \left[\frac{\lambda(A_\infty)}{D}\right]^{1/2}, \quad \Omega = \omega(A_\infty) . \tag{12.87}$$

which shows how the amplitude at infinity determines the frequency Ω.

Near $r = 0$, set

$$A(r) \sim r^c \sum_{n=0}^{\infty} a_n r^n, \quad \text{as} \quad r \to 0$$

where $a_0 \neq 0$, substitute into the first of (12.84) and equate powers of r in the usual way. The coefficient of lowest order, namely r^{c-2}, set equal to zero gives

$$c(c-1) + c - m^2 = 0 \quad \Rightarrow \quad c = \pm m .$$

For $A(r)$ to be nonsingular as $r \to 0$ we must choose $c = m$, with which

$$A(r) \sim a_0 r^m, \quad \text{as} \quad r \to 0 ,$$

where a_0 is an undetermined nonzero constant. The mathematical problem is to determine a_0 and Ω so that $A(r)$ and $\psi'(r)$ remain bounded as $r \to \infty$. From (12.81), (12.83) and the last equation, we get the behaviour of u and v near $r = 0$ as

$$\begin{pmatrix} u \\ v \end{pmatrix} \propto \begin{pmatrix} r^m \cos[\Omega t + m\theta + \psi(0)] \\ r^m \sin[\Omega t + m\theta + \psi(0)] \end{pmatrix} . \tag{12.88}$$

Koga (1982) studied phase singularities and multi-armed spirals, analytically and numerically, for the λ-ω system with

$$\lambda(A) = 1 - A^2, \quad \omega(A) = -\beta A^2 , \qquad (12.89)$$

where $\beta > 0$. Fig. 12.20 shows his computed solutions for 1-armed and 2-armed spirals.

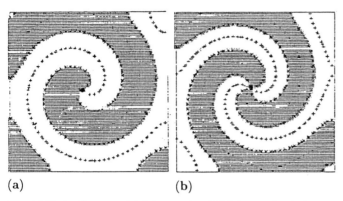

(a)   (b)

Fig. 12.20a,b. Computed (a) 1-armed and (b) 2-armed spiral wave solutions of the λ-ω system (12.79) with the λ and ω given by (12.89) with $\beta = 1$. Zero flux boundary conditions were taken on the square boundary. The shaded region is where $u > 0$. (From Koga 1982 courtesy of S. Koga)

The basic starting point to look for solutions is the assumption of the functional form for u and v given by

$$\begin{pmatrix} u \\ v \end{pmatrix} = \begin{pmatrix} A(r) \cos [\Omega t + m\theta + \psi(r)] \\ A(r) \sin [\Omega t + m\theta + \psi(r)] \end{pmatrix} . \qquad (12.90)$$

With $A(r)$ a constant and $\psi(r) \propto \ln r$ these represent rotating spiral waves as we have shown. Cohen et al. (1978) prove that for a class of $\lambda(A)$ and $\omega(A)$ the system (12.80) has rotating spiral waves of the form (12.90) which satisfy boundary conditions which asymptote to Archimedian and logarithmic spirals, that is $\psi \sim cr$ and $\psi \sim c\ln r$ as $r \to \infty$. Duffy et al. (1980) show how to reduce a general reaction diffusion system with limit cycle kinetics and *unequal* diffusion coefficients for u and v, to the case analysed by Cohen et al. (1978).

Kuramoto and Koga (1981) have studied numerically the specific λ-ω system where

$$\lambda(A) = \varepsilon - aA^2, \quad \omega(A) = c - bA^2 ,$$

where $\varepsilon > 0$ and $a > 0$. With these the system (12.80) becomes

$$w_t = (\varepsilon + ic)w - (a + ib)|w|^2 w + D\nabla^2 w .$$

We can remove the c-term by setting $w \to we^{ict}$ (algebraically the same as setting $c = 0$) and then rescale w, t and the space coordinates according to

$$w \to \left(\frac{a}{\varepsilon}\right)^{1/2} w, \quad t \to \varepsilon t, \quad \mathbf{r} \to \left(\frac{\varepsilon}{D}\right)^{1/2} \mathbf{r}$$

to get the simpler form

$$w_t = w - (1 + i\beta)|w|^2 w + \nabla^2 w , \tag{12.91}$$

where $\beta = b/a$. The space independent form of the last equation has a limit cycle solution $w = \exp(-i\beta t)$.

Kuramoto and Koga (1981) numerically investigated the spiral wave solutions of (12.91) as $|\beta|$ varies. They found that for small $|\beta|$ a steadily rotating spiral wave developed, like that in Fig. 12.20 (a) and of the form (12.88). As $|\beta|$ was increased these spiral waves became unstable and appeared to become chaotic for larger $|\beta|$. Fig. 12.21 shows the results for $|\beta| = 3.5$.

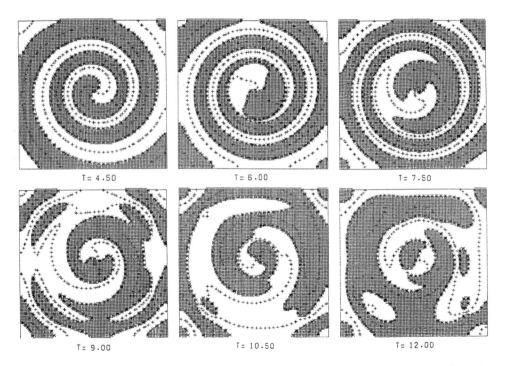

Fig. 12.21. Temporal development (time T) of chaotic patterns for the λ-ω system (12.91) for $\beta = 3.5$ and zero flux boundary conditions. (From Kuramoto and Koga 1981 courtesy of Y. Kuramoto)

Kuramoto and Koga (1981) suggest that 'phaseless' points, or black holes, such as we discussed in Chapter 8, Section 8.3, start to appear and cause the chaotic instabilities. Comparing (12.91) with (12.80) we have $\lambda = 1 - A^2$, $\omega = -\beta A^2$ and so β is a measure of how strong the local limit cycle frequency depends on the amplitude A. Since A varies with the spatial coordinate r we have a situation akin to an array of coupled, appropriately synchronized, oscillators. As $|\beta|$ increases the variation in the oscillators increases. Since stable rotating waves require a certain synchrony, increasing the variation in the local 'oscillators' tends to disrupt the synchrony giving rise to phaseless points and hence chaos. Chaos or turbulence in wavefronts in reaction diffusion mechanisms has been considered in detail by Kuramoto (1980): see other references there to this interesting problem of spatial chaos.

To conclude this section let us look at the 1-dimensional analogue of a spiral wave, namely a pulse which is emitted from the core, situated at the origin, periodically and on alternating sides of the core. If the pulses were emitted symmetrically then we would have the analogue of target patterns. Let us consider (12.82) with $\nabla^2 = \partial^2/\partial x^2$, $\lambda(A) = 1 - A^2$ and $\omega(A) = qA^2$. Now set

$$x \rightarrow \frac{x}{D^{1/2}}, \quad A = A(x), \quad \phi = \Omega t + \psi(x)$$

to get as the equations for A and ψ,

$$A_{xx} + A(1 - A^2 - \psi_x^2) = 0 \,,$$
$$\psi_{xx} + \frac{2A_x\psi_x}{A} = \Omega - qA^2 \,. \tag{12.92}$$

Boundary conditions are

$$A(x) \sim a_0 x \quad \text{as} \quad x \rightarrow 0, \quad \psi_x(0) = 0 \,,$$
$$A(x), \, \psi_x(x) \quad \text{bounded as} \quad x \rightarrow \infty \,.$$

The problem boils down to finding a_0 and Ω as functions of q so that the solution of the initial value problem to the time dependent equations is bounded. One such solution is

$$A(x) = \left(\frac{\Omega}{q}\right)^{1/2} \tanh(x/\sqrt{2}), \quad \psi_x(x) = \left(1 - \frac{\Omega}{q}\right)^{1/2} \tanh(x/\sqrt{2}) \,,$$
$$\Omega^2 + \frac{9}{2q}\Omega - \frac{9}{2} = 0 \,, \tag{12.93}$$

as can be verified. These solutions are generated periodically at the origin, alternatively on either side.

Although we touched on the stability problem for certain λ-ω solutions in Section 12.6, in general analytical determination of the stability of spiral waves is still an open question. Numerical evidence, however, suggests that many are indeed stable.

Biological waves exist which are solutions of model mechanisms other than reaction diffusion systems. For example, several of the mechanochemical models for generating pattern and form, which we discuss later in Chapter 17, also sustain travelling wave solutions. Waves which lay down a spatial pattern after passage are also of considerable importance as we shall also see later. In concluding this chapter, perhaps we should reiterate how important wave phenomena are in biology. Although this is clear just from the material in this chapter they are perhaps even more important in tissue communication during the process of embryological development. Generation of steady state spatial pattern and form is a topic of even greater importance and will be discussed at length in subsequent chapters.

Exercises

1. Consider the modified Lotka-Volterra predator-prey system in which the predator disperses via diffusion much faster than the prey: the dimensionless equations are

$$\frac{\partial u}{\partial t} = u(1 - u - v), \quad \frac{\partial v}{\partial t} = av(u - b) + \frac{\partial^2 v}{\partial x^2},$$

    where $a > 0, 0 < b < 1$ and u and v represent the predator and prey respectively. Investigate the existence of realistic travelling wavefront solutions of speed c in terms of the travelling wave variable $x + ct$, in which the wavefront joins the steady states $u = v = 0$ and $u = b, v = 1 - b$. Show that if c satisfies $0 < c < [4a(1 - b)]^{1/2}$ such wave solutions cannot exist whereas they can if $c \geq [4a(1 - b)]^{1/2}$. Further show that there is a value a^* such that for $a > a^*$ (u, v) tend to $(b, 1 - b)$ exponentially in a damped oscillatory way for large $x + ct$.

2. Consider the modified Lotka-Volterra predator prey system

$$\frac{\partial U}{\partial t} = AU \left(1 - \frac{U}{K}\right) - BUV + D_1 U_{xx},$$

$$\frac{\partial V}{\partial t} = CUV - DV + D_2 V_{xx},$$

    where U and V are respectively the prey and predator densities, A, B, C, D and K, the prey carrying capacity, are positive constants and D_1 and D_2 are the diffusion coefficients. If the dispersal of the predator is slow compared with that of the prey show that an appropriate nondimensionalization to a first approximation for $D_2/D_1 \approx 0$ results in the system

$$\frac{\partial u}{\partial t} = u(1 - u - v) + \frac{\partial^2 u}{\partial x^2}, \quad \frac{\partial v}{\partial t} = av(u - b) \,.$$

Investigate the possible existence of travelling wavefront solutions.

3. A primitive predator-prey system is governed by the model equations

$$\frac{\partial u}{\partial t} = -uv + D\frac{\partial^2 u}{\partial x^2}, \quad \frac{\partial v}{\partial t} = uv + \lambda D\frac{\partial^2 v}{\partial x^2} \,.$$

Investigate the possible existence of realistic travelling wavefront solutions in which $\lambda > 0$ and $u(-\infty, t) = v(\infty, t) = 0$, $u(\infty, t) = v(-\infty, t) = K$, a positive constant. Note any special cases. What type of situation might this system model?

4. Travelling bands of microorganisms, chemotactically directed, move into a food source, consuming it as they go. A model for this is given by

$$b_t = \frac{\partial}{\partial x}\left[Db_x - \frac{\chi b}{a}a_x\right], \quad a_t = -kb$$

where $b(x,t)$ and $a(x,t)$ are the bacteria and nutrient respectively and D, χ and k are positive constants. Look for travelling wave solutions, as functions of $z = x - ct$ where c is the wave speed, with the boundary conditions $b \to 0$ as $|z| \to \infty$, $a \to 0$ as $z \to -\infty$, $a \to 1$ as $z \to \infty$. Hence show that $b(z)$ and $a(z)$ satisfy

$$b' = \frac{b}{cD}\left[\frac{\chi kb}{a} - c^2\right], \quad a' = \frac{kb}{c} \,,$$

where the prime denotes differentiation with respect to z, and then obtain a relationship between $b(z)$ and $a(z)$.

In the special case where $\chi = 2D$ show that

$$a(z) = \left[1 + Ke^{-cz/D}\right]^{-1}, \quad b(z) = \frac{c^2}{kD}e^{-cz/D}\left[1 + Ke^{-cz/D}\right]^{-2},$$

where K is an arbitrary positive constant which is equivalent to a linear translation; it may be set to 1. Sketch the wave solutions and explain what is happening biologically.

*5. Consider the two species reduction model (12.20) for the wave front spatial variation in the Belousov Zhabotinskii reaction in which the parameters $r = b \gg 1$. That is, consider the system

$$u_t = -\frac{uv}{\varepsilon} + u(1 - u) + u_{xx}, \quad v_t = -\frac{uv}{\varepsilon} + v_{xx}, \quad 0 < \varepsilon \ll 1 \,.$$

By looking for travelling wave solutions in powers of ε

$$\begin{pmatrix} u \\ v \end{pmatrix} = \sum_{n=0}^{\infty} \begin{pmatrix} u_n(z) \\ v_n(z) \end{pmatrix} \varepsilon^n, \quad z = x - ct$$

show, by going to $O(\varepsilon)$, that the equations governing u_0, v_0 are

$$u_0 v_0 = 0, \quad u_0'' + c u_0' + u_0(1 - u_0) = v_0'' + c v_0' .$$

Hence deduce that the wave problem is split into two parts, one in which $u_0 \neq 0$, $v_0 = 0$ and the other in which $u_0 = 0$, $v_0 \neq 0$. Suppose that $z = 0$ is the point where the transition takes place. Sketch the form of the solutions you expect.

In the domain where $v_0 = 0$ the equation for u_0 is the Fisher equation, but with different boundary conditions. We must have $u(-\infty) = 1$ and $u_0(0) = 0$. Although $u_0 = 0$ for all $z \geq 0$, in general $u_0'(0)$ would not be zero. This would in turn result in an inconsistency since $u_0'(0) \neq 0$ implies there is a flux of u_0 into $z > 0$ which violates the restriction $u_0 = 0$ for all $z \geq 0$. This is called a *Stefan problem*. To be physically consistent we must augment the boundary conditions to ensure that there is no flux of u_0 into the region $z > 0$. So the complete formulation of the u_0 problem is the Fisher equation plus the boundary conditions $u_0(0) = u_0'(0) = 0$, $u(-\infty) = 1$. Does this modify your sketch of the wave? Do you think such a wave moves faster or slower than the Fisher wave? Give mathematical and physical reasons for your answer. [This asymptotic form, with $b \neq r$, has been studied in detail by Schmidt and Ortoleva (1980), who formulated the problem in the way described here and obtained analytical results for the wave characteristics.]

6. The piecewise linear model for the FHN model given by (12.29) is

$$u_t = f(u) - v + D u_{xx}, \quad v_t = bu - \gamma v ,$$
$$f(u) = H(u - a) - u ,$$

where H is the Heaviside function defined by $H(s) = 0$ for $s < 0$, $H(s) = 1$ for $s > 0$, and $0 < a < 1$, b, D and γ are all positive constants. If b and γ are small and the only rest state is $u = v = 0$ investigate qualitatively the existence, form and speed of a travelling solitary pulse.

7. The piece-wise linear FHN caricature can be written in the form

$$u_t = f(u) - v + D u_{xx}, \quad v_t = \varepsilon(u - \gamma v) ,$$
$$f(u) = H(u - a) - u ,$$

where H is the Heaviside function defined by $H(s) = 0$ for $s < 0$, $H(s) = 1$ for $s > 0$, and $0 < a < 1$, ε, D and γ are all positive constants. Sketch the null clines and determine the condition on the parameters a and γ such that three steady states exist.

When $\varepsilon \ll 1$ and a and γ are in the parameter domain such that three steady states exist investigate the existence of a threshold travelling wave front from the zero rest state to the other stable rest state if the domain is originally at the zero rest state. Determine the unique wave speed of such a front to $O(1)$ for ε small. Sketch the wave solution and make any relevant remarks about the qualitative size and form of the initial conditions which could give rise to such a travelling front. If initially u and v are everywhere at the non zero steady state discuss the possibility of a wave from it to the zero rest state. [Problems 6 and 7 have been investigated in depth analytically by Rinzel and Terman (1982).]

*13. Travelling Waves in Reaction Diffusion Systems with Weak Diffusion: Analytical Techniques and Results

*13.1 Reaction Diffusion System with Limit Cycle Kinetics and Weak Diffusion: Model and Transformed System

Although the wave phenomena we discuss in this chapter are mathematically interesting in their own right, they have been observed experimentally with Belousov-Zhabotinskii reactions by Marek and Svobodová (1975): the experimental results are the motivation for the analysis here. We first present the mathematical analysis, which follows that given by Neu (1979a), and then describe the experiments which we relate directly to the key analytical result.

We consider the general two species dimensionless reaction diffusion system in one space dimension, s, given by

$$\frac{\partial x}{\partial t} = F(x,y) + \varepsilon \frac{\partial^2 x}{\partial s^2}, \quad \frac{\partial y}{\partial t} = G(x,y) + \varepsilon \frac{\partial^2 y}{\partial s^2}, \tag{13.1}$$

where the diffusion coefficients $0 < \varepsilon \ll 1$. This means that the diffusion coupling to the reaction kinetics is weak: in the analysis below we exploit this fact. In the diffusionless situation, $\varepsilon = 0$, the reaction kinetics

$$\frac{dx}{dt} = F(x,y), \quad \frac{dy}{dt} = G(x,y) \tag{13.2}$$

possesses a stable limit cycle solution, with period T, given by

$$x = X(t), \quad y = Y(t). \tag{13.3}$$

The analysis here is similar to that given in Chapter 8, Sections *8.6–*8.9, but much less complicated algebraically, although here we must include spatial effects. We shall not duplicate all of that analysis but quote the relevant parts and give detailed references to the sections in Chapter 8 so that the reader who has followed such algebra once will be able to pass over much of the rest of this section. The development below, however, is more or less self-contained.

When $\varepsilon = 0$ every position s is undergoing a T-periodic limit cycle oscillation, which is represented by a closed trajectory γ in the (x, y) phase plane. When $\varepsilon \neq 0$, the effect of the small diffusion is to cause the orbit γ of $(x(s,t), y(s,t))$ to be displaced by a small, $O(\varepsilon)$, distance from its limit cycle trajectory given by (13.3). It is convenient in such a case to change the dependent variables from (x, y) to (A, θ), where εA is the perpendicular displacement from the original orbit γ, and θ is the phase of the oscillation. The phase parametrizes points on the limit cycle and goes from 0 to T as we make a complete circuit round γ. The transformation is given by

$$x = X(\theta) + \varepsilon A Y'(\theta), \quad y = Y(\theta) - \varepsilon A X'(\theta) . \tag{13.4}$$

(This is exactly the same transformation used and discussed in detail in Section *8.6). Fig. 13.1 illustrates the transformation motivation; the vector $(Y', -X')$ is perpendicular to the velocity vector (X', Y') which is tangent to the orbit γ at the point (X, Y).

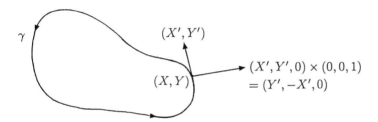

Fig. 13.1. The orbit γ is the T-period limit cycle when $\varepsilon = 0$. The vector perpendicular to γ is the vector product $(X', Y', 0) \times (0, 0, 1) = (Y', -X', 0)$.

When $\varepsilon = 0$ the limit cycle solution in terms of the phase θ is given by (13.3) as

$$x = X(\theta), \quad y = Y(\theta) \tag{13.5}$$

and in terms of the transformation (13.4) this is the situation where $A \equiv 0$ and $\theta = t + \theta_0$, where θ_0 simply gives the starting phase at $t = 0$. In the diffusionless situation, whatever θ_0 is, $d\theta/dt = 1$.

Applying the transformation (13.4) from (x, y) to (A, θ) to the original system (13.1) involves some tedious but necessary algebra and is similar to that given in Section *8.7, except here we have only one oscillator but we must include spatial effects from the diffusion terms. (The reader who wishes to dispense with the algebra may proceed immediately to (13.11) and (13.13) which give the (A, θ) reaction diffusion equations equivalent to (13.1)).

In using the change of variable (13.4) in the coupled system (13.1), we exploit the fact that $0 < \varepsilon \ll 1$ by expanding F and G in a Taylor series in ε. In the following the argument of the various functions is θ, unless otherwise stated, and primes denote differentiation with respect to θ. The first of (13.1), using (13.4), becomes

$$\frac{dx}{dt} = X'\theta_t + \varepsilon Y' A_t + \varepsilon AY'' \theta_t$$

$$= F(X,Y) + \varepsilon A[Y' F_X(X,Y) - X' F_Y(X,Y)] + \varepsilon(X''\theta_s^2 + X'\theta_{ss})$$

$$+ \varepsilon^2 (A_{ss}Y' + 2A_s Y''\theta_s + AY'''\theta_s^2 + AY''\theta_{ss}) \qquad (13.6)$$

and the second of (13.1) becomes

$$\frac{dy}{dt} = Y'\theta_t - \varepsilon X' A_t - \varepsilon AX'' \theta_t$$

$$= G(X,Y) + \varepsilon A[Y' G_X(X,Y) - X' G_Y(X,Y)] + \varepsilon[Y''\theta_s^2 + Y'\theta_{ss}]$$

$$- \varepsilon^2 (A_{ss}X' + 2A_s X''\theta_s + AX'''\theta_s^2 + AX''\theta_{ss}) . \qquad (13.7)$$

When $\varepsilon = 0$ we have, from (13.2) and (13.5),

$$X'(\theta) = F(X,Y), \quad Y'(\theta) = G(X,Y) . \qquad (13.8)$$

So, if we multiply (13.6) by X' and add to it (13.7) times Y', we get

$$(X'^2 + Y'^2)\theta_t + \varepsilon A(X'Y'' - Y'X'')\theta_t$$

$$= [X' F(X,Y) + Y' G(X,Y)] + \varepsilon A\{X'Y'[F_X(X,Y) - G_Y(X,Y)]$$

$$- X'^2 F_Y(X,Y) + Y'^2 G_X(X,Y)\}$$

$$+ \varepsilon[(X'X'' + Y'Y'')\theta_s^2 + (X'^2 + Y'^2)\theta_{ss}] + O(\varepsilon^2) . \qquad (13.9)$$

From (13.8),

$$X' F(X,Y) + Y' G(X,Y) = X'^2 + Y'^2 = R^2(\theta) , \qquad (13.10)$$

so on dividing both sides of (13.9) by $R^2 + \varepsilon A(X'Y'' - Y'X'')$ and keeping in mind the fact that ε is small, that is we expand everything as a series in ε, we get the partial differential equation for θ as

$$\frac{\partial\theta}{\partial t} = 1 + \varepsilon\left\{\gamma(\theta)A + P(\theta)\left[\frac{\partial\theta}{\partial s}\right]^2 + \frac{\partial^2\theta}{\partial s^2}\right\} + O(\varepsilon^2) , \qquad (13.11)$$

where

$$\gamma(\theta) = X'Y'(F_X - G_Y) - X'^2 F_Y + Y'^2 G_X - \frac{X'Y'' - Y'X''}{R^2} ,$$

$$P(\theta) = \frac{X'X'' + Y'Y''}{R^2} . \qquad (13.12)$$

If we now multiply (13.6) by Y' and subtract from it (13.7) times X', we get the following equation for A, on using the above θ-equation (13.11) and again remembering to expand everything in powers of ε:

$$\frac{dA}{dt} = \phi(\theta)A + Q(\theta)\left[\frac{\partial\theta}{\partial s}\right]^2 + \varepsilon U + O(\varepsilon^2),\qquad(13.13)$$

where

$$\phi(\theta) = \frac{Y'^2 F_X + X'^2 G_Y - X'Y'(F_Y + G_X) - X'X'' - Y'Y''}{R^2},$$

$$Q(\theta) = \frac{Y'X'' - X'Y''}{R^2}\qquad(13.14)$$

and U is a function of A and θ and their s derivatives. The specific forms of all these functions in the θ and A equations are not crucial for the asymptotic analysis below. However, what is important is that all of them are T-periodic in θ, and that ϕ satisfies the relation

$$\int_0^T \phi(\sigma)\, d\sigma < 0.\qquad(13.15)$$

This last relation comes from the fact that the original limit cycle oscillator is stable. This was proved in the short sub-section at the end of Section *8.7. It is the condition that perturbations A from the original limit cycle orbit γ die out as $t \to \infty$.

*13.2 Singular Perturbation Analysis: The Phase Satisfies Burgers' Equation

Equations (13.11) and (13.13) give the equations for the phase θ and the perturbation A from the original limit cycle orbit γ. The form of the equations, or rather the way ε appears in them, suggests that for small ε these ε-terms will only have an effect after a long time, $O(1/\varepsilon)$. A typical two-time procedure (see, for example, Murray 1984) suggests itself, and this we now develop for these equations (without assuming any knowledge of this two-time technique).

We look for asymptotic solutions for A and θ in the form

$$A \sim {}^0A + \varepsilon\,{}^1A, \quad \theta \sim {}^0\theta + \varepsilon\,{}^1\theta, \quad 0 < \varepsilon \ll 1,\qquad(13.16)$$

where the ${}^iA$ and ${}^i\theta$, $i = 0, 1$ depend on *two* time variables, t and $\tau = \varepsilon t$, the slow time variable, and the space variable s: now $\partial/\partial t = \partial/\partial t + \varepsilon\partial/\partial\tau$. Substituting

(13.16) into equations (13.11) and (13.13) and equating coefficients of powers of ε we get the following sequence of equations for the ^iA and $^i\theta$, $i = 0, 1$:

$$O(1): \quad \frac{\partial ^0\theta}{\partial t} = 1 \, ,$$

$$\frac{\partial ^0A}{\partial t} - \phi(^0\theta)\,^0A = Q(^0\theta) \left[\frac{\partial ^0\theta}{\partial s}\right]^2 \, , \tag{13.17}$$

$$O(\varepsilon): \quad \frac{\partial ^1\theta}{\partial t} = \gamma(^0\theta)\,^0A + P(^0\theta) \left[\frac{\partial ^0\theta}{\partial s}\right]^2 + \frac{\partial^2 \,^0\theta}{\partial s^2} - \frac{\partial ^0\theta}{\partial \tau} \, ,$$

$$\frac{\partial ^1A}{\partial t} - \phi(^0\theta)\,^1A = \left\{\phi'(^0\theta)\,^0A + Q'(^0\theta) \left[\frac{\partial ^0\theta}{\partial s}\right]^2\right\} {}^1\theta \tag{13.18}$$

$$+ 2Q(^0\theta)\left(\frac{\partial ^0\theta}{\partial s}\right)\left(\frac{\partial ^1\theta}{\partial s}\right) - \frac{\partial ^0A}{\partial \tau} + U \, ,$$

and so on for higher orders in ε.

Integrating the first of (13.17) gives

$$^0\theta = t + \psi(s, \tau) \, , \tag{13.19}$$

where ψ is a function of the slow time τ and the space variable s: ψ is undetermined at this stage as usual in such a two-time procedure. Substituting this $^0\theta$ into the second of (13.17) gives

$$\frac{\partial ^0A}{\partial t} - \phi(t + \psi)\,^0A = Q(t + \psi) \left[\frac{\partial \psi}{\partial s}\right]^2 \, . \tag{13.20}$$

This is a linear differential equation for ^0A in which the coefficients are T-periodic functions of t: the τ and s dependence only appear as parameters at this stage. For practical purposes the last equation is an ordinary differential equation for A^0 as a function of t. Its solution is directly related to the solution of

$$\frac{dz}{d\sigma} - \phi(\sigma)z = Q(\sigma) \, , \tag{13.21}$$

which is exactly the same type of equation that we studied in Chapter 8, Section *8.8, where we showed it had a unique T-periodic solution. We denote this periodic solution by $\rho(\sigma)$. There is also a decaying transient solution of (13.21), namely

$$z = h e^{V(\sigma)}, \quad V(\sigma) = \int_0^\sigma \phi(\alpha)\, d\alpha \, . \tag{13.22}$$

So, in terms of the unique T-periodic solution ρ of (13.21), the general solution ^0A of (13.20) is

$$
\begin{aligned}
^0A &= h_0(\tau)e^{V(t)} + \left[\frac{\partial\psi}{\partial s}\right]^2 \rho(t + \psi) \\
&= h_0(\tau)e^{V(t)} + \left[\frac{\partial\psi}{\partial s}\right]^2 \rho(^0\theta) .
\end{aligned}
\tag{13.23}
$$

Note that $h_0(\tau)$ is an arbitrary function of the slow time τ: from (13.19) ψ and hence $^0\theta$ are, of course, also functions of τ.

We now go on to consider the $O(\varepsilon)$ equations (13.18) with ^0A from (13.23). The $^1\theta$-equation becomes

$$
\frac{\partial^1\theta}{\partial t} = \gamma(^0\theta)h_0(\tau)e^{V(t)} + \{\gamma(^0\theta)\rho(^0\theta) + P(^0\theta)\}\left[\frac{\partial\psi}{\partial s}\right]^2 + \frac{\partial^2\psi}{\partial s^2} - \frac{\partial\psi}{\partial\tau} .
\tag{13.24}
$$

The function

$$
f(^0\theta) \equiv \gamma(^0\theta)\rho(^0\theta) + P(^0\theta)
$$

is a T-periodic function of $^0\theta$. Any periodic function can be split up into the sum of its mean plus a periodic part of zero mean. That is we can write

$$
f(^0\theta) = -\frac{k}{2} + \omega(^0\theta)
\tag{13.25}
$$

where

$$
k = -\frac{2}{T}\int_0^T f(\sigma)\,d\sigma, \qquad \int_0^T \omega(^0\theta)\,d^0\theta = 0 .
\tag{13.26}
$$

If we now integrate (13.24) we get

$$
^1\theta = \left\{\frac{\partial^2\psi}{\partial s^2} - \frac{\partial\psi}{\partial\tau} - \frac{k}{2}\left[\frac{\partial\psi}{\partial s}\right]^2\right\}t
$$

$$
+ h_0(\tau)\int^t \gamma(^0\theta)e^{V(t)}\,dt + \left[\frac{\partial\psi}{\partial s}\right]^2\int^t \omega(^0\theta)\,dt .
\tag{13.27}
$$

Now, since $\gamma(^0\theta)$ is periodic and $e^{V(t)}$ is exponentially decaying as $t \to \infty$, the first integral term is bounded. The second integral term is also bounded as $t \to \infty$ since $\omega(^0\theta)$ is periodic with zero mean: see (13.26). So, we can write the solution $^1\theta$ in (13.27) in the form

$$
^1\theta = \mu(\psi)t + B(t,\tau) ,
\tag{13.28}
$$

where $B(t,\tau)$ is a bounded function, namely the two integral terms in (13.27), and

$$
\mu(\psi) = \frac{\partial^2\psi}{\partial s^2} - \frac{\partial\psi}{\partial\tau} - \frac{k}{2}\left[\frac{\partial\psi}{\partial s}\right]^2 .
\tag{13.29}
$$

Now substitute (13.28) for $^1\theta$ into the second of (13.18) and use (13.19) for $^0\theta$ and (13.23) for ^0A, to arrive at the equation for ^1A in the form

$$\frac{\partial\,^1A}{\partial t} - \phi(^0\theta)\,^1A + C(t,\tau) = \left\{\mu(\psi)W(^0\theta)\frac{\partial\psi}{\partial s} + 2Q^0\theta\frac{\partial\mu(\psi)}{\partial s}\right\} t\frac{\partial\psi}{\partial s}\,, \tag{13.30}$$

$$W(^0\theta) = \phi'(^0\theta)\rho(^0\theta) + Q'(^0\theta)\,,$$

where

$$C(t,\tau) = [B(t,\tau) + \mu(\psi)t]\phi'(^0\theta)h_0(\tau)e^{V(t)} + \omega(^0\theta)B(t,\tau)\left[\frac{\partial\psi}{\partial s}\right]^2 - \frac{\partial\,^0A}{\partial\tau} + U\,,$$

is simply another bounded function made up of exponentially decaying or periodic terms.

The essence of two-timing comes in at this stage. The underlying and obvious assumption in seeking the asymptotic solution (13.16) is that the $O(\varepsilon)$ terms remain $O(\varepsilon)$ for all time – it is what we mean by a uniformly valid solution. So, the solution for ^1A and its time derivative from (13.30) must remain bounded for all t. Because of the explicit t-terms on the right hand side we must require the sum of them to be zero otherwise they would become unbounded. We thus arrive at the following boundedness condition for ^1A and $\partial\,^1A/\partial t$:

$$\left\{\mu(\psi)W(^0\theta)\frac{\partial\psi}{\partial s} + 2Q(^0\theta)\frac{\partial\mu(\psi)}{\partial s}\right\}\frac{\partial\psi}{\partial s} = 0\,. \tag{13.31}$$

The functions W and Q are two different functions of $^0\theta$ and so are different periodic functions of t. Thus the last equation implies that either $\partial\psi/\partial s = 0$ or $\mu(\psi) = 0$. But $\partial\psi/\partial s = 0$ means that ψ is space independent and so the solutions are also space independent to first order: this is clearly not the case. We could for example prescribe initial data so that $\partial\psi/\partial s \neq 0$. So the only other possibility gives us the necessary condition for boundedness of ^1A, namely $\mu(\psi) = 0$, which from (13.29) is

$$\mu(\psi) \equiv \frac{\partial^2\psi}{\partial s^2} - \frac{\partial\psi}{\partial\tau} - \frac{k}{2}\left(\frac{\partial\psi}{\partial s}\right)^2 = 0\,. \tag{13.32}$$

Note from (13.28) that this condition also ensures that $^1\theta$ now remains bounded since $\mu(\psi)$ is the coefficient of t.

The last equation is more familiar if we differentiate it with respect to s and set

$$u(s,\tau) = \frac{\partial\psi}{\partial s} \quad\Rightarrow\quad \frac{\partial u}{\partial t} + ku\frac{\partial u}{\partial s} = \frac{\partial^2 u}{\partial s^2}\,. \tag{13.33}$$

This is well known as *Burgers' equation* and has been widely studied: it first arose in a fluid mechanics context. This equation is discussed in detail in the book on linear and nonlinear waves by Whitham (1974). Equation (13.32) can be solved exactly using a logarithmic transformation which transforms it into the linear diffusion equation. Here with

$$\psi = -\frac{2}{k} \ln Z \quad \Rightarrow \quad \frac{\partial Z}{\partial \tau} = \frac{\partial^2 Z}{\partial s^2} \, . \tag{13.34}$$

The singular perturbation analysis in this section has produced the important and surprising result that the *phase*, when an oscillator is weakly coupled with diffusion, evolves according to Burgers' equation where the independent variables are the space s and the *slow* time τ. This has some interesting consequences as we see in the following section.

*13.3 Travelling Wavetrain Solutions for Reaction Diffusion Systems with Limit Cycle Kinetics and Weak Diffusion: Comparison with Experiment

Suppose we have a solution $\psi(s,\tau)$ of (13.32) we then get the solutions for x and y to $O(1)$ from (13.5), with $^0\theta = t + \psi(s,\tau)$ from (13.19), as

$$x \sim X(t + \psi(s,\tau)), \quad y \sim Y(t + \psi(s,\tau)) \, . \tag{13.35}$$

Before we show what these solutions mean, let us briefly digress.

Suppose $F(\xi)$, with $\xi = t + (x/c)$ and c a constant, is a periodic function of ξ with period T. Then if t increases by T or x increases by cT, $F(\xi)$ remains the same. But, as we saw in Chapter 11, Section 11.1, $F(t + (x/c))$ represents a travelling wave with wave speed c. So, with F periodic, we have a travelling wavetrain somewhat like a series of waves propagating along a string. We discussed these in detail in Chapter 12, Section 12.5. Another way of seeing this is if we go to a position x and stay there, we see an oscillating F with period T and frequency $2\pi/T$, and similarly, for a fixed time we see a wavetrain in x with wavenumber $2\pi/cT$ and hence a wavelength cT.

If we now return to the solutions x and y in (13.35), X and Y are periodic functions of t with frequency $2\pi/T$. The space variable is buried in ψ. Analogous to the last paragraph we then have $t + \psi(s,\tau)$ corresponding to $t + (x/c)$ and so the solutions x and y represent travelling wavetrain solutions with frequency $2\pi/T$ and a wavenumber $2\pi/[T\partial\psi/\partial s]$: $\partial\psi/\partial s$ corresponds to c. With these solutions, however, $\partial\psi/\partial s$ varies spatially and as a function of $\tau = \varepsilon t$, that is it varies slowly with time. From (13.33) we see therefore, that the *wavenumber* evolves slowly with time according to *Burgers' equation*.

There is a large number of known solutions for Burgers' equation, which is not surprising since they can be generated by solutions of the linear diffusion equation (13.34). (As well as these, however, there is a whole class of singular solutions which were found in by Choodnovsky and Choodnovsky (1977).) One of the exact solutions of particular interest for the application to the experiments on the Belousov-Zhabotinskii and which we discuss below, is the 'shock transition' solution of (13.33). This is like the solution for travelling wavefronts discussed in detail in Chapter 11, namely far ahead of the front/shock the solution is $u = u_1$, say, and far behind the front $u = u_2$, say. It is given by

$$u(s,\tau) = u_1 + \frac{u_2 - u_1}{1 + \exp \frac{k(u_2-u_1)(s-v\tau)}{2}} \,, \tag{13.36}$$

where $v = (u_1 + u_2)/2$ is the 'shock' speed. Clearly $u \to u_1$ as $s - v\tau \to \infty$ and $u \to u_2$ as $s - v\tau \to -\infty$. This solution is in travelling wavefront form since $u(s,\tau) = u(s - v\tau) = u(s - v\varepsilon t)$, that is $\varepsilon(u_1 + u_2)/2$ is the speed of propagation. See Whitham (1974) for a derivation of the solution of Burgers' equation or simply check that it is a solution of (13.33).

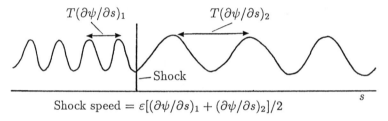

$$\text{Shock speed} = \varepsilon[(\partial\psi/\partial s)_1 + (\partial\psi/\partial s)_2]/2$$

Fig. 13.2. Solution, at a fixed time, to a reaction diffusion system with limit cycle kinetics and weak diffusion which exhibits travelling wave train solutions with a shock transition between the different wavelengths. The wavenumbers $2\pi/[T\partial\psi/\partial s]$ evolve according to Burgers' equation (13.33). The shock transition, separating regions of different wavelengths, moves at a speed $\frac{\varepsilon}{2}[(\partial\psi/\partial s)_1 + (\partial\psi/\partial s)_2]$.

What the solution (13.36) means in terms of the chemical concentrations x and y is that there is a shock transition between two different $\partial\psi/\partial s$, that is between two different wavenumbers or wavelengths. The transition, of course, takes place over an infinite distance, but with the effective transition at $s - v\tau = 0$. That is the transition between two different wavelengths moves with speed $\varepsilon[(\partial\psi/\partial s)_1 + (\partial\psi/\partial s)_2]/2$. Fig. 13.2 schematically illustrates a typical x or y as a function of s at a fixed time.

At the beginning of Section 13.1 we referred to results of experiments by Marek and Svobodova (1975) on the Belousov-Zhabotinskii reaction. Their experimental set-up consists of a continuous stirred tank reactor containing a Belousov-Zhabotinskii reaction with a given set of reaction parameters so that it oscillates with a fixed period. Attached to one end of this tank reactor is a long tubular reactor which also contains a Belousov-Zhabotinskii reaction but in which the parameters are slightly different to those in the stirred tank. Initially the concentrations in the tubular reactor are uniform in space. The two reactors are then connected. The continuous stirred tank reactor acts like a forcing function on one end of the tubular reactor and chemical concentration waves start to propagate along the tube reactor until they eventually fill its whole length. There is a clear transition region at the travelling front between the uniform state ahead of the front and the chemical concentration waves behind it. Let us now consider these results in the light of the analysis and the 'shock' transition solution discussed above. We must first try to mimic these experiments with our model.

We model the reaction going on in the tubular reactor by the dimensionless reaction diffusion equations (13.1) for two key concentrations x and y of the Belousov-Zhabotinskii reaction, typically those we considered in the two-species model in Chapter 7. We assume the diffusion coefficient $0 < \varepsilon \ll 1$. Initially x and y are spatially uniform and undergoing limit cycle oscillations with period T. To be specific let us thus take

$$x(s,0) = X(0), \quad y(s,0) = Y(0), \quad s > 0 . \tag{13.37}$$

The tank reactor at one end, say $s = 0$, provides time dependent boundary conditions. Because of the small difference between the periods of the two oscillators, it is reasonable to consider the boundary condition at $s = 0$ as

$$x(0,t) = X([1 + \varepsilon\omega]t) = X(t + \omega\tau) ,$$
$$y(0,t) = Y([1 + \varepsilon\omega]t) = Y(t + \omega\tau) . \tag{13.38}$$

So, the tubular reaction sees at one end an oscillator with a slightly different period of oscillation, namely $1/(1 + \varepsilon\omega)$.

The mathematical problem for $\psi(s,\tau)$ with these boundary and initial conditions is then, from (13.32) (that is Burgers' equation for $\partial\psi/\partial s$), (13.37) and (13.38),

$$\frac{\partial\psi}{\partial\tau} = \frac{\partial^2\psi}{\partial s^2} - \frac{k}{2}\left(\frac{\partial\psi}{\partial s}\right)^2 , \quad s > 0 \tag{13.39}$$

$$\psi(s,0) = 0, \quad \psi(0,\tau) = \omega\tau .$$

With the logarithmic transformation (13.33) this transforms to the linear problem for ϕ

$$\frac{\partial\phi}{\partial\tau} = \frac{\partial^2\phi}{\partial s^2}, \quad s > 0, \quad \phi = \exp\left(-k\psi/2\right) \tag{13.40}$$

$$\phi(s,0) = 1, \quad \phi(0,\tau) = \exp\left(-k\omega\tau/2\right) .$$

If we set $W = -k\omega > 0$, the solution ϕ of the last equation is given by (see, for example, Chapter 2 of Carslaw and Jaeger 1959)

$$\phi(s,\tau) = \mathrm{erf}\left[\frac{s}{2\sqrt{\tau}}\right] + \frac{1}{2}e^{W\tau}\left\{e^{-S\sqrt{W}}\,\mathrm{erf}\,c\left[\frac{s}{2\sqrt{\tau}} - \sqrt{W\tau}\right]\right.$$

$$\left. + e^{S\sqrt{W}}\,\mathrm{erf}\,c\left[\frac{s}{2\sqrt{\tau}} + \sqrt{W\tau}\right]\right\},$$

$$\mathrm{erf}\,z = \frac{2}{\sqrt{\pi}}\int_0^z \exp\left(-\sigma^2\right)d\sigma\,,$$

$$\mathrm{erf}\,cz = 1 - \mathrm{erf}\,z = \frac{2}{\sqrt{\pi}}\int_z^{\infty}\exp\left(-\sigma^2\right)d\sigma\,.$$

(13.41)

The error function $\mathrm{erf}\,z$ satisfies $\mathrm{erf}\,\infty = 1$, $\mathrm{erf}\,0 = 0$. We shall need the asymptotic forms of these functions for small and large z. They are, for example from Murray (1984), or on integrating by parts,

$$\mathrm{erf}\,z = \frac{2}{\sqrt{\pi}}\int_0^z \exp\left(-\sigma^2\right)d\sigma = \frac{2}{\sqrt{\pi}}\left[\int_0^{\infty} - \int_z^{\infty}\right]\exp\left(-\sigma^2\right)d\sigma$$

$$\sim 1 - \frac{1}{2z}\exp\left(-z^2\right)\quad\text{for}\quad z \gg 1$$

$$\mathrm{erf}\,cz = 1 - \mathrm{erf}\,z \sim \frac{1}{2z}\exp\left(-z^2\right)\quad\text{for}\quad z \gg 1$$

(13.42)

$$\mathrm{erf}\,z \sim \frac{2z}{\sqrt{\pi}}\quad\text{for}\quad 0 < z \ll 1$$

$$\mathrm{erf}\,cz \sim 1 - \frac{2z}{\sqrt{\pi}}\quad\text{for}\quad 0 < z \ll 1\,.$$

Now consider the solution (13.41) for $s \ll \sqrt{\tau}$ and for $s \gg \sqrt{\tau}$. For $s/(2\sqrt{\tau}) \ll 1$ and the asymptotic forms in (13.42) for z small, the solution

$$\phi(s,\tau) \sim \frac{s}{\sqrt{\pi\tau}} + \frac{1}{2}e^{W\tau}\left\{e^{-s\sqrt{W}}\,\mathrm{erf}\,c(-\sqrt{W\tau}) + e^{s\sqrt{W}}\,\mathrm{erf}\,c\sqrt{W\tau}\right\}\,,$$

$$= \frac{s}{\sqrt{\pi\tau}} + \frac{1}{2}e^{W\tau}\left\{e^{-s\sqrt{W}}[1 + \mathrm{erf}\,\sqrt{W\tau}] + e^{s\sqrt{W}}[1 - \mathrm{erf}\,\sqrt{W\tau}]\right\}\,.$$

$$\sim e^{W\tau - s\sqrt{W}}$$

for $\sqrt{W\tau}$ sufficiently large so that $\mathrm{erf}\,\sqrt{W\tau} \sim 1$. This solution is a *travelling wave form* with speed $\sqrt{W}$. For $s/\sqrt{\tau} \gg 1$, again using (13.42), we find

$$\phi(s,\tau) \sim 1 - O\left[\frac{\sqrt{\tau}}{s}\exp\left(\frac{-s^2}{4\tau}\right)\right]\,.$$

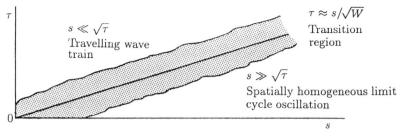

Fig. 13.3. The situation obtained when a stirred tank reactor containing an oscillating Belousov-Zhabotinskii reaction is connected, at $s = 0$ at time $t = 0$, to a tube reactor with a similar reaction but oscillating with slightly different parameters. Note the transition region between the travelling wave domain and the spatially homogeneous domain.

With these forms for ϕ, the asymptotic solution for $\psi = -(2/k)\ln\phi$, from (13.40), is then

$$\psi \sim \begin{cases} \dfrac{2}{k}(s\sqrt{W} - W\tau) \\ 0 \end{cases} \quad \text{for} \quad \begin{cases} s \ll \sqrt{\tau} \\ s \gg \sqrt{\tau} \end{cases} \tag{13.43}$$

and the corresponding wavenumber κ is proportional to $\partial\psi/\partial s$, that is

$$\kappa \propto \frac{\partial\psi}{\partial s} \sim \begin{cases} \dfrac{2\sqrt{W}}{k} \\ 0 \end{cases} \quad \text{for} \quad \begin{cases} s \ll \sqrt{\tau} \\ s \gg \sqrt{\tau} \end{cases}. \tag{13.44}$$

What these results mean as regards the chemical concentrations is that $x(s,\tau)$, for example, is like a travelling wavetrain with wavenumber

$$\frac{2\pi\kappa}{T} \propto \frac{2\pi\sqrt{W}}{kT} \quad \text{for} \quad s \ll \sqrt{\tau} ,$$

and like a spatially homogeneous temporal oscillation for $s \gg \sqrt{\tau}$: a zero wavenumber corresponds to an infinite wavelength. The transition region, or "wavefront", is at $s = \tau\sqrt{W}$. The term "wavefront" has a different sense here. It is the transition region where ahead of it the system is homogeneous and oscillating in time and behind it there is a travelling wavetrain. Fig. 13.3 illustrates the situation obtained from (13.44).

The analytical results with the assumptions implied by the boundary and initial conditions (13.37) are in qualitative agreement with the experimental observations of Marek and Svobodova (1975). For example, using Fig. 13.3, if we go to a position far ahead of the transition front (that is $s \gg \tau\sqrt{W}$) there is simply a spatially homogeneous oscillator while behind it there is a travelling wavetrain. This wavetrain is moving into the homogeneous domain with a fixed velocity, namely $\sqrt{W}$, which is directly related to the phase lag, ω, in the two oscillators at the place where they are connected.

14. Spatial Pattern Formation with Reaction/Population Interaction Diffusion Mechanisms

14.1 Role of Pattern in Developmental Biology

Embryology is that part of biology which is concerned with the formation and development of the embryo from fertilization until birth. Development of the embryo is a sequential process and follows a ground plan, which is usually laid down very early in gestation. In humans for example it is set up roughly by the 5th week. The book by Slack (1983) is a readable account of the early stages of development from egg to embryo.

Morphogenesis, the part of embryology with which we shall be concerned, is the development of pattern and form. How the developmental ground plan is established is unknown as are the mechanisms which produce the spatial patterning necessary for specifying the various organs. The following three chapters and most of this one will be devoted to mechanisms which can generate spatial pattern and form, and which have been proposed as possible pattern formation processes in a variety of morphogenetic situations. Section 14.7 will be concerned with an important ecological aspect of pattern formation, which suggests possible strategies of pest control – the mathematical analysis is different but highly relevant to many embryological situations.

Cell division starts after fertilisation. When sufficient cell division has taken place in a developing embryo the key problem is how the homogeneous mass of cells are spatially organised so that the sequential process of development can progress. Cells differentiate, in a biological sense, according to where they are in the spatial organisation. They also move around in the embryo. This latter phenomenon is an important element in morphogenesis and has given rise to a new approach to the generation of pattern and form discussed in some detail in Chapter 17.

It is impossible not to be fascinated and enthralled with the wealth, diversity and beauty of pattern in biology. Fig. 14.1 shows only four examples. How such patterns, and millions of others, were laid down is still unknown. The patterning problems posed by only Fig. 14.1 are quite diverse.

As a footnote to Fig. 14.1 (c), note the antennae on the moth. These antennae very effectively collect molecules of the chemical odorant, called a pheromone, which is extruded by the female to attract the male. The filtering efficiency of such antennae, which collect, and in effect count, the molecules, poses a very

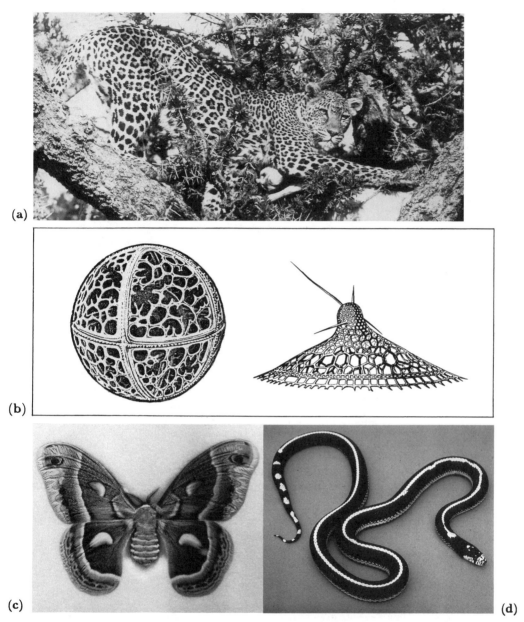

Fig. 14.1a-d. (a) Leopard (*Panthera pardus*) in the Serengetti National Park, Tanzania. Note the individual spot structure. (Photograph courtesy of Hans Kruuk) (b) Radiolarians (*Trissocyclus spaeridium* and *Eucecryphalus genbouri*). These are small marine organisms – protozoa – of the order of a millimeter across. (After Haeckel 1862, 1887) The structural architecture of radiolarians is amazingly diverse (see, for example, the plate reproductions of some of Haeckel's drawings in the Dover Archive Series, Haeckel 1974). (c) Moth (*Hyalophora cecropia*). As well as the wing patterns note the stripe pattern on the body and the structure of the antennae. (d) Californian king snake. Sometimes the pattern consists of crossbands rather than a backstripe. (Photograph courtesy of Lloyd Lemke)

different and interesting mathematical biology problem to those discussed in this book, namely how such a filter antenna should be designed to be most efficient. This specific problem is discussed in detail by Murray (1977).

The fundamental importance of pattern and form in developmental biology is self-evident. Whatever pattern we observe in the animal world it is almost certain that the process that produced it is unknown. Although the mechanism must be genetically controlled, the genes themselves cannot create the pattern. They only provide a blue-print or recipe, for the pattern generation. One of the major problems in biology is how genetic information is physically translated into the necessary pattern and form. Much of the research in developmental biology, both experimental and theoretical, is devoted to trying to determine the underlying mechanisms which generate pattern and form in early development. The detailed discussion in these next few chapters discusses some of the mechanisms which have been proposed and gives an indication of the role of mathematical modelling in trying to unravel the underlying mechanisms involved in morphogenesis.

A phenomenological concept of pattern formation and differentiation called *positional information* was proposed by Wolpert (1969, see the reviews in 1971, 1981). He suggested that cells are pre-programmed to react to a chemical (or morphogen) concentration and differentiate accordingly, into different kinds of cells such as cartilage cells. The general introductory paper by Wolpert (1977) gives a very clear and non-technical description of development of pattern and form in animals and the concepts and application of positional information.

The chemical prepattern viewpoint of embryogenesis separates the process of development into several steps; the essential first step is the creation of a morphogen concentration spatial pattern. The name 'morphogen' is used for such a chemical because it effects morphogenesis. The notion of positional information relies on a chemical pre-specification so that the cell can read out its position in the coordinates of chemical concentration, and differentiate, undergo appropriate cell shape change, or migrate accordingly. So, once the prepattern is established, morphogenesis is a slave process. Positional information is not dependent on the specific mechanism which sets up the spatial prepattern of morphogen concentration. This chapter is concerned with reaction diffusion models as the possible mechanisms for generating biological pattern. The basic chemical theory or reaction diffusion theory of morphogenesis was put forward in the classical paper by Turing (1952). Levin and Segel (1985) give a brief and readable survey of reaction diffusion theories and their generalisations.

With the complexity of animal forms the concept of positional information necessarily implies a very sophisticated interpretation of the 'morphogen map' by the cell. This need not pose any problem when we recall how immensely complex a cell is. A rough idea, for example, is given by comparing the weight per bit of information of the cell's DNA molecule, around 10^{-22}, to that of, say, imaging by an electron beam of around 10^{-10} or of a magnetic tape of about 10^{-5}. The most sophisticated and compact computer chip is simply not in the same class as a cell.

14.2 Reaction Diffusion (Turing) Mechanisms

Turing (1952) suggested that, under certain conditions, chemicals can react and diffuse in such a way as to produce steady state heterogeneous spatial patterns of chemical or morphogen concentration. In Section 9.2 in Chapter 9 we derived the governing equations for reaction diffusion mechanisms, namely (9.16), which we consider here in the form:

$$\frac{\partial \mathbf{c}}{\partial t} = \mathbf{f}(\mathbf{c}) + D\nabla^2 \mathbf{c} , \tag{14.1}$$

where $\mathbf{c}$ is the vector of morphogen concentrations, $\mathbf{f}$ represents the reaction kinetics and D is the diagonal matrix of positive constant diffusion coefficients. This chapter will mainly be concerned with models for two chemical species, $A(\mathbf{r}, t)$ and $B(\mathbf{r}, t)$ say. The equation system is then of the form

$$\frac{\partial A}{\partial t} = F(A, B) + D_A \nabla^2 A ,$$
$$\frac{\partial B}{\partial t} = G(A, B) + D_B \nabla^2 B , \tag{14.2}$$

where F and G are the kinetics, which will always be nonlinear.

Turing's (1952) idea is a simple but profound one. He said that if, in the absence of diffusion (effectively $D_A = D_B = 0$), A and B tend to a linearly stable uniform steady state then, under certain conditions, which we shall derive, spatially inhomogeneous patterns can evolve by *diffusion driven instability* if $D_A \neq D_B$. Diffusion is usually considered a *stabilising* process which is why this was such a novel concept. To see intuitively how diffusion can be destabilising consider the following, albeit urealistic, but informative analogy.

Consider a field of dry grass in which there is a large number of grasshoppers which can generate a lot of moisture by sweating if they get warm. Now suppose the grass is set alight at some point and a flame front starts to propagate. We can think of the grasshopper as an inhibitor and the fire as an activator. If there was no moisture to quench the flames the fire would simply spread over the whole field which would result in a uniform charred area. Suppose, however, that when the grasshoppers get warm enough they can generate enough moisture to dampen the grass so that when the flames reach such a pre-moistened area the grass will not burn. The scenario for spatial pattern is then as follows. The fire starts to spread – it is one of the 'reactants', the activator, with a 'diffusion' coefficient D_F say. When the grasshoppers, the inhibitor 'reactant', ahead of the flame front feel it coming they move quickly well ahead of it – that is they have a 'diffusion' coefficient, D_G say, which is much larger that D_F. The grasshoppers then sweat profusely and generate enough moisture and thus prevent the fire spreading into the moistened area. In this way the charred area is restricted to a finite domain which depends on the 'diffusion' coefficients of the reactants – fire and grasshoppers – and various 'reaction' parameters. If, instead of a single initial fire, there was a random scattering of them we can see how this process would result in a final spatially inhomogeneous steady state distribution of charred and uncharred regions in the field, since around each fire the above scenario would take place. It is clear that if the grasshoppers and flame front 'diffused' at the same speed no such spatial pattern could evolve. It is clear how to construct other analogies: another example is given in the Scientific American article by Murray (1988).

In the following section we shall describe the process in terms of reacting and diffusing morphogens and derive the necessary conditions on the reaction kinetics and diffusion coefficients. We shall also derive the type of spatial patterns we might expect. Here we briefly record for subsequent use two particularly simple hypothetical systems and one experimentally realised example, which are capable of satisfying Turing's conditions for a pattern formation system. There are now, of course, many other systems which have been used in studies of spatial patterning. These have varying degrees of experimental plausibility. With the extensive discussion of the Belousov-Zhabotinskii reaction in Chapters 7 and 12 we should particularly note it as perhaps the major experimental system.

The simplest system is the Schnakenberg (1979) reaction discussed in Chapter 6 which, with reference to the system form (14.2), has kinetics

$$F(A, B) = k_1 - k_2 A + k_3 A^2 B, \quad G(A, B) = k_4 - k_3 A^2 B , \qquad (14.3)$$

where the k's are the positive rate constants. Here A is created autocatalytically by the $k_3 A^2 B$ term in $F(A, B)$. This is one of the prototype reaction diffusion systems. The second is the activator-inhibitor mechanism suggested by Gierer and Meinhardt (1972) and discussed in Chapter 5, namely

$$F(A, B) = k_1 - k_2 A + \frac{k_3 A^2}{B}, \quad G(A, B) = k_4 A^2 - k_5 B , \qquad (14.4)$$

where here A is the activator and B the inhibitor. The $k_3 A^2/B$ term is again autocatalytic. The third, the real empirical substrate-inhibition system studied experimentally by Thomas (1975) and also described in detail in Chapter 5, has

$$F(A, B) = k_1 - k_2 A - H(A, B), \quad G(A, B) = k_3 - k_4 B - H(A, B) ,$$
$$H(A, B) = \frac{k_5 A B}{k_6 + k_7 A + k_8 A^2} . \qquad (14.5)$$

Here A and B are respectively the concentrations of the substrate oxygen and the enzyme uricase. The substrate inhibition is evident in the H-term via $k_8 A^2$. Since the H-terms are negative they contribute to reducing A and B; the rate of reduction is inhibited for large enough A.

Before commenting on the types of reaction kinetics capable of generating pattern we must nondimensionalise the systems given by (14.2) with reaction kinetics from (14.3)–(14.5). By way of example we carry out the details here for

(14.2) with F and G given by (14.3). Introduce L as a typical length scale and set

$$u = A \left(\frac{k_3}{k_2} \right)^{1/2} , \quad v = B \left(\frac{k_3}{k_2} \right)^{1/2} , \quad t^* = \frac{D_A t}{L^2} , \quad \mathbf{x}^* = \frac{\mathbf{x}}{L} ,$$

$$d = \frac{D_B}{D_A} , \quad a = \frac{k_1}{k_2} \left(\frac{k_3}{k_2} \right)^{1/2} , \tag{14.6}$$

$$b = \frac{k_4}{k_2} \left(\frac{k_3}{k_2} \right)^{1/2} , \quad \gamma = \frac{L^2 k_2}{D_A} .$$

The dimensionless reaction diffusion system becomes, on dropping the asterisks for algebraic convenience,

$$u_t = \gamma(a - u + u^2 v) + \nabla^2 u = \gamma f(u, v) + \nabla^2 u ,$$
$$v_t = \gamma(b - u^2 v) + d\nabla^2 v = \gamma g(u, v) + d\nabla^2 v , \tag{14.7}$$

where f and g are defined by these equations. We could incorporate γ into new length and time scales by setting $\gamma^{1/2}\mathbf{r}$ and γt for $\mathbf{r}$ and t respectively. This is equivalent to defining the length scale L such that $\gamma = 1$, that is $L = (D_A/k_2)^{1/2}$. We retain the specific form (14.7) for reasons which will become clear shortly as well as for the analysis in the next section and for the applications in Chapter 15.

An appropriate nondimensionalisation of the reaction kinetics (14.4) and (14.5) give (see Exercise 1)

$$f(u, v) = a - bu + \frac{u^2}{v} , \quad g(u, v) = u^2 - v ,$$
$$f(u, v) = a - u - h(u, v), \quad g(u, v) = \alpha(b - v) - h(u, v) , \tag{14.8}$$
$$h(u, v) = \frac{\rho u v}{1 + u + K u^2} ,$$

where a, b, α, ρ and K are positive parameters. If we include activator inhibition in the activator-inhibitor system in the first of these we have, for f and g,

$$f(u, v) = a - bu + \frac{u^2}{v(1 + k u^2)}, \quad g(u, v) = u^2 - v , \tag{14.9}$$

where k is a measure of the inhibition: see also Section 5.5 in Chapter 5. Murray (1982) discussed each of these systems in detail and drew conclusions as to their relative merits as pattern generators. For most pattern formation illustrations the simplest, namely (14.7), turns out to be the most robust and of course the easiest to study.

All such reaction diffusion systems can be nondimensionalised and scaled to take the general form

$$u_t = \gamma f(u, v) + \nabla^2 u, \quad v_t = \gamma g(u, v) + d\nabla^2 v , \tag{14.10}$$

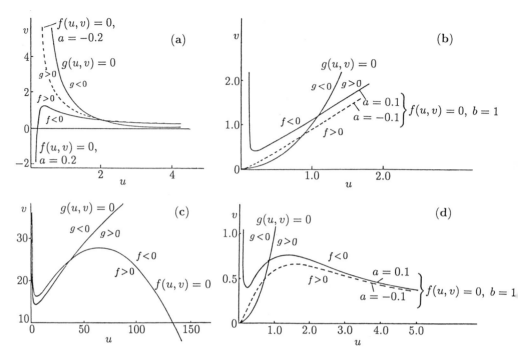

Fig. 14.2a-d. Null clines $f(u, v) = 0$, $g(u, v) = 0$: (a) The dimensionless Schnakenberg (1979) kinetics (14.7) with $a = 0.2$ and $b = 2.0$ with the dashed curve, where $a = -0.2$ and which is typical of the situation when $a < 0$. (b) The dimensionless Gierer and Meinhardt (1972) system with $a = \pm 0.1$, $b = 1$ and no activator inhibition. (c) The empirical Thomas (1975) system defined by (14.8) with parameter values $a = 150$, $b = 100$, $\alpha = 1.5$, $\rho = 13$, $K = 0.05$. (d) The kinetics in (14.9) with $a > 0$, $b > 0$ and $k > 0$, which implies activator inhibition – the dashed curve has $a < 0$.

where d is the ratio of diffusion coefficients and γ can have any of the following interpretations:

(i) $\gamma^{1/2}$ is proportional to the *linear* size of the spatial domain in one dimension. In two dimensions γ is proportional to the area. This meaning is particularly important as we shall see later in Section 14.5 and in Chapter 15.
(ii) γ represents the relative strength of the reaction terms. This means, for example, that an increase in γ may represent an increase in activity of some rate-limiting step in the reaction sequence.
(iii) An increase in γ can also be thought of as equivalent to a decrease in the diffusion coefficient ratio d.

Particular advantages of this general form are: (a) the dimensionless parameters γ and d admit a wider biological interpretation than do the dimensional parameters and (b) when we consider the domains in parameter space where particular spatial patterns appear, the results can be conveniently displayed in (γ, d) space. This aspect was exploited by Arcuri and Murray (1986).

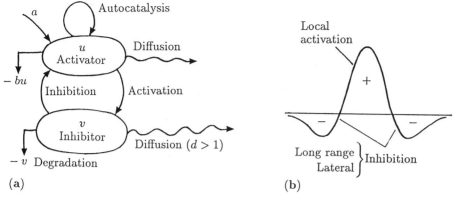

Fig. 14.3a,b. (a) Schematic representation of the activator-inhibitor system

$$u_t = a - bu - \frac{u^2}{v} + \nabla^2 u, \quad v_t = u^2 - v + d\nabla^2 v .$$

(b) Spatial representation of local activation and long range inhibition.

Whether or not the systems (14.2) are capable of generating Turing-type spatial patterns crucially depends on the reaction kinetics f and g, and the values of γ and d. The detailed form of the null clines provides essential initial information. Fig. 14.2 illustrates typical null clines for f and g defined by (14.7)–(14.9).

In spite of their different chemical motivation and derivation all of these kinetics are equivalent to some activation-inhibition interpretation and when coupled with unequal diffusion of the reactants, are capable of generating spatial patterns. The spatial activation-inhibition concept was discussed in detail in Section 9.5 in Chapter 9, and arose from an integral equation formulation: refer to equation (9.41). As we shall see in the next section the crucial aspect of the kinetics regarding pattern generation is incorporated in the form of the null clines and how they intersect in the vicinity of the steady state. There are two broad types illustrated in the last figure. The steady state neighbourhood of the null clines in Fig. 14.2 (b), (c), (d) are similar and represent one class, while that in Fig. 14.2 (a) is the other.

We should note here that there are other important classes of null clines which we do not consider, such as those in which there is more than one positive steady state: we discussed such kinetics in Chapter 6 for example. Reaction diffusion systems with such kinetics can generate even more complex spatial patterns: initial conditions here are particularly important. We also do not discuss here systems in which the diffusion coefficients are space dependent and concentration, or population, dependent: these are more important in ecological contexts. We briefly considered density dependent diffusion cases in Chapter 9.

It is often useful and intuitively helpful in model building to express the mechanism's kinetics in schematic terms with some convention to indicate auto-catalysis, activation, inhibition, degradation and unequal diffusion. If we do this, by way of illustration, with the activator-inhibitor kinetics given by the first of (14.8) in (14.10) we can adopt the convention shown in Fig. 14.3 (a).

The effect of different diffusion coefficients, here with $d > 1$, is to give the prototype spatial concept of local activation and lateral inhibition illustrated in Fig. 14.3 (b). The general concept was introduced before in Section 9.5: see Fig. 9.5. It is this generic spatial behaviour which is necessary for spatial patterning – the grasshoppers and the fire analogy is an obvious example with the fire the local activation and the grasshoppers providing the long range inhibition. It is intuitively clear that the diffusion coefficient of the inhibitor must be larger than that of the activator.

14.3 Linear Stability Analysis and Evolution of Spatial Pattern: General Conditions for Diffusion-Driven Instability

A reaction diffusion system exhibits diffusion-driven instability or *Turing instability* if the homogeneous steady state is stable to small perturbations in the absence of diffusion but unstable to small *spatial* perturbations when diffusion is present. The usual concept of instability in biology is in the context of ecology, where a uniform steady state becomes unstable to small perturbations and the populations typically exhibit some temporal oscillatory behaviour. The instability we are concerned with here is of a quite different kind. The mechanism driving the spatially inhomogeneous instability is diffusion: the mechanism determines the spatial pattern that evolves. How the pattern or mode is selected is an important aspect of the analysis.

We derive here the necessary and sufficient conditions for diffusion driven instability of the steady state and the initiation of spatial pattern for the general system (14.10). To formulate the problem mathematically we require boundary and initial conditions. These we take to be zero flux boundary conditions and given initial conditions. The mathematical problem is then defined by

$$u_t = \gamma f(u, v) + \nabla^2 u, \quad v_t = \gamma g(u, v) + d\nabla^2 v \ ,$$

$$(\mathbf{n} \cdot \nabla) \begin{pmatrix} u \\ v \end{pmatrix} = 0, \quad \mathbf{r} \text{ on } \partial B; \quad u(\mathbf{r}, 0), \ v(\mathbf{r}, 0) \text{ given} \ , \tag{14.11}$$

where ∂B is the closed boundary of the reaction diffusion domain B and $\mathbf{n}$ is the unit outward normal to ∂B. There are several reasons for choosing zero flux boundary conditions. The major one is that we are interested in self-organisation of pattern; zero flux conditions imply no external input. If we imposed fixed boundary conditions on u and v the spatial patterning could be a direct consequence of the boundary conditions as we shall see in the ecological problem below in Section 14.7. Another biologically reasonable set of conditions is periodicity: these correspond to closed three-dimensional boundaries. In Section 14.4

we carry out the analysis for a specific one- and two-dimensional situation with the kinetics given by (14.7).

The relevant homogeneous steady state (u_0, v_0) of (14.11) is the positive solution of

$$f(u, v) = 0, \quad g(u, v) = 0 . \tag{14.12}$$

Since we are concerned with *diffusion driven* instability we are interested in linear instability of this steady state that is solely *spatially* dependent. So, in the absence of any spatial variation the homogeneous steady state must be linearly stable: we first determine the conditions for this to hold. These were derived in Chapter 3 but as a reminder and for notational completeness we briefly rederive them here.

With no spatial variation u and v satisfy

$$u_t = \gamma f(u, v), \quad v_t = \gamma g(u, v) . \tag{14.13}$$

Linearising about the steady state (u_0, v_0) in exactly the same way as we did in Chapter 3, we set

$$\mathbf{w} = \begin{pmatrix} u - u_0 \\ v - v_0 \end{pmatrix} \tag{14.14}$$

and (14.13) becomes, for $|\mathbf{w}|$ small,

$$\mathbf{w}_t = \gamma A \mathbf{w}, \quad A = \begin{pmatrix} f_u & f_v \\ g_u & g_v \end{pmatrix}_{u_0, v_0} , \tag{14.15}$$

where A is the stability, or community, matrix. From now on we shall take the partial derivatives of f and g to be evaluated at the steady state unless stated otherwise. We now look for solutions in the form

$$\mathbf{w} \propto e^{\lambda t} \tag{14.16}$$

where λ is the eigenvalue. The steady state $\mathbf{w} = 0$ is linearly stable if $\operatorname{Re} \lambda < 0$ since in this case the perturbation $\mathbf{w} \to 0$ as $t \to \infty$. Substitution of (14.16) into (14.15) determines the eigenvalues λ as the solutions of

$$|\gamma A - \lambda I| = \begin{vmatrix} \gamma f_u - \lambda & \gamma f_v \\ \gamma g_u & \gamma g_v - \lambda \end{vmatrix} = 0$$

$$\Rightarrow \quad \lambda^2 - \gamma(f_u + g_v)\lambda + \gamma^2(f_u g_v - f_v g_u) = 0 , \tag{14.17}$$

so

$$\lambda_1, \lambda_2 = \frac{1}{2}\gamma \left[(f_u + g_v) \pm \{(f_u + g_v)^2 - 4(f_u g_v - f_v g_u)\}^{1/2} \right] . \tag{14.18}$$

Linear stability, that is $\operatorname{Re} \lambda < 0$, is guaranteed if

$$\operatorname{tr} A = f_u + g_v < 0, \quad |A| = f_u g_v - f_v g_u > 0 . \tag{14.19}$$

Since (u_0, v_0) are functions of the parameters of the kinetics, these inequalities thus impose certain constraints on the parameters. Note that for all cases in Fig. 14.2 in the neighbourhood of the steady state, $f_u > 0$, $g_v < 0$, and for Fig. 14.2 (a) $f_v > 0$, $g_u < 0$ while for Fig. 14.2 (b)–(d) $f_v < 0$, $g_u > 0$. So tr A and $|A|$ could be positive or negative: here we are only concerned with the conditions and parameter ranges which satisfy (14.19).

Now consider the full reaction diffusion system (14.11) and again linearise about the steady state, which with (14.14) is $\mathbf{w} = 0$, to get

$$\mathbf{w}_t = \gamma A\mathbf{w} + D\nabla^2\mathbf{w}, \quad D = \begin{pmatrix} 1 & 0 \\ 0 & d \end{pmatrix} . \tag{14.20}$$

To solve this system of equations subject to the boundary conditions (14.11) we first define $\mathbf{W}(\mathbf{r})$ to be the time independent solution of the spatial eigenvalue problem defined by

$$\nabla^2\mathbf{W} + k^2\mathbf{W} = 0, \quad (\mathbf{n} \cdot \nabla)\mathbf{W} = 0 \quad \text{for} \quad \mathbf{r} \text{ on } \partial B , \tag{14.21}$$

where k is the eigenvalue. For example, if the domain is one-dimensional, say $0 \le x \le a$, $\mathbf{W} \propto \cos(n\pi x/a)$ where n is an integer: this satisfies zero flux conditions at $x = 0$ and $x = a$. The eigenvalue in this case is $k = n\pi/a$. So $1/k = a/n\pi$ is a measure of the wave-like pattern: the eigenvalue k is called the *wavenumber* and $1/k$ is proportional to the wavelength ω_c; $\omega_c = 2\pi/k = 2a/n$ in this example. From now on we shall refer to k in this context as the wavenumber. With finite domains there is a discrete set of possible wavenumbers.

Let $\mathbf{W}_k(\mathbf{r})$ be the eigenfunction corresponding to the wavenumber k. Each eigenfunction $\mathbf{W}_k$ satisfies zero flux boundary conditions. Because the problem is linear we now look for solutions $\mathbf{w}(\mathbf{r}, t)$ of (14.20) in the form

$$\mathbf{w}(\mathbf{r}, t) = \sum_k c_k e^{\lambda t}\mathbf{W}_k(\mathbf{r}) . \tag{14.22}$$

where the constants c_k are determined by a Fourier expansion of the initial conditions in terms of $\mathbf{W}_k(\mathbf{r})$. λ is the eigenvalue which determines temporal growth. Substituting this form into (14.20) with (14.21) and cancelling $e^{\lambda t}$, we get, for each k,

$$\lambda\mathbf{W}_k = \gamma A\mathbf{W}_k + D\nabla^2\mathbf{W}_k$$
$$= \gamma A\mathbf{W}_k - Dk^2\mathbf{W}_k .$$

We require nontrivial solutions for $\mathbf{W}_k$ so the λ are determined by the roots of the characteristic polynomial

$$|\lambda I - \gamma A + Dk^2| = 0 .$$

Evaluating the determinant with A and D from (14.15) and (14.20) we get the eigenvalues $\lambda(k)$ as functions of the wavenumber k as the roots of

$$\lambda^2 + \lambda[k^2(1+d) - \gamma(f_u + g_v)] + h(k^2) = 0 \ ,$$
$$h(k^2) = dk^4 - \gamma(df_u + g_v)k^2 + \gamma^2|A| \ . \tag{14.23}$$

The steady state (u_0, v_0) is linearly stable if both solutions of (14.23) have $\mathrm{Re}\,\lambda < 0$. We have already imposed the constraints that the steady state is stable in the absence of any spatial effects, that is $\mathrm{Re}\,\lambda(k^2 = 0) < 0$. The quadratic (14.23) in this case is (14.17) and the requirement that $\mathrm{Re}\,\lambda < 0$ gave conditions (14.19). For the steady state to be unstable to *spatial* disturbances we require $\mathrm{Re}\,\lambda(k) > 0$ for some $k \neq 0$. This can happen if either the coefficient of λ in (14.23) is negative, or if $h(k^2) < 0$ for some $k \neq 0$. Since $(f_u + g_v) < 0$ from conditions (14.19) and $k^2(1+d) > 0$ for all $k \neq 0$ the coefficient of λ, namely

$$[-\gamma(f_u + g_v) + k^2(1+d)] > 0 \ ,$$

so the only way $\mathrm{Re}\,\lambda(k^2)$ can be positive is if $h(k^2) < 0$ for some k. This is immediately clear from the solutions of (14.23), namely

$$2\lambda = [k^2(1+d) - \gamma(f_u + g_v)] \pm \{[k^2(1+d) - \gamma(f_u + g_v)]^2 - 4h(k^2)\}^{1/2} \ .$$

Since we required the determinant $|A| > 0$ from (14.19) the only possibility for $h(k^2)$ in (14.23) to be negative is if $(df_u + g_v) > 0$. Since $(f_u + g_v) < 0$ from (14.19) this implies that $d \neq 1$ and f_u and g_v must have opposite signs. So, a further requirement to those in (14.19) is

$$df_u + g_v > 0 \quad \Rightarrow \quad d \neq 1 \ . \tag{14.24}$$

With the reaction kinetics giving the null clines in Fig. 14.2 we noted that $f_u > 0$ and $g_v < 0$, so the first condition in (14.19) and the last inequality require that the diffusion coefficient ratio $d > 1$. For example, in terms of the activator-inhibitor mechanism (14.8) this means that the inhibitor must diffuse faster than the activator as we noted above.

The last inequality is necessary but not sufficient for $\mathrm{Re}\,\lambda > 0$. For $h(k^2)$ to be negative for some non-zero k, the minimum $h_{\min}$ must be negative. From (14.23), elementary differentiation with respect to k^2 shows that

$$h_{\min} = \gamma^2 \left[|A| - \frac{(df_u + g_v)^2}{4d}\right], \quad k^2 = k_m^2 = \gamma \frac{df_u + g_v}{2d} \ . \tag{14.25}$$

Thus the condition that $h(k^2) < 0$ for some $k^2 \neq 0$ is

$$\frac{(df_u + g_v)^2}{4d} > |A| \ . \tag{14.26}$$

At bifurcation, when $h_{\min} = 0$, we require $|A| = (df_u + g_v)^2/4d$ and so for fixed kinetics parameters this defines a critical diffusion coefficient ratio $d_c(> 1)$ as the appropriate root of

$$d_c^2 f_u^2 + 2(2f_v g_u - f_u g_v)d_c + g_v^2 = 0 \ . \tag{14.27}$$

The critical wavenumber k_c is then given by

$$k_c^2 = \gamma \frac{d_c f_u + g_v}{2d_c} = \gamma \left[\frac{|A|}{d_c}\right]^{1/2} = \gamma \left[\frac{f_u g_v - f_v g_u}{d_c}\right]^{1/2} . \tag{14.28}$$

Fig. 14.4 (a) shows how $h(k^2)$ varies as a function of k^2 for various d.

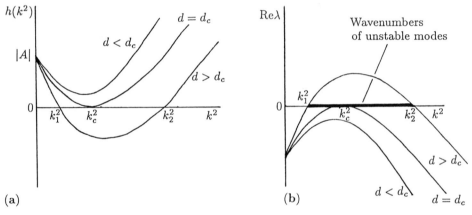

(a)   (b)

Fig. 14.4a,b. (a) Plot of $h(k^2)$ defined by (14.23) for typical kinetics illustrated in Fig. 14.2. When the diffusion coefficient ratio d increases beyond the critical value d_c, $h(k^2)$ becomes negative for a finite range of $k^2 > 0$. (b) Plot of the largest of the eigenvalues $\lambda(k^2)$ from (14.23) as a function of k^2. When $d > d_c$ there is a range of wavenumbers $k_1^2 < k^2 < k_2^2$ which are linearly unstable.

Whenever $h(k^2) < 0$, (14.23) has a solution λ which is positive for the same range of wavenumbers that make $h < 0$. From (14.23) with $d > d_c$ the range of unstable wavenumbers $k_1^2 < k^2 < k_2^2$ is obtained from the zeros k_1^2 and k_2^2 of $h(k^2) = 0$ as

$$k_1^2 = \gamma \frac{(df_u + g_v) - \{(df_u + g_v)^2 - 4d|A|\}^{1/2}}{2d} < k^2$$

$$< \gamma \frac{(df_u + g_v) + \{(df_u + g_v)^2 - 4d|A|\}^{1/2}}{2d} = k_2^2 . \tag{14.29}$$

Fig. 14.4 (b) plots a typical $\lambda(k^2)$ against k^2. The expression $\lambda = \lambda(k^2)$ is called a *dispersion relation*. We discuss the importance and use of dispersion relations in more detail in the next two sections. Note that, within the unstable range, $\text{Re}\,\lambda(k^2) > 0$ has a maximum for the wavenumber k_m obtained from (14.25) with $d > d_c$. This implies that there is a fastest growing mode in the summation (14.22) for $\mathbf{w}$; this is an attribute we shall now exploit.

If we consider the solution $\mathbf{w}$ given by (14.22), the dominant contributions as t increases are those modes for which $\text{Re}\,\lambda(k^2) > 0$ since all other modes tend to zero exponentially. From Fig. 14.4, or analytically from (14.29), we determine the range, $k_1^2 < k^2 < k_2^2$ where $h(k^2) < 0$, and hence $\text{Re}\,\lambda(k^2) > 0$, and so from (14.22)

$$\mathbf{w}(\mathbf{r}, t) \sim \sum_{k_1}^{k_2} c_k \exp\left[\lambda(k^2)t\right] \mathbf{W}_k(\mathbf{r}) \quad \text{for large } t \,. \tag{14.30}$$

An analysis and graph of the dispersion relation are thus extremely informative in that they immediately say which eigenfunctions, that is which spatial patterns, are linearly unstable and grow exponentially with time. We must keep in mind that, with finite domain eigenvalue problems, the wavenumbers are discrete and so only certain k in the range (14.29) are of relevance: we discuss the implications of this later.

The key assumption, and what in fact happens, is that these linearly unstable eigenfunctions in (14.30) which are growing exponentially with time will eventually be bounded by the nonlinear terms in the reaction diffusion system of equations and an ultimate steady state spatially inhomogeneous solution will emerge. A key element in this assumption is the existence of a confined set for the kinetics (see Chapter 3). We would intuitively expect that if a confined set exists for the kinetics, the same set would also contain the solutions when diffusion is included. This is indeed the case and can be rigorously proved; see Smoller (1983). So, part of the analysis of a specific mechanism involves the demonstration of a confined set within the *positive* quadrant. A general nonlinear analysis for the evolution to the finite amplitude steady state spatial patterns is still lacking but singular perturbation analyses for d near the bifurcation value d_c have been carried out and a non-uniform spatially heterogeneous solution is indeed obtained (see, for example, Lara and Murray 1983; an early nonlinear analysis of a predator-prey system was given by Segel and Levin 1976). There have been many spatially inhomogeneous solutions evaluated numerically using a variety of specific reaction diffusion mechanisms. The results presented in the next chapter illustrate some of the richness of pattern which can be generated.

To recap, we have now obtained conditions for the generation of spatial patterns by *two*-species reaction diffusion mechanisms of the form (14.11). For convenience we reproduce them here. Remembering that all derivatives are evaluated at the steady state (u_0, v_0), they are, from (14.19), (14.24), (14.26),

$$\begin{aligned} f_u + g_v < 0, \quad f_u g_v - f_v g_u > 0 \,, \\ df_u + g_v > 0, \quad (df_u + g_v)^2 - 4d(f_u g_v - f_v g_u) > 0 \,. \end{aligned} \tag{14.31}$$

The derivatives f_u and g_v must be of opposite sign: with the reaction kinetics exhibited in Fig. 14.2, $f_u > 0$, $g_v < 0$ so the first and third of (14.31) imply that the ratio of diffusion coefficients $d > 1$. If the conditions (14.31) are satisfied there is a scale(γ)-dependent range of patterns, with wavenumbers defined by (14.29), which are linearly unstable. The spatial patterns which initially grow (exponentially) are those eigenfunctions $\mathbf{W}_k(\mathbf{r})$ with wavenumbers k_1 and k_2 determined by (14.29); namely those in (14.30). Note that the scale parameter γ plays a crucial role in these expressions, a point we consider further in the next section. Generally we would expect the kinetics and morphogen diffusion coefficients to be fixed. The only natural variable parameter is then γ which reflects the size of the embryo or rather the embryonic domain we are considering.

Diffusion Driven Instability in Infinite Domains: Continuous Spectrum of Eigenvalues

In a finite domain the possible wavenumbers k and corresponding spatial wavelengths of allowable patterns are discrete and depend in part on the boundary conditions. In developmental biology the size of the embryo during the period of spatial patterning is often sufficiently large, relative to the pattern to be formed, that the 'boundaries' cannot play a major role in isolating specific wavelengths, as for example in the generation of patterns of hair, scale and feather primordia (see Chapter 17). Thus, for practical purposes the pattern formation domain is effectively infinite. Here we describe how to determine the spectrum of unstable eigenvalues for an infinite domain – it is considerably easier than for a finite domain.

We start with the linearised system (14.20) and look for solutions in the form

$$\mathbf{w}(\mathbf{r}, t) \propto \exp\left[\lambda t + i\mathbf{k} \cdot \mathbf{r}\right] ,$$

where $\mathbf{k}$ is the wave vector with magnitude $k = |\mathbf{k}|$. Substitution into (14.20) again gives

$$|\lambda I - \gamma A + Dk^2| = 0$$

and so the dispersion relation giving λ in terms of the wavenumbers k is again given by (14.23). The range of eigenvalues for which $\text{Re}\,\lambda(k^2) > 0$ is again given by (14.29). The crucial difference between the situation here and that for a finite domain is that there is always a spatial pattern if, in (14.29), $0 < k_1^2 < k_2^2$ since we are not restricted to a discrete class of k^2 defined by the eigenvalue problem (14.21). So at bifurcation when k_c^2, given by (14.28), is linearly unstable the mechanism will evolve to a spatial pattern with the critical wavelength $\omega_c = 2\pi/k_c$. Thus the wavelength with the maximum exponential growth in Fig. 14.4(b) will be the pattern which generally emerges. In the next chapter on biological applications we shall see that the difference between a finite domain and an effectively infinite one has important biological implications: finite domains put considerable restrictions on the allowable patterns.

14.4 Detailed Analysis of Pattern Initiation in a Reaction Diffusion Mechanism

Here we consider a specific two-species reaction diffusion system and carry out the detailed analysis. We lay the groundwork in this section for the subsequent applications to real biological pattern formation problems. We calculate the eigenfunctions, obtain the specific conditions on the parameters necessary to initiate spatial patterns and determine the wavenumbers and wavelengths of the spatial disturbances which initially grow exponentially.

We study the simplest reaction diffusion mechanism (14.7), first in one space dimension, namely

$$u_t = \gamma f(u, v) + u_{xx} = \gamma(a - u + u^2 v) + u_{xx} ,$$
$$v_t = \gamma g(u, v) + d v_{xx} = \gamma(b - u^2 v) + d v_{xx} . \tag{14.32}$$

The kinetics null clines $f = 0$ and $g = 0$ are illustrated in Fig. 14.2 (a). The uniform positive steady state (u_0, v_0) is

$$u_0 = a + b, \quad v_0 = \frac{b}{(a + b)^2}, \quad b > 0, \quad a + b > 0 \tag{14.33}$$

and, at the steady state,

$$f_u = \frac{b - a}{a + b}, \quad f_v = (a + b)^2 > 0, \quad g_u = \frac{-2b}{a + b},$$
$$g_v = -(a + b)^2 < 0, \quad f_u g_v - f_v g_u = (a + b)^2 > 0 . \tag{14.34}$$

Since f_u and g_v must have opposite signs we must have $b > a$. With these expressions, conditions (14.31) require

$$f_u + g_v < 0 \quad \Rightarrow \quad 0 < b - a < (a + b)^3 ,$$
$$f_u g_v - f_v g_u > 0 \quad \Rightarrow \quad (a + b)^2 > 0 ,$$
$$d f_u + g_v > 0 \quad \Rightarrow \quad d(b - a) > (a + b)^3 , \tag{14.35}$$
$$(d f_u + g_v)^2 - 4d(f_u g_v - f_v g_u) > 0$$
$$\Rightarrow \quad [d(b - a) - (a + b)^3]^2 > 4d(a + b)^4 .$$

These inequalities define a domain in (a, b, d) parameter space, called the *Turing space*, within which the mechanism will be unstable to certain spatial disturbances of given wavenumbers k, which we now determine.

Consider the related eigenvalue problem (14.21) and let us choose the domain to be $x \in (0, p)$ with $p > 0$. We then have

$$\mathbf{W}_{xx} + k^2 \mathbf{W} = 0, \quad \mathbf{W}_x = 0 \text{ for } x = 0, p \tag{14.36}$$

the solutions of which are

$$\mathbf{W}_n(x) = \mathbf{A}_n \cos(n\pi x/p), \quad n = \pm 1, \pm 2, \ldots \qquad (14.37)$$

where the $\mathbf{A}_n$ are arbitrary constants. The eigenvalues are the *discrete* wavenumbers $k = n\pi/p$. Whenever (14.34) are satisfied and there is a range of wavenumbers $k = n\pi/p$ lying within the bound defined by (14.29), then the corresponding eigenfunctions $\mathbf{W}_n$ are linearly unstable. Thus the eigenfunctions (14.37) with wavelengths $\omega = 2\pi/k = 2p/n$ are the ones which initially grow with time like $\exp\{\lambda([n\pi/p]^2)t\}$. The band of wavenumbers from (14.29), with (14.34), is given by

$$\gamma L(a,b,d) = k_1^2 < k^2 = \left(\frac{n\pi}{p}\right)^2 < k_2^2 = \gamma M(a,b,d)$$

$$L = \frac{([d(b-a)-(a+b)^3] - \{[d(b-a)-(a+b)^3]^2 - 4d(a+b)^4\}^{1/2}}{2d(a+b)},$$

$$M = \frac{([d(b-a)-(a+b)^3] + \{[d(b-a)-(a+b)^3]^2 - 4d(a+b)^4\}^{1/2}}{2d(a+b)}.$$

$$(14.38)$$

In terms of the wavelength $\omega = 2\pi/k$, the range of unstable modes $\mathbf{W}_n$ have wavelengths bounded by ω_1 and ω_2 where

$$\frac{4\pi^2}{\gamma L(a,b,d)} = \omega_1^2 > \omega^2 = \left(\frac{2p}{n}\right)^2 > \omega_2^2 = \frac{4\pi^2}{\gamma M(a,b,d)}. \qquad (14.39)$$

Note in (14.38) the importance of scale, quantified by γ. The smallest wavenumber is π/p, that is $n = 1$. For fixed parameters a, b and d, if γ is sufficiently small (14.38) says that there is *no allowable* k in the range, and hence no mode $\mathbf{W}_n$ in (14.37), which can be driven unstable. This means that all modes in the solution $\mathbf{w}$ in (14.30) tend to zero exponentially and the steady state is stable. We discuss this important role of scale in more detail below.

From (14.30) the spatially heterogeneous solution which emerges is the sum of the unstable modes, namely

$$\mathbf{w}(x,t) \sim \sum_{n_1}^{n_2} \mathbf{C}_n \exp\left[\lambda\left(\frac{n^2\pi^2}{p^2}\right)t\right] \cos\frac{n\pi x}{p} \qquad (14.40)$$

where λ is given by the positive solution of the quadratic (14.23) with the derivatives from (14.34), n_1 is the smallest integer greater than or equal to pk_1/π, n_2 the largest integer less than or equal to pk_2/π and $\mathbf{C}_n$ are constants which are determined by a Fourier series analysis of the initial conditions for $\mathbf{w}$. Initial conditions in any biological context involve a certain stochasticity and so it is inevitable that the Fourier spectrum will contain the whole range of Fourier modes, that is the $\mathbf{C}_n$ are non-zero. We can therefore assume at this stage that

γ is sufficiently large to ensure that allowable wavenumbers exist in the unstable range of k. Before discussing the possible patterns which emerge let us first obtain the corresponding two-dimensional result.

Consider the two-dimensional domain defined by $0 < x < p$, $0 < y < q$ whose rectangular boundary we denote by ∂B. The spatial eigenvalue problem in place of that in (14.36) is now

$$\nabla^2 \mathbf{W} + k^2 \mathbf{W} = 0, \quad (\mathbf{n} \cdot \nabla)\mathbf{W} = 0 \quad \text{for} \quad (x, y) \text{ on } \partial B \qquad (14.41)$$

the eigenfunctions of which are

$$\mathbf{W}_{p,q}(x, y) = \mathbf{C}_{n,m} \cos \frac{n\pi x}{p} \cos \frac{m\pi y}{q}, \quad k^2 = \pi^2 \left(\frac{n^2}{p^2} + \frac{m^2}{q^2} \right), \qquad (14.42)$$

where n and m are integers. The two-dimensional modes $\mathbf{W}_k(x, y)$ which are linearly unstable are those with wavenumbers k, defined by the last equation, lying within the unstable band of wavenumbers defined in terms of a, b and d by (14.38). We again assume that γ is sufficiently large so that the range of unstable wavenumbers contains at least one possible mode. Now the unstable spatially patterned solution is given by (14.30) with (14.42) as

$$\mathbf{w}(x, y, t) \sim \sum_{n,m} \mathbf{C}_{n,m} \exp\left[\lambda(k^2)t\right] \cos \frac{n\pi x}{p} \cos \frac{m\pi y}{q},$$

$$\gamma L(a, b, d) = k_1^2 < k^2 = \pi^2 \left(\frac{n^2}{p^2} + \frac{m^2}{q^2} \right) < k_2^2 = \gamma M(a, b, d),$$

(14.43)

where the summation is over all pairs (n, m) which satisfy the inequality, L and M are defined by (14.38) as before and $\lambda(k^2)$ is again the positive solution of (14.23) with the expressions for the derivatives of f and g given by (14.34). As t increases a spatial pattern evolves which is initially made up of the modes in (14.43).

Now consider the type of spatial patterns we might expect from the unstable solutions in (14.40) and (14.43). Suppose first that the domain size, as measured by γ, is such that the range of unstable wavenumbers in (14.38) admits only the wavenumber $n = 1$: the corresponding dispersion relation for λ in terms of the wavelengths $\omega = 2p/n$, is illustrated in Fig. 14.5 (a) below. The only unstable mode, from (14.37) is then $\cos(\pi x/p)$ and the growing instability is given by (14.40) as

$$\mathbf{w}(x, t) \sim \mathbf{C}_1 \exp\left[\lambda\left(\frac{\pi^2}{p^2}\right)t\right] \cos \frac{\pi x}{p},$$

where λ is the positive root of the quadratic (14.23) with f_u, f_v, g_u and g_v from (14.34) and with $k^2 = (\pi/p)^2$. Here all other modes decay exponentially with time. We can only determine the $\mathbf{C}_1$ from initial conditions. To get an intuitive understanding for what is going on, let us simply take $\mathbf{C}_1$ as $(\varepsilon, \varepsilon)$ for some small positive ε and consider the morphogen u; that is from the last equation and the definition of $\mathbf{w}$ from (14.14),

$$u(x,t) \sim u_0 + \varepsilon \exp\left[\lambda\left(\frac{\pi^2}{p^2}\right)t\right]\cos\frac{\pi x}{p}. \qquad (14.44)$$

This unstable mode, which is the dominant solution which emerges as t increases, is illustrated in Fig. 14.5 (b). In other words, this is the pattern predicted by the dispersion relation in Fig. 14.5 (a).

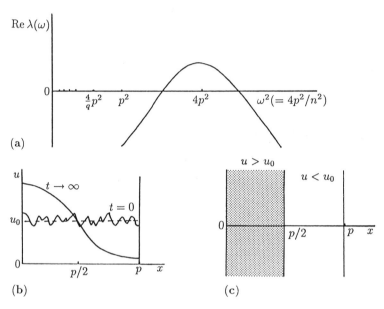

(a)

(b)   (c)

Fig. 14.5a-c. (a) Typical dispersion relation for the growth factor $\operatorname{Re}\lambda$ as a function of the wavelength ω obtained from a linearisation about the steady state. The only mode which is linearly unstable has $n = 1$; all other modes have $\operatorname{Re}\lambda < 0$. (b) The temporally growing linear mode which eventually evolves from random initial conditions into a finite amplitude spatial pattern such as shown in (c), where the shaded area corresponds to a concentration higher than the steady state u_0 and the unshaded area to a concentration lower than the steady state value.

Clearly if the exponentially growing solution was valid for all time it would imply $u \to \infty$ as $t \to \infty$. For the mechanism (14.32) the kinetics has a confined set, within the positive quadrant, which bounds the solution. So the solution in the last equation must be bounded and lie in the positive quadrant. We hypothesise that this growing solution eventually settles down to a spatial pattern which is similar to the single cosine mode shown in Fig. 14.5 (b). As mentioned before, singular perturbation analyses in the vicinity of the bifurcation in one of the parameters, for example near the critical domain size for γ such that a single wavenumber is just unstable, or when the critical diffusion coefficient ratio is near d_c, bear this out as do the many numerical simulations of the full nonlinear equations. Fig. 14.5 (c) is a useful way of presenting spatial patterned results for reaction diffusion mechanisms – the shaded region represents a concentration above the steady state value while the unshaded region represents concentrations below the steady state value. As we shall see, this simple way of presenting the results is very useful in the application of chemical prepattern theory to patterning problems in developmental biology, where it is postulated that cells differentiate when one of the morphogen concentrations is above (or below) some threshold level.

Let us now suppose that the domain size is doubled, say. With the definition of γ chosen to represent scale this is equivalent to multiplying the original γ by 4 since in the one-dimensional situation $\sqrt{\gamma}$ is proportional to size, that is length, of the domain. This means that the dispersion relation and the unstable range are simply moved along the k^2-axis or along the ω^2-axis. Suppose the original $\gamma = \gamma_1$. The inequalities (14.38) determine the unstable modes as those with wavelengths $\omega(= 2\pi/k)$ determined by (14.39), namely

$$\frac{4\pi^2}{\gamma_1 L(a, b, d)} > \omega^2 > \frac{4\pi^2}{\gamma_1 M L(a, b, d)} . \tag{14.45}$$

Let this be the case illustrated in Fig. 14.5 (a) and which gave rise to the pattern in Fig. 14.5 (c). Now let the domain double in size. We consider exactly the same domain as in Fig. 14.5 but with an increased γ to $4\gamma_1$. This is equivalent to having the same γ_1 but with a domain 4 times that in Fig. 14.5. We choose the former means of representing a change in scale. The equivalent dispersion relation is now illustrated in Fig. 14.6 (a) – it is just the original one of Fig. 14.5 (a) moved along so that the wavelength of the excited or unstable mode now has $\omega = p$, that is $n = 2$. The equivalent spatial pattern is then as in Fig. 14.6 (b). As we shall see in the applications chapter which follows, it is a particularly convenient way, when presenting spatial patterned solutions, to incorporate scale solely via a change in γ.

We can thus see with this example how the patterning process works as regards domain size. There is a basic wavelength picked out by the analysis for a given $\gamma = \gamma_1$, in this example that with $n = 1$. As the domain grows it eventually can incorporate the pattern with $n = 2$ and progressively higher modes the larger the domain, as shown in Fig. 14.6 (c). In the same way if the domain is sufficiently small there is clearly a $\gamma = \gamma_c$ such that the dispersion relation, now moved to the right in Fig. 14.6 (a), will not even admit the wavelength with $n = 1$. In this case no mode is unstable and so no spatial pattern can be generated. The concept of a critical domain size for the existence of spatial pattern is an important one both in developmental biology, and in spatially dependent ecological models as we shall see later.

Note in Fig. 14.6 (b) the two possible solutions for the same parameters and zero flux boundary conditions. Which of these is obtained depends on the bias in the inital conditions. Their existence poses certain conceptual difficulties

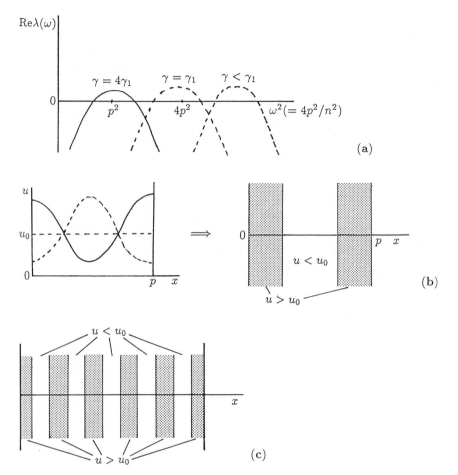

Fig. 14.6a-c. (a) Dispersion relation $\operatorname{Re}\lambda$ as a function of the wavelength ω when the single mode with $n = 2$ is unstable for a domain size $4\gamma_1$; the dashed curves are those with $\gamma = \gamma_1$ and $\gamma < \gamma_c < \gamma_1$ where γ_c is the scale value of the domain that will not admit any heterogeneous pattern. (b) The spatial pattern in the morphogen u predicted by the dispersion relation in (a). The dashed line, the mirror image about $u = u_0$ is also an allowable form of this solution. The initial conditions determine which pattern is obtained. (c) The spatial pattern obtained when the domain is sufficiently large to fit in the number of unstable modes equivalent to $n = 10$: the shaded regions represent morphogen levels $u > u_0$, the uniform steady state.

from a developmental biology point of view within the context of positional information. If cells differentiate when the morphogen concentration is larger than some threshold then the differentiated cell pattern is obviously different for each of the two possible solutions. Development, however, is a sequential process and so a previous stage generally cues the next. In the context of reaction diffusion models this implies a bias in the initial conditions towards one of the patterns.

Now consider the two-dimensional problem with a dispersion relation such that the unstable modes are given by (14.43). Here the situation is not so

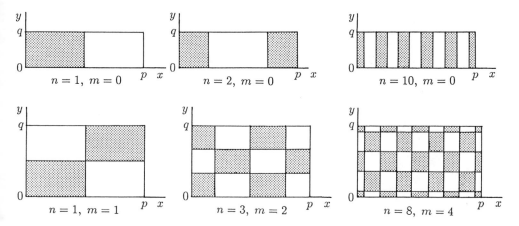

Fig. 14.7. Typical two-dimensional spatial patterns indicated by the linearly unstable solution (14.43) when various wavenumbers are in the unstable range. The shaded regions are where $u > u_0$, the uniform steady state.

straightforward since for a given γ, representing the *scale*, the actual modes which are unstable now depend on the domain *geometry* as measured by the length p and the width q. Referring to (14.43), first note that if the width is sufficiently small, that is q is small enough, even the first mode with $m = 1$ lies outside the unstable range. The problem is then equivalent to the above one-dimensional situation. As the width increases, that is q increases, genuine two-dimensional modes with $n \neq 0$ and $m \neq 0$ become unstable since $\pi^2(n^2/p^2 + m^2/q^2)$ lies in the range of unstable wavenumbers. Fig. 14.7 illustrates typical temporally growing spatial patterns indicated by (14.43) with various non-zero n and m.

Regular Planar Tesselation Patterns

The linear patterns illustrated in the last figure arise from the simplest two-dimensional eigenfunctions of (14.41). Less simple domains require the solutions of

$$\nabla^2 \psi + k^2 \psi = 0, \quad (\mathbf{n} \cdot \nabla)\psi = 0 \quad \text{for} \quad \mathbf{r} \text{ on } \partial B . \qquad (14.46)$$

Except for simple geometries the analysis quickly becomes quite complicated. Even for circular domains the eigenvalues have to be determined numerically. Surprisingly, there are some elementary solutions for symmetric domains which tessellate the plane, namely squares, hexagons, rhombi and, by subdivision, triangles: these were found by Christopherson (1940). In other words we can cover the complete plane with, for example, regular hexagonal tiles. [The basic symmetry group of regular polygons are hexagons, squares and rhombi, with, of course, triangles, which are subunits of these.] Thus we want solutions ψ where the unit cell, with zero flux conditions on its boundary, is one of the regular tesselations which can cover the plane. That is we want solutions which are *cell* periodic: the

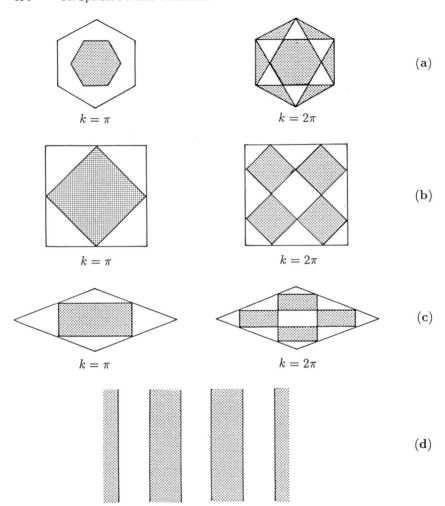

Fig. 14.8a-d. (a) Patterns which are obtained with the solution (14.47) with $k = \pi$ and $k = 2\pi$. The shaded region is where $\psi > 0$ and the unshaded region where $\psi < 0$. (b) Patterns generated by the solution (14.48) for a square tesselation with $k = \pi$ and $k = 2\pi$. (c) Rhombic patterns from (14.49) with $k = \pi$ and $k = 2\pi$. (d) One-dimensional roll patterns from (14.50).

word 'cell' is, of course, here meant as the unit of tesselation.

The solution of (14.46) for a hexagon is

$$
\begin{aligned}
\psi(x,y) &= \frac{\cos k \left(\frac{\sqrt{3}y}{2} + \frac{x}{2} \right) + \cos k \left(\frac{\sqrt{3}y}{2} - \frac{x}{2} \right) + \cos kx}{3} \\
&= \frac{\cos \left\{ kr \sin \left(\theta + \frac{\pi}{6} \right) \right\} + \cos \left\{ kr \sin \left(\theta - \frac{\pi}{6} \right) \right\} + \cos \left\{ kr \sin \left(\theta - \frac{\pi}{2} \right) \right\}}{3} .
\end{aligned}
$$

$$(14.47)$$

From (14.46), ψ is independent to the extent of multiplication by an arbitrary constant: the form chosen here makes $\psi = 1$ at the origin. This solution satisfies zero flux boundary conditions on the hexagonal symmetry boundaries if $k = n\pi$, $n = \pm 1, \pm 2, \ldots$ Fig. 14.8 (a) shows the type of pattern the solution can generate.

The polar coordinate form shows the invariance to hexagonal rotation, that is invariance to rotation by $\pi/3$, as it must. That is

$$\psi(r, \theta) = \psi(r, \theta + \frac{\pi}{3}) = H\psi(r, \theta) = \psi(r, \theta) \,,$$

where H is the hexagonal rotation operator.

The solution for the square is

$$
\begin{aligned}
\psi(x, y) &= \frac{\cos kx + \cos ky}{2} \\
&= \frac{\cos (kr \cos \theta) + \cos (kr \sin \theta)}{2} \,,
\end{aligned}
\tag{14.48}
$$

where $k = \pm 1, \pm 2 \ldots$ and $\psi(0,0) = 1$. This solution is square rotationally invariant since

$$\psi(r, \theta) = \psi(r, \theta + \frac{\pi}{2}) = S\psi(r, \theta) = \psi(r, \theta) \,,$$

where S is the square rotational operator. Typical patterns are illustrated in Fig. 14.8 (b).

The solution for the rhombus is

$$
\begin{aligned}
\psi(x, y) &= \frac{\cos kx + \cos\{k(x \cos \phi + y \sin \phi)\}}{2} \\
&= \frac{\cos \{kr \cos \theta\} + \cos \{k(r \cos(\theta - \phi)\}}{2} \,,
\end{aligned}
\tag{14.49}
$$

where ϕ is the rhombus angle and again $k = \pm 1, \pm 2, \ldots$ This solution is invariant under a rhombic rotation, that is

$$\psi(r, \theta; \phi) = \psi(r, \theta + \pi; \phi) = R\psi(r, \theta; \phi) \,,$$

where R is the rhombic rotation operator. Illustrative patterns are shown in Fig. 14.8 (c).

A further cell periodic solution is the one-dimensional version of the square; that is there is only variation in x. The solutions here are of the form

$$\psi(x, y) = \cos kx, \quad k = n\pi, \quad n = \pm 1, \pm 2, \ldots \tag{14.50}$$

and represent rolls with patterns as in Fig. 14.8 (d). These, of course, are simply the one-dimensional solutions (14.37).

When the full nonlinear equations are solved numerically with initial conditions taken to be small random perturbations about the steady state, linear theory turns out to be a good predictor of the ultimate steady state in the one-dimensional situation, particularly if the unstable modes have large wavelengths, that is small wavenumbers. With larger wavenumbers the predictions are less reliable – and even more so with two-dimensional structures. Since the equations we have studied are linear and invariant when multiplied by a constant, we can have equivalent solutions which are simply mirror images in the line $u = u_0$: refer to Fig. 14.6 (b). Thus the pattern that evolves depends on the initial conditions and the final pattern tends to be the one closest to the initial conditions. There is, in a sense, a basin of attraction for the spatial patterns as regards the initial conditions. Once again near bifurcation situations singular perturbation analysis indicates nonlinear patterns closely related to the linear predictions. In general, however, away from the bifurcation boundaries linear predictions are much less reliable: see the computed patterns exhibited in the next chapter. Except for the simplest patterns, we should really use linear theory for two and three dimensions only as a guide to the wealth of patterns which can be generated by pattern formation mechanisms. Linear theory does however determine the parameter ranges for pattern generation.

The application of reaction diffusion pattern generation to specific developmental biology problems is usually within the context of a prepattern theory whereby cells differentiate according to the level of the morphogen concentration. If the spatial pattern is quite distinct, that is with relatively large gradients, less sensitive tuning is required of the cells in order to carry out their assigned roles than if the pattern variation or the concentration gradients are small. It is useful therefore to try to get a quantitative measure of spatial heterogeneity, which is meaningful biologically, so as to compare different mechanisms. Another biologically relevant method will be discussed in the next section.

Berding (1987) introduced a 'heterogeneity' function for the spatial patterns generated by reaction diffusion systems with zero flux boundary conditions. Suppose the general mechanism (14.10), in one space variable, is diffusionally unstable and the solutions evolve to the spatially inhomogeneous steady state solutions $U(x)$ and $V(x)$ as $t \to \infty$. With the definition of γ in (14.6) proportional to the square of the domain length we can measure domain size by γ and hence take x to be in $(0,1)$. Then (U, V) satisfy the dimensionless equations

$$U'' + \gamma f(U, V) = 0, \quad dV'' + \gamma g(U, V) = 0 ,$$
$$U'(0) = U'(1) = V'(0) = V'(1) = 0 . \tag{14.51}$$

The non-negative heterogeneity function is defined by

$$H = \int_0^1 (U'^2 + V'^2) \, dx \geq 0 , \tag{14.52}$$

which depends only on the parameters of the system and the domain scale γ. H is an 'energy function'. If we now integrate by parts, using the zero flux boundary conditions in (14.51),

$$H = - \int_0^1 (UU'' + VV'') \, dx$$

which, on using (14.51) for U'' and V'', becomes

$$H = \frac{\gamma}{d} \int_0^1 [dU f(U, V) + V g(U, V)] \, dx \ . \tag{14.53}$$

If there is no spatial patterning, U and V are simply the uniform steady state solutions of $f(U, V) = g(U, V) = 0$ and so $H = 0$, as also follows, of course, from the definition (14.52).

From (14.53) we see how the scale parameter and diffusion coefficient ratio appear in the definition of heterogeneity. For example, suppose the domain is such that it sustains a single wave for $\gamma = \gamma_1$, in dimensional terms a domain length $L = L_1$ say. If we then double the domain size to $2L_1$ we can fit in two waves and so, intuitively from (14.52), H must increase as there is more heterogeneity. Since $\gamma \propto L^2$, H from (14.53) is simply quadrupled. From an embryological point of view, for example, this means that as the embryo grows we expect more and more structure. An example of this increase in structure in a growing domain is illustrated in Fig. 14.15 below. Berding (1987) discusses particular applications and compares specific reaction diffusion mechanisms as regards their potential for heterogeneity.

14.5 Dispersion Relation, Turing Space, Scale and Geometry Effects in Pattern Formation in Morphogenetic Models

We first note some general properties about the dispersion relation and then exploit it further with the specific case we analysed in the last section. The formation of spatial patterns by any morphogenetic model is principally a nonlinear phenomenon. However, as we noted, a good indication of the patterns in one dimension can be obtained by a simple linear analysis. For spatial patterns to form, we saw that two conditions must hold simultaneously. First, the spatially uniform state must be stable to small perturbations, that is all $\lambda(k^2)$ in (14.22) have $\mathrm{Re}\, \lambda(k^2 = 0) < 0$, and second, only patterns of a certain spatial extent, that is patterns within a definite range of wavelengths k, can begin to grow, with $\mathrm{Re}\, \lambda(k^2 \neq 0) > 0$. These conditions are encapsulated in the dispersion relation in either the (λ, k^2) or (λ, ω^2) forms such as in Fig. 14.4 (b) and Fig. 14.5 (a). The latter, for example, also says that if the spatial pattern of the disturbances have k^2 large, that is, very small wavelength disturbances, the steady state is again linearly stable. A dispersion relation therefore immediately gives the initial rate of growth or decay of patterns of various sizes. Dispersion relations are obtained from the general evolution equations of the pattern formation mechanism. A general and non-technical biologically oriented discussion of pattern formation models is given by Oster and Murray (1989).

Since the solutions to the linear eigenfunction equations such as (14.36) are simply sines and cosines, the 'size' of various spatial patterns is measured by the wavelength of the trigonometric functions; for example $\cos\left(n\pi x/p\right)$ has a wavelength $\omega = 2p/n$. So, the search for growing spatial patterns comes down to seeing how many sine or cosine waves can 'fit' into a domain of a given size. The two-dimensional situation is similar, but with more flexibility as to how they fit together.

A very important use of the dispersion relation is that it shows immediately whether patterns can grow, and if so, what the size of the patterns are. The curves in Fig. 14.4 (b) and Fig. 14.5 (a) are the prototype – no frills or 'vanilla' – dispersion relation for generating spatial patterns. We shall see later that other forms are possible and imply different pattern formation phenomena. However, these are less common and little is known at the present time about the patterns which evolve from them. The mechanochemical models discussed in detail in Chapter 17 can in fact generate a surprisingly rich spectrum of dispersion relations (see Murray and Oster 1984) most of which cannot be generated by two or three-species reaction diffusion models.

The prototype dispersion relation has the two essential characteristics mentioned above: (i) the spatially featureless state ($k = 0$, $\omega = \infty$) is stable: that is, the growth rate of very large wavelength waves is negative, and (ii) there is a small band, or window, of wavelengths which can grow (that is, a finite band of unstable 'modes', $\cos\left(n\pi x/L\right)$, for a finite number of integers n). Of these growing modes, one grows fastest; the one closest to the peak of the dispersion curve. This mode, k_m say, is the solution of

$$\frac{\partial \lambda}{\partial k^2} = 0 \quad \Rightarrow \quad \max\left[\operatorname{Re}\lambda\right] = \operatorname{Re}\lambda(k_m^2) .$$

Strictly k_m may not be an allowable mode in a finite doamin situation. In this case it is the possible mode closest to the analytically determined k_m.

Thus the dispersion curve shows that while the spatially homogeneous state is stable, the system will amplify patterns of a particular spatial extent, should they be excited by random fluctuations, which are always present in a biological system, or by cues from earlier patterns in development. Generally, one of the model parameters is 'tuned' until the dispersion curve achieves the qualitative shape shown. For example, in Fig. 14.4 (b) if the diffusion ratio d is less than the critical d_c, $\operatorname{Re}\lambda < 0$ for all k^2. As d increases, the curve rises until $d = d_c$ after which it pushes its nose above the axis at some wavenumber k_c, that is wavelength $\omega_c = 2\pi/k_c$, whereupon a cosine wave of that wavelength can start to grow, assuming it is an allowable eigenfunction. This critical wavenumber is given by (14.28) and, with $d = d_c$ from (14.27), we thus have

$$\omega_c = \frac{2\pi}{k_c} = 2\pi \left\{\frac{d_c}{\gamma^2(f_u g_v - f_v g_u)}\right\}^{1/4} . \tag{14.54}$$

With the illustrative example (14.32) there are 4 dimensionless parameters: a and b, the kinetics parameters, d, the ratio of diffusion coefficients and γ, the scale parameter. We concentrated on how the dispersion relation varied with d and showed how a bifurcation value d_c existed when the homogeneous steady state became unstable, with the pattern 'size' determined by k_c or ω_c given by the last equation. It is very useful to know the parameter space, involving all the parameters, wherein pattern forms and how we move into this pattern forming domain by varying whatever parameter we choose, or indeed when we vary more than one parameter. Clearly the more parameters there are the more complicated is this corresponding parameter or Turing space. Let us now determine the parameter space for the model (14.32) by extending the parametric method we described in Chapter 6, Section 6.4 for determining the space in which oscillatory solutions were possible. The technique was developed and applied to several reaction diffusion models by Murray (1982).

The conditions on the parameters a, b and d for the mechanism (14.32) to generate spatial patterns, if the domain is sufficiently large, are given by (14.35) with γ coming into the picture via the possible unstable modes determined by (14.38). Even though the inequalities (14.35) are probably the simplest realistic set we could have in any reaction diffusion mechanism they are still algebraically quite messy to deal with. With other than extremely simple kinetics it is not possible to carry out a similar analysis analytically. So let us start with the representation of the steady state used in Section 6.4, namely (6.24), with u_0 as the nonnegative parametric variable; that is v_0 and b are given in terms of a and u_0 from (6.24), or (14.33) above, as

$$v_0 = \frac{u_0 - a}{u_0^2}, \quad b = u_0 - a . \tag{14.55}$$

The inequalities (14.35), which define the conditions on the parameters for spatial patterns to grow, involve, on using the last expressions,

$$f_u = -1 + 2u_0v_0 = 1 - \frac{2a}{u_0}, \quad f_v = u_0^2 ,$$

$$g_u = -2u_0v_0 = -\frac{2(u_0 - a)}{u_0}, \quad g_v = -u_0^2 . \tag{14.56}$$

We now express the conditions for diffusion-driven instability given by (14.31) as inequalities in terms of the parameter u_0: these define boundary curves for domains in parameter space. With the first,

$$f_u + g_v < 0 \quad \Rightarrow \quad 1 - \frac{2a}{u_0} - u_0^2 < 0$$

$$\Rightarrow \quad a > \frac{u_0(1 - u_0^2)}{2}, \quad b < u_0 - a = \frac{u_0(1 + u_0^2)}{2} , \tag{14.57}$$

$$\Rightarrow \quad b = \frac{u_0(1 + u_0^2)}{2}$$

as the boundary curve where, since we are interested in the boundary curve, the $b = u_0 - a$ comes from the steady state definition (14.55) and where we replace a by its expression from the inequality involving only u_0 and a. These define a domain parametrically in (a, b) space as we let u_0 take all positive values; if the inequality is replaced by an equality sign, (14.57) define the boundary curve parametrically. We now do this with each of the conditions in (14.31).

The second condition of (14.31), using (14.56), requires

$$f_u g_v - f_v g_u > 0 \quad \Rightarrow \quad u_0^2 > 0 , \tag{14.58}$$

which is automatically satisfied. The third condition, requires

$$df_u + g_v > 0$$
$$\Rightarrow \quad a < \frac{u_0(d - u_0^2)}{2d}, \quad b = u_0 - a = \frac{u_0(d + u_0^2)}{2d} , \tag{14.59}$$
$$\Rightarrow \quad b = \frac{u_0(d + u_0^2)}{2d}$$

as the boundary curve.

The fourth condition in (14.31) is a little more complicated. Here

$$(df_u + g_v)^2 - 4d(f_u g_v - f_v g_u) > 0$$
$$\Rightarrow \quad [u_0(d - u_0^2) - 2da]^2 - 4du_0^4 > 0$$
$$\Rightarrow \quad 4a^2 d^2 - 4adu_0(d - u_0^2) + [u_0^2(d - u_0^2)^2 - 4u_0^4 d] > 0$$

which, on factorizing the left hand side, implies

$$a < u_0 \frac{1 - \frac{2u_0}{\sqrt{d}} - \frac{u_0^2}{d}}{2} \quad \text{or} \quad a > u_0 \frac{1 + \frac{2u_0}{\sqrt{d}} - \frac{u_0^2}{d}}{2} .$$

Thus this inequality results in *two* boundary curves, namely

$$a = u_0 \frac{1 - \frac{2u_0}{\sqrt{d}} - \frac{u_0^2}{d}}{2}, \quad b = u_0 - a = u_0 \frac{1 + \frac{2u_0}{\sqrt{d}} + \frac{u_0^2}{d}}{2} ,$$
$$a = u_0 \frac{1 + \frac{2u_0}{\sqrt{d}} - \frac{u_0^2}{d}}{2}, \quad b = u_0 - a = u_0 \frac{1 - \frac{2u_0}{\sqrt{d}} + \frac{u_0^2}{d}}{2} . \tag{14.60}$$

The curves, and the enclosed domains, defined parametrically by (14.57)–(14.60), define the parameter space or *Turing space* (see Murray 1982), where the steady state can be diffusionally driven unstable and hence create spatial patterns. As we noted in Section 14.4 the first and third conditions in (14.35) require f_u and g_v to have opposite signs which require $b > a$ and hence $d > 1$.

It is now a straightforward plotting exercise to obtain the curves defined by (14.57)–(14.60); we simply let u_0 take on a range of positive values and calculate the corresponding a and b for a given d. In general, with inequalities (14.57)–(14.60), five curves are involved in defining the boundaries. Here, as is often the case, several are redundant in that they are covered by one of the others. For example, in the first of (14.60),

$$a < u_0 \frac{1 - \frac{2u_0}{\sqrt{d}} - \frac{u_0^2}{d}}{2} < u_0 \frac{1 - \frac{u_0^2}{d}}{2} \, ,$$

since we are considering $u_0 > 0$, so (14.59) is automatically satisfied if we satisfy the first condition in (14.60). Also, since $d > 1$,

$$u_0 \frac{1 - \frac{u_0^2}{d}}{2} > u_0 \frac{1 - u_0^2}{2}$$

so the curve defined by (14.57) lies below the curve defined by (14.59): the former is a lower limiting boundary curve, so a suitable domain is defined if we use the first of (14.60). Furthermore, since

$$u_0 \frac{1 - \frac{u_0^2}{d}}{2} < u_0 \frac{1 + \frac{2u_0}{\sqrt{d}} - \frac{u_0^2}{d}}{2}$$

there can be no domain satisfying (14.59) and the second curve in (14.60).

Finally, therefore, for this mechanism we need only two parametric curves, namely those defined by (14.57) and the first of (14.60), and the Turing space is determined by

$$a > u_0 \frac{1 - u_0^2}{2}, \quad b = u_0 \frac{1 + u_0^2}{2} \, ,$$

$$a < u_0 \frac{1 - \frac{2u_0}{\sqrt{d}} - \frac{u_0^2}{d}}{2}, \quad b = u_0 \frac{1 + \frac{2u_0}{\sqrt{d}} + \frac{u_0^2}{d}}{2} \, . \tag{14.61}$$

We know that when $d = 1$ there is no Turing space, that is there is no domain where spatial patterns can be generated. The curves defined by (14.61) with $d = 1$ contradict each other and hence no Turing space exists. Now let d take on values greater than 1. For a critical d, d_c say, a Turing space starts to grow for $d > d_c$. Specifically $d = d_c = 3 + 2\sqrt{2}$, calculated from (14.61) by determining the d such that both curves give $a = 0$ at $b = 1$ and at this value the two inequalities are no longer contradictory. The space is defined, in fact, by two surfaces in (a, b, d) space. Fig. 14.9 shows the cross-sectional regions in (a, b) parameter space where the mechanism (14.32) can generate spatial patterns.

Even if a and b, for a given $d > 1$, lie within the Turing space this does not guarantee that the mechanism will generate spatial patterns, because scale and geometry play a major role. Depending on the size of γ and the actual spatial domain in which the mechanism operates, the unstable eigenfunctions, or modes,

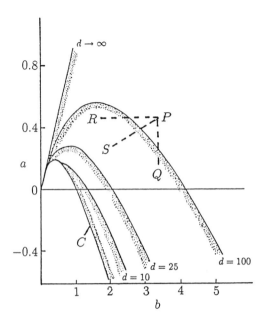

Fig. 14.9. Turing space for (14.32); that is the parameter space where spatial patterns can be generated by the reaction diffusion mechanism (14.32). For example, if $d = 25$ any values for a and b lying within the domain bounded by the curves marked C (that is $d = 1$) and $d = 25$ will result in diffusion driven instability. Spatial pattern will evolve if the domain (γ) is sufficiently large for allowable k^2, defined by (14.38) and (14.43).

may not be allowable solutions. It is here that the detailed form of the dispersion relation comes in again. To be specific let us consider the one-dimensional finite domain problem defined by (14.36). The eigenvalues, that is the wavenumbers, $k = n\pi/p$, $n = \pm 1, \pm 2 \ldots$ are *discrete*. So, referring to Fig. 14.4 (b), unless the dispersion relation includes in its range of unstable modes at least one of these discrete values no structure can develop. We must therefore superimpose on the Turing space in Fig. 14.9 another axis representing the scale parameter γ. If γ is included in the parameters of the Turing space it is not necessarily simply connected since, if the dispersion relation, as γ varies, does not include an allowable eigenfunction in its unstable modes, no pattern evolves. Let us consider this aspect and use of the dispersion relation in more detail.

The Turing space involves only dimensionless parameters which are appropriate groupings of the dimensional parameters of the model. The parameters a, b and d in the last figure are, from (14.6),

$$a = \frac{k_1}{k_2} \left(\frac{k_3}{k_2} \right)^{1/2} , \qquad b = \frac{k_4}{k_2} \left(\frac{k_3}{k_2} \right)^{1/2} , \qquad d = \frac{D_B}{D_A} .$$

Suppose, for example, $d = 100$ and a and b have values associated with P in Fig. 14.9, that is the mechanism is not in a pattern formation mode. There is no unique way to move into the pattern formation domain; we could decrease either a or b so that we arrive at Q or R respectively. In dimensional terms we can reduce a, for example, by appropriately changing k_1, k_2 or k_3 – or all of them. Varying other than k_1 will also affect b, so we have to keep track of b as well. If we only varied k_2 the path in the Turing space is qualitatively like that from P to

S. If d can vary, which means either of D_A or D_B can vary, we can envelope P in the pattern formation region by simply increasing d. Interpreting the results from a biological point of view, therefore, we see that it is the *orchestration of several effects which produce pattern*, not just one, since we can move into the pattern formation regime by varying one of several parameters. Clearly we can arrive at a specific point in the space by one of several paths. The concept of equivalent effects, via parameter variation, producing the same pattern is an important one in the interpretation and design of relevant experiments associated with any model. It is not a widely appreciated concept in biology. We shall discuss some important biological applications of the practical use of dimensionless groupings in subsequent chapters.

To recap briefly, the dispersion relation for the general reaction diffusion sytem (14.10) is given by the root $\lambda(k^2)$ of (14.23) with the larger real part. The key to the existence of unstable spatial modes is whether or not the function

$$h(k^2) = dk^4 - \gamma(df_u + g_v)k^2 + \gamma^2(f_u g_v - f_v g_u) \tag{14.62}$$

is negative for a range of $k^2 \neq 0$: see Fig. 14.4 (a). Remember that the f and g derivatives are evaluated at the steady state (u_0, v_0) where $f(u_0, v_0) = g(u_0, v_0) = 0$, so $h(k^2)$ is a quadratic in k^2 whose coefficients are functions only of the parameters of the kinetics, d the diffusion coefficient ratio and the scale parameter γ. The minimum, $h_{\min}$, at $k = k_m$ corresponds to the λ with the maximum Re λ and hence the mode with the largest growth factor $\exp[\lambda(k_m^2)t]$. From (14.25), or simply from the last equation, $h_{\min}$ is given by

$$h_{\min} = h(k_m^2) = -\gamma^2 \frac{df_u^2 + \frac{g_v^2}{d} - 2(f_u g_v - 2f_v g_u)}{4},$$
$$k_m^2 = \gamma \frac{df_u + g_v}{2d}. \tag{14.63}$$

The bifurcation between spatially stable and unstable modes is when $h_{\min} = 0$. When this holds there is a critical wave number k_c which, from (14.28) or again simply derived from (14.62), is when the parameters are such that

$$(df_u + g_v)^2 = 4d(f_u g_v - f_v g_u) \quad \Rightarrow \quad k_c^2 = \gamma \frac{df_u + g_v}{2d}. \tag{14.64}$$

As the parameters move around the Turing space we can achieve the required equality, the first of the equations in (14.64), by letting one or other of the parameters pass through its bifurcation values, all other parameters being kept fixed. In the last section, and in Fig. 14.4 (b) for example, we chose d as the parameter to vary and for given a and b we evaluated the bifurcation value d_c. In this situation, just at bifurcation, that is, when $h_{\min}(k_c^2) = 0$, a single spatial pattern with wavenumber k_c is driven unstable, or excited, for $d = d_c + \varepsilon$, where $0 < \varepsilon \ll 1$. This critical wavenumber from (14.64) is proportional to $\sqrt{\gamma}$ and so

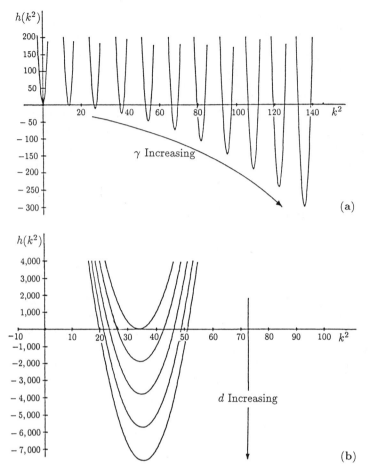

Fig. 14.10a,b. (a) Isolation of unstable modes (that is $h(k^2) < 0$ in (14.23)) by setting the diffusion ratio $d = d_c + \varepsilon$, $0 < \varepsilon \ll 1$ and varying the scale γ for the Thomas (1975) kinetics (14.8) with $a = 150$, $b = 100$, $\alpha = 1.5$, $\rho = 13$, $K = 0.05$, $d = 27.03$: the critical $d_c = 27.02$. (b) The effect of increasing d with all other parameters fixed as in (a). As $d \to \infty$ the range of unstable modes is bounded by $k^2 = 0$ and $k^2 = \gamma f_u$.

we can vary which spatial pattern is initiated by varying γ. This is called *mode selection* and is crucial in applications as we shall see later.

In the case of finite domains we can isolate a specific mode to be excited, or driven unstable, by choosing the width of the band of unstable wavenumbers to be narrow and centred round the desired mode. Let us take the parameters in the kinetics to be fixed and let $d = d_c + \varepsilon$, $0 < \varepsilon \ll 1$ we then get from (14.64) the appropriate γ for a specified k as

$$\gamma \approx \frac{2d_c k^2}{d_c f_u + g_v}, \qquad (14.65)$$

where the kinetics parameters at bifurcation, sometimes called the *marginal kinetics state*, satisfy the first of (14.64). So, by varying γ we can isolate whatever mode we wish to be excited. Fig. 14.10 (a) shows a typical situation. Arcuri and Murray (1986) have carried out an extensive Turing space analysis for the Thomas (1975) mechanism in such a case. Note in Fig. 14.10 (a) that as γ increases $h_{\min}$ becomes more negative, as is indicated by (14.63).

Suppose now we keep γ and the kinetics parameters fixed, and let d increase from its bifurcation value d_c. From (14.63) $h_{\min} \sim -(\gamma f_u)^2 d/4$ for d large and so $\lambda \to \infty$ with d. The width of the band of unstable modes has wavenumbers bounded by the zeros k_1 and k_2 of $h(k^2)$ in (14.62). These are given by (14.29), or immediately from (14.62) as

$$
\begin{aligned}
k_1^2 &= \gamma \frac{(df_u + g_v) - \{(df_u + g_v)^2 - 4d(f_u g_v - f_v g_u)\}^{1/2}}{2d}, \\
k_2^2 &= \gamma \frac{(df_u + g_v) + \{(df_u + g_v)^2 - 4d(f_u g_v - f_v g_u)\}^{1/2}}{2d},
\end{aligned}
\tag{14.66}
$$

from which we get

$$
k_1^2 \sim 0, \quad k_2^2 \sim \gamma f_u \quad \text{as} \quad d \to \infty.
\tag{14.67}
$$

So, for a fixed scale there is an upper limit for the excited mode wavenumber and hence a lower limit for the possible wavelengths of the spatial patterns. Fig. 14.10 (b) illustrates a typical case for the Thomas (1975) system given by (14.8).

With all kinetics parameters fixed, each parameter pair (d, γ) defines a unique parabola $h(k^2)$ in (14.62), which in turn specifies a set of unstable modes. We can thus consider the (d, γ) plane to be divided into regions where specific modes or a group of modes are diffusively unstable. When there are several unstable modes, because of the form of the dispersion relation, such as in Fig. 14.4 (b), there is clearly a mode with the largest growth rate since there is a maximum $\text{Re }\lambda$ for some k_m^2 say. From (14.23), the positive eigenvalue $\lambda_+(k^2)$ is given by

$$
2\lambda_+(k^2) = \gamma(f_u + g_v) - k^2(1 + d) + \{[\gamma(f_u + g_v) - k^2(1 + d)]^2 - 4h(k^2)\}^{1/2}
$$

which has a maximum for the wavenumber k_m given by

$$
k^2 = k_m^2 = \frac{\gamma}{d-1} \left\{ (d+1) \left[-\frac{f_v g_u}{d} \right]^{1/2} - f_u + g_v \right\}.
\tag{14.68}
$$

As we have noted the prediction is that the fastest growing k_m-mode will be that which dominates and hence will be the mode which evolves into the steady state nonlinear pattern. This is only a reasonable prediction for the lower modes. The probable reason is that with the higher modes the interaction caused by the nonlinearities is more complex than when only the simpler modes are linearly unstable. Thus using (14.68) we can map the regions in (d, γ) space where a specific mode, and hence pattern, will evolve: see Arcuri and Murray (1986). Fig. 14.11 (a), (b) show the mappings for the Thomas (1975) system in one space dimension calculated from the linear theory and the full nonlinear system, while Fig. 14.11 (c) shows the corresponding spatial morphogen patterns indicated by Fig. 14.11 (b).

An important use of such parameter spaces is the measure of the robustness of the mechanism under consideration. With Fig. 14.11 (b), for example, suppose the biological conditions result in a (d, γ) parameter pair giving P, say, in the region which evolves to the 4-mode. A key property of any model is how sensitive it is to the inevitable random perturbations which exist in the real world. From Fig. 14.11 (b) we see what leeway there is if a 4-mode pattern is required in the developmental sequence. This (d, γ) space is but one of the relevant spaces to consider, of course, since any mechanism involves other parameters. So, in assessing robustness, or model sensitivity, we must also take into account the size and shape of the Turing space which involves all of the kinetics parameters. Probably (d, γ) spaces will not be too different qualitatively from one reaction diffusion system to another. What certainly is different, however, is the size and shape of the Turing space, and it is this space which provides another useful criterion for comparing relevant robustness of models. Murray (1982) studied this specific problem and compared various specific reaction diffusion mechanisms with this in mind. He came to certain conclusions as to the more robust mechanisms – both the Thomas (1975) and Schnakenberg (1979) systems, given respectively by (14.7) and (14.8), have relatively large Turing spaces, whereas that of the activator-inhibitor model of Gierer and Meinhardt (1972), given by (14.9) is quite small and implies a considerable sensitivity of pattern to small parameter variation. In the next chapter on specific pattern formation problems in biology we touch on other important aspects of model relevance which are implied by the form of the dispersion relation and the nondimensionalisation used.

The parameter spaces designating areas for specific patterns were all obtained with initial conditions taken to be random perturbations about the uniform steady state. Even in the low modes the polarity can be definitively influenced by biased initial conditions. We can, for example, create a single hump pattern with a single maximum in the centre of the domain or with a single minimum in the centre: see Fig. 14.6 (b). So even though specific modes can be isolated, initial conditions can strongly influence the polarity. When several of the modes are excitable, and one is naturally dominant from the dispersion relation, we can still influence the ultimate pattern by appropriate initial conditions. If the initial conditions include a mode within the unstable band and whose amplitude is sufficiently large, then this mode can persist through the nonlinear region so as to dominate the other unstable modes and the final pattern qualitatively often has roughly that wavelength. We discuss this in more detail in the next section. These facts also have highly relevant implications for biological applications.

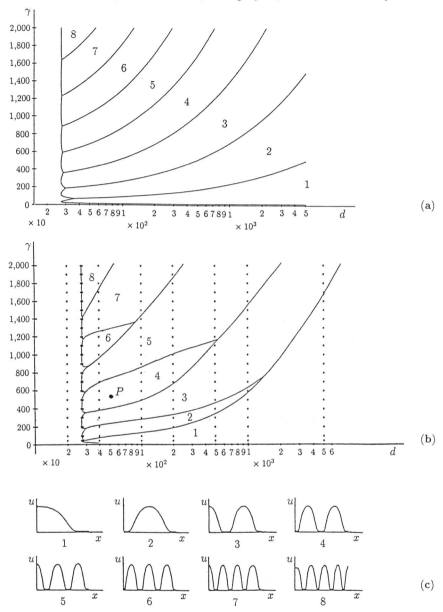

Fig. 14.11a-c. (a) Predicted solution space, based on linear theory, showing the regions with the fastest growing modes for the Thomas (1975) system (14.8) with parameter values as in Fig. 14.10 and zero flux boundary conditions. (b) A typical space as evaluated from the numerical simulation of the full nonlinear Thomas system (14.8) with the same parameter values and zero flux boundary conditions as in (a). Each (γ, d) point marked with a period represents a specific simulation of the full nonlinear system. (c) The corresponding spatial morphogen concentration patterns obtained with parameters d and γ in the regions indicated in (b). Spatial patterns can be visualised by setting a threshold u^* and shading for $u > u^*$. The first two morphogen distributions, for example, correspond to the first two patterns in Fig. 14.7. (From Arcuri and Murray 1986)

14.6 Mode Selection and the Dispersion Relation

Consider a typical 'vanilla' or simplest dispersion relation giving the growth factor λ as a function of the wavenumber or wavelength ω such as shown in Fig. 14.4 (b) where a band of wavenumbers is linearly unstable. Let us also suppose the domain is finite so that the spectrum of eigenvalues is discrete. In the last section we saw how geometry and scale played key roles in determining the particular pattern predicted from linear theory, and this was borne out by numerical simulation of the nonlinear system: see also the results presented in the next chapter. We pointed out that initial conditions can play a role in determining, for example, the polarity of a pattern or whether a specific pattern will emerge. If the initial conditions consist of small random perturbations about the uniform steady state then the likely pattern to evolve is that with the largest linear growth. In many developmental problems, however, the trigger for pattern initiation is scale – there are several examples in the following chapter. In other developmental situations a perturbation from the uniform steady state is initiated at one end of the spatial domain and the spatial pattern develops from there, eventually spreading throughout the whole domain. The specific pattern that evolves for a given mechanism therefore can depend critically on how the instability is initiated. In this section we investigate this further so as to suggest what patterns will evolve from which initial conditions, for given dispersion relations, as key parameters pass through bifurcation values. The problem of which pattern will evolve, namely mode selection, is a constantly recurring one. The following discussion, although motivated by reaction diffusion pattern generators, is quite general and applies to any pattern formation model which produces a similar type of dispersion relation.

Consider a basic dispersion relation $\lambda(\omega^2)$ where the wavelength $\omega = 2\pi/k$ with k the wavenumber, such as in Fig. 14.12 (a). Now take a one-dimensional domain and consider in turn the three possible ways of initiating pattern as shown in Fig. 14.12 (b), (c), (d).

Consider first the case in Fig. 14.12 (b). Here the initial perturbation has all modes present in its expansion in terms of the eigenfunctions and so all modes in the unstable band of wavelengths in Fig. 14.12 (a) are stimulated. The mode with the maximum λ, ω_2, is the one with the fastest growth and it ultimately dominates. The steady state inhomogeneous pattern that persists is then that with wavelength ω_2.

In Fig. 14.12 (c) we envisage the domain to be growing at a rate that is slow compared with the time to generate spatial pattern. For small $L(t)$ the domain is such that it cannot contain any wave with wavelengths in the unstable band. When it reaches L_c, the critical domain size for pattern, it can sustain the smallest wavelength pattern, namely that with wavelength ω_1. In the time it takes $L(t)$ to grow sufficiently to allow growth of the next wavenumber, that with wavelength ω_1 is sufficiently established to dominate the nonlinear stage. So the final pattern that emerges is that with the base wavelength ω_1.

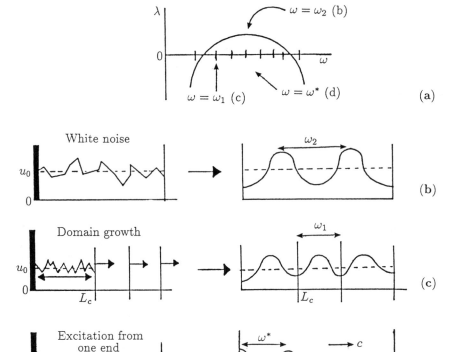

Fig. 14.12a-d. (a) Typical basic, or 'vanilla', dispersion relation giving the growth coefficient λ as a function of the wavelength ω of the spatial pattern. (b) Here the initial disturbance is a random perturbation (white noise) about the uniform steady state u_0. The pattern which evolves corresponds to ω_2 in (a), the mode with the largest growth rate. (c) Pattern evolution in a growing domain. The first unstable mode to be excited, ω_1, remains dominant. (d) Here the initial disturbance is at one end and it lays down a pattern as the disturbance moves through the domain. The pattern which evolves has a wavelength ω^* somewhere within the band of unstable wavelengths.

Travelling Wave Initiation of Pattern

Consider the situation, as in Fig. 14.12 (d), where the pattern is initiated at one end of the domain. We expect the final pattern to have a wavelength somewhere within the unstable band predicted by the dispersion relation. To see how to calculate the wavenumber in general let us start with an infinite one-dimensional domain and a general linear system

$$\mathcal{J}\mathbf{w} = 0, \quad \mathbf{w}(x,t) \propto \exp\left(ikx + \lambda t\right) \quad \Rightarrow \quad \lambda = \lambda(k) , \tag{14.69}$$

where $\mathcal{J}$ is a linear operator such as associated with the linear form of reaction diffusion equations and the dispersion relation $\lambda(k)$ is like that in Fig. 14.4 (b) or in the last figure with ω replaced by k; in other words the classic form. The general solution $\mathbf{w}$ of the linear system in (14.69) is

$$\mathbf{w}(x,t) = \int \mathbf{A}(k) \exp\left[ikx + \lambda(k)t\right] dk , \qquad (14.70)$$

where the $\mathbf{A}(k)$ are determined by a Fourier transform of the initial conditions $\mathbf{w}(x,0)$. Since we are concerned with the final structure and not the transients we do not need to evaluate $\mathbf{A}(k)$ here.

Suppose the initial conditions $\mathbf{w}(x,0)$ are confined to a small finite domain around $x = 0$ and the pattern propagates out from this region. We are interested in the wave-like generation of pattern as shown in the second figure in Fig. 14.12 (d). This means that we should look at the form of the solution well away from the origin. In other words, we should focus our attention on the asymptotic form of the solution for x and t large but such that x/t is $O(1)$, which means we move with a velocity $c = x/t$ and so are in the vicinity of the 'front' – roughly where the arrow is in the second figure in Fig. 14.12 (d). We write (14.70) in the form

$$\mathbf{w}(x,t) = \int \mathbf{A}(k) \exp\left[\sigma(k)t\right] dk, \quad \sigma(k) = ikc + \lambda(k), \quad c = \frac{x}{t} . \qquad (14.71)$$

The asymptotic evaluation of this integral for $t \to \infty$ is given by analytically continuing the integrand into the complex k-plane and using the method of steepest descents (see Murray's (1984) book, Chapter 3) which gives

$$\mathbf{w}(x,t) \sim \mathbf{J}(k_0) \left[\frac{2\pi}{t|\sigma''(k_0)|}\right]^{1/2} \exp\left\{t[ik_0c + \lambda(k_0)]\right\}$$

where $\mathbf{J}$ is a constant and k_0 (now complex) is given by

$$\sigma'(k_0) = ic + \lambda'(k_0) = 0 . \qquad (14.72)$$

The asymptotic form of the solution is thus

$$\mathbf{w}(x,t) \sim \mathbf{K} t^{-1/2} \exp\left\{t[ick_0 + \lambda(k_0)]\right\} \qquad (14.73)$$

where $\mathbf{K}$ is a constant.

For large t the wave 'front' is roughly the point between the pattern forming tail and the leading edge which initiates the disturbances, that is, where $\mathbf{w}$ neither grows nor decays. This is thus the point where

$$\mathrm{Re}\left[ick_0 + \lambda(k_0)\right] \approx 0 . \qquad (14.74)$$

At the 'front' the wavenumber is $\mathrm{Re}\,k_0$ and the solution frequency of oscillation ω is

$$\omega = \mathrm{Im}\left[ick_0 + \lambda(k_0)\right] .$$

Denote by k^* the wavenumber of the pattern laid down behind the 'front'. We now assume there is conservation of nodes across the 'front' which implies

$$k^*c = \omega = \text{Im}\left[ick_0 + \lambda(k_0)\right] . \tag{14.75}$$

The three equations (14.72), (14.74) and (14.75) now determine k_0 and the quantities we are interested in, namely c and k^*, are respectively the speed at which the pattern is laid down and the steady state pattern wavenumber. This technique has been used by Dee and Langer (1983) for a reaction diffusion mechanism. They simulate the dynamics of the pattern generation for a specific system.

   Becker and Field (1985) obtained spatial patterns from a numerical simulation of the Field-Noyes model, with diffusion, for the Belousov-Zhabotinskii reaction discussed in Chapter 7. They also found such patterns experimentally. Stationary spatial patterns in the BZ reaction were also obtained numerically and experimentally by Varek et al. (1979).

Dynamics of Pattern Formation in Growing Domains

The time evolution of patterns in growing domains can be quite complex, particularly if the domain growth is comparable with the generation time of the spatial pattern and there are two or more space dimensions. The form of the dispersion relation as the scale γ increases can have highly pertinent biological implications as we shall see in Chapter 17 when we consider cartilage formation in the developing limb. Here we introduce the phenomenon and discuss some of the implications of two specific classes of dispersion relation behaviour as γ increases.

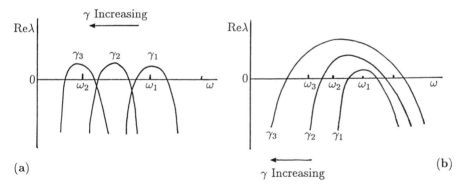

Fig. 14.13a,b. (a) As the scale γ increases from γ_1 to γ_3 the dispersion relation isolates specific modes interspersed with gaps during which no pattern can form. (b) Here as γ increases the number of unstable modes increases: the mode with maximum growth varies with γ. Unstable modes exist for all $\gamma \geq \gamma_1$.

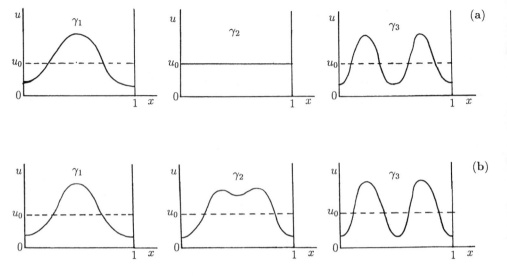

Fig. 14.14a,b. (a) Development of spatial patterns with a dispersion relation dependence on scale, via γ, as shown in Fig. 14.13 (a). (b) Sequential development of pattern as γ varies according to the dispersion relation in Fig. 14.13 (b).

In Fig. 14.6 (a) we saw that as the scale γ increased the dispersion curve was moved along the axis where it successively excited modes with smaller wavelengths. Fig. 14.13 (a) is a repeat example of this behaviour. Fig. 14.13 (b) is another possible behaviour of a dispersion relation as the scale γ increases. They imply different pattern generation scenarios for growing domains.

Consider first the situation in Fig. 14.13 (a). Here for $\gamma = \gamma_1$ the mode with wavelength ω_1 is excited and starts to grow. As the domain increases we see that for $\gamma = \gamma_2$ no mode lies within the unstable band and so the pattern decays to the spatially uniform steady state. With further increase in scale, to $\gamma = \gamma_3$ say, we see that a pattern with wavelength ω_2 is created. So the pattern formation is effectively a discrete process with successively more structure created as γ increases but with each increase in structure interspersed with a regime of spatial homogeneity. Fig. 14.14 (a) illustrates the sequence of events as γ increases in the way we have just described.

Consider now the behaviour implied by the dispersion relation dependence on scale implied by Fig. 14.13 (b). Here the effect of scale is simply to increase the band of unstable modes. The dominant mode changes with γ so there is a continuous evolution from one mode, dominant for $\gamma = \gamma_1$ say, to another mode as it becomes dominant for $\gamma = \gamma_3$ say. This dynamic development of pattern is illustrated Fig. 14.14 (b).

When comparing different models with experiment it is not always possible to choose a given time as regards pattern generation to carry out the experiments. When it is possible, then perhaps similarity of pattern is a sufficient first step in

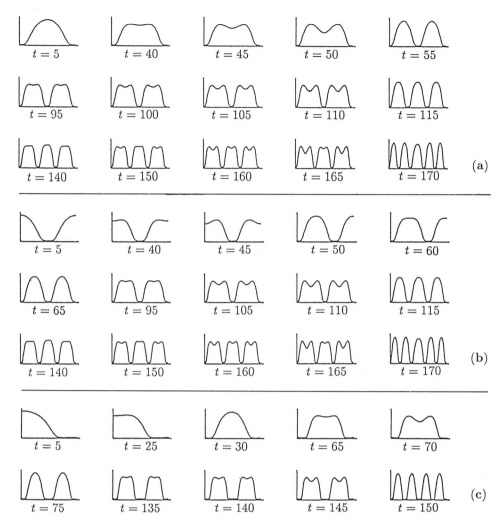

Fig. 14.15a-c. Sequence of one-dimensional spatial patterns numerically simulated with the mechanism (14.10) with kinetics (14.8). Zero flux boundary conditions were used and the growth in scale is $\gamma(t) = s + 0.1t^2$ with s fixed, which simulates a linear rate of growth since $\gamma \propto (\text{length})^2$. Parameter values for the kinetics are as in Fig. 14.10 (a) except for d. (a) and (b) have $d = 30$ ($d_c \approx 27$), $s = 100$ and two different sets of initial random perturbations. Note how the two sets of patterns converge as time t increases. (c) has $d = 60$, $s = 50$. As d increases more modes are missed in the pattern sequence and there is a distinct tendency towards frequency doubling. (After Arcuri and Murray 1986)

comparison with theory. When it is not possible, the dynamic form of the pattern can be important and can be the key step in deciding which mechanism is the more appropriate. We shall recall these comments later in Chapter 17.

A computed example of dynamic pattern formation as the scale γ is increased is shown in Fig. 14.15.

In these simulations the mechanism's pattern generation time is smaller than a representative growth time since the sequence of patterns clearly form before breaking up to initiate the subsequent pattern. This is an example of a dispersion relation behaviour like that in Fig. 14.13 (b); that is there is no regime of spatial homogeneity. The tendency to period doubling indicated by Fig. 14.15 (c) is interesting and as yet unexplored. Arcuri and Murray (1986) consider this and other aspects of pattern formation in growing domains.

14.7 Pattern Generation with Single Species Models: Spatial Heterogeneity with the Spruce Budworm Model

We saw above that if the domain size is not large enough, that is γ is too small, reaction diffusion models with zero flux boundary conditions cannot generate spatial patterns. Zero flux conditions imply that the reaction diffusion domain is isolated from the external environment. We now consider different boundary conditions which take into account the influence of the region exterior to the reaction diffusion domain. To be specific, consider the single reaction diffusion equation in the form

$$u_t = f(u) + D\nabla^2 u , \tag{14.76}$$

and think of the model in an ecological setting; that is, u denotes the population density of a species. Here $f(u)$ is the species' dynamics and so we shall assume $f(0) = 0$, $f'(0) \neq 0$, $f(u_i) = 0$ for $i = 1$ if there is only one (positive) steady state or $i = 1, 2, 3$ if there are three. Later we shall consider the population dynamics $f(u)$ to be those of the spruce budworm, which we studied in detail in Chapter 1, Section 1.2 and which has three steady states as in Fig. 1.4. The diffusion coefficient D is a measure of the dispersal efficiency of the relevant species.

We consider in the first instance the one-dimensional problem for a domain $x \in (0, L)$, the exterior of which is completely hostile to the species. This means that on the domain boundaries $u = 0$. The mathematical problem we consider is then

$$u_t = f(u) + Du_{xx} ,$$
$$u(0, t) = 0 = u(L, t), \quad u(x, 0) = u_0(x) ,$$
$$f(0) = 0, \quad f'(0) > 0, \quad f(u_2) = 0, \quad f'(u_2) > 0 , \tag{14.77}$$
$$f(u_i) = 0, \quad f'(u_i) < 0, \quad i = 1, 3 ,$$

where u_0 is the initial population distribution. The question we want to answer is whether or not such a model can sustain spatial patterns.

In the spatially homogeneous situation $u = 0$ and $u = u_2$ are unstable and u_1 and u_3 are stable steady states. In the absence of diffusion the dynamics imply that u tends to one or other of the stable steady states and which it is depends on the initial conditions. In the spatial situation, therefore, we would expect $u(x, t)$ to try to grow from $u = 0$ except at the boundaries. Because $u_x \neq 0$ at the boundaries the effect of diffusion implies that there is a flux of u out of the domain $(0, L)$. So for u small there are two competing effects, the growth from the dynamics and the loss from the boundaries. As a first step we examine the linear problem obtained by linearising about $u = 0$. The relevant formulation is, from (14.77),

$$u_t = f'(0)u + Du_{xx} ,$$
$$u(0, t) = u(L, t) = 0, \quad u(x, 0) = u_0(x) . \tag{14.78}$$

We look for solutions in the form

$$u(x, t) = \sum_n a_n e^{\lambda t} \sin (n\pi x / L) ,$$

which by inspection satisfy the boundary conditions at $x = 0, L$. Substitution of this into (14.78) and equating coefficients of $\sin (n\pi x / L)$ determines λ as $\lambda = [f'(0) - D(n\pi/L)^2]$ and so the solution is given by

$$u(x, t) = \sum_n a_n \exp \left\{ \left[f'(0) - D \left(\frac{n\pi}{L} \right)^2 \right] t \right\} \sin \frac{n\pi x}{L} , \tag{14.79}$$

where the a_n are determined by a Fourier series expansion of the initial conditions $u_0(x)$. We do not need a_n in this analysis. From (14.79) we see that the dominant mode in the expression for u is that with the largest λ, namely that with $n = 1$, since

$$\exp \left[f'(0) - D \left(\frac{n\pi}{L} \right)^2 \right] t < \exp \left[f'(0) - D \left(\frac{\pi}{L} \right)^2 \right] t, \quad \text{for all} \quad n \geq 2 .$$

So, if the dominant mode tends to zero as $t \to \infty$, so then do all the rest. We thus get as our condition for the linear stability of $u = 0$

$$f'(0) - D \left(\frac{\pi}{L} \right)^2 < 0 \quad \Rightarrow \quad L < L_c = \pi \left[\frac{D}{f'(0)} \right]^{1/2} . \tag{14.80}$$

In dimensional terms D has units cm^2s^{-1} and $f'(0)$ units s^{-1} since it is the linear birth rate (for u small $f(u) \approx f'(0)u$) which together give L_c in centimetres. Thus if the domain size L is less than the critical size L_c, $u \to 0$ as $t \to \infty$ and no spatial structure evolves. The larger the diffusion coefficient the larger is the critical domain size; this is in keeping with the observation that as D increases so also does the flux out of the region.

The scenario for spatial structure in a growing domain is that as the domain grows and L just passes the bifurcation length L_c, $u = 0$ becomes unstable and the first mode

$$a_1 \exp \left[f'(0) - D \left(\frac{\pi}{L} \right)^2 t \right] \sin \frac{\pi x}{L}$$

starts to grow with time. Eventually the nonlinear effects come into play and $u(x, t)$ tends to a steady state spatially inhomogeneous solution $U(x)$, which, from (14.77), is determined by

$$DU'' + f(U) = 0, \quad U(0) = U(L) = 0 , \tag{14.81}$$

where the prime denotes differentiation with respect to x. Because $f(U)$ is nonlinear we cannot, in general, get an explicit solution for U.

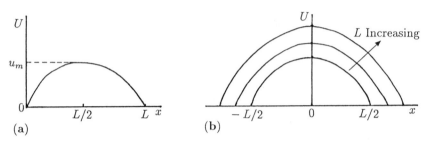

(a)

(b)

Fig. 14.16a,b. (a) Steady state pattern in the population u governed by (14.77) when the domain length $L > L_c$, the critical size for instability in the zero steady state. Note the symmetry about $L/2$. (b) Schematic steady state solution with the origin at the symmetry point where $u = u_m$ and $u_x = 0$.

From the spatial symmetry in (14.77) and (14.81) – setting $x \to -x$ leaves the equations unchanged – we expect the solutions to be symmetric in x about the mid-point $x = L/2$. Since $u = 0$ at the boundaries we assume the mid-point is the maximum, u_m say, where $U' = 0$: it is helpful now to refer to Fig. 14.16 (a). If we multiply (14.81) by U' and integrate with respect to x from 0 to L we get

$$\frac{1}{2} DU'^2 + F(U) = F(u_m), \quad F(U) = \int_0^U f(s) \, ds \tag{14.82}$$

since $U = u_m$ when $U' = 0$. It is convenient to change the origin to $L/2$ so that $U'(0) = 0$ and $U(0) = u_m$: that is set $x \to x - L/2$. Then

$$\left(\frac{D}{2}\right)^{1/2} \frac{dU}{dx} = [F(u_m) - F(U)]^{1/2}$$

which integrates to give

$$|x| = \left(\frac{D}{2}\right)^{1/2} \int_{U(x)}^{u_m} [F(u_m) - F(w)]^{-1/2} \, dw , \tag{14.83}$$

which gives the solution $U(x)$ implicitly: typical solutions are illustrated schematically in Fig. 14.16 (b). The boundary conditions $u = 0$ at $x = \pm L/2$ and the last equation give

$$L = (2D)^{1/2} \int_0^{u_m} [F(u_m) - F(w)]^{-1/2} \, dw \quad \Rightarrow \quad u_m = u_m(L) . \qquad (14.84)$$

We thus obtain, implicitly, u_m as a function of L. The actual determination of the dependence of u_m on L has to be carried out numerically. Note the singularity in the integrand when $w = u_m$, but because of the square root it is integrable. Typically u_m increases with L as illustrated in Fig. 14.16 (b).

Spatial Patterning of the Spruce Budworm

Now consider the model for the spruce budworm, the dynamics for which we derived in Section 1.2 in Chapter 1. Here, using (1.8) for $f(u)$, (14.77) becomes

$$u_t = ru \left(1 - \frac{u}{q} \right) - \frac{u^2}{1 + u^2} + Du_{xx} = f(u) + Du_{xx} , \qquad (14.85)$$

where the positive parameters r and q relate to the dimensionless quantities associated with the dimensional parameters in the model defined by (1.7); q is proportional to the carrying capacity and r is directly proportional to the linear birth rate and inversely proportional to the intensity of predation. The population dynamics $f(u)$ is sketched in Fig. 14.17 (a) when the parameters are in the parameter domain giving three positive steady states u_1, u_2 and u_3, the first and third being linearly stable and the second unstable. With $F(u)$ defined by (14.82) and substituted into (14.84) we have u_m as a function of the domain size L. This was evaluated numerically by Ludwig et al. (1979) the form of which is shown in Fig. 14.17 (b): there is another critical length, L_0 say, such that for $L > L_0$ more than one solution exists. We analyse this phenomenon below.

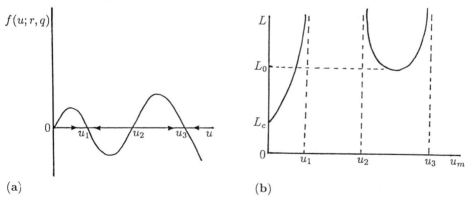

(a) (b)

Fig. 14.17a,b. (a) Typical dynamics $f(u; r, q)$ for the spruce budworm as defined by (14.85). (b) The maximum population u_m as a function of the domain size L. For $u_m < u_1$ the population is in the refuge range, whereas $u_m > u_2$ for $L > L_0$, which is in the outbreak regime.

From an ecological viewpoint we would like to know the critical domain size L_0 when the maximum population can be in the outbreak regime; that is $u_m > u_2$ in Fig. 14.17 (a). This is determined from numerical integration of (14.84) and is shown in Fig. 14.17 (b). When $L > L_0$ we see from Fig. 14.17 (b) that there are three possible solutions with different u_m. The ones with u in the refuge and outbreak regimes are stable and the other, the middle one, is unstable. Which solution is obtained depends on the initial conditions. Later we shall consider possible ecological uses of this model in the control of the budworm. Before doing so we describe a useful technique for determining approximate values for L_0 analytically.

Analytical Method for Determining Critical Domain Sizes and Maximum Populations

The numerical evaluation of $u_m(L)$ when there are three possible u_m for a given L is not completely trivial. Since the critical domain size L_0, which sustains an outbreak, is one of the important and useful quantities we require for practical applications, we now derive an *ad hoc* analytical method for obtaining it by exploiting an idea described by Lions (1982).

The steady state problem is defined by (14.81). Let us rescale the problem so that the domain is $x \in (0, 1)$ by setting $x \to x/L$ so that the equivalent $U(x)$ is now determined from

$$DU'' + L^2 f(U) = 0, \quad U(0) = U(1) = 0 . \tag{14.86}$$

From Fig. 14.16 the solution looks qualitatively like a sine. With the rescaling so that $x \in (0, 1)$ the solution is thus qualitatively like $\sin(\pi x)$. This means that $U'' \approx -\pi^2 U$ and so the last equation implies

$$-D\pi^2 U + L^2 f(U) \approx 0 \quad \Rightarrow \quad \frac{D\pi^2 U}{L^2} \approx f(U) . \tag{14.87}$$

We are interested in the value of L such that the last equation has three roots for U: this corresponds to the situation in Fig. 14.17 (b) when $L > L_0$. Thus all we need do to determine an approximate L_0 is simply to plot the last equation as in Fig. 14.18 and determine the value L such that three solutions exist.

For a fixed dispersal coefficient D we see how the solutions U vary with L. As L increases from $L \approx 0$ the first critical L, L_c, is given when the straight line $D\pi U/L^2$ intersects $f(U)$; that is when $D\pi^2/L^2 = f'(0)$, as given by (14.80). As L increases further we can determine the critical L_0 when $D\pi^2 U/L_0^2$ is tangent to the curve $f(U)$, at P in the Fig. 14.18. It is just a matter of determining L which gives a double positive root of

$$\frac{D\pi^2 U}{L^2} = f(U) .$$

It is left as an exercise (Exercise 7) to determine L_0 as a function of r, q and D when $f(U)$ is given by (14.85). For any given L the procedure also determines,

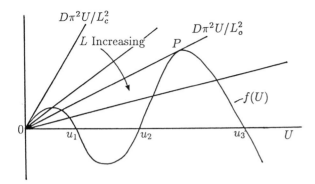

Fig. 14.18. Approximate analytic procedure for determining the critical domain sizes L_c and L_0 which can sustain respectively a refuge and an outbreak in the species population where the dynamics is described by $f(U)$. L_c is the value of L when $D\pi^2 U/L^2$ is tangent to $f(U)$ at $U = 0$. L_0 is given by the value of L when $D\pi U/L^2$ is just tangent to $F(U)$ at P.

approximately, the maximum U. From Fig. 14.18 we clearly obtain by this procedure a similar figure to that in Fig. 14.17 (b). This procedure is quite general for determining critical domain sizes, both for structure bifurcating from the zero steady state and for domains which can sustain larger populations arising from population dynamics with multiple positive steady states.

14.8 Spatial Patterns in Scalar Population Interaction-Reaction Diffusion Equations with Convection: Ecological Control Strategies

In practical applications of such models the domains of interest are usually two-dimensional and so we must consider (14.76). Also, with insect pests in mind, the exterior region is not generally completely hostile, so $u = 0$ on the boundaries is too restrictive a condition. Here we briefly consider a one and two-dimensional problem in which the exterior domain is not completely hostile and there is a prevailing wind. This is common in many insect dispersal situations; it can modify the spatial distribution of the population in a major way.

Suppose, for algebraic simplicity, that the two-dimensional domain is a rectangular region B defined by $0 \le x \le a$, $0 \le y \le b$ having area A. The completely hostile problem is then given by

$$u_t = f(u) + D\left(\frac{\partial^2 u}{\partial x^2} + \frac{\partial^2 u}{\partial y^2}\right),$$

$$u = 0 \quad \text{for} \quad (x, y) \text{ on } \partial B .$$

(14.88)

Following the same procedure as in the last section for u small we get the solution of the linearised problem to be

$$u(x,y,t) = \sum_{n,m} a_{mn} \exp\left\{\left[f'(0) - D\pi^2\left(\frac{n^2}{a^2} + \frac{m^2}{b^2}\right)\right]t\right\} \sin\frac{n\pi x}{a} \sin\frac{m\pi y}{b}$$

(14.89)

So the critical domain size, which involves both a and b, is given by any combination of a and b such that

$$\frac{a^2 b^2}{a^2 + b^2} = \frac{D\pi^2}{f'(0)} \ .$$

Since

$$a^2 + b^2 > 2ab = 2A \quad \Rightarrow \quad \frac{a^2 b^2}{a^2 + b^2} < \frac{A}{2}$$

we get an inequality estimate for spatial patterning to exist, namely

$$A > \frac{2D\pi^2}{f'(0)} \ .$$

(14.90)

Estimates for general two-dimensional domains have been obtained by Murray and Sperb (1983). Clearly the mathematical problem is that of finding the smallest eigenvalue for the spatial domain considered.

In all the scalar models considered above the spatial patterns obtained have only a single maximum. With completely hostile boundary conditions these are the only type of patterns that can be generated. With two-species reaction diffusion systems, however, we saw that more diverse patterns could be generated. It is natural to ask whether there are ways in which similar multi-peak patterns could be obtained with single species models in a one-dimensional context.

Suppose now that there is a constant prevailing wind $\mathbf{w}$ which contributes a convective flux $(\mathbf{w} \cdot \nabla)u$ to the conservation equation for the population $u(\mathbf{r}, t)$. Also suppose that the exterior environment is not completely hostile in which case appropriate boundary conditions are

$$(\mathbf{n} \cdot \nabla)u + hu = 0, \quad \mathbf{r} \text{ on } \partial B \ ,$$

(14.91)

where $\mathbf{n}$ is the unit normal to the domain boundary ∂B. The parameter h is a measure of the hostility: $h = \infty$ implies a completely hostile exterior, whereas $h = 0$ implies a closed environment, that is zero flux boundaries. We briefly consider the latter case later. The mathematical problem is thus

$$u_t + (\mathbf{w} \cdot \nabla)u = f(u) + D\nabla^2 u \ ,$$

(14.92)

with boundary conditions (14.91) and given initial distribution $u(\mathbf{r}, 0)$. Here we consider the one-dimensional problem and follow the analysis of Murray and Sperb (1983), who also deal with the two-dimensional analogue and more general aspects of such problems.

The problem we briefly consider is the one-dimensional system which defines the steady state spatially inhomogeneous solutions $U(x)$. From (14.91) and (14.92), since

$$(\mathbf{w} \cdot \nabla)u = w_1 u_x \,,$$

$$(\mathbf{n} \cdot \nabla)u + hu = 0 \quad \Rightarrow \quad u_x + hu = 0, \ x = L; \quad u_x - hu = 0, \ x = 0 \,,$$

where w_1 is the x-component of the wind $\mathbf{w}$, the mathematical problem for $U(x)$ is

$$DU'' - w_1 U' + f(U) = 0 \,,$$
$$U'(0) - hU(0) = 0, \quad U'(L) + hU(L) = 0 \,. \tag{14.93}$$

We study the problem using phase plane analysis by setting

$$U' = V, \quad DV' = w_1 V - f(U) \quad \Rightarrow \quad \frac{dV}{dU} = \frac{w_1 V - f(U)}{DV} \,, \tag{14.94}$$

and we look for phase plane trajectories which, from the boundary conditions in (14.93), join any point on one of the following lines to any point on the other line:

$$V = hU, \quad V = -hU \,. \tag{14.95}$$

The phase plane situation is illustrated in Fig. 14.19 (a), (b) as we shall now show.

Refer first to Fig. 14.19 (a). From (14.94) we get the sign of dV/dU at any point (U, V). On the curve $V = f(U)/w_1$, $dV/dU = 0$ with dV/dU positive and negative when (U, V) lies respectively above (if $V > 0$) and below it. So, if we start on the boundary line $V = hU$ at say P the trajectory will qualitatively be like T_1 since $dV/dU < 0$ everywhere on it. If we start at S, say, although the trajectory starts with $dV/dU < 0$ it intersects the $dV/dU = 0$ line and passes through to the region where $dV/dU > 0$ and so the trajectory turns up. The trajectories T_2, T_3 and T_4 are all possible scenarios depending on the parameters and where the solution trajectory starts. T_3 and T_4 are not solution trajectories satisfying (14.94) since they do not terminate on the boundary curve $V = -hU$. T_1 and T_2 are allowable solution paths and each has a single maximum U where the trajectory crosses the $V = 0$ axis.

We now have to relate the corresponding domain length L to these solution trajectories. To be specific let us focus on the trajectory T_2. Denote the part of the solution with $V > 0$ by $V^+(U)$ and that with $V < 0$ by $V^-(U)$. If we now integrate the first equation in (14.94) from U_Q to $U_{Q'}$, that is the U-values at either end of the T_2 trajectory, we get the corresponding length of the domain for the solution represented by T_2 as

$$L = \int_{U_Q}^{U_m} [V^+(U)]^{-1} \, dU + \int_{U_m}^{U_{Q'}} [V^-(U)]^{-1} \, dU \,. \tag{14.96}$$

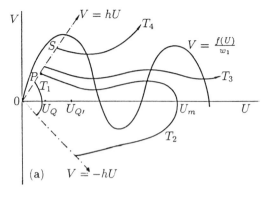

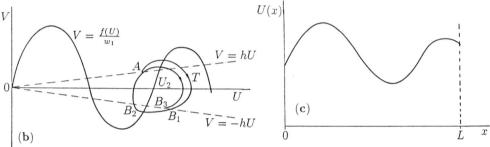

Fig. 14.19a-c. (a) With h sufficiently large the possible trajectories from $V = hU$ to $V = -hU$ admit solution trajectories like T_1 and T_2 with only a single maximum U_m. (b) For small enough h it is possible to have more complex patterns as indicated by the specimen trajectory T. (c) A typical solution $U(x)$ for a phase trajectory like T in (b).

So, for each allowable solution trajectory we can obtain the corresponding size of the solution domain. The qualitative form of the solution $U(x)$ as a function of x can be deduced from the phase trajectory since we know U and U' everywhere on it and from the last equation we can calculate the domain size. With the situation represented by Fig. 14.19 (a) there can only be a single maximum in $U(x)$. Because of the wind convection term, however, there is no longer the solution symmetry of the solutions as in the last Section 14.7.

Now suppose the exterior hostility decreases, that is h in (14.95) decreases, so that the boundary lines are now as illustrated in Fig. 14.19 (b). Proceeding in the same way as for the solution trajectories in Fig. 14.19 (a) we see that it is possible for a solution to exist corresponding to the trajectory T. On sketching the corresponding solution $U(x)$ we see that here there are two maxima in the domain: see Fig. 14.19 (c). In this situation however we are in fact patching several possible solutions together. Referring to Fig. 14.19 (b) we see that a possible solution is represented by that part of the trajectory T from A to B_1. It has a single maximum and a domain length L_1 given by the equivalent of (14.96). So if we restrict the domain size to be L_1 this is the relevant solution. However, if we allow a larger L the continuation from B_1 to B_2 is now possible and so the trajectory AB_1B_2 corresponds to a solution of (14.93). Increasing L further we can include the rest of the trajectory to B_3. It is thus possible to have multi-humped solutions if the domain is large enough. The length L corresponding to the solution path T is obtained in exactly the same way as above, using the equivalent of (14.96).

So, for small enough values of h it is possible to have more and more structure as the trajectory winds round the point u_2 in the (U, V) phase plane. For such solutions to exist, of course, it is essential that $w_1 \neq 0$. If $w_1 = 0$ the solutions are symmetric about the U-axis and so no spiral solutions are possible. Thus a prevailing wind is essential for complex patterning. It also affects the critical domain size for patterns to exist. General results and further analysis are given by Murray and Sperb (1983).

An Insect Pest Control Strategy

Consider now the problem of insect pest control. The forest budworm problem is very much a two-dimensional spatial problem. As we pointed out in Chapter 1, Section 1.2, a good control strategy would be to maintain the population at a refuge level. As we also showed in Section 1.2 it would be strategically advantageous if the dynamics parameters r and q in (14.85) could be changed so that only a single positive steady state exists. This is not really ecologically feasible. With the more realistic spatial problem, however, we have a further possible means of keeping the pest levels within the refuge range by ensuring that their spatial domains are of a size that do not permit populations in the outbreak regime. The arguments go through for two-dimensional domains, but for illustrative purposes let us consider first the one-dimensional situation.

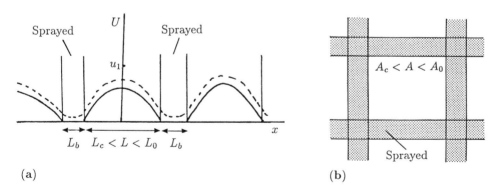

(a)   (b)

Fig. 14.20a,b. (a) A possible control strategy to contain the insect pest in a refuge rather than an outbreak environment. Strips – insect 'breaks' - are sprayed to maintain an effective domain size $L < L_0$, the critical size for an outbreak. The broken line is more a typical situation in practice. (b) Equivalent two-dimensional analogue where $A > A_c$ is a typical domain which can sustain a pest refuge population but which is not sufficient to sustain an outbreak, that is $A < A_0$.

Refer to Fig. 14.17 (b). If the spatial region was divided up into regions with size $L < L_0$, that is so that the maximum u_m was always less than u_1, the refuge population level, we would have achieved our goal. So, a possible strategy is to spray the region in strips so that the non-sprayed regions impose an effective $L < L_0$ as in Fig. 14.20 (a): the solid vertical lines separating the sprayed regions are the boundaries to a completely hostile exterior.

Of course it is not practical to destroy all pests that stray out of the unsprayed region, so a more realsitic model is that with boundary conditions (14.91) where some insects can survive outside the untreated domain. The key mathematical problem to be solved then is the determination of the critical width of the insect 'break' L_b. This must be such that the contributions from neighbouring untreated areas do not contribute a sufficient number of insects, which diffuse through the break, to initiate an outbreak in the neighbouring patches even though $L < L_0$, the critical size in isolation. A qualitative population distribution would typically be as shown by the dashed line in Fig. 14.20 (a).

The two-dimensional analogue is clear but the solution of the optimisation problem is more complicated. First the critical domain A_0 which can sustain an insect pest outbreak has to be determined for boundary conditions (14.91). Then the width of the sprayed strips have to be determined. It is not a trivial problem to solve, but certainly a possible one. A preliminary investigation of these problems has been carried out by Ben-Yu, Mitchell and Sleeman (1986).

Although we have concentrated on the budworm problem the techniques and control strategies are equally applicable to other insect pests. The field of insect dispersal presents some very important ecological problems, such as the control of killer bees now sweeping up through the western United States (see, for example Taylor (1977)) and locust plagues in Africa. Levin (see for example, 1974, 1976, 1981a,b) has made realistic and practical studies of these and other problems associated with spatially heterogeneous ecological models. The concept of a break control strategy to prevent the spatial spread of a disease epidemic will be discussed in some detail later in Chapter 20 when we discuss the spatial spread of rabies.

*14.9 Nonexistence of Spatial Patterns in Reaction Diffusion Systems: General and Particular Results

Nonexistence of Stable Spatial Patterns for Scalar Equations in One Dimension with Zero Flux Boundary Conditions

Since we found more complex patterns in the last section as the hostility parameter h in (14.91) decreased, it is natural to ask if *a single reaction diffusion equation with zero prevailing wind, that is $\mathbf{w} = 0$, and zero flux boundary conditions, that is $h = 0$, can sustain* stable *spatial patterns*. Here the question of stability is important since zero flux conditions effectively allow for free movement of the population density on the boundaries. So, it is always a possibility that the solution will just evolve to the uniform steady state. Here we show that *a scalar* reaction diffusion equation cannot sustain a spatial pattern if we restrict it to one space dimension.

Consider the following problem:

$$u_t = f(u) + u_{xx}, \quad x \in (0,1), \quad t > 0$$
$$u_x(0,t) = u_x(1,t) = 0, \quad t > 0 .$$
$$(14.97)$$

Steady state spatially nonuniform solutions $U(x)$ satisfy

$$U'' + f(U) = 0, \quad U'(0) = U'(1) = 0 .$$
$$(14.98)$$

We shall prove that if $U(x)$ with $U'(x)$ not identically zero is a solution of (14.98) then $U(x)$ is unstable.

It is sufficient to prove that U is linearly unstable so let

$$u(x,t) = U(x) + w(x,t), \quad |w| \ll |U|_{max} ,$$
$$w(x,t) = y(x)e^{\lambda t} .$$
$$(14.99)$$

The solution U is stable if $\mathrm{Re}\,\lambda < 0$. Substitution into (14.97) and retaining only linear terms gives

$$\lambda y = y'' + f'(U(x))y, \quad y'(0) = y'(1) = 0 .$$
$$(14.100)$$

We realistically assume that $f'(U)$ is bounded, by $K(> 0)$ say. If we consider the eigenvalue problem with K replacing $f'(U)$ we have

$$y'' + (K - \lambda)y = 0, \quad y'(0) = y'(1) = 0$$

which has a discrete set of eigenvalues for $\lambda < K$ and no non-trivial solutions for all $\lambda > K$. This implies that with f' bounded, the discrete eigenvalues λ of (14.100) must be bounded above and so there must be a largest eigenvalue, λ_0 say.

With λ_0 the largest eigenvalue of the set $\lambda_0, \lambda_1, \lambda_2, \ldots$ of the problem defined by (14.100), we have

$$\lambda_0 > \lambda_1 \geq \lambda_2 \ldots$$

Now consider the eigenvalue problem

$$\mu y = y'' + f'(U(x))y, \quad y(0) = y(1) = 0 ,$$
$$(14.101)$$

that is with fixed boundary values. Note that $z = U'(x)$ is an eigenfunction of (14.101) if $\mu = 0$ since

$$0 = U''' + f'(U)U' \qquad z'' + f'(U)z = 0 ,$$
$$\Rightarrow$$
$$U'(0) = U'(1) = 0 \qquad z(0) = z(1) = 0 .$$

But this implies that $\mu = 0$ is an eigenvalue of (14.101) which implies that the largest eigenvalue $\mu_0 \geq 0$. If $f'(U)$ is large we certainly require $\mu > 0$ whereas if it is very small $y'' \approx \mu y$ which requires $\mu < 0$.

Now consider the two eigenvalue problems defined by (14.100) and (14.101), that is with zero flux and zero boundary conditions on y. We have the general result that $\lambda_0 > \mu_0$; see, for example, Courant and Hilbert (1961) and Sperb (1981). But $\mu_0 \geq 0$ so we have $\lambda_0 > 0$. That is, the largest eigenvalue of (14.100) is positive and so, from (14.99), w grows exponentially with time and hence $U(x)$ is linearly unstable. Thus there are no stable spatially patterned solutions of (14.97). The only solutions $u(x,t)$ of (14.97) are thus the homogeneous steady states given by solutions of $f(u) = 0$.

This result does *not* carry over completely to scalar equations in more than one space dimension as has been shown by Matano (1979) in the case where $f(u)$ has two linearly stable steady states. The spatial patterns that can be obtained however, depend on specific domain boundaries, non-convex to be specific. For example, a dumb bell shaped domain with a sufficiently narrow neck is an example. The pattern depends on the difficulty of diffusionally transporting enough flux of material through the neck to effect a change from one steady state to another to thus achieve homogeneity.

Large Diffusion Prevents Spatial Patterning in Reaction Diffusion Mechanisms with Zero Flux Boundary Conditions

We showed in Section 14.3 and Section 14.4 how reaction diffusion systems with zero flux boundary conditions could generate a rich spectrum of spatial patterns if the parameters and kinetics satisfied appropriate conditions: crucially the diffusion coefficients had to be different. Here we show that for general multi-species systems patterning can be destroyed if the diffusion is sufficiently large, which is intuitively what we might expect, although it is not obvious if the diffusion coefficients are unequal.

Before discussing the multi-species multi-dimensional theory it is pedagogically helpful to consider first the general one-dimensional two-species reaction diffusion system

$$u_t = f(u,v) + D_1 u_{xx}, \quad v_t = g(u,v) + D_2 v_{xx} \tag{14.102}$$

with zero flux boundary conditions and initial conditions

$$u_x(0,t) = u_x(1,t) = v_x(0,t) = v_x(1,t) = 0$$
$$u(x,0) = u_0(x), \quad v(x,0) = v_0(x), \tag{14.103}$$

where $u_0'(x)$ and $v_0'(x)$ are zero on $x = 0, 1$. Define an energy integral E by

$$E(t) = \frac{1}{2} \int_0^1 (u_x^2 + v_x^2)\, dx . \tag{14.104}$$

This is, except for the 1/2, the heterogeneity function introduced in (14.52). Differentiate E with respect to t to get

$$\frac{dE}{dt} = \int_0^1 (u_x u_{xt} + v_x v_{xt}) \, dx$$

and substitute from (14.102), on differentiating with respect to x, to get, on integrating by parts,

$$\frac{dE}{dt} = \int_0^1 \Big[u_x(D_1 u_{xx})_x + u_x(f_u u_x + f_v v_x)$$

$$+ \, v_x(D_2 v_{xx})_x + v_x(g_u u_x + g_v v_x) \Big] \, dx \; ,$$

$$= \Big[u_x D_1 u_{xx} + v_x D_2 v_{xx} \Big]_0^1 - \int_0^1 (D_1 u_{xx}^2 + D_2 v_{xx}^2) \, dx$$

$$+ \int_0^1 \Big[f_u u_x^2 + g_v v_x^2 + (f_v + g_u) u_x v_x \Big] \, dx \; .$$

Because of the zero flux conditions the integrated terms are zero.

Now define the quantities d and m by

$$d = \min(D_1, D_2), \quad m = \max_{u,v} (f_u^2 + f_v^2 + g_u^2 + g_v^2)^{1/2} \tag{14.105}$$

where $\max\limits_{u,v}$ means the maximum for u and v taking all possible solution values. If we want we could define m by some norm involving the derivatives of f and g: it is not crucial for our result. From the equation for dE/dt, with these definitions, we then have

$$\frac{dE}{dt} \le -d \int_0^1 (u_{xx}^2 + v_{xx}^2) \, dx + 4m \int_0^1 (u_x^2 + v_x^2) \, dx$$

$$\le (4m - 2\pi^2 d) E \; , \tag{14.106}$$

where we have used the result

$$\int_0^1 u_{xx}^2 \, dx \ge \pi^2 \int_0^1 u_x^2 \, dx \tag{14.107}$$

with a similar inequality for v: see Appendix 4 for a derivation of (14.107).

From the inequality (14.106) we now see that if the minimum diffusion coefficient d, from (14.105), is large enough so that $(4m - 2\pi^2 d) < 0$ then $dE/dt < 0$, which implies that $E \to 0$ as $t \to \infty$ since $E(t) \ge 0$. This implies, with the definition of E from (14.104), that $u_x \to 0$ and $v_x \to 0$ which implies spatial homogeneity in the solutions u and v as $t \to \infty$. The result is not precise since there are many appropriate choices for m; (14.105) is just one example. The purpose of the result is simply to show that it is possible for diffusion to dampen *all* spatial heterogeneities. We comment briefly on the biological implication of this result below.

We now prove the analogous result for general reaction diffusion systems. Consider

$$\mathbf{u}_t = \mathbf{f}(\mathbf{u}) + D\nabla^2 \mathbf{u} \,, \tag{14.108}$$

where $\mathbf{u}$, with components u_i, $i = 1, 2, \ldots, n$, is the vector of concentrations or populations, and D is a diagonal matrix of the positive diffusion coefficients D_i, $i = 1, 2, \ldots, n$ and $\mathbf{f}$ is the nonlinear kinetics. The results we shall prove are also valid for a diffusion matrix with certain cross-diffusion terms, but for simplicity here we only deal with (14.108). Zero flux boundary and inital conditions for $\mathbf{u}$ are

$$(\mathbf{n} \cdot \nabla)\mathbf{u} = 0 \ \mathbf{r} \text{ on } \partial B, \quad \mathbf{u}(\mathbf{r}, 0) = \mathbf{u}_0(\mathbf{r}) \,, \tag{14.109}$$

where $\mathbf{n}$ is the unit outward normal to ∂B, the boundary of the domain B. As before we shall assume that all solutions $\mathbf{u}$ are bounded for all $t \geq 0$. Practically this is effectively assured if a confined set exists for the reaction kinetics.

We now generalise the previous analysis: it will help to refer to the equivalent steps in the above. Define the energy $E(t)$ by

$$E(t) = \frac{1}{2} \int_B \|\nabla \mathbf{u}\|^2 \, d\mathbf{r} \,, \tag{14.110}$$

where the norm

$$\|\nabla \mathbf{u}\|^2 = \sum_{i=1}^{n} |\nabla u_i|^2 \,.$$

Let d be the smallest eigenvalue of the matrix D, which in the case of a diagonal matrix is simply the smallest diffusion coefficient of all the species. Now define

$$m = \max_{\mathbf{u}} \|\nabla_{\mathbf{u}} \mathbf{f}(\mathbf{u})\| \tag{14.111}$$

where $\mathbf{u}$ takes on all possible solution values and $\nabla_{\mathbf{u}}$ is the gradient operator with respect to $\mathbf{u}$.

Differentiating $E(t)$ in (14.110), using integration by parts, the boundary conditions (14.109) and the original system (14.108) we get, with $\langle \mathbf{a}, \mathbf{b} \rangle$ denoting the inner product of $\mathbf{a}$ and $\mathbf{b}$,

$$\frac{dE}{dt} = \int_B \langle \nabla \mathbf{u}, \nabla \mathbf{u}_t \rangle \, d\mathbf{r}$$

$$= \int_B \langle \nabla \mathbf{u}, \nabla D\nabla^2 \mathbf{u} \rangle \, d\mathbf{r} + \int_B \langle \nabla \mathbf{u}, \nabla \mathbf{f} \rangle \, d\mathbf{r}$$

$$= \int_{\partial B} \langle \nabla \mathbf{u}, D\nabla^2 \mathbf{u} \rangle \, d\mathbf{r} - \int_B \langle \nabla^2 \mathbf{u}, D\nabla^2 \mathbf{u} \rangle \, d\mathbf{r} + \int_B \langle \nabla \mathbf{u}, \nabla_{\mathbf{u}} \mathbf{f} \cdot \nabla \mathbf{u} \rangle \, d\mathbf{r}$$

$$\leq -d \int_B |\nabla^2 \mathbf{u}|^2 \, d\mathbf{r} + mE \ . \tag{14.112}$$

In Appendix 4 we show that when $(\mathbf{n} \cdot \nabla)\mathbf{u} = 0$ on ∂B

$$\int_B |\nabla^2 \mathbf{u}|^2 \, d\mathbf{r} \geq \mu \int_B \|\nabla \mathbf{u}\|^2 \, d\mathbf{r} \ , \tag{14.113}$$

where μ is the least positive eigenvalue of

$$\nabla^2 \phi + \mu\phi = 0, \quad (\mathbf{n} \cdot \nabla)\phi = 0 \quad \mathbf{r} \text{ on } \partial B \ ,$$

where ϕ is a scalar. Using the result (14.113) in (14.112) we get

$$\frac{dE}{dt} \leq (m - 2\mu d)E \quad \Rightarrow \quad \lim_{t \to \infty} E(t) = 0 \text{ if } m < 2\mu d \tag{14.114}$$

and so, once again, if the smallest diffusion coefficient is large enough this implies that $\nabla \mathbf{u} \to 0$ and so all spatial patterns tend to zero as $t \to \infty$.

Othmer (1977) has pointed out that the parameter m defined by (14.105) and (14.111) is a measure of the sensitivity of the reaction rates to changes in $\mathbf{u}$ since $1/m$ is the shortest kinetic relaxation time of the mechanism. On the other hand $1/(2\mu d)$ is a measure of the longest diffusion time. So the result (14.114), which is $1/m > 1/(2\mu d)$, then implies that if the shortest relaxation time for the kinetics is greater than the longest diffusion time then all spatial patterning will die out as $t \to \infty$. The mechanism will then be governed solely by kinetics dynamics. Note that the solution of the latter can include limit cycle oscillations.

Suppose we consider the one-dimensional situation with a typical embryological domain of interest, say $L = O(1\,\text{mm})$. With $d = O(10^{-6}\text{cm}^2\text{s}^{-1})$ the result (14.114) then implies that homogeneity will result if the shortest relaxation time of the kinetics $1/m > L^2/(2\pi^2 d)$, that is a time of $O(500\,\text{s})$.

Consider the general system (14.108) rescaled so that the length scale is 1 and the diffusion coefficients are scaled relative to D_1 say. Now return to the formulation used earlier, in (14.10), for instance, in which the scale γ appears with the kinetics in the form $\gamma \mathbf{f}$. The effect of this on the condition (14.114) now produces $\gamma m - 2\mu < 0$ as the stability requirement. We immediately see from this form that there is a critical γ, proportional to the domain area, which in one dimension is $(\text{length})^2$, below which no structure can exist. This is of course a similar result to the one we found in Section 14.3 and Section 14.4.

We should reiterate that the results here give qualitative bounds and not estimates for the various parameters associated with the model mechanisms. The evaluation of an appropriate m is not easy. In Sections 14.3 and 14.4 we derived specific quantitative relations between the parameters, when the kinetics were of a particular class, to give spatially structured solutions. The general results in this section, however, apply to all types of kinetics, oscillatory or otherwise, as long as the solutions are bounded.

In this chapter we have dealt primarily with reaction or population interaction kinetics which, in the absence of diffusion, do not exhibit oscillatory behaviour in the restricted regions of parameter space which we have considered. We may ask whether it is possible to have any spatial structure when oscillatory kinetics is coupled with diffusion. We saw in Chapters 12 and 13 that such a combination could give rise to travelling wave trains when the domain was infinite. If the domain is finite we could anticipate a kind of regular sloshing around within the domain which is a reflexion of the existence of spatially and temporally unstable modes. This can in fact occur but it is not always so. One case to point is the classical Lotka-Volterra system with equal diffusion coefficients for the species. Murray (1975) showed that in a finite domain all spatial heterogeneities must die out (see Exercise 11).

In the next chapter we shall be discussing several specific practical biological pattern formation problems. In later chapters we shall describe other mechanisms which can generate spatial patterns. An important system which has been widely studied is the reaction-diffusion-chemotaxis mechanism for generating aggregation patterns in the slime mold amoebae, one model for which we derived in Chapter 9, Section 9.4. Using exactly the same kind of analysis we discussed above for diffusion driven instability we can show how spatial patterns in the amoebae can arise and the conditions on the parameters under which this will happen (see Exercise 2).

Exercises

1. Determine the appropriate nondimensionalisation for the reaction kinetics in (14.4) and (14.5) which result in the forms (14.8).

2. An activator-inhibitor reaction diffusion system in dimensionless form is given by

$$u_t = \frac{u^2}{v} - bu + u_{xx}, \quad v_t = u^2 - v + dv_{xx} ,$$

where b and d are positive constants. Which is the activator and which the inhibitor? Determine the positive steady states and show, by an examination of the eigenvalues in a linear stabiltiy analysis of the diffusionless situation, that the reaction kinetics cannot exhibit oscillatory solutions if $b < 1$.

Determine the conditions for the steady state to be driven unstable by diffusion. Show that the parameter domain for diffusion driven instability is given by $0 < b < 1$, $db > 3 + 2\sqrt{2}$ and sketch the (b, d) parameter space in which diffusion driven instability occurs. Further show that at the bifurcation to such an instability the the critical wave number k_c is given by $k_c^2 = (1 + \sqrt{2})/d$.

*3. An activator-inhibitor reaction diffusion system with activator inhibition is modelled by

$$u_t = a - bu + \frac{u^2}{v(1 + Ku^2)} + u_{xx} ,$$

$$v_t = u^2 - v + dv_{xx} ,$$

where K is a measure of the inhibition and a, b and d are constants. Sketch the null clines for positive b, various $K > 0$ and positive or negative a.

Show that the (a, b) Turing space for diffusion driven instability is defined parameterically by

$$a = bu_0 - (1 + Ku_0^2)^2$$

combined with

$$b > 2[u(1 + Ku_0^2)]^{-1} - 1, \quad b > 0, \quad b > 2[u(1 + Ku_0^2)]^{-2} - \frac{1}{d} ,$$

$$b < 2[u(1 + Ku_0^2)]^{-2} - 2\sqrt{2}[du(1 + Ku_0^2)]^{-1/2} + \frac{1}{d} ,$$

where the parameter u_0 takes on all values in the range $(0, \infty)$. Sketch the Turing space for (i) $K = 0$ and (ii) $K \neq 0$ for various d. [Murray 1982]

4. Determine the relevant axisymmetric eigenfunctions $\mathbf{W}$ and eigenvalues k^2 for the circular domain bounded by R defined by

$$\nabla^2 \mathbf{W} + k^2 \mathbf{W} = 0, \quad \frac{d\mathbf{W}}{dr} = 0 \text{ on } r = R .$$

Given that the linearly unstable range of wavenumbers k^2 for the reaction diffusion mechanism (13.7) is given by

$$\gamma L(a, b, d) < k^2 < \gamma M(a, b, d) ,$$

where L and M are defined by (13.38), determine the critical radius R_c of the domain below which no spatial pattern can be generated. For R just greater than R_c sketch the spatial pattern you would expect to evolve.

5. Consider the reaction diffusion mechanism given by

$$u_t = \gamma \left(\frac{u^2}{v} - bu \right) + u_{xx}, \quad v_t = \gamma(u^2 - v) + dv_{xx} ,$$

where γ, b and d are positive constants. For the domain $0 \leq x \leq 1$ with zero flux conditions determine the dispersion relation $\lambda(k^2)$ as a function of the wavenumbers k of small spatial perturbations about the uniform steady state. Is it possible with this mechanism to isolate successive modes by judicious variation of the parameters? Is there a bound on the excitable modes as $d \to \infty$ with b and γ fixed?

6. Suppose fishing is regulated within a zone H km from a country's shore (taken to be a straight line) but outside of this zone over-fishing is so excessive that the population is effectively zero. Assume that the fish reproduce logistically, disperse by diffusion and within the zone are harvested with an effort E. Justify the following model for the fish population u(x,t):

$$u_t = ru \left(1 - \frac{u}{K}\right) - Eu + Du_{xx} \ ,$$

$$u = 0 \text{ on } x = H, \quad u_x = 0 \text{ on } x = 0 \ ,$$

where r, K, $E(< r)$ and D are positive constants.

   If the fish stock is not to collapse show that the fishing zone H must be greater than $[D/(r - E)]^{1/2}$ km. Briefly discuss any ecological implications.

7. Use the approximation method described in Section 14.7 to determine analytically the critical length L_0 as function of r, q and D such that an outbreak can exist in the spruce budworm population model

$$u_t = ru \left(1 - \frac{u}{q}\right) - \frac{u^2}{1 + u^2} + Du_{xx}, \quad u = 0 \text{ on } x = 0, L \ .$$

Determine the maximum population u_m when $L = L_0$.

8. Suppose that a two species reaction diffusion mechanism in u and v generates steady state spatial patterns $U(x)$, $V(x)$ in a one-dimensional domain of size L with zero flux boundary conditions $u_x = v_x = 0$ at both boundaries $x = 0$ and $x = L$. Consider the heterogeneity functions defined by

$$H_G(w) = L^{-1} \int_0^L w_x \, dx, \quad H_S(w) = L^{-1} \int_0^L [w_x - H_G(w)]^2 \, dx \ .$$

Biologically the first of these simply measures the gradient while the second measures the deviation from the simple gradient. Show that the heterogeneity or energy integral

$$H(t) = L^{-1} \int_0^L (U'^2 + V'^2) \, dx = [H_G(U)]^2 + [H_G(V)]^2 + H_S(U) + H_S(V) \ .$$

[Berding 1987]

9. Show that the reaction diffusion mechanism

$$u_t = f(u) + D\nabla^2 u \ ,$$

where the concentration vector u has n components, D is a diagonal diffusion matrix with elements d_i, $i = 1, 2, \ldots, n$ and f is the nonlinear kinetics, linearises about a positive steady state to

$$\mathbf{w}_t = A\mathbf{w} + D\nabla^2\mathbf{w} \,,$$

where A is the Jacobian matrix of $\mathbf{f}$ at the steady state.

Let k be the eigenvalue of the problem defined by

$$\nabla^2\mathbf{W} + k^2\mathbf{W} = 0, \quad (\mathbf{n}\cdot\nabla)\mathbf{W} = 0 \quad \mathbf{r} \text{ on } \partial B \,.$$

On setting $\mathbf{w} \propto \exp[\lambda t + i\mathbf{k}\cdot\mathbf{r}]$ show that the dispersion relation $\lambda(k^2)$ is given by the solutions of the characteristic polynomial

$$P(\lambda) = |A - k^2 D - \lambda I| = 0 \,.$$

Denote the eigenvalues of $P(\lambda)$, with and without diffusion, by λ_i^+ and λ_i^- respectively. Diffusion driven instability occurs if $\operatorname{Re}\lambda_i^- < 0$, $i = 1, 2, \ldots, n$ and at least one $\operatorname{Re}\lambda_i^+ > 0$ for some $k^2 \neq 0$.

From matrix algebra there exists a transformation T such that

$$|A - \lambda I| = |T^{-1}(A - \lambda I)T| = \prod_{i=1}^{n}(\lambda_i^- - \lambda) \,.$$

Use this result and the fact that $\operatorname{Re}\lambda_i^- < 0$ to show that if $d_i = d$ for all i then $\operatorname{Re}\lambda_i^+ < 0$ for all i and hence that a necessary condition for diffusion driven instability is that at least one diffusion coefficient is different from the rest.

10. The linearisation of a reaction diffusion mechanism about a positive steady state is

$$\mathbf{w}_t = A\mathbf{w} + D\nabla^2\mathbf{w} \,,$$

where A is the Jacobian matrix of the reaction kinetics evaluated at the steady state.

If the matrix $A + A^T$, where T denotes the transpose, is stable this means that all of its eigenvalues λ are real and negative. Show that $\mathbf{w}\cdot A\mathbf{w} < -\delta\mathbf{w}\cdot\mathbf{w}$, for some $\delta > 0$.

[Hint: By considering $\mathbf{w}_t = A\mathbf{w}$ first show that $(\mathbf{w}^2)_t = 2\mathbf{w}A\mathbf{w}$. Then show that $\mathbf{w}_t^T\cdot\mathbf{w} = \mathbf{w}^T A^T\mathbf{w}$ and $\mathbf{w}^T\cdot\mathbf{w}_t = \mathbf{w}^T A\mathbf{w}$ to obtain $(\mathbf{w}^2)_t = \mathbf{w}^T(A + A^T)\mathbf{w}$. Thus deduce that $\mathbf{w}A\mathbf{w} = 1/2\mathbf{w}^T(A^T + A)\mathbf{w} < -\delta\mathbf{w}\cdot\mathbf{w}$ for some $\delta > 0$.]

Let k^2 be the eigenvalues of the eigenvalue problem

$$\nabla^2\mathbf{w} + k^2\mathbf{w} = 0 \,.$$

By considering dE/dt, where

$$E(t) = \int_B \mathbf{w}\cdot\mathbf{w}\,d\mathbf{r}$$

with B the spatial domain, show that $\mathbf{w}^2 \to 0$ as $t \to \infty$ and hence that such reaction diffusion systems cannot generate spatial patterns if the Jacobian matrix is of this particular form.

11. Consider the Lotka-Volterra predator-prey system (see Chapter 3, Section 3.1) with diffusion given by

$$u_t = u(1 - v) + Du_{xx}, \quad v_t = av(u - 1) + Dv_{xx}$$

in the domain $0 \le x \le 1$ with zero flux boundary conditions. By multiplying the first equation by $a(u - 1)$ and the second by $(v - 1)$ show that

$$S_t = DS_{xx} - D\sigma^2 ,$$

$$S = au + v - \ln(u^a v), \quad \sigma^2 = a\left(\frac{u_x}{u}\right)^2 + \left(\frac{v_x}{v}\right)^2 \ge 0 .$$

Determine the minimum S for all u and v. Show that necessarily $\sigma \to 0$ as $t \to \infty$ by supposing σ^2 tends to a non-zero bound the consequences of which are not possible. Hence deduce that no spatial patterns can be generated by this model in a finite domain with zero flux boundary conditions.
[This result can also be obtained rigorously, using maximum principles: the detailed analysis is given by Murray (1975).]

12. The amoebae of the slime mold *Dictyostelium discoideum*, with density $n(x, t)$, secrete a chemical attractant, cyclic-AMP, and spatial aggregations of amoebae start to form. One of the models for this process (and discussed in Section 9.4) gives rise to the system of equations, which in their one-dimensional form, are

$$n_t = D_n n_{xx} - \chi(na_x)_x, \quad a_t = hn - ka + D_a a_{xx} ,$$

where a is the attractant concentration and h, k, χ and the diffusion coefficients D_n and D_a are all positive constants. Nondimensionalise the system.

Consider (i) a finite domain with zero flux boundary conditions and (ii) an infinite domain. Examine the linear stability about the steady state (which introduces a further parameter here), derive the dispersion relation and discuss the role of the various parameter groupings. Hence obtain the conditions on the parameters and domain size for the mechanism to initiate spatially heterogeneous solutions.

Experimentally the chemotactic parameter χ increases during the life cycle of the slime mold. Using χ as the bifurcation parameter determine the critical wave length when the system bifurcates to spatially structured solutions in an infinite domain. In the finite domain situation examine the bifurcating instability as the domain is increased.

Briefly describe the physical processes operating and explain intuitively how spatial aggregation takes place.

15. Animal Coat Patterns and Other Practical Applications of Reaction Diffusion Mechanisms

In this chapter we discuss some real biological pattern formation problems and show how the modelling discussed earlier in the book, particularly the last chapter, can be applied. As an applications chapter of theories developed earlier, it contains considerably more biology than mathematics. Since all models for spatial pattern generation are necessarily nonlinear, practical applications require numerical solutions since no non-trivial analytical solutions are available, nor are likely to be in the near future. A preliminary linear analysis however is usually very useful, generally a necessity in fact. In each of the applications the biological modelling is discussed in detail. Most of the finite amplitude patterns reproduced are numerical solutions of the model equations and are applied directly to the specific biological situation.

In Section 15.1 we shall show how the pattern of animal coat markings such as on the zebra, leopard and so on, could be generated using a reaction diffusion mechanism. In the other sections we shall describe two other pattern formation problems, namely butterfly wing patterns in Section 15.2, and patterns which presage hairs in whorls during regeneration in *Acetabularia*, an important marine alga, in Section 15.3.

Reaction diffusion theory has now been applied to quite a large number of biological situations. For example, Kauffman et al. (1978) (see also Kauffman 1981) presented one of the first practical applications to the early segmentation of the embryo of the fruit fly *Drosophila*. The paper by Bunow et al. (1980) is specifically related to this work but has important general implications: they discuss pattern sensitivity among other things. The book by Meinhardt (1982) is primarily concerned with activation-inhibition reaction diffusion model applications.

A key problem with the application of Turing's (1952) theory of morphogenesis is the identification of the morphogens and this has been a major obstacle to its acceptance as one of the essential processes in development. The fact that certain chemicals are essential for development does not necessarily mean they are morphogens. Identification of their role in the *patterning* process is necessary for this. It is partially for this reason that we discuss in Section 15.3 hair initiation in *Acetabularia*, where calcium is proposed as an example of a real morphogen. Theoretical and experimental evidence will be presented to back up the hypothesis.

15.1 Mammalian Coat Patterns – 'How the Leopard Got Its Spots'

Mammals exhibit a rich and varied spectrum of coat patterns: Fig. 15.1 shows some typical markings. The beautifully illustrated (all drawings) multi-volume (7) series of books, *East African Mammals*, produced since 1971 by Jonathan Kingdon (see, for example, the volume on carnivores, 1978, and large mammals, 1979) give, among other things, the most comprehensive and accurate survey of the wealth and variety of animal coat patterns. The book by Portmann (1952) has some interesting observations on animal forms and patterns. However, as with almost all biological pattern generation problems the mechanism involved has not yet been determined. Murray (1979, 1981a,b) studied this particular pattern formation problem in some depth and it is mainly this work we discuss here: see also the general article in *Scientific American* by Murray (1988). Among other things he suggested that a *single* mechanism could be responsible for generating practically all of the common patterns observed. Murray's theory was based on a chemical concentration hypothesis by Searle (1968) who mentioned the potential of a Turing mechanism. Murray (1979, 1981) took a reaction diffusion system, which could be diffusively driven unstable, as the possible mechanism responsible for laying down most of the spacing patterns; these are the morphogen prepatterns for the animal coat markings. The subsequent differentiation of the cells to produce melanin simply reflects the spatial pattern of morphogen concentration.

In the last chapter we showed how such reaction diffusion mechanisms can generate spatial patterns. In this section we (i) present results of numerical simulations of a specific reaction diffusion system with geometries relevant to the zoological problem; (ii) compare the patterns with those observed in many animals and finally; (iii) highlight the circumstantial evidence to substantiate the hypothesis that a single mechanism is all that is possibly required. Bard (1981) and Young (1984) also investigated animal coat patterns from a reaction diffusion point of view. Cocho et al. (1987) proposed a quite different model based on cell-cell interaction and energy considerations: it is essentially a cellular automata approach.

Although the development of the colour pattern on the integument, that is the skin, of mammals occurs towards the end of embryogenesis, we suggest that it reflects an underlying pre-pattern that is laid down much earlier. In mammals the pre-pattern is formed in the early stages of embryonic development – in the first few weeks of gestation. In the case of the zebra, for example, this is around 21–35 days; the gestation period is about 360 days. To create the colour patterns certain genetically determined cells, called melanoblasts, migrate over the surface of the embryo and become specialised pigment cells, called melanocytes, which lie in the basal layer of the epidermis. Hair colour comes from the melanocytes generating melanin, within the hair follicle, which then passes into the hair. As a result of graft experiments, it is generally agreed that whether or not a melanocyte produces melanin depends on the presence of a chemical although we

(a)

Fig. 15.1a-d. Typical animal coat markings on the (a) Leopard (*Panthera pardus*); (b) Zebra (*Equus grevyi*); (c) Giraffe (*Giraffa camelopardalis tippelskirchi*) (Photographs courtesy of Hans Kruuk); (d) Tiger (*Felis tigris*).

(b)

(c)   (d)

do not yet know what it is. In this way the observed coat colour pattern reflects an underlying chemical prepattern, to which the melanocytes are reacting to produce melanin.

For any pattern formation mechanism to be applicable the scale of the actual size of the patterns has to be large compared to the cell diameter. For example, the number of cells in a leopard spot, at the time of laying down the pattern, is probably of the order of 0.5 mm, that is of the order of 100 cells. Since we do not know what reaction diffusion mechanism is involved, and since all such systems are effectively mathematically equivalent as we saw in the previous chapter, all we need at this stage is a specific system to study numerically. Solely for illustrative purposes Murray (1979, 1981a) chose the Thomas (1976) system, the kinetics of which is given in (14.5) in the last chapter: it is a real experimental system with parameters associated with real kinetics. The nondimensional system is given by (14.7) with (14.8), namely

$$\frac{\partial u}{\partial t} = \gamma f(u, v) + \nabla^2 u, \quad \frac{\partial v}{\partial t} = \gamma g(u, v) + d\nabla^2 v$$

$$f(u, v) = a - u - h(u, v), \quad g(u, v) = \alpha(b - v) - h(u, v) \qquad (15.1)$$

$$h(u, v) = \frac{\rho u v}{1 + u + K u^2} .$$

Here a, b, α, ρ and K are positive parameters. The ratio of diffusion coefficients, d, must be greater than one for diffusion driven instability to be possible. Recall from the last chapter that the scale factor γ is a measure of the domain size.

With the integument of the mammalian embryo in mind the domain is a closed surface and appropriate conditions for the simulations are periodic boundary conditions with relevant initial conditions – random perturbations about a steady state. We envisage the process of pattern formation to be activated at a specific time in development, which implies that the reaction diffusion domain size and geometry is prescribed. The initiation switch could, for example, be a wave progressing over the surface of the embryo which effects the bifurcation parameter in the mechanism which in turn activates diffusion-driven instability. What initiates the pattern formation process, and how it is initiated, are not the problems we address here. We wish to consider only the pattern formation potential of the mechanism and see whether or not the evidence for such a system is borne out when we compare, the patterns generated by the mechanism and observed animal coat markings with similar geometrical constraints to those in the embryo.

In Sections 14.4 and 14.5 in the last chapter we saw how crucially important the scale and geometry of the reaction diffusion domain were in governing the actual spatial patterns of morphogen concentration which started to grow when the parameters were in the space where the system was diffusionally unstable. Refer also to Figs. 14.6 and 14.7. (It will be helpful in the following to have the analyses and discussions in Sections 14.1–14.6 in mind.)

To investigate the effects of geometry and scale on the type of spatial patterns generated by the full nonlinear system (15.1) we chose for numerical simulation a series of two-dimensional domains which reflect the geometric constraints of an embryo's integument.

Let us first consider the typical markings found on the tails and legs of animals, which we can represent as tapering cylinders, the surface of which is the reaction diffusion domain. From the analysis in Sections 14.4 and 14.5 when the mechanism undergoes diffusion driven instability, linear theory gives the range of unstable modes, k^2, in terms of the parameters of the model system: in two space dimensions with domain defined by $0 < x < p$, $0 < y < q$, these are given by (14.43) as

$$\gamma L = k_1^2 < k^2 = \pi^2 \left(\frac{n^2}{p^2} + \frac{m^2}{q^2} \right) < k_2^2 = \gamma M \tag{15.2}$$

where L and M are functions only of the kinetics parameters of the reaction diffusion mechanism. With zero flux boundary conditions, the solution, which involves exponentially growing modes about the uniform steady state, is given by (14.43) as

$$\sum_{n,m} \mathbf{C}_{n,m} \exp\left[\lambda(k^2)t\right] \cos \frac{n\pi x}{p} \cos \frac{m\pi y}{q} ,$$

$$\text{where} \quad k^2 = \pi^2 \left(\frac{n^2}{p^2} + \frac{m^2}{q^2} \right) \tag{15.3}$$

where the $\mathbf{C}$ are constants and summation is over all pairs (n,m) satisfying (15.2).

Now consider the surface of a tapering cylinder of length s with $0 \le z \le s$ and with circumferential variable θ. The linear eigenvalue problem equivalent to that in (14.41) requires the solutions $\mathbf{W}(\theta, z; r)$ of

$$\nabla^2 \mathbf{W} + k^2 \mathbf{W} = 0 , \tag{15.4}$$

with zero flux conditions at $z = 0$ and $z = s$ and periodicity in θ. Since we are only concerned here with the surface of the tapering cylinder as the domain, the radius of the cone, r, at any point is essentially a 'parameter' which reflects the thickness of the cylinder at a given z. The equivalent solution to (15.3) is

$$\sum_{n,m} \mathbf{C}_{n,m} \exp\left[\lambda(k^2)t\right] \cos\left(n\theta\right) \cos \frac{m\pi z}{s} ,$$

$$\text{where} \quad k^2 = \frac{n^2}{r^2} + \frac{m^2\pi^2}{s^2} \tag{15.5}$$

where the summation is over all pairs (n,m) satisfying the equivalent of (15.2), namely

$$\gamma L = k_1^2 < k^2 = \frac{n^2}{r^2} + \frac{m^2\pi^2}{s^2} < k_2^2 = \gamma M \tag{15.6}$$

Note that r appears here as a parameter.

Now consider the implications as regards the linearly growing spatial patterns, which we know for simple patterns usually predict the finite amplitude spatial patterns which are ultimately obtained. If the tapering cylinder is everywhere very thin this means r is small. In turn this implies that the first circumferential mode with $n = 1$, and all others with $n > 1$, in (15.5) lie outside the unstable range defined by (15.6). In this case the unstable modes involve only z-variations. In other words it is equivalent to the one-dimensional situation with only one-dimensional patterns as in Fig. 14.6: see also Fig. 15.2 (a). If, however, r is large enough near one end so that $n \neq 0$ is in the unstable range defined by (15.6), θ-variations appear. We thus have the situation in which there is a gradation from a two-dimensional pattern in z and θ at the thick end to the one-dimensional pattern at the thin end. Fig. 15.2 shows some numerically computed solutions (using a finite element procedure) of (15.1) for various sizes of a tapering domain. In Fig. 15.2 (a), (b) the only difference is the scale parameter γ, which is the bifurcation parameter here. In fact in all the numerical simulations of the nonlinear system (15.1) reproduced in Figs. 15.2–15.4, the mechanism parameters were kept fixed and *only* the scale and geometry varied. Although the bifurcation parameter is scale (γ), geometry plays a crucial role.

The tail patterns illustrated in Fig. 15.2 are typical of many spotted animals, particularly the cats (*Felidae*). The cheetah, jaguar and genet are good examples of this pattern behaviour. In the case of the leopard (*Panthera pardus*) the spots almost reach the tip of the tail, whereas with the cheetah there is always a distinct striped part and the genet has a totally striped tail. This is consistent with the embryonic tail structure of these animals around the time we suppose the pattern formation mechanism is operative. The genet embryo tail has a remarkably uniform diameter which is relatively quite thin: in the photograph of the fully grown genet in Fig. 15.2 (h) the hair is typically fluffed up. The pre-natal leopard tail sketched in Fig. 15.2 (g) is sharply tapered and relatively short: the adult leopard tail (see Fig. 15.1 (a) and Fig. 15.2 (g)) is long but it has the same number of vertebrae. Thus the fact that the spots go almost to the tip is consistent with a rapid taper, with stripes, often incomplete, only appearing at the tail tip, if at all. This post-natal stretching is also reflected in the larger spots further down the tail as compared to those near the base or the body generally; refer to Fig. 15.1 (a) and Fig. 15.2 (g).

Consider now the typical striping on the zebra as in Fig. 15.1 (b) and Fig. 15.3 (a), (b). From the simulations reproduced in Fig. 15.2 (a) we see that reaction diffusion mechanisms can generate stripes easily. Zebra striping was investigated in detail by Bard (1977) who argued that the pattern was laid down around the 3rd to 5th week through gestation. He did not discuss any actual patterning mechanism but from the results in this section this is not a problem. The different species of zebra have different stripe patterns and he suggested that the stripes were thus laid down at different times in gestation. Fig. 15.3 shows the hypothesised patterning on embryos at different stages in gestation and schematically shows the effect of growth.

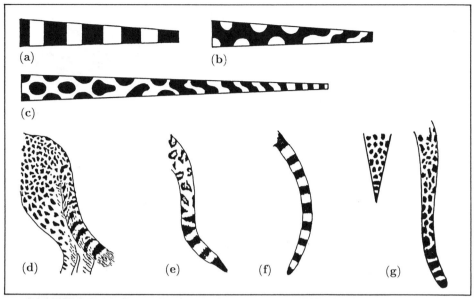

Fig. 15.2a-h. Computed solutions of the nonlinear reaction diffusion system (15.1) with zero flux boundary conditions and initial conditions taken as random perturbations about the steady state. The dark regions represent concentrations in the morphogen u above the steady state u_s. Parameter values: $\alpha = 1.5$, $K = 0.1$, $\rho = 18.5$, $a = 92$, $b = 64$ (these imply a steady state $u_s = 10$, $v_s = 9$), $d = 10$. With the same geometry, in (a) the scale factor $\gamma = 9$ and in (b) $\gamma = 15$. Note how the pattern bifurcates to more complex patterns as γ increases. In (c) the scale factor $\gamma = 25$ and a longer domain is used to illustrate clearly the spot-to-stripe transition: here the dark regions have $u < u_s$. (d) Typical tail markings from an adult cheetah (*Acinonyx jubatis*). (e) Typical adult jaguar (*Panthera onca*) tail pattern. (f) Just pre-natal tail markings in a male genet (*Genetta genetta*). (After Murray 1981a,b) (g) Typical markings on the tail of an adult leopard. Note how far down the tail the spots are with only a few stripes near the tip. See also the photograph in Fig. 15.1 (a) where the leopard's tail is conveniently draped so as to demonstrate this trait clearly. The pre-natal leopard tail is very much shorter and shows why the adult pattern is as shown. (h) A common genet (*Genetta genetta*) showing the distinctly striped tail emerging from a spotted body. (Photograph courtesy of Hans Kruuk)

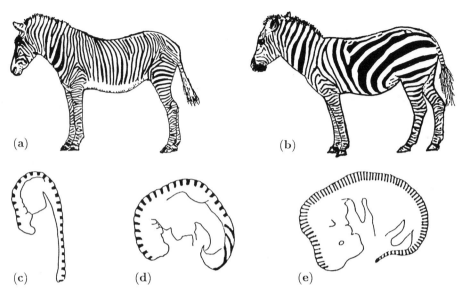

Fig. 15.3a-e. Typical zebra patterns: (a) *Equus grevyi*; (b) *Equus burchelli*. Proposed stripe pattern 0.4 mm apart superimposed on two zebra embryos: (c) 21 day embryo; (d) The effect of 3–4 days of the pattern in (c). (e) A similar stripe pattern laid down on a 5 week old embryo. ((c)-(e) redrawn after Bard 1977)

By noting the number of adult stripes, and how they had been distorted by growth if laid down as a regular stripe array, Bard (1977) deduced that the distance between the stripes when they were laid down was about 0.4 mm. He also deduced the time in gestation when they were created. Fig. 15.3 (c), with pattern distortion with growth as shown in Fig. 15.3 (d), is consistent with the stripe pattern on the zebra *Equus burchelli* as in Fig. 15.3 (b): see also the photograph in Fig. 15.1 (b). Grevy's zebra, *Equus grevyi*, in Fig. 15.3 (a) has many more stripes and these are laid down later in gestation, around 5 weeks, as in Fig. 15.3 (e) where again they are taken to be 0.4 mm apart.

Fig. 15.4a,b. (a) Typical examples of scapular stripes on the foreleg of zebra (*Equus zebra zebra*). (b) Pedicted spatial pattern from the reaction diffusion mechanism: see also Fig. 15.8 (e) below. (After Murray, 1979, 1981a)

If we now look at the scapular stripes on the foreleg of zebras as illustrated in Fig. 15.4 (a), we have to consider an actual pattern formation mechanism as done by Murray (1980, 1981). Here we see that the mathematical problem is that of the junction between a linear striped domain joined at right angles to another striped domain: Fig. 15.4 (b) is the pattern predicted by the reaction diffusion mechanism for such a domain. The experimentally obtained pattern displayed below in Fig. 15.8 (e) confirms this mathematical prediction.

The markings on zebras are extremely variable yet remain within a general stripe theme. Animals which are almost completely black with lines of white spots as well as those almost completely white have been seen: see, for example, Kingdon (1979).

If we now consider the usual markings on the tiger (*Felis tigris*) as in Fig. 15.1 (d) we can see how its stripe pattern could be formed by analogy to the zebra. The gestation period for the tiger is around 105 days. We anticipate the pattern to be laid down quite early on, within the first few weeks, and that the mechanism generates a regularly spaced stripe pattern at that time. Similar remarks to those made for the zebra regarding growth deformation on the stripe pattern equally apply to tiger stripes. Many tigers show similar distortions in the adult animal.

Let us now consider the giraffe, which is one of the largest animals that still exhibits a spotted pattern. Fig. 15.5 (a) is a sketch of a giraffe embryo 35–45 days old: it already has a clearly recognizable giraffe shape, even though the gestation period is about 457 days. The prepattern for the giraffe coat pattern has almost certainly been laid down by this time. Fig. 15.5 (b) is a sketch of typical neck markings on the reticulated giraffe. Figs. 15.5 (c)–(e) are tracings, on approximately the same scale, of trunk spots from the major giraffe species. Fig. 15.5 (f) shows a typical pattern computed from the mechanism (15.1) with the same kinetics parameter values as for Fig. 15.2.

We arbitrarily chose the homogeneous steady state as the threshold for melanocytes to produce melanin, represented by the dark regions in the figure. It is possible, of course, that the threshold which triggers melanogenesis is either lower or larger than the homogeneous steady state. For example, if we choose a lower threshold we get a different pattern: Fig. 15.5 (g) is an example in which we chose a lower threshold in the simulations which gave Fig. 15.5 (f). This produces larger areas of melanin. We can thus see how the markings on different species of giraffe could be achieved simply, if the melanocytes are programmed to react to a lower morphogen concentration. The giraffe photographs in Murray (1988) illustrate this particularly clearly.

The dramatic effect of scale is clearly demonstrated in Fig. 15.6 where only scale varies from one picture to the other – as indicated by the different values for γ. It is not suggested that this is necessarily the typical shape of the integument at the time of pre-pattern formation, it is only a non-trivial specimen shape to illustrate the results and highlight the striking effect of scale on the patterns generated. If the domain size (γ) is too small, then no spatial pattern can be generated. We discussed this in detail in the last chapter, but it is clear from

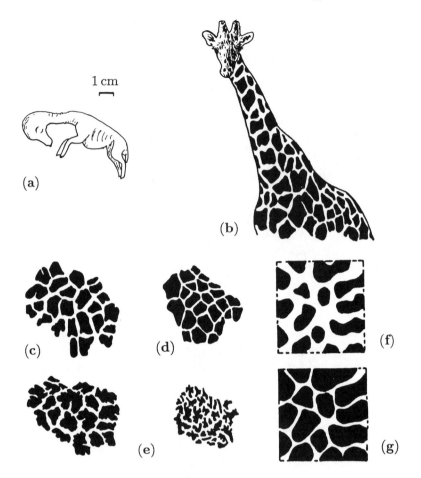

Fig. 15.5a-g. (a) Giraffe (*Giraffa camelopardalis*): 35–45-day embryo. (b) Typical neck spots on the reticulated giraffe (*Giraffa camelopardalis reticulata*). (c)-(e) Tracings (after Dagg 1968) of trunk spots (to the same scale) of giraffe, *Giraffa camelopardalis* (c) *rothschildi*, (d) *reticulata*, (e) *tippelskirchi*. (f) Spatial patterns obtained from the model mechanism (15.1) with kinetics parameter values as in Fig. 15.2. (g) Spatial pattern obtained when a lower threshold than in (f) is considered to initiate melanogenesis in the same simulations which gave (f). (From Murray 1981b, 1988)

the range of unstable modes, m and n in (15.6) for example. With a small enough domain, that is small enough γ, even the lowest non-zero m and n lie outside the unstable range. This implies that in general very small animals can be expected to be uniform in colour; most of them are. As the size increases, γ passes through a series of bifurcation values and different spatial patterns are generated. However, for very large domains as in Fig. 15.6 (g), the morphogen concentration distribution is again almost uniform: the structure is very fine. This might appear, at first sight, somewhat puzzling. It is due to the fact that for large domains, large γ, the linearly unstable solutions derived from (15.3)

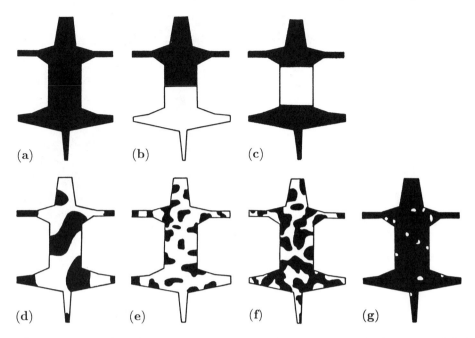

Fig. 15.6a-g. Effect of body surface scale on the spatial patterns formed by the reaction diffusion mechanism (15.1) with parameter values $\alpha = 1.5$, $K = 0.125$, $\rho = 13$, $a = 103$, $b = 77$ (steady state $u_s = 23$, $v_s = 24$), $d = 7$. Domain dimension is related directly to γ. (a) $\gamma < 0.1$; (b) $\gamma = 0.5$; (c) $\gamma = 25$; (d) $\gamma = 250$; (e) $\gamma = 1250$; (f) $\gamma = 3000$; (g) $\gamma = 5000$. (From Murray 1980, 1981a)

have the equivalent of large m and n, which implies a very fine scale pattern; so small, in fact, that essentially no pattern can be seen. This suggests that most very large animals, such as elephants, should be almost uniform in colour, as indeed most are.

Consider now the first bifurcation from a uniform coat pattern as implied by Fig. 15.6 (b). Fig. 15.7 are sketches and a photograph of two striking examples of the half-black, half-white pattern in Fig. 15.6 (b), namely the ratel, or honey badger, and the Valais goat. The next bifurcation for a longer and still quite thin embryo is elegantly illustrated in Fig. 15.7 (d) which relates to the pattern in Fig. 15.6 (c).

In all the numerical simulations with patterns other than the simplest, such as in Figs. 15.6 (a), (b), (c) the final patterns were dependent on the initial conditions. However, for a given set of parameters, geometry and scale, the patterns for all initial conditions are *qualitatively* similar. From the point of view of the applicability of such mechanisms for generating animal coat patterns, this dependence on initial conditions is a very positive attribute of such models. The reason is that the initial random conditions for each animal are unique to that animal and hence so is its coat pattern, but each lies within its own general class. So, all leopards have a spotted pattern yet each has a unique distribution

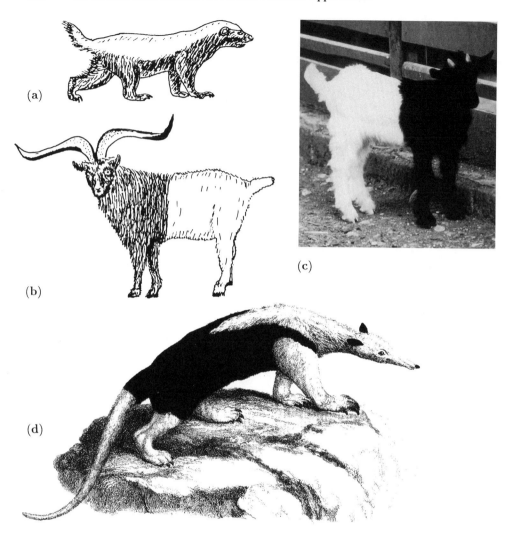

Fig. 15.7a-d. Examples of the simplest coat patterns found in animals: (a) ratel or honey badger (*Mellivora capensis*). (b) Adult valais goat (*Capra aegagrus hircus*). (After Herán 1976). (c) Young valais goat. (Photograph courtesy of Avi Baron and Paul Munro) (d) The next pattern bifurcation is dramatically and elegantly illustrated in an early 19th century print of the anteater (*Tamandua tetradactyl*).

of spots. On tigers and zebras, for example, the stripe patterns can be quite diverse while still adhering to a general theme.

Although we have considered only a few specific coat markings (see Murray 1981a,b, 1988 for further discussion) we see that there is a striking similarity between the patterns that the model generates and those found on a wide variety of animals. Even with the restrictions we imposed on the parameters for our simulations the wealth of possible patterns is remarkable. The patterns depend strongly on the geometry and scale of the reaction domain although later growth may distort the initial pattern.

To summarize we hypothesize that almost all animal coat patterns can be generated by a single mechanism. Any reaction diffusion mechanism capable of generating diffusion-driven spatial patterns would be a plausible model. The pattern which evolves is determined by the time the mechanism is activated since this relates directly to the geometry and scale of the embryo's integument. The time of the activation wave (such as that illustrated in Fig. 14.12 (d)), or activation switch, is inherited. With most small animals with short gestation periods we would expect uniformity in color, which is generally the case. For larger surface integument at the time of activation, the first bifurcation produces patterns where animals can be half black and half white: see Fig. 15.7. For progressively larger domains at activation more and more pattern structures emerge, with a progression through certain anteaters, zebras onto the large cats and so on. The simpler patterns are remarkably stable, that is they are quite insensitive to conditions at the time the mechanism is activated. At the upper end of the size scale we have more variability within a class as in the close spotted giraffes. As mentioned we expect very large animals to be uniform in colour again, which indeed is generally the case, with elephants, rhinoceri and hippopotami being typical examples.

As mentioned above, we expect the time of activation of the mechanism to be inherited and so, at least in animals where the pattern is important for survival, pattern formation is initiated when the embryo is a given size. Of course, the conditions on the embryo's surface at the time of activation naturally exhibit a certain randomness which produces patterns which depend uniquely on the initial conditions, the geometry and the scale. A very important aspect of this type of mechanism is that, for a given geometry and scale, the patterns found for a variety of random initial conditions are qualitatively similar. For example, with a spotted pattern it is essentially only the distribution of spots which varies. The resultant individuality is important for both kin and group recognition. Where the pattern is of little importance to the animals survival, as with domestic cats, the activation time need not be so carefully controlled and so pattern polymorphism, or variation, is much greater.

This model also offers possible explanations for various pattern anomalies on some animals. Under certain circumstances a change in the value of one of the parameters can result in a very marked change in the pattern obtained. An early activation for example in a zebra would result in an all black animal. A delay in activating the mechanism would give rise to spots on the underlying black field. Examples of both of these have been observed and recorded, for example, by Kingdon (1979).

Whether a parameter change affects the pattern markedly depends on how close the parameter value is to a bifurcation value (recall the discussion in Section 14.5 and Fig. 14.11). The fact that a small change in a parameter near a bifurcation boundary can result in relatively large changes in pattern has important implications for evolutionary theory as we shall see later in Chapter 18.

It is an appealing idea that a single mechanism could generate all of the observed mammalian coat patterns. Reaction diffusion models, and the new and powerful mechanochemical models discussed later in Chapter 17, have many of the attributes such a pattern formation mechanism must have. The latter in fact have a pattern generation potential even richer than that of reaction diffusion mechanisms. The considerable circumstantial evidence which comes from comparing the patterns generated by the model mechanism with specific animal pattern features is encouraging. The fact that many general and specific features of mammalian coat patterns can be explained by this simple theory does not, of course, mean that it is correct, but so far they have not been explained satisfactorily by any other theory. The above results nevertheless support a single all-encompassing mechanism for pattern formation for animal coat markings.

As an interesting mathematical footnote, the initial stages of spatial pattern formation by reaction diffusion mechanisms (when departures from uniformity are small) poses the same type of mathematical eigenvalue problem as that describing the vibration of thin plates or drum surfaces. The vibrational modes are also governed by (15.4) except that $\mathbf{W}$ now represents the amplitude of the vibration. So, we can highlight experimentally how crucial geometry and scale are to the patterns by examining analogous vibrating drum surfaces. If the size of the surface is too small it will simply not sustain vibrations; that is the disturbances simply die out very quickly. Thus a minimum size is required before we can excite any sustainable vibration. If we consider a domain similar to that in Fig. 15.6 for the drum surface, which in our model is the reaction diffusion domain, we get a set of increasingly complicated modes of possible vibration as we increase the size.

Although it is not easy to use the same boundary conditions for the vibrations that we used in the reaction diffusion simulations, the general features of the patterns exibited must be qualitatively similar from mathematical considerations. The equivalent of γ in the vibrating plate problem is the frequency of the forcing vibration. So if a pattern forms on a plate vibrating at a given frequency then the pattern formed on a larger but similar plate is the same as on the original plate vibrated at a proportionally larger frequency. According to linear vibration theory, a doubling of the plate size, for example, is equivalent to keeping the original plate size and doubling the frequency. These experiments were carried out for geometries similar to those in Fig. 15.2 (c), Fig. 15.4 and Fig. 15.6 and the results are shown in Fig. 15.8 (see Xu, Vest and Murray 1983 for further details).

15.2 A Pattern Formation Mechanism for Butterfly Wing Patterns

The variety of different patterns, as well as their spectacular colouring, on butterfly and moth wings is astonishing. Fig. 15.9 (a), (b) show but two examples:

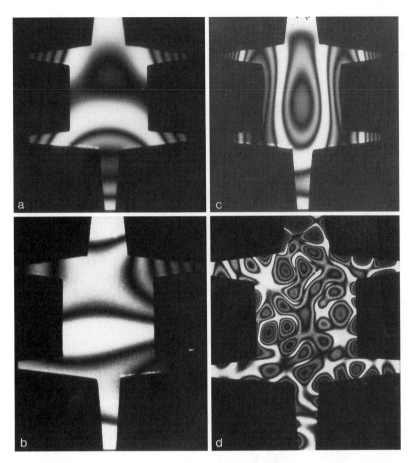

Fig. 15.8a-f. Sequence of time-average holographic interferograms on a plate excited by sound waves of increasing frequency from (a) to (d). The vibrational patterns are broadly in line with the patterns shown in Fig. 15.6. Increasing frequency is equivalent to vibration at a constant frequency and increasing the plate size. (e) This shows vibrations very similar to the predicted pattern in Fig. 15.4 (b) while in (f) the spot to stripe transition in a tapering geometry is clearly demonstrated. (From Xu, Vest and Murray 1983)

see also Fig. 15.19 below. There are close to a million different types of butterflies and moths. The study of butterfly wing colours and patterns has a long history, often carried out by gifted amateur scientists, particularly in the 19th century. In this century there has been a burgeoning of scientific activity. A good and succinct review of the major elements in lepidopteran wing patterns is given by Nijhout (1978); see also Nijhout (1985a). Although the spectrum of different wing patterns is at first sight bewildering, Schwanwitsch (1924) and Suffert (1927) showed that in the case of the Nymphalids there are relatively few pattern elements: see, for example, the review by Nijhout (1978). Fig. 15.9 (c) shows the basic groundplan for the wing patterns in the Nymphalids; each pattern has a

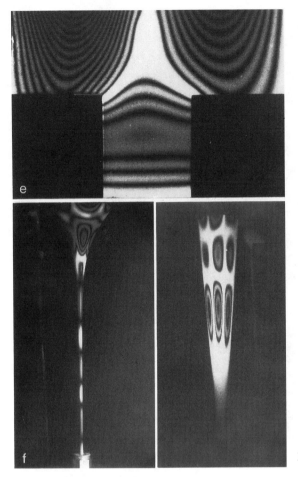

Fig. 15.8e,f

specific name. In this section we shall discuss a possible model mechanism for generating some of these regularly recurring patterns and compare the results with specific butterfly patterns and experiments.

As with the development of the coat patterns on mammals the patterns on the wings of lepidoptera (butterflies and moths) appear towards the end of morphogenesis but they reflect an underlying prepattern that was laid down much earlier. The prepattern in lepidoptera is probably laid down during the early pupal stage or in some cases it perhaps starts just before (Nijhout 1980a).

Here we shall describe and analyse a possible model mechanism for wing patterns proposed by Murray (1981b). We apply it to various experiments concerned with the effect on wing patterns of cautery at the pupal stage in the case of the 'determination stream hypothesis' (Kuhn and von Englehardt 1933), and on transplant results associated with the growth of ocelli or eyespots (Nijhout 1980a) all of which will be described below. As in the last section a major feature of the model is the crucial dependence of the pattern on the geometry and scale

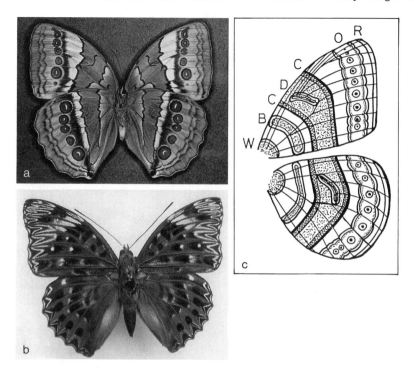

Fig. 15.9a-c. Examples of the varied and complex patterns on butterfly wings: (a) *Stichophthalma camadeva*. (Photograph courtesy of H.F. Nijhout) (b) *Dichorragia nesimachus*. (c) Basic groundplan of the pattern elements in the forewing and backwings in the Nymphalids (after Schwanwitsch 1924, Suffert 1927). The letters denote: marginal bands (R), border ocelli (O), central symmetry bands (C), discal spots (D), basal bands (B), wing root bands (W). The butterfly in (a) exhibits almost all of the basic pattern elements. The arrowhead patterns in (b) pose a particular challenge to any pattern formation mechanism.

of the wing when the pattern is laid down. Although the diversity of wing patterns might indicate that several mechanisms are required, among other things we shall show here how seemingly different patterns can be generated by the same mechanism.

As just mentioned, the formation of wing pattern can be made up by a combination of relatively few pattern elements. Of these, the *central symmetry patterns* (refer to Fig. 15.9 (c)) are common, particularly so in moth wings, and roughly consist of mirror image patterns about a central anterior-posterior axis across the middle of the wing (see, for example, Fig. 15.9 (a)). They were studied extensively by Kuhn and von Engelhardt (1933) in an attempt to understand the pattern formation on the wings of the small moth *Ephestia kuhniella*. They proposed a phenomenological model in which a 'determination stream' emanates from sources at the anterior and posterior edges of the wing and progresses as a wave across the wing to produce anterior-posterior bands of pigment: see Fig. 15.10 (b) below. They carried out microcautery experiments on the pupal wing and their results were consistent with their phenomenological hypothesis. Work by Henke (1943) on 'spreading fields' in *Lymantria dispar* (see Fig. 15.12 (g) below) also supports this hypothesis. The results from the model mechanism discussed in this section will also be related to his experiments. The model relies on scale-forming stem cells in the epithelium reacting to underlying patterns laid down during the pupal or just pre-pupal stage. Goldschmidt (1920) suggested that primary patterns may be laid down before the pattern is seen: this seems to be borne out by more recent experimental studies.

Eyespots or ocelli are important elements in many butterfly wings: see the examples in Fig. 15.9 (a). Nijhout (1980a,b) presents evidence, from experiments on the nymphalid butterfly *Precis coenia*, that the foci of the eyespots are the influencing factors in their pattern formation. The foci generate a morphogen, the level of which activates a colour-specific enzyme. Colour production, that is melanogenesis, in *Precis coenia* involves melanins which are not all produced at the same time (Nijhout 1980b). In another survey Sibatani (1980) proposes an alternative model based on the existence of an underlying prepattern and suggests that the ocellus-forming process involves several interacting variables. These two models are not necessarily mutually exclusive since a 'positional information' (Wolpert 1969) model relies on cells reacting in a specified manner to the concentration level of some morphogen.

The cautery work of Kuhn and von Engelhardt (1933) suggests that there are at least two mechanisms in the pattern formation in *Ephestia kuhniella*, since different effects are obtained depending on the time after pupation at which the cauterization occurs. There are probably several independent pattern-formation systems operating, as was suggested by Schwanwitsch (1924) and Suffert (1927). However, the same mechanism, such as that discussed here, could simply be operating at different times, which could imply different parameter values and different geometries and scale to produce quite different patterns. It would also not be unreasonable to postulate that the number of melanins present indicates the minimum number of mechanisms, or separate runs of the same mechanism.

Although the main reason for studying wing pattern in lepidoptera is to try to understand their formation with a view to finding a pattern generation mechanism (or mechanisms), another is to show evidence for the existence of diffusion fields greater than about 100 cells (about 0.5 mm), which is about the maximum found so far. With butterfly wing patterns, fields of $O(5\,\mathrm{mm})$ seem to exist. From a modelling point of view an interesting aspect is that the evolution of pattern looks essentially two-dimensional, so we must again consider the roles of both geometry and scale. As in the last section we shall see that seemingly different patterns can be generated with the same mechanism simply by its activation at different times on different geometries and on different scales.

Model Mechanism: Diffusing Morphogen – Gene Activation System

We first briefly discuss central symmetry patterns (see Fig. 15.9 (c)) since it is the experimental work on these which motivates the model mechanism. These

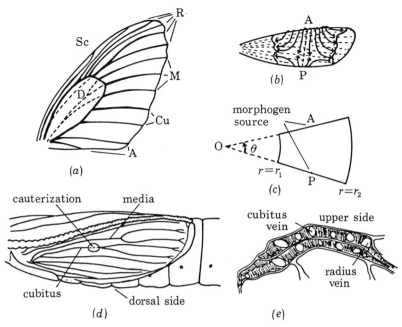

Fig. 15.10a-e. (a) Forewing of a generalized lepidopteran with the basic venation nomenclature: A, anal; Cu, cubitus; M, media; R, radius; Sc, subcosta, D, discal. The regions between veins are wing cells. Dotted lines represent veins that exist at the pupal stage, but later atrophy. (b) Hypothesized 'determination stream' for central symmetry pattern formation (after Kuhn and von Engelhardt 1933). (c) Idealized pupal wing with A and P the anterior and posterior sources of the determination stream (morphogen). (d) Schematic representation of the right pupa wing approximately 6–12 hours old (after Kuhn and von Engelhardt 1933). (e) Schematic cross section through the wing vertically through the cauterized region, showing the upper and lower epithelia and veins. (After Kuhn and von Engelhardt 1933)

crossbands of pigment generally run from the anterior to the posterior of the wings and are possibly the most prevalent patterns. Dislocation of these bands along wing cells, namely regions bounded by veins and a wing edge, can give rise to a remarkably wide variety of patterns (see, for example, Nijhout 1978, 1985a): Fig. 15.9 (a) displays a good example. Fig. 15.10 (a) is a diagram of the forewing of a generalized lepidopteran, and illustrates typical venation including those in the discal cell (D) where the veins later atrophy and effectively disappear.

Kuhn and von Engelhardt (1933) carried out a series of experiments, using microcautery at the pupal stage, to try to see how central symmetry patterns arose on the forewing of the moth *Ephestia kuhniella*. Some of their results are illustrated in Fig. 15.12 (a)–(c) below. They seem consistent with a 'determination stream' or wave emanating from sources on the anterior and posterior edges of the wing, namely at A and P in Fig. 15.10 (b). The front of this wave is associated with the position of the crossbands of the central symmetry system. The work of Schwartz (1962) on another moth tends to confirm the existence of such a determination stream for central symmetry systems. Here we develop a possible mechanism, which we suggest operates just after pupation, for generating this specific pattern (as well as others) and we compare the results with experiment.

We assume that there are sources of a morphogen, with concentration S, situated at A and P on the anterior and posterior edges of the wing, which for simplicity (not necessity) in the numerical calculations is idealized as shown in Fig. 15.10 (c) to be a circular sector of angle θ bounded by radii r_1 and r_2. At a given time in the pupal stage, which we assume to be genetically determined, a given amount of morphogen S_0 is released and it diffuses across the wing. The wing has an upper and lower epithelial surface layer of cells and vein distribution such as illustrated in Fig. 15.10 (d), (e). The pattern on the upper and lower sides of the wing are determined independently. As the morphogen diffuses we assume it is degraded via first order kinetics. The diffusion field is the wing surface and so we have zero flux boundary conditions for the morphogen at the wing edges. The governing equation for the morphogen concentration $S(r, \theta, t)$ is then

$$\frac{\partial S}{\partial t} = D \left(\frac{\partial^2 S}{\partial r^2} + r^{-1} \frac{\partial S}{\partial r} + r^{-2} \frac{\partial^2 S}{\partial \theta^2} \right) - KS \tag{15.7}$$

where $D(\mathrm{cm^2 s^{-1}})$ is the diffusion coefficient and $K(\mathrm{s^{-1}})$ the degradation rate constant.

As S diffuses across the wing surface, suppose the cells react in response to the local morphogen level, and a gene G is activated by S to produce a product g. We assume that the kinetics of the gene product exhibits a biochemical switch behaviour such as we discussed in Chapter 1 and in more detail in Chapter 5 (note specifically Exercise 3): see also Fig. 15.11 (c), (d) below. Such a mechanism can effect a permanent change in the gene product level as we shall see. (Alternatively, a model, with similar kinetics, in which the morphogen activates a colour-specific enzyme that depends on the local morphogen level is another possible mechanism.) There are now several such biochemically plausible switch mechanisms (see, for example, Edelstein 1972; Babloyantz and Hiernaux 1975). It is not important at this stage which switch mechanism we use for the gene product g, but, to be specific we use the one proposed by G. Mitchison and used by Lewis et al. (1977), namely

$$\frac{dg}{dt} = K_1 S + \frac{K_2 g^2}{K_4 + g^2} - K_3 g \ , \tag{15.8}$$

where the K's are positive parameters. Here g is activated linearly by the morphogen S, by its own product in a nonlinear positive feedback way and linearly degraded proportional to itself. $g(t; r, \theta)$ is a function of position through S.

The model involves S_0 of morphogen released on the wing boundaries at A and P as in Fig. 15.10 (c) in the idealized wing geometry we consider. The morphogen satisfies (15.7) within the domain defined by

$$r_1 \leq r \leq r_2, \quad 0 \leq \theta \leq \theta_0 \ , \tag{15.9}$$

and S satisfies zero flux boundary conditions. S_0 is released from $A(r = r_A, \theta = \theta_0)$ and $P(r = r_P, \theta = 0)$ as delta functions at $t = 0$: initially $S = 0$ everywhere. We take the gene product to be initially zero, that is $g(0; r, \theta) = 0$. The appropriate boundary and initial conditions for the mathematical problem are then

$$S(r, \theta, 0) = 0, \quad r_1 < r < r_2, \quad 0 < \theta < \theta_0,$$

$$S(r, 0, 0) = S_0 \delta(r - r_P), \quad S(r, \theta_0, 0) = S_0 \delta(r - r_A),$$

$$\frac{\partial S}{\partial r} = 0, \quad 0 \leq \theta \leq \theta_0, \quad r = r_1, \quad r = r_2,$$

$$\frac{\partial S}{\partial \theta} = 0, \quad r_1 < r < r_2, \quad \theta = 0, \quad \theta = \theta_0,$$

$$g(0; r, \theta) = 0,$$

(15.10)

where $\delta(t)$ is the Dirac delta function. Equations (15.7) and (15.8) with (15.10) uniquely determine S and g for all $t > 0$.

As always, it is useful to introduce nondimensional quantities to isolate the key parameter groupings and indicate the relative importance of different terms in the equations. Let $L(\text{cm})$ be a standard reference length and $a(\text{cm})$, for example $r_2 - r_1$, a relevant length of interest in the wing. Introduce dimensionless quantities by

$$\gamma = \left(\frac{a}{L}\right)^2, \quad S^* = \frac{S}{S_0}, \quad r^* = \frac{r}{a}, \quad t^* = \frac{D}{a^2}t,$$

$$k = \frac{KL^2}{D}, \quad k_1 = \frac{K_1 S_0 L^2}{D\sqrt{K_4}}, \quad k_2 = \frac{K_2 L^2}{D\sqrt{K_4}},$$

(15.11)

$$k_3 = \frac{K_3 L^2}{D}, \quad g^* = \frac{g}{\sqrt{K_4}}$$

and the model system (15.7), (15.8) with (15.10) becomes, on dropping the asterisks for notational simplicity,

$$\frac{\partial S}{\partial t} = \frac{\partial^2 S}{\partial r^2} + r^{-1}\frac{\partial S}{\partial r} + r^{-2}\frac{\partial^2 S}{\partial \theta^2} - \gamma k S,$$

$$\frac{dg}{dt} = \gamma\left(k_1 S + \frac{k_2 g^2}{1 + g^2} - k_3 g\right) = \gamma f(g; S).$$

(15.12)

The last equation defines $f(g; S)$.

Recall, from the last section, the reason for introducing the scale parameter γ, namely the convenience in making scale changes easier. If our 'standard' wing has $a = L$, that is $\gamma = 1$, then for the *same* parameters a similar wing but twice the size has $a = 2L$, that is $\gamma = 4$, but it can be represented diagrammatically by the same-sized figure as for $\gamma = 1$. (Recall Fig. 15.6 which exploited this aspect.)

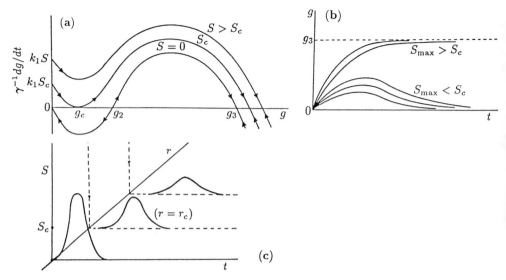

Fig. 15.11a-c. (a) Biochemical switch mechanism with typical bistable kinetics such as from (15.12). The graph shows $\gamma^{-1}dg/dt$ against g for appropriate k_1, k_2, k_3 and several values of S. The critical S_c is defined as having two stable steady states for $S < S_c$ and one, like $g = g_3$, for $S > S_c$. (b) Schematic behaviour of g as a function of t from (15.12) for various pulses of S which increase from $S = 0$ to a maximum $S_{\max}$ and then decrease to $S = 0$ again. The lowest curve is for the pulse with the smallest $S_{\max}$. The final state of g, for large time, changes discontinuously from $g = 0$ to $g = g_3$ if $S_{\max}$ passes through a critical threshold $S_{th}(> S_c)$. (c) Schematic solution, (15.16) below, for the morphogen S as a function of position r measured from the release point of the morphogen.

The initial and boundary conditions (15.10) in nondimensional form are algebraically the same except that now

$$S(r, 0, 0) = \delta(r - r_P), \quad S(r, \theta_0, 0) = \delta(r - r_A) \,. \tag{15.13}$$

The switch and threshold nature of the gene kinetics mechanism in the second of (15.12) can be seen by considering the schematic graph of $\gamma^{-1}dg/dt$ as a function of g, as in Fig. 15.11 (a), for various constant values for S and appropriate k's.

To determine a range of k's in (15.12) so that the kinetics exhibit a switch mechanism, consider first $f(g; 0)$ from (15.12): refer also to the last figure. We simply require $\gamma^{-1}dg/dt = f(g; 0)$ to have 2 positive steady states, that is the solutions of

$$f(g; 0) = 0 \quad \Rightarrow \quad g = 0, \quad g_1, g_2 = \frac{k_2 \pm (k_2^2 - 4k_3^2)^{1/2}}{2k_2} \tag{15.14}$$

must all be real. This is the case if $k_2 > 2k_3$. For $S > 0$ the curve of $f(g; S)$ is simply moved up and, for small enough S, there are 3 steady states, two of which are stable. The S-shape plot of $\gamma^{-1}dg/dt$ against g is typical of a switch mechanism.

Now suppose that at a given time, say $t = 0$, $g = 0$ everywhere and a pulse of morphogen S is released. It activates the gene product since with $S > 0$, $dg/dt > 0$ and so, at each position, g increases with time typically as in Fig. 15.11 (b), which are curves for g as a function of time for a given S. If $\dot{S}$ never reaches the critical threshold $S_{th}(> S_c)$, then as S decreases again to zero after a long time so does g. However, if S exceeds S_c for a sufficient time, then g can increase sufficiently so that it tends, eventually, to the local steady-state equivalent to g_3, thus effecting a switch from $g = 0$ to $g = g_3$. That S must reach a threshold is intuitively clear. What is also clear is that the detailed kinetics in the second of (15.12) are not critical as long as they exhibit the threshold characteristics illustrated in Fig. 15.11.

Even the linear problem for S posed by (15.12) with the relevant boundary and initial conditions (15.10) is not easily solved analytically. We know S qualitatively looks like that in Fig. 15.11 (c) with S reaching a different maximum at each r. Also for a given $S(r,t)$ the solution for g has to be found numerically. The critical threshold S_{th} which effects the switch is also not trivial to determine analytically: Kath and Murray (1985) present a singular perturbation solution to this problem for fast switch kinetics. Later, when we consider eyespot formation and what are called dependent patterns, we shall derive some approximate analytical results. For central symmetry patterns however we shall rely on numerical simulations of the model system.

Intuitively we can see how the mechanism (15.12), in which a finite amount of morphogen S_0 is released from A and P in Fig. 15.10 (c), can generate a spatial pattern in gene product (or colour-specific enzyme). The morphogen pulse diffuses and decays as it spreads across the wing surface, and as it does so it activates the gene G to produce g. If over a region of the wing $S > S_{th}$, then g increases sufficiently from $g = 0$ to move towards g_3 so that when S finally decreases g continues to move towards g_3 rather than returning to $g = 0$. The growth in g, governed by (15.12), is not instantaneous and so the critical S_{th} is larger than S_c in Fig. 15.11 (a). The coupling of the two processes, diffusion and gene transcription, in effect introduces a time lag. Thus as the pulse of morphogen diffuses across the wing as a quasi-wave (see Fig. 15.11 (c)), it generates a domain of permanently non-zero values of g, namely g_3, until along some curve on the wing, S has decreased sufficiently ($S < S_{th}$) so that g returns to $g = 0$ rather than continuing to increase to $g = g_3$. We are interested in determining the switched-on or activated domain: Kath and Murray (1985) also did this for fast switch kinetics. We now apply the mechanism to several specific pattern elements.

Central Symmetry Patterns; Scale and Geometry Effects;
Comparison with Experiments

We first consider how the model may apply to central symmetry patterns and specifically to the experiments of Kuhn and von Engelhardt (1933). We assume that the morphogen S emanates from morphogen sources at A and P on the wing edges, as in Fig. 15.10 (c). The morphogen 'wave' progresses and decays as it moves across the wing until the morphogen level S is reduced to the critical concentration S_{th}, below which the gene-activation kinetics cannot generate a permanent non-zero product level as described above. Now relate the spatial boundary between the two steady-state gene-product levels, the threshold front, with the determination front of Kuhn and von Englehardt (1933). The cells, which manifest the ultimate pigment distribution, are considered to react differentially in the vicinity of this threshold front. The idea that cells react differentially at marked boundaries of morphogen concentrations has also been suggested by Meinhardt (1986) in early segmentation in the fruit fly *Drosophila* embryo.

We require the ultimate steady state solution for the gene product g which requires the solution of the full nonlinear time and space-dependent problem (15.12) with (15.13) and the dimensionless form of (15.10). The numerical results below were obtained using a finite difference scheme. The main parameters that can be varied are the k's and γ. The qualitative behaviour of the pattern formation mechanism and the critical roles played by the geometry and scale can best be highlighted by choosing an appropriate set of values for the parameters and keeping them fixed for *all* of the calculations. The results are shown in Figs. 15.9–12. The parameter values did not have to be carefully selected. In all of the simulations the same amount of morphogen was released at the sources A and P in the middle of the anterior and posterior wing edges.

Consider first the experiments on the moth *Ephestia kuhniella*: its wing is, in fact, quite small, the actual size is about that of the nail on one's little finger. Fig. 15.12 (a) illustrates a normal wing with typical markings while Fig. 15.12 (b), (c) show the results of thermal microcautery (Kuhn and von Engelhardt 1933). Fig. 15.12 (d) is the idealized normal wing: the shaded region is the residual non-zero gene product left behind the determination wave of morphogen.

When a hole, corresponding to thermal cautery, is inserted in the idealised wing we assume that the morphogen level in the hole is zero. That is we set $S = 0$ on the hole boundary on the assumption that any morphogen which diffuses into the hole is destroyed. The numerical results corresponding to the geometry of the experiments are shown in Fig. 15.12 (e), (f), which relate respectively to the experimental results in Fig. 15.12 (b), (c). Fig. 15.12 (g) is another example with a larger cauterization, while Fig. 15.12 (h) is of a comparably cauterized wing of *Limantria dispar* (Henke 1943). Fig. 15.12 (i) is the model's prediction if cauterization removes the source of morphogen at the posterior edge of the wing. No such experiments appear to have been done to establish where the sources of the determination stream are.

Let us now consider the effect of geometry and scale. Even with such a simple model the variety of patterns that can be generated is impressive. For the same values for the kinetics parameters k, k_1, k_2 and k_3 in (15.12), Fig. 15.13 (a)-(c) illustrate, for a *fixed geometry*, some of the effects of *scale* on the spatial patterns. These, of course, are qualitatively as we would expect intuitively. As

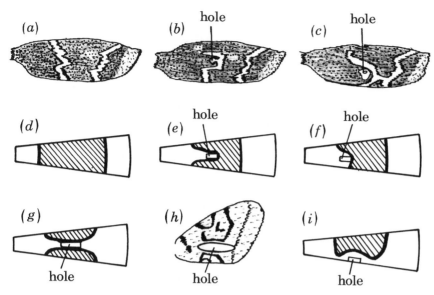

Fig. 15.12a-i. Effect of cauterization on the central symmetry pattern. (a)-(c) are from the experimental results of Kuhn and von Engelhardt (1933) on the moth *Ephestia kuhniella* during the first day after the pupation: (a) normal wing, (b) and (c) cauterized wings with the hole as indicated. (d) Idealized model normal wing in which the 'determination stream' has come from morphogen sources at A and P on the anterior and posterior wing edges: the hatched region represents a steady-state non-zero gene product. (e), (f), (g) and (i) are computed solutions from the model mechanism with cauterized holes as indicated. (h) Effect of cauterization, during the first day after pupation, on the cross-bands of the forewing of *Lymantria dispar* (after Henke 1943). Simulation 'experiments' on the model for comparison; the correspondence is (a)-(d), (b)-(e), (c)-(f), (g)-(h). If cauterization removes the determination stream's source of morphogen at the posterior edge, the pattern predicted by the model is as shown in (i). Parameter values used in the calculations for (15.12) for the idealized wing in Fig. 15.10 (c) for all of (d)-(g), and (i): $k_1 = 1.0 = k_3$, $k_2 = 2.1$, $k = 0.1$, $\gamma = 160$ and unit sources of S at $\theta = 0$ and $\theta = 0.25$ radians, $r_1 = 1$, $r_2 = 3$. (From Murray 1981b)

mentioned above, central symmetry patterns are particularly common in moth wings. Fig. 15.13 (d), (e) show just two such examples, namely the chocolate chip (*Psodos coracina*) and black mountain (*Clostera curtula*) moths respectively: compare these with Fig. 15.13 (a), (b).

The effect of geometry is also important and again we can intuitively predict its general effect with this model when the morphogen is released at the same points on the wing edges. The patterns illustrated in Fig. 15.14 were obtained by simply varying the angle subtended by the wing edges.

The comparison between the experimental results and the results from solving the model system's equations for appropriate domains is encouraging. The solutions generate a region where the morphogen has effected a switch from a zero to a non-zero steady state for the gene product g. If the process was repeated with a different gene product and with a slightly smaller morphogen release it is clear that we could generate a single sharp band of differentiated cells.

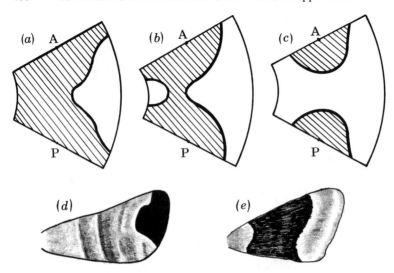

Fig. 15.13a-e. Some effects of scale on spatial patterns generated by the mechanism (15.12) when morphogen is released from sources at A and P. (a)-(c) have the same set of parameter values as in Fig. 15.12, namely $k_1 = 1.0 = k_3$, $k_2 = 2.1$, $k = 0.1$, with the wing defined by $r_1 = 1$, $r_2 = 3$, $\theta = 1.0$ radian, for different domain sizes: (a) $\gamma = 2$; (b) $\gamma = 6$; (c) $\gamma = 40$. A wing with $\gamma = \gamma_2(> \gamma_1)$ has linear dimensions $\sqrt{\gamma_2/\gamma_1}$ larger than that with $\gamma = \gamma_1$. The shaded region has a non-zero gene product. (d) *Psodos coracina* and (e) *Clostera curtula* are examples of fairly common patterns on moth wings. (From Murray 1981b)

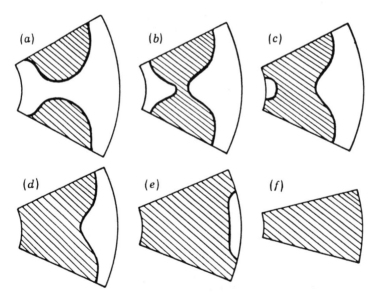

Fig. 15.14a-f. Simple effects of geometry on the spatial patterns. The morphogen is released in the same way as for Fig. 15.13 with the same k-parameter values and a fixed scale parameter $\gamma = 10$ with $r_1 = 1$, $r_2 = 3$ taken for all simulations of (15.12). Geometry is changed by simply varying the angle, in radians, of the sector: (a) $\theta = 1.0$; (b) $\theta = 0.975$; (c) $\theta = 0.95$; (d) $\theta = 0.9$; (e) $\theta = 0.8$; (f) $\theta = 0.5$.

Dependent Patterns

Consider now dependent patterns, which are also very common, in which pigment is restricted to the vicinity of the veins. The pattern depends on the position in the wing of the veins, hence the name for these patterns. Here we consider the morphogen to be released from the boundary veins of the wing cells and so a non-zero gene product g is created near the veins and it is this pattern which is reflected by the pigment-generating cells. If we consider the wing cell to be modelled by the sector of a circle, just as the wing in the situations discussed above, we now have the morphogen released all along the cell boundaries except the outer edge. In this case if we consider the wing cell to be very long so that the problem is quasi-one-dimensional we can derive, given S_{th}, an analytical expression for the width of the gene product spatial pattern.

Consider the one-dimensional problem in which a given amount of morphogen is released at $x = 0$. The idealised mathematical problem is defined by

$$\frac{\partial S}{\partial t} = \frac{\partial^2 S}{\partial x^2} - \gamma k S ,$$

$$S(x,0) = \delta(x), \quad S(\infty, t) = 0 .$$

(15.15)

with solution

$$S(x,t) = (4\pi t)^{-1/2} \exp\left[-\gamma k t - \frac{x^2}{4t}\right], \quad t > 0 .$$

(15.16)

This is qualitatively like that sketched in Fig. 15.11 (c). For a given x the maximum S, $S_{\max}$ say, is given at time t_m where

$$\frac{\partial S}{\partial t} = 0 \quad \Rightarrow \quad t_m = \frac{-1 + \{1 + 4\gamma k x^2\}^{1/2}}{4\gamma k}$$

(15.17)

which on substitution in (15.16) gives

$$S_{\max}(x) = \left[\frac{\gamma k}{\pi(z-1)}\right]^{1/2} \exp\left[\frac{-z}{2}\right] \quad \text{where} \quad z = (1 + 4\gamma k x^2)^{1/2} .$$

(15.18)

Now from the kinetics mechanism (15.12), $S_{\max} = S_{th}$ is the level which effects a switch from $g = 0$ to $g = g_3$ in Fig. 15.11 (a). Substituting in (15.8) we can calculate the distance x_{th} from the vein where $g = g_3$ and hence, in our model, the domain of a specific pigmentation. Thus x_{th} is the solution of

$$S_{th} = \left\{\frac{\gamma k}{\pi(z_{th}-1)}\right\}^{1/2} \exp\left[\frac{-z_{th}}{2}\right], \quad x_{th} = \left[\frac{z_{th}^2 - 1}{4\gamma k}\right]^{1/2}$$

(15.19)

An alternative form of the equation for x_{th} is

$$z_{th} + \ln\left[\frac{S_{th}^2 \pi (z_{th}-1)}{\gamma k}\right] = 0, \quad x_{th} = \left[\frac{z_{th}^2 - 1}{4\gamma k}\right]^{1/2} .$$

(15.20)

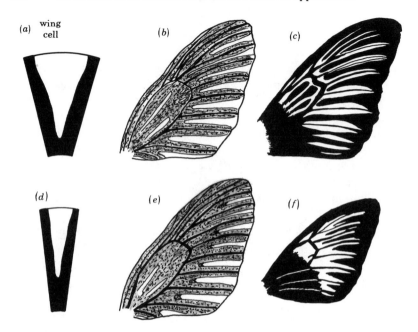

Fig. 15.15a-f. Examples of dependent patterns. (a),(d) Computed patterns from the mechanism (15.12) for a wing cell with parameter values: $k = 0.1$, $k_1 = 1.0 = k_3$, $k_2 = 2.1$, $\gamma = 250$ with the morphogen source strength ρ in the anterior and posterior veins: (a) $\rho = 0.075$, (d) $\rho = 0.015$, and $\rho = 0$ on the cross veins. (b),(e) Schematic predicted pattern from the wing cell patterns in (a), (d) applied to the generalized wing of Fig. 15.10 (a): shaded regions have a non-zero gene product g. (c),(f) Specific examples of dependent patterns on the forewing of two Papilionidae: (c) *Troides hypolitus*, (f) *Troides haliphron*.

In dimensional terms, from (15.11), the critical distance $x_{th}(\text{cm})$ is thus given in terms of the morphogen pulse strength S_0, the diffusion coefficient $D(\text{cm}^2\text{s}^{-1})$ and the rate constant $K(\text{s}^{-1})$. The role of the kinetics parameters in (15.12) comes into the determination of the critical S_{th}. The analytical evaluation of S_{th} is given by Kath and Murray (1985).

Let us now consider dependent patterns with the mechanism operating in a domain which is an idealised wing cell with a given amount of morphogen released in a pulse from the bounding veins: refer to Fig. 15.15. From the above analysis we expect a width of the non-zero gene product g on either side of the vein. As in the central symmetry patterns we expect geometry and scale to play major roles in the final pattern obtained for given values of the model parameters, and we can now intuitively predict the qualitative behaviour of the solutions.

Now apply the model mechanism (15.12) to the idealised wing cell with a given amount, ρ, of morphogen released per unit length of the bounding veins (refer also to Fig. 15.10 (a)). With the same parameter values as for Fig. 15.13 and Fig. 15.14 the equations were again solved using a finite difference scheme with zero flux conditions along the boundaries after a pulse of morphogen had been released from the three edges representing the veins. Fig. 15.15 (a), (d) show

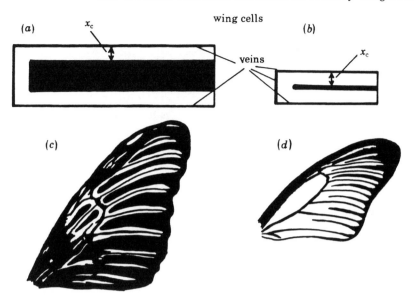

Fig. 15.16a-d. (a),(b) Idealized wing cells based on the analytical solution (15.20) for the critical switched on domain. These clearly illustrate the effect of scale. The pattern (unshaded in this case) width is fixed for given parameter values. (c),(d) Examples of dependent patterns from two Papilionidae: (c) *Troides prattorum*; (d) *Iterus zalmoxis*.

examples of the computed solutions with Fig. 15.15 (b), (e) the approximate resulting wing patterns generated on a full wing. The role of geometry in the patterns is as we would now expect. Fig. 15.12 (c), (f) are specific, but typical, examples of the forewing of *Troides hypolitus* and *Troides haliphron* respectively. Such dependent patterns are quite common in the Papilionidae.

Now consider scale effects. Fig. 15.16 (a), (b) directly illustrate these schematically when the veins are approximately parallel. Fig. 15.16 (c), (d) show examples of the forewings of *Troides prattorum* and *Iterus zalmoxis*. The distance from the vein of the pigmented pattern depends in a nonlinear way on the parameters and the amount of morphogen released. If these values are fixed, the distance from the vein is *independent* of scale. That is, the mechanism shows that the *intravenous* strips between pigmented regions vary according to how large the wing cell is. This is in agreement with the observations of Schwanwitsch (1924) and the results in Fig. 15.16 exemplify this.

These results are consistent with the observation of Schwanwitsch (1924) on nymphalids and certain other families. He noted that although the width of intravenous stripes (in our model the region between the veins where $g = 0$) is species-dependent, the pigmented regions in the vicinity of the veins are the same size. In several species the patterns observed in the discal cell (D in Fig. 15.10 (a)) reflect the existence of the veins that subsequently atrophy: see Fig. 15.15 (c) and Fig. 15.16 (c) of the forewing of the female *Troides prattorum*.

Eyespot or Ocelli Patterns

Eyespot patterns are very common: see, for example, Fig. 15.9 (a). Nijhout (1980b) performed transplant experiments on the buckeye butterfly (*Precis coenia*) wherein he moved an incipient eyespot from one position on the wing to another where normally an eyespot does not form. The result was that an eyespot formed at the new position. This suggests that there is possibly a source of some morphogen at the eyespot centre from which the morphogen diffuses outwards and activates the cells to produce the circular patterns observed. So, once again it seems reasonable to investigate the application of the above mechanism to these results.

We assume that the eyespot centre emits a pulse of morphogen in exactly the same way as for the central symmetry patterns. The idealised mathematical problem for the morphogen from the first of (15.12) in plane axisymmetric polar coordinates, with initial and boundary conditions from (15.10) and (15.13), is

$$\frac{\partial S}{\partial t} = \frac{\partial^2 S}{\partial r^2} + r^{-1}\frac{\partial S}{\partial r} - \gamma k S \,,$$

$$S(r,0) = \delta(r), \quad S(\infty,t) = 0 \,. \tag{15.21}$$

The solution is

$$S(r,t) = (4\pi t)^{-1}\exp\left[-\gamma k t - \frac{r^2}{4t}\right], \quad t > 0 \,, \tag{15.22}$$

which is like the function of time and space sketched in Fig. 15.11 (c).

As before we can calculate the size of the gene-activated region given the critical threshold concentration $S_{th}(> S_c)$ which effects the transition from $g = 0$ to $g = g_3$ in Fig. 15.11 (a). In exactly the same way as we used to obtain (15.20) the maximum S_{max} for a given r from (15.22) is

$$S_{max} = \frac{\gamma k}{2\pi(z-1)}e^{-z}, \quad z = [1 + \gamma k r^2]^{1/2}$$

and so the radius of the activated domain, r_{th} say, is given by the last equation with $S_{max} = S_{th}$, as

$$z_{th} + \ln\left[\frac{2\pi(z_{th}-1)S_{th}}{\gamma k}\right] = 0, \quad z_{th} = [1 + \gamma k r_{th}^2]^{1/2} \,. \tag{15.23}$$

Many eyspots have several concentric ring bands of colour. With the above experience we now know what the patterns will be when we solve the system (15.12) with conditions that pertain to this eyespot situation. If, instead, we let the mechanism run twice with slightly different amounts of morphogen released we shall obtain two separate domains which overlap. Fig. 15.17 (a) shows the numerical result of such a case together with the predicted pattern in Fig. 15.17 (b)

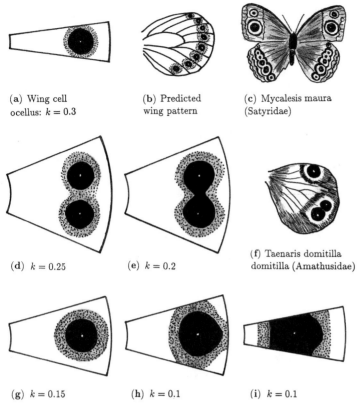

(a) Wing cell
ocellus: $k = 0.3$

(b) Predicted
wing pattern

(c) Mycalesis maura
(Satyridae)

(d) $k = 0.25$

(e) $k = 0.2$

(f) Taenaris domitilla
domitilla (Amathusidae)

(g) $k = 0.15$

(h) $k = 0.1$

(i) $k = 0.1$

Fig. 15.17a-i. (a) A patterned eyespot generated within a wing cell by two emissions of morphogen each with its own gene product. With the same parameter values, the dark region had less morphogen injected than that with the shaded domain. The k-parameter values are as in Fig. 15.9-13 except for $k = 0.3$. **(b)** The predicted overall wing pattern if an eyespot was situated in each wing cell with **(c)** a typical example (*Mycalesis maura*). **(d)** and **(e)** illustrate the effect of different degradation constants k which result in coalescing eyspots with **(f)** an actual example (*Taenaris domitilla*). **(g)-(i)** demonstrate the effect of different geometries.

if an eyespot is situated in each distal wing cell, and an example of a specific butterfly which exhibits this result is Fig. 15.17 (c).

The simple model proposed in this section can clearly generate some of the major pattern elements observed on lepidopteran wings. As we keep reiterating in this book. this is not sufficient to say that such a mechanism is that which necessarily occurs. The evidence from comparison with experiment is, however, suggestive of a diffusion based model. From the material discussed in detail in Chapter 14 we could also generate such patterns by appropriately manipulating a reaction diffusion system capable of diffusion-driven pattern generation. What is required at this stage if such a model mechanism is indeed that which operates, is an estimation of parameter values and how they might be varied under controlled experimental conditions. We thus consider how the model might apply

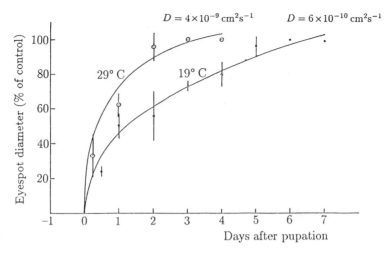

Fig. 15.18. The diameter of a growing eyespot in the buckeye butterfly (*Precis coenia*) as a function of time after pupation. The experiments by Nijhout (1980a) were carried out at two different temperatures, 19°C and 29°C. The continuous curves are best fits from the analytical expression (15.25), which is derived from the simple morphogen diffusion model, the first of (15.7).

quantitatively to the experimental results of Nijhout (1980a) who measured the diameter of a growing eyespot as a function of time: the results are reproduced in Fig. 15.18.

   Let us now relate the model analysis to the experiments. The solution for $S(r,t)$ is given by (15.22). We want the value of r such that $S = S_{th}$. Denote this by R; it is a function of t. From (15.22) with $S = S_{th}$ we have

$$S_{th} = (4\pi t)^{-1} \exp\left[-\gamma kt - \frac{R^2}{(4t)}\right], \quad t > 0$$

which gives

$$R^2(t) = -4t[\gamma kt + \ln(4\pi t S_{th})] . \tag{15.24}$$

For comparison with the experiments we require the diameter $d(= 2R)$ in dimensional terms. We consider a single eyespot with the standard length a in the nondimensionalisation (15.11) to be the diameter of the control in the experiment. Since we are interested in the growth of the eyespot to its normal size this means that $L = a$ and hence $\gamma = 1$. Thus the time varying diameter $d(t)$ in Fig. 15.18 is simply $2R(t)$ which, on using (15.11) and (15.24) gives

$$d^2(t) = -16Dt\left\{Kt + \ln\left[\frac{4\pi S_{th}tD}{S_0 a^2}\right]\right\} \tag{15.25}$$

$$= -16Dt[Kt + \ln t + C] ,$$

where $C = \ln\left[4\pi S_{th}D/(S_0a^2)\right]$ is simply a constant and $D(\text{cm}^2\text{s}^{-1})$ and $t(\text{sec})$ are now dimensional. Note from (15.25) that

$$d(t) \sim O([t\ln t]^{1/2}) \quad \text{as} \quad t \to 0. \tag{15.26}$$

The maximum diameter d_m is obtained in the same way as above for the gene-activated domain size for dependent patterns, specifically (15.20). Here it is given by d_m, the solution of

$$z + \ln\left(\frac{z-1}{2K}\right) + C = 0, \quad z = \left(1 + \frac{Kd_m^2}{4D}\right)^{1/2}. \tag{15.27}$$

If we now use (15.25) and the experimental points from Fig. 15.18, we can determine D, k and C from a best fit analysis. From the point of view of experimental manipulation it is difficult to predict any variation in the degradation constant K since we do not know what the morphogen is. There is, however, some information as to how diffusion coefficients vary with temperature. Thus the parameter whose value we can deduce, and which we can potentially use at this stage, is the diffusion coefficient D. From the experimental results in Fig. 15.18 and the best fit with (15.25), we obtained values of $D = 4 \times 10^{-9}\,\text{cm}^2\text{s}^{-1}$ at $29°C$ and $D = 6 \times 10^{-10}\,\text{cm}^2\text{s}^{-1}$ at $19°C$. Although we cannot independently measure the diffusion coefficient of a morphogen that we cannot yet identify, the order of magnitude of these values and how D varies with temperature seems reasonable.

From (15.25) we obtain the velocity of spread, $v(t)$, of the eyespot as $dd(t)/dt$, and so

$$v(t) = -2D\frac{\{[Kt + \ln t + C] + t\left(K + \frac{1}{t}\right)\}}{\{-Dt[Kt + \ln t + C]\}^{1/2}} \tag{15.28}$$

from which we deduce that

$$v(t) \sim 2(-Dt^{-1}\ln t)^{1/2} \quad \text{as} \quad t \to 0. \tag{15.29}$$

Nijhout (1980a) found the average wave speed to be $0.27\,\text{mm/day}$ at $29°C$ and $0.12\,\text{mm/day}$ at $19°C$. With the best fit values of the parameters, (15.28) gives the velocity of spread as a function of t with (15.29) showing how quickly the initial growth rate is. With the diffusion coefficient estimates deduced above, the ratio of the initial wave speeds from (15.29) at $29°C$ and $19°C$ is $(D_{29}/D_{19})^{1/2} = (4 \times 10^{-9}/6 \times 10^{-10})^{1/2} \approx 2.58$. This compares favorably with the ratio of the average wave speeds found experimentally, namely $0.27/0.12 = 2.25$. This adds to the evidence for such a diffusion controlled pattern formation mechanism as above.

If mechanisms such as we have discussed in this section are those which operate, the dimension of the diffusion field of pattern formation is of the order of several millimetres. This is much larger than any so far found in other embryonic situations. One reason for assuming they do not occur is that development of pattern via diffusion would, in general, take too long if distances were larger than a millimetre; over this time enough growth and development would take place, to imply considerable sensitivity in pattern formation. In pupal wings, however, this is not so, since pattern can develop over a period of days during which the scale and geometry vary little. With the experience from the last few chapters, the original misgiving is no longer valid since if we include reaction diffusion mechanisms, not only can complex patterns be formed but biochemical messages can also be transmitted very much faster than pure diffusion. A final point regarding eyespots is that the positioning of the centres can easily be achieved with reaction diffusion models and the emission of the morphogen triggered by a wave sweeping over the wing or by nerve activation: we discuss neural models in the next chapter.

It is most likely that several independent mechanisms are operating, possibly at different stages, to produce the diverse patterns on butterfly wings (Schwanwitsch 1924; Suffert 1927). It is reasonable to assume, as a first modelling step, that the number of mechanisms is the same as the number of melanins present. In the case of the nymphalid *Precis coenia* there are four differently coloured melanins (Nijhout 1980b).

With the relatively few pattern elements (in comparison with the vast and varied number of patterns that exist) in Suffert's (1927) groundplan, it seems worthwhile to explore further the scope of pattern formation possibilities of plausible biochemical diffusion models such as that discussed here. Fig. 15.19 however shows a few more of the complex wing patterns which have yet to be generated with such reaction diffusion mechanisms. By introducing anisotropy in the diffusion (that is diffusion depends on the direction) it may be possible to generate further patterns which are observed, such as the arrowhead in Fig. 15.9 (b) and in the last figure.

Perhaps we should turn the pattern formation question around and ask: 'What patterns *cannot* be formed by such simple mechanisms?' As a pattern generation problem, butterfly wing patterns seem particularly appropriate to study since it appears that pattern in the wings is developed comparatively late in development and interesting transplant experiments (Nijhout 1980a) and cautery-induced colour patterns (Nijhout 1985b) are feasible, as are the colour pattern modifications induced by temperature shocks (see, for example, Nijhout 1984).

15.3 Modelling Hair Patterns in a Whorl in *Acetabularia*

The green marine alga *Acetabularia*, a giant unicellular organism – see the beautiful photograph in Fig. 15.20 - is a fascinating plant which constitutes a link in the marine food chain (Bonotto 1985). The feature of particular interest to us here is its highly efficient self-regenerative properties which allow for laboratory controlled regulation of its growth. *Acetabularia* has been the subject of several meetings: see, for example, the proceedings edited by Bonotto et al. (1985). In this section we describe a model, proposed by Goodwin, Murray and Baldwin

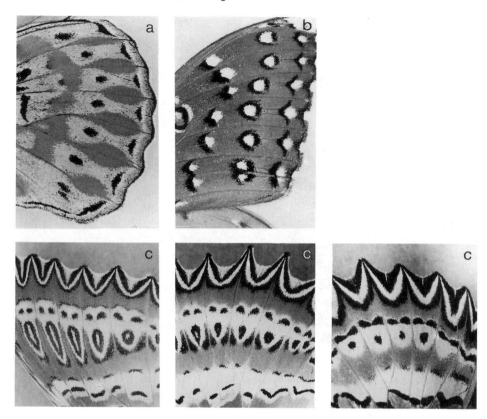

Fig. 15.19a-c. Photographs of examples of butterfly wing patterns the formation of which have not yet been modelled. (a) *Crenidomimas cocordiae*; (b) *Hamanumida daedalus*; (c) Three examples from the genus *Cethosia*. Note here the topological similarity of these three patterns from the elongated pattern on the left to the much flatter form on the right. (Photographs courtesy of H.F. Nijhout)

(1985), for the mechanism which controls the periodic hair spacing in the whorl of a regenerating head of *Acetabularia*. Experimental evidence is presented, not only to corroborate the analytical quantitative results of the mechanism, but more importantly to suggest that the initiation of hairs is controlled by calcium, possibly the elusive morphogen in *Acetabularia*. Fuller biological details are given in Goodwin et al. (1985).

The alga consists of a narrow stalk around 4–5 cm long on the top of which is a round cap about 1 cm across: see Fig. 15.21 (a). The stalk is a thin cylindrical shell of cytoplasm. Free calcium, Ca^{2+}, plays a crucial role in the regeneration, after amputation, of the periodic distribution of the whorl hairs and eventually the cap. There are various stages in regeneration as schematically shown in Fig. 15.21 (b)-(d). After amputation there is an extension of the stalk, then a tip flattening and finally the formation of a whorl. Further extension of the stalk can take place with formation of other whorls. Fig. 15.21 (e) is a schematic cross section of the stalk at the growth region and is the relevant spatial domain in our model.

Fig. 15.20. The marine algae *Acetabularia ryukyuensis*. (Photograph courtesy of I. Shihira-Ishikawa)

The model we develop and the mechanism we analyse is specifically concerned with the spatial pattern which cues the periodic distribution of hairs on a whorl. Experiments (see Goodwin et al. 1985) show that there are definite limits to the concentration of Ca^{2+} in the external medium, within which whorl formation will take place. Fig. 15.22 shows the experimental results: below about 2 mM and above about 60 mM external calcium, whorls do not form. The normal value in artificial sea water is 10 mM Ca^{2+}. With about 5 mM only one whorl is produced after which the cap forms.

The experimental results suggest that the rate of movement of calcium from the external medium through the outer wall of the plant is intimately involved in growth determination and the initiation of a whorl of hairs. It is for this reason that calcium is proposed as a true morphogen in *Acetabularia*. If it is indeed a morphogen then it should play a role in the distribution of hairs or rather the mean distance between them, the wavelength of hair distribution. Experiments were conducted to determine the effect of the external free calcium concentration on the hair wavelength: see Fig. 15.22 as well as Fig. 15.25 below. Analysis of the model mechanism, which we discuss below, also corroborates the spacing hypothesis.

Let us consider some of the evidence for a reaction diffusion mechanism. First recall from Chapter 14 that if we have a spatial structure generated by

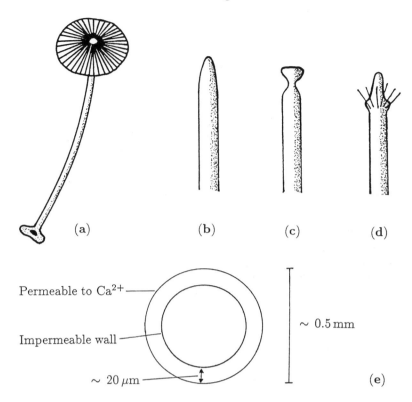

Fig. 15.21a-e. (a) Typical mature *Acetabularia*. (b)-(d) the various stages in the growth of a whorl: (b) extension, (c) flattening of the tip, (d) formation of the whorl. (e) Transverse cross section of the growth region of the stalk: note the typical dimensions.

a Turing-type reaction diffusion system the number of structures is not scale invariant. For example if we have a given one-dimensional domain with several waves in morphogen concentration, a domain twice the size will have twice the number of waves as long as the parameters are kept fixed. This is an intrinsic characteristic of the spatial properties of reaction diffusion models of the kind discussed in the last chapter. We should perhaps note here the model reaction diffusion mechanisms proposed by Pate and Othmer (1984) which *are* scale invariant with regard to pattern formation. This problem of size invariance has also been addressed, for example, by Babloyantz and Bellemans (1985) and Hunding and Sørensen (1988). Any model can be made to display size adaption if the parameters vary appropriately.

There is considerable variability in the number of hairs in a whorl; they vary from about 5 to 35. Experiments show that for plants maintained under the same conditions the hair spacing, w say, is almost constant and that the number of hairs is proportional to the radius of the stalk. The mechanism thus regulates the hair *spacing* irrespective of the size of the plant. This relation between scale and pattern number is a property of reaction diffusion systems as

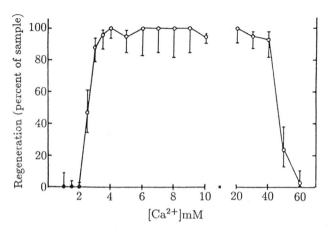

Fig. 15.22. Experimental results on the effect of external calcium Ca^{2+} in the surrounding medium in the formation of whorls in *Acetabularia* after amputation. Note the lower and upper limits below and above which no regeneration occurs. (From Goodwin, Murray and Baldwin 1985)

we demonstrated in the last chapter. In fact it is a property of other pattern formation mechanisms which involve the space variables in a similar way as we shall see later in Chapter 17 when we discuss mechanochemical pattern generators, which are quite different; so such a property is by no means conclusive evidence.

Harrison et al. (1981) showed that the spacing, w, of hairs depends on the ambient temperature, T, according to $\ln w \propto 1/T$. This Arrhenius-type of temperature variation suggests a chemical reaction kinetics factor, again in keeping with a reaction diffusion theory. In other words the spacing depends on the kinetics parameters.

The model we now develop is for the generation of the spatial distribution of a morphogen, identified with calcium, which is reflected in the spatial distribution of the hairs in the whorl. We assume initiation is governed by the overall reactions of two species u and v, the latter considered to be the concentration of Ca^{2+}, with u the other morphogen, as yet unknown. The spatial domain we consider is the annular cross-section of the stalk as illustrated in Fig. 15.21 (e). The available evidence is not sufficient for us to suggest any specific reaction kinetics for the reaction diffusion system so we choose the simplest two-species mechanism, the Schnackenberg (1979) system we considered in some detail in Section 6.4 in Chapter 6, specifically the dimensionless form (14.10). It is

$$u_t = \gamma(a - u + vu^2) + \nabla^2 u = \gamma f(u,v) + \nabla^2 u , \qquad (15.30)$$

$$v_t = \gamma(b - vu^2) + d\nabla^2 v = \gamma g(u,v) + d\nabla^2 v , \qquad (15.31)$$

which define $f(u,v)$ and $g(u,v)$ and where a, b, γ and d are positive parameters. With the annular domain, u and v are functions of r, θ and t with the domain defined by

$$R_i \leq r \leq R_0, \quad 0 \leq \theta < 2\pi , \tag{15.32}$$

where R_i and R_0 are the dimensionless inside and outside radii of the annulus respectively, and the Laplacian

$$\nabla^2 = \frac{\partial^2}{\partial r^2} + r^{-1}\frac{\partial}{\partial r} + r^{-2}\frac{\partial^2}{\partial \theta^2} . \tag{15.33}$$

The scale parameter γ is proportional to R_i^2 here.

We introduce further nondimensional quantities and redefine the already dimensionless variable r by

$$r^* = \frac{r}{R_i}, \quad \delta = \frac{R_0}{R_i}, \quad R^2 = R_i^2\gamma \tag{15.34}$$

and the system (15.30)–(15.32) becomes, on dropping the asterisks for notational convenience,

$$u_t = R^2(a - u + vu^2) + \frac{\partial^2 u}{\partial r^2} + r^{-1}\frac{\partial u}{\partial r} + r^{-2}\frac{\partial^2 u}{\partial \theta^2} , \tag{15.35}$$

$$v_t = R^2(b - vu^2) + d\left(\frac{\partial^2 v}{\partial r^2} + r^{-1}\frac{\partial v}{\partial r} + r^{-2}\frac{\partial^2 v}{\partial \theta^2}\right) , \tag{15.36}$$

with the reaction diffusion domain now given by

$$1 \leq r \leq \delta, \quad 0 \leq \theta < 2\pi . \tag{15.37}$$

Biologically the inner wall of the stalk is impermeable to calcium so we assume zero flux conditions for both u and v on $r = 1$. There is a net flux of calcium into the annulus. However the intracellular concentration level of calcium is $O(10^{-4}\,\text{mM})$ compared with the external level of $1\,\text{mM}$ to $100\,\text{mM}$. Thus the influx of calcium is essentially independent of the internal concentration. The spatial dimensions of the annulus give values for δ of about 1.05 to 1.1 which implies that it is sufficiently thin for the geometry to be considered quasi one-dimensional. We can thus reflect the inward flux of calcium by the source term b in the v (that is calcium) equation (15.35). We can then take zero flux conditions at the outer boundary $r = \delta$ as well as on $r = 1$. We are thus concerned with the system (15.35) and (15.36) in the domain (15.37) with boundary conditions

$$u_r = v_r = 0 \quad \text{on} \quad r = 1, \delta . \tag{15.38}$$

In the last chapter we discussed in detail the diffusion driven spatial patterns generated by such reaction diffusion mechanisms and obtained the various conditions the parameters must satisfy. Here we only give a brief sketch of the analysis, which in principle is the same. We consider small perturbations about the uniform steady state (u_0, v_0) of (15.35) and (15.36), namely

$$u_0 = a + b, \quad v_0 = \frac{b}{(a+b)^2} , \tag{15.39}$$

by setting

$$\mathbf{w} = \begin{pmatrix} u - u_0 \\ v - v_0 \end{pmatrix} \propto \psi(r, \theta) e^{\lambda t} \tag{15.40}$$

where $\psi(r, \theta)$ is an eigenfunction of the Laplacian on the annular domain (15.37) with Neumann boundary conditions (15.38). That is

$$\nabla^2 \psi + k^2 \psi = 0, \quad (\mathbf{n} \cdot \nabla) \psi = 0 \quad \text{on} \ r = 1, \delta , \tag{15.41}$$

where the possible k are the wavenumber eigenvalues which we must determine. In the usual way of Chapter 14 we are interested in wavenumbers k such that $\text{Re} \, \lambda(k^2) > 0$. The only difference between the analysis here and that in the last chapter is the different analysis required for the eigenvalue problem.

The Eigenvalue Problem

Since the dimensions of the relevant annular region in *Acetabularia* implies $\delta \sim 1$ we can negelect the r-variation as a first approximation and the eigenvalue problem is one-dimensional and periodic in θ with $\psi = \psi(\theta)$. So in (15.41) the r-variation is ignored and the eigenvalue problem becomes

$$\frac{d^2 \psi}{d\theta^2} + k^2 \psi = 0, \quad \psi(0) = \psi(2\pi), \quad \psi'(0) = \psi'(2\pi) \tag{15.42}$$

which has solutions

$$k = n, \quad \psi(\theta) = a_n \sin n\theta + b_n \cos n\theta \quad \text{for integers} \ n \geq 1 . \tag{15.43}$$

where the a_n and b_n are constants. The exact problem from (15.41) is

$$\frac{\partial^2 \psi}{\partial r^2} + r^{-1} \frac{\partial \psi}{\partial r} + r^{-2} \frac{\partial^2 \psi}{\partial \theta^2} + k^2 \psi = 0 \tag{15.44}$$

with

$$\psi_r(1, \theta) = \psi_r(\delta, \theta) = 0$$
$$\psi(r, 0) = \psi(r, 2\pi), \quad \psi_\theta(r, 0) = \psi_\theta(r, 2\pi) . \tag{15.45}$$

Solve (15.44) by separation of variables by setting

$$\psi(r, \theta) = R_n(r)[a_n \sin n\theta + b_n \cos n\theta] \tag{15.46}$$

which on substituting into (15.44) gives

$$R_n'' + r^{-1}R_n' + \left(k^2 - \frac{n^2}{r^2}\right)R_n = 0, \quad R_n'(1) = R_n'(\delta) = 0 . \tag{15.47}$$

The solution is

$$R_n(r) = J_n(k_n r)Y_n'(k_n) - J_n'(k_n)Y_n(k_n r) , \tag{15.48}$$

where the J_n and Y_n are the nth order Bessel functions and the eigenvalues $k^2 = k_n^2$ are determined by the boundary conditions. The form (15.48) automatically satisfies the first of (15.45) while the second requires

$$J_n(k_n\delta)Y_n'(k_n) - J_n'(k_n)Y_n(k_n\delta) = 0 . \tag{15.49}$$

For each n in the last equation there is an infinity of solutions k_n^j, $j = 1, 2, \ldots$ These values have been evaluated numerically by Bridge and Angrist (1962). We know, of course, that as $\delta \to 1$ the problem becomes one-dimensional and the eigenvalues $k \to n$, so we expect $k_n^j(\delta) \to n$ as $\delta \to 1$. (In fact, this can be shown analytically by setting $\delta = 1 + \varepsilon$ in (15.49) and carrying out a little asymptotic analysis as $\varepsilon \to 0$.) This completes our discussion of the eigenvalue problem.

In the last chapter, specifically Section 14.5, we discussed the role of the dispersion relation in pattern creation and obtained the Turing space of the parameters wherein spatial perturbations of specific wavenumbers about the uniform steady state (u_0, v_0) in (15.39) could be driven unstable. That is, in (15.40), Re $\lambda(k^2) > 0$ for a range of wavenumbers. The range of wavenumbers is obtained from the general expressions in (14.66) in Section 14.5: there is a slight change in notation, with R^2 for γ used in (15.35) and (15.36). With the notation here the range is given by

$$K_1^2 < k^2 < K_2^2$$
$$K_2^2, K_1^2 = R^2\frac{(df_u + g_v) \pm \{(df_u + g_v)^2 - 4\alpha(f_u g_v - f_v g_u)\}^{1/2}}{2d} \tag{15.50}$$

where here, and in the rest of this section, the derivatives are evaluated at the steady state (u_0, v_0) given by (15.39). In the quasi one-dimensional situation with eigenvalue problem (15.42), the eigenvalues k are simply the positive intergers $n \geq 1$. From the last equation and (15.34) the range of spatial patterns which are linearly unstable is thus proportional to the radius of the annulus R_i.

For each eigenvalue k satisfying (15.50) there is a corresponding Re $\lambda(k^2) > 0$ and among all these (discrete) k's there is one which gives a maximum Re $\lambda =$ Re $\lambda_M =$ Re $\lambda(k_M^2)$. k_M is again obtained as in Section 14.5, specifically equation

(14.68) which in the notation here gives

$$
\begin{aligned}
k_M^2 &= \frac{R^2}{d-1} \left\{ (d+1) \left[\frac{-f_v g_u}{d} \right]^{1/2} - f_u + g_v \right\} \\[2mm]
&= \frac{R^2}{d-1} \left\{ -\frac{b-a}{b+a} - (b+a)^2 + (d+1) \left[\frac{2b(b+a)}{d} \right]^{1/2} \right\}
\end{aligned}
\tag{15.51}
$$

on evaluating the derivatives at (u_0, v_0) from (15.39) (or simply getting them from (14.34)).

As we also discussed in Chapter 14, at least in a one-dimensional situation the fastest growing mode is a good indicator of the ultimate finite amplitude steady state spatial pattern. That is the pattern wavelength w in the quasi one-dimensional situation is given by the dimensionless length $w = 2\pi/k_M$, with k_M from (15.51). If we now choose the basic length to be the radius r_i of the annulus then in dimensional terms from (15.34) we see that the dimensional wavenumber $k_{Md} = k_M/r_i$ and so the dimensional wavelength

$$
w_d = r_i w = \frac{r_i 2\pi}{r_i k_{Md}} = \frac{2\pi}{k_{Md}},
$$

which is independent of the radius r_i. Thus, in our model, hair *spacing* is *independent* of the stalk radius. With the experience from Chapter 14, this, of course, is exactly as we should expect.

For qualitative comparison with the experimental results in Fig. 15.22 we must consider the effect on the pattern formed as the external calcium concentration varies, that is as b varies. The major experimental facts (see Goodwin et al. 1985) are: (i) There is a range of external calcium concentrations within which whorls will form. That is if b is too high or too low no hairs are initiated. (ii) Within this range, the hair spacing decreases as the calcium concentration increases, quickly at first but then becoming more gradual. (iii) The amplitude of the pattern decreases to zero as the concentration of Ca^{2+} approaches the upper and lower limits. We now want to derive relevant quantities from the model to compare with these basic experimental facts.

We must derive some analytical measure of the amplitude of the pattern which is formed by the mechanism. In practical terms only a finite amount of time is available to generate required patterns. In reaction diffusion models the steady state pattern is obtained, from a mathematical viewpoint, only as $t \to \infty$. Linear theory, however, provides information on the fastest growing mode which generally dominates the patterning, thus giving a good prediction of the final qualitative picture of steady state morphogen concentrations. It is quite likely, if a morphogen theory obtains, that differentiation to initiate a hair takes place when the morphogen level reaches some threshold value. So, it is reasonable to suppose that the maximum linear growth rate $\mathrm{Re}\,\lambda(k_M^2)$ gives some indication of the actual morphogen amplitude observed – certainly if $\lambda_M = 0$ the amplitude must be zero. We thus use $\text{Re}\,\lambda(k_M^2)$ as our amplitude measure which we get by substituting k_M from (15.51) into the expression for λ (the larger of the two solutions of (14.23)) namely

$$2\lambda_M = \gamma(f_u + g_v) - \frac{k_M^2}{d+1} + \left\{\left[\gamma(f_u + g_v) - \frac{k_M^2}{d+1}\right]^2 - 4h(k_M^2)\right\}^{1/2},$$

$$h(k_M^2) = dk_M^4 - \gamma(df_u + g_v)k_M^2 + \gamma(f_u g_v - f_v g_u).$$

With the kinetics from (15.35) and (15.36) and k_M from (15.51), a little tedious algebra gives the maximum growth rate as

$$\lambda_M = \lambda(k_M^2) = (d-1)^{-1}\left\{d\frac{b-a}{b+a} + (b+a)^2 - [2bd(b+a)]^{1/2}\right\}. \qquad (15.52)$$

Consider now the (a, b) Turing space for the system (15.30) and (15.31) given in Fig. 14.9 for various values of the diffusion ratio d. We reproduce one of the curves for reference in Fig. 15.23 (a) and relate the parameter b to the external calcium concentration. Referring to Fig. 15.23 (a) if we consider a fixed a, $a_1(> a_m)$ say, then, as we increase the calcium concentration b from zero, we see that no pattern is formed until it reaches the lower threshold value $b_{\min}$. Further increase in b moves the parameters into the parameter space for pattern formation. Fig. 15.23 (b) shows a typical computed pattern obtained numerically in the quasi one-dimensional situation. This is the case when the stalk wall is sufficiently thin, that is $\delta \approx 1$ in (15.37), and r-variations in (15.35) and (15.36) can be neglected. Fig. 15.23 (c) shows the corresponding pattern on the annulus where the shaded region is above a concentration threshold. We assume that when this happens a hair is initiated. If the annular region is wider, that is δ is larger (approximately $\delta > 1.2$) so that r-variations have to be considered, the spatial pattern generated takes on a more two-dimensional aspect, an example of which is shown in Fig. 15.23 (d); here $\delta = 1.5$.

As b increases beyond $b_{\max}$ the parameters move out of the Turing space and the mechanism can no longer create a spatial pattern. This is in keeping with the experimental fact (i) above and illustrated in the quantitative experimental results in Fig. 15.22. Note from Fig. 15.23 (a) that this qualitative behaviour only happens if the fixed a is greater than a_m and is not too large. For a fixed $a_0 < a < a_m$ we see that as b increases from zero there are two separate domains where pattern can be generated.

In Fig. 15.23 (a), if $a = a_1$, for example, the maximum linear growth rate λ_M is zero at $b = b_{\min}$ and $b = b_{\max}$: this can be derived analytically from (15.52). For $b_{\min} < b < b_{\max}$ the growth rate $\lambda_M > 0$. Using the analytical expression for the maximal growth rate λ_M in (15.52) we computed its variation with b as b took increasing values from $b_{\min}$ to $b_{\max}$. As discussed above we relate the maximal growth rate with the amplitude of the resulting pattern: Fig. 15.24 displays the results.

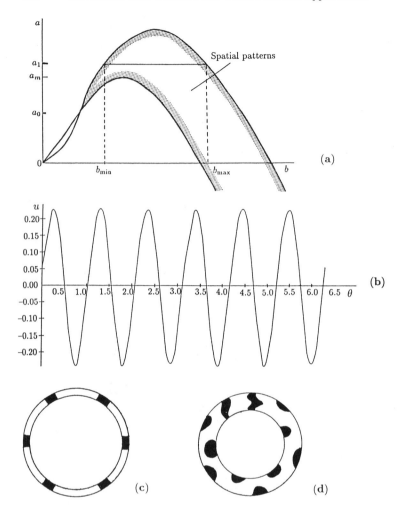

Fig. 15.23a-d. (a) Typical Turing space for the mechanism (15.30) and (15.31) for a fixed d. Spatial patterns can be generated when a and b lie within the region indicated. (b) Computed solution structure for u graphed relative to the steady state from (15.35)–(15.37) as a function of θ in the quasi one-dimensional situation where $\delta \approx 1$ (that is we set $\partial/\partial r \equiv 0$): parameter values $a = 0.1$, $b = 0.9$, $d = 9$, $R = 3.45$ (steady state $u_0 = 1.9$, $v_0 \approx 0.25$). (c) The equivalent pattern on the stalk: shaded regions represent high concentration levels of Ca^{2+} above a given threshold. (d) As the width of the annular region increases the pattern generated becomes more two-dimensional and less regular.

Since the experiments also measure the effect on the wavelength w of varying the external Ca^{2+} concentration we also examine the predicted behaviour of w from the above analysis as b varies. For fixed a and d, (15.51) gives the dependence of k_M on b and hence of the pattern wavelength $w = 2\pi/k_M$. We find from (15.51), with appropriate $a(> a_m)$, that the wavelength decreases with b as we move through the pattern formation region in Fig. 15.23 (a). Fig. 15.25 illustrates

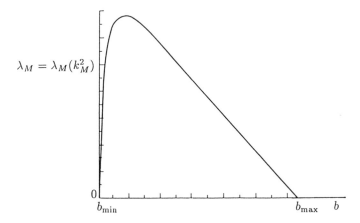

$$\lambda_M = \lambda_M(k_M^2)$$

Fig. 15.24. Typical computed maximal growth λ_M from (15.52) as a function of b when $a(> a_m)$ and b lie within the parameter range giving spatial structure in Fig. 15.23 (a). λ_M relates directly to the amplitude of the steady state standing wave in calcium Ca^{2+}. Compare this form with the experimental results shown in Fig. 15.22: b_{min} and b_{max} are equivalent to the 2 mM and 60 mM points respectively.

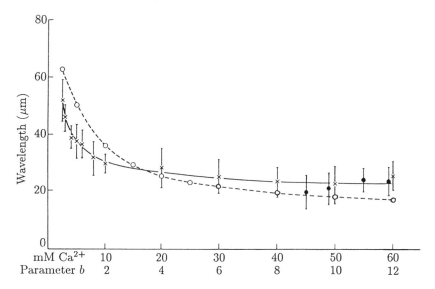

Fig. 15.25. Experimental ($\times$) and theoretical ($\circ$) results for the variation of the hair spacing wavelength (distance in microns (μm) between hairs) on a regenerating whorl of *Acetabularia* as a function of the external calcium concentration. This relates to the parameter b in the model mechanism (15.35) and (15.36). The bars show standard deviations from the average distance between hairs on groups of plants and not on individual hairs. The solid ($\bullet$) period denotes where whorls were formed but, with extensive gaps where hairs failed to form, although the mean hair spacing where they formed normally was the same as in plants with complete whorls, for that calcium concentration. (From Goodwin, Murray and Baldwin 1985)

the computed behaviour using the dimensional wavelength obtained from k_M in (15.51) as compared with the experimental results from Goodwin et al. (1985). The parameters a, d and R were fitted to give a best fit, but nevertheless the quantitative comparison is reasonable when b is varied.

The material presented here is an example of how a model mechanism and an experimental programme can be directly related and developed together. The hypothesis that calcium could be one of the morphogens in a reaction diffusion system was explored and a specific mechanism suggested which satisfied some of the required conditions dictated by experiment, such as a window of external calcium concentration where hair patterns could be formed. Certainly not all reaction diffusion mechanisms exhibit this behaviour.

We chose a simple two-species mechanism which incorporated key biological facts and identified one of the morphogens with calcium. The question arises as to what the other morphogen, u, could be. One candidate proposed was cyclic-AMP (cAMP) which is important in cellular metabolism. cAMP induces the release of calcium from mitochondria while calcium inhibits cAMP production. However the conditions for spatial structure require $d > 1$ which means cAMP has to diffuse faster than calcium which, with cAMP's larger molecular weight, is not the case. Another candidate is the proton H^+ since there is some evidence of a close connection between calcium and the proton pump activity and pH in the morphogenesis of *Acetabularia*. With the present state of knowledge, however, the identity of either morphogen must still be speculative.

The formation of a spatial pattern in calcium concentration is viewed as the prepattern for hair initiation. Actual hair growth with its mechanical deformation of the plant is a subsequent process which uses and reflects the prepattern. It is possible that calcium is directly coupled to the mechanical properties of the cytoplasm, the shell of the stalk. Such a coupling could be incorporated into the mechanochemical theory of morphogenesis discussed in detail later in Chapter 17. In fact the mechanisms proposed there do not need a prepattern prior to hair initiation; the whole process takes place simultaneously.

16. Neural Models of Pattern Formation

Perhaps the most obvious, ubiquitous, important and complex spatial patterning processes are those associated with the nervous system such as pattern recognition and the transmission of visual information to the brain. This is a vast field of study. In this chapter we give only an introduction to some of the models, involving nerve cells, which have been proposed as pattern generators. Basic to the concept of neural activity is the nerve cell, or neuron. The neuron consists of a cell body with its dendrites, axon and synapses. It is a bit like a tree with the roots the dendrites, the base the cell body, the trunk the axon and the numerous branches the synapses. The cell receives information, from other cells, through its dendrites, passes messages along the axon to the synapses which in turn pass signals on to the dendrites of other cells. This neuronal process is central to brain functioning. The axons are the connectors and make up the white matter in the brain with the dendrites and synapses making up the grey matter.

We start pedagogically in Section 16.1 with a simple scalar model to introduce the basic concepts and show how spatial pattern evolves from a neural type of model. In Section 16.2 we derive a model for stripe formation, the ocular dominance stripes, in the visual cortex based on neural activation while in Section 16.3 we discuss a theory of hallucination patterns. Finally in Section 16.4 we describe and analyse an application of a neural activity theory to the formation of mollusc shell patterns.

16.1 Spatial Patterning in Neural Firing with a Simple Activation-Inhibition Model

Nerve cells can fire spontaneously, that is they show a sudden burst of activity. They can also fire repeatedly at a constant rate. Whether or not a cell fires, depends on its autonomous firing rate and the excitatory and inhibitory input it gets from neighbouring cells: this input can be from other than nearest neighbours, that is long range interaction. Such input can be positive, which induces activity, or negative which inhibits it. In Section 9.5 in Chapter 9 we briefly discussed an integral equation, namely (9.41), incorporating similar concepts. We showed how it related to a diffusion-type model and used it to introduce the idea of long-range diffusion. Here we discuss it again and in more detail since

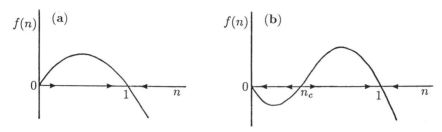

Fig. 16.1a,b. Functional form for the rate of change of the firing rate $f(n)$: (a) Situation with a single stable positive steady rate of firing, a representative $f(n) = rn(1 - n)$; (b) Example with bistable threshold kinetics with linearly stable steady state rates $n = 1$ and the quiescent state $n = 0$. A perturbation from $n = 0$ to $n > n_c$ implies $n \to 1$: a representative $f(n) = rn(n - n_c)(1 - n)$.

it introduces the key ingredient for the models we discuss later in this chapter. Although this model is primarily for pedagogical purposes and is not a model for any specific phenomenon it turns out to be closely related, mathematically, to a special case of the neural model we discuss in the following section.

Consider the one-dimensional situation in which the cells are functions only of x and t. Denote by $n(x, t)$ the firing rate of the cells. In the absence of any neigbourhood influences we assume the cells can be in a quiescent state or fire autonomously at a uniform rate which we normalise to 1. If the cells' firing rate is perturbed we assume it evolves according to

$$\frac{dn}{dt} = f(n) \,, \tag{16.1}$$

where $f(n)$ has zeros at $n = 0$ and $n = 1$, the steady state rates, with a functional form as in Fig. 16.1 (a). Here the only steady state is $n = 1$. $f(n)$ might exhibit bistable threshold kinetics whereby there is an unstable threshold steady state, n_c say, such that if $n > n_c$, $n \to 1$ and if $n < n_c$, $n \to 0$. A typical bistable form for $f(n)$ is illustrated in Fig. 16.1 (b). The kinetics of the firing dynamics (16.1) determines the subsequent firing rate given any initial rate n. In the case of bistable kinetics if $n(x, 0) = 0$, input from neighbouring cells could temporarily raise the firing rate to $n > n_c$ in which case the cells could eventually fire at a constant rate $n = 1$.

Let us now include spatial variation and incorporate the effect on the firing rate of cells at position x, of neighbouring cells at position x'. We assume the effect of close neighbours is greater than that from more distant ones; the spatial variation is incorporated in a weighting function w which is a funcion of $|x - x'|$. We must integrate the effect of all neighbouring cells on the firing rate and we model this by a convolution integral involving an influence kernel.

To be specific let us take $f(n)$ as in Fig. 16.1 (a) where the positive steady state is $n = 1$. Now modify (16.1) so that if $w > 0$ there is a positive contribution to the firing rate from neighbours if $n > 1$ and a negative one if $n < 1$. Thus a possible first attempt at modelling the mechanism is the integro-differential equation

$$\frac{\partial n}{\partial t} = f(n) + \int_D w(x - x')[n(x',t) - 1]\, dx' \qquad (16.2)$$
$$= f(n) + w * (n - 1) \, ,$$

where D is the spatial domain over which the influence kernel $w(x)$ is defined and $*$ denotes the convolution defined by this equation. The form (16.2) ensures that $n = 1$ is a solution. To complete the formulation of the model we must specify the influence kernel w, which we shall assume to be symmetric and so

$$w(|x - x'|) = w(x - x') = w(x' - x)$$

A non-symmetric kernel could arise as the result of some superimposed gradient. To ensure that n is always non-negative we must ensure that $n_t > 0$ for small n. This requires

$$\int_D w(|x - x'|)\, dx' < 0 \qquad (16.3)$$

since $f(n) \to 0$ with n and (16.2) reduces to

$$\frac{\partial n}{\partial t} \sim - \int_D w(|x - x'|)\, dx' > 0$$

This condition is satified with the typical kernels we envisage in the practical applications: see, for example, the analysis of the illustrative example in (16.13) below.

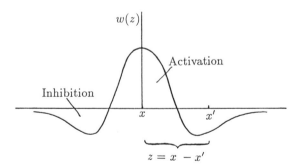

Fig. 16.2. Typical kernel $w(z)$ which exhibits local activation – long range inhibition. $w(z)$ measures the effect of cells at position x' on cells at position x; here $z = x - x'$.

In our model we envisage a cell to have a short range activation effect and a long range inhibitory effect. This is typical of the cell behaviour in the models discussed in the following sections. Pattern formation concepts based on local activation and lateral, or long range, inhibition have been discussed fully and in a pedagogical way by Oster and Murray (1989). Such a cell-cell influence is incorporated in a kernel of the form illustrated in Fig. 16.2. On the infinite domain, $w(z)$ is a continuous symmetric function of the variable $z = x - x'$ such that

$$w(z) \to 0 \quad \text{as} \quad |z| \to \infty; \quad z = x - x' . \tag{16.4}$$

We can see intuitively how the mechanism (16.2) with $w(z)$ as in Fig. 16.2 can start to create spatial patterns. To be specific consider $f(n)$ as in Fig. 16.1 (a) so $n = 1$ is the spatially uniform steady state solution of (16.2). Now impose small spatially heterogeneous perturbations about $n = 1$. If, in a small region about x where $w > 0$, we have $n > 1$, the effect of the integral term is to increase n autocatalytically in this region while at the same time inhibiting the nearest neighbourhood where $w < 0$. Thus small perturbations above the steady state will start to grow while those less than the steady state will tend to decrease further thus enhancing the heterogeneity. It is the classic activation-inhibition situation for generating spatial pattern. What we want, of course, is to quantify analytically the growth and pattern wavelength in terms of the model parameters.

Consider the infinite domain and linearize (16.2) about the positive steady state $n = 1$ by setting

$$u = n - 1, \quad |u| \ll 1$$
$$\Rightarrow \quad u_t = -au + \int_{-\infty}^{\infty} w(|x - x'|)u(x',t) \, dx', \quad a = |f'(1)| . \tag{16.5}$$

Now look for solutions in the form

$$u(x,t) \propto \exp\left[\lambda t + ikx\right] , \tag{16.6}$$

where k is the wavenumber and λ the growth factor. Substituting into (16.5), setting $z = x - x'$ in the integral and cancelling $\exp\left[\lambda t + ikx\right]$ gives λ as a function of k, that is the dispersion relation, from (16.5) as

$$\lambda = -a + \int_{-\infty}^{\infty} w(z) \exp\left[ikz\right] dz = -a + W(k) , \tag{16.7}$$

where $W(k)$, defined by the last equation, is simply the Fourier transform of the kernel $w(z)$. Remember that $w(z) = w(|z|)$ with the kernels we consider. We could of course solve (16.5) in general by taking the Fourier transform and using the convolution theorem on the integral term. This is equivalent to summing u in (16.6) over all k.

A simple symmetric kernel of the form illustrated in Fig. 16.2 can, for example, be constructed from a combination of exponentials of the form

$$\exp\left[-bx^2\right], \quad b > 0 \tag{16.8}$$

which has Fourier transform

$$\int_{-\infty}^{\infty} \exp\left[-bx^2 + ikx\right] dx = \left(\frac{\pi}{b}\right)^{1/2} \exp\left[\frac{-k^2}{4b}\right] . \tag{16.9}$$

This exponential and its transform are sketched in Fig. 16.3 (a).

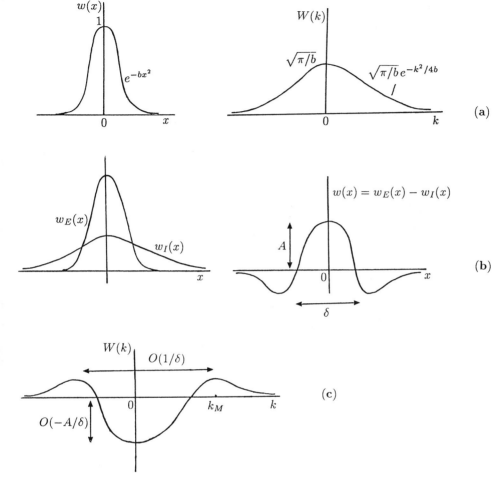

Fig. 16.3a-c. (a) Simple exponential kernel and its Fourier transform. Note how the the widths and height in x- and k-space relate to each other. (b) Typical activation (w_E) and inhibition (w_I) kernels which give a composite kernel $w = w_E - w_I$ with a local activation ($w > 0$) and long range inhibition ($w < 0$). (c) Sketch of $W(k) = W_E(k) - W_I(k)$, the Fourier transform of $w_E - w_I$. Note again how the height and width of the kernel in (b) and its transform are related.

Two particularly relevant properties of Fourier transforms for our purposes are that the taller and narrower the function is in x-space the shorter and broader is its transform in k-space. Fig. 16.3 (b) shows how to construct a short range activation and long range inhibition kernel from two separate kernels while Fig. 16.3 (c) is a sketch of its Fourier transform. The transform kernel $W(k)$ is similar in shape to the original kernel $w(x)$ but upside down.

Now consider the dispersion relation (16.7) giving the growth factor $\lambda = \lambda(k)$. From the full discussion of dispersion relations in Chapter 14, specifically Section 14.5, we can determine much of the pattern generation potential of the model. From the form of the transform of the activation-inhibition kernel in Fig. 16.3 (c) we can plot λ as a function of k as shown in Fig. 16.4. We see, from (16.7) that if the parameter a is large enough (and it is clearly a stabilising factor in equation (16.5)), $\lambda < 0$ for all wavenumbers k and the uniform solution $n = 1$ is linearly stable to all spatial disturbances. As the parameter a decreases, a critical bifurcation value a_c is reached where $\lambda = 0$ with $\lambda > 0$ for a finite range of k when $a < a_c$. The critical bifurcation a_c depends on the structure of the kernel: it is obtained from Fig. 16.3 (c), which defines k_M from (16.7) as $a_c = W(k_M)$, with k_M from (16.10). When $a < a_c$, for each $k_1 < k < k_2$ in Fig. 16.4 $\lambda(k) > 0$ and the solution $u(x,t)$ in (16.6) is exponentially growing with time-like $\exp[\lambda(k)t]$. There is a fastest growing mode, with wavenumber k_M from (16.10), which dominates the linear solution. Eventually linear theory no longer holds and the nonlinear terms in (16.2) bound the solution and it evolves to a spatially heterogeneous steady state. The final steady state finite amplitude structure is generally closely related to the wavelength of the fastest growing linear mode as was the case with reaction diffusion models discussed in Chapter 14 in the one-dimensional situation.

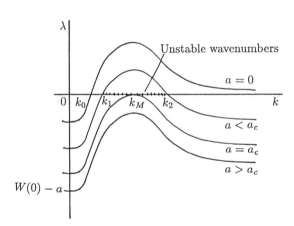

Fig. 16.4. Dispersion relation $\lambda(k)$ as a function of the wavenumber k, from (16.7), for various values of the bifurcation parameter a. $W(k)$, sketched in Fig. 16.3 (c), defines $W(k_M)$ and k_M.

Suppose $0 < a < a_c$. Then as t increases the dominant part of the solution $u(x,t)$ of (16.5) is given by the sum of all exponentially growing modes of the form (16.6), that is those with wavenumbers bounded by k_1 and k_2 in Fig. 16.4, namely

$$u(x,t) \sim \int_{k_1}^{k_2} A(k)\exp\left[\lambda(k)t + ikx\right] dk ,$$

where $A(k)$ is determined from the initial conditions. All other modes tend to zero exponentially since $\lambda(k) < 0$. Using Laplace's method the asymptotic expansion of the integral (see Murray 1984) gives

$$u(x,t) \sim A(k_M) \left\{ \frac{-2\pi}{t\lambda''(k_M)} \right\}^{1/2} \exp\left[\lambda(k_M)t + ik_M x\right]$$

The maximum λ is given by $\lambda(k_M)$ where $\lambda'(k_M) = 0$ and k_M is the maximum of $W(k)$, that is

$$\lambda_M = \lambda(k_M), \quad W'(k_M) = 0, \quad W''(k_M) < 0 . \tag{16.10}$$

The appearance of spatial pattern according to the dispersion relation in Fig. 16.4 has almost, but crucially not quite, the no-frills 'vanilla' dispersion relation form discussed in Chapter 14. It differs in a major way as $a \to 0$. From (16.7), $\lambda = W(k)$ when $a = 0$, which, as is clear from Fig. 16.3 (c), gives an *infinite* range of unstable wavenumbers to the right of k_0 where $W(k) = 0$. That is disturbances with very large wavenumbers – very small wavelengths – are all unstable. The ultimate steady state spatial structures in this situation probably depend critically on the initial conditions. From the numerical simulations of the models in the following sections, even in two dimensions, irregular stripe patterns possibly dominate. The ultimate pattern here is strictly a nonlinear phenomenon.

In Chapter 9 we related the integral formulation (16.2) to the differential equation reaction diffusion formulation in which the spatial interaction is via diffusion. To relate our analysis to the discussion in Chapter 9 we start with (16.2) on an infinite domain, with w symmetric, and set $z = x' - x$ to get

$$n_t = f(n) + \int_{-\infty}^{\infty} w(z)[n(x+z,t) - 1]\,dz . \tag{16.11}$$

If the kernel's influence is restricted to a narrow neighbourhood around $z = 0$ we can expand $n(x - z)$ in a Taylor series to get

$$n_t = f(n) + [n(x,t) - 1]\int_{-\infty}^{\infty} w(z)\,dz + n_x \int_{-\infty}^{\infty} zw(z)\,dz$$

$$+ \frac{n_{xx}}{2}\int_{-\infty}^{\infty} z^2 w(z)\,dz + \frac{n_{xxx}}{3!}\int_{-\infty}^{\infty} z^3 w(z)\,dz + \dots$$

The integrals are the moments of $w(z)$. Exploiting the symmetry property of $w(z)$, the odd moments of the kernel, that is those with odd powers of z in the integrand, are zero, so

$$n_t = f(n) + w_0(n-1) + w_2 n_{xx} + w_4 n_{xxxx} + \dots$$

$$w_{2m} = \frac{1}{(2m)!}\int_{-\infty}^{\infty} z^{2m} w(z)\,dz, \quad m = 0, 1, 2, \dots \tag{16.12}$$

where w_{2m} is the $2m$-th moment of the kernel w. The signs of the moments depend critically on the form of w.

As an example, suppose we choose the kernel to be

$$w(z) = b_1 \exp\left[-\frac{z^2}{d_1}\right] - b_2 \exp\left[-\frac{z^2}{d_2}\right], \quad b_1 > b_2, \quad d_1 < d_2 \qquad (16.13)$$

with b_1, b_2, d_1 and d_2 positive parameters: this form is of the type in Fig. 16.3 (b). Its transform (compare with (16.9)) is

$$W(k) = \int_{-\infty}^{\infty} w(z) \exp[ikz]\, dz$$

$$= \sqrt{\pi} \left\{ b_1 d_1 \exp\left[-\frac{(d_1 k)^2}{4}\right] - b_2 d_2 \exp\left[-\frac{(d_2 k)^2}{4}\right] \right\}, \qquad (16.14)$$

which is of the type in Fig. 16.3 (c) if

$$W(0) < 0 \quad \Rightarrow \quad b_1 d_1 - b_2 d_2 < 0. \qquad (16.15)$$

In this case k_M in Fig. 16.3 (c) and (16.10) exists and is given by the non-zero solutions of $W'(k) = 0$ namely

$$k_M^2 = 4(d_2 - d_1)^{-1} \ln\left[\frac{b_2}{b_1}\left(\frac{d_2}{d_1}\right)^3\right] > 0 \qquad (16.16)$$

since $d_2/d_1 > 1$ and $b_2 d_2/b_1 d_1 > 1$.

The moments w_{2m}, defined by (16.12), can be evaluated exactly for the kernel (16.13) by noting that if

$$I(b) = \int_{-\infty}^{\infty} \exp\left[-bz^2\right] dz = \left(\frac{\pi}{b}\right)^{1/2}. \qquad (16.17)$$

differentiating successively with respect to b immediately gives the even moments. For example

$$I'(b) = -\frac{1}{2}\left(\frac{\pi}{b^3}\right)^{1/2} = -\int_{-\infty}^{\infty} z^2 \exp\left[-bz^2\right] dz.$$

With $w(z)$ in (16.13) we thus determine the moments as

$$w_0 = \sqrt{\pi}(b_1 d_1 - b_2 d_2), \quad w_2 = \frac{\sqrt{\pi}(b_1 d_1^3 - b_2 d_2^3)}{4}, \dots \qquad (16.18)$$

and so on: they can be either positive or negative depending on the parameters. The decision as to where the expansion should be terminated in (16.12) depends on the relative magnitude of the moments. The wider the kernel's spatial influence the more terms required in the expansion.

If (16.15) holds for (16.13), $w_0 < 0$ and $w_2 < 0$. With $w_0 < 0$ this kernel satisfies the criterion (16.3) which ensures that the solution for n in (16.2) is always non-negative. It is also clear from (16.12) in the space-independent situation for n small since

$$f(n) + w_0(n - 1) \sim -w_0 > 0 \ .$$

If $w_2 < 0$, that is negative diffusion, this term is *destabilising* while if $w_4 < 0$, that is negative long range diffusion, the 4th order term is *stabilising*. This situation arises in an offshoot of the models discussed in Section 16.4 below. The evolution of spatially structured solutions of such higher order equations like (16.12) with $w_2 < 0$, $w_4 < 0$ has been analysed in detail by Cohen and Murray (1981) whose method of analysis can also be applied to the model mechanism (16.2).

Now that we have introduced the basic ideas let us consider some specific practical neural models.

16.2 A Mechanism for Stripe Formation in the Visual Cortex

Visual information is transmitted via the optic nerve from the retinal cells in each eye through a kind of relay station, called the lateral geniculate nucleus, to the visual cortex. The inputs from each eye are relayed separately and are distributed within a specific layer – layer IVc – of the cortex. Experiments involving electrophysiological recordings from the visual cortex of cats and monkeys have shown that the spatial pattern of nerve cells – neurons – in the visual cortex that are stimulated by the right eye form spatial bands, which are interlaced with bands of neurons which can be stimulated by the left eye. As mentioned, these neurons branch and form synapses (regions where nervous impulses move from one neuron to another). By moving the electrode within the cortex, and relating the response to the input from the eyes, the pattern of stripes are found to be about $350\mu m$ wide and about the same apart. These are the *ocular dominance stripes* the explanation for which is associated with the names of D.H. Hubel and T.N. Wiesel and for which they were awarded a Nobel Prize in 1981. Fig. 16.5 (a) shows a reconstruction of the spatial pattern of these ocular dominance stripes obtained from a macaque monkey.

The appearance of ocular dominance stripes is not universal in mammals; mice, rats and American monkeys, for example, do not seem to have them. It is possible that they are not necessary for the visual process but are associated with some other process. Even where they exist they are not always so clearly delineated as is the case with the cat and macaque monkey.

It is difficult not to be struck by the resemblance of ocular dominance stripes to many others in the natural world, such as the stripes on zebras as in Fig. 16.5 (b) and fingerprints as in Fig. 16.5 (c). However, we must resist reading too much into this similarity. Although it is possible that a neural-type model might be responsible for animal coat markings it is unlikely that it is the case for fingerprints. The mechanochemical mechanism discussed later in Chapter 17 is

(a) 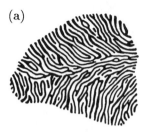   (b) 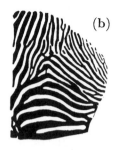   (c)

Fig. 16.5a-c. (a) Spatial pattern of ocular dominance stripes in the visual cortex of a macaque monkey. The dark bands are areas which receive input from one eye while the unshaded regions receive input from the other eye. (After Hubel and Wiesel 1977). (b) Stripe pattern on the rear flank of a Grevy's zebra (see also Fig. 15.3 (a)). (c) Typical human fingerprint pattern.

a good candidate. In any attempt at modelling fingerprint formation the police manual by Cherrill (1954), which has some interesting historical material, the book by Loesch (1983) and the papers by Elsdale and Wasoff (1976), Green and Thomas (1978) provide very useful background information with which to start. A general introductory account to dermatoglyphs is given by Cummins and Midlo (1943). The short general topological discussion of fingerprints by Penrose (1979) is also of interest and has implications for the type of model required.

In the case of the monkey it seems that the initial patterning process starts before birth, and is apparent just before birth. The pattern formation process is complete by six weeks of age. The actual stripe pattern can be altered as long as it has not been completely formed (Hubel et al. 1977). For example, removal or closure of one eye for about 7 weeks after birth causes the width of the bands, which reflect input from that eye, to become narrower while those from the other eye get broader. If both eyes are blindfolded there seems to be little effect on the stripes. There seems to be a critical period for producing such monocular vision since blindfolding seems to have no effect after the stripes have been formed. If these experiments are carried out after about 2 months of age, when the pattern generation is complete, there seems to be no effect on the pattern.

Hubel et al. (1977) suggested that the stripes are formed during development as a consequence of competition between the eye terminals in the cortex. That is if a region is dominated by input synapses associated with the right eye it inhibits the establishment of left eye synapses, at the same time enhancing the establishment of right eye synapses. This is like an activation-inhibition mechanism involving two species. Intuitively we can see how this could produce spatial patterns in an initial uniform distribution of unspecified synapses on the cortex. It is also an attractive idea since it is in keeping with the experimental results associated with eye closure during development. For example, if one eye is closed there is no inhibition of the other eye's terminals and so there are no stripes. Swindale (1980) proposed a model mechanism based on this idea of Hubel et al. (1977) and is the model we discuss in detail in this section.

Model Mechanism for Generating Ocular Dominance Stripes

The model is effectively an extension of that discussed in the last section except that here we consider two different classes of cells. Consider the layer (IVc) in the visual cortex, the layer on which the ocular dominance stripes are formed, to be a two-dimensional domain D and denote by $n_R(\mathbf{r}, t)$ and $n_L(\mathbf{r}, t)$ the left and right eye synapse densities on this cortical surface at position $\mathbf{r}$ at time t. The stimulus function, which measures the effect of neighbouring cells (neurons), is made up of positive and negative parts which respectively enhance and inhibit the cells' growth. As before this is represented by the sum of the product of a weighted kernel (w) and the cell density (n). The stimulus, both positive and negative, from neighbouring synapses on the rate of growth of right eye synapses n_R, for example, is taken to be the sum, s_R, of the appropriate convolutions; that is

$$s_R = w_{RR} * n_R + w_{LR} * n_L \, ,$$

$$w_{RR} * n_R = \int_D n_R(|\mathbf{r} - \mathbf{r}^*|) w_{RR}(\mathbf{r}^*) \, d\mathbf{r}^*, \qquad (16.19)$$

$$w_{LR} * n_L = \int_D n_L(|\mathbf{r} - \mathbf{r}^*|) w_{LR}(\mathbf{r}^*) \, d\mathbf{r}^* \, .$$

The stimulus function s_L for the left synapses n_L is defined similarly. It is reasonable to assume that under normal developmental conditions the activation and inhibition effects from each type of synapse is symmetric, that is

$$w_{RR} = w_{LL} = w_a, \quad w_{RL} = w_{LR} = w_i \, .$$

We envisage these weighting kernels to exhibit short range activation with long range, or lateral, inhibition such as illustrated in Fig. 16.2, where the activation region is of the order of half the stripe width, say $200 \, \mu$m, with the inhibition region from about $200 \, \mu$m to $600 \, \mu$m. With the above idea of right synapses locally enhancing growth of right synapses and laterally inhibiting left synapses, and vice versa, the qualitative forms of w_a and w_i are illustrated in Fig. 16.6.

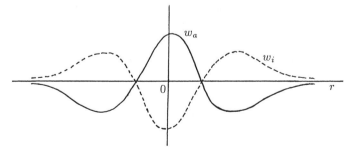

Fig. 16.6. Typical activation, w_a, and inhibition, w_i, kernels, as functions of r, for ocular dominance stripe formation. The activation region around the origin is of the order of $200 \, \mu$m with the inhibition region of order $400 \, \mu$m.

The cell densities n_R and n_L must, of course, be non-negative and bounded above by some cell density, N say. We must include such restraints in the model formulation and do so with density dependent multiplicative factors on the stimulation functions. We thus propose as the model mechanism

$$\frac{\partial n_R}{\partial t} = f(n_R)[w_a * n_R + w_i * n_L]$$

$$\frac{\partial n_L}{\partial t} = f(n_L)[w_a * n_L + w_i * n_R] \,, \tag{16.20}$$

where $f(n)$ has zeros at $n = 0$ and $n = N$ with $f'(0) > 0$ and $f'(N) < 0$; a logistic form for $f(n)$ is a reasonable one to take at this stage.

Analysis

The mechanism (16.20) with $f(n)$ qualitatively as described, clearly limits the growth of synapses and reflects, for example, the existence of some factor which reduces the rate of synapse growth with increasing density of that synapse density. The model also implies that if the synapses from a given eye type ever disappear from a region of the cortex they can never reappear.

The last property is the same as saying that the steady states $n_R = 0$, $n_L = N$ and $n_L = 0$, $n_R = N$ are stable. For example let us consider n_R and look at the stability of $n_R = 0$, $n_L = N$ from the first of (16.20). To be specific let us take $f(n) = n(N - n)$. For small n_R in a given region, any growth is inhibited by the neighbouring presence of n_L, that is $w_i * n_L < 0$ in the first of (16.20). So in this region, retaining only first order terms in n_R, (16.20) gives

$$\frac{\partial n_R}{\partial t} \approx N n_R w_i * N < 0$$

and so $n_R = 0$, $n_L = N$ is stable. Symmetry arguments show that $n_L = 0$, $n_R = N$ is also linearly stable.

Many of the numerical simulations were carried out with the initial distribution such that the total density at any point, namely $n_R + n_L$, is a constant and equal to N. This would be the case if there was a fixed number of post-synaptic sites on the cortex which are always occupied by one or other eye synapse. This assumption implies that $\partial n_R / \partial t = -\partial n_L / \partial t$. The two equations in (16.20) when written in terms of one of the densities must be the same. This requires $w_a = -w_i$. This simplifies the model considerably, which then reduces to

$$n_R = f(n_R)[w_a * (2n_R - N)] = f(n_R)[2w_a * n_R - K] \,, \tag{16.21}$$

$$n_L = N - n_R, \quad K = N \int_D w_a(|\mathbf{r} - \mathbf{r}^*|) \, d\mathbf{r}^* \,,$$

namely a scalar equation for n_R with certain common features to that considered in detail in the last section: $w_a(r)$, the solid line kernel in Fig. 16.6, is similar to that in Fig. 16.2.

Equation (16.21) has 3 steady states:

$$n_R = 0, \; n_L = N; \quad n_R = N, \; n_L = 0; \quad n_R = n_L = \frac{N}{2} \, . \tag{16.22}$$

We have already shown that the first two are linearly stable.

Consider now the stability of the third steady state in which the synapses from each eye are equally distributed over the cortex. The model intuitively implies this state is unstable. Consider a small perturbation $u(\mathbf{r}, t)$ about the steady state, substitute into the first equation in (16.21) and retain only first order terms in u to obtain

$$u_t = \frac{1}{4} N^2 [w_a * u] \, . \tag{16.23}$$

This equation is similar to that in (16.5) if $a = 0$. Now look for solutions in a similar way to that in (16.6), but now it is in two space dimensions, namely

$$u(\mathbf{r}, t) \propto \exp\left[\lambda t + i\mathbf{k} \cdot \mathbf{r}\right] \tag{16.24}$$

where $\mathbf{k}$ is the eigenvector with wavelength $2\pi/k$ where $k = |\mathbf{k}|$. Substitution into (16.23) gives the dispersion relation as

$$\lambda = \frac{1}{4} N^2 W_a(k), \quad W_a(k) = \int_D w_a(|\mathbf{r} - \mathbf{r}^*|) \exp\left[i\mathbf{k} \cdot \mathbf{r}^*\right] d\mathbf{r}^* \, . \tag{16.25}$$

$W_a(k)$ is simply the Fourier transform of the kernel w_a over the domain of the cortex.

If we consider the one-dimensional situation, $W_a(k)$ is similar in form to $W(k)$ in Fig. 16.3 (c) since w_a is similar to $W(k)$ there. Referring to Fig. 16.4 we see that there is an infinite range of unstable wavenumbers. Although there is a wavenumber k_M giving a with maximum growth rate $\lambda = \lambda_M$ it is not clear that it will dominate when nonlinear effects are included. In fact numerical simulations seem to indicate a strong dependence on initial conditions, as appears to be the case whenever there is an infinite range of unstable modes. The situation in the two-dimensional situation is similar in that there is an infinite range of wavevectors $\mathbf{k}$ which give linearly unstable modes in (16.24). The dominant solution is given by an integral of the modes in (16.24), with $\lambda(k)$ from (16.25), over the $\mathbf{k}$-space of all unstable wavenumbers, namely where $W_a(k) > 0$: it is infinite in extent. The instability evolves to a spatially heterogeneous steady state.

Numerical simulation of the model mechanism confirms this and generates spatial patterns such as in Fig. 16.7 (a), which shows the result of one such computation.

The development of ocular stripes takes place over a period of several weeks during which considerable growth of the visual cortex takes place. This can have some effect on the stripe pattern formed by the model mechanism. Numerical

(a)   (b)

Fig. 16.7a,b. (a) Numerical solution of the model with constant total synapse density N with antisymmetric kernels ($w_a = -w_i$); that is (16.21). The kernel chosen, similar to that in (16.13), was $w_a = A \exp\left[-r^2/d_1\right] - B \exp\left[-(r - h)^2/d^2\right]$ with parameter values $A = 0.3$, $B = 1.0$, $d_1 = 5$, $d_2 = 1.4$, $h = 3.7$. The dark and light regions correspond to the final steady state synapse densities n_R and n_L. (b) Simulation giving the pattern formed when the domain is subjected to unidirectional growth of 20% during pattern formation: here n_R and n_L could vary independently. The same kernel as in (a) was used for w_a and w_i, with parameter values $A = 1$, $B = 0.9$, $d_1 = 3$, $d_2 = 10$, $h = 0$ for w_a and $A = -1$, $B = -1$, $d_1 = 4$, $d_2 = 10$, $h = 0$ for w_i. (Redrawn from Swindale 1980)

simulations indicate that the stripes tend to run in the direction of growth as shown in Fig. 16.7 (b).

With the model developed here it is possible to carry out a variety of 'experiments' such as monocular deprivation. This can be done by restricting the input from one eye for a period of time during development. Swindale (1980) carried out simulations in this situation and found certain critical periods where deprivation played a major role: these results are in keeping with many of the real experimental results.

It is interesting to note that this, and other neural models seem to have a preference for stripe pattern formation whereas reaction diffusion models in large domains seem to favour a more spotted pattern. It would be extremely illuminating to know mathematically why this should be the case. From a linear analysis we see that there is an *infinite* range of unstable modes with the eigenvalues bounded away from $k = 0$. From a linear instability point of view this seems to be the main difference between the dispersion relation here and those in Chapter 14. It is tempting to speculate that a model with a dispersion relation with an unbounded range of unstable wavenumbers, $0 < k < \infty$, has a preference for irregular stripes as opposed to isolated individual patches.

16.3 A Model for the Brain Mechanism Underlying Visual Hallucination Patterns

Hallucinations occur in a wide variety of situations such as with migraine headaches, epilepsy, advanced syphilis and, particularly since the 1960's, as a result of external stimulus by drugs such as the extremely dangerous LSD and

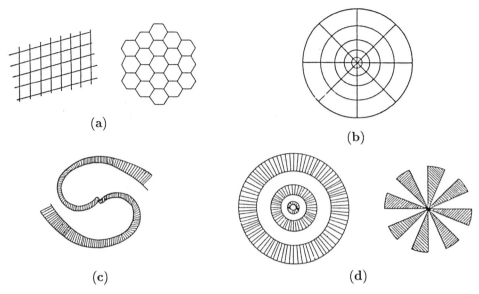

Fig. 16.8a-d. Typical examples of the four basic pattern types observed by hallucinating subjects: (a) lattice; (b) cobweb; (c) spiral; (d) tunnel and funnel. (After Ermentrout and Cowan 1979)

mescaline (derived from the peyote cactus). A general description is given by Oster (1970).

Hallucinogenic drugs acquired a certain mystique since users felt it could alter their perception of reality. From extensive studies of drug induced hallucinations by Klüver (1967), it appears that in the early stages the subject sees a series of simple geometric patterns which can be grouped into four pattern types. These four categories (Klüver 1967) are: (i) lattice, network, grating honeycomb; (ii) cobweb; (iii) spiral; (iv) tunnel, funnel, cone. Fig. 16.8 shows typical examples of these pattern types. In Fig. 16.8 (a) the fretwork type is characterised by regular tesselation of the plane by a repeating unit: that in Fig. 16.8 (b), the spiderweb, is a kind of distorted Fig. 16.8 (a).

The hallucinations are independent of peripheral input: for example, experiments showed that LSD could produce visual hallucinations in blind subjects. These experiments, and others such as those in which electrodes in the subcortical regions generated visual experiences, suggest that the hallucinogenic patterns are generated in the visual cortex. Ermentrout and Cowan's (1979) seminal paper is based on the assumption that the hallucinations are cortical in origin and proposed and analysed a neural net model for generating the basic patterns: see also the discussion on large scale nervous activity by Cowan (1982) and the less technical, more physiological, exposition by Cowan (1987). Ermentrout and Cowan (1979) suggest the patterns arise from instabilities in neural activity in the visual cortex: we discuss their model in detail in this section.

The Geometry of the Basic Patterns in the Visual Cortex

A visual image in the retina is projected conformally onto the cortical domain. The retinal image, which is described in polar coordinates (r, θ), is distorted in the process of transcription to the cortical image where it is described in (x, y) Cartesian coordinates. It is a mechanism for the creation of these cortical projection patterns that we need to model. The packing of retinal ganglion cells (the ones that transmit the image via the lateral geniculate nucleus which relays it to the visual cortex) decreases with distance from the centre and so small objects in the centre of the visual field are much bigger when mapped onto the cortical plane. So, a small area $dxdy$ in the cortical plane corresponds to $Mrdrd\theta$ in the retinal disk, where M is the cortical magnification parameter which is a function of r and θ. Cowan (1977) deduced the specific form of the visuo-cortical transformation from physiological measurements; it is defined by

$$x = \alpha \ln \left[\beta r + (1 + \beta^2 r^2)^{1/2} \right], \quad y = \alpha \beta r \theta (1 + \beta^2 r^2)^{-1/2}, \tag{16.26}$$

where α and β are constants. Close to the centre (the fovea) of the visual field, that is r small, the transformation is approximately given by

$$x \sim \alpha \beta r, \quad y \sim \alpha \beta r \theta, \quad r \ll 1, \tag{16.27}$$

whereas for r far enough away from the centre (roughly greater than a solid angle of $1°$)

$$x \sim \alpha \ln [2\beta r], \quad y \sim \alpha \theta. \tag{16.28}$$

Thus, except very close to the fovea, a point on the retina denoted by the complex coordinate z is mapped onto the point with complex coordinate w in the visual cortex according to

$$w = x + iy = \alpha \ln [2\beta r] + i\alpha \theta = \alpha \ln [z], \quad z = 2\beta r \exp [i\theta]. \tag{16.29}$$

This is the ordinary complex logarithmic mapping. It has been specifically discussed in connection with the retino-cortical magnification factor M. Fig. 16.9 shows typical patterns in the retinal plane and their corresponding shapes in the cortical plane as a result of the transformation (16.29): see any complex variable book which discusses conformal mappings in the complex plane or simply apply (16.29) to the various shapes such as circles, rectangles and so on. We can thus summarise, from Fig. 16.9, the cortical patterns which a mechanism must be able to produce as: (i) cellular patterns of squares and hexagons, and (ii) roll patterns along some constant direction. All of these patterns tesselate the plane and belong to the class of doubly-periodic patterns in the plane. In Section 14.4 in Chapter 14 we saw that reaction diffusion mechanisms can generate similar patterns at least near the bifurcation from homogeneity to heterogeneity.

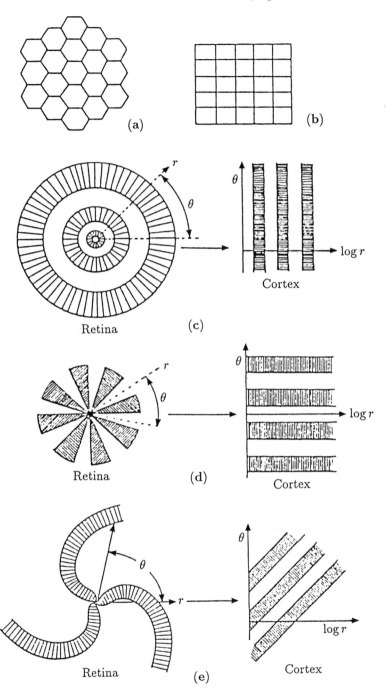

Fig. 16.9a-e. Corresponding patterns under the visuo-cortical transformation. (a) The lattice patterns in Fig. 16.8 (a), except for distortions, are effectively unchanged. The other visual field hallucination patterns are on the left with their corresponding cortical images on the right: (c) tunnel; (d) funnel; (e) spiral. (After Ermentrout and Cowan 1979)

Model Neural Mechanism

The basic assumption in the model is that the effect of drugs, or any of the other causes of hallucinations, is to cause instabilities in the neural activity in the visual cortex and these instabilities result in the visual patterns experienced by the subject. Ermentrout and Cowan's (1979) model considers the cortical neurons, or nerve cells, to be of two types, excitatory and inhibitory, and assume that they influence each others activity or firing rate (recall the discussion in Section 16.1). We denote the continuum spatially distributed neural firing rates of the two cell types by $e(\mathbf{r}, t)$ and $i(\mathbf{r}, t)$ and assume that cells at position $\mathbf{r}$ and time t influence themselves and their neighbours in an excitatory and inhibitory way much as we described in the last section with activation and inhibition kernels.

Here the activity at time t strictly depends on the time history of previous activity and so in place of the dependent variables e and i we introduce the time course grained activities

$$
\begin{pmatrix} E(\mathbf{r}, t) \\ I(\mathbf{r}, t) \end{pmatrix} = \int_{-\infty}^{t} h(t - \tau) \begin{pmatrix} e(\mathbf{r}, t) \\ i(\mathbf{r}, t) \end{pmatrix} d\tau , \tag{16.30}
$$

where $h(t)$ is a temporal response function which incorporates decay and delay times: $h(t)$ is a decreasing function with time which is typically approximated by a decaying exponential $\exp[-at]$ with $a > 0$.

There is physiological evidence (see, for example, Ermentrout and Cowan 1979) that suggests the activity depends on the self-activation through E and inhibition through I. The activity of E and of I also decay exponentially with time, so the model mechanism can be written as

$$
\begin{aligned}
\frac{\partial E}{\partial t} &= -E + S_E(\alpha_{EE} w_{EE} * E - \alpha_{IE} w_{IE} * I) , \\
\frac{\partial I}{\partial t} &= -I + S_I(\alpha_{EI} w_{EI} * E - \alpha_{II} w_{II} * I) ,
\end{aligned} \tag{16.31}
$$

where, from physiological evidence, the functions S_E and S_I are a typical threshold functions of their argument, such as the S shown in Fig. 16.10 (a) and the α's are constants related to the physiology and, for example, drug dosage: note that S is bounded for all values of its argument with $S(0) = 0$. The convolutions are taken over the two-dimensional cortical domain and the kernels here are non-negative, symmetric and decaying with distance, as illustrated in Fig. 16.10 (b): a symmetric decaying exponential such as $\exp[-(x^2 + y^2)]$ is an example. The argument in the interaction functions, S_E for example, represents the difference between the weighted activation of the local excitation and the local inhibition due to the presence of inhibitors. The inhibitors are enhanced through the argument of the S_I function, in the I equation, via the w_{EI} convolution. The inhibitors also inhibit their own production via the w_{II} convolution. There are similarities with the model discussed in the last section except there the activation and inhibition was included in each kernel.

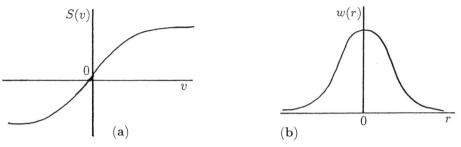

Fig. 16.10a,b. (a) Typical threshold response function $S(v)$; (b) Typical kernel function $w(\mathbf{r})$ in the model mechanism (16.31).

Stability Analysis

Let us now examine the linear stability of the spatially uniform steady state of (16.31), namely $E = I = 0$; that is the rest state. The nonlinearity in the system is in the functions S so the linearised form of (16.31), where now E and I are small, is

$$\frac{\partial E}{\partial t} = -E + S'_E(0)(\alpha_{EE} w_{EE} * E - \alpha_{IE} w_{IE} * I)$$

$$\frac{\partial I}{\partial t} = -I + S'_I(0)(\alpha_{EI} w_{EI} * E - \alpha_{II} w_{II} * I) ,$$

(16.32)

where, because of the forms in Fig. 16.10 (a), the derivatives $S'_E(0)$ and $S'_I(0)$ are positive constants. We now look for spatially structured solutions in a similar way to that used in the last two sections except that here we are dealing with a system rather than single equations ((16.5) and (16.23)), by setting

$$\begin{pmatrix} E(\mathbf{r}, t) \\ I(\mathbf{r}, t) \end{pmatrix} = \mathbf{V} \exp\left[\lambda t + i\mathbf{k} \cdot \mathbf{r}\right] = \mathbf{V} \exp\left[\lambda t + ik_1 x + ik_2 y\right]$$

(16.33)

where $\mathbf{k}$ is the wave vector with wavenumbers k_1 and k_2 in the (x, y) coordinate directions; λ is the growth factor and $\mathbf{V}$ the eigenvector. If $\lambda > 0$ for certain $\mathbf{k}$, these eigenfunctions are linearly unstable in the usual way.

Substituting (16.33) into the linear system (16.32) gives a quadratic equation for $\lambda = \lambda(k)$, the dispersion relation, where $k = |\mathbf{k}| = (k_1^2 + k_2^2)^{1/2}$. For example, with (16.33)

$$w_{EE} * E = \int_D w_{EE}(|\mathbf{r} - \mathbf{r}^*|) \exp\left[\lambda t + i\mathbf{k} \cdot \mathbf{r}^*\right] d\mathbf{r}^*$$

$$= \exp\left[\lambda t\right] \int_D w_{EE}(|\mathbf{u}|) \exp\left[i\mathbf{k} \cdot \mathbf{u} + i\mathbf{k} \cdot \mathbf{r}\right] d\mathbf{u}$$

$$= \exp\left[\lambda t + i\mathbf{k} \cdot \mathbf{r}\right] \int_D w_{EE}(|\mathbf{u}|) \exp\left[i\mathbf{k} \cdot \mathbf{u}\right] d\mathbf{u}$$

$$= W_{EE}(\mathbf{k}) \exp\left[\lambda t + i\mathbf{k} \cdot \mathbf{r}\right]$$

(16.34)

where $W_{EE}(\mathbf{k})$ is the two-dimensional Fourier transform of $w_{EE}(\mathbf{r})$ over the cortical domain D. A typical qualitative form for the w-kernels and its transform is:

$$w(\mathbf{r}) = \exp\left[-b(x^2 + y^2)\right]$$
$$\Rightarrow \quad W(\mathbf{k}) = \frac{\pi}{b}\exp[-k^2/4b], \quad k^2 = k_1^2 + k_2^2 . \tag{16.35}$$

Setting (16.33) into (16.32) and cancelling $\exp\left[\lambda t + i\mathbf{k}\cdot\mathbf{r}\right]$, the quadratic for λ is given by the characteristic polynomial

$$\begin{vmatrix} -\lambda - 1 + S'_E\alpha_{EE}W_{EE} & -S'_E\alpha_{IE}W_{IE} \\ S'_I\alpha_{EI}W_{EI} & -\lambda - 1 - S'_I\alpha_{II}W_{II} \end{vmatrix} = 0 . \tag{16.36}$$

For algebraic convenience let us incorporate the derivatives $S'_E(0)$ and $S'_I(0)$ into the α-parameters. In anticipation of a bifurcation to spatially structured solutions as some parameter, p say, is increased, again for simplicity let us assume that the mechanism is modulated by p multiplying the α-parameters and so we shall write pa for α. In the case of drug-induced hallucinations p could be associated with drug dosage. With this notation, namely

$$S'_E\alpha_{EE} = pa_{EE}, \quad S'_E\alpha_{IE} = pa_{IE} ,$$
$$S'_I\alpha_{II} = pa_{II}, \quad S'_I\alpha_{EI} = pa_{EI} , \tag{16.37}$$

equation (16.35) for λ is

$$\lambda^2 + L(\mathbf{k})\lambda + M(\mathbf{k}) = 0 ,$$
$$L(\mathbf{k}) = 2 - pa_{EE}W_{EE}(\mathbf{k}) + pa_{II}W_{II}(\mathbf{k}) ,$$
$$M(\mathbf{k}) = 1 + p^2 a_{IE}a_{EI}W_{IE}(\mathbf{k})W_{EI}(\mathbf{k}) - pa_{EE}W_{EE}(\mathbf{k}) \tag{16.38}$$
$$- p^2 a_{EE}a_{II}W_{EE}(\mathbf{k})W_{II}(\mathbf{k}) + pa_{II}W_{II}(\mathbf{k}) .$$

Solutions of the type (16.33) are linearly stable in the usual way if $\mathrm{Re}\,\lambda < 0$ and unstable if $\mathrm{Re}\,\lambda > 0$. Since here we are interested in the spatial patterns which arise from spatially structured instabilities we require the space independent problem to be stable, that is $\mathrm{Re}\,\lambda(0) < 0$: recall the related discussion in Section 14.3 in Chapter 14. Here this is the case if

$$\mathrm{Re}\,\lambda(k = 0) < 0 \quad \Rightarrow \quad L(0) > 0, \quad M(0) > 0 , \tag{16.39}$$

which impose conditions on the parameters in (16.38).

Let us now look at the dispersion relation $\lambda = \lambda(k)$, the solution of (16.38). From the typical kernel forms, such as in (16.35), we see that the $W(\mathbf{k})$ tend to zero as $k \to \infty$ in which case (16.38) becomes

$$\lambda^2 + 2\lambda + 1 \approx 0 \quad \Rightarrow \quad \lambda < 0 \quad \text{for} \quad k \to \infty$$

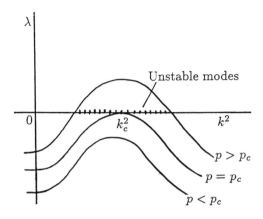

Fig. 16.11. Basic dispersion relation giving the growth rate $\lambda(k)$ as a function of the wavenumber $k = |\mathbf{k}|$. The bifurcation parameter is p (for example a measure of drug dosage): at bifurcation, where $p = p_c$, from spatially uniform to spatially heterogeneous solutions the critical wavenumber is $k_c = (k_{1c}^2 + k_{2c}^2)^{1/2}$.

and so large wavenumber solutions, that is those with $k \gg 1$ and hence small wavelengths, are linearly stable. This plus condition (16.39) points to the basic spatial pattern formation type of dispersion relation like that illustrated in Fig. 16.11. Here we have used the parameter p as the bifurcation parameter which we specifically consider related to the hallucinogenic drug dosage.

The mechanism for spatial pattern creation is then very like that with all other pattern formation mechanisms we have so far discussed. That is the pattern is generated when a parameter passes through a bifurcation value, p_c say, and that for larger p there is a finite range of unstable wavenumbers which grow exponentially with time, $O(\exp[\lambda(k)t])$, where $\lambda(k) > 0$ for a finite range of k. For $p = p_c + \varepsilon$, where $0 < \varepsilon \ll 1$, the spatially patterned solutions are approximately given by the solutions of the linear system (16.32) just like the linear eigenvalue problems we had to solve in Chapter 14, such as that posed by (14.46). The asymptotic procedure to show this is now fairly standard: for example Lara Ochoa and Murray (1983) used it in the equivalent reaction diffusion situation while Lara Ochoa (1984) carried out the analysis for the two-dimensional higher order diffusion equation derived in Section 9.6.

Let us now consider the type of patterns we can generate with the linear system (16.32) and relate them to the hallucinogenic patterns illustrated in Fig. 16.9. Near bifurcation, spatially heterogenous solutions are constructed from the exponential form in (16.33), namely

$$\mathbf{V} \exp[k_1 x + k_2 y], \quad k_1^2 + k_2^2 = k_c^2, \tag{16.40}$$

where $\mathbf{V}$ is the eigenvector corresponding to the eigenvalue $\mathbf{k}_c$ and k_1 and k_2 are the wavenumbers in the coordinate directions. We are specifically interested in solution units which can tesselate the plane, such as rolls, hexagons and so on. It will be helpful to recall (or revise) the latter part of Section 14.4 where we examined such solutions and their relation to the basic symmetry groups of hexagon, square and rhombus. All such solutions can be constructed from

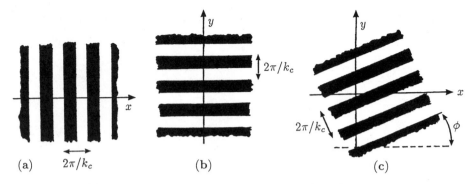

Fig. 16.12a-c. (a) Steady state solutions for E and I from (16.42) which give vertical stripes, that is rolls. (b) and (c) respectively illustrate horizontal rolls and those at an angle ϕ obtained from the solutions (16.43) and (16.44). Note the comparison between these and the visual cortex patterns in Fig. 16.9 and their related visual images.

combinations of the specific basic units

$$\exp\left[k_c x\right], \quad \exp\left[k_c y\right], \quad \exp\left[k_c(y\cos\phi \pm x\sin\phi)\right] \tag{16.41}$$

which are respectively periodic with period $2\pi/k_c$ in the x-direction, y-direction and perpendicular to the lines $y\cos\phi \pm x\sin\phi$ which make angles of $\pm\phi$ with the x-axis.

Let us first consider the simplest periodic structures in the visual cortex, namely the right hand forms in Fig. 16.9 (c), (d), (e). The small amplitude steady state expressions for E and I are of the form

$$\begin{pmatrix} E(x,y) \\ I(x,y) \end{pmatrix} = \mathbf{V}(p_c, k_c^2)\cos\left(a + k_c x\right) \tag{16.42}$$

in the case of vertical stripes, where a is a constant which simply fixes the origin. The patterns have period $2\pi/k_c$ and are illustrated in Fig. 16.12 (a) and correspond to those on the right in Fig. 16.9 (c). E and I are constant along lines $x = $ constant. In a similar way

$$\begin{pmatrix} E(x,y) \\ I(x,y) \end{pmatrix} = \mathbf{V}(p_c, k_c^2)\cos\left(a + k_c y\right) \tag{16.43}$$

represents the horizontal striping shown in Fig. 16.12 (b), which in turn corresponds to those on the right in Fig. 16.9 (d) while those corresponding to Fig. 16.9 (e) are reproduced in Fig. 16.12 (c) and given by

$$\begin{pmatrix} E(x,y) \\ I(x,y) \end{pmatrix} = \mathbf{V}(p_c, k_c^2)\cos\left(a + k_c x\cos\phi + k_c y\sin\phi\right), \tag{16.44}$$

where a is again a constant.

Let us now consider unit solutions with hexagonal symmetry, that is with the patterns in Fig. 16.9 (a) in mind. We wish to construct solutions for E and I which are invariant under the hexagonal rotation operator H. In polar coordinates (r, θ) this means that

$$H[E(r, \theta)] = E(r, \theta + \frac{\pi}{3}) = E(r, \theta) . \tag{16.45}$$

Such hexagonal solutions involve the specific exponential forms

$$\exp\left[ik_c\left(\frac{\sqrt{3}y}{2} \pm \frac{x}{2}\right)\right]; \quad k_{1c} = \pm\frac{k_c}{2}, \quad k_{2c} = \frac{k_c\sqrt{3}}{2}$$

$$\exp[ik_c x]; \quad k_{1c} = k_c, \quad k_{2c} = 0$$

and the relevant E and I are given by (see also (14.47) in Section 14.4)

$$\begin{pmatrix} E(x, y) \\ I(x, y) \end{pmatrix} = \mathbf{V}(p_c, k_c^2)\left\{ \cos\left[a + k_c\left(\frac{\sqrt{3}y}{2} + \frac{x}{2}\right)\right] \right.$$

$$\left. + \cos\left[b + k_c\left(\frac{\sqrt{3}y}{2} - \frac{x}{2}\right)\right] + \cos\left[c + k_c x\right] \right\} \tag{16.46}$$

where a, b and c are constants. In polar coordinates, a form which makes it clear that the solution is invariant under the hexagonal rotation (16.45), is

$$\begin{pmatrix} E(x, y) \\ I(x, y) \end{pmatrix} = \mathbf{V}(p_c, k_c^2)\left\{ \cos\left[a + k_c r \sin\left(\theta + \pi/6\right)\right] \right.$$

$$\left. + \cos\left[b + k_c(b + k_c r \sin\left(\theta - \pi/6\right)\right] + \cos\left[c + k_c r \cos\left(\theta - \pi/6\right)\right] \right\} \tag{16.47}$$

where a, b and c are constants. Fig. 16.13 illustrates the hexagonal tesselation from this solution for two sets of values for a, b and c: there are infinitely many patterns parameterized by a, b and c but they are all hexagonal (see also Fig. 14.8 (a)).

The procedure for generating the other patterns, the square and rhombus, is clear. In the case of the square lattice associated with Fig. 16.9 (b) the relevant solution is (compare also (14.48) in Section 14.4)

$$\begin{pmatrix} E(x, y) \\ I(x, y) \end{pmatrix} = \mathbf{V}(p_c, k_c^2)\{\cos\left[a + k_c x\right] + \cos\left[b + k_c y\right]\} \tag{16.48}$$

while the rhombus tesselation solution (compare with (14.49)) is

$$\begin{pmatrix} E(x, y) \\ I(x, y) \end{pmatrix} = \mathbf{V}(p_c, k_c^2)\{\cos\left[a + k_c x\right] + \cos\left[b + k_c(x \cos\phi + y \sin\phi)\right]\} , \tag{16.49}$$

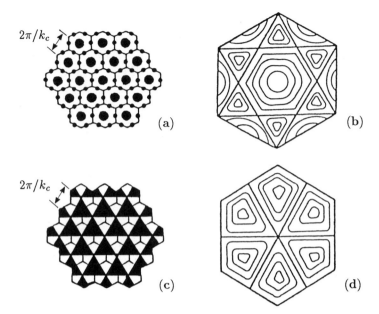

Fig. 16.13a-d. Hexagonal patterns from the near bifurcation solution (16.46) with two different parameter sets for a, b and c. The shaded areas have $E > 0$ and the unshaded $E < 0$; on the contours lines both $I(x,y)$ and $E(x,y)$ are constant. (a) $a = b = c = 0$ with the contours of a single hexagonal cell shown in (b). (c) $a = \pi/2$, $b = c = 0$ with (d) showing the corresponding contour lines for a single cell. Compare these patterns with the hallucinogenic pattern in Fig. 16.9 (a). (After Ermentrout and Cowan 1979)

where again a and b are constants. Patterns from these solutions are illustrated in Fig. 16.14.

The above linear analysis of the neural net model (16.31) for generating spatial patterns in the density of excitatory and inhibitory nerve cell synapses in the visual cortex shows that the mechanism can generate the required patterns which correspond to the basic hallucinogenic patterns in Fig. 16.9. These patterns are initiated when a physiological parameter p passes through a bifurcation value in the usual way of creating spatial patterns with a dispersion relation like that in Fig. 16.11. The modelling here is similar in many respects to that in the last section. This suggests that physiologically we can would expect actual pattern dimensions associated with hallucinogenic patterns to be comparable, roughly 2 mm. Further discussion on the applications and physiological implications is given by Cowan (1987).

If the model mechanism is to be an explanation for hallucinogenic patterns, the solutions we have described must be stable. The question of stability is not an easy one and in general, at this stage anyway, the best indication seems to be from the numerical simulation of the full nonlinear equations. Bifurcation and asymptotic analyses, however, provide strong indications. This has been done by Ermentrout and Cowan (1979).

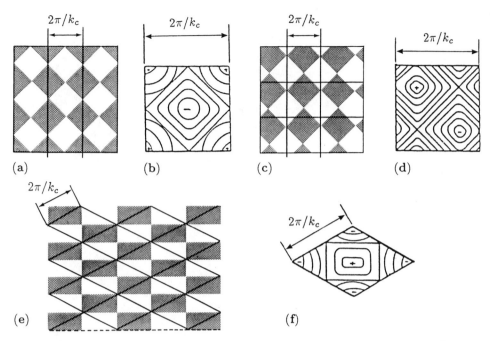

Fig. 16.14a-f. Lattice patterns which tesselate the domain with square and rhombic cell units. The shaded areas have $E > 0$ and the unshaded $E < 0$; on the contours lines, $I(x, y)$ and $E(x, y)$ are constant. (a) Solution (16.48) with $a = b = 0$ with (b) showing the contours for a single cell. (c) $b = -a = \pi/2$ with the contours of a single cell shown in (d). Compare these patterns with the hallucinogenic pattern in Fig. 16.9 (b). (e) Rhombus solution (16.49) with $a = b = \pi$ with (f) showing the corresponding contour lines for a single rhombic cell. (After Ermentrout and Cowan 1979)

16.4 Neural Activity Model for Shell Patterns

The intricate and colourful patterns on mollusc shells are almost as dramatic as those on butterfly wings: see, for example, Fig. 16.15. The reasons for these patterns, unlike the case with butterflies, is somewhat of a mystery since many of these species with spectacular patterns spend their life buried in mud.

Ermentrout, Campbell and Oster (1986) suggest that, since these markings probably do not appear to serve any adaptive purpose, this is why there are so many extreme polymorphic patterns observed in certain species. The novel model proposed by Ermentrout et al. (1986), which is the one we discuss in mathematical detail in this section, combines elements of discrete time models (refer to Chapters 2 and 4) and continuous spatial variation as in the previous sections. Several other modelling attempts have been made to reproduce some of the observed mollusc shell patterns: Waddington and Cowe (1969) and Wolfram (1984), for example, used a cellular automata phenomenolical approach while Meinhardt and Klingler (1987) employed an activator-inhibitor reaction diffusion model. All of these models can mimic many of the commoner shell patterns and

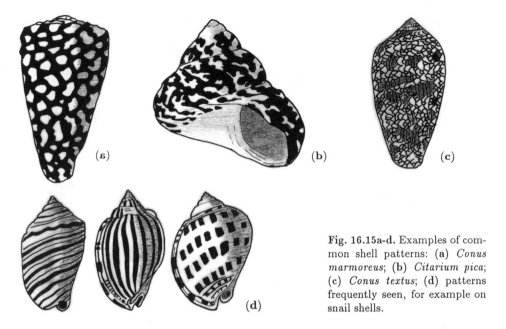

Fig. 16.15a-d. Examples of common shell patterns: (a) *Conus marmoreus*; (b) *Citarium pica*; (c) *Conus textus*; (d) patterns frequently seen, for example on snail shells.

so, as we keep reiterating, the only way to determine which mechanism pertains must be through the different experiments that each suggests. Having said that however, shell patterns are formed over several years and it would be surprising if a reaction diffusion system could sustain the necessary coherence over such a long period. The nervous system, on the otehr hand, is an integral part of the mollusc's physiology throughout its life. As regards cellular automata models, they make no connection with any of the underlying biological processes involved in the mollusc's growth and development.

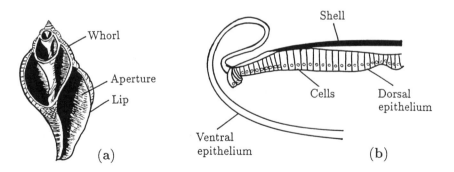

Fig. 16.16a,b. (a) Typical spiral shell structure. (b) Basic anatomical elements of the mantle region on a shell. (Drawn after Ermentrout et al. 1986)

A typical shell is a conical spire made up of tubular whorls which house the visceral mass of the animal: see Fig. 16.16 (a). These whorls are laid down sequentially and wind round a central core and eventually terminating at the aperture: see Fig. 16.16 (a). A readable introductory survey of the biology of molluscs is given in the textbook by Barnes (1980). Just under the shell is the mantle with epithelial cells which secrete the material for shell growth. Fig. 16.16 (b) is a simplified diagram of the anatomy of the mantle region. The basic assumption in the Ermentrout et al. (1986) model is that the secretory activity of the epithelial cells is controlled by nervous activity and that the cells are enervated from the central ganglion (concentrated masses of nerve cells – a kind of brain). The cells are activated and inhibited by the neural network which joins the secretory cells to the ganglion.

The Neural Model for Shell Pattern Formation

The specific assumptions of the model, and which we shall enlarge upon below, are:

(i)   Cells at the mantle edge secrete material intermittently.
(ii) The secretion depends on (*a*) the neural stimulation, S, from surrounding regions of the mantle, and (*b*) the accumulation of an inhibitory substance, R, present in the secretory cell.
(iii) The net neural stimulation of the secretory cells consists of the difference between the excitatory and inhibitory inputs from the surrounding tissue (recall the discussions in the previous two sections).

It is known that the shell is laid down intermittently. At the start of each secretory period the assumption is that the mantle aligns with the previous pattern and extends it. The alignment is effected through *a* sensing (a kind of tasting) of the pigmented (or unpigmented) areas of the previous secretion session. Alternatively it is possible that the pigmented shell laid down in the previous session stimulates the mantle neurons locally to continue with that pattern.

We now incorporate these assumptions into a model mechanism with discrete time and continuous space. It will be helpful to refer to Fig. 16.17. Consider the edge of the mantle to consist of a line of secretory cells, or mantle pigment cells, with coordinate x measured along the line. Let $P_t(x)$ be the amount of pigment secreted by a cell at position x at time t (the unit of period will be taken as 1). Let $A_t(x)$ be the average activity of the mantle neural net, $R_t(x)$ the amount of inhibitory substance produced by the cell and the functional $S[P]$, the net neural stimulation, which depends on P_{t-1}, the pigment secreted during the previous session. Although the final patterns are two-dimensional, the way they are laid down lets us consider, in effect, a one-dimensional model.

The model equation for the neural activity is taken to be

$$A_{t+1}(x) = S[P_t(x)] - R_t(x) \qquad (16.50)$$

which simply says that the neural activity is stimulated by the net neural stim-

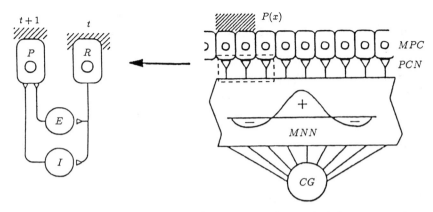

Fig. 16.17. Schematic model for neural activation and control of pigment secretion: see text for an explanation. Here *MPC* denotes the mantle pigment cells, *PCN* the pigment cell neurons, *MNN* the mantle neural net and *CG* the central ganglion. *R* denotes the receptor cells sensing pigment laid down in the time period t, P denotes the pigment cells secreting pigment in time period $t+1$, E denotes the excitatory neurons and I the inhibitory neurons. (After Ermentrout et al. 1986)

ulation in the earlier session $(S[P_t])$ and inhibited by the previous session's inhibitory substance (R_t). We assume the inhibitory material, R_t, depends linearly on the amount of pigment secreted during the previous period while at the same time degrading at a constant linear rate δ. The governing conservation equation for R_t is then

$$R_{t+1}(x) = \gamma P_t(x) + \delta R_t(x) , \qquad (16.51)$$

where the rate of increase $\gamma < 1$ and the rate of decay $\delta < 1$ are positive parameters. Because this is a discrete equation (recall Chapter 2) note that the decay term is positive. For example, if $\gamma = 0$, R_t decreases with each time step if $0 < \delta < 1$: it increases if $\delta > 1$.

We now assume that pigment will be secreted only if the mantle activity is stimulated above some threshold, A^* say, and thus write

$$P_t(x) = H(A - A^*) \qquad (16.52)$$

where H is the Heaviside function: $H = 0$ when $A < A^*$ and $H = 1$ when $A > A^*$.

It is reasonable, at this stage of the modelling, to assume that the pigment secretion P_t is simply proportional to the activity A_t and subsume the threshold behaviour into the stimulation function $S[P_t]$. With this, the simpler model is then

$$P_{t+1}(x) = S[P_t(x)] - R_t(x) \qquad (16.53)$$

$$R_{t+1}(x) = \gamma P_t(x) + \delta R_t(x) . \qquad (16.54)$$

which we shall now study in detail.

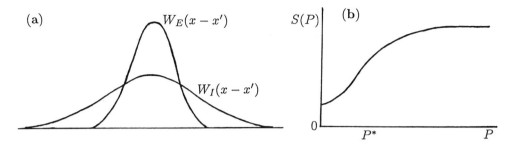

Fig. 16.18a,b. (a) Schematic form of the activation, w_E, and inhibition, w_I, kernels in (16.55). The width of the inhibition kernel is greater than that of the activation kernel; compare with Fig. 16.3 (b). (b) Typical threshold form for the stimulation function $S[P]$ in the model equation (16.53): P^* is the approximate threshold value equivalent to the activation threshold A^* in (16.52).

Consider now the neural stimulation functional $S[P_t]$. This consists of excitatory and inhibitory effects. Although secretion at a given time $t+1$ depends only on the excitation during the time period t to $t+1$, each period's excitation depends on the stimulation from sensing the previous period's pigment pattern. It is reasonable to assume that the time constants for neural interactions are much faster than those for shell growth and so we use an average neural firing rate in the mantle. We thus define the excitation $E_{t+1}(x)$ and inhibition $I_{t+1}(x)$ by the convolution integrals

$$E_{t+1}(x) = \int_\Omega w_E(|x' - x|)P_t(x')\,dx' = w_E * P_t$$

$$(16.55)$$

$$I_{t+1}(x) = \int_\Omega w_I(|x' - x|)P_t(x')\,dx' = w_I * P_t \, ,$$

where Ω is the mantle domain and may be circular, or a finite length, depending on the shell we are considering. The excitatory and inhibitory kernels w_E and w_I are measures of the effect of neural contacts between cells at x' and those at x. They represent nonlocal spatial effects which we are now familiar with from the previous sections of this chapter. Fig. 16.18 (a) illustrates the general form of these kernels while Fig. 16.17 sketches their combined effect, namely a positive excitation (+) and a negative inhibition (–), in the mantle neural net (*MNN*).

From the analysis in Sections 16.1–16.3 (and intuitively with the experience we now have) we require the width of the inhibition kernel w_I to be greater than that of the activation kernel w_E; see Fig. 16.18 (a). For illustrative numerical simulation purposes only, Ermentrout et al. (1986) chose the following kernels:

$$w_j = 0 \quad \text{for} \quad |x| > \sigma_j, \quad j = E, I$$

$$w_j = q_j \left\{ 2^p - [1 - \cos{(\pi x/\sigma_j)}]^p \right\} \quad \text{for} \quad |x| \le \sigma_j, \quad j = E, I$$

$$(16.56)$$

where q_j are chosen such that

$$\int_\Omega w_j(x)\,dx = \alpha_j, \quad j = E, I\,. \tag{16.57}$$

The parameters σ_j measure the range of the kernels; $\sigma_I > \sigma_E$ in our model. The parameter p, which of course could also be a p_j, controls the sharpness of the cut-off. For p small the kernels are sharply peaked while for p large they are almost rectangular. The amplitude of the excitatory and inhibition kernel functions is controlled by the α_j where, in our system, we expect $\alpha_E > \alpha_I$.

The stimulation functional S is composed of the difference between the excitatory and inhibitory elements. For analysis a fairly general form as in Fig. 16.18 (b) suffices but for numerical simulation a specific form for S which exhibits a threshold-type behaviour is required and Ermentrout et al. (1986) chose

$$S[P_t(x)] = S_E[E_t(x)] - S_I[I_t(x)]$$
$$S_j(u) = \{1 + \exp[-\nu_j(u - \theta_j)]\}^{-1}, \quad j = E, I \tag{16.58}$$

where the ν_j controls the sharpness of the threshold switch which is located at θ_j. The sharpness $(S'_j(\theta_j))$ increases with ν_j: θ_j corresponds to P^* in Fig. 16.18 (b).

The complete model system used in the numerical simulations is given by (16.53)–(16.58): it has 11 parameters, or 12 if we have different p in (16.56). Rescaling reduces the number while other parameters appear only as products. The influence-function parameters are the α, σ and p, those associated with the firing threshold are the ν and θ, while the dynamic or refractory parameters, in (16.53) and (16.54), are δ and γ. We now examine analytically the way spatial patterns are generated by the model.

Linear Stability Analysis

Equations (16.53) and (16.54) can be combined into the single scalar equation by iterating (16.53) once in time and using (16.54) and (16.53) again to get

$$P_{t+2} = S[P_{t+1}] - \gamma P_t - \delta(S[P_t] - P_{t+1})$$
$$= S[P_{t+1}] + \delta P_{t+1} - \delta S[P_t] - \gamma P_t\,. \tag{16.59}$$

For notational simplicity we have written P_t for $P_t(x)$ and so on.

In the usual linear stability way, we now look for the homogeneous steady state, P_0 say, perturb it linearly and analyse the resulting linear equation. From (16.59)

$$P_0 = S[P_0] + \delta P_0 - \delta S[P_0] - \gamma P_0$$
$$\Rightarrow \quad P_0 = \frac{(1 - \delta)S[P_0]}{1 + \gamma - \delta} \tag{16.60}$$

with the form for $S[P]$ in Fig. 16.18 (b) at least one positive steady state solution exists: simply draw each side as a function of P_0 to see this. Now linearise about P_0 by writing

$$P_t(x) = P_0 + u_t(x), \quad |u_t| \text{ small} . \tag{16.61}$$

On substitution into (16.59) and retaining only linear terms in u, we get

$$u_{t+2} - L_0[u_{t+1}] - \delta u_{t+1} + \delta L_0[u_t] + \gamma u_t = 0 , \tag{16.62}$$
$$L_0[u] = S_E'(P_0) w_E * u - S_I'(P_0) w_I * u ,$$

where L_0 is a linear integral (convolution) operator. The eigenfunctions of L_0 on a periodic domain of length L are $\exp[2\pi i n x / L]$, $n = 1, 2, \ldots$. Since we consider the finite linear size L of the domain, that is the length of the shell lip, to be very much larger than the range of the kernels, these eigenfunctions are approximately those for L.

Since (16.62) is a linear equation, discrete in time and continuous in x, we look for solutions in the form

$$u_t \propto \lambda^t \exp[ikx] , \tag{16.63}$$

which on substituting into (16.62) gives

$$\lambda^{t+2} \exp[ikx] - \lambda^{t+1} L_0[\exp[ikx]] - \delta \lambda^{t+1}$$
$$+ \delta \lambda^t L_0[\exp[ikx]] + \gamma \lambda^t \exp[ikx] = 0 . \tag{16.64}$$

Here

$$L_0[\exp[ikx]] = S_E'(P_0) \int_\Omega w_E(|x' - x|) \exp[ikx'] \, dx'$$
$$- S_I'(P_0) \int_\Omega w_I(|x' - x|) \exp[ikx'] \, dx' ,$$
$$= \exp[ikx]\{S_E' W_E(k) - S_I' W_I(k)\}$$
$$= \exp[ikx] L^*(k) , \tag{16.65}$$

which defines $L^*(k)$, where $W_E(k)$ and $W_I(k)$, the Fourier transforms of $w_E(x)$ and $w_I(x)$ over the domain Ω, are defined by

$$W_j(k) = \int_\Omega W_j(z) \exp[ikz] \, dz, \quad j = E, I . \tag{16.66}$$

Using (16.65) in (16.64) and cancelling $\lambda^t \exp[ikx]$ we get the characteristic equation for λ as

$$\lambda^2 - \lambda(L^*(k) + \delta) + (\delta L^*(k) + \gamma) = \lambda^2 + a(k)\lambda + b(k) = 0 , \tag{16.67}$$

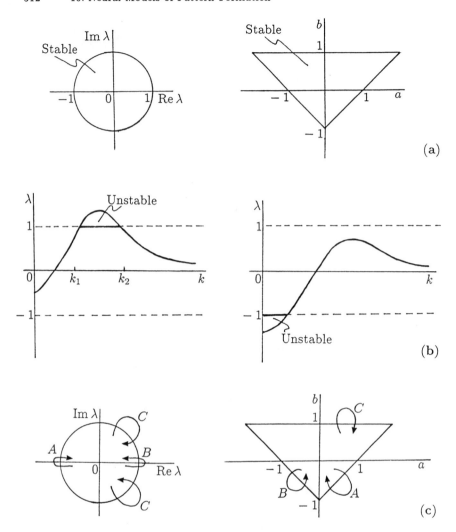

Fig. 16.19a-c. (a) The unit disk in the λ-plane and the stability domain in the (a, b) parameter plane. (b) Typical dispersion relations $\lambda(k)$ for spatial instability: since the solution (16.63) is discrete in time, instability occurs if $|\lambda| > 1$. For illustrative purposes these forms have λ real: see Fig. 16.20 (b) below for complex λ. (c) Instability path in the λ-plane for each type of bifurcation from the stability triangle. These are obtained using the solutions for λ in (16.68) and considering the signs of a and b.

which define $a(k)$ and $b(k)$. The solutions $\lambda = \lambda(k)$ give the dispersion relations for the linear problem (16.62).

Unlike previous models, the time variation in the solutions is discrete and so linear stability here requires $|\lambda(k)| < 1$ (recall the analysis in Chapter 2). That is, for the solutions (16.62) to be stable for all k, $\lambda(k)$ from (16.67) must lie within the unit circle in the complex λ-plane. From (16.67)

$$\lambda = \frac{-a \pm (a^2 - 4b)^{1/2}}{2} \tag{16.68}$$

and so we deduce, using a little elementary algebra, that the condition $|\lambda| < 1$ is satisfied if $a(k)$ and $b(k)$ lie within the triangle in the (a, b) plane shown in Fig. 16.19 (a).

Let us now consider the kind of spatial instabilities which appear as $|\lambda|$ passes through the bifurcation value 1. This means that the solutions λ must move across the unit circle in the complex λ-plane. The pattern evolution from such dispersion relations is a little more subtle than those we have considered up to now, with the novel aspects directly related to the discrete time element in the model. Typical dispersion relations are illustrated in Fig. 16.19 (b), one for $\lambda > 1$ and one for $\lambda < -1$. With the latter, the instability (16.63) increases with time in an oscillatory way with fastest growth for modes with wavenumbers $k = 0$: in other words there is no spatial homogeneity. We come back to this case below.

The model parameters, which of course include the shape of the activation and inhibition kernels w_E and w_I, define a point, (a_0, b_0) say, in (a, b) parameter space using the definitions of a and b in (16.67). From Fig. 16.19 (a), if (a_0, b_0) lies within the triangular domain, $|\lambda| < 1$ and the steady state P_0 is stable since from (16.63) $u_t(x) \to 0$ as $t \to \infty$. Instability is initiated as a parameter (or parameters) varies so that the related point in the (a, b) plane moves out of the stability triangle. Recalling the models and analyses in the previous sections, the total effect of the excitation and inhibition kernels introduced here is like a single kernel, $w(x)$ say, with a local, or short range, activation and a long range, or lateral, inhibition. $w(x)$ and its Fourier transform $W(k)$ have qualitatively similar properties, respectively, to a composite kernel in the linear convolution $L_0[u(x)]$ defined by (16.62) and its Fourier transform $L^*(k)$ defined by (16.65). Fig. 16.20 (a) is a sketch of a typical composite kernel, $L(x)$ say, equivalent to the two contributions in $L_0(x)$ together with its Fourier transform $L^*(k)$. Recall the similar forms in Fig. 16.3 (b) and Fig. 16.3 (c).

Now consider what type of pattern is created as a parameter passes through a critical bifurcation value, thus moving λ out of the unit disc in Fig. 16.19 (a). There are basically three different ways λ crosses the unit circle, namely bifurcation through: (i) $\lambda = 1$, (ii) $\lambda = -1$ and (iii) $\lambda = \exp[i\phi]$, $\phi \neq 0$, $\phi \neq \pi$. We consider the pattern evolution implications of each of these in turn.

(i) *Bifurcation through* $\lambda = 1$. From Fig. 16.19 (c), which was obtained from an analysis of (16.68), we see that this occurs if both $a(k) < 0$ and $b(k) < 0$ and the point (a, b) crosses the bifurcation line in the (a, b) plane in the 3rd quadrant. From the definitions in (16.67) this means that for a range of wavenumbers k,

$$\delta > -L^*(k) > \frac{\gamma}{\delta} \quad \Rightarrow \quad \delta^2 > \gamma . \tag{16.69}$$

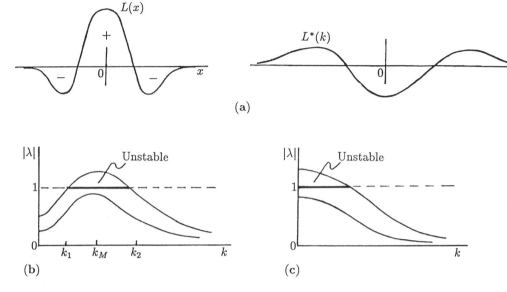

Fig. 16.20a-c. (a) Qualitative shape of the composite kernel $L(x)$ in the convolution $L_0(x)$ in (16.62), that is the difference between excitatory and inhibitory convolutions, and its Fourier transform $L^*(k)$: compare with Fig. 16.3 (c). (b),(c) Typical dispersion relations which result in spatial patterns. In (b) eigenfunctions (16.63) for a finite range of wavenumbers $0 < k_1 < k < k_2$ are unstable while in (c) the unstable modes include $k = 0$ as the fastest growing eigenfunction.

Referring to the right hand figure in Fig. 16.20 (a) this means that the possible range of wavenumbers k which can satisfy (16.69) is in the vicinity of $k = 0$.
From (16.61) and (16.62)

$$P_{t+1}(x) - P_0 \propto \lambda^t \exp[ikx] \tag{16.70}$$

which grows with time for all eigenfunctions with wavenumbers k in the range in which (16.69) is satisfied. The fastest growing solution is dominated by the minimum k. If $k = 0$ we have

$$P_{t+1}(x) - P_0 \propto \lambda^t(0) . \tag{16.71}$$

This in fact does form a spatial pattern on the shell consisting of regularly spaced horizontal stripes – homogeneous stripes parallel to the shell edge. These are incremental lines and are simply homogeneous lines of pigment laid down in each time step: see Fig. 16.21 (a) and also the middle shell pattern in Fig. 16.15 (d). If $k \neq 0$ then spatial heterogenity arises as in the next case.

(ii) *Bifurcation through* $\lambda = -1$. From Fig. 16.19 (c) we see that this situation arises if $a(k) > 0$ and $b(k) < 0$ and the bifurcation line in the (a, b) plane is

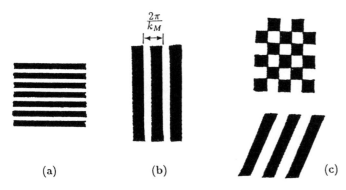

Fig. 16.21a-c. (a) Horizontal stripe pigment pattern arising from a bifurcation at $\lambda = +1$ and generated by solutions (16.71). (b) Spatial pattern of vertical stripes from (16.74) which arises when the bifurcation is at $\lambda = -1$. (c) Patterns which arise from complex bifurcations with solutions from (16.76).

crossed in the 4th quadrant. From the definitions in (16.67) this requires

$$-L^*(k) > \max\left[\delta, \frac{\gamma}{\delta}\right] \quad \text{for} \quad 0 < k_1 < k < k_2 . \tag{16.72}$$

Now from (16.61) and (16.62)

$$P_{t+1}(x) - P_0 \propto \lambda^t \exp[ikx] \tag{16.73}$$

which grows with time for eigenfunctions with wavenumbers k in the range bounded by k_1 and k_2 with a maximum λ, $\lambda_M = \lambda(k_M)$: refer also to Fig. 16.20 (b). So, after some time the dominant solution is

$$P_{t+1}(x) - P_0 \propto \lambda_M^t \exp[ik_M x] . \tag{16.74}$$

Since λ_M is real and greater than 1, λ_M^t is always positive so the pigment laid down at each step is in line with the pigment laid down at the previous step. Thus (16.74) forms a vertical, regularly spaced, stripe pattern with wave length $2\pi/k_M$ as shown in Fig. 16.21 (b): these are basic longitudinal bands.

(iii) *Bifurcation through* $\lambda = \exp[i\phi]$, $\phi \neq 0$, $\phi \neq \pi$. We see from Fig. 16.19 (c) that in this case (a, b) has to leave the stability triangle in the 1st or 2nd quadrant, so $b(k) > 0$ and $a(k)$ can be positive or negative. To be specific take the case $a(k) > 0$. From (16.67) this requires

$$-L^*(k) > \delta, \quad \delta L^*(k) + \gamma > 0 \quad \Rightarrow \quad \delta^2 < \gamma . \tag{16.75}$$

In this type of bifurcation

$$P_{t+1}(x) - P_0 \propto \lambda^t \exp[ikx] . \tag{16.76}$$

Since here λ is complex, each time step moves the pattern along the x-axis by an amount equal to $\arg \lambda$. Thus this solution generates stripe patterns, which are at a fixed angle, or checkered patterns, such as illustrated in Fig. 16.21 (c). An oblique stripe pattern is one of the basic shell patterns.

The values of the parameters along with the scale determine which patterns are created. One example of the role of parameters, for example, is immediately obtained from the restrictions on δ and γ in (16.69) and (16.75). The former generates incremental lines, as in Fig. 16.21 (a), while the latter creates oblique stripes as in Fig. 16.21 (c). δ and γ are respectively the degradation and production rates of the inhibitory material R_t in the model (16.53) and (16.54). There is thus a bifurcation line in (δ, γ) space, namely $\delta^2 = \gamma$, across which the pattern which is generated changes from incremental lines to oblique and checkered patterns. Suppose the shell pattern which is evolving is a checkered one, that is $\delta^2 < \gamma$. If there is a sudden reduction in the production of the inhibitory substance (R_t), that is γ is reduced, the pattern formed afterwards is a horizontal striped one. Many shells exhibit such abrupt pattern changes: see the example in Fig. 16.22 (d).

Spatial Patterns Generated by the Neural Model

Of course the pattern predictions in Fig. 16.21 are based on a linear theory and the final stable patterns come from the full nonlinear model and are obtained numerically (Ermentrout et al. 1986). The linear analysis however is a good predictor for the finite amplitude patterns formed by the full nonlinear system. In the simulations presented below the kernels w in (16.56) and stimulation functions S in (16.58), with the same ν, were used. The mechanism was solved in the form of the single second order difference equation (16.59) with random initial conditions.

The patterns depend on the parameter values. In parameter space there are bifurcation lines and surfaces across which pattern bifurcation takes place. Fig. 16.22 (a)-(c) shows examples of longitudinal bands, checkered patterns and oblique stripes in one species while the middle shell in Fig. 16.15 (d) is a good example of incremental lines. If, during development, a bifurcation surface is crossed then there is an abrupt change of pattern: Fig. 16.22 (d) shows an example.

The direction of the oblique stripes also depends on the parameter values. In the model we assumed that the parameters were constant as the mantle grew. It is not difficult to imagine a situation where there is a gradient in parameter values in the mantle and this can give a bias to the stripe direction. This gradient could change its direction as growth proceeds thus causing a change in the stripe direction. These give rise to divaricate patterns, examples of which are shown in Fig. 16.23 together with simulations which mimic them.

The model can generate a wide spectrum of patterns which are commonly observed (see Ermentrout et al. 1986 for examples of other patterns not discussed here). They include, for example wavy stripes, checks, irregularly stripe patterns,

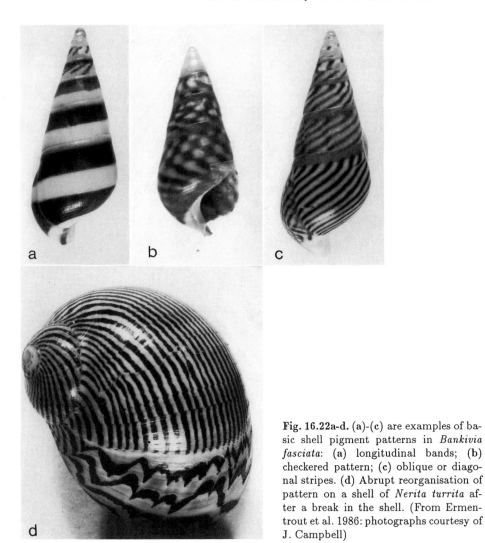

Fig. 16.22a-d. (a)-(c) are examples of basic shell pigment patterns in *Bankivia fasciata*: (a) longitudinal bands; (b) checkered pattern; (c) oblique or diagonal stripes. (d) Abrupt reorganisation of pattern on a shell of *Nerita turrita* after a break in the shell. (From Ermentrout et al. 1986: photographs courtesy of J. Campbell)

tents and so on. Fig. 16.24 (a) shows an example of irregular wandering stripes while Fig. 16.24 (b) shows typical tent patterns common in the courtly cones. The latter simulation has somewhat longer range kernels. From simulations it appears that the excitation threshold and kinetics parameters in the equation for the refractory substance R_t play particularly important roles in the patterns created.

The model (16.53) and (16.54) under certain limiting circumstances can be shown (Ermentrout et al. 1986) to reduce to a cellular automata mechanism which has been studied by Wolfram (1984) and which exhibits chaotic and tent patterns. It is to be expected that similar chaotic pattern behaviour can be generated by the system here: the patterns in Fig. 16.24 could be examples. The

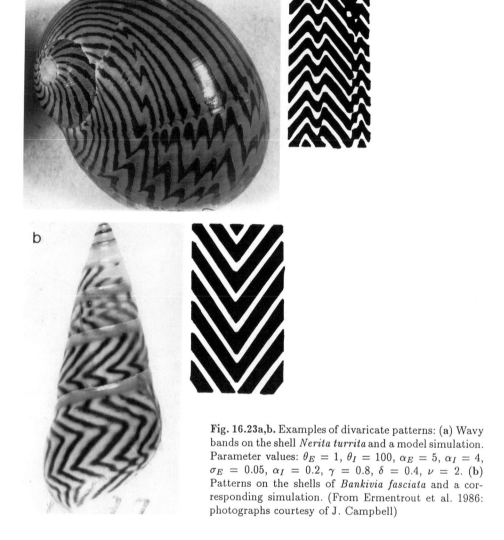

Fig. 16.23a,b. Examples of divaricate patterns: (a) Wavy bands on the shell *Nerita turrita* and a model simulation. Parameter values: $\theta_E = 1$, $\theta_I = 100$, $\alpha_E = 5$, $\alpha_I = 4$, $\sigma_E = 0.05$, $\alpha_I = 0.2$, $\gamma = 0.8$, $\delta = 0.4$, $\nu = 2$. (b) Patterns on the shells of *Bankivia fasciata* and a corresponding simulation. (From Ermentrout et al. 1986: photographs courtesy of J. Campbell)

basic pattern formation potential of such discrete time neural models has only been explored here for basic patterns. If such models are those which govern shell pattern formation then the shell gives a hard copy print-out of the neural activity of the mollusc's mantle and how it interacts with the shell geometry.

As mentioned in the introduction similar patterns can be generated with reaction diffusion models. This is not surprising since, as we showed before, this type of mechanism can be couched in terms of short range activation and long range inhibition. The mechanochemical models for pattern formation to be discussed in the following chapter are also capable of generating similar patterns

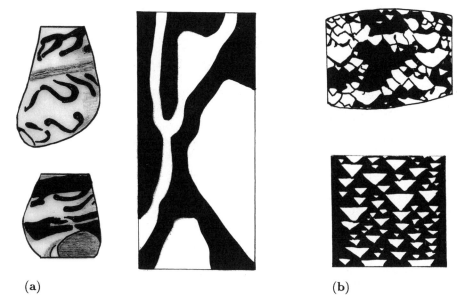

(a)   **(b)**

Fig. 16.24a,b. (a) Two examples of irregular wandering stripe patterns in *Bankivia fasciata* and a model simulation. Parameter values: $\theta_E = 4.5$, $\theta_I = 0.32$, $\alpha_E = 15$, $\alpha_I = 0.5$, $\sigma_E = 0.1$, $\alpha_I = 0.15$, $\gamma = 0.1$, $\delta = 0.8$, $\nu = 8$. (b) Typical tent characteristics of the textile and courtly cones (the section shown is from a *Conus episcopus*: see also Fig. 16.15 (c)) and a related simulation. Parameter values: $\theta_E = 5.5$, $\theta_I = 5.5$, $\alpha_E = 10$, $\alpha_I = 4$, $\sigma_E = 0.1$, $\alpha_I = 0.2$, $\gamma = 0.3$, $\delta = 0.2$, $\nu = 8$. (Drawn from Ermentrout et al. 1986)

To see what connection the discrete time model has with a continuous time model we now briefly consider the analogue we get from a continuous approximation to the discrete model we have just studied.

Continuous Time Model Analogue

If we subtract P_t and R_t from each side of (16.53) and (16.54) respectively we have

$$P_{t+1}(x) - P_t(x) = S[P_t(x)] - R_t(x) - P_t(x)$$

$$R_{t+1}(x) - R_t(x) = \gamma P_t(x) + \delta R_t(x) - R_t(x)$$

which suggests an analogous continuous time model of the form

$$\frac{\partial P}{\partial t} = S[P] - R - P \,, \tag{16.77}$$

$$\frac{\partial R}{\partial t} = \gamma P - (1 - \delta)R \,, \tag{16.78}$$

where $R(x, t)$ and $P(x, t)$ are now functions of continuous space and time, and $S[P]$ has a sigmoid form as a function of its argument as in Fig. 16.18 (b). Remember from the original formulation that $\gamma < 1$ and $\delta < 1$. If we take the excitatory and inhibitory functional forms S_E and S_I in (16.58) to be the same, then with (16.55), we can take

$$S[P] = S(w_E * P - w_I * P) = S(w * P) \qquad (16.79)$$

where $w(x)$ is a typical local-activation/lateral-inhibition kernel such as illustrated in Fig. 16.20 (a). The continuous time model mechanism, given by (16.77)–(16.79), is a integro-differential equation system.

If we now consider the kernel's influence to be restricted to a small neighbourhood around x we can use the same procedure as in Section 16.1 to expand the integral in (16.79). This gives

$$\int w(|x' - x|) P(x') \, dx' = \int w(z) P(x + z) \, dz$$
$$\approx M_0 P + M_2 \frac{\partial^2 P}{\partial x^2} + M_4 \frac{\partial^4 P}{\partial x^4} + \cdots \qquad (16.80)$$

where we have used the symmetry properties of the kernel and where the moments

$$M_{2m} = [(2m)!]^{-1} \int z^{2m} w(z) \, dz, \quad m = 1, 2, \ldots \qquad (16.81)$$

If the kernel is very narrow then higher order moments $|M_{2m}|$, $m \geq 2$ are small compared with $|M_0|$ and $|M_2|$ and expanding S in (16.79) in a Taylor series, (16.77) and (16.78) reduce to a familiar reaction diffusion form, namely

$$\frac{\partial P}{\partial t} = S(M_0 P) - R - P + D \frac{\partial^2 P}{\partial x^2} = f(P, R) + D \frac{\partial^2 P}{\partial x^2} , \qquad (16.82)$$

$$\frac{\partial R}{\partial t} = \gamma P - (1 - \delta) R = g(P, R) , \qquad (16.83)$$

which define the kinetics functions f and g and where the diffusion coefficient $D = M_2$. With the form of S from Fig. 16.18 (b) the nullclines are schematically as shown in Fig. 16.25: at least one positive steady state exists.

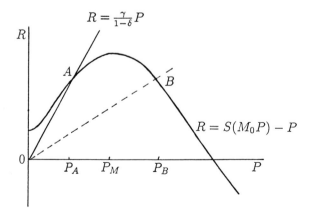

Fig. 16.25. Schematic nullclines $f(P, R) = 0$, $g(P, R) = 0$ in the phase plane for the reaction diffusion system (16.82) and (16.83). As γ decreases, the steady state moves from A to the right, towards B.

We analysed reaction diffusion systems in depth in the Chapter 14 and know that under appropriate circumstances spatial patterns can be generated: see Section 14.3 in Chapter 14. We briefly investigate here whether or not the system (16.82) and (16.83) can be diffusion driven unstable and hence produce spatial patterns. We can say at the outset that if the single steady state is at B the system cannot generate spatial patterns under any circumstances (see Chapter 14). In the following we consider the steady state to be at a point A, where $P_A < P_M$ in Fig. 16.25.

It is left as a revision exercise to show that the dispersion relation $\lambda = \lambda(k)$ about the steady state A in Fig. 16.25 is given by the characteristic equation

$$\lambda^2 + a(k)\lambda + b(k) = 0 , \tag{16.84}$$

where

$$a(k) = Dk^2 - (f_P + g_R) ,$$
$$b(k) = -g_R Dk_2 + (f_P g_R - f_R g_P) , \tag{16.85}$$

where the derivatives of $f(P,R)$ and $g(P,R)$ are evaluated at the uniform steady state. From (16.83) $g_R = -(1-\delta) < 0$ and f_P can be positive or negative. Since we want the system to generate spatial patterns in a Turing sense we require (see Section 14.3)

$$f_P + g_R < 0, \quad f_P g_R - f_R g_P > 0 \tag{16.86}$$

which makes $a(k) > 0$ and, since $g_R < 0$, $b(k) > 0$ for all k. In this case the solutions of (16.84) have Re $\lambda < 0$ and so this specific system cannot form spatial patterns. This is not, however, the end of the matter.

The usual reaction diffsuion form was obtained by terminating the integral expansion in (16.80) at the second moment M_2. If the range of the excitation and inhibition is less local then we must include higher order moments. Not only that, in (16.83) we tacitly assumed that $D = M_2$ is positive. This need not be the case, it depends on the form of the activation-inhibition kernel. If the lateral inhibition is longer range and more severe then it is possible for M_2 in (16.81) to be negative. If the kernel is such that this is the case then we must include at least one higher moment term, namely $M_4 \partial^4 P/\partial x^4$, and more if M_4 is positive. The reason for this is that if $D = M_2 < 0$ then diffusion in (16.82) is destabilising (recall the discussion in Section 9.5 in Chapter 9) whereas if we include long range diffusion which is associated with M_4 it is stabilising if $M_4 < 0$. If we include some long range diffusion then, in place of (16.82) and (16.83) we obtain

$$\frac{\partial P}{\partial t} = S(M_0 P) - R - P - D_1 \frac{\partial^2 P}{\partial x^2} - D_2 \frac{\partial^4 P}{\partial x^4} \tag{16.87}$$

$$\frac{\partial R}{\partial t} = \gamma P - (1-\delta)R = g(P,R) \tag{16.88}$$

where now $D_1 = -M_2 > 0$, $D_2 = -M_4 > 0$ represent diffusion coefficients, with the higher 4th order operator representing the non-local, or long-range, diffusion element.

Carrying out a linear analysis for (16.87) and (16.88) the dispersion relation is again given by (16.84) but with

$$a(k) = D_2 k^4 - D_1 k^2 - (f_P + g_R) ,$$
$$= D_2 k^4 - D_1 k^2 - (f_P - 1 + \delta) ,$$
$$b(k) = -D_2 g_R k^4 + g_R D_1 k^2 + (f_P g_R - f_R g_P) ,$$
$$= D_2 (1 - \delta) k^4 - (1 - \delta) D_1 k^2 - [(1 - \delta) f_P + \gamma f_R] .$$

(16.89)

If either of $a(k)$ or $b(k)$ become negative for a range of non-zero wavenumbers k then $\mathrm{Re}\,\lambda > 0$ and the steady state is spatially unstable. Both of $a(k)$ and $b(k)$ are quadratic in k^2 with the coefficient of k^2 negative. Fig. 16.26 sketches $a(k)$ by way of illustration: $b(k)$ behaves qualitatively the same. It is clear that if D_1 is large enough, that is the destabilising term, the system becomes linearly unstable.

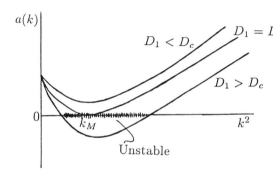

Fig. 16.26. Sketch of $a(k)$ from (16.89). When $a(k) < 0$ the dispersion relation $\lambda(k)$ for the system (16.87) and (16.88) has $\mathrm{Re}\,\lambda > 0$ for a range of wavenumbers and the modes with these wavenumbers are linearly unstable. When $a(k) < 0$, or $b(k) < 0$, $\mathrm{Re}\,\lambda > 0$ and spatial patterns evolve.

From (16.89), since $g_R < 0$ the minimum of $a(k)$ and $b(k)$ is at $k_m^2 = D_1/2D_2$ and

$$a(k_m) = -\frac{D_1^2}{4D_2} - (f_P - 1 + \delta) ,$$
$$b(k_m) = -(1 - \delta)\frac{D_1^2}{4D_2} - [f_P(1 - \delta) + \gamma f_R] .$$

(16.90)

So, bifurcation of the steady state occurs when $D_1 = D_c$ where

$$D_c^2 = \min \left[-4D_2(f_P - 1 + \delta), \quad 4D_2 \left(f_P + \frac{\gamma f_R}{1 - \delta} \right) \right] .$$

(16.91)

Because of (16.86) both terms on the right in (16.91) are positive. For $D_1 > D_c$ there is thus a range of unstable modes with wavenumbers k such that $\mathrm{Re}\,\lambda(k) > 0$ for $k \neq 0$ since $a(k) < 0$: see Fig. 16.26. The system (16.87) and (16.88) is thus capable of generating spatial patterns. With this continuous-time analogue we could envisage the shell patterns to evolve as the shell grew because of a slow change in the parameters, such as D_1, the bifurcation parameter we used for illustration here.

All of the models discussed in this chapter rely on a local activation and longer range inhibition element. In many ways an integral formulation is preferable to a traditional differential equation formulation since the former is intuitively more easily related to the biological concepts. We can also see intuitively how spatial pattern can arise. The mechanisms proposed here, in Sections 16.2–16.4, all have some physiological justification but at this stage need considerable more experimental studies to justify their acceptance. In the following chapter we introduce a new approach to pattern formation in which the mechanisms are based on known and accepted properties of cells and tissues.

Exercises

1. Consider the integral equation model

$$\frac{\partial n}{\partial t} = f(n) + \int_{-\infty}^{\infty} w(x - x')[n(x', t) - 1]\, dx' \, .$$

where $n = 1$ is a zero of $f(n) = 0$ with $f'(1) < 0$. Construct activation-inhibition kernels w from (i) simple square waves and (ii) exponentials of the form $\exp[-|x|/a]$. Sketch the resulting dispersion relation $\lambda(k)$ and determine the critical parameter values for spatially structured solutions. Find the wavelength of the fastest growing mode in a linearly unstable solution.

2. Consider a model for ocular dominance stripes with left and right eye synapse densities n_R and n_L given by the system

$$\frac{\partial n_R}{\partial t} = f(n_R)[w_{RR} * n_R + w_{LR} * n_L] \, ,$$

$$\frac{\partial n_L}{\partial t} = f(n_L)[w_{LL} * n_L + w_{RL} * n_R] \, ,$$

where $*$ denotes the convolution and $f(n) = n(N - n)$ where N is the constant total synapse density; that is $n_R + n_L = N$ and the interaction kernels w are as illustrated in Fig. 16.7.

Show first that conservation of total synapse density implies

$$w_{RR} = -w_{RL}, \quad w_{LL} = -w_{LR} \, .$$

Setting

$$u = n_R - n_L, \quad w = w_{RR} + w_{LL}, \quad K = N * (w_{RR} - w_{LL})$$

show that the model mechanism reduces to the single convolution equation

$$\frac{\partial u}{\partial t} = (N^2 - u^2)(w * u + K).$$

(When there is eye symmetry $w_{RR} = w_{LL}$ and so $K = 0$.)
If initially $n_R = n_L$ then for small t, u is small and hence

$$\frac{\partial u}{\partial t} = N^2(w * u + K).$$

If the domain is one-dimensional, $x \in (-\infty, \infty)$, show that spatial patterned solutions for $u(x, t)$ will evolve with time by determining and analysing the dispersion relation.

In the case of symmetric eye inputs determine the wavelength of the fastest growing mode in terms of the parameters when the kernel w is given by (16.13).

3. A model for shell patterns is described by the system (16.53) and (16.54). Choose Heaviside functions for the stimulation functions S in (16.56) and rectangular forms for the excitation (w_E) and inhibition (w_I) kernels in (16.55): keep in mind the requirement that the inhibition is of longer range than the excitation. Hence derive the piece-wise linear system governing the pigmentation pattern $P_t(x)$ and the refractory substance $R_t(x)$.

Starting with this piecewise linear model derive the equation governing $P_t(x)$ for large t when the build-up of R_t by P_t is zero; that is the parameter $\gamma = 0$.

17. Mechanical Models for Generating Pattern and Form in Development

17.1 Introduction and Background Biology

Development of spatial pattern and form is one of the central issues in embryology. The formation of structure in embryology is known as morphogenesis. Pattern generation models are generally grouped together as morphogenetic models. These models provide the embryologist with possible scenarios as to how pattern is laid down and how the embryonic form might be created. Although genes of course play a crucial role in the control of pattern formation, genetics says nothing about the actual *mechanisms* involved nor how the vast range of pattern and form that we see evolves from a homogeneous mass of dividing cells.

Broadly speaking the two prevailing views of pattern generation that have dominated the thinking of embryologists in the past few years are the long standing Turing chemical pre-pattern approach, which we have discussed at length in previous chapters, and the more recent mechanochemical approach developed by G.F. Oster and J.D. Murray and their colleagues (for example, Odell et al. 1981, Murray, Oster and Harris 1983, Oster, Murray and Harris 1983, Murray and Oster 1984a, Murray and Oster 1984b, Oster, Murray and Maini 1985). General descriptions have been given by Murray and Maini (1986), Oster and Murray (1989) and Murray, Maini and Tranquilo (1988). In this chapter we discuss this new mechanochemical approach, which, among other things, considers the role that mechanical forces play in the process of morphogenetic pattern formation, and apply it to several specific developmental problems of current widespread interest in embryology. A clear justification for the need for such a mechanical approach to the development of pattern in cellular terms is inferred from the following quote from Wolpert (1977): "It is clear that the egg contains not a description of the adult, but a program for making it, and this program may be simpler than the description. Relatively simple cellular forces can give rise to complex changes in form; it seems simpler to specify how to make complex shapes than to describe them."

The two approaches are basically quite different. In the chemical pre-pattern approach, pattern formation and morphogenesis take place sequentially. First the chemical concentration pattern is laid down, then the cells interpret this pre-pattern and differentiate accordingly. So, in this approach, morphogenesis is essentially a slave process which is determined once the chemical pattern has been established. Mechanical shaping of form which occurs during embryogenesis is not addressed in the chemical theory of morphogenesis. The elusiveness of these chemical morphogens is proving a considerable drawback in the acceptance of such a theory of morphogenesis. There is, however, no question but that chemicals play crucially important roles in development.

In the mechanochemical approach, pattern formation and morphogenesis is considered to go on simultaneously as a single process. Here the chemical patterning and the form-shaping movements of the cells and the embryological tissue interact continuously to produce the observed spatial pattern. Another important aspect of this approach is that the models associated with it are formulated in terms of measurable quantities such as cell densities, forces, tissue deformation and so on. This focuses attention on the morphogenetic process itself and in principle is more amenable to experimental investigation. The principal use of any theory is in its predictions and, even though each theory might be able to create similar patterns, they are mainly distinquished by the different experiments they suggest. We discuss some of the experiments associated with the mechanical theory later in this chapter and also in the next.

A particularly telling point in favour of simultaneous development is that such mechanisms have the potential for self correction. Embryonic development is usually a very stable process with the embryo capable of adjusting to many outside disturbances. The process whereby a prepattern exists and then morphogenesis takes place is effectively an open loop system. These are potentially unstable processes and make it difficult for the embryo to make the necessary adjustment to such disturbances as development proceeds.

In this chapter we discuss morphogenetic processes which involve coordinated movement or patterning of populations of cells. The two types of early embryonic cells we are concerned with are fibroblast, or dermal or mesenchymal cells and epidermal, or epthelial cells. Fibroblast cells are capable of independent movement, due to long finger-like protrusions called filopodia or lamellapodia which grab onto adhesive sites and pull themselves along (you can think of such a cell as something like a minute octopus): spatial aggregation patterns in these appear as spatial variations in cell number density. Fibroblasts can also secrete fibrous material which helps to make up the extracellular matrix (ECM) tissue within which the cells move. Epidermal cells, on the other hand, in general do not move but are packed together in sheets and spatial patterns in their population are manifested by cell deformations. Fig. 17.1 schematically sets out some of the key properties of the two types of cells. A good description of the types of cells, their movement properties and characteristics and their role in embryogenesis, is given in the textbook by Walbot and Holder (1987). The definitive text exclusively on the cell is by Alberts et al. (1983). These books are particularly relevant to the material in this chapter.

We first consider mesenchymal (fibroblast) cell pattern formation in early embryogenesis. In animal development the basic body plan is more or less laid down in the first few weeks, such as the first 4 weeks in man, where gestation is about 280 days, and not much more in the case of a giraffe, for example,

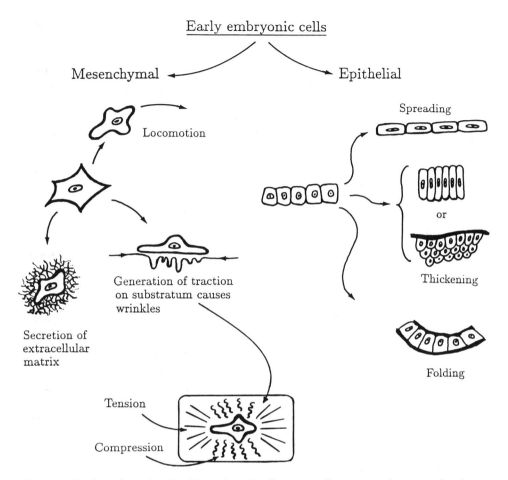

Fig. 17.1. Early embryonic cells. Mesenchymal cells are motile, generate large traction forces and can secrete extracellular matrix which forms part of the tissue within which the cells move. When these cells are placed on a thin silicon rubber substratum their traction forces deform the rubber sheet: see the photograph in Fig. 17.2. Epthelial cells do not move about but can spread or thicken when subjected to forces; this affects cell division (see, for example, Folkman and Moscona 1978).

which has a gestation period of nearly 460 days. It is during this crucial early period that we expect pattern and form generating mechanisms, such as we propose here, to be operative. The models we discuss here take into account considerably more biological facts than those we have hitherto considered. Not surprisingly this makes the models more complicated. It is essential, however, for mathematical biologists genuinely concerned with real biology to appreciate the complexity of biology. So, it is appropriate that we should now discuss the modelling of mechanisms for some of the more complex but realistic aspects of development of pattern and form and to which experimentalists can specfically and concretely relate. All of the models we propose in this chapter are firmly based on macroscopic experimentally measurable variables and on generally accepted properties of embryonic cells.

In the following section we derive a fairly general model and subsequently deduce simpler versions. This is rather different to the approach we have adopted up to now and reflects, in part, the complexity of real modelling in embryology and in part on the assumption that the readers are now more sophisticated in their approach.

We should add here that these models pose numerous challenging mathematical, both analytical and numerical, and biological modelling problems which have not yet been investigated in any depth.

17.2 Mechanical Model for Mesenchymal Morphogenesis

Several factors affect the movement of embryonic mesenchymal cells. As mentioned before, a good description of cells in early embryogenesis and which is particularly relevant to the material in this chapter, is given in the textbooks by Walbot and Holder (1987) and Alberts et al. (1983). Among these factors are: (i) convection, whereby cells may be passively carried along on a deforming substratum; (ii) chemotaxis, whereby a chemical gradient can direct cell motion both up and down a concentration gradient; (iii) contact guidance in which the substratum on which the cells crawl suggest a preferred direction; (iv) contact inhibition by the cells whereby a high density of neighbouring cells inhibits motion; (v) haptotaxis, which we describe below, where the cells move up an adhesive gradient; (vi) diffusion, where the cells move randomly but generally down a cell density gradient; (vii) galvanotaxis where movement from the field generated by electric potentials, which are known to exist in embryos, provides a preferred direction of motion. These effects are all well documented from experiment.

The model field equations we propose in this section encapsulate the key features which affect cell movement within its extracellular environment. We shall not include all of the affects just mentioned but it will be clear how they can be incorporated and their effect quantified. The subsequent analysis of the field equations will show how regular patterned aggregates of cells come about. In Sections 17.5 and 17.6 we shall describe two real applications of the model, in particular to the highly organised patterns on skin such as the primordia which become feathers and scales, and the condensation of cells which mirror the cartilage pattern in developing limbs.

The basic mechanical model hinges on two key experimentally determined properties of mesenchymal cells *in vivo*: (i) cells migrate within a tissue substratum made up of fibrous extracellular matrix, which we shall often refer to as the ECM, and other cells (Hay 1981); (ii) cells can generate large traction forces (Harris, Stopak and Wild 1981). Fig. 17.2 is a photograph of cells on a thin silicone substratum: the tension and compression lines they generate are

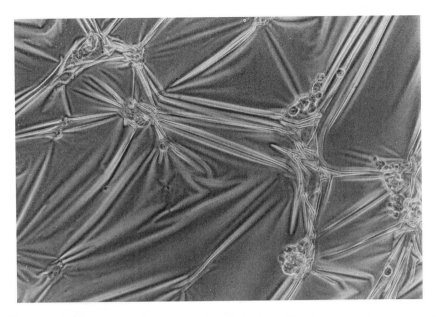

Fig. 17.2. Mesenchymal cells on an elastic substratum. The strong tractions generated deform the substratum and create compression and tension wrinkles. The tension wrinkles can extend several hundreds of cell diameters. (Photograph courtesy of A.K. Harris)

clearly seen: see also Harris, Ward and Stopak (1980). The basic mechanism we shall develop, models the mechanical interaction between the motile cells and the elastic substratum, within which they move.

Mesenchymal cells move by exerting forces on their surroundings, consisting of the elastic fibrous ECM and the surface of other cells. They use their cellular protrusions, the filopodia or lamellapodia, which stretch out from the cell in all directions, gripping whatever is available and pulling. (The biology of these protrusions is discussed by Trinkaus (1980): see also the book by Trinkaus (1984) which is useful background reading for morphogenetic modelling. Oster (1984) specifically discusses the mechanism of how an individual cell crawls.) As the cells move through the ECM they deform it by virtue of their traction forces. These deformations in the ECM induce anisotropy effects which in turn affect the cell motion. The resulting coordination of the various effects, such as we have just mentioned, result in spatially organised cell aggregations. The basic model is essentially that proposed by Murray et al. (1983), Murray and Oster (1984a,b), with a detailed biological description by Oster et al. (1983).

The model, a continuum one, consists of three equations governing (i) the conservation equation for the cell population density, (ii) the mechanical balance of the forces between the cells and the ECM, and (iii) the conservation law governing the ECM. Let $n(\mathbf{r}, t)$ and $\rho(\mathbf{r}, t)$ denote respectively the cell density (the number of cells per unit volume) and ECM density at position $\mathbf{r}$ and time t. Denote by $\mathbf{u}(\mathbf{r}, t)$ the displacement vector of the ECM, that is a material point in the matrix initially at position $\mathbf{r}$ undergoes a displacement to $\mathbf{r} + \mathbf{u}$. We derive forms for each of these equations in turn.

Cell Conservation Equation

The general form of the conservation equation is (recall Chapter 9)

$$\frac{\partial n}{\partial t} = -\nabla \cdot \mathbf{J} + M , \qquad (17.1)$$

where $\mathbf{J}$ is the flux of cells, that is the number crossing a unit area in unit time, and M is the mitotic or cell proliferation rate: the specific form is not important at this stage. For simplicity we shall take a logistic model for the cell growth, namely $rn(N - n)$ where r is the initial proliferative rate and N is the maximum cell density in the absence of any other effects. We include in $\mathbf{J}$ some of the factors mentioned above which affect cell motion.

Convection. With $\mathbf{u}(\mathbf{r}, t)$ the displacement vector of the ECM, the convective flux contribution $\mathbf{J}_c$ is

$$\mathbf{J}_c = n\frac{\partial \mathbf{u}}{\partial t} . \qquad (17.2)$$

Here the velocity of deformation of the matrix is $\partial \mathbf{u}/\partial t$ and the amount of cells transported is simply n times this velocity. It is likely that the convective flux is the most important contribution to cell transport.

Random Dispersal. Cells tend to disperse randomly when in a homogeneous isotropic medium. Classical diffusion (see Chapter 9) contributes a flux term $-D_1\nabla n$ which models the random motion in which the cells respond to *local* variations in the cell density and tend to move down the density gradient. This results in the usual diffusion contribution $D_1\nabla^2 n$ to the conservation equation and represents local, or *short range* random motion.

In developing embryos the cell densities are relatively high and classical diffusion, which applies to dilute systems is not, perhaps, sufficiently accurate. The long filopodia extended by the cells can sense density variations beyond their nearest neighbours and so we must include a *nonlocal* effect on diffusive dispersal since the cells sense more distant densities and so respond to neighbouring *averages* as well. Fig. 17.3 schematically illustrates why this long range sensing can be important.

The Laplacian operator acting on a function reflects the difference between the value of the function at position $\mathbf{r}$ and its local average as can be seen on writing it in its simplest finite difference approximation. Alternatively the Laplacian can be written in the form

$$\nabla^2 n \propto \frac{n_{av}(\mathbf{r}, t) - n(\mathbf{r}, t)}{R^2}, \quad \text{as} \quad R \to 0 \qquad (17.3)$$

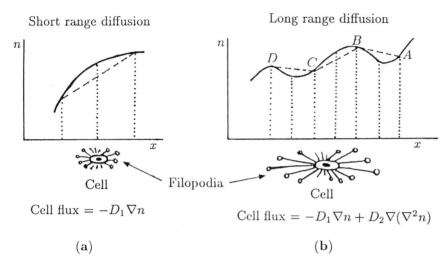

Short range diffusion

Cell flux $= -D_1 \nabla n$

(a)

Long range diffusion

Cell flux $= -D_1 \nabla n + D_2 \nabla(\nabla^2 n)$

(b)

Fig. 17.3a,b. In (a) the filopodia only sense the immediate neighbouring densities to determine the gradient (the broken line) and hence disperse in a classical random manner giving a flux of $-D_1 \nabla n$. With the situation in (b) the long filopodia can sense not only neighbouring densities but also neighbouring averages which contribute a long range diffusional flux term $D_2 \nabla(\nabla^2 n)$. This contributes to directed dispersal which is not necessarily in the same direction as indicated by short range (again denoted by broken lines) diffusion. Long range diffusion suggests general movement of cells from A to D whereas short range diffusion implies movement from D to C, B to C and B to A.

where n_{av} is the average cell concentration in a sphere of radius R about $\mathbf{r}$ defined by

$$n_{av}(\mathbf{r}) = \frac{4\pi R^3}{3} \int_V n(\mathbf{r} + \mathbf{s}) \, ds \ . \tag{17.4}$$

where V is the volume of the sphere. If the integrand in (17.4) is expanded in a Taylor's series and n_{av} substituted in (17.3) the proportionality factor is 10/3. Again recall the full discussion and analysis in Chapter 9.

The flux of cells is thus given by

$$\mathbf{J}_D = -D_1 \nabla n + D_2 \nabla(\nabla^2 n) \ , \tag{17.5}$$

where $D_1 > 0$ is the usual Fickian diffusion coefficient and $D_2 > 0$ is the long range diffusion coefficient. The long range contribution gives rise to a biharmonic term in (17.1). In the morphogenetic situations we consider, we expect the effect of diffusion to be relatively small. Nonlocal diffusive dispersal was considered by Othmer (1969); his work is particularly apposite to the cell situation. Cohen and Murray (1981) derived and considered a related model in an ecological context.

Recall from Section 9.5 (Chapter 9) that this long range diffusion has a stabilising effect if $D_2 > 0$. We can see this immediately if we consider the long range diffusion equation, obtained by substituing (17.4) into (17.1) and, omitting the mitotic term M, to get

$$\frac{\partial n}{\partial t} = -\nabla \cdot \mathbf{J}_D = D_1 \nabla^2 n - D_2 \nabla^4 n \ .$$

We now look for solutions of the form $n(r, t) \propto \exp[\lambda t + i\mathbf{k} \cdot \mathbf{r}]$ where $\mathbf{k}$ is the usual wave vector. Substituting this into the last equation gives the dispersion relation as $\lambda = -D_2 k^4 - D_1 k^2 < 0$ for all wavenumbers $k(= |\mathbf{k}|)$. So $n \to 0$ as $t \to \infty$, which implies $n = 0$ is stable. If the biharmonic term had $D_2 < 0$, $n = 0$ would be unstable for wavenumbers $k^2 > -D_1/D_2$.

Haptotaxis or Mechanotaxis. The traction exerted by the cells on the matrix generates gradients in the matrix density $\rho(\mathbf{r}, t)$. We associate the density of matrix with the density of adhesive sites for the cell lamellapodia to get a hold of. Cells free to move in an adhesive gradient tend to move up it since the cells can get a stronger grip on the denser matrix. This results in a net flux of cells *up* the gradient which, on the simplest assumption, is proportional to $n\nabla\rho$. It is very similar to chemotaxis (recall Chapter 9, Section 9.4). Because of the physical properties of the matrix and the nonlocal sensing properties of the cells we should perhaps also include a long range effect, similar to that which gave the biharmonic term in (17.5). In this case the haptotactic flux is given by

$$\mathbf{J}_h = n(a_1 \nabla \rho - a_2 \nabla^3 \rho) \tag{17.6}$$

where $a_1 > 0$ and $a_2 > 0$.

The cell conservation equation (17.1), with the flux contributions to $\mathbf{J}$ from (17.2), (17.5) and (17.6) with the illustrative logistic form for the mitosis M, becomes

$$\frac{\partial n}{\partial t} = -\nabla \cdot \underbrace{\left[n \frac{\partial \mathbf{u}}{\partial t} \right]}_{\text{convection}} + \underbrace{\nabla \cdot [D_1 \nabla n - D_2 \nabla (\nabla^2 n)]}_{\text{diffusion}}$$

$$- \underbrace{\nabla \cdot n[a_1 \nabla \rho - a_2 \nabla^3 \rho]}_{\text{haptotaxis}} + \underbrace{rn(N - n)}_{\text{mitosis}} \ , \tag{17.7}$$

where D_1, D_2, a_1, a_2, r and N are positive parameters.

We have not included galvanotaxis nor chemotaxis in (17.7) but we can easily deduce what such contributions would look like. If ϕ is the electric potential then the galvanotaxis flux can be written as

$$\mathbf{J}_G = gn\nabla\phi \ , \tag{17.8}$$

where the parameter $g > 0$. If c is the concentration of a chemotactic chemical, a chemotactic flux can then be of the form

$$\mathbf{J}_C = \chi n \nabla c \ ,$$

where $\chi > 0$ is the chemotaxis parameter. Another effect which could be important but which we shall also not include here, is the guidance cues which come from the directional cues in the ECM. For example, there is experimental evidence that matrix strain results in aligned fibres which encourages movement along the directions of strain as opposed to movement across the strain lines. This effect can be incorporated in the equation for n by making the diffusion and haptotactic coefficients functions of the elastic strain tensor (see, for example, Landau and Lifshitz 1970) of the ECM defined by

$$\varepsilon = \frac{1}{2}(\nabla \mathbf{u} + \nabla \mathbf{u}^T) \, . \tag{17.9}$$

In principle the qualitative form of the dependence of, for example, $D_1(\varepsilon)$ and $D_2(\varepsilon)$ on ε can be deduced from experiments. In the following, however, we shall take D_1, D_2, a_1 and a_2 to be constants.

In (17.7) we modelled the mitotic, or cell proliferation rate, by a simple logistic growth with linear growth rate r. The detailed form of this term is not critical as long as it is qualitatively similar. It is now well known from experiment (for example, Folkman and Mascona 1978) that the mitotic rate is dependent on cell shape. So, within our continuum framework, r should depend on the displacement $\mathbf{u}$. A brief review of the ECM and its effect on cell shape, proliferation and differentiation is given by Watt (1986). At this stage, however, we shall also not include this potentially important effect.

One of the purposes of this section is to show how possible effects can easily be incorporated in the model. Although the conservation equation (17.7) is clearly not the most general possible, it suffices to show what can be expected in more realistic model mechanisms for biological pattern generation.

The analysis of such models allows us to compare the various effects as to their pattern formation potential and hence to come up with the simplest realistic system which can generate pattern and which is experimentally testable. Simpler systems are discussed later in Section 17.4. Perhaps we should mention here that only the inclusion of convection in the cell conservation equation is essential. Intuitively this is what we might have expected at least as regards transport effects.

Cell-Matrix Mechanical Interaction Equation

The composition of the fibrous extracellular matrix, the ECM, within which the cells move is complex and moreover, its constituents change as development proceeds. Its mechanical properties have not yet been well characterised. Here, however, we are interested only in the mechanical interaction between the cells and the matrix. Also the mechanical deformations are small so, as a reasonable first approximation, we take the composite material of cells plus matrix to be modelled as a linear, isotropic viscoelastic continuum with stress tensor $\sigma(\mathbf{r}, t)$.

The time scale of embryonic motions during development is very long (hours) and the spatial scale is very small (less than a millimetre or two). We are thus in a very low Reynolds number regime (cf. Purcell 1977) and so can ignore inertial effects in the mechanical equation for the cell-ECM interaction. Thus we assume that the traction forces generated by the cells are in mechanical equilibrium with the elastic restoring forces developed in the matrix and any external forces present. The mechanical cell-matrix equation is then (see, for example, Landau and Lifshitz 1970)

$$\nabla \cdot \boldsymbol{\sigma} + \rho \mathbf{F} = 0 , \tag{17.10}$$

where $\mathbf{F}$ is the external force acting on the matrix (per unit matrix) and $\boldsymbol{\sigma}$ is the stress tensor. (This equation applied to a spring loaded with a weight, simply says the applied force is balanced by the elastic force from the extended spring.) We must now model the various contributions to $\boldsymbol{\sigma}$ and $\mathbf{F}$.

Consider first the stress tensor $\boldsymbol{\sigma}$. It consists of contributions from the ECM and the cells and we write

$$\boldsymbol{\sigma} = \boldsymbol{\sigma}_{\text{ECM}} + \boldsymbol{\sigma}_{\text{cell}} . \tag{17.11}$$

The usual expression for a linear viscoelastic material (Landau and Lifshitz 1970) gives the stress-strain constitutive relation as

$$\boldsymbol{\sigma}_{\text{ECM}} = \underbrace{[\mu_1 \boldsymbol{\varepsilon}_t + \mu_2 \theta_t \mathbf{I}]}_{\text{viscous}} + \underbrace{E'[\boldsymbol{\varepsilon} + \nu' \theta \mathbf{I}]}_{\text{elastic}} , \tag{17.12}$$

$$\text{where} \quad E' = E/(1+\nu), \quad \nu' = \nu/(1-2\nu) .$$

The subscript t denotes partial differentiation, $\mathbf{I}$ is the unit tensor, μ_1 and μ_2 are the shear and bulk viscosities of the ECM, $\boldsymbol{\varepsilon}$ is the strain tensor defined above in (17.9), $\theta (= \nabla \cdot \mathbf{u})$ is the dilation and E and ν are the Young's modulus and Poisson ratio respectively.

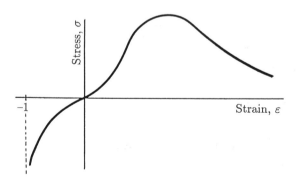

Fig. 17.4. The effect of straining the extracellular matrix is to align the fibres and stiffen the material. If we think of a one-dimensional situation the strain from (17.9) is $\varepsilon = \partial u/\partial x$ and the dilation $\theta = \partial u/\partial x$. The effective elastic modulus E is the gradient of the stress-strain curve. It increases with strain until the yield point whereupon it levels off and drops for large enough strains as the material tears. The ECM is in compression when $\varepsilon < 0$. Because a given amount of material (cells + matrix) cannot be squeezed to zero, there is a lower limit of $\varepsilon = -1$ where the stress tends to $-\infty$.

The assumption of isotropy is certainly a major one. While the ECM may be isotropic in the absence of cell tractions (and even this is doubtful) it is probably no longer isotropic when subjected to cellular forces. Although we shall not specifically consider a non-isotropic model, we should be aware of the kind of anisotropy that might be included in a more sophisticated model. When a fibrous material is strained the fibres tend to align in the directions of the principal stresses and the effective elastic modulus in the direction of strain increases. With the main macroscopic effect of fibre alignment being to strengthen the material in the direction of strain we can model this by making the elastic modulus E an increasing function of the dilation θ, at least for small θ. It does not of course increase indefinitely since eventually the material would yield. Fig. 17.4 is a typical stress-strain curve. It is possible that ν is also a function of θ; here, however, we take it to be constant.

Fibrous materials are also characterised by nonlocal elastic interactions since the fibres can transmit stress between points in the ECM quite far apart. By arguments analogous to those which lead to the biharmonic term in the cell conservation equation (17.7) we should include long range effects in the elsstic stress for the composite material. The anisotropic effect discussed in the last paragraph and this nonlocal effect can be modelled by writing in place of the elastic contribution in (17.12)

$$\boldsymbol{\sigma}_{\text{ECM}}]_{\text{elastic}} = E'(\theta)[\varepsilon + \beta_1\nabla^2\varepsilon + \nu'(\theta + \beta_2\nabla^2\theta)\mathbf{I}] \,,$$
$$\text{where}\quad E' = E(\theta)/(1+\nu), \quad \nu' = \nu/(1-2\nu) \tag{17.13}$$

and the β's are parameters which measure long range effects. However at this stage of modelling it is reasonable to take $\beta_1 = \beta_2 = 0$ and $E(\theta)$ to be a constant.

Now consider the contribution to the stress tensor from the cell tractions, that is $\boldsymbol{\sigma}_{\text{cell}}$. The more cells there are the greater the traction force. There is, however, experimental evidence indicating cell-cell contact inhibition with the traction force decreasing for large enough cell densities. This can be simply modelled by assuming that the cell traction forces, $\tau(n)$ per unit mass of matrix, initially increase with n but eventually decrease with n for large enough n. Here we simply choose

$$\tau(n) = \frac{\tau n}{1 + \lambda n^2} \,, \tag{17.14}$$

where τ (dyne-cm/gm) is a measure of the traction force generated by a cell and λ is a measure of how the force is reduced because of neighbouring cells: we come back to this below. Typical experimental values for τ are in of the order of 10^{-3} dyne/μm of cell edge, which is a very substantial force (Harris et al. 1981). The actual form of the force generated per cell, that is $\tau(n)/n$, as a function of cell density can be determined experimentally.

If the filopodia, with which the cells attach to the ECM, extend beyond their immediate neighbourhood, as they probably do, it is again reasonable to include a nonlocal effect analogous to the long range diffusion effect we included in the cell conservation equation. We thus take the contribution $\boldsymbol{\sigma}_{\text{cell}}$ to the stress tensor to be

$$\boldsymbol{\sigma}_{\text{cell}} = \tau(1 + \lambda n^2)^{-1} n(\rho + \gamma\nabla^2\rho)\mathbf{I} \,, \tag{17.15}$$

where $\gamma > 0$ is the measure of the nonlocal long range cell-ECM interactions. The long range effects here are probably more important than the long range diffusion and haptotaxis effects in the cell conservation equation.

If the cells are densely packed the nonlocal effect would primarily be between the cells and in this case a more appropriate form for (17.15) would be

$$\boldsymbol{\sigma}_{\text{cell}} = \tau(1 + \lambda n^2)^{-1} \rho(n + \gamma\nabla^2 n)\mathbf{I} \,. \tag{17.16}$$

Finally let us consider the body force $\mathbf{F}$ in (17.10). With the applications we have in mind, and discussed below, the matrix material is attached to a substratum of underlying tissue, or the epidermis, by what can perhaps best be described as being similar to guy ropes. We model these restraining forces as body forces proportional to the density of the ECM and the displacement of the matrix from its unstrained position and thus take

$$\mathbf{F} = -s\mathbf{u} \,, \tag{17.17}$$

where $s > 0$ is an elastic parameter characterizing the substrate attachments.

In the model we analyse we shall not include all the effects we have discussed but only those we feel are the more essential at this stage. So, the force equation we take for the mechanical equilibrium between the cells and the ECM is (17.10), with (17.11)–(17.17), which gives

$$\nabla \cdot \left[\underbrace{\mu_1\boldsymbol{\varepsilon}_t + \mu_2\theta_t\mathbf{I}}_{\text{viscous}} + \underbrace{E'(\boldsymbol{\varepsilon} + \nu'\theta\mathbf{I})}_{\text{elastic}} \right.$$

$$\left. + \underbrace{\tau n(1 + \lambda n^2)^{-1}(\rho + \gamma\nabla^2\rho)\mathbf{I}}_{\text{cell traction}} \right] - \underbrace{s\rho\mathbf{u}}_{\substack{\text{external} \\ \text{forces}}} = 0 \,, \tag{17.18}$$

$$\text{where} \quad E' = \frac{E}{1 + \nu} \,, \quad \nu' = \frac{\nu}{1 - 2\nu} \,. \tag{17.19}$$

Matrix Conservation Equation

The conservation equation for the matrix material, $\rho(\mathbf{r}, t)$, is

$$\frac{\partial\rho}{\partial t} + \nabla \cdot (\rho\mathbf{u}_t) = S(n, \rho, \mathbf{u}) \,. \tag{17.20}$$

where matrix flux is taken to be mainly via convection and $S(n, \rho, \mathbf{u})$ is the rate of secretion of matrix by the cells. Secretion and degradation is thought to play a role in certain situations involving mesenchymal cell organisation, and it certainly does in wound healing, an important possible application briefly described in Section 17.10. However, on the time scale of cell motions that we shall be considering here we can neglect this effect and shall henceforth assume $S = 0$. Experimental evidence (Hinchliffe and Johnson 1980) also indicates that $S = 0$ during chondrogenesis and pattern formation of skin organ primordia.

Equations (17.7), (17.18) and (17.20) with $S = 0$ constitute the field equations for our model pattern formation mechanism for fibroblast cells. The three dependent variables are the density fields $n(\mathbf{r}, t)$, $\rho(\mathbf{r}, t)$ and the displacement field $\mathbf{u}(\mathbf{r}, t)$. The model involves 14 parameters, namely D_1, D_2, a_1, a_2, r, N, μ_1, μ_2, τ, λ, γ, s, E and ν, all of which are in principle measurable and some of which are currently being investigated experimentally.

As usual, to assess the relative importance of the various effects, and to simplify the analysis, we nondimensionalize the equations. We use general length and time scales L and T, a uniform initial matrix density ρ_0 and set

$$\mathbf{r}^* = \frac{\mathbf{r}}{L}, \quad t^* = \frac{t}{T}, \quad n^* = \frac{n}{N}, \quad \mathbf{u}^* = \frac{\mathbf{u}}{L}, \quad \rho^* = \frac{\rho}{\rho_0},$$

$$\nabla^* = L\nabla, \quad \theta^* = \theta, \quad \varepsilon^* = \varepsilon, \quad \gamma^* = \frac{\gamma}{L^2}, \quad r^* = rNT,$$

$$s^* = \frac{s\rho_0 L^2(1+\nu)}{E}, \quad \lambda^* = \lambda N^2, \quad \tau^* = \frac{\tau\rho_0 N(1+\nu)}{E}, \quad (17.21)$$

$$a_i^* = \frac{a_i\rho_0 T}{L^2}, \quad \mu_i^* = \frac{\mu_i(1+\nu)}{TE}, \quad i = 1, 2$$

$$D_1^* = \frac{D_1 T}{L^2}, \quad D_2^* = \frac{D_2 T}{L^4}.$$

The nondimensionalisation has reduced the 14 parameters to 12 parameter groupings. Depending on what time scale we are particularly concerned with we can reduce the set of 12 parameters further. For example, if we choose T as the mitotic time $1/rN$, then $r^* = 1$: this means we are interested in the evolution of pattern on the mitotic time scale. Alternatively we could choose T so that $\gamma^* = 1$ or $\mu_i = 1$ for $i = 1$ or $i = 2$. Similarly we can choose a relevant length scale and further reduce the number of groupings.

With the nondimensionalisation (17.21) the model mechanism (17.7), (17.18) and (17.20), with matrix secretion $S = 0$ (although it would be interesting to study the effect of a matrix source term S on the subsequent analysis), becomes, on dropping the asterisks for notational simplicity,

$$n_t = D_1\nabla^2 n - D_2\nabla^4 n - \nabla \cdot [a_1 n\nabla\rho - a_2 n\nabla(\nabla^2\rho)]$$
$$- \nabla \cdot (n\mathbf{u}_t) + rn(1-n), \quad (17.22)$$

$$\nabla \cdot \left\{ (\mu_1\varepsilon_t + \mu_2\theta_t\mathbf{I}) + (\varepsilon + \nu'\theta\mathbf{I}) + \frac{\tau n}{1+\lambda n^2}(\rho + \gamma\nabla^2\rho)\mathbf{I} \right\} = s\rho\mathbf{u}, \quad (17.23)$$

$$\rho_t + \nabla \cdot (\rho\mathbf{u}_t) = 0. \quad (17.24)$$

Note that the dimensionless parameters, all of which are positive, are divided into those associated with the cell properties, namely a_1, a_2, D_1, D_2, r, τ, λ and those related to the matrix properties, namely μ_1, μ_2, ν', γ and s.

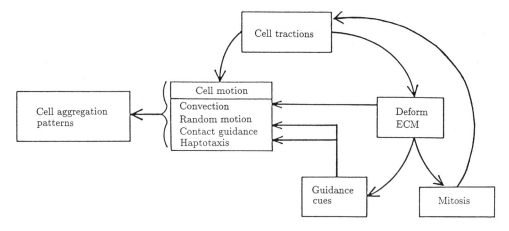

Fig. 17.5. Conceptual framework for the mechanical models. Cell tractions play a central role in orchestrating pattern formation.

Although the model system (17.22)–(17.24) is analytically formidable the model's conceptual framework is quite clear, as illustrated in Fig. 17.5. As we have noted, this model does not include all the effects that might be relevant, particularly the effect of matrix secretion. Although later we shall derive considerably simpler systems we should have some idea of what a model looks like that incorporates many of the features that biologists might feel are important. As we also said above, one of the major roles of such modelling and subsequent analysis is to indicate what features are *essential* for pattern formation. So, in the initial linear analysis that follows we shall retain all of the terms in the model (17.22)–(17.24) and only set various parameters to zero in the general results to see what effects may be redundant or dwarfed by others.

17.3 Linear Analysis, Dispersion Relation and Pattern Formation Potential

In order to model spatial aspects observed in embryonic development the equation system (17.22)–(17.24) must admit spatially inhomogeneous solutions. Considering their complexity and with the experience gained from the study of the pattern forming models in Chapters 14–16 we have little hope, at this stage, of finding useful analytical solutions to such nonlinear systems. We know, however, that much of the pattern formation potential is predicted by a linear analysis about uniform steady state solutions. We also now know that such linear predictions are not infallible and must be backed up by numerical simulations if finite amplitude structures far from homogeneity are required. (It will be helpful in the following to recall in detail the material and discussions relating to spatial pattern formation in Chapter 14, particularly Sections 14.3–14.6.)

Before carrying out a linear analysis let us note that one of the applications of this theory will be to the pattern formation process that accompanies the formation of skin organ primordia for feathers, scales and teeth, for example: see Section 17.5. The initial cell aggregations which appear in the dermis, that is the layer just under the epidermis on which the scales and feathers start, only differ in cell density from the surrounding tissue by fairly small amounts. Therefore it is worthwhile from the practical application viewpoint to carry out a detailed linear analysis of the field equations, not only as a first analytical step to indicate spatial pattern potentialities and guide numerical work, but also because the patterns themselves may involve solutions that effectively fall within the linear regime. The latter are often effectively those from a nonlinear theory close to bifurcation from uniformity. We come back to the biological applications in more detail below.

The uniform steady state solutions of (17.22)–(17.24) are

$$n = \mathbf{u} = \rho = 0, \quad n = 1, \quad \mathbf{u} = \rho = 0, \quad n = \rho = 1, \quad \mathbf{u} = 0 . \tag{17.25}$$

The first two solutions are not relevant as $\rho = 0$ is not relevant in the biological situation. The third solution is relevant (ρ here is normalised to 1 by the nondimensionalisation)) and the linear stability of this solution is found in the usual way (recall, in particular, Section 14.3 in Chapter 14) by seeking solutions of the linearized equations from (17.22)–(17.24). We thus consider $n - 1$, $\rho - 1$ and $\mathbf{u}$ to be small and, on substituting into the nonlinear system and retaining only linear terms in $n - 1$, $\rho - 1$ and $\mathbf{u}$ and their derivatives, we get the following linear system, where for algebraic convenience we have written n and ρ for $n - 1$ and $\rho - 1$ respectively:

$$n_t - D_1 \nabla^2 n + D_2 \nabla^4 n + a_1 \nabla^2 \rho - a_2 \nabla^4 \rho + \theta_t + rn = 0 , \tag{17.26}$$

$$\nabla \cdot \left[(\mu_1 \varepsilon_t + \mu_2 \theta_t \mathbf{I}) \right.$$

$$\left. + (\varepsilon + \nu' \theta \mathbf{I}) + \{\tau_1 \rho + \tau_2 n + \tau_1 \gamma \nabla^2 \rho)\mathbf{I} \right] - s\mathbf{u} = 0 , \tag{17.27}$$

$$\rho_t + \theta_t = 0 , \tag{17.28}$$

where

$$\tau_1 = \frac{\tau}{1+\lambda}, \quad \tau_2 = \frac{\tau(1-\lambda)}{(1+\lambda)^2} . \tag{17.29}$$

Note that if $\lambda > 1$, $\tau_2 < 0$; λ, which is non-negative, is a measure of the cell-cell contact inhibition.

We now look for solutions to these linearized equations by setting

$$(n, \rho, \mathbf{u}) \propto \exp\left[\sigma t + i\mathbf{k} \cdot \mathbf{r}\right] \tag{17.30}$$

where $\mathbf{k}$ is the wavevector and σ is the linear growth factor (not to be confused with the stress tensor). In the usual way (cf. Chapter 14) substitution of (17.30) into (17.26)–(17.28) gives the dispersion relation $\sigma = \sigma(k^2)$ as solutions of the polynomial in σ given by the determinant

$$\begin{vmatrix} \sigma + D_1 k^2 + D_2 k^4 + r & -a_1 k^2 - a_2 k^4 & ik\sigma \\ ik\tau_2 & ik\tau_1 - ik^3\tau_1\gamma & -\sigma\mu k^2 - (1+\nu')k^2 - s \\ 0 & \sigma & ik\sigma \end{vmatrix} = 0 .$$

where $k = |\mathbf{k}|$. A little algebra gives $\sigma(k^2)$ as the solutions of

$$\sigma[\mu k^2\sigma^2 + b(k^2)\sigma + c(k^2)] = 0 ,$$
$$b(k^2) = \mu D_2 k^6 + (\mu D_1 + \gamma\tau_1)k^4 + (1 + \mu r - \tau_1 - \tau_2)k^2 + s ,$$
$$c(k^2) = \gamma\tau_1 D_2 k^8 + (\gamma\tau_1 D_1 - \tau_2 D_2 + D_2 - a_2\tau_1)k^6 \qquad (17.31)$$
$$\quad + (D_1 + sD_2 - \tau_1 D_1 + \gamma\tau_1 r - a_1\tau_2)k^4$$
$$\quad + (r + sD_1 - r\tau_1)k^2 + rs .$$

Here we have set $\mu = \mu_1 + \mu_2$ and τ_1, τ_2, μ and s replace $\tau_1/(1+\nu')$, $\tau_2/(1+\nu')$, $\mu/(1+\nu')$ and $s/(1+\nu')$ respectively. The dispersion relation is the solution of (17.31) with the largest $\mathrm{Re}\,\sigma \geq 0$, so

$$\sigma(k^2) = \frac{-b(k^2) + \{b^2(k^2) - 4\mu k^2 c(k^2)\}^{1/2}}{2\mu k^2} ,$$
$$\sigma(k^2) = \frac{-b(k^2)}{2\mu k^2}, \quad \text{if} \quad c(k^2) \equiv 0 . \qquad (17.32)$$

Spatially heterogeneous solutions of the linear system are characterized by a dispersion $\sigma(k^2)$ which has $\mathrm{Re}\,\sigma(0) \leq 0$ but which exhibits a range of unstable modes with $\mathrm{Re}\,\sigma(k^2) > 0$ for $k^2 \neq 0$. From (17.31), if $k^2 = 0$, the spatially homogeneous case, we have $b(0) = s > 0$ and $c(0) = rs > 0$ since all the parameters are positive. So $\sigma = -c/b < 0$ and hence stability obtains. Thus we require conditions for $\mathrm{Re}\,\sigma(k^2) > 0$ to exist for at least some $k^2 \neq 0$. All the solutions (17.30) with these k's are then linearly unstable and grow exponentially with time. In the usual way we expect these unstable heterogeneous linear solutions will evolve into finite amplitude spatially structured solutions. Heuristically we see from the nonlinear system (17.22) that such exponentially growing solutions will not grow unboundedly – the quadratic term in the logistic growth prevents this. In models where the mitotic rate is not set to zero, the contact inhibition term (17.23) ensures that solutions are bounded. Numerical simulations of the full system by Perelson et al. (1986) bear this out.

The linearly unstable solutions have a certain predictive ability as to the qualitative character of the finite amplitude solutions. The predictability again seems to be limited to only small wavenumbers in a one-dimensional situation. As we saw in Chapters 14 and 15 this was usually, but not always, the case with reaction diffusion systems.

From the dispersion relation (17.32), the only way a solution with $\mathrm{Re}\,\sigma(k^2) >$ 0 can exist is if $b(k^2) < 0$ or $c(k^2) < 0$ or both. Since the only negative terms involve the traction parameter τ, occurring in τ_1 and τ_2, a necessary condition for the mechanism to generate spatially heterogeneous solutions is that the cell traction $\tau > 0$. Note from (17.29) that it is possible that τ_2 can be negative. It is also clear heuristically from the mechanism that τ must be positive since the cell traction forces are the only contribution to the aggregative process in the force-balance equation (17.23). So, with $\tau > 0$, sufficient conditions for spatial structured solutions to exist are when the parameters ensure that $b(k^2) < 0$ and/or $c(k^2) < 0$ for some $k^2 > 0$. Because of the central role of the cell traction we shall use τ as the bifurcation parameter. There is also a biological reason for choosing τ as the bifurcation parameter. It is known that, *in vitro*, the traction generated by a cell can increase with time (for a limited period) typically as illustrated in Fig. 17.6.

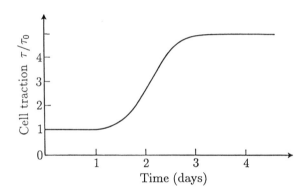

Fig. 17.6. Qualitative *in vitro* behaviour of fibroblast cell traction with time after placing the cells on a dish. τ_0 is a base value, typically of the order of $10^{-2}\,\mathrm{Nm}^{-1}$ of cell edge.

We can deduce the qualitative effects of some of the various terms in the model, as regards their pattern formation potential, by simply looking at the expressions for $b(k^2)$ and $c(k^2)$. Since we require b or c to be negative before spatial pattern will evolve from random initial cell densities, we see, for example, that if the tethering, quantified by s, is increased it tends to stabilise the solutions since it tends to make both b and c more positive. The long range effect of the cells on the matrix, quantified by the parameter γ, is also a stabilising influence. So also is the viscosity. On the other hand long range haptotaxis, via a_2, is always destabilizing. Considerable quantitative information can be obtained simply from the polynomial coefficients in the dispersion relation. Much of it, of course, is intuitively clear. Where the parameters combine, however, further analysis is necessary to draw biological implications; examples are given in the following section.

The expressions for $\sigma(k^2)$ in (17.32), and $b(k^2)$ and $c(k^2)$ in (17.31), determine the domains in parameter space where spatially inhomogeneous linearly unstable solutions exist. They also give the bifurcation surfaces in parameter space, that is the surfaces which separate homogeneous from inhomogeneous solutions. It is algebraically very complicated to determine these surfaces in general. In any case, because of the dimensionality of the parameter space, it would be of little conceptual help in understanding the basic features of the pattern formation process. It is more instructive to consider various special cases whereby we assume one or more of the various factors affecting cell motion and matrix deformation to be negligible. One result of this is to produce several very much simpler model mechanisms which are all capable of generating spatial patterns. Which mechanism is most appropriate for a given biological situation must be determined by the biology.

With the polynomial complexity of $b(k^2)$ and $c(k^2)$ in the dispersion relation $\sigma(k^2)$ in (17.32) we can expect complex linear growth behaviour. In the following section we consider some particular models, all of which are capable of generating spatial patterns. We also display the remarkable variety of dispersion relations which mechanical models can produce from relatively simple mechanisms.

17.4 Simple Mechanical Models Which Generate Spatial Patterns with Complex Dispersion Relations

In this section we consider some special cases of (17.22)–(17.24) where one or more of the factors affecting cell motion or the mechanical equilibrium are assumed to be negligible; each highlights something new. These are deduced by simply setting various parameters to zero and examining the resultant dispersion relation $\sigma(k^2)$ from (17.32). It is not, of course, a haphazard procedure: we examine the effect on $b(k^2)$ and $c(k^2)$ and determine, *a priori*, the likely outcome.

(i) $D_1 = D_2 = a_1 = a_2 = 0$: no cell diffusion and no haptotaxis, $r = 0$: no cell division.

From the general model (17.22)–(17.24) the mechanism becomes

$$n_t + \nabla \cdot (n\mathbf{u}_t) = 0 ,$$

$$\nabla \cdot \left\{ (\mu_1 \varepsilon_t + \mu_2 \theta_t \mathbf{I}) + (\varepsilon + \nu' \theta \mathbf{I}) + \frac{\tau n}{1 + \lambda n^2}(\rho + \gamma \nabla^2 \rho)I \right\} = s\rho\mathbf{u} , \qquad (17.33)$$

$$\rho_t + \nabla \cdot (\rho\mathbf{u}_t) = 0 .$$

The implication of the simple conservation equations for n and ρ is that the cells and matrix are simply convected by the matrix. As mentioned in Section 17.3 this is believed to be the major transport process. The one-dimensional version of the model mechanism is

$$n_t + (n u_t)_x = 0 ,$$

$$\mu u_{xxt} + u_{xx} + [\tau n(1 + \lambda n^2)^{-1}(\rho + \gamma \rho_{xx})]_x = s\rho u , \qquad (17.34)$$

$$\rho_t + (\rho u_t)_x = 0 ,$$

where we have set $\mu = (\mu_1 + \mu_2)/(1 + \nu')$, $\tau = \tau(1 + \nu')$, $s = s/(1 + \nu')$. This system linearises to

$$n_t + u_{tx} = 0 \; ,$$

$$\mu u_{xxt} + u_{xx} + [\tau_1 n + \tau_2 \rho + \tau_1 \gamma \rho_{xx}]_x = su \; , \qquad (17.35)$$

$$\rho_t + u_{tx} = 0 \; ,$$

where, from (17.24) and (17.31),

$$\tau_1 = \frac{\tau}{1 + \lambda}, \quad \tau_2 = \frac{\tau(1 - \lambda)}{(1 + \lambda)^2} \; ,$$

with, as we just said, τ replacing $\tau/(1 + \nu')$.

For the system (17.33) we have from (17.31), $c(k^2) \equiv 0$ and so the dispersion relation from (17.32) is

$$\sigma(k^2) = \frac{-b(k^2)}{2\mu k^2} \; , \qquad (17.36)$$

$$b(k^2) = \gamma \tau_1 k^4 + (1 - \tau_1 - \tau_2)k^2 + s \; .$$

The only way we can have $\operatorname{Re} \sigma > 0$, and here, of course, σ is real, is if $b(k^2) < 0$ for some $k^2 > 0$. This requires $\tau_1 + \tau_2 > 1$ and from the second of (17.36)

$$b_{\min} = s - \frac{(\tau_1 + \tau_2 - 1)^2}{4\gamma \tau_1} < 0 \; . \qquad (17.37)$$

In terms of τ, λ, γ and s this becomes

$$\tau^2 - \tau(1 + \lambda)^2[1 + \gamma s(1 + \lambda)] + \frac{(1 + \lambda)^4}{4} > 0 \; , \qquad (17.38)$$

which implies that spatial patterns will evolve only if

$$\tau > \tau_c = (1 + \lambda)^2 \frac{1 + \gamma s(1 + \lambda) + \{[1 + \gamma s(1 + \lambda)]^2 - 1\}^{1/2}}{2} \; . \qquad (17.39)$$

The other root is not relevant since it implies $\tau_1 + \tau_2 < 1$ and so from (17.36) $b(k^2) > 0$ for all k. The surface $\tau = \tau_c(\lambda, \gamma, s)$ is the bifurcation surface between spatial homogeneity and heterogeneity. In view of the central role of cell traction, and the form of the traction versus time curve in Fig. 17.6, it is natural to take τ as the bifurcation parameter. As soon as τ increases beyond the critical value τ_c, the value which first makes $b(k^2)$ zero, the uniform steady state bifurcates to a spatially unstable state. The natural groupings are $\tau/(1 + \lambda)^2$ and $\gamma s(1 + \lambda)$ and the bifurcation curve $\tau/(1 + \lambda)^2$ versus $\gamma s(1 + \lambda)$ is a particularly simple monotonic one.

When (17.39) is satisfied, the dispersion relation (17.36) is a typical basic dispersion relation (cf. Fig. 14.4 (b)) which initiates spatial patterns as illustrated in Fig. 17.8 (a) which is given towards the end of this section. All wavenumbers k in the region where $\sigma(k^2) > 0$ are linearly unstable: here that is the range of k^2 where $b(k^2) < 0$, which from (17.36) is given by

$$k_1^2 < k^2 < k_2^2$$

$$k_1^2, k_2^2 = \frac{(\tau_1 + \tau_2 - 1) \pm \{(\tau_1 + \tau_2 - 1)^2 - 4s\gamma\tau_1\}^{1/2}}{2\gamma\tau_1} \tag{17.40}$$

$$\tau_1 = \frac{\tau}{1 + \lambda}, \quad \tau_2 = \frac{\tau(1 - \lambda)}{(1 + \lambda)^2},$$

where τ, γ, λ and s must satisfy (17.39). There is a fastest growing linear mode which again predicts, in the one-dimensional model with random initial conditions, the ultimate nonlinear spatial pattern (cf. Chapter 14, Section 14.6). Other ways of initiating the instability results in different preferred modes. We discuss nonlinear aspects of the models later and present some simulations of a full nonlinear model in the context of a specific biological application.

Before considering another example, note the form of the cell conservation equation in (17.35). With no cell mitosis there is no natural cell density which we associated with the maximum logistic value N in (17.7). Here we can use N (or n_0 to highlight the different situation) for the nondimensionalisation in the usual way but now it comes in as another arbitrary parameter which can be varied. It appears, of course, in several dimensionless groupings defined by (17.21) and so offers more potential for experimental manipulation to test the models. Later we shall describe the results from experiments when the cell density is reduced. With (17.21), we can thus determine how dimensionless parameters which involve N vary, and hence predict the outcome, from a pattern formation point of view, by investigating the dispersion relation. All models without any cell proliferation have this property. Another property of such models is that the final solution will depend on the initial conditions, that is, as cell density is conserved, different initial conditions (and hence different total cell number) will give rise to different patterns. However, as the random perturbations are small, the differences in final solution will be small. This is biologically realistic as no two patterns are *exactly* alike.

(ii) $D_2 = \gamma = 0$: no long range diffusion and no cell-ECM interactions. $a_1 = a_2 = 0$: no haptotaxis. $r = 0$: no cell proliferation.

From the general model (17.22)–(17.24) the cell equation involves diffusion and convection and the model mechanism is now

$$n_t = D_1\nabla^2 n - \nabla \cdot (n u_t)$$

$$\nabla \cdot \left\{ (\mu_1\varepsilon_t + \mu_2\theta_t \mathbf{I}) + (\varepsilon + \nu'\theta\mathbf{I}) + \frac{\tau\rho n\mathbf{I}}{1 + \lambda n^2} \right\} = s\rho\mathbf{u}, \tag{17.41}$$

$$\rho_t + \nabla \cdot (\rho\mathbf{u}_t) = 0,$$

From (17.31) and (17.32) the dispersion relation for the system is

$$\sigma(k^2) = \frac{-b \pm [b^2 - 4\mu k^2 c]^{1/2}}{2\mu k^2} \,,$$

$$b(k^2) = \mu D_1 k^4 + (1 - \tau_1 - \tau_2)k^2 + s \,,$$

$$c(k^2) = D_1 k^2 [k^2(1 - \tau_2) + s] \,. \tag{17.42}$$

As noted above, in this model the homogeneous steady state cell density $n = N$ where N is now simply another parameter.

The critical value of τ, with τ_1 and τ_2 defined by (17.29), which makes the minimum of $b(k^2)$ zero is

$$b_{\min} = 0 \quad \Rightarrow \quad \tau_1 + \tau_2 = 1 + 2(\mu s D_1)^{1/2}$$

$$\Rightarrow \quad \tau_{b=0} = \tau_c = (1 + \lambda)^2 \left[\frac{1}{2} + (\mu s D_1)^{1/2} \right] \tag{17.43}$$

and $c(k^2)$ becomes negative for a range of k^2 if $\tau_1 > 1$, that is

$$c(k^2) < 0 \quad \Rightarrow \quad \tau_{c=0} = \tau_c = 1 + \lambda \,. \tag{17.44}$$

(The special case when $\lambda = 1$ makes $\tau_2 = 0$ exactly: such specific cases are unlikely to be of biological interest.) Now, as τ increases, whether b or c becomes zero first depends on other parameters groupings. In the case of the minimum of b becoming zero first, this occurs at the critical wavenumber which is obtained from the expression for $b(k^2)$ in (17.42) with (17.43) as

$$[k]_{b=0} = \left(\frac{s}{\mu D_1} \right)^{1/4} \,. \tag{17.45}$$

If τ is such that $c(k^2)$ becomes zero first, then from (17.42)

$$c(k^2) < 0 \quad \text{for all} \quad k^2 > \frac{s}{\tau_1 - 1} \,. \tag{17.46}$$

The linear *and* nonlinear solution behaviour depends critically on whether $b(k^2)$ or $c(k^2)$ becomes zero first. That is whether $\tau_{c=0}$ is greater than or less than $\tau_{b=0}$. Suppose that, as τ increases from zero, $b = 0$ first. From the dispersion relation (17.42), σ is complex at $\tau_c = \tau_{b=0}$ and so for τ just greater than τ_c, the solutions

$$n, \rho, \mathbf{u} \sim O(\exp [\text{Re}\,\sigma(k_{b=0}^2)t + i\text{Im}\,\sigma(k_{b=0}^2)t + i\mathbf{k}_{b=0} \cdot \mathbf{r}]) \,. \tag{17.47}$$

These solutions represent exponentially growing *travelling* waves and the prediction is that no steady state finite amplitude solutions will evolve.

On the other hand if $\tau_{c=0}$ is reached first then σ remains real, at least near the critical k^2, and it would seem that spatial structures would evolve in the usual way. In any simulation related to a real biological application, a very careful analysis of the dispersion relation and the possible parameter values which are to be used is an absolutely essential part of the process since there can be transitions from normal evolving spatial patterns through unstable travelling waves to spatial patterning again. With the non-standard type of partial differential equation we are dealing with such behaviour might be thought to be an artifact of the numerical simulations.

An even simpler version of this model, which still exhibits spatial structure, has $D_1 = 0$. The system in this case is, from (17.41),

$$n_t + \nabla \cdot (nu_t) = 0$$

$$\nabla \cdot \left\{ (\mu_1 \varepsilon_t + \mu_2 \theta_t I) + (\varepsilon + \nu' \theta I) + \frac{\tau \rho n I}{1 + \lambda n^2} \right\} = s\rho \mathbf{u} , \qquad (17.48)$$

$$\rho_t + \nabla \cdot (\rho u_t) = 0 .$$

Here, from (17.31), $c(k^2) \equiv 0$ and

$$\sigma(k^2) = -\frac{b}{2\mu k^2}, \quad b(k^2) = (1 - \tau_1 - \tau_2)k^2 + s ,$$

Thus we require τ and λ to satisfy

$$\tau_1 + \tau_2 > 1 \quad \Rightarrow \quad \tau > \frac{(1+\lambda)^2}{2}$$

$$\Rightarrow \quad \sigma(k^2) > 0 \quad \text{for all} \quad k^2 > \frac{s}{2\tau(1+\lambda)^{-2} - 1} . \qquad (17.49)$$

The dispersion relation here, and illustrated in Fig. 17.9 (a) given at the end of this section, is fundamentally different to that in Fig. 17.8 (a) below: there is an *infinite* range of unstable wave numbers. That is, perturbations with very large wavenumbers, which correspond to very small wavelengths, are unstable. This is because the version (17.48) of the model has no long range effects included: such effects tend to smooth out small wavelength patterns. It is not clear what pattern will evolve from random initial conditions. It is likely that the ultimate spatial structure depends intimately on the initial conditions. Asymptotic analyses are still lacking for systems with such dispersion relations.

(iii) $D_1 = D_2 = 0$: no cell diffusion. $a_1 = a_2 = 0$: no haptotaxis. $\mu_1 = \mu_2 = 0$: no viscoelastic effects in the ECM.

Here the system (17.22)–(17.24) reduces to

$$n_t + \nabla \cdot (nu_t) = rn(1 - n) ,$$

$$\nabla \cdot \left\{ (\varepsilon + \nu' \theta I) + \frac{\tau n}{1 + \lambda n^2} (\rho + \gamma \nabla^2 \rho) I \right\} = s\rho \mathbf{u} , \qquad (17.50)$$

$$\rho_t + \nabla \cdot (\rho u_t) = 0 ,$$

which, in one space dimension, is

$$n_t + (nu_t)_x = rn(1 - n)$$
$$[u_x + \tau n(1 + \lambda n^2)^2(\rho + \gamma\rho_{xx})]_x = s\rho u \qquad (17.51)$$
$$\rho_t + (\rho u_t)_x = 0 .$$

where again we have incorporated $(1 + \nu')$ in τ and s. The dispersion relation in this case is, using (17.31) and (17.32),

$$\sigma(k^2) = -\frac{c(k^2)}{b(k^2)} ,$$

$$b(k^2) = \gamma\tau_1 k^4 + (1 - \tau_1 - \tau_2)k^2 + s , \qquad (17.52)$$

$$c(k^2) = \gamma\tau_1 r k^4 + r(1 - \tau_1)k^2 + rs .$$

Here, as τ increases $b(k^2)$ becomes zero first, so the bifurcation traction value is given by τ_c where $b(k^2) = 0$. This expression for $b(k^2)$ is the same as that in (17.36) and so the critical τ_c is given by (17.39). Here, however, $c(k^2)$ is not identically zero and so the dispersion relation is quite different as τ increases beyond τ_c. $c(k^2)$ can also become negative. Denote the critical τ_c when $b(k^2)$ and $c(k^2)$ first become zero by $\tau_c^{(b)}$ and $\tau_c^{(c)}$ respectively: here $\tau_c^{(b)} < \tau_c^{(c)}$. In this case the $\sigma(k^2)$ behaviour as τ increases is illustrated in Fig. 17.7, which exhibits a fundamentally different dispersion relation to what we have found and discussed before.

First note in Fig. 17.7 that the range of unstable wave numbers is finite and that there are two bifurcation values for the traction parameter τ. The pattern formation potential of a system with such a dispersion relation is much richer than is possible with the standard dispersion form in Fig. 17.8 (a) below. Linear theory, of course, is not valid where the linear growth is infinite. So, from an analytical point of view, we must include other effects which effectively round off the discontinuities in $\sigma(k^2)$. This in turn implies the existence of a singular perturbation problem. We do not consider such problems here but we can see intuitively that such dispersion relations, with large linear growth rates, imply a 'fast focusing' of modes with preferred wavenumbers. For example, in Fig. 17.7 (e) we would expect the modes with wavenumbers at the lower end of the lower band and the upper end of the upper band to be the dominant modes. Which mode eventually dominates in the nonlinear theory will depend critically on the initial conditions.

With large linear growth we can see how necessary it is to have some cell-cell inhibition in the full nonlinear system. If there is fast focusing there is the possibility of unlimited growth in, for example, the cell density. This in turn implies the appearance of spike-like solutions. The effect of the inhibition terms, as measured by the parameter λ, is thus essential.

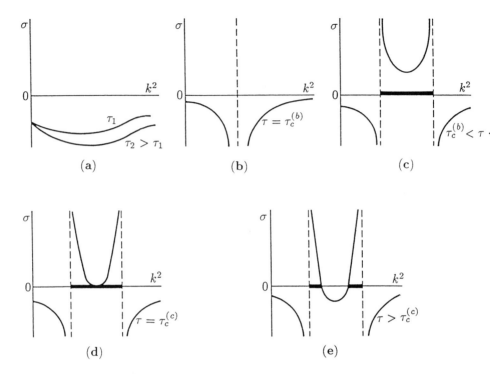

Fig. 17.7a–e. Qualitative variation in the dispersion relation $\sigma(k^2)$ in (17.52) for the model system (17.50) (and (17.51)) as the traction parameter τ increases. The bifurcation values $\tau_c^{(b)}$ and $\tau_c^{(c)}$ denote the values of τ where $b(k^2) = 0$ and $c(k^2) = 0$ respectively. The wave numbers of the unstable modes are denoted by the heavy line on the k^2-axis.

It is now clear how to investigate various simpler models derived from the more complicated basic model (17.22)–(17.24). Other examples are left as exercises.

Fig. 17.8 and Fig. 17.9 demonstrate the richness of dispersion relation types which exist for the class of mechanical models (17.22)–(17.24). Fig. 17.8 shows only some of the dispersion relations which have finite ranges of unstable modes while Fig. 17.9 exhibits some of the possible forms with infinite ranges of unstable modes. A nonlinear analysis in the vicinity of bifurcation to spatial heterogeneity, such as has been done by Maini and Murray (1988), can be used on the mechanisms which have a dispersion relation of the form illustrated in Fig. 17.8 (a). A nonlinear theory for models with dispersion relations with an infinite range of unstable modes, such as those in Fig. 17.9, is, as we noted, still lacking as is that for dispersion relations which exhibit infinite growth modes (Fig. 17.8 (c)-(g)). Although we anticipate that the pattern will depend more critically on initial conditions than in the finite range of unstable mode situations, this has also not been established.

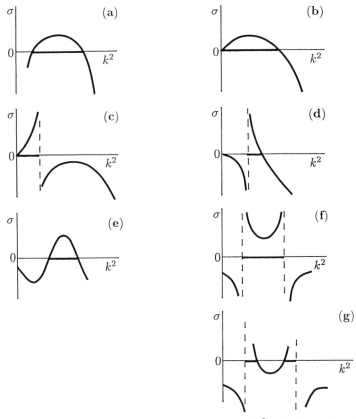

Fig. 17.8a-g. Examples of dispersion relations $\sigma(k^2)$, obtained from (17.32) with (17.31) for mechanical models based on the mechanism (17.22)–(17.24). The various forms correspond to the specific conditions listed in Table 17.1. Realistic models for those with infinite growth must be treated as singular perturbation problems, with small values for the appropriate parameters in terms which have been omitted so as to make the linear growth finite although large.

Table 17.1. Mechanical models, derived from the basic system (17.22)–(17.24) with positive non-zero parameters denoted by •, which have dispersion relations with a finite range of unstable wavenumbers. The corresponding dispersion relation forms are given in Fig. 17.8, with $\lambda = 0$, $\tau_1 = \tau_2 = \tau$.

Fig. 17.8	D_1	D_2	a_1	a_2	r	μ	s	γ	λ	Condition on τ
(a)	○	○	○	○	○	•	•	•	○	$\tau > \{1 + \gamma s + [(1 + \gamma s)^2 - 1]^{1/2}\}/2$
(b)	•	○	○	○	○	•	○	•	○	$1 > \tau > 1/2$
(c)	•	○	○	○	○	○	○	•	○	$1 > \tau > 1/2$
(d)	•	○	○	○	○	○	○	•	○	$1 < \tau;\ \ D_1 > (2\tau - 1)/(\tau - 1)$
(e)	•	•	•	•	○	○	•	○	○	$1/2 > \tau;\ \ [\tau(D_1 + a_1) - D_1 - sD_2]^2$
		○		•						$> 4sD_1[D_2 + \tau(a_1 a_2 - D_2)]$
(f)	○	○	○	○	•	○	•	•	○	$1/2 < \tau < 1;\ \ (2\tau - 1)^2 > 4s\gamma\tau$
(g)	○	○	○	○	•	○	•	•	○	$1 < \tau < 1;\ \ (\tau - 1)^2 > 4s\gamma\tau$

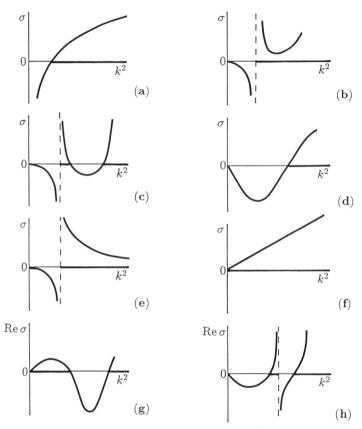

Fig. 17.9a–h. Examples of dispersion relations $\sigma(k^2)$, obtained from (17.32) with (17.31) for mechanical models based on the mechanism (17.22)–(17.24), with an infinite range of unstable modes. The conditions listed in Table 17.2 relate the models to specific forms. In (g) and (h) the imaginary part of σ is nonzero.

Table 17.2. Mechanical models, derived from the basic system (17.22)–(17.24) with positive non-zero parameters denoted by •, which have dispersion relations with an infinite range of unstable wavenumbers. The corresponding dispersion relation forms are given in Fig. 17.9, with $\lambda = 0$, $\tau_1 = \tau_2 = \tau$.

Fig. 17.9	D_1	D_2	a_1	a_2	r	μ	s	γ	λ	Condition on τ
(a)	○	○	○	○	○	•	•	○	○	$\tau > 1/2$
(b)	•	○	○	○	○	○	•	○	○	*$\tau < 1/2$
	•	○	•	○	○	○	•	○	○	$D_1/(D_1 + a_1) > \tau > 1/2$
(c)*	•	•	•	•	○	○	•	○	○	$\tau > 1/2$, plus quadratic condition on τ
(d)	•	○	•	○	○	○	•	○	○	$D_1/(D_1 + a_1)\tau 1/2$
(e)	○	○	○	○	•	○	•	○	○	$1/2 < \tau < 1$
(f)	•	○	○	○	○	○	○	○	○	$1/2 < \tau < 1$

* This is only one of the possibilities, depending on $c(k^2)$ and its zeros.

Table 17.3. Mechanical models, derived from the basic system (17.22)–(17.24) with positive non-zero parameters denoted by •, which have dispersion relations with an infinite range of unstable wavenumbers and which admit temporal oscillations. The corresponding dispersion relation forms are given in Fig. 17.9, with $\lambda = 0$, $\tau_1 = \tau_2 = \tau$.

Fig. 17.9	D_1	D_2	a_1	a_2	r	μ	s	γ	λ	Condition on τ are very complicated
(g)	•	○	○	○	○	•	•	○	○	
(h)	○	○	○	○	•	•	•	○	○	

Mechanical models, as we noted above, are also capable of generating travelling waves: these are indicated by dispersion relations with complex σ. Table 17.3 gives examples of models which admit such solutions.

From a biological applications viewpoint two and three-dimensional patterns are naturally of great interest. With the experience gained from the study of reaction diffusion models in Chapters 14 and 15, and in neural models in Chapter 16, we expect the simulated patterns for the full nonlinear models here to reflect some of the qualitative features of the linearized analysis of the basic model equations (17.26)–(17.28). This is motivation for looking at possible symmetries in the solutions. We do this by taking the divergence of the linear force balance equation (17.27) which, with (17.26) and (17.28), gives, using the identity

$$\operatorname{div} \varepsilon = \operatorname{grad}(\operatorname{div} \mathbf{u}) - \frac{1}{2}\operatorname{curl}\operatorname{curl}\mathbf{u} ,$$

$$n_t - D_1\nabla^2 n + D_2\nabla^4 n + a_1\nabla^2\rho - a_2\nabla^4\rho + \theta_t + rn = 0 , \tag{17.53}$$

$$\nabla^2[(\mu\theta_t + (1+\nu')\theta + \tau_1 n + \tau_2\rho + \tau_1\gamma\nabla^2\rho)] - s\theta = 0 , \tag{17.54}$$

$$\rho_t + \theta_t = 0 , \tag{17.55}$$

where $\tau_1 = \tau/(1+\lambda)$, $\tau_2 = \tau(1-\lambda)/(1+\lambda)^2$ and $\mu = \mu_1 + \mu_2$. To determine a reasonably full spectrum of relevant solutions of this set of equations is not trivial. However, we can look for periodic solutions, which tesselate the plane, for example. Such solutions (cf. Chapters 14 and 16) satisfy

$$\Gamma(\mathbf{r} + m\boldsymbol{\omega}_1 + l\boldsymbol{\omega}_2) = \Gamma(\mathbf{r}) , \tag{17.56}$$

where $\Gamma = (n, \mathbf{u}, \rho)$, m and l are integers and $\boldsymbol{\omega}_1$ and $\boldsymbol{\omega}_2$ are independent vectors. A *minimum* class of such periodic solutions of the linear system (17.53)–(17.55) include at least the eigenfunctions of

$$\nabla^2\psi + k^2\psi = 0, \quad (\mathbf{n}\cdot\nabla)\psi = 0 \quad \text{for} \quad \mathbf{r} \text{ on } \partial B , \tag{17.57}$$

where $\mathbf{n}$ is the unit normal vector on the boundary ∂B of the domain B. With these boundary conditions the solutions are periodic. As we saw in Chapter 14, Section 14.4 and Chapter 16 regular plane periodic tesselation has the basic symmetry group of the hexagon, square (which include rolls) and rhombus solutions given respectively by equations (14.47), (14.48) and (14.49). For convenience the solutions in polar coordinate form (r, ϕ) are reproduced again here:

Hexagon: $\quad \psi(r, \theta) = \left[\cos\left\{kr\sin\left(\phi + \frac{\pi}{6}\right)\right\} + \cos\left\{kr\sin\left(\phi - \frac{\pi}{6}\right)\right\}\right.$

$$\left. + \cos\left\{kr\sin\left(\phi - \frac{\pi}{2}\right)\right\}\right]/3 . \qquad (17.58)$$

Square: $\quad \psi(r, \phi) = \left[\cos\left\{kr\cos\phi\right\} + \cos\left\{kr\sin\phi\right\}\right]/2 \qquad (17.59)$

Rhombus: $\quad \psi(r, \phi; \delta) = \left[\cos\left\{kr\cos\phi\right\} + \cos\left\{kr\cos(\phi - \delta)\right\}\right]/2 , \quad (17.60)$

where δ is the rhombic angle. Such symmetric solutions are illustrated in Fig. 17.10.

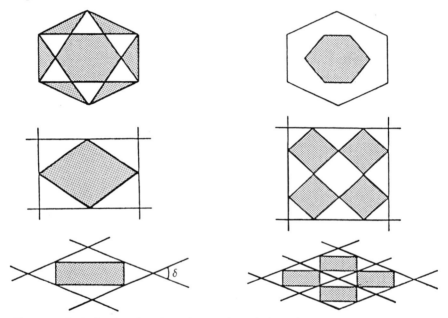

Fig. 17.10. Periodic eigenfunction solutions (17.58)–(17.60) which tesselate the plane. Here the dark regions, for example, represent higher densities.

Small Strain Approximation: A Caricature Mechanical Model for Two-dimensional Patterns

It seems that in many embryological situations the strain, cell density and ECM density changes during the pattern formation process are small. Such assumptions can lead to the linear model system (17.26)–(17.28). Linear systems, however, pose certain problems regarding long term stability. We can exploit the small strain approximation to derive a simple scalar equation model which retains certain key nonlinearities which allow us to carry out a nonlinear analysis in the two-dimensional situation and obtain stable nonlinear solutions of biological relevance.

To illustrate this let us consider the nonlinear dimensionless system (17.33), the nontrivial steady state of which we take as $n = \rho = 1$, $\mathbf{u} = 0$. Because of the small strains we linearise the cell and matrix conservation equations, the first and third of (17.33), to get

$$n_t + \nabla \cdot \mathbf{u}_t = 0 \quad \Rightarrow \quad n_t + \theta_t = 0 \;,$$

$$\rho_t + \nabla \cdot \mathbf{u}_t = 0 \quad \Rightarrow \quad \rho_t + \theta_t = 0 \;,$$

since the dilation $\theta = \nabla \cdot \mathbf{u}$. Integrating with respect to t and using the fact that when $\theta = 0$, $n = \rho = 1$ we get

$$n(\mathbf{r}, t) = 1 - \theta(\mathbf{r}, t) = \rho(\mathbf{r}, t) \;. \tag{17.61}$$

Since we consider θ small, certainly $\theta < 1$, n and ρ remain positive, as is necessary of course.

Because of (17.61) we now replace the external force $s\rho\mathbf{u}$ in the force balance equation, the second of (17.33), by its linear approximation $s\mathbf{u}$. Now substitute the linear forms relating n and ρ to the dilation from (17.61) into the second of (17.33), take the divergence of the resulting equation and use the tensor identity

$$\nabla \cdot \varepsilon = \operatorname{grad} \operatorname{div} \mathbf{u} - \frac{1}{2} \operatorname{curl} \operatorname{curl} \mathbf{u} \;.$$

This yields the following scalar equation for the dilation θ :

$$\mu \nabla^2 \theta_t + \nabla^2 \theta + \tau \nabla^2[(1 - \theta)^2 - \gamma(1 - \theta)\nabla^2 \theta] - s\theta = 0 \;, \tag{17.62}$$

where we have incorporated $1 + \nu'$ into redefinitions for μ, τ and s and for algebraic simplicity taken $\lambda = 0$. The effect of $\lambda \neq 0$ is simply to introduce a multiplicative term $[1 + 2\lambda\theta/(1 + \lambda)]$ in the square bracket in (17.62) and have $\tau/(1 + \lambda)$ in place of τ.

Maini and Murray (1988) carried out a nonlinear analysis on the caricature model (17.62) and obtained roll and hexagonal solutions. The significance of the latter will be discussed in the following section on a biological application to skin organ morphogenesis.

Perhaps it should be mentioned here that the spectrum of spatial patterns possible with the mechanism (17.22)–(17.24) and its numerous simplifications is orders of magnitude greater than with a reaction diffusion system – even three species systems. The implications of a paper by Penrose (1979) are that tensor systems have solutions with a wider class of singularities than vector systems. Since the cell-matrix equation is a tensor equation, its solutions should thus include a wider class of singularities than reaction diffusion vector systems. Even with the linear system (17.26)–(17.28) the analytical and numerical studies have only just started.

In the following two sections we consider two biologically important and widely studied pattern formation problems using mechanical models of pattern generation. As always the actual mechanism is not yet known but a mechanical mechanism we suggest is certainly a strong candidate.

17.5 Periodic Patterns of Feather Germs

Generation of regular patterns occurs in many situations in early embryogenesis. These are particularly evident in skin organ morphogenesis such as in the formation of feather and scale primordia and are widely studied (see, for example, Sengel 1976, Davidson 1983). Feather formation has much in common with scale formation during early development of the primordia. Here we shall concentrate on feather germ formation with particular reference to the chick, and fowl in general. Feather primordial structures are distributed across the surface of the animal in a characteristic and regular hexagonal fashion. The application of the Oster-Murray mechanical theory to feather germ primordia was first put forward by Murray et al. (1983) and Oster et al. (1983) and it is their scenario we describe here. We first present the biological background which suggests using a mechanical model.

Vertebrate skin consists essentially of two layers; an epithelial epidermis overlays a much thicker mesenchymal dermis and is separated from it by a fibrous basal lamina. The layer of epithelial cells, which in general do not move, can deform as we described in Section 17.1. Dermal cells are loosely packed and motile and can move around in the extracellular matrix, the ECM, as we described earlier. The earliest observable developmental stages of feather and scale germs begin the same way. We shall concentrate here on the initiation and subsequent appearance of feather rudiments in the dorsal pteryla – the feather forming region on the chick back.

In the chick the first feather rudiments become visible about 6 days after egg fertilization. Each feather germ, or primordium, consists of a thickening of the epidermis with one or more layers of columnar cells, called a placode, beneath which is an aggregation of dermal (mesenchymal) cells, called a papilla. Excellent pictures of papillae and placodes are given by Davidson (1983). The dermal condensations are largely the result of cell migration, with localised proliferation playing a secondary role. Whether or not the placodes form prior to the dermal papillae is a controversial issue. There is considerable experimental work going on to determine the order of appearance or, indeed, whether the interaction between the epidermis and dermis producee the patterns simultaneously. The dermis seems to determine the spatial patterning – as shown by epidermal-dermal recombination experiments (Rawles 1963, Dhouailly 1975). The model we discuss here is for the formation of dermal papillae. Subsequent development, however, is a coordinated process involving both the epidermal and dermal layers (Wessells 1977, Sengel 1976). The book by Sengel (1976) gives a full discussion of experimental work on the morphogenesis of skin.

Davidson (1983) demonstrates that chick feather primordia appear sequentially. A central column of dermal cells forms on the dorsal pteryla and subsequently breaks up into a row of papillae. As the papillae form, tension lines develop joining the cell aggregation centres. With the above mechanical models this is consistent with the cells trying to align the ECM. Now lateral rows of papillae form sequentially from the central column outwards – in the ventral direction – but these are interdigitated with the papillae in the preceding row: see Fig. 17.11 (a)-(f). These lateral rows spread out from the central midline almost like a wave of pattern initiation. Experiments by Davidson (1983) tend to confirm this wave theory – later we show how these results can be explained by our model and we present corroborative numerical results.

These observations suggest that it is reasonable first to model the pattern formation process for the initial row of papillae by a one-dimensional column of cells and look for the conditions for spatial instability which generate a row of papillae. This is stage 1 and is represented by the sequential process illustrated in Fig. 17.11 (a)-(d).

We have seen in the previous sections that the mechanical model (17.22)–(17.24), and simpler models derived from it as in Section 17.4, that as the cell traction parameter τ increases beyond some critical value τ_c the uniform steady state becomes spatially unstable. With the standard 'vanilla' dispersion relation as in Fig. 17.8 (a), the mode (17.30) with a specific wavenumber k_c, that is with wavelength $2\pi/k_c$, first becomes unstable and a spatial pattern starts to evolve: this generates a regular pattern of dermal papillae.

Simulations of the one-dimensional version of the full nonlinear mechanical model (17.22)–(17.24) with negligible long range haptotaxis, $a_2 = 0$, and, reflecting the biological situation of small cell mitosis, $r = 0$, have been carried out by Perelson et al. (1986). They particularly addressed the problem of mode selection in models with many parameters, and proposed a simple scheme for determining parameter sets to isolate and 'grow' a specific wavelength pattern. Fig. 17.12 (a) shows a typical steady state pattern of cell aggregations (the papillae), ECM displacement and density variations. As we would expect intuitively, the cell aggregations are in phase with the ECM density variation ρ, and both are out of phase with the ECM displacement u. The reason is that the cell aggregations pull the matrix towards the areas of higher cell density thus stretching the matrix between them: Fig. 17.11 (e) illustrates what is going on physically.

Patterns of the type illustrated in Fig. 17.12 (a) occur only if the cell traction parameter is above a certain critical value (see Section 17.4). Thus a possible scenario for the formation of the pattern along the dorsal mideline is that there is a wave of initiation that sweeps down the column and this could be related to tissue age; in this case, *in vitro* experiments show that the cell traction parameter increases: refer to Fig. 17.6. As the cells become stronger τ passes through the critical value τ_c and pattern is initiated. Note that in the model, one-dimensional pattern develops simultaneously whereas the experiments suggest sequential development. This is reminiscent of the mode of pattern formation illustrated in Fig. 14.12 (d) in Chapter 14.

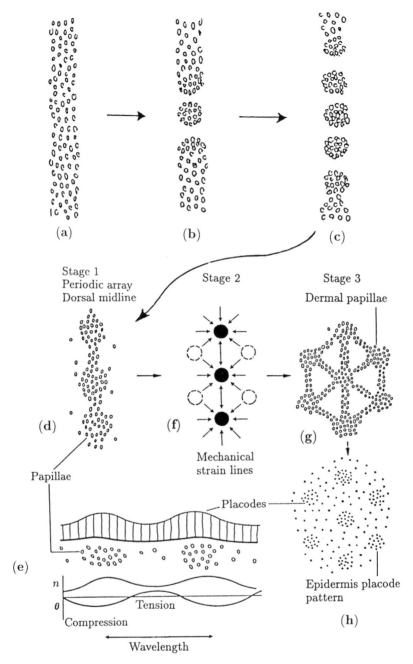

Fig. 17.11a-h. (a)-(d) These show the predicted sequential break up of a uniform distribution of mesenchymal (dermal) motile cells into regular cell condensations with a wavelength determined by the parameters of the model mechanism (stage 1). These cell aggregations are the primordial papillae for feathers and scales. (e) Vertical cross section qualitatively showing the feather germ primordia. The placodes in the epidermis are underlain by the papillae which create the stress field. (**f**) Subsequent aggregations form laterally. The prestressed strain field from the first line of condensations induces a bias so that the neighbouring line of papillae interdigitate with the first line (stage 2). The resulting periodic array is thus hexagonal, the basic unit of which is illustrated in (**g**) (stage 3). (**h**) The epidermal placode pattern that mirrors the dermal pattern.

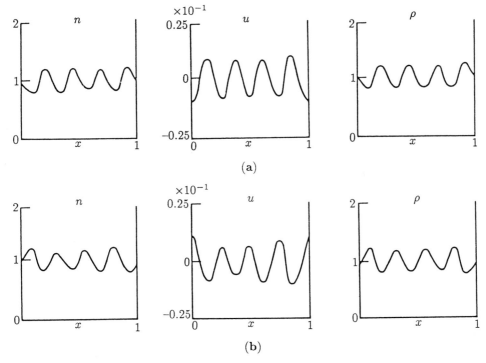

Fig. 17.12a,b. Steady state solutions, for the cell density n, ECM displacement u and ECM density ρ of the nonlinear one-dimensional version of the mechanical model (17.22)–(17.24). **(a)** Periodic boundary conditions were used and initial conditions were random perturbations about the uniform steady state $n = \rho = 1$, $u = 0$. **(b)** Heterogeneous steady state solutions with the initial displacement pattern in **(a)**. Parameter values: $D_1 = a_1 = \gamma = 10^{-3}$, $\lambda = 0.12$, $\tau = 1.65$, $s = 400$, $\mu = 1$. Note that n and ρ are in phase and both are out of phase with u.

Let us now consider the formation of the distinctive hexagonal two-dimensional pattern of papillae. We described above how a wave of pattern initiation seems to spread out from the dorsal midline. This suggests that the pattern of matrix strains set up by the initial row of papillae biases the formation of the secondary condensations at positions displaced from the first line by half a wavelength. Fig. 17.12 (b) shows the appropriate numerical simulation based on such a scenario: note how the patterns are out of phase with those in Fig. 17.12 (a). If we now look at Fig. 17.11 (f), (g) we see how this scenario generates a regular hexagonal pattern in a sequential way like a wave emanating from the central dorsal midline.

This 'wave' is, however, not a wave in the usual sense since if the dermal layer is cut along a line parallel to the dorsal midline the wave simply starts up again beyond the cut *ab initio*. This is consistent with Davidson's (1983) experimental observations who specifically investigated the qualitative effect on spacing of stretching and cutting the epidermis.

This quasi-one-dimensional scenario, although suggested by linear theory, was to a certain extent validated by the nonlinear simulations. A better verification would be from a simulation of the two-dimensional model. However, using our scenario, it is possible to make predictions as to the change in wavelength as the experimental parameters are changed. For example, one version of the models predicts a spacing that increases as the total number of cells N decreases. This agrees with experimental observation (Davidson, personal communication 1983).

One of the most useful aspects of a nondimensional analysis, with resulting nondimensional groupings of the parameters, is that it is possible to assess how different physical effects, quantified by the parameters in the nondimensionalisation, trade off against one another. For example, with the nondimensional groupings (17.21) we see from the definition of the dimensionless traction, namely $\tau^* = \tau \rho_0 N (1 + \nu)/E$, that with the model (17.22)–(17.24) the effect of a reduction in the cell traction τ is the same as a reduction in the cell density or an increase in the elastic modulus E. To see clearly the overall equivalence the bifurcation surfaces in parameter space must be considered. An important caveat in interpreting results from experimental manipulation is that quite different cell or matrix alterations can produce compensating and thus equivalent results. Although such a caveat is applicable to any model mechanism it is particularly apposite to experimentation with mechanical models since the morphogenetic variables are unquestionably real.

We should perhaps mention here that an alternative model based on a reaction diffusion theory was proposed by Nagorcka (1986), Nagorcka and Mooney (1985) for the initiation and development of scale and feather primordia, and by Nagorcka and Mooney (1982) for the formation of hair fibres.

The modelling here does not cast light on the controversy regarding the *order* of formation of placodes and papillae. However, since the traction forces generated by the dermal cells can be quite large the model lends support to the view that the dermis controls the pattern even if it does not initiate it. Current thinking tends towards the view that initiation requires tissue interaction between the dermis and epidermis. It is well known that mechanical deformations effect mitosis and so tissue interaction seems natural with mechanical models: see also the discussion in Oster et al. (1983). Nagorcka, Manoranjan and Murray (1987) investigated a tissue interaction mechanism specifically with the complex patterns of scales in mind. We describe some of these complex patterns and their model in Section 17.10. A recent experimental paper by Nagawa and Nakanishi (1988) confirms the importance of mechanical influences of the mesenchyme on epithelial branching morphology.

17.6 Cartilage Condensations in Limb Morphogenesis

The vertebrate limb is one of the most widely and easily studied developmental systems and such studies have played a major role in embryology (see, for example, Hinchliffe and Johnson 1980, Thorogood 1983). We shall discuss some evolutionary aspects of limb development in the next chapter. Here we propose a mechanical model for generating the pattern of cell condensations which evolve in a developing limb bud and which eventually become cartilage: it was first put forward by Murray et al. (1983) and Oster et al. (1983): a related mechanochemical model was later proposed by Oster, Murray and Maini (1985).

The pattern in developing limb buds which determines the final cartilage patterns, which later ossify into bones, involves aggregations of chondrocyte cells, which are mesenchymal cells such as we have been considering. The basic evolution of chondrocyte patterns takes place sequentially as the limb bud grows, which it does from the distal end. Fig. 17.13 gives an explanation of how, with geometry and scale as bifurcation parameters, chondrogenesis could proceed. The actual sequence of patterns for the developing chick limb is illustrated in Fig. 17.13 (c): Fig. 17.13 (d) is a photograph of a normal adult limb. The detailed explanation of the process based on a mechanical mechanism is the following.

As the limb bud grows, through cell proliferation in the apical ectodermal ridge, which is at the distal end, the cross section of the tissue domain, which includes the ECM and mesenchymal cells, is approximately circular but with an elliptical bias. Let us consider this to be the two-dimensional domain for our mechanical model with zero flux boundary conditions for the cells n and matrix ρ. The condition for $\mathbf{u}$ is an imposed restraining force which comes from the epidermis – the sleeve of the limb bud. The dispersion relation for the mechanism with such a domain is reminiscent of that for reaction diffusion mechanisms with similar geometry, as discussed in detail in Chapter 14, particularly Section 14.5. Let us suppose that as the cells age the traction increases as in Fig. 17.6 and eventually passes through the critical value τ_c. The detailed form of the dispersion relation is such that in the appropriate parameter space this first bifurcation produces a single central aggregation of cells recruited from the surrounding tissue.

The axial cell aggregations are influenced by the cross-sectional shape as shown in Fig. 17.13 (a). As the cells condense into a single aggregation they generate a strong centrally directed stress as in Fig. 17.13 (b). This radial stress deforms the already slightly elliptical cross section to make it even more elliptical. This change in geometry in turn induces a secondary bifurcation to two condensations because of the changed flatter geometry of the cross section. An aerofoil section gives rise to two condensations of different size as in Fig. 17.13 (a), path 3: these we associate with, for example, the radius and ulna in forelimbs as in the photograph in Fig. 17.13 (d).

We should interject, here, that this behaviour is directly equivalent to the situation with the patterns generated by reaction diffusion mechanisms as illustrated in Fig. 14.11 (c) and Fig. 14.14. A similar scenario for sequential laying down of a pre-pattern for cartilage formation by reaction diffusion models equally applies. However, importantly with a mechanical model, the condensation of cells influence the *shape* of the domain and can actually *induce* the sequence of bifurcations shown in Fig. 17.13 (c).

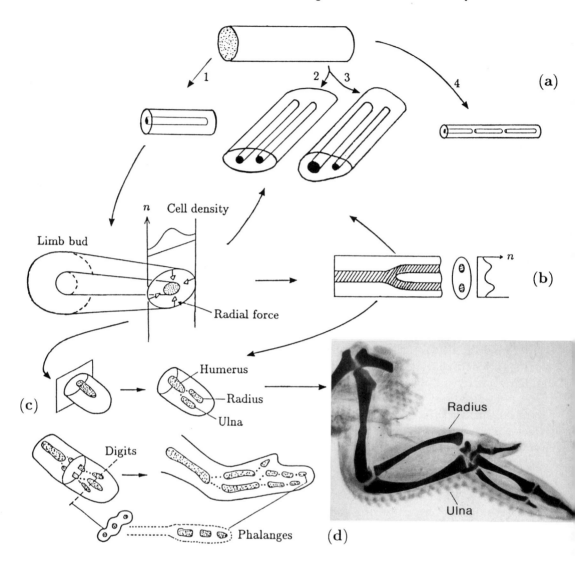

Fig. 17.13a-d. (a) The type of axial condensation is influenced by the cross-sectional shape of the limb. Initially a single condensation, path 1, will be produced (for example, the humerus in (c)). A more elliptical cross section allows two aggregations to form with an aerofoil-shaped domain producing unequal condensations, paths 2 and 3 (for example, the radius and ulna in (c)). In a long thin cylinder the axial condensations form segmental units, path 4 (for example, the phalanges in (c)). (b) This shows how the mechanical mechanism influences cross-sectional form and hence induces the required sequence of chondrogenic patterns. As the cells form the central condensation their tractions deform the limb thus making it more elliptical. At a critical ellipticity the pattern bifurcates to two condensations. How three condensations are formed is important and explained in the text: refer also to Fig. 17.16 (d). (c) The schematic bifurcation sequence of chondrocyte (mesenchymal) cell aggregations which presage cartilage formation in the developing chick limb. (d) Photograph of the normal cartilage pattern in the limb of a 10 day chick. Bar, 1 mm. (Photograph courtesy of L. Wolpert and A. Hornbruch)

After a two-condensation state has been obtained, further growth and flattening can generate the more distal patterns. By the time the limb bud is sufficiently flat, cell recruitment effectively isolates patterning of the digits. Now subsequent growth induces longitudinal or segmental bifurcations with more condensations simply fitted in as the domain, effectively linearly now, increases and we get the simple laying down of segments, for example, the phalanges in Fig. 17.13 (c), as predicted by Fig. 17.13 (a), path 4.

It is important to reiterate here, that the sequence of cell pattern bifurcations need not be generated by a changing geometry; it can result from a variation of other parameters in the model. Also asymmetric condensations can result from a spatial variation or asymmetry in a parameter across the limb cross-section. There is well documented experimental evidence for asymmetric properties, which, of course, are reflected in the different bone shapes and sizes in the limb such as the radius and ulna in Fig. 17.13 (d). Whatever triggers the bifurcations as we move from the proximal to distal part of the limb, the natural sequence is from a single condensation, to two condensations and then to several as in Fig. 17.13 (c): see also Fig. 17.16 below.

Much of the extensive experimental work on chondrogenesis has been to investigate the chondrogenic patterns which result from tissue grafts. The major work on chick limbs was initiated and carried out by Lewis Wolpert and his co-workers and it has been the principal stimulus for much of the current research in this line, on other animals as well as chicks: see, for example, Wolpert and Hornbruch (1987), Smith and Wolpert (1981) and the more review-type article by Wolpert and Stein (1984). One set of experiments involves taking a piece of tissue from one part of a limb bud and grafting it onto another as in Fig. 17.14 (a). The region from the donor limb is referred as the zone of polarizing activity – the ZPA. This results in the double limb as in Fig. 17.14 (b).

Consider now the double limb in Fig. 17.14 (b) in the light of our model and let us examine how this can arise. Let us be specific and take geometry and scale as the bifurcation parameters. The effect of the tissue graft is to increase the width of the limb cross section by increasing the cell division in the apical ectodermal ridge – the distal edge of the wing bud (Smith and Wolpert 1981). This means that at each stage, after the graft, the domain is sufficiently wide for a double set of cell condensations to form and thus generate a double limb as shown in Fig. 17.14 (b). Not all grafts result in a double limb. Different double patterns which appear, are obtained and which depends on where and when in development the graft is inserted. Fig. 17.14 (c) shows an example of a natural double hand. It is not uncommon for people to have six fingers on a hand, often an inherited trait. (Ann Boleyn, one of Henry VIII's wives had six fingers – unfortunately she had the wrong appendage cut off!) An experimental prediction of the model then is that if the limb bud with a graft is geometrically constrained within a scale commensurate with a single limb, it would not be able to undergo the double bifurcation sequence necessary for a double limb to form. The book by Walbot and Holder (1987) gives a good description of such graft experiments

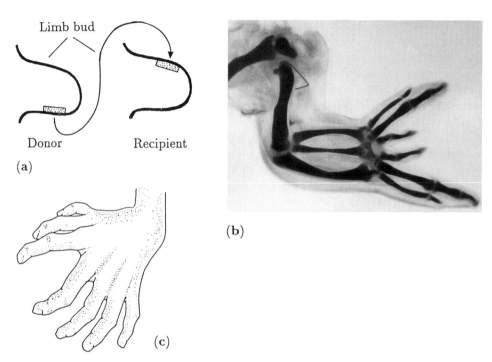

Limb bud

Donor   Recipient

(a)

(b)

(c)

Fig. 17.14a-c. (a) Graft experiments involve taking a small piece of tissue from one limb bud and grafting it onto another. The effect of such a graft is to induce increased cell proliferation and hence increase the subsequent size of limb. The result is to induce growth commensurate with a domain in which multiple cell condensations can be fitted in at each stage of growth and hence result in double limbs. (b) Photograph of a double limb in a 10 day chick following an anterior graft of tissue from the posterior region, the zone of polarizing activity (ZPA), of another limb as in (a). The grafted tissue creates the appropriate symmetry which results in a mirror image limb. (From Wolpert and Hornbruch 1987: photograph courtesy of L. Wolpert and A. Hornbruch) (c) A natural example of a double hand of a Boston man: note the lack of thumb and the mirror symmetry. (After Walbot and Holder 1987)

and discusses the results in terms of a positional information approach with, in effect, a chemical prepattern background.

Some extremely interesting experiments have been carried out which show that many of the results of graft experiments can be achieved by subjecting the limb bud, during development, to doses of retinoic acid sequestered in small beads inserted into the limb. The retinoic acid is slowly released in the tissue. A quantitative analysis of the effects are given by Tickle et al. (1985). Disruption of the chondrogeneic process by chemicals and drugs is well known – the thalidomide affair is a tragic example. This work highlights one of the important uses of any theory which might help in unravelling the mechanism involved in chondrogeneis. Until we understand the process it is unlikely we shall be able to understand how drugs, chemicals and so on will disrupt the process during development.

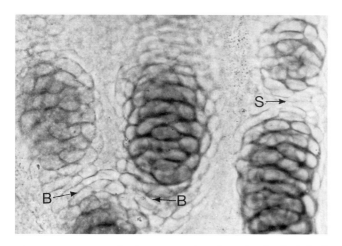

Fig. 17.15. Longitudinal cross section through the limb bud of a salamander *Ambystoma mexicanum*: note the branching bifurcation of cell condensation. At a later stage the branches of the *Y* separate from the main stem through cell recruitment at the ends, these are marked *B*. A segmental bifurcation can be seen starting at *S*. (Photograph courtesy of P. Alberch)

When we look at the cartilage patterns after the initial pattern has been laid down we see, as in Fig. 17.13 (c), that there is a gap between the bifurcations, for example, between the humerus and the radius and ulna. Whether or not there is a gap in cell condensation as the pattern bifurcates from a single to a double aggregation depends on the dispersion relation: these are the same possibilities we discussed regarding Fig. 14.14 (a), (b) in Chapter 14. It has been shown from experiment in the case of cartilage patterning in the developing limb for a large number of animals that the bifurcation is a clear branching process as seen in Fig. 17.15. In fact, with the mechanical model the bifurcation is continuous. The separation comes from subsequent recruitment of cells to form the observed gaps. This bifurcation patterning puts certain constraints on allowable dispersion relations which, in turn, imply certain constraints which any model mechanism must satisfy.

Morphogenetic Rules for Cartilage Morphogenesis in the Limb

With a completely symmetric geometry and tissue isotropy it is possible to move through the bifurcation space of parameters from one aggregation, to two, to three and so on. With reaction diffusion mechanisms, such as we considered in Chapter 14, it is possible to choose a path in some parameter space to achieve this: refer to Fig. 14.11 (b) for a specific example. It is also possible to do this with mechanical models. However, with the natural anisotropy in embryological tissue such isotropy does not exist. The question then arises as to how the pattern sequence from a double to a triple condensation is effected. We believe that for all practical purposes the process must be that in which one branch of the double condensation itself undergoes a branching bifurcation while near the other branch either a focal condensation appears or it undergoes a segmental bifurcation (see Fig. 17.16). Let us now note another experimentally observed fact, namely that during chondrogenesis there appears to be little cell division (Hinchliffe and Johnson 1980). This implies that condensations principally form through recruitment of cells from neighbouring tissue. Thus, as the limb bud grows the pattern bifurcation that takes place following a branching bifurcation is as illustrated in Fig. 17.16 (c). Fig. 17.16 (a), (b) show the other two basic condensation elements in setting up a cell condensation pattern in a developing limb.

If we now take the bifurcating pattern elements in Fig. 17.16 (a)-(c) as the three allowable types of cell condensations we can see how to construct any limb cartilage pattern by repeated use of the basic condensation elements in Fig. 17.16 (a)-(c). Fig. 17.16 (e), which is the forelimb of a salamander, is just one example to illustrate the process. So, even without considering any specific mechanism, we hypothesize an important *set of morphogenetic rules* for the patterning sequence of cartilage in the development of the vertebrate limb. This hypothesis, encapsulated in the theory put forward by Murray et al. (1983) and Oster et al. (1983), has recently been exploited by Oster et al. (1988) who present extensive experimental evidence for its validity. In the next chapter, where further examples are given of the application of these general rules, we shall see how it introduces developmental constraints and makes practical predictions, which are of considerable importance in evolutionary biology.

In the above discussion we had in mind a mechanical model for pattern formation in mesenchymal cells. The morphogenetic rules which we deduced equally apply to reaction diffusion models of pattern formation. In fact we believe they are model independent, or rather any model mechanism for chondrogenic pattern formation must be capable of generating such a sequence of bifurcating patterns. In that sense we cannot yet say which mechanism obtains.

Experimental evidence from amphibians suggests that osmotic properties of the ECM may be important in morphogenesis. Hyaluronate is a principle component of the ECM and can exist in a swollen osmotic state. As the condensation of chondrocytes start the cells secrete an enzyme, hyaluronidase, which degrades the hyaluronate. This could lead to the osmotic collapse of the matrix thus bringing the cells into close enough contact to initiate active contractions and thus generate cell aggregations. Cell motility is probably not important in this scenario. A modification of the mechanical model to incorporate these chemical aspects and the added forces caused by osmotic pressure has been proposed and analysed by Oster, Murray and Maini (1985). They showed that such a mechanochemical model would generate similar chondrogenic patterning for the developing limb.

A major role of theory in morphogenesis is to suggest possible experiments to distinguish different models each of which can generate the appropriate sequence of patterns observed in limb chondrogensis. Mechanical models lend themselves to experimental scrutiny more readily than reaction diffusion models because of the elusivenesss of chemical morphogens. In the next chapter we shall describe in detail certain predictions and experimental results based on them: they have important evolutionary implications regarding vertebrate limb development.

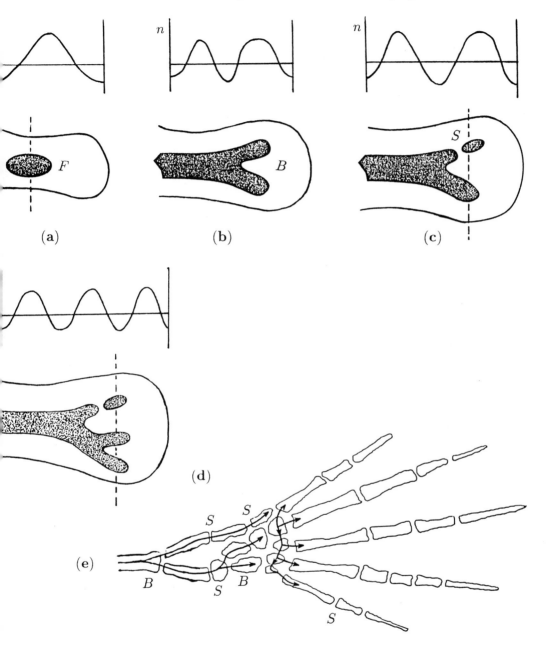

Fig. 17.16a-e. (a)-(c) The three basic types of cell condensations which generate cartilage patterns in the developing vertebrate limb. These are postulated as the morphogenetic rules for cartilage pattern generation for all vertebrate limbs. (a) Focal condensation, F; (b) Branching bifurcation, B; (c) Segmentation condensation, S. (d) Formation of more patterns is by further branching or independent foci. (e) An example of a branching sequence showing how the cartilage patterns in the limb of a salamander can be built up from a sequence of F, B and S bifurcations.

17.7 Mechanochemical Model for the Epidermis

The models we considered in earlier sections were concerned with internal tissue, the dermis and mesenchyme. The epithelium, an external tissue of epidermal cells, is another major tissue system in the early embryo which plays an important role in regulating embryogenesis. We briefly alluded to the interaction between the dermis and the epidermis in the formation of skin organ primordia. Many major organs rely on tissue interactions so their importance cannot be overemphasized: see, for example, Nagawa and Nakanishi (1988) who specifically investigate mechanical effects. As a necessary prerequisite in understanding dermal-epidermal tissue interactions we must thus have a model for the epidermis. Here we shall describe a model for the epithelium. The cells which make it up are quite different to the fibroblast cells we considered above. They do not actively migrate but are arranged in layers or sheets, which can bend and deform during embryogenesis: see Fig. 17.1. During these deformations the cells tend to maintain contact with their nearest neighbours and importantly, unlike dermal cells, they cannot generate traction forces under normal conditions. There are noted exceptions, one of which we shall briefly discuss later when we consider a model for wound healing.

Odell et al. (1981) modelled the epithelium as a sheet of discrete cells adhering to a basal lamina. Using a model for the contractile mechanism within the cell they showed how many morphogenetic movements of epithelial sheets could result from the mechanical interactions between the constituent cells. Basic to their model was the mechanochemistry of the cytogel, the interior of the cell, which provides an explanation for the contractile properties of the cells. The continuum model we describe in this section was proposed by Murray and Oster (1984a) and is based on the discrete model of Odell et al. (1981).

A different but related model for epithelial movement is proposed by Mittenthal and Mazo (1983). They model the epithelium as a fluid elastic shell which allows cell rearrangement and such spatial heterogeneity creates tensions which can alter the shell shape.

Some Biological Facts about Cytogel

The cell cytoplasm consists largely of a viscoelastic gel which is a network of macromolecular fibres mostly composed of actin linked by myosin crossbridges, the same elements involved in muscle contraction. This network has a number of complex responses and is a dynamic structure. The cell can contract actively by regulating the assembling and disassembling of the crosslinking of the fibres and carry out a variety of shape changes. When the fibres are strongly linked the cytoplasm tends to gel whereas when they are weakly linked it solates. Here we shall focus on just two key mechanical properties which subsume the complex process into a mechanochemical constitutive relation.

Chemical control of the cell's contractility, related to the sol-gel transition or degree of actomyosin crosslinking, is mainly due to the local concentration of free

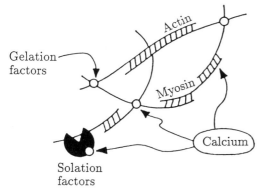

Fig. 17.17. Microspcopic description. The cytogel consists of actin and myosin which generate the traction forces. Solation and gelation enzymes control the connectivity of the gel and thus its viscosity and elasticity. Although there is a complex of chemicals involved, in the model, we consider free calcium to control the activity of the solation and gelation factors and hence the contractile apparatus. We model the macroscopic properties of the cytogel as a viscoelastic continuum with a viscosity, an elastic modulus and an active traction force τ.

calcium in the cytogel. Calcium regulates the activity of the solation and gelation factors and the contractile machinery. Fig. 17.17 gives a cartoon summary of the principal components of the contractile apparatus although we shall not deal with it at this level of detail.

At low concentration levels calcium encourages crosslinking in the gel and, on a sliding filament concept (as in muscles), more crossbridges become operative and this implies that the fibres tend to shorten and hence become stronger. It is similar to the increased strength of muscle in a contracted state and a stretched state – you can lift a much heavier object with your arm bent than with it outstretched. So as free calcium concentration goes up the gel first starts to contract actively. If the concentration gets too high, however, the gel becomes solated (the network begins to break apart) and cannot support any stress. There is thus a 'window' of calcium concentration which is optimal for contractile activity. Thus the concentration of free calcium and the mechanical forces associated with the cytogel must be key variables in our mechanical model.

Force Balance Equation for Cytogel Contractility

We model the epithelial sheet of cells as a viscoelastic continuum of cytogel. As with the model for mesenchymal cells, inertial forces are negligible, so the force balance mechanical equation can be taken to be

$$\nabla \cdot (\boldsymbol{\sigma}_V + \boldsymbol{\sigma}_E) + \rho \mathbf{F} = 0 , \qquad (17.63)$$

where $\mathbf{F}$ represents the external body forces per unit density of cytogel, ρ is the cytogel density, assumed constant, and the stress is the sum of a viscoelastic stress σ_V and an elastic stress σ_E given by (cf. (17.12))

$$\sigma_V = \mu_1 \varepsilon_t + \mu_2 \theta_t \mathbf{I} \,,$$

$$\sigma_E = \underbrace{E(1+\nu)^{-1}(\varepsilon + \nu'\theta\mathbf{I})}_{\substack{\text{elastic} \\ \text{stress}}} - \underbrace{\tau\mathbf{I}}_{\substack{\text{active} \\ \text{contraction} \\ \text{stress}}} \,, \qquad (17.64)$$

where τ is the contribution of the active cytogel traction to the elastic stress and ε, θ, $\mathbf{I}$, μ_1, μ_2, E, ν and ν' have the same meaning as in the cell-matrix equation (17.18) in the dermal model in Section 17.2. Here, however, the nonlinear dependence of the parameters on the dependent variables is different.

There is a relationship between the two models. In the mesenchymal model we considered the cell-matrix material as an elastic continuum in which were imbedded motile contractile units – the dermal cells. In the model here the elastic continuum cytogel also has contractile units – the actomyosin crossbridges. However, we do not need to account for any motion of these contractile units except for the deformation of the gel sheet itself. In our model for the epithelium the role of the cells is now played by the chemical trigger for contraction, namely the free calcium concentration which we denote by $c(\mathbf{r}, t)$. Thus we model the constitutive parameters μ_1, μ_2, E and τ in the stress tensor in (17.64) as follows. In fact, as we did in Section 17.2, we shall incorporate more generality than we shall subsequently study, not only for completeness but because the full model poses interesting and as yet unsolved mathematical problems.

(i) *Viscosity Parameters μ_1 and μ_2.* The severing of the gel network effected by the calcium results in a precipitous drop in the apparent viscosity and so we model $\mu_i(c)$, $i = 1,2$ by a typical sigmoidal curve as illustrated in Fig. 17.18 (a).

(ii) *Elasticity.* A characteristic property of the actomyosin fibrils is that as the amount of overlap of the actin fibres increases, so do the number of number of crossbridges, and so the fibre gets stronger as it contracts. Also, when a fibrous material is stretched the fibres tend to align in the direction of the stress and the effective elasticity increases. This means, as in the full dermal model, that the material is anisotropic. These are nonlinear effects which we can model by taking the elastic modulus E as a function of the dilation θ. Choosing $E(\theta)$ as a decreasing function of the dilation, as illustrated in Fig. 17.18 (b), is a reasonable form to start with.

(iii) *Active Traction.* The fibrous material of the cytogel starts to generate contractile forces once the actomyosin machinery is triggered to contract. The onset of contraction occurs when the free calcium is in the micromolar range. When the calcium level gets too high the fibrous material can no longer exert any contractile stress. We thus model the active stress contribution $\tau(c)$ as a function of calcium as illustrated in Fig. 17.18 (c).

(iv) *Body Force.* Movement of the epithelial layer is inhibited by its attachment to the basal lamina, which separates it from the mesenchyme, by restraining tethers,

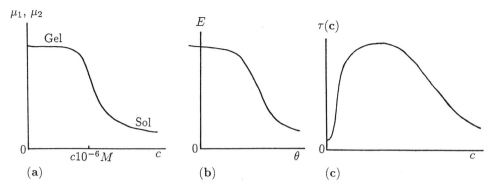

Fig. 17.18a-c. Typical nonlinear dependence on the dependent variables of the constitutive parameters in the stress tensor in (17.64). In (a) and (b) the precipitous drop in the viscosities and actomyosin network elastic modulus occurs when the gel solates. (c) Dependence of actomyosin traction as a function of c.

equivalent to the 'guy lines' in the dermal model. We assume this restraining force per unit cytogel density, $\mathbf{F} = s\mathbf{u}$, where s is a factor reflecting the strength of the attachments and $\mathbf{u}(\mathbf{r}, t)$ is the displacement of a material point of the cytogel. The form is similar to that used in the dermal model.

We incorporate these effects into the stress tensor (17.64) and the force balance equation (17.63) for the cytogel which takes the form

$$\nabla \cdot \left\{ \mu_1 \varepsilon_t + \mu_2 \theta_t \mathbf{I} + E(1 + \nu)^{-1}[\varepsilon + \nu'\theta \mathbf{I} - \tau(c)\mathbf{I}] \right\} - s\rho\mathbf{u} = 0 \ . \qquad (17.65)$$

For algebraic simplicity, or rather less complexity, in the analysis that follows, we shall take the viscosities μ_i, $i = 1, 2$ and E to be constant. An analysis including variable viscosities and elastic modulus would be interesting, particularly on the wave propagation aspects of the model which we discuss later.

Conservation Equation for Calcium

We must first describe some of the chemical aspects of the cytogel. Calcium is sequestered in membranous vesicles dispersed throughout the cytogel. It is released from the vesicles by an autocatalytic process known as calcium-stimulated calcium release (CSCR). This means that if the free calcium outside the vesicles exceeds a certain threshold value it causes the vesicles to release their store of calcium (it is like the toilet flush principle). We can model this aspect by a threshold kinetics. If we assume the resequestration of calcium is governed by first order kinetics, we can combine the processes in a kinetics function $R(c)$ where

$$R(c) = \frac{\alpha c^2}{1 + \beta c^2} - \delta c \ , \qquad (17.66)$$

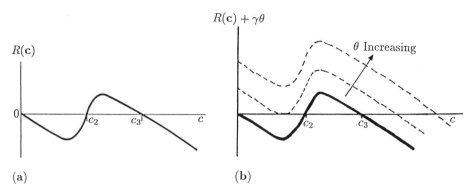

Fig. 17.19a,b. (a) Qualitative form of the calcium-stimulated calcium release kinetics function $R(c)$. A release of calcium can be triggered by an increase of calcium: this is a switch from the zero steady state to $c = c_3$. (b) A strain-induced calcium release, that is stretch activation, based on the kinetics function $R(c) + \gamma\theta$ in (17.67), if $\gamma\theta$ exceeds a certain threshold where θ is the dilation.

where α, β and δ are positive constants. The form of $R(c)$ is typically S-shaped as shown in Fig. 17.19 (a): if $4\beta\delta^2 < \alpha^2$ there are two linearly stable steady states at $c = 0$ and $c = c_3$ and an unstable steady state at $c = c_2$.

The release of calcium can also be triggered by straining the cytogel, a phenomenon known as 'stretch activation'. We can model this by including in the kinetics $R(c)$ a term $\gamma\theta$ where γ is the release per unit strain and θ is the dilation. Fig. 17.19 (b) shows the effect of such a term and how it can trigger calcium release if it exceeds a certain threshold strain. (Certain insect flight muscles exhibit this phenomenon in that stretching induces a contraction by triggering a local calcium release.)

Calcium, of course, also diffuses so we arrive at a model conservation equation for calcium given by

$$\frac{\partial c}{\partial t} = D\nabla^2 c + R(c) + \gamma\theta$$

$$= D\nabla^2 c + \frac{\alpha c^2}{1 + \beta c^2} - \delta c + \gamma\theta \ . \tag{17.67}$$

where D is the diffusion coefficient of the calcium. We have already discussed this equation in detail in Chapter 15, Section 15.2 (cf. also Chapter 5, Exercise 3) and have shown it gives rise to excitable kinetics. We should emphasize here that the kinetics in (17.67) is simply a model which captures the qualitative features of the calcium kinetics. The biochemical details of the process are not yet completely understood.

The mechanochemical model for the cytogel consists of the mechanical equilibrium equation (17.65), and the calcium conservation equation (17.67). They are coupled through the calcium induced traction term $\tau(c)$ in (17.65) and the strain-activation term $\gamma\theta$ in (17.67). In the subsequent analysis we shall take $E(\theta)$, the viscosities μ_i, $i = 1, 2$ and the density ρ to be constants.

We nondimensionalize the equations by setting

$$r^* = \frac{r}{L}, \quad t^* = \delta t, \quad c^* = \frac{c}{c_3}, \quad \mathbf{u}^* = \frac{\mathbf{u}}{L},$$

$$\theta^* = \theta, \quad s^* = \frac{\rho s L^2(1+\nu)}{E}, \quad \mu_i^* = \frac{\mu_i \delta(1+\nu)}{E}, \quad (i = 1, 2),$$

$$\varepsilon^* = \varepsilon, \quad \tau^*(c^*) = \frac{(1+\nu)\tau(c)}{E}, \quad R^*(c^*) = \frac{R(c)}{\delta c_3},$$

$$\alpha^* = \frac{\alpha c_3}{\delta}, \quad \beta^* = \beta c_3^2, \quad \gamma^* = \frac{\gamma}{\delta c_3}, \quad D^* = \frac{D}{L^2 \delta},$$

(17.68)

where L is some appropriate characteristic length scale and c_3 is the largest zero of $R(c)$ as in Fig. 17.19 (b). Substituting these into (17.65) and (17.67) and omitting the asterisks for notational simplicity, we have the dimensionless equations for the cytogel continuum as

$$\nabla \cdot \{\mu_1 \varepsilon_t + \mu_2 \theta_t \mathbf{I} + \varepsilon + \nu'\theta\mathbf{I} - \tau(c)\mathbf{I}\} = s\mathbf{u},$$

$$\frac{\partial c}{\partial t} = D\nabla^2 c + \frac{\alpha c^2}{1 + \beta c^2} - c + \gamma\theta = D\nabla^2 c + R(c) + \gamma\theta.$$

(17.69)

The boundary conditions depend on the biological problem we are considering. These are typically zero flux conditions for the calcium and periodic or stress-free conditions for the mechanical equation.

Linear Stability Analysis and Spatial Pattern Generation

The linear stability of the homogeneous steady state solutions of (17.69) can be carried out in the usual way and is left as an exercise: here we only give the main results. The homogeneous steady states are

$$\mathbf{u} = \theta = 0, \quad c = c_i, \quad i = 1, 2, 3,$$

(17.70)

where c_i are the zeros of $R(c)$. The dispersion relation $\sigma = \sigma(k^2)$, where here σ (not to be confused with the stress tensor) is the exponential temporal growth factor and k is the wavenumber of the linear perturbation, is given by

$$\mu k^2 \sigma^2 + b(k^2)\sigma + d(k^2) = 0,$$

$$b(k^2) = \mu D k^4 + (1 + \nu' - \mu R_i')k^2 + s,$$

$$d(k^2) = D(1+\nu')k^4 + [sD - \gamma\tau_i' - (1+\nu')R_i']k^2 - sR_i',$$

(17.71)

where

$$\mu = \mu_1 + \mu_2, \quad R_i' = R'(c_i), \quad \tau_i' = \tau'(c_i),$$

(17.72)

and prime denotes differentiation with respect to c (except that on ν').

Let us assume that the precipitous drop in the traction $\tau(c)$ occurs at $c \geq c_2$, then, from Fig. 17.18 (c) and Fig. 17.19 (b), we have

$$\tau_1' > 0, \quad \tau_i' < 0, \quad i = 2, 3; \quad R_2' > 0, \quad R_i' < 0, \quad i = 1, 3 . \tag{17.73}$$

For the steady state $c = c_1 = \mathbf{u} = 0$, $b(k^2) > 0$ and $d(k^2) > 0$, unless γ is large, for all k^2 in which case $\mathrm{Re}\,\sigma(k^2) < 0$ and this steady state is stable. The steady state $c = c_2$, $\mathbf{u} = 0$ is unstable for $k^2 = 0$ and so is generally unstable. The steady state $c = c_3$, $\mathbf{u} = 0$ has $b(k^2) > 0$ for all k^2 but it is possible for $d(k^2)$ to become negative for $k^2 > 0$ if γ is sufficiently large. This situation gives rise to the simplest dispersion relation like that in Fig. 17.8 (a) whereby a finite band of wavenumbers spatial perturbations are unstable, one of which has a fastest linear growth in the usual way.

With tissue interactions in mind, we can see how this model can be modified to incorporate a mechanical influence from the mechanical dermal model. This would appear as a dermal input in the calcium equation proportional to θ_D where θ_D is a dilation contribution from the dermis. If the strain activation parameter γ is not sufficiently large for the steady state $c = c_3$, $\mathbf{u} = 0$ to be unstable for $k^2 > 0$, the effect of the dermal input θ_D could initiate an epidermal instability since it enhances the $\gamma\theta$ term in the calcium equation, the second of (17.69). Inclusion of such a term also changes the possible steady states since these are now given by

$$R(c) + \theta_D = 0, \quad \mathbf{u} = 0 .$$

The qualitative effect of θ_D, when it is constant for example, can be easily seen from Fig. 17.19 (b): a threshold effect for large enough θ_D is evident.

It is clear how tissue interaction is a natural consequence in these mechanical models. A nonuniform dermal cell distribution can trigger the epithelial sheet to form placodes. This scenario would indicate that the papillae preceed the placodes. On the other hand the epidermal model can also generate spatial patterns on its own and in turn affect the dermal mechanism by a transferred strain and hence effect dermal patterns. Also the epidermal model could be triggered to disrupt its uniform state by an influx of calcium (possibly from a reaction diffusion system). So, at this stage we can draw no conclusions as to the order of appearance of placodes and papillae solely from the study of these models without further experimental input. However, it is an attractive feature of the models that tissue interaction between dermis and epidermis can be so naturally incorporated.

17.8 Travelling Wave Solutions of the Cytogel Model

One of the interesting features of the model of Odell et al. (1981) was its ability to propogate contraction waves in the epithelium. Intuitively we expect the continuum model here to exhibit similar behaviour. The appearance of contraction waves is a common phenomenon during embryogenesis: recall the discussion in Section 11.6 on post-fertilization waves on eggs. In this section we investigate such travelling waves in a one-dimensional context. This problem can be solved numerically but to determine the solution behaviour as a function of the parameters we consider a simplified form of the equations, which retains the key qualitative behaviour but for which we can obtain analytical solutions.

The one-dimensional version of (17.69) is

$$\mu u_{xxt} + u_{xx} - \tau'(c)c_x - su = 0 \; ,$$
$$c_t - Dc_{xx} - R(c) - \gamma u_x = 0 \; ,$$
(17.74)

where $\mu = \mu_1 + \mu_2$ and where we have incorporated $(1 + \nu')$ into the redefined μ and s. We look for travelling wave solutions (cf. Chapter 11, Section 11.1) of the form

$$u(x,t) = U(z), \quad c(x,t) = C(z); \quad z = x + Vt$$
(17.75)

where V is the wave speed: with $V > 0$ these represent waves travelling to the left. Substitution into (17.74) gives

$$\mu V U''' + U'' - \tau'(C)C' - sU = 0 \; ,$$
$$V C' - DC'' - R(C) - \gamma U' = 0 \; ,$$
(17.76)

where $R(C)$ and $\tau(C)$ are qualitatively as shown in Fig. 17.19 (a) and Fig. 17.18 (c) respectively. This system gives a fourth order phase space the solutions of which would have to be found numerically for given $R(C)$ and $\tau(C)$.

To proceed analytically we use a technique proposed by Rinzel and Keller (1973) for waves in the Fitzhugh-Nagumo system (cf. Chapter 12, Section 12.4) which involves a piece-wise linearisation of the nonlinear functions $R(C)$ and $\tau(C)$. This method retains the key qualitative features of the full nonlinear problem but reduces it to a linear one with a different linear system for different ranges in z. The procedure involves patching the solutions together at the region boundaries.

Here we shall be interested in a travelling wave which joins the low calcium state $c = 0$ to the high one $c = c_3$; that is we wish to consider a wave which triggers calcium release. Let us further assume that during the transition the active traction drops from its high value to zero as the concentration of calcium passes through a threshold value c_T, in other words we are considering 'solation' waves. By a suitable rescaling we can normalize the traction so that $\tau(C)$ goes from $\tau = 1$ to $\tau = 0$: we also take $c_T = c_2$, the unstable steady state of $R(c)$. We could retain a distinct and separate c_T but this unecessarily complicates the analysis. Lane, Manoranjan and Murray (1987) investigated the effect of separating c_T from c_2 and showed that it was not of qualitative significance. Again for simplicity we normalize the steady state $c = c_3 = 1$. We now introduce the piecewise linear functions for $\tau(C)$ and $R(C)$ in (17.76) by:

$$\tau(C) = 1 - H(C - c_2) ,$$
$$R(C) = -C + H(C - c_2) ,$$

(17.77)

where $H(C - c_2)$ is the Heaviside function; $H = 0$ if $C < c_2$ and $H = 1$ if $C > c_2$. These functional forms, illustrated in Fig. 17.20, preserve the major qualitative feature of the nonlinear forms of $R(c)$ and the 'solation' end of the traction force $\tau(c)$. Strictly we should choose $\tau(C)$ as a 'top hat', with

$$\tau(C) = H(C) - H(C - c_2) .$$

(17.78)

The following analytical proceedure would still go through but the resulting algebraic complexity would obscure the main points of the technique.

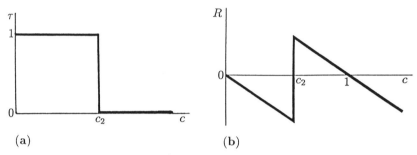

(a)   (b)

Fig. 17.20a,b. (a) Piecewise linear form of the traction force $\tau(c)$. This reflects the drop in τ for high calcium levels, that is the part of the curve in Fig. 17.18 (c) where τ drops from its maximum to low values for large c. (b) Piecewise linear form for $R(c)$: compare this with Fig. 17.19 (a).

With (17.77) the nonlinear system (17.76) is reduced to a set of linear equations. We look for wave solutions satisfying

$$U(-\infty) = C(-\infty) = 0, \quad U(\infty) = 0, \quad C(\infty) = 1 .$$

(17.79)

With $V > 0$ in (17.75) these boundary conditions imply we have a wave*front* solution in $C(z)$, which increases as the wave passes, accompanied by a wave *pulse* in $U(z)$. We determine the analytical solution by patching together the solutions of the linear equations which apply in the intervals $[-\infty, 0]$ and $[0, \infty]$ of the z-axis. Since we are investigating travelling waves, we can take $C = c_2$ to be at $z = 0$ without loss of generality. We use this condition below. With (17.77) the system (17.76) reduces to

$$\mu V U''' + U'' + \delta(C - c_2)C' - sU = 0 ,$$
$$V C' - DC'' + C - H(C - c_2) - \gamma U' = 0 ,$$

(17.80)

where $\delta(\cdot)$, the derivative of the Heaviside function, is the Dirac delta function.

Note that if a solution of (17.80) is given by

$$(C, U, c_2, \mathbf{p}, V) \quad \text{then} \quad (1 - C, -U, 1 - c_2, \mathbf{p}, V) \tag{17.81}$$

is also a solution, where $\mathbf{p} = (\mu, s, D, \gamma)$. This means that if we solve for a wavefront solution in C and consider $V > 0$, we immediately obtain, because of (17.81), a waveback solution in C. A waveback solution is where C decreases as it passes a given point.

System (17.80) in the z-interval $z \leq 0$ becomes

$$\mu V U''' + U'' - sU = 0, \quad z \neq 0$$
$$VC' - DC'' + C - \gamma U' = 0, \quad z < 0 \tag{17.82}$$

while in the interval $z \geq 0$

$$\mu V U''' + U'' - sU = 0, \quad z \neq 0$$
$$VC' - DC'' + C - 1 - \gamma U' = 0. \quad z > 0 \tag{17.83}$$

Recall the definition of the δ-function in Chapter 9: this is why it does not appear in (17.82) and (17.83) when we exclude $z = 0$. These have to be solved with boundary conditions (17.79), which are appropriate for a pulse in U and a front connecting $C = 0$ to $C = 1$. From (17.82) and (17.83) we can also specify continuity for U, C and U' at $z = 0$. Two other conditions arise from integration of the equations across $z = 0$; the first of (17.80) gives a discontinuity in U'' at $z = 0$, while the second leads to continuity in C' at $z = 0$. To see this we first introduce the following notation for the solution parts:

$$U_-(z), \ C_-(z) \ \text{for} \ z \leq 0; \quad U_+(z), \ C_+(z) \ \text{for} \ z \geq 0 . \tag{17.84}$$

Integrating (17.80) across $z = 0$, that is from $z = 0-$ to $z = 0+$, gives

$$\left[\mu V U'' \right]_{0-}^{0+} + \left[U' \right]_{0-}^{0+} + \int_{0-}^{0+} \delta(C - c_2) C' \, dz - s \int_{0-}^{0+} U \, dz = 0 ,$$

$$\left[VC \right]_{0-}^{0+} - \left[DC' \right]_{0-}^{0+} + \int_{0-}^{0+} C \, dz - \int_{0-}^{0+} H(C - c_2) \, dz - \gamma \left[U \right]_{0-}^{0+} = 0 .$$

Because of continuity of C, U and U' the latter shows that C' is also continuous at $z = 0$ while the former gives, in the notation of (17.84),

$$[U_+'{}'(0) - U_-'{}'(0)] = -\frac{1}{\mu V} \tag{17.85}$$

since

$$\int \delta(C - c_2) C' \, dz = \int \delta(C - c_2) \, dC = 1 .$$

Thus the boundary conditions for (17.82) and (17.83) in terms of $U_\pm(z)$, $C_\pm(z)$ are

$$\left.\begin{array}{c} U_-(z) \to 0 \\ C_-(z) \to 0 \end{array}\right\} z \to -\infty \qquad \left.\begin{array}{c} U_+(z) \to 0 \\ C_+(z) \to 1 \end{array}\right\} z \to +\infty$$

$$U_-(0) = U_+(0); \quad U''_+(0) - U''_-(0) = \frac{1}{\mu V}; \quad U'_-(0) = U'_+(0) \qquad (17.86)$$

$$C'_-(0) = C'_+(0); \quad C_-(0) = C_+ = (0) \ .$$

With these conditions the linear equations (17.82) and (17.83) can be solved. Note that use of a Heaviside function for τ results in the U-equation being decoupled from the C-equation.

We look for solutions of (17.82) and (17.83) proportional to $\exp[\lambda z]$. The auxiliary equation for the U-equation is

$$V\mu\lambda^3 + \lambda^2 - s = 0 \ , \qquad (17.87)$$

which has one positive real root and two roots with negative real parts, as can be seen by plotting the left hand side as a function of λ. The analytical solution for $U(z)$ is given as the sum of exponentials, $\exp[\lambda_i z]$ where λ_i, $i = 1, 2, 3$ are the three solutions of (17.87), and can be one of two cases, either real or complex, depending on whether (17.87) has three real or one real and two complex solutions. This occurs for $V < V_{\mathrm{crit}}$ and $V > V_{\mathrm{crit}}$ respectively, where

$$V_{\mathrm{crit}} = \frac{2}{3\mu\sqrt{3s}} \ . \qquad (17.88)$$

We get this by differentiating (17.87) as a function of λ and finding the condition for the local maximum to intersect the λ-axis. The two cases for U correspondingly produce two cases for C which are obtained on substituting U in the C-equations, the second of (17.82) and (17.83). We must now consider the two cases where the wave velocity $V < V_{\mathrm{crit}}$ and $V > V_{\mathrm{crit}}$.

Solutions for $V < V_{\mathrm{crit}}$. Here there is one positive root, λ_1, and two negative roots, λ_2 and λ_3, of (17.87) and so, using the conditions (17.86) as $z \to \pm\infty$, the U-pulse is of the form

$$\begin{aligned} U_-(z) &= A_1 \exp[\lambda_1 z], \quad z \le 0 \\ U_+(z) &= A_2 \exp[\lambda_2 z] + A_3 \exp[\lambda_3 z], \quad z \ge 0 \end{aligned} \qquad (17.89)$$

where A_i, $i = 1, 2, 3$ are arbitrary constants.

The solution $C(z)$ for the calcium wavefront is now obtained by solving the second of each of (17.82) and (17.83) after substituting for $U(z)$ from (17.89). After a little algebra we get

$$\begin{aligned} C_-(z) &= B_2 \exp[\eta_2 z] + F_1 \exp[\lambda_1 z], \quad z \le 0 \\ C_+(z) &= 1 + B_1 \exp[\eta_1 z] + F_2 \exp[\lambda_2 z] + F_3 \exp[\lambda_3 z], \quad z \ge 0 \end{aligned} \qquad (17.90)$$

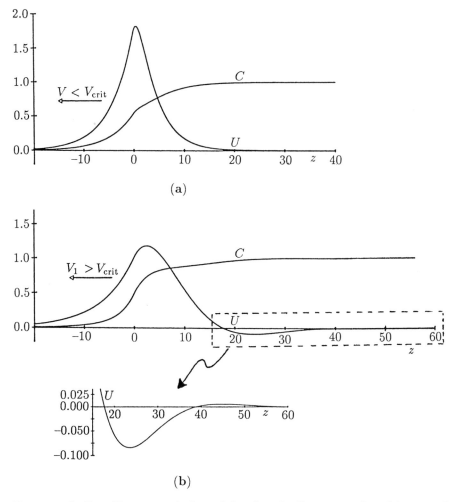

Fig. 17.21a,b. Travelling wave solutions of the piecewise linear cytogel model system (17.80) computed from the analytical solutions. Parameter values $\mu = 10.0$, $s = 0.058$, $D = 9.0$, $\gamma = 1.0$, which imply the critical velocity $V_{\text{crit}} = 0.16$. **(a)** Here the uniform steady state $c_2 = 0.537$ (cf. Fig. 17.20 (b)) which gives $V = 0.1 < V_{\text{crit}}$. **(b)** $c_2 = 0.488$, giving $V = 1.0 > V_{\text{crit}}$. Note the different form of the deformation U-pulse; there is now a compression region and a tension region. (After Lane, Murray and Manoranjan 1987)

where B_i, $i = 1, 2$ are arbitrary constants, F_i, $i = 1, 2, 3$ are functions of the A_i in (17.89), and η_i, $i = 1, 2$ are the roots of the auxiliary polynomial for the $C(z)$ equation, given by

$$\eta_1 = \frac{V - (V^2 + 4D)^{1/2}}{2D}, \quad \eta_2 = \frac{V + (V^2 + 4D)^{1/2}}{2D}.$$

On using the continuity and jump conditions (17.86), the arbitrary constants A_i and B_i, and hence F_i, are now determined in terms of the model parameters $\mathbf{p} = (\mu, s, D, \gamma)$ and the, as yet, undetermined velocity V. At this stage we use the condition we assumed above, namely that at $z = 0$, $C(0) = c_2$. After some tedious algebra this gives a consistency relation

$$C(0) = F(\mathbf{p}, V) = c_2 , \tag{17.91}$$

where F is a known, but algebraically complicated, function. This consistency function provides the relationship which determines the speed V of the solutions. The final determination of the speed depends on the form of this function. It was investigated in detail by Lane et al. (1987) who showed that the solutions are quite sensitive to variations in c_2.

In principle this completely determines the pulse solution $U(z)$ and the wavefront solution $C(z)$. Fig. 17.21 (a) illustrates the solutions for a given set of model parameters.

Solutions for $V > V_{\mathrm{crit}}$. In this case the U-pulse is not monotonic, because the roots of the auxiliary equation (17.87) are complex, and it passes through positive and negative values for $z > 0$. In the same way as we obtained the solution when $V < V_{\mathrm{crit}}$ we find the solution here is

$$
\begin{aligned}
U_-(z) &= A_1 \exp\left[\lambda_1 z\right] , \\
U_+(z) &= \left[A_2 \cos bz + A_3 \sin bz\right] \exp\left[az\right] ,
\end{aligned}
\tag{17.92}
$$

where λ_1 and $a \pm ib$, $a < 0$, are the roots of (17.87) and the A_i, $i = 1, 2, 3$ have to be determined. The C_+ solution, for example, is obtained as

$$C_+(z) = 1 + B_1 \exp\left[\eta_1 z\right]) + \left[J \cos bz + K \sin bz\right] \exp\left[az\right] , \tag{17.93}$$

where η_1 is the appropriate solution of the auxiliary equation of the C-equation and J and K are constants. Fig. 17.21 (b) is a computed example of the solutions where $V > V_{\mathrm{crit}}$; we again had to use the consistency relation (17.91).

Depending on the form of the consistency function quite complex wave solution behaviour can be exhibited by this relatively simple one-dimensional cytogel model. Lane et al. (1987) present further details based on a numerical investigation of $F(\mathbf{p}, V)$. They also investigated the stability properties of the solutions. One highly relevant aspect of their study was to show that the piecewise linear caricature was a good predictor of the behaviour of the solutions to the full nonlinear model.

Let us briefly consider the dimensional wave speed using (17.68). It is given by

$$V_{\mathrm{dimensional}} = \delta L V .$$

If, for example, we choose the length L to be associated with the basal lamina attachment, that is $s^* = 1$ in (17.68), then

$$L^2 = \frac{E}{\rho s(1 + \nu)} \quad \Rightarrow \quad V_{\text{dimensional}} = \delta V \left[\frac{E}{\rho s(1 + \nu)} \right]^{1/2} .$$

Thus we see that the stronger the basal attachment the slower is the propagation speed, as might be expected. If we look at the phenomenon on a diffusion length scale this gives another qualitative behaviour, now as a function of the dimensional diffusion coefficient D

$$L = \left(\frac{D}{\delta} \right)^{1/2} \quad \Rightarrow \quad V_{\text{dimensional}} = (\delta D)^{1/2} V .$$

Thus the propagation speed increases with diffusion. The qualitative dependence of the propagation speed with variation in the other parameters is not always so clear since they appear in more than one dimensionless grouping and hence in the evaluation of V.

We should briefly note here that we have investigated only travelling wave *fronts*. As we said above, the analysis can be used to consider wave *back* solutions. We can thus see how to construct, in principle, wave pulse solutions: it is an algebraically formidable problem.

One of the interesting applications of this model was to the post-fertilization waves on vertebrate eggs which we discussed in some detail in Chapter 11, Section 11.6. There, we were specifically concerned with the calcium waves that swept over the egg. We noted that these were accompanied by deformation waves and alluded to the work in this chapter. Lane et al. (1987) took the cytogel model described here in a spherical geometry, representing the egg's surface, and investigated surface waves with post-fertilization waves in mind: see Fig. 11.10. It is informative to re-read Section 11.6 in the light of the full mechanochemical cytogel model and the analysis we have presented in this section.

17.9 Formation of Microvilli

Micrographs of the cellular surface frequently show populations of microvilli, which are foldings on the cell membrane, arrayed in a regular hexagonal packing as shown in the photograph in Fig. 17.22. Oster, Murray and Odell (1985) have proposed a modification of the model in Section 17.7 to explain these patterns and it is their model we now discuss.

The model is based on the following sequence of events. First, the cytogel is triggered to contract, probably by an increase in the level of calcium, and as it does so spatial patterns are formed as a result of the instability of the uniform steady state. They proposed that the hexagonal patterns observed are essentially the hexagonal periodic solutions for the tension patterns which are generated.

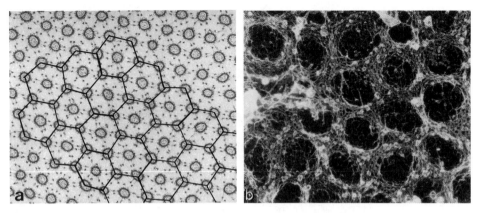

Fig. 17.22a,b. (a) Micrograph of the hexagonal array on a cellular surface after the microvilli (foldings) have been sheared off: The photograph has been marked to highlight the hexagonal array. (Photograph courtesy of A.J. Hudspeth) (b) View of a field of microvilli from the cytoplasmic side, the inside of the cell. Note the bands of aligned actin fibres enclosing the regions of sparse actin density. (Micrograph courtesy of D. Begg)

The pattern established in this way creates arrays of lacunae, or spaces, which are less dense in actomyosin. At this stage, osmotic pressure expands these regions outward to initiate the microvilli. One of the new elements in the model is the inclusion of an osmotic pressure.

The biological assumptions underlying the model are: (i) The subcortical region beneath the apical membrane of a cell consists largely of an actin-dense gel. (ii) The gel can contract by a sliding filament mechanism involving myosin cross-bridges linking the actin fibers as in the above model for the epithelium: see Fig. 17.17. We shall show that these assumptions are sufficient to ensure that an actin sheet will not necessarily remain spatially homogeneous, but can form a periodic array of actin fibres. This arrangement of actin fibres could be the framework for the extrusion of the microvilli by osmotic forces.

We should interject here, that Nagawa and Nakanishi (1988) comment that dermal cells have the highest gel-contraction activity.

Here we briefly describe the model and give only the one-dimensional analysis. The extension to more dimensions is the same as in the models in earlier sections. We consider a mechanochemical model for the cytogel which now involves the sol-gel and calcium kinetics, all of which satisfy conservation equations, and an equation for the mechanical equilibrium of the various forces which are acting on the gel. The most important difference with this model, however, is the extra force from the osmotic pressure; this is fairly ubiquitous during development.

We consider the cytogel to consist of a viscoelastic continuum involving two components, the sol, $S(x,t)$, and the gel, $G(x,t)$, whose state is regulated by calcium, $c(x,t)$. There is a reversible transition from gel to sol. The actomyosin gel is made up of crosslinked fibrous components while the sol is made up of the non-crosslinked fibres. The state of cytogel is specified by the sol, gel and calcium concentration distribution plus the mechanical state of strain, $\varepsilon(= u_x(x,t))$, of the gel.

We consider the sol, gel and calcium to diffuse, although the diffusion coefficient of the gel is very much smaller, because of its crosslinks, than that of the sol and calcium. We also reasonably assume there is a convective flux contribution (recall Section 17.2) to the conservation equations for the gel and sol.

With the experience gained from the study of the previous mechanochemical models the conservation equations for S, G and c are taken to be

$$\text{Sol:} \qquad S_t + (Su_t)_x = D_S S_{xx} - F(S,G,\varepsilon) \,, \qquad (17.94)$$

$$\text{Gel:} \qquad G_t + (Gu_t)_x = D_G G_{xx} + F(S,G,\varepsilon) \,, \qquad (17.95)$$

$$\text{Calcium:} \quad c_t = D_c c_{xx} + R(c,\varepsilon) \,, \qquad (17.96)$$

where the D's are diffusion coefficients and $R(c,\varepsilon)$ is qualitatively similar to the kinetics $R(c)$ in (17.66) plus the strain activation term (cf. Fig. 17.19). The specific form of the kinetics terms in (17.94) and (17.95) reflects the conservation of the sol-gel system; for example a loss in sol is directly compensated by a gain in gel. These two equations can be collapsed into one differential equation and one algebraic, namely $G + S = $ constant. The function $F(S,G,\varepsilon)$ incorporates the details and strain dependence of the sol-gel reaction kinetics which we take to be

$$F(S,G,\varepsilon) = k_+(\varepsilon)S - k_-(\varepsilon)G \,. \qquad (17.97)$$

The schematic forms of the rates, $k_+(\varepsilon)$ and $k_-(\varepsilon)$, are shown in Fig. 17.23 (a). These forms are consistent with the sol-gel behaviour in which if the gel is dilated, that is ε increases, its density decreases and the mass action rate of gelation increases. Conversely when the gel contracts the gel density goes up and the solation rate increases. Thus the gel concentration may increase with increasing strain: Fig. 17.23 (b) illustrates the sol-gel equilibrium states.

In the mechanical force balance the osmotic pressure is a major contributory force in gel and is a decreasing function of strain. The definition of strain implies $\varepsilon \geq -1$. (For example, if we consider a strip of gel of unstretched length L_0 and disturbed length L, the strain is $(L - L_0)/L_0$: the absolute minimum of L is zero so the minimum strain is -1.) As $\varepsilon \to -1$ the osmotic pressure becomes infinitely large (we cannot squeeze a finite amount of gel into no space), while for large strains the osmotic effects are small. We thus model the osmotic forces qualitatively by a stress tensor contribution

$$\sigma_0 = \frac{\pi}{1 + \varepsilon} \,,$$

where π is a positive parameter.

The elastic forces not only involve the classical linear stress strain law but also, because of the long strand-like character of the gel, they involve long range effects (cf. Section 17.2). We model this by a stress tensor

$$\sigma_E = GE(\varepsilon - \beta\varepsilon_{xx}) \,,$$

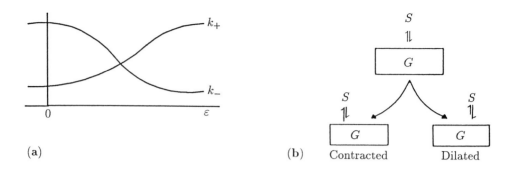

Fig. 17.23a,b. (a) Qualitative form of the gelation rate $k_+(\varepsilon)$ and solation rate $k_-(\varepsilon)$. (b) In the unstressed state the gel (G) is in chemical equilibrium with the sol (S). When the gel contracts its density goes up and the chemical equilibrium shifts towards the sol phase, whereas when the gel is dilated the gel density decreases so that the equilibrium is shifted towards the gel phase. Thus the equilibrium gel fraction is an increasing function of strain. The size of the arrows indicate the relative rate changes under contraction and dilation.

where E is the elastic modulus and $\beta > 0$ is a measure of the long range elastic effect: recall the discussion of long range effects in Section 17.2. The inclusion of G is because the elastic force acts through the fibres of the gel and so the more gel the stronger the force. The elastic forces oppose the osmotic pressure.

The contribution from the active contraction of the gel depends on the calcium concentration and the strain. We model this by

$$\sigma_A = \frac{G\tau(c)}{1 + \varepsilon^2} ,$$

where $\tau(c)$ is a measure of the active traction strength which depends on the calcium concentration c, increasing with increasing c for at least low values of c. The ε-dependence is suggested by the fact that the contractile force is smaller if the gel is dilated, that is larger ε. Again we require the multiplicative G since traction also acts through the gel fibres.

Finally the gel has an effective viscosity, which results in a viscous force

$$\sigma_V = G\mu\varepsilon_t ,$$

where μ is the viscosity. This force again acts through the gel fibres.

All of the forces are in equilibrium and so we have the continuum mechanical force balance equation

$$\sigma_x = [\sigma_0 + \sigma_E + \sigma_A + \sigma_V]_x = 0 ,$$

$$\sigma = \underbrace{\frac{\pi}{1 + \varepsilon}}_{\text{osmotic}} - \underbrace{GE(\varepsilon - \beta\varepsilon_{xx})}_{\text{elastic}} - \underbrace{\frac{G\tau(c)}{1 + \varepsilon^2}}_{\substack{\text{active} \\ \text{stress}}} - \underbrace{G\mu\varepsilon_t}_{\text{viscous}} . \tag{17.98}$$

Here there are no external body forces restraining the gel. Note that only the osmotic pressure tends to dilate the gel.

Equations (17.94)–(17.96), (17.98) together with constitutive relations for $F(S, G, \varepsilon)$ and $R(c, \varepsilon)$ along with appropriate boundary and initial conditions, constitute this mechanochemical model for the cytogel sheet.

Simplified Model System

This system of nonlinear partial differential equations can be analyzed on a linear basis, similar to what we have now done many times, to demonstrate the pattern formation potential. This is algebraically quite messy and unduly complicated if we simply wish to demonstrate the powerful pattern formation capabilities of the mechanism. The full system also poses a considerable numerical simulation challenge. So, to highlight the model's potential we consider a simplified model which retains the major physical features.

Suppose that the diffusion time scale of calcium is very much faster than that of the gel and of the gel's viscous response (that is $D_c \gg D_G, \mu$). Then, from (17.96) we can take c to be constant (because D_c is relatively large). Thus c now appears only as a parameter and we replace $\tau(c)$ by τ. If we assume the diffusion coefficients of the sol and gel are the same, (17.94) and (17.95) imply $S + G = S_0$, a constant and so $S = S_0 - G$.

If we integrate the force balance equation (17.98) we get,

$$G \mu \varepsilon_t = GE\beta \varepsilon_{xx} + H(G, \varepsilon) , \tag{17.99}$$

where

$$H(G, \varepsilon) = \frac{\pi}{1 + \varepsilon} - \frac{G\tau}{1 + \varepsilon^2} - GE\varepsilon - \sigma_0 , \tag{17.100}$$

where the constant stress σ_0 is negative, in keeping with the convention we used in (17.98).

Equation (17.99) is in the form of a 'reaction diffusion' equation where the strain, ε, plays the role of 'reactant' with the 'kinetics' given by $H(G, \varepsilon)$. The 'diffusion' coefficient depends on the gel concentration and the elastic constants E and β.

With $S = S_0 - G$ the gel equation (17.95) becomes

$$G_t + (Gu_t)_x = k_+(\varepsilon)S_0 - [k_+(\varepsilon) + k_-(\varepsilon)]G + D_G G_{xx} . \tag{17.101}$$

Now introduce nondimensional quantities

$$G^* = \frac{G}{S_0}, \quad \varepsilon^* = \varepsilon, \quad x^* = \frac{x}{\sqrt{\beta}}, \quad t^* = \frac{tE}{\mu} ,$$

$$k_+^* = \frac{k_+ \mu}{E}, \quad k_-^* = \frac{k_- \mu}{E}, \quad \sigma_0^* = \frac{\sigma_0}{S_0 E} , \tag{17.102}$$

$$D^* = \frac{D_G \mu}{\beta E}, \quad \tau^* = \frac{\tau}{E}, \quad \pi^* = \frac{\pi}{S_0 E} ,$$

with which (17.99), (17.100) and (17.101) become, on dropping the asterisks for notational simplicity,

$$G\varepsilon_t = G\varepsilon_{xx} + f(G,\varepsilon) \ ,$$
$$G_t + (Gu_t)_x = DG_{xx} + g(G,\varepsilon) \ , \qquad (17.103)$$

where

$$f(G,\varepsilon) = -\sigma_0 + \frac{\pi}{1+\varepsilon} - \frac{G\tau}{1+\varepsilon^2} - G\varepsilon \ ,$$
$$g(G,\varepsilon) = k_+(\varepsilon) - [k_+(\varepsilon) + k_-(\varepsilon)]G \ , \qquad (17.104)$$

where the qualitative forms of $k_+(\varepsilon)$ and $k_-(\varepsilon)$ are shown in Fig. 17.23 (a).

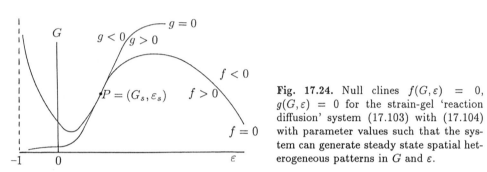

Fig. 17.24. Null clines $f(G,\varepsilon) = 0$, $g(G,\varepsilon) = 0$ for the strain-gel 'reaction diffusion' system (17.103) with (17.104) with parameter values such that the system can generate steady state spatial heterogeneous patterns in G and ε.

The 'reaction diffusion' system (17.103) with (17.104) is similar to those studied in depth in Chapters 12–15 and so we already know the wide range of pattern formation potential. Although in most of these analyses there was no convection, its presence simply enhances the steady state and wave pattern formation capabilities of the system. Typical nullclines $f = 0$ and $g = 0$ are illustrated in Fig. 17.24. Note that there is a nontrivial steady state (ε_s, G_s). Note also that the strain 'reactant' ε can be negative; it is bounded below by $\varepsilon = -1$.

If we linearize the system about the steady state as usual by writing

$$(w,v) = (G - G_s, \varepsilon - \varepsilon_s) \propto \exp[\lambda t + ikx] \ , \qquad (17.105)$$

and substitute into the linearized system from (17.103), namely

$$G_s v_t = G_s v_{xx} + f_G w + f_\varepsilon v \ ,$$
$$w_t + G_s v_t = Dw_{xx} + g_G w + g_\varepsilon v \ , \qquad (17.106)$$

where the partial derivatives of f and g are evaluated at the steady state (G_s, ε_s), we get the dispersion relation $\lambda(k^2)$ as a function of the wavenumber k. It is given by the roots of

$$G_s \lambda^2 + b(k)\lambda + d(k) = 0 \ , \qquad (17.107)$$

where

$$b(k) = G_s(1+D)k^2 - [f_\varepsilon + G_s g_G - G_s f_G] \,,$$
$$d(k) = G_s D k^4 - [Df_\varepsilon + G_s g_G]k^2 + [f_\varepsilon g_G - f_G g_\varepsilon] \,. \tag{17.108}$$

To get spatially heterogeneous structures we require

$$\text{Re}\,\lambda(0) < 0 \quad \Rightarrow \quad b(0) > 0, \quad d(0) > 0$$
$$\text{Re}\,\lambda(k) > 0 \quad \Rightarrow \quad b(k) < 0 \quad \text{and/or} \quad d(k) < 0 \quad \text{for some} \quad k \neq 0 \,. \tag{17.109}$$

The first of these in terms of the f and g derivatives at the steady state requires

$$-[f_\varepsilon + G_s g_G - G_s f_G] > 0, \quad f_\varepsilon g_G - f_G g_\varepsilon > 0 \,. \tag{17.110}$$

From Fig. 17.24 we see that

$$f_\varepsilon > 0, \quad f_G < 0, \quad g_\varepsilon > 0, \quad g_G < 0 \,, \tag{17.111}$$

and so (17.110) gives specific conditions on the parameters in f and g in (17.104). For spatial instability we now require the second set of conditions in (17.109) to hold. Because of the first of (17.110) it is not possible for $b(k)$ in (17.108) to be negative, so the only possibility for pattern is if $d(k)$ can become negative. This can happen only if the coefficient of k^2 is negative and the minimum of $d(k)$ is negative. This gives the conditions on the parameters for spatially unstable modes as

$$Df_\varepsilon + G_s g_G > 0 \,,$$
$$(Df_\varepsilon + G_s g_G)^2 - 4G_s D(f_\varepsilon g_G - f_G g_\varepsilon) > 0 \,, \tag{17.112}$$

together with (17.110). The forms of f and g as functions of G and ε are such that these conditions can be satisfied.

From Chapter 14 we know that such reaction diffusion systems can generate a variety of one-dimensional patterns and in two dimensions, hexagonal structures. Although here we have only considered the one-dimensional model, the two-dimensional space system can indeed generate hexagonal patterns. Another scenario for generating hexagonal patterns, and one which is, from the viewpoint of generating a regular two-dimensional pattern, a more stable process, is if the patterns are formed sequentially as they did in the formation of feather germs in Section 17.5, with each row being displaced half a wavelength.

Let us now return to the formation of the microvilli and Fig. 17.22. The tension generated by the actomyosin fibres aligns the gel along the directions of stress. Thus the contracting gel forms a tension structure consisting of aligned fibres in a hexagonal array. Now between the dense regions the gel is depleted and less able to cope with the osmotic swelling pressure, which is always present in the cell interior. The suggestion, as mentioned at the beginning of this section, then is that this pressure pushes the sheet, at these places, into incipient microvilli, patterns of which are illustrated in Fig. 17.22. These are the steady state patterned solutions of the model mechanism.

17.10 Other Applications of Mechanochemical Models

There are many potential applications, some of which have not yet been examined in depth, others not at all. Here we mention a few, but only very briefly. Some of these have already been modelled by reaction diffusion models but we mention them since they could equally well be modelled from a mechanochemical viewpoint.

(i) *Hair Patterning in Acetabularia.* In Chapter 15, Section 15.3 we proposed a reaction diffusion model for the hair patterning in the marine organism *Acetabularia.* During the process of regeneration, after head amputation, a variety of events takes place, one of which is the emergence of a regular pattern of hair primordia around the stem. As we noted, calcium plays a crucial role in the process. For example if too little free calcium is present in the surrounding medium no pattern evolves. On the other hand if too much calcium is present then again no pattern evolves. Associated with this regular pattern are wrinkles in the stem near the tip where the pattern appears. It is not unreasonable to relate the mechanical model for the cytogel in Section 17.7 to this model and to examine its regular pattern forming potential in a cylindrical domain with a view to explaining such mechanical deformation effects.

(ii) *Animal Coat Patterns.* We saw in Chapter 15, Section 15.1 that many of the patterns observed on mammalian coat patterns could be generated with a morphogen based reaction diffusion mechanism. The patterns thus generated were considered the chemical prepatterns to which the melanoblast cells, the precursors of melanocytes, the pigment forming cells respond. The evidence presented for such a theory was based on observational comparisons and on certain developmental constraints which were dictated by the geometry and scale of the animal's surface when the pre-pattern is laid down. These melanoblast cells migrate from the neural crest early in development. Since the model discussed in Section 17.2 deals specifically with such migratory cells it is possibly directly applicable to patterns found on mammalian coats. In view of the evident richness of patterns which our mechanical models can generate, it is clear that we can obtain not only similar patterns to those from a reaction diffusion mechanism but others which the latter cannot exhibit. Since the partial differential equations (17.22)–(17.24) for example are also domain and boundary dependent the same kind of patterns and developmental constraints mentioned in Section 17.2 will be found.

(iii) *Wound Healing.* The disfiguring scars which result from cutaneous wounds, particularly those associated with burns, are mainly due to the large traction forces generated by the fibroblast cells which recolonize the wound area. (They can also be responsible for the impaired use of joints.) The shape of the boundary is known to affect the amount of scarring. In some cases the wound contraction can be advantageous. The mechanochemical models provide a framework for understanding how events in wound healing are organised and also a means for enhancing or mitigating contraction effects. The models can be used to simulate wound closure. The geometry of the 'wound' boundary can be varied and the subsequent force field determined with a view to designing an optimal wound boundary for minimal scarring. In this application we should include a matrix secretion term as in (17.20) as well as cell proliferation. Preliminary work on the application of mechanical models to wound healing is reported in the review of the mechanical approach by Murray, Maini and Tranquilo (1988). Wound healing is a complicated, important and fascinating process which poses a considerable modelling challenge. The potential benefits unquestionably justify any effort to do so.

(iv) *Rejection of Artificial Joints.* One of the problems with artificial hip joints is that the cement for fixing them inside the femur does not form a good bond with the living tissue. The concepts used in setting up the cell motility mechanisms suggest that a way to effect a more suitable bond might be to try and produce an adhesive which is sufficiently porous to allow movement of cells into it, by virtue of the cell tractions.

(v) *Complex Pattern Generation from Tissue Interaction.* In many reptiles and animals there are complex spatial patterns of epidermal scales and underlying osteoderms, which are bony ossified-like dermal plates, in which there is no simple one to one size correspondence although the patterns are still highly correlated: Fig. 17.25 shows some specific examples. There are also scale patterns whereby a regular pattern appears to be made up from a superposition of two patterns with different basic wavelengths as the illustrative examples in Fig. 17.26 show.

Numerous epidermal-dermal tissue recombinant studies (for example, Rawles 1963, Dhouailly 1975) clearly demonstrate that the instructions needed to initiate skin organ primordia are common to at least three different zoological classes, namely mammals, birds and reptiles.

In Section 17.5 we proposed a mechanical mechanism for generating dermal papillae and it is also possible to generate the spatial patterns we associate with placodes with the model in Section 17.7. Nagorcka (1987) proposed a reaction diffusion mechanism for the initiation and development of priomordia (see also Nagorcka, 1986; Nagorcka and Mooney, 1985) as well as for the formation of the appendages themselves such as hair fibres (Nagorcka and Mooney, 1982).

None of these models have yet been shown to have the capacity to produce complex spatial patterns such as illustrated in Fig. 17.25 and Fig. 17.26. (Making the parameters space dependent is not an acceptable way – this is simply putting the pattern in first.) The patterns in Fig. 17.26 may be viewed as a superposition of two patterns whose wavelengths differ by a factor of at least two as can be seen by comparing the distance between neighbouring small scales denoted λ_s, and neighbouring large scales denoted by λ_l.

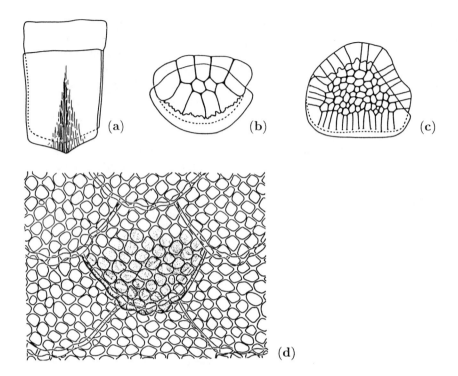

Fig. 17.25a-d. Examples of the different relationship between the osteoderms (bony dermal plates) and the overlying horny epidermal scales (after Otto 1908). (a) The dorsal region of the girdle-tailed lizard *Zonurus cordylus*. (b) The dorsal caudal (tail) region in the skink (a small lizard) *Chalcides (Gongylus) ocellatus* of the family *Scincidae*. (c) The region near the cloaca (anus) of the apotheker or "pharmacist" skink *Scincus officinalis*. (d) The ventral region of the common gecko, *Tarentola mauritanica*: here we have shaded one of the large epidermal scales. The small structures are osteoderms. (From Nagorcka, Manoranjan and Murray 1987)

These complex patterns suggest the need to explore the patterns which can be formed by an interactive mechanism which combines the mechanisms for the epidermis and dermis respectively. The work of Nagawa and Nakashini (1988) substantiates this.

Nagorcka et al. (1987) consider the pattern forming properties of an integrated mechanism consisting of a dermal cell traction model and a reaction diffusion mechanism of epidermal origin, which interact with each other. Preliminary results from their composite 'tissue interaction' mechanism suggest that spatial patterns similar to those in Fig. 17.26 can be produced as a single pattern by the integrated mechanism.

The dispersion relation of integrated mechanisms can be thought of as involving two independent dispersion relations, each with a set of unstable wavenumbers but of different ranges. If the coupling is weak it is reasonable to think of the composite mechanism as being unstable to perturbations with wavelengths approximately equal to λ_s and λ_l, which characterize the two mechanisms indepen-

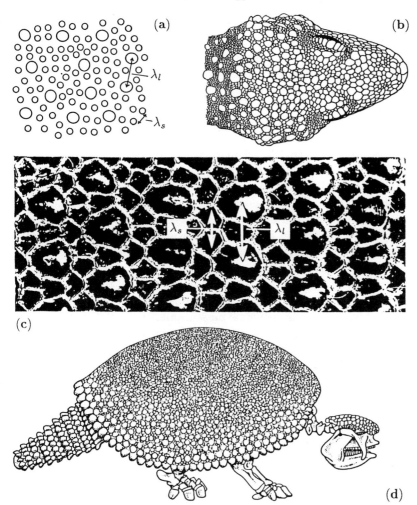

Fig. 17.26a-d. (a) An example of a feather pattern composed of two basic units, one of small diameter and one of large diameter, seen in the skin area under the beak of a species of common coot, *Fulic atra*, after 12 days of incubation. Associated with the pattern are two wavelengths λ_s and λ_l, namely the distances between neighbouring small and large feather follicles, respectively. (After Gerber 1939) (b) Typical small and large scales in the dorsal head region of lizards, here *Cyrtodactylus fedtschenkoi* of the *Gekkonidae* family. (After Leviton and Anderson 1984) The regional variation in the arrangement could be quantified by the ratio λ_s/λ_l. (c) In this example the small and large epidermal scales are in one to one correspondence with the underlying bony scutes (osteoderms) forming the secondary dermal armour in, at least, some species of armadillo, such as in *Dasypus novemcinctus* shown here. (d) The bony scutes seen in the carapace of the *Glyptodon* (after Romer 1962), an ancestor of the armadillo. (From Nagorcka, Manoranjan and Murray 1987)

dently. The patterns obtained are similar to those in Fig. 17.26 provided a large difference exists between the mechanisms' intrinsic wavelengths. The observed patterns may also be produced by any two mechanisms whose dispersion relations are characterized by two separated ranges of unstable wavenumbers. The mechanical model gives such dispersion relations: see, for example, Fig. 17.7 (e).

We can not conclude from the results briefly described here, that an integrated mechanism is responsible for the initiation of skin organ primordia. The results do, however, indicate the potential of such mechanisms for generating complex spatial patterns in the skin of animals of different species from several zoological classes, and warrant an extensive programme of investigation of spatial pattern generation by integrated systems.

There can be no doubt that mechanochemical processes are involved in development. The models we have described here represent a new approach and the concepts suggest that mechanical forces could be the principal guiding elements in producing the correct sequence of tissue patterning and shape changes which are found in the developing embryo. Whereas in reaction diffusion models a chemical morphogen prepattern is set up, which is *then* read and interpreted by the cell as required, in a mechanochemical framework pattern formation and morphogenesis are one and the same.

The models simply reflect the laws of mechanics as applied to tissue cells and their environment, and are based on known biological and biochemical facts: all of the parameters involved are in principle measurable.

We should add that these models are first attempts, and considerable mathematical analysis is required to investigate their potentialities to the full. In turn this will suggest model modifications in the usual way of realistic biological modeling. At this stage the analysis has only just been started but is sufficient to indicate a wealth of wide-ranging patterns and mathematically challenging problems. The models have already been applied realistically to a variety of morphogenetic problems of current major interest. The results and basic ideas have initiated considerable experimental investigation and new ways of looking at a wide spectrum of embryological problems.

Exercises

1. A mechanical model for pattern formation consists of the following equations for the $n(x,t)$, the cell density, $u(x,t)$, the matrix displacement and $\rho(x,t)$ the matrix density:

$$n_t + (nu_t)_x = 0 \ ,$$

$$\mu u_{xxt} + u_{xx} + \tau[n\rho + \gamma\rho_{xx}]_x - s\rho u = 0 \ ,$$

$$\rho_t + (\rho u_t)_x = 0 \ ,$$

where μ, τ, γ and s are positive parameters. Briefly describe what mechanical effects are included in this model mechanism. Show, from first principles, that the dispersion relation $\sigma(k^2)$ about the uniform nontrivial steady state is given by

$$\sigma(k^2) = -\frac{\gamma\tau k^4 + (1 - 2\tau)k^2 + s}{\mu k^2} \, ,$$

where k is the wave number.

Sketch the dispersion relation as a function of k^2 and determine the critical traction or tractions τ_c when the uniform steady state becomes linearly unstable. Determine the wavelength of the unstable mode at bifurcation to heterogeneity.

Calculate and sketch the space in the $(\gamma s, \tau)$ plane within which the system is linearly unstable. Show that the fastest growing linearly unstable mode has wavenumber $(s/\gamma\tau)^{1/4}$. Thus deduce that for a fixed value of γs the wavelength of the fastest growing unstable mode is inversely proportional to the root of the strength of the external tethering.

2. The mechanical model mechanism governing the patterning of dermal cells of density n in an extracellular matrix of density ρ, whose displacement is measured by u, is represented by

$$n_t = D_1 n_{xx} - D_2 n_{xxxx} - a(n\rho_x)_x - (nu_t)_x + rn(1 - n) \, ,$$

$$\mu u_{xxt} + u_{xx} + \tau[n(\rho + \gamma\rho_{xx})]_x = su\rho \, ,$$

$$\rho_t + (\rho u_t)_x = 0 \, ,$$

where all the parameters are non-negative. Explain what each term represents physically.

Show that the trivial steady state is unstable and determine the dispersion relation $\sigma(k^2)$ for the nontrivial steady state $n = \rho = 1$, $u = 0$.

Sketch the dispersion relation σ as a function of k^2 for various values of τ in the situation where viscous effects are negligible. Briefly discuss the implications from a spatial pattern generating point of view. Now sketch the dispersion relation when the viscosity parameter $0 < \mu \ll 1$ and point out any crucial differences with the $\mu = 0$ case.

3. Consider the dimensionless equations for the cytogel continuum given by

$$\nabla \cdot \{\mu_1 \varepsilon_t + \mu_2 \theta_t \mathbf{I} + \varepsilon + \nu'\theta\mathbf{I} + \tau(c)\mathbf{I}\} = s\mathbf{u} \, ,$$

$$\frac{\partial c}{\partial t} = D\nabla^2 c + \frac{\alpha c^2}{1 + \beta c^2} - c + \gamma\theta = D\nabla^2 c + R(c) + \gamma\theta \, .$$

where ε and θ are the strain tensor and dilation respectively, $\mathbf{u}$ is the cytogel displacement of a material point and c is the concentration of free calcium. The active traction function $\tau(c)$ is as illustrated in Fig. 17.18 and μ_1, μ_2, ν', α, D, γ and β are constants.

Investigate the linear stability of the homogeneous steady state solutions c_i, $i = 1, 2, 3$ and show that the dispersion relation $\sigma = \sigma(k^2)$, where here

σ is the exponential temporal growth factor and k is the wavenumber of the linear perturbation, is given by

$$\mu k^2 \sigma^2 + b(k^2)\sigma + d(k^2) = 0 \, ,$$
$$b(k^2) = \mu D k^4 + (1 + \nu' - \mu R_i')k^2 + s \, ,$$
$$d(k^2) = D(1 + \nu')k^4 + [sD + \gamma \tau_i' - (1 + \nu')R_i']k^2 - sR_i' \, ,$$

where

$$\mu = \mu_1 + \mu_2, \quad R_i' = R'(c_i), \quad \tau_i' = \tau'(c_i), \quad i = 1, 2, 3$$

Show that it is possible for the model to generate spatial structures if γ is sufficiently large. Determine the critical wavenumber at bifurcation.

18. Evolution and Developmental Programmes

18.1 Evolution and Morphogenesis

We shall never fully understand the process of evolution until we know how the environment affects the mechanisms that produce pattern and form in embryogenesis. Natural selection must act on the developmental programmes to effect change. We require, therefore a morphoglogical view of evolution, which goes beyond the traditional level of observation to a morphological explanation of the observed diversity. Later in this chapter we shall discuss some specific examples whereby morphogenesis has been experimentally influenced to produce early embryonic forms; early, that is, from an evolutionary point of view. This chapter has no mathematics *per se* and is more or less a stand-alone biological chapter. However, the concepts developed and their practical applications are firmly based on the models, and their analysis, presented and elaborated in earlier chapters, particularly Chapters 14 and 17.

Natural selection is the process of evolution in which there is preferential survival of those who are best adapted to the environment. There is enormous diversity and within species such diversity arises from random genetic mutations and recombination. We must therefore ask why there is not a continuous spectrum of forms, shapes and so on, even within a single species. The implication is that the development programmes must be sufficiently robust to withstand a reasonable amount of random input. From the extensive genetic research on the fruit fly *Drosophila*, it seems that only a finite range of mutations are possible, relatively few in fact.

The general belief is that evolution never moves backward, although it might be difficult to provide a definition of what we mean by direction. If evolution takes place in which a vertebrate limb moves from being three-toed to four-toed, from a morphoglogical view of evolution there is no reason, if conditions are appropriate, that there cannot be a transition 'back' from the four-toed to the three-toed variety. From our study of pattern formation mechanisms this simply means that the sequential bifurcation programme is different. In Section 18.3 we shall see an example where an experimentally induced change in the parameters of the mechanism of morphogenesis results in embryonic forms which, with the excepted direction of evolutionary change, means that evolution has moved backwards.

If we take the development of the vertebrate limb we saw in Section 17.6 in the last chapter that development was sequential in that the humerus preceded the formation of the radius and ulna and these preceded the formation of the subsequent cartilage patterns such as the phalanges. As a specific example, we argued that the formation of the humerus could cue the next bifurcation by influencing the geometry of the limb bud. We also saw how graft experiments could alter the pattern sequence and we showed how the result was a natural consequence when examined from a mechanistic viewpoint.

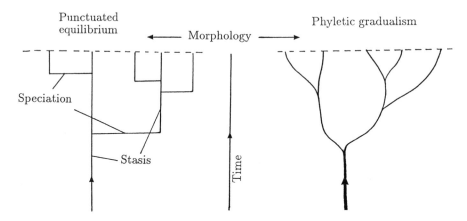

Fig. 18.1. Punctuated equilibrium implies that, as we move through geological time, changes in speciation occur very quickly (on geological time) as compared with stasis, the period between speciation events. Phyletic gradualism says that speciation and diversification are gradual evolutionary processes.

So, intimately associated with the concept of bifurcation programmes, are discrete events whereby there is a discrete change from one pattern to another as some parameter passes through a bifurcation value. The possibility of discrete changes in a species as opposed to gradual changes is at the root of a current controversy in evolution, between what is called *punctuated equilibrium* and *phyletic gradualism*, which has raged for about the past 15 years. (Neo-Darwinism is the term which has been used for punctuated equilibrium.) Put simply, punctuated equilibrium is the view that evolutionary change, or speciation and morphological diversification, takes place effectively instantaneously on geological time, whereas gradualism implies a more gradual evolution to a new species or a new morphology. The arguments for both must come from the fossil records and different sets of data are used to justify each view – sometimes even the same set of data is used! Fig. 18.1 schematically shows the two extremes.

From a strictly observational approach to the question we would require a much more extensive fossil record than currently exists, or is ever likely to. From time to time recently discovered sites are described which provide fine-scaled palaeontological resolution of speciation events. For example, Williamson (1981) describes one of these in northern Kenya for molluscs and uses it to argue for a punctuated equilibrium view of evolution. On the other hand Sheldon (1987) gathered fossil data, from sites in mid-Wales, on trilobites (crab-like marine creatures that vary in size from a few millimetres to tens of centimetres) and on the basis of his study argues for a gradualist approach. From an historical point of view, the notion of punctuated equilibrium was very clearly put by Darwin (1873) himself in the 6th and later editions of his book, *On the Origin of Species*, in which he said (see the summary at the end of Chapter XI, p. 139), "although each species must have passed through numerous transitional stages, it is probable that the periods, during which each underwent modification, though many and long as measured by years, have been short in comparison with the periods during which each remained in an unchanged condition." (The corresponding passage in the first edition is in the summary of Chapter X, p.139.)

From our study of pattern formation mechanisms in earlier chapters the controversy seems artificial. We have seen, particulary from Chapters 14–17, that a slow variation in a parameter can affect the final pattern in a continuous and discrete way. For example, consider the mechanism for generating butterfly wing patterns in Section 15.2. A continuous variation in one of the parameters, when applied, say, to forming a wing eyespot, results in a continuous variation in the eyespot size. The expression (equation (15.24), for example) for the radius of the eyespot shows a continuous dependence on the paramaters of the model mechanism. In the laboratory, for example, the varying parameter could be temperature. Such a continuous variation falls clearly within the gradualist view of evolutionary change.

On the other hand suppose we consider Fig. 14.11 (b), (c) which we reproduce here as Fig. 18.2 (a), (b) for convenience. It encapsulates the correspondence between discrete patterns and two of the mechanism's dimensionless parameters. Although Fig. 18.2 is the bifurcation space for a specific reaction diffusion mechanism we can obtain comparable bifurcation spaces for the mechanochemical models in the last chapter. In Section 17.6 in the last chapter we noted that the effect of a tissue graft on the cartilage patterns in the developing limb was to increase cell proliferation and hence the size of the actual limb bud. Let us, for illustrative purposes, focus on the development of the vertebrate limb. In Fig. 18.2 (a), if we associate cell number with domain size γ we see that as γ continuously increases for a fixed d, say $d = 100$, we have bifurcation values in γ when the pattern changes abruptly from one pattern in Fig. 18.2 (b) to another. So, a continuous variation in a parameter here effects discontinuous changes in the final spatial pattern. This pattern variation clearly falls within a punctuated equilibrium approach to evolution.

Thus, depending on the mechanism and the specific patterning feature we focus on, we can have a gradual or discontinuous change in form. So to reiterate our comment above it is clear that to understand how evolution takes place we must understand the morphogenetic processes involved.

Although the idea that morphogenesis is important in understanding species diversity goes back to the mid-19th century, it is only relatively recently that it

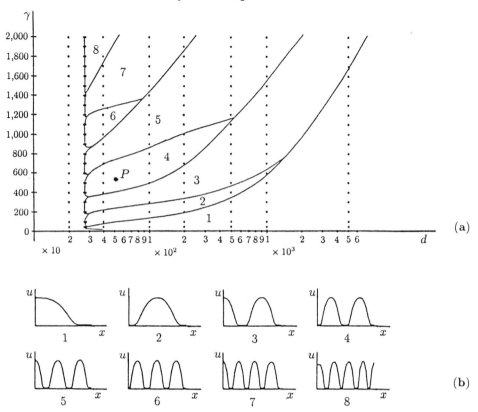

Fig. 18.2a.b. (a) Solution space for a reaction diffusion mechanism (system (14.8) in Chapter 14) with domain size, γ, and morphogen diffusion coefficient ratio, d. (b) The spatial patterns in morphogen concentration with d, γ parameter values in the regions indicated in (a). (From Arcuri and Murray 1986)

has been raised again in a more systematic way by, for example, Alberch (1980) and Oster and Alberch (1982): we briefly describe some of their ideas below. More recently Oster et al. (1988) have presented a detailed study of vertebrate limb morphology, which is based on the notion of the morphological rules described in the last chapter. The latter paper presents experimental evidence to justify their morphogenetic view of evolutionary change: later in the chapter we describe their ideas and some of the supporting evidence.

Morphogenesis is a complex dynamic process in which development takes place in a sequential way with each step following, or bifurcating, from a previous one. Alberch (1980, 1982), and Oster and Alberch (1982) suggest that development can be viewed as involving a only a small set of rules of cellular and mechanochemical interactions which, as we have seen from previous chapters, can generate complex morphologies. Irrespective of the actual mechanisms, they see developmental programmes as increasingly complex interactions between cell populations and their gene activity. Each level of the patterning process has its

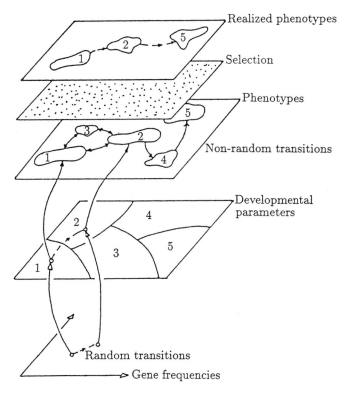

Realized phenotypes

Selection

Phenotypes

Non-random transitions

Developmental parameters

Random transitions

Gene frequencies

Fig. 18.3. A schematic diagram showing how random genetic mutations can be filtered out to produce a stable phenotype. For example, here random genetic mutations affect the size of the various developmental parameters. With the parameters in a certain domain, 1 say, the mechanisms create the specific pattern 1 at the next level up: this is a possible phenotype. Depending on the size of the random mutations we can move from one parameter domain to another and end up with a different phenotype. There are thus a finite number of realisable forms. At the next stage selection takes place and the final result is a number (reduced) of realized phenotyes. (From Oster and Alberch 1982)

own dynamics (mechanism) and it in turn imposes certain constraints on what is possible. This is clear from our studies on pattern formation models wherein the parameters must lie in specific regions of parameter space to produce specific patterns: see, for example, Fig. 18.2. Alberch (1982) and Oster and Alberch (1982) encapsulate their ideas of a developmental programme and developmental bifurcations in the diagram shown in Fig. 18.3.

If the number or size of the mutations is sufficiently large, or sufficiently close to a bifurcation boundary, there can be a qualitative change in morphology. From our knowledge of pattern formation mechanisms, with Fig. 18.3, we can see how different stability domains correspond to different phenotypes and how certain genetic mutations can result in a major morphological change and others do not. Not only that, we can see how transitions between different morphologies are constrained by the topology of the parameter domains for a given morphology.

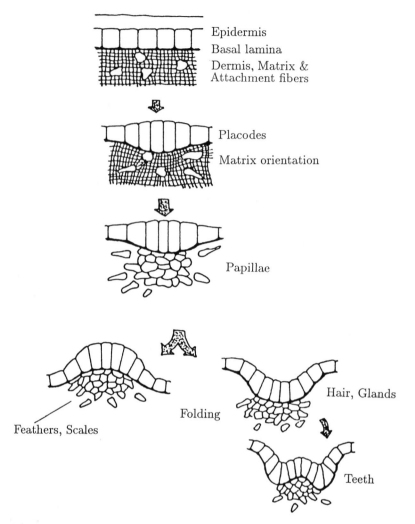

Fig. 18.4. Key mechanical events in the dermis and epidermis in development of skin organ promordia. (After Oster and Alberch 1982)

For example, a transition between state 1 and 2 is more likely than between 1 and 5 and furthermore, to move from 1 to 5 intervening states have to be traversed. An important point to note is that existing morphological forms depend crucially on the history of their past forms. The conclusion therefore is that the appearance of novel phenotypic forms is *not* random, but can be discontinuous. As Alberch (1980) notes: "We need to view the organism as an integrated whole, the product of a developmental program and constrained by developmental and functional interactions. In evolution, selection may decide the winner of a given game but development non-randomly defines the players."

Developmental Constraints

In previous chapters we have shown that, for given morphogenetic mechanisms, geometry and scale impose certain developmental constraints. For example, in Chapter 15, Section 15.1 we noted that a spotted animal could have a striped tail but not the other way round. In the case of pattern formation associated with skin organ primordia, as discussed in the last chapter, we have mechanical examples which exhibit similar developmental constraints. Holder (1983) carried out an extensive observational study of 145 hands and feet of four classes of tetrapod vertebrates. He concluded that developmental constraints were important in the evolution of digit patterns.

Fig. 18.4 shows some of the key mechanical steps in the early development of certain skin organs such as feathers, scales and teeth. In Section 17.6 we addressed the problem of generating cell condensation patterns which we associated with the papillae. In the model for epithelial sheets, discussed in Section 17.8, we saw how spatially heterogeneous patterns could be formed and even initiated by the dermal patterns. Odell et al. (1981) showed how buckling of sheets of discrete cells such as in Fig. 18.4, could arise. From the sequential view of development we might ask whether it is possible to move onto a different developmental pathway by disrupting a mechanical event. There is experimental evidence that a transition can be effected from the scale pathway to the feather pathway (for example, Dhouailly et al. 1980) by treating the skin organ primordia with retinoic acid. In their experiments feathers were formed on chick foot scales.

18.2 Evolution and Morphogenetic Rules in Cartilage Formation in the Vertebrate Limb

In Section 17.6 in the last chapter we showed how a mechanical model could generate the cartilage patterns in the vertebrate limb. There we proposed a simple set of general morphogenetic construction rules for how the major features of limb cartilage patterns are established. Here we shall use these results and draw on comparative studies of limb morphology and experimental embryological studies of the developing limb to support our general theory of limb morphogenesis. We shall then put the results in an evolutionary context. The following is mainly based on the work of Oster, Shubin, Murray and Alberch (1988).

Since the limb is one of the most morphologically diversified of the vertebrate organs and one of the more easily studied developmental systems it is not surprising it is so important in both embryology and evolutionary biology. Coupled with this, is a rich fossil record documenting the evolution of limb diversification (see, for example, Hinchliffe and Johnson 1981 for a comprehensive discussion).

Although morphogenesis appears deterministic on a macroscopic scale, on a microscopic scale cellular activities during the formation of the limb involve considerable randomness. Order emerges as an average outcome with some high probability. We argued in Section 17.6 that some morphogenetic events are ex-

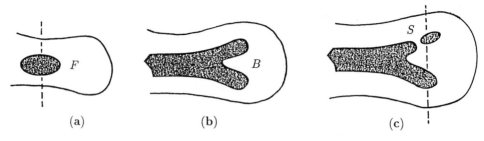

Fig. 18.5a-c. Morphogenetic rules: the three basic cell condensation types, namely a single or focal condensation, F, as in (a), a branching bifurcation, B, as in (b) and a segmental condensation, S, as in (c). More complicated patterns can be built up from a combination of these basic bifurcations: see Fig. 18.8, Fig. 18.10 (cf. Fig. 17.16 (e)) and Fig. 18.11 below.

tremely unlikely, such as trifurcations from a single chondrogenic condensation. Mathematically, of course, they are not strictly forbidden by the pattern formation process, be it mechanochemical or reaction diffusion, but are highly unlikely since they correspond to a delicate choice of conditions and parameter tuning. This is an example of a 'developmental constraint' although the term 'developmental bias' would be more appropriate.

Let us recall the key results in Section 17.6 regarding the 'morphogenetic rules' for limb cartilage patterning. These are summarized in Fig. 17.16 (a)-(c) the key parts of which we reproduce for convenience in Fig. 18.5.

The morphogenetic process starts with a uniform field of mesenchymal cells from which a precartilagenous focal condensation of mesenchymal cells forms in the proximal region of the limb bud. With the mechanical model discussed in Chapter 17, this is the outcome of a model involving the cells, the extracellular matrix (ECM) and its displacement. With the model of Oster et al. (1985), various mechano*chemical* processes are also involved. Subsequent differentiation of the mesenchymal cells is intimately tied to the process of condensation. It seems that differentiation and cartilage morphogenesis are frequently interrelated phenomena. An alternative cell-chemotaxis model with cell differentiation whereby condensation and morphogenesis takes place simultaneously has been proposed by Oster and Murray (1989).

There is a zone of recruitment created around the chondrogenic focus. That is, an aggregation of cells autocatalytically enhances itself while depleting cells in the surrounding tissue. This is effectively setting up a lateral inhibitory field against further aggregation. Because nearby foci compete for cells this leads to almost cell-free regions between foci. In other words, a condensation focus establishes a 'zone of influence' within which other foci are inhibited from forming.

As the actual cartilagenous element develops, the cells seem to separate into two regions;, the outer region consists of flattened cells concentrically arranged, while the cells in the inner region are rounded. The outer cells differentiate to form the perichondrium which sheaths the developing bone. As proposed by Archer, Rodney and Wolpert (1983) and Oster, Murray and Maini (1985), the

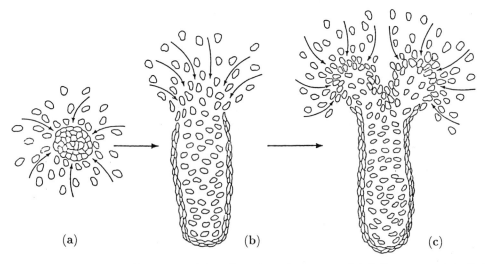

Fig. 18.6a-c. Schematic illustration of the cell condensation process. (a) Cells aggregate initally into a central focus. Development of the cartilagenous element restricts cell recruitment to the distal end of the condensation. (c) When conditions are appropriate the aggregation undergoes a Y-bifurcation. (After Shubin and Alberch 1986)

perichondrium constrains the lateral growth of cartilage and forces its elongation. It also restricts the lateral recruitment of additional cells, so that cells are added to this initial condensation primarily by adding more mesenchymal cells at the distal end thus affecting linear growth as illustrated in Fig. 18.6, which also shows the general features of the condensation process.

As we noted in Section 17.6, limb morphogenetic patterns are usually laid down sequentially, and not simultaneously over an entire tissue (Hinchliffe and Johnson 1981). The latter method would be rather unstable. Theoretical models show that sequential pattern generation is much more stable and reproducible. Recall the simulations associated with the formation of animal coat patterns in Chapter 15, Section 15.1, where the final pattern was dependent on the initial conditions, as compared with the robust formation of hexagonal feather germ and scale arrays in birds, discussed in Section 17.5 in the last chapter, and the supporting evidence from the model simulations by Perelson et al. (1986).

Although most of the pattern formation sequence proceeds in a proximo-distal direction, the differentiation of the digital arch (see Fig. 18.10 below) occurs sequentially from anterior to posterior. The onset of the differentiation of the digital arch is correlated with the sudden broadening and flattening of the distal region of the limb bud into a paddle-like shape. From the typical dispersion relations for pattern generation mechanisms (recall, for example, the detailed discussion in Chapter 14, Section 14.5) such a change in geometry can initiate independent patterns and is the key to understanding this apparent exception to the sequential development rule. Physically, this means that where the domain is large enough, an independent aggregation arises and is far enough away from the

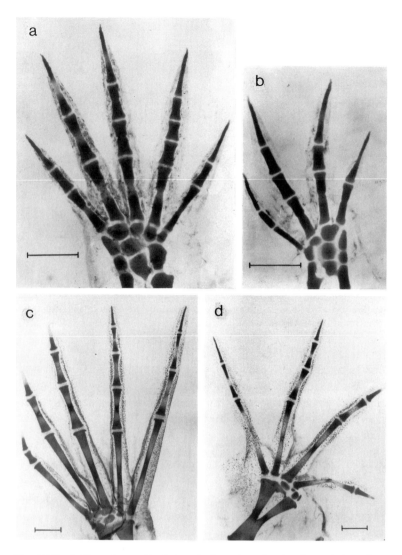

Fig. 18.7a-d. Experimentally induced alterations in the foot of the salamander *Ambystoma mexicanum* and the frog *Xenopus laevis* through treatment of the limb bud with colchicine. (a) Normal right foot of the salamander and (b) the treated left foot. (c) Normal right foot of the frog with (d) the treated left foot. (From Alberch and Gale 1983: photographs courtesy of P. Alberch)

other aggregations that it can recruit cells to itself without being dominated by the attractant powers of its larger neighbours. Of course other model parameters are also important elements in the ultimate pattern and its sequential generation and initiation. The key point is that, irrespective of whether reaction diffusion or mechanochemical models create the chondrogenic condensations, the model parameters, which include the size and shape of the growing limb bud, are crucially important in controlling pattern. Experimental manipulations clearly confirm this importance.

Alberch and Gale (1983, 1985) treated a variety of limb buds with the mitotic inhibitor colchicine. This chemical reduces the dimensions of the limb by reducing cell proliferation. As we predicted, from our knowledge of pattern generation models and their dispersion relations, such a reduction in tissue size reduces the number of bifurcation events, as illustrated in Fig. 18.7.

Note that a possibility that cannot be ruled out is that colchicine affects the timing and number of bifurcations by altering some other developmental parameter, such as cell traction or motility, in addition to the size of the recruitment domains. This alteration, of course, is still consistent with the theory. At this stage further experiments are required to differentiate between the various possibilities. The main point is that these experiments confirm the principle that alterations in developmental parameters (here tissue size) can change the normal sequence of bifurcation events, with concomitant changes in limb morphology that are significant.

Using the basic ideas of cartilage pattern formation in Oster et al. (1983), Shubin and Alberch (1986) carried out a series of comparative studies with amphibians, reptiles, birds and mammals, and confirmed the hypothesis that tetrapod limb development consists of iterations of the processes of focal condensation, segmentation and branching. Furthermore, they showed that the patterns of precartilage cell condensation display several striking regularities in the formation of the limb pattern. Fig. 18.8 presents just some of these results: other examples are also given in Oster et al. (1988).

Condensation, branching and segmentation are an intrinsic property of cartilage forming tissue, although where and when condensation occurs depends on several factors. The stability and reproducibility of a condensation pattern depends crucially on its sequential formation. Patou (1973), in some interesting experiments, removed and disaggregated the tissue and cells involved in cartilage formation from the leg buds of duck and chick embryos. The two populations were then mixed and repacked into the empty limb bud sleeves. The resulting cartilage patterns were highly abnormal and did not display the characteristics of either species, as seen from Fig. 18.9. In all cases, however, the condensation patterns were generated by iterations of the three basic processes of condensation, branching and segmentation, as shown in Fig. 18.5. The results support the theoretical conclusion that branching, segmentation and *de novo* condensation events are reflections of the basic cellular properties of cartilage forming tissue.

Certain Patterns of Chondrogenesis are Unlikely

The study of theoretical models shows that there are considerable restrictions as to the possible patterns of chondrogenesis, as we noted in the last chapter. For example, it is highly unlikely that a trifurcation is possible, that is, a branching of *one* element into three elements, or one into many. Even though subsequent growth may present the appearance of a 1-to-3 splitting, the theory suggests

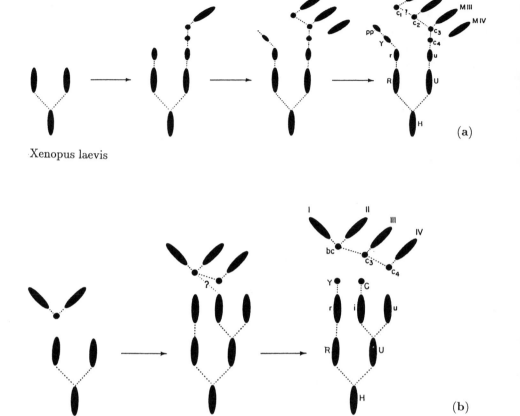

Xenopus laevis

(a)

Ambystoma

(b)

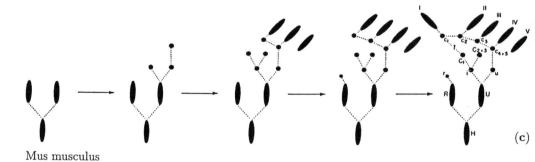

Mus musculus

(c)

Fig. 18.8a-c. Comparative examples of the branching and segmentation in cartilage patterning. (a) Amphibian. The foreleg of the frog *Xenopus laevis*; (b) Reptile. Forelimb of the salamander *Ambystoma mexicanum*; (c) Mammal. Limb of the house mouse *Mus musculus*. These are all constructed from repetitive use of the three basic morphogenetic rules displayed in Fig. 18.5.

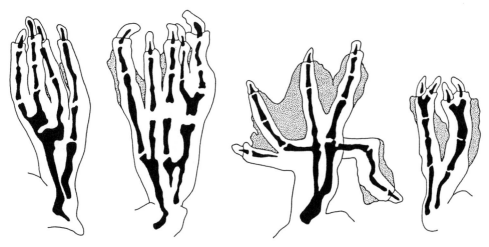

Fig. 18.9. Cartilage patterns obtained after the tissue and cells from the limb buds of a duck and chick embryo were extracted, mixed together and then repacked into the limb bud sleeves. (From Patou 1973) The patterns are highly irregular but are still generated by iterations of the basic morphogenetic rules (Fig. 18.5).

that all branchings are initially binary. This is because a trifurcation is possible only under a very narrow set of parameter values, particularly when we include asymmetries. Such a delicate combination of parameter requirements almost always leads to an unstable pattern.

We now see how the study of pattern formation mechanisms can define more precisely the notion of a 'developmental constraint'. The above discussion together with that in Section 17.6 on limb morphogenesis is only one example. It is based on a pattern formation sequence for laying down the cell aggregation pattern reflected in the final limb architecture.

Alberch (1989) has applied the notion of the unlikelihood of trifurcations to other examples of internal constraints in development. He argues that this is the reason we do not see any three-headed monsters. There are numerous examples of two-headed snakes and other reptiles, siamese twins and so on. Three-headed monsters, of which a few have been reported, although often of questionable veracity, are constructed from a bifurcation followed by a subsequent bifurcation of one of the branches.

The study of monsters – teratology - has a long history. Art – that of Hieronymus Bosch is a particularly good source – and mythology are replete with splendid monsters and new morphologies. One mediaeval description of a three-headed human is that it was born with one head human, the other a wolf's and the third a bloody mass without skin. It finally died after it appeared before the city Senate and made a series of dire predictions! There is much interesting 19th century writing on monsters. A CIBA symposium in 1947 was specifically concerned with monsters in nature (Hamburger 1947) and art (Born 1947).

18.3 Developmental Constraints, Morphogenetic Rules and the Consequences for Evolution

Variation and selection are the two basic components of an evolutionary process. Genetic mutations generate novelties in the population, while natural selection is limited by the amount of variability present although it is usually quite high. There is generally no direct correspondence between genetic and morphological divergence. This lack of correspondence suggested looking for constraints on final phenotype in the mapping from genes to phenotypes, such as occurs in developmental processes. We should remember that genes do not specify patterns or structures; they change the construction recipe by altering the molecular structures or by regulating other genes that specify cellular behaviour. So it is well to reiterate what we said at the begining of the chapter, namely that only with an understanding of developmental mechanisms can we address the central question of how genes can produce ordered anatomical structures.

There has been considerable recent interest in the role of constraints in evolution which has led to the widespread usage of the term 'developmental constraint' (for example, see the recent review by Levinton, 1986). Unfortunately the concept of developmental constraint, as mentioned before, is often used loosely to describe a phenomenological pattern (Williamson 1982). From the above discussion and application of the morphogenetic rules for generating limb cartilage architecture, we can see how it might be varied during evolution and thus give a more precise operational definition of 'developmental constraints' on morphological evolution. From this perspective we can resolve certain puzzles concerning the evolutionary homology (that is the phenomenon of having the same phylogenetic origin but not the same final structure) of bone structures which appear to be geometrically related in the adult skeleton.

Fig. 18.7 shows that the application of mitotic inhibitors to developing limbs produces a smaller limb with a regular reduction in the number of digits and the number of tarsal elements. Perturbing development in this way produced morphologies characterized by the absence of certain elements. These lost elements resulted from the inability of a growing cartilage focus to undergo segmentation or branching events. The patterns produced in these experiments paralleled much of the variation of the species' limbs found in nature.

An extreme experimental variant in the salamander (*Ambystoma*) limb (when treated with mitotic inhibitors) showed a striking resemblance to the pattern of limb evolution in the paedomorphic, or early embryological, form *Proteus* as shown in Fig. 18.10. This suggests that *Proteus* and *Ambystoma* share common developmental mechanisms and hence a common set of developmental constraints. The similarity can be explained in terms of the bifurcation properties, or rather failure of aggregations to bifurcate, of the pattern formation process which restricts the morphogenetic events to the three types of spatial condensation we enumerated in our catalogue of morphogenetic rules.

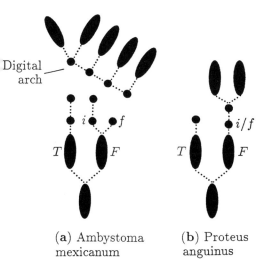

Digital arch

(a) Ambystoma mexicanum

(b) Proteus anguinus

Fig. 18.10a,b. Comparison of the embryonic precartilage connections in the salamnders *Ambystoma mexicanum* (a) and *Proteus anquinus* (b). In the former the fibula (F) branches into the fibulare (f) and intermedium (i). In *Proteus*, however, this branching event has been replaced by a segmentation of F into the single element, either the fibulare or the intermedium, although into which is neither answerable nor a proper question. (After Shubin and Alberch 1986)

It is dangerous to relate geometrically similar elements without knowledge of the underlying developmental programme, for the processes that created the elements may not correspond. For example, the loss of a digit as in Fig. 18.7 may result from the failure of a branching bifurcation; then it is not sensible to ask 'which' digit was lost, since the basic sequence has been altered: see also Fig. 18.10 and Fig. 18.11. In the latter figure Alberch (1989) uses the morphogenetic 'rules' outlined in Fig. 18.5 to show how the differences can be explained.

With these examples we see that it is not easy to compare the cartilage elements themselves but rather the morphogenetic processes that created them. Thus the development of limb elements can be compared using the bifurcation patterns of pre-cartilage condensations and evolutionary changes can be resolved into iterations of condensation, branching and segmentation events.

At the beginning of this chapter we mentioned the possibility of evolution moving backward. It is clearly possible when we consider evolution of form as simply variations in mechanical parameters. Fig. 18.11 is an unequivocal example where this has happened solely through changing the morphogenetic processes – evolutionary change certainly need not always have the same direction. A change in environmental conditions can obviously affect mechanism parameters and hence pattern.

Finally, we should note that the construction rules based on bifurcation sequences suggests that descriptions of limb diversity using D'Arcy Thompson's 'grid deformations' can be misleading (Thompson 1917). He showed, for example, that by superimposing a grid on a fish shape, different species could be 'derived' by simply deforming the grid structure. This gives only a phenomenological description. This kind of comparison between geometries is good only for 'topological' deformations, and precludes branching and *de novo* condensations, because these bifurcations destroy the underlying assumption of continuity upon which the method rests.

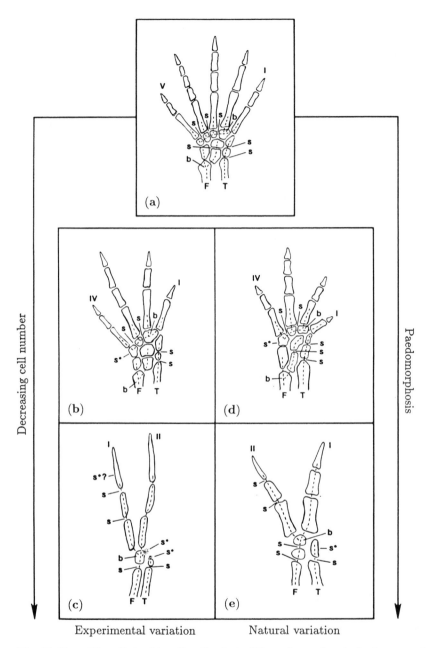

Decreasing cell number

Paedomorphosis

Experimental variation   Natural variation

Fig. 18.11a-e. The effect of treating the foot of the salamander *Ambystoma mexicanum* with the mitotic inhibitor colchicine is to reduce the number of skeletal elements, for example from the four-toed (b) to the two-toed (c). The effect of the colchicine is to reduce the cell number in the limb (and hence the size). The dotted lines show the type of bifurcations during development; here *b* and *s* denoting respectively *Y*-bifurcations and segmental bifurcations as in Fig. 18.5 (b), (c) respectively. The loss of elements in the developmentally disturbed, or evolutionary derived, species is usually caused by a failure of the foci to undergo a branching bifurcation. An asterisk marks the failure of a segmentation event. (From Oster et al. 1988)

Conclusions

The specific set of construction rules encapsuled in Fig. 18.5 provide a scenario for the construction of the vertebrate limb. Such a scheme allows us to be more precise about what we mean by a developmental constraint. It is not suggested for a moment that this list is complete or definitive; nevertheless, such rules point to a new approach to the study of limb morphology, and suggest how the shape of the tetrapod limb has been constrained during evolution. They also show how extremely useful the study of pattern formation mechanisms can be in suggesting practical 'laws' of development.

Limbs develop initially within thin, tubular boundaries. This initial domain shape defines two specific developmental constraints: (i) Limb development must be largely sequential; (ii) Proximal development is initiated by a single focal condensation. Subsequent distal development must proceed from this focus by branching and segmentation using the basic elements in Fig. 18.5.

The growth pattern of the limb bud frequently produces a wide distal paddle. Within the paddle region there is room for focal condensation and extensive branching and segmentation into carpal and tarsal elements. This branching and segmentation is staggered since branching inhibits colateral branching or segmentation because of the competition for cells. If the paddle region is sufficiently large, independent foci – the digital arch – can arise. Subsequent development is from anterior to posterior, that is from the thickest part of the paddle towards the posterior 'open field'.

Given that the limb has a characteristic growing period, which limits its final size, the number of branching bifurcations is limited. This probably accounts for the fact that limbs generally have at most 5 or 6 terminal digits. As we saw in Section 17.6 in the last chapter, grafting experiments which result in duplicate limbs (cf. Fig. 17.14 (a), (b)) require a much larger tissue mass to sustain the supernumerary digits.

As we learn more about the process of morphogenesis, and not only for the limb, we shall be able to add to our small list of construction rules. Each of these will add further constraints on the evolution of limb morphology. We feel that the mechanistic viewpoint we have espoused here will provide a more concrete definition of 'developmental constraints' in evolution.

19. Epidemic Models and the Dynamics of Infectious Diseases

The study of epidemics has a long history with a vast variety of models and explanations for the spread and cause of epidemic outbreaks. Even today they are often attributed to evil spirits or displeased gods. AIDS (acquired immunodeficiency syndrome), *the* epidemic of the 1980's and probably of the 20th century, has been ascribed by many as a punishment sent by God. Hippocrates (459–377 BC), in his essay on 'Airs, Waters and Localities' wrote that one's temperament, personal habits and environment were important factors – not unreasonable even today. Somewhat less relevant, but not without its moments of humour, is Alexander Howe's (1865) book in which he sets out his 'Laws of Pestilence' in 31 propositions of which the following, proposition 2, is typical: 'The length of the interval between successive periodic visitations corresponds with the period of a single revolution of the lunar node, and a double revolution of the lunar apse time'.

In this chapter and the following, we shall describe some models for the population dynamics of disease agents and the spatio-temporal spread of infections. We can then try to exploit them in the control, or ideally the eradication, of the disease or infection we are considering. The practical use of such models must rely heavily on the realism put into the models. As usual, this does not mean the inclusion of all possible effects, but rather the incorporation in the model mechanisms, in as simple a way as possible, what appear to be the major components. Like most models they generally go through several versions before qualitative phenomena can be explained or predicted with any degree of confidence. Great care must be exercised before practical use is made of any epidemic models. However, even simple models should, and frequently do, pose important questions with regard to the underlying process and possible means of control of the disease or epidemic. One such case study is the model proposed by Capasso and Paveri-Fontana (1979) for the 1973 cholera epidemic in Bari in southern Italy.

An interesting early mathematical model, involving a nonlinear ordinary differential equation, by Bernoulli (1760), considered the effect of cow-pox innoculation on the spread of smallpox. The article has some interesting data on child mortality at the time. It is probably the first time that a mathematical model was used to assess the practical advantages of a vaccination control programme.

Models can be extremely useful in giving reasoned estimates for the level of vaccination for the control of directly transmitted infectious diseases: see, for example, Anderson and May (1982a, 1985, 1986). The theoretical papers on epidemic models by Kermack and McKendrick (1927, 1932, 1933) have had a major influence in the development of mathematical models: we describe one of these in Section 19.1. The modelling literature is now extensive. A good introduction and survey of the variety of problems and models for the spread and control of infectious diseases are given, for example, by the books by Bailey (1975) and Hoppensteadt (1975) on mathematical models, the survey by Wickwire (1977) and the collection of articles on the population dynamics of infectious diseases edited by Anderson (1982).

In this chapter we discuss several models which incorporate some general aspects of epidemiological modelling of disease transmission and the time development of epidemics. In the following chapter we consider the geographic spread of infectious diseases and describe in detail a practical model for the spatial spread of rabies and a possible means of its control.

There are basically two broad types of model. In one the total population is taken to be approximately constant with, for example, the population divided into susceptible, infected and immune groups: other groupings are also possible, depending on the disease. We discuss models in this category in Section 19.1. In the other, the population size is affected by the disease via the birth rate, mortality and so on. Host-parasite interacting populations often come into this category: we discuss one such model later in the chapter.

19.1 Simple Epidemic Models and Practical Applications

In the models we consider here the total population is taken to be constant. If a small group of infected individuals are introduced into a large population, a basic problem is to describe the spread of the infection within the population as a function of time. Of course this depends on a variety of circumstances, including the actual disease involved, but as a first attempt at modelling directly transmitted diseases we make some not unreasonable general assumptions.

Consider a disease which, after recovery, confers immunity (which includes deaths: dead individuals are still counted). The population can then be divided into three distinct classes; the susceptibles, S, who can catch the disease; the infectives, I, who have the disease and can transmit it; and the removed class, R, namely those who have either had the disease, or are recovered, immune or isolated until recovered. The progress of individuals is schematically described by

$$S \longrightarrow I \longrightarrow R.$$

Such models are often called *SIR* models.

The assumptions made about the transmission of the infection and incubation period are crucial in any model. With $S(t)$, $I(t)$ and $R(t)$ as the number of individuals in each class we assume here that: (i) The gain in the infective class is at a rate proportional to the number of infectives and susceptibles, that is rSI, where $r > 0$ is a constant. The susceptibles are lost at the same rate. (ii) The rate of removal of infectives to the removed class is proportional to the number of infectives, that is aI where $a > 0$ is a constant. (iii) The incubation period is short enough to be negligible; that is a susceptible who contracts the disease is infective right away.

We now consider the various classes as uniformly mixed: that is every pair of individuals has equal probability of coming into contact with one another. The model mechanism is then

$$\frac{dS}{dt} = -rSI , \tag{19.1}$$

$$\frac{dI}{dt} = rSI - aI , \tag{19.2}$$

$$\frac{dR}{dt} = aI . \tag{19.3}$$

where $r > 0$ is the infection rate and $a > 0$ the removal rate of infectives. This is the classic Kermack-McKendrick (1927) model. We are, of course, only interested in non-negative solutions for S, I and R. This is a primitive model but, even so, we can make some highly relevant general comments about epidemics and, in fact, adequately describe some specific epidemics with such a model.

The constant population size is built into the system (19.1)–(19.3) since, on adding the equations,

$$\frac{dS}{dt} + \frac{dI}{dt} + \frac{dR}{dt} = 0 \quad \Rightarrow \quad S(t) + I(t) + R(t) = N , \tag{19.4}$$

where N is the total size of the population. Thus, S, I and R are all bounded above by N. The mathematical formulation of the epidemic problem is completed given initial conditions such as

$$S(0) = S_0 > 0, \quad I(0) = I_0 > 0, \quad R(0) = 0 . \tag{19.5}$$

A key question in any epidemic situation is, given r, a, S_0 and the initial number of infectives I_0, whether the infection will spread or not, and if it does how it develops with time, and of course when it will start to decline. From (19.2)

$$\left[\frac{dI}{dt}\right]_{t=0} = I_0(rS_0 - a) \underset{<}{\overset{>}{_{}}} 0 \quad \text{if} \quad S_0 \underset{<}{\overset{>}{_{}}} \frac{a}{r} = \rho . \tag{19.6}$$

Since, from (19.1), $dS/dt \leq 0$, $S \leq S_0$ we have, if $S_0 < a/r$,

$$\frac{dI}{dt} = I(rS - a) \leq 0 \quad \text{for all} \quad t \geq 0 , \tag{19.7}$$

in which case $I_0 > I(t) \to 0$ as $t \to \infty$ and so the infection dies out: that is, no epidemic can occur. On the other hand if $S_0 > a/r$ then $I(t)$ initially increases and we have an epidemic. The term 'epidemic' means that $I(t) > I_0$

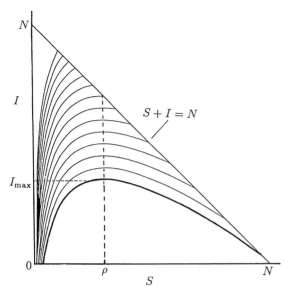

Fig. 19.1. Phase trajectories in the susceptibles (S)-infectives (I) phase plane for the SIR model epidemic system (19.1)–(19.3). The curves are determined by the initial conditions $I(0) = I_0$ and $S(0) = S_0$. With $R(0) = 0$, all trajectories start on the line $S + I = N$ and remain within the triangle since $0 < S + I \leq N$ for all time. An epidemic situation formally exists if $I(t) > I_0$ for any time $t > 0$: this always occurs if $S_0 > \rho(= a/r)$ and $I_0 > 0$.

for some $t > 0$: see Fig. 19.1 above. We thus have a *threshold phenomenon*. If $S_0 > S_c = a/r$ there is an epidemic while if $S_0 < S_c$ there is not. The critical parameter $\rho = a/r$ is sometimes called the *relative removal rate* and its reciprocal $\sigma(= r/a)$ the infection's *contact rate*.

We write

$$R_0 = \frac{rS_0}{a},$$

where R_0 is the basic *reproduction rate* of the infection, that is, the number of secondary infections produced by one primary infection in a wholly susceptible population. Here $1/a$ is the average infectious period. If more than one secondary infection is produced from one primary infection, that is $R_0 > 1$, clearly an epidemic ensues. The whole question of thresholds in epidemics is obviously important. A mathematical introduction to the subject is given by Waltman (1974).

We can derive some other useful analytical results from this simple model. From (19.1) and (19.2)

$$\frac{dI}{dS} = -\frac{(rS - a)I}{rSI} = -1 + \frac{\rho}{S}, \quad \rho = \frac{a}{r}, \quad (I \neq 0).$$

The singularities all lie on the $I = 0$ axis. Integrating the last equation gives the (I, S) phase plane trajectories as

$$I + S - \rho \ln S = \text{constant} = I_0 + S_0 - \rho \ln S_0 \,, \tag{19.8}$$

where we have used the initial conditions (19.5). The phase trajectories are sketched in Fig. 19.1. Note that with (19.5), all initial values S_0 and I_0 satisfy $I_0 + S_0 = N$ since $R(0) = 0$ and so for $t > 0$, $0 \le S + I < N$.

If an epidemic exists we should like to know how severe it will be. From (19.7) the maximum I, $I_{\max}$, occurs at $S = \rho$ where $dI/dt = 0$. From (19.8), with $S = \rho$,

$$
\begin{aligned}
I_{\max} &= \rho \ln \rho - \rho + I_0 + S_0 - \rho \ln S_0 \\
&= I_0 + (S_0 - \rho) + \rho \ln \left(\frac{\rho}{S_0} \right) \\
&= N - \rho + \rho \ln \left(\frac{\rho}{S_0} \right) .
\end{aligned} \tag{19.9}
$$

For any initial values I_0 and $S_0 > \rho$, the phase trajectory starts with $S > \rho$ and we see that I increases from I_0 and hence an epidemic ensues. It may not necessarily be a severe epidemic as is the case if I_0 is close to $I_{\max}$. It is also clear from Fig. 19.1 that if $S_0 < \rho$ then I decreases from I_0 and no epidemic occurs.

Since the axis $I = 0$ is a line of singularities, on all trajectories $I \to 0$ as $t \to \infty$. From (19.1), S decreases since $dS/dt < 0$ for $S \ne 0, I \ne 0$. From (19.1) and (19.3)

$$
\begin{aligned}
\frac{dS}{dR} &= -\frac{S}{\rho} \\
\Rightarrow \quad S &= S_0 \exp\left[-R/\rho\right] \ge S_0 \exp\left[-N/\rho\right] > 0 \\
\Rightarrow \quad 0 &< S(\infty) \le N \,.
\end{aligned} \tag{19.10}
$$

In fact from Fig. 19.1, $0 < S(\infty) < \rho$. Since $I(\infty) = 0$, (19.4) implies that $R(\infty) = N - S(\infty)$. Thus, from (19.10)

$$S(\infty) = S_0 \exp\left[-\frac{R(\infty)}{\rho}\right] = S_0 \exp\left[-\frac{N - S(\infty)}{\rho}\right]$$

and so $S(\infty)$ is the positive root $0 < z < \rho$ of the transcendental equation

$$S_0 \exp\left[-\frac{N - z}{\rho}\right] = z \,. \tag{19.11}$$

We then get the total number of susceptibles who catch the disease in the course of the epidemic as

$$I_{\text{total}} = I_0 + S_0 - S(\infty) \,, \tag{19.12}$$

where $S(\infty)$ is the positive solution z of (19.11). An important implication of this analysis, namely that $I(t) \to 0$ and $S(t) \to S(\infty) > 0$, is that the disease dies out from a lack of *infectives* and *not* from a lack of susceptibles.

The threshold result for an epidemic is directly related to the relative removal rate ρ – if $S_0 > \rho$ an epidemic ensues whereas it does not if $S_0 < \rho$. For a given disease, the relative removal rate varies with the community and hence determines whether an epidemic may occur in one community and not in another. The number of susceptibles S_0 also plays a role, of course. For example, if the density of susceptibles is high and the removal rate, a, of infectives is low (through ignorance, lack of medical care, inadequate isolation and so on) then an epidemic is likely to occur. Expression (19.9) gives the maximum number of infectives while (19.12) gives the total number who get the infection in terms of $\rho(= a/r)$, I_0, S_0 and N.

In most epidemics it is difficult to determine how many new infectives there are each day since only those that are removed, for medical aid or whatever, can be counted. Public Health records generally give the number of infectives per day, week or month. So, to apply the model to actual epidemic situations in general, we need to know the number removed per unit time, namely dR/dt as a function of time.

From (19.10), (19.4) and (19.3) we get an equation for R alone, namely

$$\frac{dR}{dt} = aI = a(N - R - S) = a\left(N - R - S_0 \exp\left[-\frac{R}{\rho}\right]\right), \quad R(0) = 0 , \quad (19.13)$$

which can only be solved analytically in a parametric way: the solution in this form however is not very convenient. Of course, if we know a, r, S_0 and N it is a simple matter to compute the solution numerically. Usually we do not know all the parameters and so we have to carry out a best fit procedure assuming, of course, the epidemic is reasonably described by such a model. In practice, however, it is often the case that if the epidemic is not large, R/ρ is small; certainly $R/\rho < 1$. Following Kermack and McKendrick (1927) we can then approximate (19.13) by

$$\frac{dR}{dt} = a\left[N - S_0 + \left(\frac{S_0}{\rho} - 1\right)R - \frac{S_0 R^2}{2\rho^2}\right] .$$

Factoring the right hand side quadratic in R, we can integrate this equation to get, after some elementary but tedious algebra, the solution

$$R(t) = \frac{\rho^2}{S_0}\left[\left(\frac{S_0}{\rho} - 1\right) + \alpha \tanh\left(\frac{\alpha a t}{2} - \phi\right)\right] \quad (19.14)$$

$$\alpha = \left[\left(\frac{S_0}{\rho} - 1\right)^2 + \frac{2S_0(N - S_0)}{\rho^2}\right]^{1/2} , \quad \phi = \frac{\tanh^{-1}\left(\frac{S_0}{\rho} - 1\right)}{\alpha} .$$

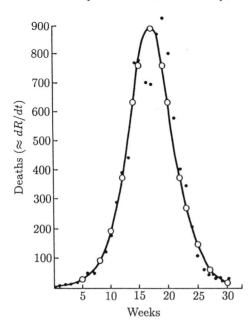

Fig. 19.2. Bombay plague epidemic of 1905–6. Comparison between the data (•) and theory (○) from the (small) epidemic model and where the number of deaths is approximately dR/dt given by (19.16). (After Kermack and McKendrick 1927)

The removal rate is then given by

$$\frac{dR}{dt} = \frac{a\alpha^2\rho^2}{2S_0}\,\text{sech}^2\left(\frac{\alpha a t}{2} - \phi\right),\qquad(19.15)$$

which involves only 3 parameters, namely $a\alpha^2\rho^2/(2S_0)$, αa and ϕ. With epidemics which are not large, it is this function of time which we should fit to the Public Health records. On the other hand, if the disease is such that we know the actual number of the removed class then it is $R(t)$ in (19.14) we should use. If R/ρ is not small, however, we must use the differential equation (19.13) to determine $R(t)$.

We now apply the model to two very different epidemic situations.

Bombay Plague Epidemic 1905–6

This plague epidemic lasted for almost a year. Since most of the victims who got the disease died, the number removed per week, that is dR/dt, is approximately equal to the number of deaths per week. On the basis that the epidemic was not severe (relative to the population size), Kermack and McKendrick (1927) compared the actual data with (19.15), determined the best fit for the three parameters and got

$$\frac{dR}{dt} = 890\,\text{sech}^2(0.2t - 3.4).\qquad(19.16)$$

This is illustrated in Fig. 19.2 together with the actual epidemic data.

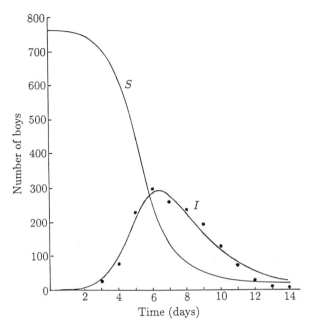

Fig. 19.3. Influenza epidemic data (•) for a boys boarding school as reported in British Medical Journal, 4th March 1978. The continuous curves for the infectives (I) and susceptibles (S) were obtained from a best fit numerical solution of the *SIR* system (19.1)–(19.3): parameter values $N = 763$, $S_0 = 762$, $I_0 = 1$, $\rho = 202$, $r = 2.18 \times 10^{-3}$/day. The conditions for an epidemic to occur, namely $S_0 > \rho$ is clearly stisfied and the epidemic is severe since R/ρ is not small.

Influenza Epidemic in an English Boarding School 1978

In the 4th March 1978 issue of the British Medical Journal there is a report with detailed statistics of a flu epidemic in a boys boarding school with a total of 763 boys. Of these 512 were confined to bed during the epidemic, which lasted from 22nd January to 4th February 1978. It seems that one infected boy initiated the epidemic. This situation has many of the requirements assumed in the model derivation. Here however, the epidemic is severe and the full system has to be used. Here, when a boy was infected he was put to bed and so we have $I(t)$ directly from the data. Since in this case we have no analytical solution for comparison with the data, a best fit numerical technique was used directly on the equations (19.1)–(19.3) for comparison of the data. Fig. 19.3 illustrates the resulting time evolution for the infectives, $I(t)$, together with the epidemic statistics. The R-equation (19.3) is uncoupled: the solution for $R(t)$ is simply proportional to the area under the $I(t)$ curve.

Raggett (1982) applied the *SIR* model (19.1)–(19.3) to the outbreak of plague in the village of Eyam in England from 1665–66. In this remarkable altruistic incident, the village sealed itself off when plague was discovered, so as to prevent it spreading to the neighbouring villages, and it was successful. By the end of the epidemic only 83 of the original population of 350 survived. Thus here,

$S(\infty) = 83$ out of an inital $S_0 = 350$. This is another example, like the school flu epidemic, where the epidemic was severe. Raggett (1982) shows how to determine the parameters from the available data and knowledge of the etiology of the disease. He reiterates the view that although the initial form was proabaly bubonic plague, the pneumonic form (see the following chapter, Section 20.2) most likely became prevalent; the latter form can be transmitted from the cough of a victim. The comparison between the solutions from the determinsitc model and the Eyam data is very good. The comparison is much better than that obtained from the corresponding stochastic model, which Raggett (1982) also considered. We discuss a model for the spatial spread of plague in Section 20.2 in the following chapter.

If a disease is *not* of short duration then (19.1), the equation for the susceptibles, should include birth and death terms. Mortality due to natural causes should also be included in equation (19.2) for the infectives and in (19.3) for the removed class. The resulting models can be analysed in a similar way to that used here and in Chapter 3 on interacting populations: they are all just systems of ordinary differential equations. It is not surprising, therefore, that oscillatory behaviour in disease epidemics is common: these are often referred to as epidemic waves. Here they are *temporal* waves. *Spatial* epidemic waves appear as an epidemic spreads geographically. The latter are also common and we consider them in the next chapter.

Many diseases have a latent or incubation period when a susceptible has become infected but is not yet infectious. Measles, for example, has an 8–13 day latent period. The incubation time for AIDS, on the other hand, is anything from a few months to years after the patient has been shown to have antibodies to the human immunodeficiency virus (HIV). We can, for example, incorporate this as a delay effect, or by introducing a new class, $E(t)$ say, in which the susceptible remains for a given length of time before moving into the infective class. Such models give rise to integral equation formulations and they can exhibit oscillatory behaviour as might be expected from the inclusion of delays. Some of these are described by Hoppensteadt (1975). Nonlinear oscillations in such models have been studied by Hethcote, Stech and van den Driessche (1981). An alternative approach recently used in modelling AIDS will be discussed below in Section 19.4. Finally age, a, is often a crucial factor in disease susceptibility and infectiousness. The models then become partial differential equations with independent variables (t, a): we consider one such model in Section 19.6.

There are many modifications and extensions which can and often must be incorporated in epidemic models; these depend critically on the disease. In the following sections we discuss a few more general models to illustrate different but important points. The books and references already cited describe numerous models and go into them in considerable detail.

19.2 Modelling Venereal Diseases

The increasing incidence of sexually transmitted diseases (STD), such as gonorrhea, chlamydia, syphilis and, of course, AIDS, is a major health problem in both developed and developing countries. In the U.S.A., for example, it is believed that more than 2 million people contract gonorrhea annually. Among reportable communicable diseases its incidence is far greater than the combined totals of syphilis, measles, tuberculosis, hepatitis plus others. World wide it does not, of course, exceed in numbers such diseases as malaria, hookworm, trachoma and so on which run into many millions: we discuss a parasite model below in Section 19.5.

STDs have certain characteristics which are different from other infections, such as measles or rubella (German measles). One difference is that they are mainly restricted to the sexually active community. Another is that often the carrier is asymptomatic (that is the carrier shows no overt symptoms) until quite late on in the development of the infection. A third crucial difference is that STDs induce little or no acquired immunity following an infection. Equally important in virus infections is the lack of present knowledge of the parameters which characterize the transmission dynamics.

Although gonorrhea, syphilis and AIDS are well known, with the latter growing alarmingly, one of the STDs which has far outstripped gonorrhea is the little known *Chlamydia trachomatis*, which in the mid-1980s struck of the order of 4 million Americans annually. It can produce sterility in women without their ever showing any overt symptoms. Diagnostic techniques have only recently been sufficiently refined to make diagnosis more accurate and less expensive. The asymptomatic character of this disease among women is serious. Untreated it causes pelvic inflammatory disorders (PID) which is often accompanied by chronic pain, fever and sterility. With pregnancy, PID, among other complications, can aften cause premature delivery, ectopic pregnancies (that is the fertilized egg is implanted outside the womb) which are life threatening. Untreated gonorrhea, for example, can also cause, blindness, PID, heart failure and ultimately death. STDs are a major cause of sterility in women. The consequences of untreated STDs in general are very unpleasant.

The vertical transmission of STDs from mother to newborn children is another of the threats and tragedies of many STDs. Another problem is the appearance of new strains: in connection with AIDS, a new virus, HIV2 has now been found. With gonorrhea, for example, a new strain, *Neisseria gonorrhea*, was discovered in the 1970s which proved resistant to penicillin.

In this section we present a simple classical epidemic model which incorporates some of the basic elements in the heterosexual spread of venereal diseases. We have in mind such diseases as gonorrhea: AIDS we discuss separately later in Section 19.4. The monograph by Hethcote and Yorke (1984) gives a very good survey of models used for the spread and control of gonorrhea. They show how models and data can be used to advantage: the conclusions they arrive at are specifically aimed at public health workers.

For the model here we assume there is uniformly promiscuous behaviour in the population we are considering. As a simplification we consider only heterosexual encounters. The population consists of two interacting classes, males and females, and infection is passed from a member of one class to the other. It is a criss-cross type of disease in which each class is the disease host for the other. In all of the models we have assumed homogeneous mixing between certain populations subgroups. Recently Dietz and Hadeler (1988) have considered epidemic models for STDs in which there is heterogeneous mixing. For example, the pairing of two susceptibles confers temporary immunity.

Criss-cross infection is similar in many ways to what goes on in malaria and bilharzia, for example, where two criss-cross infections occur. In bilharzia it is between humans and a particular type of snail. Bilharzia, or schistosomiasis, has been endemic in Africa for a very long time. Very young male children who contract the disease start to pass blood in the urine around the age of puberty. (Bilharzia was so common that the ancient Egyptians believed that this passing of blood was the male equivalent of menstruation. Those boys who did not contract the disease were so unique that they were believed to have been chosen by the gods and hence should become priests! Since this perhaps implies they had some immunity they tended to live longer. Perhaps this is why so many of the mummies of older important people are of priests. There are, of course, other possible explanations.)

Since the incubation period for venereal diseases is usually quite short – in gonorrhea for example it is 3–7 days – when compared to the infectious period, we use an extension of the simple epidemic model in Section 19.1. We divide the promiscuous male population into susceptibles S, infectives I and a removed class R: the similar female groups we denote by S^*, I^* and R^*. If we do not include any transition from the removed class to the susceptible group, the infection dynamics is schematically

$$
\begin{array}{ccccc}
S & \longrightarrow & I & \longrightarrow & R \\
& \times & & & \\
S^* & \longrightarrow & I^* & \longrightarrow & R^*
\end{array}
\tag{19.17}
$$

Here I^* infects S and I infects S^*.

As we noted above, the contraction of gonorrhea does not confer immunity and so an individual removed for treatment becomes susceptible again after recovery. In this case a better dynamics flow diagram for gonorrhea is

$$
\begin{array}{ccccc}
S & \longrightarrow & I & \longrightarrow & R \\
& \times & & & \\
S^* & \longrightarrow & I^* & \longrightarrow & R^*
\end{array}
\tag{19.18}
$$

An even simpler version involving only susceptibles and infectives is

$$S \underset{}{\overset{}{\longrightarrow}} I$$

$$(19.19)$$

$$S^* \longrightarrow I^*$$

which, by way of illustration, we now analyse. It is a criss-cross SI model.

We take the total number of males and females to be constant and equal to N and N^* respectively. Then, for (19.19)

$$S(t) + I(t) = N, \quad S^*(t) + I^*(t) = N^* \qquad (19.20)$$

As before we now take the rate of decrease of male susceptibles to be proportional to the male susceptibles times the infectious female population with a similar form for the female rate. We assume that once infectives have recovered they rejoin the susceptible class. A model for (19.19) is then (19.20) together with

$$\frac{dS}{dt} = -rSI^* + aI, \qquad \frac{dS^*}{dt} = -r^*S^*I + a^*I^*$$

$$\frac{dI}{dt} = rSI^* - aI, \qquad \frac{dI^*}{dt} = r^*S^*I - a^*I^* \qquad (19.21)$$

where r, a, r^* and a^* are positive parameters. We are interested in the progress of the disease given initial conditions

$$S(0) = S_0, \quad I(0) = I_0, \quad S^*(0) = S_0^*, \quad I^*(0) = I_0^* . \qquad (19.22)$$

Although (19.21) is a 4th order system, with (19.20) it reduces to a 2nd order system in either S and S^* or I and I^*. In the latter case we get

$$\frac{dI}{dt} = rI^*(N - I) - aI, \qquad \frac{dI^*}{dt} = r^*I(N^* - I^*) - a^*I^* , \qquad (19.23)$$

which can be analysed in the (I, I^*) phase plane in the standard way (cf. Chapter 3). The equilibrium points, that is the steady states of (19.23), are $I = 0 = I^*$ and

$$I_s = \frac{NN^* - \rho\rho^*}{\rho + N^*}, \qquad I_s^* = \frac{NN^* - \rho\rho^*}{\rho^* + N} ,$$

$$(19.24)$$

$$\rho = \frac{a}{r}, \qquad \rho^* = \frac{a^*}{r^*} .$$

Thus non-zero positive steady state levels of the infective populations exist only if $NN^*/\rho\rho^* > 1$: this is the *threshold condition* somewhat analagous to that found in Section 19.1.

With the experience gained from Chapter 2, we now expect that, if the positive steady state exists then the zero steady state is unstable. This is indeed the case. The eigenvalues λ for the linearization of (19.23) about $I = 0 = I^*$ are given by

$$\begin{vmatrix} -a - \lambda & rN \\ r^*N^* & -a^* - \lambda \end{vmatrix} = 0$$

$$\Rightarrow \quad 2\lambda = -(a + a^*) \pm \left[(a + a^*)^2 + 4aa^* \left(\frac{NN^*}{\rho\rho^*} - 1\right)\right]^{1/2} .$$

So, if the threshold condition $NN^*/\rho\rho^* > 1$ holds, $\lambda_1 < 0 < \lambda_2$ and the origin is a saddle point in the (I, I^*) phase plane. If the threshold condition is not satisfied, that is $NN^*/\rho\rho^* < 1$, then the origin is stable since both $\lambda < 0$. In this case positive I_s and I_s^* do not exist.

If I_s and I_s^* exist, meaning in the context here that they are positive, then linearizing (19.23) about it, the eigenvalues λ satisfy

$$\begin{vmatrix} -a - rI_s^* - \lambda & rN \\ r^*N^* & -a^* - r^*I_s - \lambda \end{vmatrix} = 0$$

that is,

$$\lambda^2 + \lambda[a + a^* + rI_s^* + r^*I_s] + [a^*rI_s^* + ar^*I_s + rr^*I_sI_s^* + aa^* - rr^*NN^*] = 0 ,$$

the solutions of which have $\operatorname{Re}\lambda < 0$ and so the positive steady state (I_s, I_s^*) in (19.24) is stable.

The threshold condition for a non-zero steady state infected population is $NN^*/\rho\rho^* = (rN/a)(r^*N^*/a^*) > 1$. We can interpret each term as follows. If every male is susceptible then rN/a is the average number of males contacted by a female infective during her infectious period: a reciprocal interpretation holds for r^*N^*/a^*. These quantities, rN/a and r^*N^*/a^*, are the maximal male and female *contact rates* respectively.

Although parameter values for contacts during an infectious stage are somewhat unreliable from individual questionaires, what is abundantly clear from the statistics since 1950 is that an epidemic has occurred in a large number of countries and so $NN^*/\rho\rho^* > 1$. From data given by a male and female infective, in the U.S.A. in 1973, regarding the number of contacts during a period of their infectious state, figures of maximal contact rates of $N/\rho \approx 0.98$ and $N^*/\rho^* \approx 1.15$ were calculated for the male and female respectively which give $NN^*/\rho\rho^* \approx 1.127$. If we take, by way of example, the total male and female promiscuous populations to be, say, 20 million, each then the steady state for the male infectives from (19.24) is $I_s = 1.12$ million and for the female infectives $I_s^* = 1.21$ million. It must be kept in mind that this model is a very simple one and principally pedagogical.

19.3 Multi-group Model for Gonorrhea and Its Control

Although the *SI* model in the last section is a particularly simple one, it is not too unrealistic. In the case of gonorrheal infections it neglects many relevant factors. For example, a large proportion of females, although infected and infectious, show no obvious symptoms; that is, they form an asymptomatic group. There are, in fact, various population subgroups. For example, we could reasonably have susceptible, symptomatic, treated infective and untreated infective groups. Lajmanovich and Yorke (1976) proposed and analysed an 8-group model for gonococcal infections consisting of sexually (i) very active and (ii) active females (males) who are asymptomatic when infectious and (iii) very active and (iv) active females (males) who are symptomatic when infectious.

If the total populations of active male and female are N and N^*, assumed constant, we can normalize the various group populations as fractions of N and N^*. Denote the groups of women with indices 1,3,5,7 and the men with 2,4,6,8. Then if N_i, $i = 1, 2, \ldots, 8$ denote the *normalized* populations

$$N_1 + N_3 + N_5 + N_7 = 1, \quad N_2 + N_4 + N_6 + N_8 = 1 . \tag{19.25}$$

Since neither immunity nor resistance is acquired in gonococcal infections we consider only two class, susceptibles and infectives. If $I_i(t)$, $i = 1, 2, \ldots, 8$ denote the fractions infecious at any time t, the fractional numbers of susceptibles at that time are then $1 - I_i(t)$, $i = 1, 2, \ldots, 8$.

We again assume homogeneous mixing. For each group let D_i be the mean length of time (in months) of the infection in group i. Then, there is a $1/D_i$ chance of an infective recovering each month. This implies that the removal rate per month is I_i/D_i.

Let L_{ij} be the number of effective contacts per month of an infective in group j with an individual in group i. Since the model here considers only *hetero*sexual contacts we have

$$L_{ij} = 0 \quad \text{if } i + j \text{ even} .$$

The matrix $[L_{ij}]$ is called the *contact matrix*. Although there are seasonable variations in the L_{ij} we take them to be constant here. Then the average number of susceptibles infected per unit time (month) in group i by group j is $L_{ij}(1 - I_i)$. Thus the differential equation model system is

$$\underbrace{\frac{d(N_i I_i)}{dt}}_{\substack{\text{rate of new} \\ \text{infectives}}} = \underbrace{\sum_{j=1}^{8} L_{ij}(1 - I_i)N_j I_j}_{\substack{\text{rate of new} \\ \text{infectives (incidence)}}} - \underbrace{\frac{N_i I_i}{D_i}}_{\substack{\text{recovery rate} \\ \text{of infectives}}} \tag{19.26}$$

with given initial conditions $I_i(0) = I_{i0}$.

By considering the linearization about the non-zero steady state the effect of varying the parameters can be assessed and hence the effects of various control strategies. This model is analysed in detail by Lajmanovich and Yorke (1976).

Major aims in control include of course the reduction in incidence and an increase in detection, each of which affects the long term progress of the spread of the disease. So, screening, detection and treatment of infectives is the major first step in control. The paper by Hethcote, Yorke and Nold (1982) compares various control methods for gonorrhea: it also has references to many of the models which have been proposed.

As an example, suppose C is a parameter proportional to the number of women screened and CR_i is the rate at which infected women are detected in group i. Let EP_i be the general supplementary detection rate where E is a measure of the effort put in and P_i is the population of a group i: E depends on the control strategy. Then, in place of (19.26) we have the control model

$$\frac{d(N_i I_i)}{dt} = \sum_{j=1}^{8} L_{ij}(1 - I_i)N_j I_j - \frac{N_i I_i}{D_i} - CR_i - EP_i . \tag{19.27}$$

Different control methods imply different R_i and P_i.

Suppose there is general screening of women (the major control procedure in the U.S.A.). On the basis that the number of infected women detected is directly proportional to the number infected and the supplementary programme is general screening of women population, we have

$$P_i = R_i = I_i N_i, \quad i = 1, 3, 5, 7; \qquad P_i = R_i = 0, \quad i = 2, 4, 6, 8 . \tag{19.28}$$

If the programme is for men, the odd and even number range is interchanged.

These and other control procedures are discussed in the paper by Hethcote et al. (1982): see also Hethcote and Yorke (1984). They also discuss the important problem of parameter estimation and finally carry out a comparison of various control strategies. The cost and social range of screening are not negligible factors in the practical implementation of such programmes. The political and sociological considerations can also be rather sensitive.

It should be emphasized again, that venereal disease models, which are to be used in control programmes, must have a realistic validation, which can only come from a comparison of their solutions and predictions with actual data. This should apply to all disease control models.

19.4 AIDS: Modelling the Transmission Dynamics of the Human Immunodeficiency Virus (HIV)

The human immunodeficiency virus, HIV, leads to acquired immunodeficiency syndrome – AIDS. When antibodies to HIV are detected the patient is infected and said to be seropositive or HIV positive. The virulence of AIDS and the rate of spread of the epidemic are alarming. The prognosis is that it will be the most serious pandemic, that is a world epidemic, of at least this century and many liken it to the Black Death of the mid-14th century (see Chapter 20).

After antibodies to HIV are detected there is a latent period before the patient exhibits full-blown AIDS, the end-stage disease. How long this period is likely to be in individual cases is not known. Documented evidence shows it can be from months to years. It is also not known what proportion of the population is seropositive. In fact, not very much is known about the basic epidemiological parameters associated with the spread of the virus. The ridiculous 'head-in-the-sand' attitude and the 'there-is-no-AIDS-in-our-country' attitude of many countries particularly in the 1980s is astonishing and highly irresponsible. Although there are social problems associated with gathering data on the number of people who have the HIV, it is unlikely that the epidemic will be contained if this information is not available in the near future. The lack of knowledge creates enormous difficulties in designing effective control programmes, not to mention health-care facilities. Education programmes as to how it can spread are the minimum requirement. Although in developed countries AIDS is primarily associated with the homosexual community this is not the case in many of the underdeveloped communities, where heterosexual spread of the disease is prevalent. We must expect, therefore, that it will spread into the heterosexual community in developed countries as current signs indicate.

A review of some of the more reliable epidemiological data is given, for example, by Anderson et al. (1986) and May and Anderson (1987) who also present some first models of various aspects of the virus transmission. The statistics from the U.S.A. and the U.K. give a very clear picture of the seriousness and magnitude of the epidemic. As more data becomes available, the models will be able to become more sophisticated. In this section we describe two simple models from Anderson et al. (1986) to give a flavour of the modelling problems. These models are partially pedagogical and do not include many of the factors which could and should be included in more realistic models. For example, we shall not consider the possibility of new viral strains appearing, which clearly will have very important consequences.

A major problem with AIDS is the variable length of the incubation period from the time the patient is diagnosed as seropositive until he exhibits the symptoms of AIDS. This has major consequences for the spread of the virus. So, the first model we consider is for the time evolution of the disease between those infected and those with AIDS.

Consider a population in which all of the people are infected with HIV at time $t = 0$. Denote by $y(t)$ the fraction of the population who have AIDS at time t, and by $x(t)$ the fraction who are seropositive but do not yet have AIDS. We thus have $x(t) = 1 - y(t)$. Let $v(t)$ be the rate of conversion from infection to AIDS. A simple model for the dynamics with relevant initial conditions is then

$$\frac{dx}{dt} = -v(t)x, \quad \frac{dy}{dt} = v(t)x ,$$

$$x(0) = 1, \quad y(0) = 0 ,$$

$$\tag{19.28}$$

where $x + y = 1$.

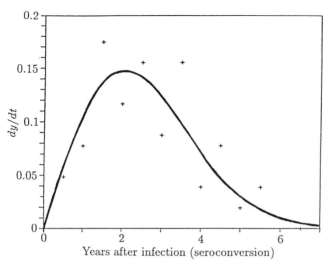

Fig. 19.4. The rate of change in the proportion of the population who develop AIDS who were infected with HIV (through blood transfusion) at time $t = 0$. The data, from Peterman, Drotman and Curran (1985), provide a best-fit value of $a = 0.237\,\mathrm{yr}^{-1}$ for the model solution (19.30). (After Anderson et al. 1986)

This model assumes that all infected people develop AIDS, which is not necessarily the case. If we assume that the patient's immune system to opportunistic diseases, such as cancer, is progressively impaired from the time of infection, then $v(t)$ is an increasing function of time. Let us take a linear dependence with

$$v(t) = at \ , \tag{19.29}$$

where $a > 0$ is a constant. The solution of (19.28) is then given by

$$x(t) = \exp\left[-\frac{at^2}{2}\right], \quad y(t) = 1 - \exp\left[-\frac{at^2}{2}\right] \ . \tag{19.30}$$

Peterman, Drotman and Curran (1985) present data on 194 cases of blood transfusion-associated AIDS. This data is shown in Fig. 19.4. The solution (19.30) was applied to this data and the parameter a determined from a best fit to the data: the continuous curve in Fig. 19.4 shows the result for the rate of increase, dy/dt, in AIDS patients as a function of time. The comparison between theory and data is good.

Epidemic Model

Here we are interested in the devlopment of an AIDS epidemic in a homosexual population. Let us assume there is a constant immigration rate B of susceptible males into a population of size $N(t)$. Let $X(t)$, $Y(t)$, $A(t)$ and $Z(t)$ denote respectively the number of susceptibles, infectious males, AIDS patients and

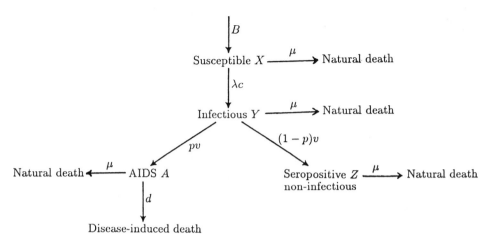

Fig. 19.5. The flow diagram of the disease as modelled by the system (19.31)–(19.34). B represents the recruitment of susceptibles into the homosexual community. The rate of transferral from the susceptible to the infectious class is λc. A proportion of the infectious class become non-infectious with the rest developing AIDS. Natural (non-AIDS induced) death is also included in the model.

the number of seropositives who are non-infectious. We assume susceptibles die naturally at a rate μ: if there were no AIDS the steady state population would then be $N^* = B/\mu$. We assume AIDS patients die at a rate d: typically $1/d$ is of the order of 9 to 12 months. Fig. 19.5 is a flow diagram of the disease on which we base our model.

As in previous models we consider uniform mixing. A reasonable first model system, based on the flow diagram Fig. 19.5, is then

$$\frac{dX}{dt} = B - \mu X - \lambda cX, \quad \lambda = \frac{\beta Y}{N}, \tag{19.31}$$

$$\frac{dY}{dt} = \lambda cX - (v + \mu)Y, \tag{19.32}$$

$$\frac{dA}{dt} = pvY - (d + \mu)A, \tag{19.33}$$

$$\frac{dZ}{dt} = (1 - p)vY - \mu Z, \tag{19.34}$$

$$N(t) = X(t) + Y(t) + Z(t) + A(t). \tag{19.35}$$

Here B is the recruitment rate of susceptibles, μ is the natural (non-AIDS related) death rate, λ is the probability of acquiring infection from a randomly chosen partner ($\lambda = \beta Y/N$, where β is the transmission probability), c is the number of sexual partners, d is the AIDS-related death rate, p is the proportion of seropositives who are infectious and v, defined before, is the rate of conversion from infection to AIDS but here taken to be constant. (Actually λ here is more appropriately $\beta Y/(X + Y + Z)$ but A is considered small in comparison with N). With v constant, $1/v$, equal to D say, is then the average incubation time of the disease. Note that in this model the total population $N(t)$ is not constant, as was the case in the conventional epidemic models in Section 19.1. If we add equations (19.31)–(19.34) we get

$$\frac{dN}{dt} = B - \mu N - dA .$$ (19.36)

An epidemic ensues if the basic reproductive rate $R_0 > 1$, that is the number of secondary infections which arise from a primary infection is greater than one. In (19.32) if, at $t = 0$, an infected individual is introduced into an otherwise infection free population of susceptibles, we have initially $X \approx N$ and so near $t = 0$,

$$\frac{dY}{dt} \approx (\beta c - v - \mu)Y \approx v(R_0 - 1)Y$$ (19.37)

since the average incubation time, $1/v$, from infection to development of the disease, is very much shorter than the average life expectancy, $1/\mu$, of a susceptible; that is $v \gg \mu$. Thus the approximate threshold condition for an epidemic to start is, from the last equation,

$$R_0 \approx \frac{\beta c}{v} > 1 .$$ (19.38)

Here the basic reproductive rate R_0 is given in terms of the number of sexual partners c, the transmission probability β and the average incubation time of the disease $1/v$.

When an epidemic starts, the system (19.31)–(19.35) evolves to a steady state given by

$$X^* = \frac{(v + \mu)N^*}{c\beta}, \quad Y^* = \frac{(d + \mu)(B - \mu N^*)}{pvd}$$

$$Z^* = \frac{(1 - p)(d + \mu)(B - \mu N^*)}{pd\mu}, \quad A^* = \frac{B - \mu N^*}{d},$$ (19.39)

$$N^* = \frac{B\beta[\mu(v + d + \mu) + vd(1 - p)]}{[v + \mu][\beta(d + \mu) - pv]} .$$

If we linearize about this steady state it can be shown that (X, Y, Z, A) tends to (X^*, Y^*, Z^*, A^*) in a damped oscillatory manner with a period of oscillation given in terms of the model parameters: the algebra is messy. With typical current values for the parameters the period of epidemic outbreaks is of the order of 30–40 years. It is highly unlikely though, that the parameters characterizing social behaviour associated with the disease would remain unchanged over that time span.

We can get some interesting information from an analysis of the system during the early stages of an epidemic. Here the population consists of almost all susceptibles and so $X \approx N$ and the equation for the growth of the infectious, that is seropositive, Y class is approximated by (19.37), the solution of which is

$$Y(t) = Y(0) \exp\left[v(R_0 - 1)t\right] = Y(0) \exp\left[rt\right] , \qquad (19.40)$$

where R_0 is the basic reproductive rate, $1/v$ is the average infectious period and $Y(0)$ is the initial number of infectious people introduced into the susceptible population. The intrinsic growth rate, $r = v(R_0 - 1)$, is positive only if an epidemic exists $(R_0 > 1)$. From (19.40) we can obtain the doubling time for the epidemic, that is the time t_d when $Y(t_d) = 2Y(0)$, as

$$t_d = r^{-1} \ln 2 = \frac{\ln 2}{v(R_0 - 1)} . \qquad (19.41)$$

We thus see that the larger the basic reproductive rate R_0 the shorter the doubling time.

If we substitute (19.40) into equation (19.33) for the AIDS patients, we get

$$\frac{dA}{dt} = pvY(0)\exp\left[rt\right] - (d + \mu)A .$$

Early on in the epidemic there are no AIDS patients, that is $A(0) = 0$, and so the solution is given by

$$A(t) = pvY(0)\frac{\exp\left[rt\right] - \exp\left[(d + \mu)t\right]}{r + d + \mu} . \qquad (19.41)$$

Estimates for the parameter r have been calculated by Anderson and May (1986) from data from 6875 homosexual and bisexual men who attended a clinic in San Francisco over the period 1978 to 1985: the average value is $0.88\,\mathrm{yr}^{-1}$. Crude estimates (Anderson and May 1986, Anderson et al. 1986) for the other parameter values are $R_0 = 3\text{--}4$, $d + \mu \approx d \approx d = 1 - 1.33\,\mathrm{yr}^{-1}$, $p = 10\%\text{--}30\%$ (probably higher), $v \approx .22\,\mathrm{yr}^{-1}$, $c = 2\text{--}6$ partners per month. With these estimates we then get an approximate doubling time for the seropositive class as roughly 9 months.

Numerical simulations of the model system of equations (19.31)–(19.34) give a clear picture of the epidemic development after the introduction of HIV into a susceptible homosexual population. Fig. 19.6 shows one such simulation: the model predicts that seropositive incidence reaches a maximum around 12–15 years after the introduction of the virus into the population.

In spite of the simplicity of the models the results are in line with observation in homosexual communities. More complex models have been proposed, for example, by Anderson et al. (1986) and Anderson (1988). A review of some of the current mathematical models for the transmission dynamics of HIV infection and AIDS is given by Isham (1988). With the accumulation of more data and information of the epidemic, even more sophisticated models will be required. A

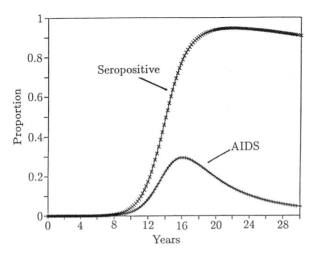

Fig. 19.6. Numerical solution of the model system (19.31)–(19.34) with initial conditions $A(0) = Z(0) = 0$, $S(0) + Y(0) = N(0) = 100,000$. Parameter values $B = 13333.3 \, \text{yr}^{-1}$, $v = 0.2 \, \text{yr}^{-1}$, $\mu = 1/32 \, \text{yr}^{-1}$, $d = 1 \, \text{yr}^{-1}$, $p = 0.3$, the basic reproductive rate of the epidemic $R_0 \approx \beta c/v = 5.15$. The graphs give the proportion of seropositives and the proportion who develop AIDS. Compare the AIDS curve with that in Fig. 19.4. (After Anderson et al. 1986)

practical use of simple models at this stage is that, among other things, it poses questions which can guide data collection and focus on what useful information can be obtained from sparse data. The crude estimates of epidemic severity doubling time, and so on, are in themselves of considerable interest and use. Finally, when the epidemic moves in a major way into the heterosexual community, as it already has in some countries, even more complex models will be required.

19.5 Modelling the Population Dynamics of Acquired Immunity to Parasite Infection

Gastrointestinal nematode parasites infections in man are of immense medical importance throughout the developing world. An estimated 800–1000 million people are infected with *Ascaris lumbricoides*, 700–900 million with the hookworms *Ancylostoma duodenale* and *Nector americanus* and 500 million with the whipworm *Trichuris trichiura* (Walsh and Warren 1979). To design optimal control policies, we must have an understanding of the factors which regulate parasite abundance and influence the size and stability of helminth populations. So, in this section we present a model for the immunological response by the host against gastrointestinal parasites which was proposed and studied by Berding et al. (1986). We shall see that such simple modelling can have highly significant implications for real world control programmes.

Parasites invoke extremely complex immunological responses from their mammalian hosts. We still do not know exactly how these come about but current

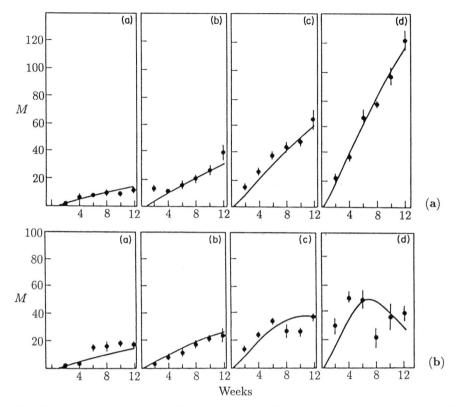

Fig. 19.7a,b. Change in mean adult worm burden, M, in mice hosts fed on a protein diet for a repeated infection over a 12-week period: **(a)** low protein diet; **(b)** high protein diet. The infection rates are (a) 5, (b) 10, (c) 20, (d) 40 larvae/mouse/2 weeks. The periods are the experimental points from Slater and Keymer (1986). The continuous lines are solutions of the mathematical model: how these were obtained will be described below in the subsection on the population dynamics model and analysis. (From Berding et al. 1986)

experimental research provides some important pointers which form the basis for the mathematical model.

Let us first summarize the relevant biological facts starting with a brief review of key experiments. Laboratory experiments in which mice are repeatedly exposed to parasite infection at constant rates can provide a suitable test for mathematical models of helminth population dynamics. Experiments relevant for our model (Slater and Keymer 1986) involve two groups of 120 mice, which are fed on artificial diets containing either 2% ('low protein') or 8% ('high protein') weight for weight protein. Both groups were subdivided into 4 groups of 30 mice, which we denote by (a), (b), (c) and (d), which are subjected to repeated infection with larvae of the nematode *Heligmosoides polygyrus*. The subgroups are infected at different rates: group (a) with 5 larvae/mouse/two weeks, group (b) with 10 larvae/mouse/two weeks, group (c) with 20 larvae/mouse/two weeks, and group (d)

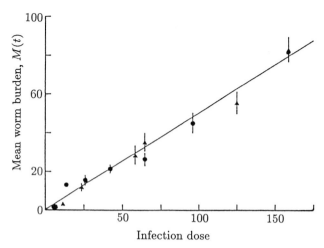

Fig. 19.8. These experimental results show the worm survival after a single infection of larvae. There is a linear relationship between larval dose and adult worm burden: these results are after 14 days from the infection. The circles represent mice fed on a low protein diet and the triangles are for mice fed on a high protein diet. The solid line is a best fit linear description of the data: the gradient is 0.64, from which we deduce that 64% of the larvae survive. (From Berding et al. 1986)

(d) with 40 larvae/mouse/two weeks. So, we have a total of 8 subgroups of 30 mice differing either in their infection rates or in the protein diets they are fed on. It is known that protein deprivation impairs the function of the immune system so this scheme lets us compare parasite population dynamics, under various infectious conditions, in the presence and the absence of an acquired immune response.

The most important exeperimental observations were the temporal changes in the mean worm burden, M, namely the total number of adult worms divided by the total number of hosts. Every two weeks throughout the experiment a sample of 5 mice from each group was examined for the presence of adult parasites. The number of parasites present in each mouse was determined by post mortem examination of the small intestine. The main experimental results are shown in Fig. 19.7 which display the mean worm burden as a function of time for the low and the high protein groups, respectively. The letters (a), (b), (c) and (d) refer to infection rates of 5, 10, 20 and 40 larvae per two weeks.

Other experiments were carried out to quantify parasite establishment and survival in primary infection. Here only a single dose of larvae was given unlike the repeated infection in Fig. 19.7. The results shown in Fig. 19.8 gives the mean worm burden as a function of the infection dose. From this figure we estimate that approximately 64% of the infective larvae survive to become adult worms.

The survival of adult worms in a single infection is summarized in Fig. 19.9: it again shows the mean worm burden as a function of time. In this situation the worm population remains free from effects of the host's immune system: we use this figure to estimate the natural death rate of the adult worms.

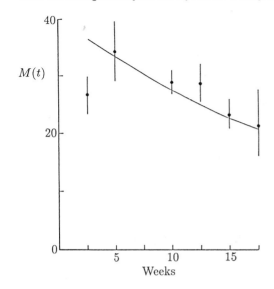

Fig. 19.9. Survival of adult worms which follows a single infection of 50 larvae/-mouse on day zero. The results are for mice fed on a low protein diet. The continuous line is a best fit for an exponential survival model with a constant death rate $\delta = 5.6 \times 10^{-3}\,\mathrm{day}^{-1}$: that is we assume the worms die proportional to their mean population (see equation (19.45) below). (From Berding et al. 1986)

So as to be able to construct a realistic model, let us summarize these and related experimental observations:

(A1) Infective parasite larvae, after ingestion by the host, develop into tissue dwelling larvae which become adult worms found within the lumen of the alimentary canal. This invokes a distinct immunological response from the host. Typically, the tissue dwelling larvae are in the most immunogenic stage in the parasite life cycle. We thus assume that the immune system is triggered according to the larval burden experienced by the host.

(A2) Many experiments point to the presence of delay, that is memory, effects in immune response. Some of these effects can be accounted for by including delay in the models.

(B1) The experimental results shown in Figs. 19.7 suggest, importantly, that the strength of the immune response is very much dependent on the nutritional status of the host. We shall interpret the differences in the dynamics of infection in mice feeding on low and high protein diets, that is Figs. 19.7 (a) and Fig. 19.7 (b) respectively, as a consequence of a relationship between the nutritional status and immunological competence.

(B2) The similarity between the Figs. 19.7 (a) $(a),(b)$ and Fig. 19.7 (b) $(a),(b)$ on the one hand, and the differences between Figs. 19.7 (a) $(c),(d)$ and Fig. 19.7 (b) $(c),(d)$ on the other, clearly indicate a threshold behaviour of the immune system. Biologically this means that the full activation of the immune system requires a certain threshold of exposure to parasite infection.

(B3) Available evidence on the effectiveness of the immune response, which we define here as the per capita rate of limitation in parasite establishment and survival, in relation to its stimulus, that is increased exposure to infection, suggests the following scenario. After an initial increase, the activity of the immune response to the parasites saturates at a maximum level. Further stimulation does not seem to increase the subsequent effectiveness of acquired immunity. So, we assume here that the activity of immune response saturates at a defined maximum level.

(C) The immunological response may act against several stages in the parasite life cycle. In some strains of mice it directly kills tissue dwelling larvae. However, in others the immune response is not capable of preventing larval development. In these, larvae subjected to immunological attack emerge as stunted adults, with a correspondingly high mortality rate. So, to reflect these experimental findings we model immunological competence by an increased mortality rate of the adult parasite.

Let us now construct the model on the basis of these assumptions, firmly based on experimental observations, in three main steps:

(i) We introduce a variable, E, for the immune system, which takes into account assumptions (A1) and (A2), by

$$E = \int_{t-T}^{t} L(t') \, dt' \, , \tag{19.42}$$

where $L(t')$ denotes the mean number of tissue dwelling larvae in a host at time t' with T the time-span over which the immune system retains memory of past infections. So, E is a measure of the number of larvae in the host during the time interval $(t - T, t)$. Note that with the form (19.42) different situations (for example, a small infection persistent for long time and a large infection persistent for a short time) can lead to the same values of E.

(ii) To account in a simple way for the biological facts in (B1), (B2) and (B3) we introduce an expression to describe the immune system's activity, namely

$$I \equiv I_{\alpha\beta}(E) = \frac{\alpha E^2}{\beta + E^2} \, , \tag{19.43}$$

where E is the input variable (19.42), α is the maximum functional activity of the host's immune response and β provides a measure of the sensitivity of the immune system. (Recall the predation response in the budworm model dynamics in Chapter 1.) According to (B1) α also reflects the nutritional status of the host being considered: we can think of α as a monotonic increasing function of the nutritional status. β also may be host specific since it seems likely that β also has a direct biological interpretation in genetic terms since different strains of mice differ in their immune response against parasitic infections.

(iii) Finally we have to incorporate (19.42) with (19.43) into a dynamical model for the complete host-parasite community. According to assumption (C), and independent of the specific dynamical situation under consideration, the activity of the host's immune system simply leads to an increase in the mortality of the adult parasites. This requires an additional loss term in the dynamical equations for the mean worm burden $M(t)$ of the form $-IM(t) < 0$, where I, the strength of immunological response, plays the role of a death rate for parasites; it depends on the level of infection.

Population Dynamics Model and Analysis

From the above, mice fed on low protein diets appear to have little or no immune response: we shall refer to these as the low protein diet group (LPG) and investigate the dynamics of their mean worm burden by a simple immigration-death model. On the other hand, hosts feeding on a high protein diet are expected to show an immune response.

Low Protein Model

Let us start by considering the parasite dynamics of the LPG. The parasites, harboured by a host population of constant size, are subdivided into 2 categories: larvae in the wall of the small intestine, and adult worms in the gut lumen. We model the dynamics of the mean number of larvae, L, per host by

$$\frac{dL}{dt} = \lambda_i - \mu DL, \quad i = 1, 2, 3, 4 \tag{19.44}$$

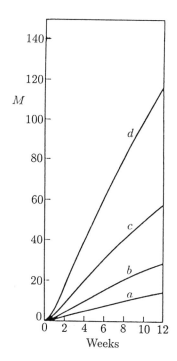

Fig. 19.10. Mean worm burden $M(t)$ for mice on the low protein diet (LPD) obtained from the analytical solution (19.46) of the model (19.44) and (19.45). The curves correspond to the different larvae infection rates λ_i, $i = 1, 2, 3, 4$: (a) 5, (b) 10, (c) 20, (d) 40 larvae per mouse per 2 weeks. Parameter values: $\mu = 0.125\,\mathrm{day}^{-1}$, $D = 1.56$, $\delta = 5.6 \times 10^{-3}\,\mathrm{day}^{-1}$. These curves correspond to those superimposed on Fig. 19.7 (a).

where λ_i, $i = 1, 2, 3, 4$, refer to the experimentally controlled infection rates, for example, 5, 10, 20 and 40 larvae per mouse per two weeks as in the experiments recorded in Fig. 19.7. Here $1/D = C_L$ denotes the proportion of larvae developing into adult worms after a developmental time delay t_L, here denoted by $1/\mu$. For the parasite *Heligmosoides polygyrus*, $t_L = 1/\mu \approx 8$ days and from Fig. 19.8 we estimate $C_L = 0.64$. We can now evaluate the net loss rate of the larval population per host as $\mu D \approx 0.195\, \mathrm{day}^{-1}$ which implies (i) an effective life span of a larval worm of $1/(\mu D) \approx 5.12$ days, and (ii) the natural larval mortality rate $\mu_0 = \mu(D - 1) \approx 0.07\, \mathrm{day}^{-1}$.

We model the dynamics of the mean adult worm burden, M, by

$$\frac{dM}{dt} = \mu L - \delta M , \tag{19.45}$$

where δ denotes the natural death rate of the adult worms in the absence of competitive or immunological constraints. We estimate $\delta = 5.6 \times 10^{-3}\, \mathrm{day}^{-1}$ from the experimental results of a single infection shown in Fig. 19.9. which implies an adult worm life-span of approximately 25 weeks.

Solutions of the linear equations (19.44) and (19.45), with the initial conditions $L(0) = M(0) = 0$, are simply

$$L(t) = \frac{\lambda_i}{\mu D}[1 - \exp(-\mu D t)]$$

$$M(t) = \frac{\lambda_i}{D}\Big\{\delta^{-1}[1 - \exp(-\delta t)] \tag{19.46}$$

$$+ (\mu D - \delta)^{-1}[\exp(-\mu D t) - \exp(-\delta t)]\Big\}, \quad i = 1, 2, 3, 4.$$

Fig. 19.10 plots $M(t)$ for $i = 1, 2, 3, 4$ for the first twelve weeks using the above estimates for the parameter values. These are the curves which are superimposed on the experimental resuts in Fig. 19.7 (a): there is very good quantitative agreement.

High Protein Model

With this diet the host's immune system comes into play and so we have to incorporate its action into the dynamical equation (19.45) for the worm burden. In line with the observation (iii) above, this equation now takes the form

$$\frac{dM}{dt} = \mu L - (\delta + I)M , \tag{19.47}$$

$$I = \frac{\alpha E^2}{\beta + E^2}, \quad E = \int_{t-T}^{t} L(t')\, dt' , \tag{19.48}$$

where I is the cumulative effect of increased mortality of the worms by the immune response.

The larvae equation is still taken to be (19.44) since we assume the immune response does not principally alter the larvae dynamics. The infection pattern in the laboratory situation is then given by (19.46) as

$$
L(t) = \begin{cases} 0, & t < 0 \\ \dfrac{\lambda_i}{\mu D}(1 - \exp[-\mu Dt]), & t > 0 \end{cases} \tag{19.49}
$$

where λ_i, $i = 1, 2, 3, 4$ are the different larval infection rates. This generates the immune sytem input function E given by (19.48): integration gives

$$
E(t) = \begin{cases} \dfrac{\lambda_i}{\mu D}\left\{t - (\mu D)^{-1}(1 - \exp[-\mu Dt])\right\}, & 0 < t < T \\ \dfrac{\lambda_i}{\mu D}\left\{T - (\mu D)^{-1}\exp(-\mu Dt)(1 - \exp[\mu DT])\right\}, & t > T \end{cases} \tag{19.50}
$$

which, as $t \to \infty$, asymptotes to the constant $\lambda_i T/\mu D$.

The high protein model consists of (19.44) and (19.47) and to solve it we must first obtain estimates for the immune system parameters T, α, and β, respectively the memory time from past infections, the maximum mortality contribution from the immune system and the worm burden the immune response is switched on. Accurate estimates of immunological memory time T are not available. Some data (Rubin et al. 1971) indicate that some mice retain active immunity against *Heligmosoides polygyrus* for at least 30 weeks after infection. On the basis of this we assume T is at least larger than the experimental duration time of 12 weeks of experiments: see Fig. 19.7.

Consider now the parameter α, which characterizes maximum functional activity of the host immune response and also reflects the nutritional status of the host. We can estimate it from the asymptotic steady state value $M(\infty) = M_\infty$ of the worm burden. Let us consider the highest infection rate λ_4, then from (19.50) in the limit $t \to \infty$, we have

$$
E = \frac{\lambda_4 T}{\mu D} ,
$$

which on substituting into (19.48) gives

$$
I \approx \alpha \quad \text{for} \quad T \gg \frac{\mu D\sqrt{\beta}}{\lambda_4} \tag{19.51}
$$

The experimentally observed saturation, as described in (B3), ensures the validity of this assumption on T. We can use (19.51) with (19.47) at the steady state to determine α, to get

$$
\alpha M_\infty = \mu L(\infty) - \delta M_\infty \quad \Rightarrow \quad \alpha = \frac{\lambda_4}{DM_\infty} - \delta . \tag{19.52}
$$

Since, within the experimental observation time, the system does not reach its final steady state, we use (19.52) to predict M_∞ as a *function of α* , namely

$$M_\infty = \frac{\lambda_4}{D(\alpha + \delta)} . \tag{19.53}$$

Finally we use the experimental data given in Fig. 19.7 (b) (d), which corresponds to the highest rate of infection, to determine the sensitivity of the immune system as measured by β. Note there that the mean adult worm burden rises to a maximum value M^* at a time t^* and then declines under the influence of host immunity, despite continual reinfection, to settle at the asymptotic steady state value M_∞. For the maximum point (M^*, t^*), $M^* = M(t^*)$, equations (19.47) and (19.48) gives

$$0 = \mu L - \delta M^* - \frac{\alpha E^2 M^*}{\beta + E^2} . \tag{19.54}$$

Since, in the laboratory situation, t^* satisfies

$$\frac{1}{\mu D} \ll t^* < T , \tag{19.55}$$

we use the first of (19.50) to get E and then solve (19.54) for β to get

$$\beta = \frac{E^2(\alpha M^* - \mu L + \delta M^*)}{\mu L - \delta M^*} , \tag{19.56}$$

where $L(t)$ and $E(t)$ are their values at $t = t^*$; as before $M^* = M(t^*)$. We estimate the values for $M^*(\approx 50$ worms) and $t^*(\approx 7$ weeks) from the experimental data in Fig. 19.7 (b) (d) and in turn use (19.56) in subsequent calculations to determine the sensitivity β, which is measured in worm^{2}day^2, as a function of α. So, we have used the experimental data and the fact that the laboratory situation is in the regime $t < T$ to determine the respective parameters α and β by using (19.52) and (19.56).

We can now analyze the complete nonlinear immigration-death model

$$\frac{dL}{dt} = \lambda_i - \mu DL, \quad i = 1, 2, 3, 4 ,$$
$$\frac{dM}{dt} = \mu L - (\delta + I)M , \tag{19.57}$$

where the acquired immune response function I is given in terms of E and L by (19.48). Numerical integration of (19.57) gives the solution for the mean adult worm burden as a function of time: the results are plotted in Fig. 19.11 for the first twelve weeks. The different curves again represent the different infection rates.

Here we have chosen α equal to $0.5\,\mathrm{day}^{-1}$, which implies $\beta \approx 6.1 \times 10^6$ worms^{2}day^2. With these, the solutions give a very satisfactory fit to the

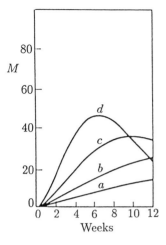

Fig. 19.11. The time evolution in the mean worm burden, $M(t)$, in mice hosts fed on a high protein diet for a 12-week period of repeated infection, obtained from a numerical integration of equations (19.57), which govern the population dynamics in the presence of host immune response. These curves are the ones used to compare with the experimental data in Fig. 19.7 (b). Parameter values: $\mu = 0.125\,\mathrm{day}^{-1}$, $D = 1.56$ and $\delta = 5.6 \times 10^{-3}\,\mathrm{day}^{-1}$ as in Fig. 19.10, and $\alpha = 0.5$, which implies $\beta \approx 0.757$, for the maximum functional activity of the immune system.

experimental data in Fig. 19.7. Since α is related to M_∞ by (19.53), we thus predict that a continuation of the present experimental setting eventually leads to an asymptotic steady state of $M_\infty = 4$ worms. If we restrict ourselves to the same genetic type of hosts and the same dietary conditions, the model can also be used to investigate more realistic situations, for example that in which the hosts are subjected to natural infection. We briefly discuss this below.

For even more general applications of (19.57), such as to arbitrary nutritional conditions or different strains of mice, further experiments are necessary to clarify: (i) the detailed functional dependence of the maximum functional activity α on the nutritional status of the hosts, (ii) the specific relationship of the sensitivity β to various strains of mice, and (iii) the size of the memory time (T). With these the system (19.67) can be used to predict the time-evolution and the final steady state of the mean worm burden dependence on the nutritional status and the genetic properties of the hosts being considered.

Among the goals of any mathematical modelling in epidemiology are: (i) to provide a proper mechanistic description of the field situation and (ii) to provide a sound basis for making practical predictions. Usually, however, a major difficulty is the reasonable estimation of the many parameters which are involved in the models. Controlled laboratory experiments, which study particular aspects of the complete dynamics, while keeping all other parts of the system under experimental control, have proved very useful in this respect. The experiments described here have specifically highlighted the role of the immune response. As a result we have been able to develop and exploit a simple but realistic mathematical model, which admits a full *quantitative* description of the population dynamics in the presence of host immune response.

At this point a few cautionary remarks should be made. First, the model as it stands does not, nor was it intended to, give a full picture of the underlying delicate biochemical and biocellular processes. It does, however, provide a quantitative picture of the macroscopic features of immune response: the per capita rate of limitation in parasite survival can be related quantitatively to the antigenic stimulus (that is, the exposure to infection). Secondly, The choice of the input function E for the immune system in (19.42) and in particular (19.43) is, of course, not unique; it seems, however, the most plausible one in view of the biological observations listed. In fact the qualitative features of the experimental data are reproduced even with a linear function $I(E)$ in place of the immune activity function in (19.48). However, numerical simulations show that this latter model assumption gives a more satisfactory, *simultaneous* fit of the four graphs corresponding to the four different infection rates, (Fig. 19.7 (b) (a)-(d)), than a linear version of (19.43). In summary then, the model is supported by the following facts: (i) it is in keeping with the biological observations, (ii) it provides a quantitative fit for the experimental data used to test it, and (iii) the parameters introduced are biologically meaningful and can be estimated.

The importance of an acquired immune response in human infection with several species of helminth parasites have also been shown, for example, in the recent immunological and epidemiological studies of Butterworth et al. (1985). They describe the immune response of 'resistant' and 'susceptible' Kenyan schoolchildren to infection with the blood fluke *Schistosoma mansoni*. The role of human immunity in controlling other worm infection is similarly well established. There is an urgent need for fieldwork studies: simple mathematical models of the type described and used here can be of enormous help in their design and interpretation. In addition, extension of the modelling technique to the 'real world' can provide a cheap and effective way of testing the efficiency of various parasite control programmes, without resort to lengthy and expensive field trials. Further modelling on the lines described in this section have been carried out by Berding et al. (1987) for further laboratory studies in which there is a genetically heterogeneous host population and in which there is natural transmission of the parasite. As before the mice populations had different protein diets. They also discuss the significance of the results from a real world medical viewpoint.

What is already abundantly clear is that in real world practical terms, the nutritional status of the host is an important factor in the population dynamics of a parasite infections, and clearly must not be ignored in the design of optimal health control policies.

*19.6 Age Dependent Epidemic Model and Threshold Criterion

In many diseases the chronological age of the individual is an important factor in assessing his vulnerability and infectiousness. For example, the data quoted by Bernoulli (1760) on the incidence and severity of smallpox with age is a vivid illustration: vulnerability and mortality go down markedly with age. A variety of age dependent models are discussed in in the book by Hoppensteadt (1975). Dietz

(1982), for example, proposes such a model for river blindness (onchocersiasis) and uses it to compare various possible control strategies.

Age may also be interpreted as the time from entry into a particular population class such as the susceptibles, infectives or the removed group. The two interpretations of age are often the same. With the specific case we analyse in Section 19.7, on a drug use epidemic model, age within a class, the users, is the relevant interpretation.

Consider the population to be divided into susceptibles, $S(t)$, and infectives, $I(a,t)$, where a is the age from exposure to the disease. The number of susceptibles decreases through exposure to the disease. The removal rate of susceptibles is taken to be

$$\frac{dS}{dt} = -\left[\int_0^\tau r(a')I(a',t)\,da'\right]S, \quad S(0) = S_0 . \tag{19.58}$$

That is, the removal due to infectives is weighted with a function $r(a)$ which is a measure of the infectiousness of the infectives. Since the infective is only infectious for a limited time, τ, this is the upper limit in the integral.

To get the equation for the infective population $I(a,t)$ we use a conservation approach. In a time Δ there is an advance in chronological age and infective class age from (t,a) to $(t+\Delta, a+\Delta)$. Conservation then says that the change in the number of infectives in a time Δ must be balanced by the number removed. We thus have, in time Δ,

$$I(a+\Delta, t+\Delta) - I(a,t) = -\lambda(a)I(a,t)\Delta$$

where $\lambda(a)$ is the age dependent removal factor. In the limit as $\Delta \to 0$ we then get, on expanding in a Taylor series, the partial differential equation

$$\frac{\partial I}{\partial t} + \frac{\partial I}{\partial a} = -\lambda(a)I . \tag{19.59}$$

At time $t = 0$ there is some given age distributed class of infectives $I_0(a)$. At $a = 0$ there is recruitment from the susceptible class into the infectives. Since *all* new infectives come from the susceptibles, the 'birth rate' $I(0,t)$ is equal to $-dS/dt$. Thus the boundary conditions for (19.59) are

$$I(a,0) = I_0(a), \quad I(0,t) = -\frac{dS}{dt}, \quad t > 0 . \tag{19.60}$$

The model integro-differential equation system now consists of (19.58)-(19.60) where $I_0(a)$ and S_0 are given. The functions $r(a)$ and $\lambda(a)$ are known, at least qualitatively, for the disease and in control procedures can often be manipulated.

An infection will not spread if the number of susceptibles expected to be infected by each infective drops below one. If the number exceeds one then the infection will spread and we have an epidemic. The number γ of initial susceptibles expected to be infected by each infective is

$$\gamma = S_0 \int_0^\tau r(a) \exp\left[-\int_0^a \lambda(a')\,da'\right] da \,. \tag{19.61}$$

As in (19.58), $r(a)$ here is the infective capability of an infective. It is weighted with an exponential function which is the probability of an initial infective surviving to age a: $\lambda(a)$ is the same as in (19.59). The threshold value for an epidemic is $\gamma = 1$ above which the infection spreads. We shall now show how the severity of the epidemic, as measured by the ratio $S(\infty)/S_0$, depends on γ. Clearly from (19.58) since $dS/dt \leq 0$, $S(t) \to S(\infty)$ where $0 \leq S(\infty) \leq S_0$.

We solve the mathematical problem (19.58)–(19.60) using characteristics. (A similar proceedure was used in Chapter 1, Section 1.7, in the single population model with age distribution.) The characteristics of (19.59) are the straight lines.

$$\frac{dt}{da} = 1 \quad \Rightarrow \quad a = t + a_0, \quad a > t \tag{19.62}$$
$$= t - t_0, \quad a < t$$

where a_0 and t_0 are respectively the age of an individual at time $t = 0$ in the given original population and the time of birth of an infective: see Fig. 19.12.

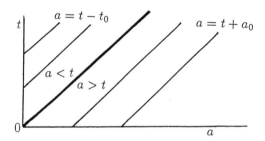

Fig. 19.12. Characteristics for the infectives equation (19.59). On $t = 0$, $I(a,0) = I_0(a)$, which is given, and on $a = 0$, $I(0,t) = -dS/dt$, $t > 0$.

The characteristics form of (19.59) is

$$\frac{dI}{da} = -\lambda(a)I \quad \text{on} \quad \frac{dt}{da} = 1 \,,$$

and so, with Fig. 19.12 in mind, integrating the last equation we get

$$I(a,t) = I_0(a_0) \exp\left[-\int_{a_0}^a \lambda(a')\,da'\right], \quad a > t$$
$$= I(0,a_0) \exp\left[-\int_0^a \lambda(a')\,da'\right], \quad a < t \,.$$

Thus, from (19.62),

$$I(a,t) = I_0(a-t) \exp\left[-\int_{a-t}^a \lambda(a') \, da'\right], \quad a > t$$

$$= I(0, a-t) \exp\left[-\int_0^a \lambda(a') \, da'\right], \quad a < t.$$

(19.63)

From (19.58) the solution $S(t)$ is

$$S(t) = S_0 \exp\left[-\int_0^t \left\{\int_0^\tau r(a)I(a,t') \, da\right\} dt'\right].$$

(19.64)

Using (19.63) for $I(a,t)$, in the ranges $a < t$ and $a > t$,

$$\int_0^\tau r(a)I(a,t') \, da = \int_0^t r(a)I(0, t'-a) \exp\left[-\int_0^a \lambda(a') \, da'\right] da$$

$$+ \int_t^\tau r(a)I_0(a-t') \exp\left[-\int_{a-t'}^a \lambda(a') \, da'\right] da$$

(19.65)

Since the time of infectiousness is τ, the last integral vanishes if $t > \tau$: we can think of it in terms of $r(a) = 0$ if $a > \tau$. For $S(t)$ in (19.64) we have, using (19.60) and (19.65),

$$\int_0^t \int_0^\tau r(a)I(a,t') \, da \, dt'$$

$$= -\int_0^t \int_0^{t'} r(a) \exp\left[-\int_0^a \lambda(a') \, da'\right] \frac{dS(t'-a)}{dt'} \, da \, dt'$$

$$+ \int_0^t \int_{t'}^\tau r(a)I_0(a-t') \exp\left[-\int_{a-t'}^a \lambda(a') \, da'\right] da \, dt'.$$

Interchanging the order of integration in the first integral on the right hand side we get

$$\int_0^t \int_0^\tau r(a)I(a,t') \, da \, dt$$

$$= -\int_0^t r(a) \exp\left[-\int_0^a \lambda(a') \, da'\right] (S(t-a) - S_0) \, dt + m(t),$$

(19.66)

where

$$m(t) = \int_0^t \int_{t'}^\tau r(a)I_0(a-t') \exp\left[-\int_{a-t'}^a \lambda(a') \, da'\right] da \, dt'.$$

(19.67)

Substituting (19.66) into (19.64) we then get

$$S(t) = S_0 \exp\left\{-m(t) + \int_0^t r(a) \exp\left[-\int_0^a \lambda(a')\,da'\right](S(t-a) - S_0)\,da\right\}.$$
(19.68)

If we now let $t \to \infty$, remembering that $r(a) = 0$ for $a > \tau$, we get, using γ defined in (19.61),

$$F = \exp[-m(\infty) + \gamma(F - 1)], \quad F = \frac{S(\infty)}{S_0}.$$
(19.69)

We are interested in the severity of the epidemic as measured by F, that is the fraction of the susceptible population that survives the epidemic, and how it varies with γ. For given $r(a)$, $I_0(a)$ and $\lambda(a)$, (19.67) gives $m(t)$ and hence $m(\infty)$. If $0 < m(\infty) = \varepsilon \ll 1$, Fig. 19.13 shows how F varies with γ. For each value of γ there are two roots for F but, since $S(\infty) \le S_0$, only the root $F = S(\infty)/S_0 \le 1$ is relevant. Note how the severity of the epidemic is small for ε small as long as $\gamma < 1$ but it increases dramatically, that is $S(\infty)/S_0 < 1$, for $\gamma > 1$. For example if $0 < \varepsilon \ll 1$ and $\gamma \approx 1.85$, $S(\infty)/S_0 \approx 0.25$.

Suppose a single infective is introduced into a susceptible population of size S_0. We can approximate this by writing $I_0(a) = \delta(a)$, the Dirac delta function, then

$$\int_0^\tau I_0(a)\,da = 1.$$

In this case, from (19.67),

$$\begin{aligned}
m(t) &= \int_0^t \int_{t'}^\tau r(a)I_0(a - t')\exp\left[-\int_{a-t'}^a \lambda(a')\,da'\right]da\,dt' \\
&= \int_0^t \int_{t'}^\tau r(a)\delta(a - t')\exp\left[-\int_{a-t'}^a \lambda(a')\,da'\right]da\,dt' \\
&= \int_0^\infty r(t')\exp\left[-\int_0^{t'} \lambda(a')\,da'\right]dt' \\
&= \frac{\gamma}{S_0}
\end{aligned}$$

from (19.61). Thus (19.69) becomes

$$F = \exp\left[\gamma\left(F - 1 - \frac{1}{S_0}\right)\right], \quad F = \frac{S(\infty)}{S_0}.$$
(19.70)

Since $1/S_0 \ll 1$ in general the solutions for F in terms of γ are typically as given in Fig. 19.13. Thus $\gamma > 1$ need *not* be large for a severe epidemic to occur. Therefore, it is the estimation of the parameter γ in (19.61) that is critical in the epidemiology of age dependent models. This we do in the following section

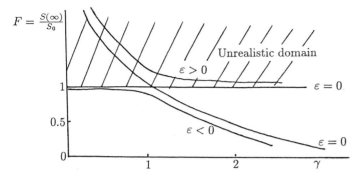

Fig. 19.13. Dependence of the epidemic severity $F = S(\infty)/S_0$, namely the fraction of the susceptible population who survive the epidemic, on the threshold parameter γ from (19.69), namely $F = \exp\left[-\varepsilon + \gamma(F - 1)\right]$. The only realistic values are $F = S(\infty)/S_0 \leq 1$.

for a very simple and primitive model of drug use: the model and anlysis was presented by Hoppensteadt and Murray (1981).

19.7 Simple Drug Use Epidemic Model and Threshold Analysis

The spread of the use of self-administered drugs, therapeutic and illicit, is in some cases a result of the enthusiastic proselytizing by a user in the initial stages of use. We describe here a simple illustrative model discussed by Hoppensteadt and Murray (1981) for the etiology of such a drug and show how to determine the threshold parameter γ. This entails the evaluation of the infectiousness which we relate to the response of the user to the drug. The novel feature of the epidemic model studied here is the inclusion of the user's personal response to the drug. The model is a pedagogical one: we do not have a specific drug in mind.

Suppose the drug is introduced into the blood stream in dosages $d(t)$ and let it be removed at a rate proportional to $c(t)$, the drug concentration in the blood; that is a first order kinetics removal. The governing equation for the blood concentration $c(t)$ is then

$$\frac{dc}{dt} = d(t) - kc, \quad c(0) = 0 , \tag{19.71}$$

where $k > 0$ is constant and $t = 0$ is the time the individual is first recruited as a user. In drug abuse the dosage $d(t)$ tends to be oscillatory or approximately periodic with a progressively decreasing period. The solution of (19.71) is

$$c(t) = e^{-kt} \int_0^t e^{kt'} d(t') \, dt' . \tag{19.72}$$

For many drugs the body has specific sites and it is the binding of these sites which evokes a response in the user. Denote the number of free sites, that is active or unbound, by $A(t)$, the number of bound, that is inactive, sites by $B(t)$ and the total number by N. We assume here that no new sites are being created so $A(t) + B(t) = N$. We take as a site binding model

$$\varepsilon \frac{dA}{dt} = \alpha B - \beta cA, \quad A(0) = N \,,$$
$$\varepsilon \frac{dB}{dt} = \beta cA - \alpha B, \quad B(0) = 0 \,,$$

$$(19.73)$$

where α, β and ε are positive constants. The inclusion of ε here is for later algebraic convenience. We are thus assuming that the rate of binding of active sites is proportional to the amount of the drug $c(t)$ in the body and the number of active sites available, that is $\beta cA/\varepsilon$. There is also a replenishment of the active sites proportional to the number of bound sites, that is $\alpha B/\varepsilon$. With $A + B = N$ the equation for B is given in (19.73).

Suppose now that the reaction $r(t)$ to the drug is proportional to the blood concentration and the number of free sites. We thus take it to be

$$r(t) = Rd(t)A(t) \,,$$

$$(19.74)$$

where $R > 0$ is a measure of the individual's response to the drug.

If the rate of binding is very fast, that is α and β are $O(1)$ and $0 < \varepsilon \ll 1$ in (19.73), the number of free and bound receptors reach equilibrium very quickly. Then, using $A + B = N$,

$$B = \frac{\beta cA}{\alpha} \quad \Rightarrow \quad A = \frac{\alpha N}{\alpha + \beta c}, \quad B = \frac{\beta Nc}{\alpha + \beta c} \,,$$

$$(19.75)$$

and the individual's response is

$$r = \frac{R\alpha Nc}{\alpha + \beta c} \,,$$

$$(19.76)$$

which is a Michaelis-Menten (cf. Chapter 5, Section 5.2) type of response which saturates to $r_{\max} = R\alpha N/\beta$ for large blood concentration levels c. Note that with B as in (19.75) the response $r = R\alpha B/\beta$, that is the response is proportional to the number of bound sites.

If ε in (19.73) is $O(1)$ we can incorporate it into the α and β; this is equivalent to setting $\varepsilon = 1$. Now with $B = N - A$ the equation for $A(t)$ from (19.73), with $\varepsilon = 1$, is

$$\frac{dA}{dt} = \alpha N - A(\alpha + \beta c), \quad A(0) = N$$

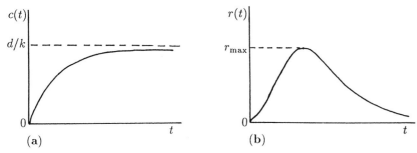

Fig. 19.14a,b. (a) The blood concentration $c(t)$ of the drug: from (19.78) it saturates to d/k after a long time. (b) The body's response to the drug from (19.79). Note the initial increase before it tails off with continuous drug use.

which has solution

$$A(t) = N \exp\left[-\int_0^t \{\alpha + \beta c(t')\}\, dt'\right]$$
$$+ \alpha N \int_0^t \exp\left[-\int_{t'}^t \{\alpha + \beta c(\tau)\}\, d\tau\right] dt' , \tag{19.77}$$

with $c(t)$ from (19.72).

If $d(t)$ is known we can carry out the integrations explicitly to get $c(t)$ and $A(t)$: it is algebraically rather complicated for even a simple periodic $d(t)$. Since the algebraic details in such a case initially tend to obscure the key elements we consider here the special case $d(t) = d$, a constant, and assume that the recovery rate of active sites from their bound state is very small, that is $\alpha \approx 0$. Then, from (19.72) giving $c(t)$ and the last equation giving $A(t)$, we have

$$c(t) = \frac{d(1 - e^{-kt})}{k} ,$$
$$A(t) = N \exp\left[-\frac{\beta d}{k}\left\{t + \frac{1}{k}(e^{-kt} - 1)\right\}\right] \tag{19.78}$$

and the response $r(t)$ from (19.74) is

$$r(t) = RcA = \frac{RNd}{k}(1 - e^{-kt}) \exp\left[-\frac{\beta d}{k}\left\{t + \frac{1}{k}(e^{-kt} - 1)\right\}\right] . \tag{19.79}$$

Fig. 19.14 illustrates the form of $c(t)$ and $r(t)$ from (19.78) and (19.79).

It is interesting to note that even with this very simple illustrative model, the response of an individual does not just increase with dosage: after an initial stage of increasing response it actually decreases with time.

Now consider the possibility of an epidemic of drug use appearing in a population S_0 of non-users after the introduction of a single user. We assume

$1/S_0 \ll 1$, as is reasonable, and so $F = S(\infty)/S_0$ is given by the solution $F < 1$ in Fig. 19.13 for the appropriate γ, which we now evaluate.

Here age is measured from the first time of using the drug. There is no time limit for infectiousness so in the definition (19.61) for γ we set $\tau = \infty$. From Fig. 19.14 (b) the response $r(t) \to 0$ as $t \to \infty$: that is the infectiousness, or proselytizing fervour, becomes less effective with time. For simplicity we assume the probability factor in (19.61) has λ constant and so

$$\gamma = S_0 \int_0^\infty r(t) e^{-\lambda t} \, dt . \tag{19.80}$$

We can now evaluate γ for various limiting situations in terms of the parameters α, β, γ and k in the user model (19.71)–(19.74).

In the case $d(t) = d$, a constant, we get Table 19.1 (from Hoppensteadt and Murray 1981) after some elementary algebra. It gives the user's response $r(t)$ and the corresponding epidemiological parameter γ. For example, in the case $0 < \varepsilon \ll 1$, (19.76) holds if $\alpha \ll \beta$, $r(t) \approx RN\alpha/\beta$, a constant, and (19.80) gives $\gamma \approx S_0 RN\alpha/(\alpha\beta)$. On the other hand if $0 < \varepsilon \ll 1$ and $\beta \gg \alpha$ then, from (19.75), $r(t) = RNc(t)$ and, with $c(t)$ from (19.78), γ is given, from (19.80), by

$$\gamma = S_0 \int_0^\infty \frac{RNd(1 - e^{-kt})}{k} e^{-\lambda t} \, dt = \frac{S_0 RNd}{\lambda(\lambda + k)} .$$

A similar type of asymptotic approach results in the other forms in Table 19.1.

Table 19.1

Case			$r(t)$	γ/S_0
(i)	$\varepsilon \ll 1,$	$\alpha \ll \beta$	$RN\alpha/\beta$	$RN\alpha/\lambda\beta$
(ii)	$\varepsilon \ll 1,$	$\beta \gg \alpha$	$RNc(t)$	$RND/[\lambda(\lambda + k)]$
(iii)	$\varepsilon \ll 1,$	$k \gg 1$	RND/k	$RND/k\lambda$
(iv)	$\varepsilon \ll 1,$	$k \ll 1, \quad \alpha/\beta D \ll 1$	$RNDt/[1 + (\beta Dt/\alpha)] \sim RN\alpha/\beta$	$RN\alpha/\beta\lambda$
(v)	$\varepsilon = 1,$	$k \ll 1$	$NDrt \exp[-2D\beta t]$	$RND/(2D\beta + \lambda)^2$
(vi)	$\varepsilon = 1,$	$k \gg 1$	$RD \exp[-D\beta t/k]/k$	$RND/k\lambda$

In the case of most self-administered drugs $0 < \varepsilon \ll 1$, that is the response is very fast. The possibility of an epidemic depends on relative magnitude of the various parameters in a simple way. This case is covered by (i)–(iv) in Table 19.1. For example if the rate of freeing of bound sites is much slower than the binding rate, $\beta \gg \alpha$ (case (ii)) then, since most sites will be bound, the user's reaction is small. This reduces the user's 'infectiousness' and hence the epidemic risk.

If we increase the cure rate, that is increase λ, there is a reduction in γ. Decreasing the individual's response, such as by education or chemotherapy, also reduces γ and hence reduces the possibility of a severe epidemic. The results are in line with a heuristic common sense approach.

If we define the critical population S_c by

$$S_c = \left\{ \int_0^\infty r(a) \exp\left[-\int_0^a \lambda(a') \, da' \right] da \right\}^{-1}, \qquad (19.81)$$

then if $S_0 > S_c$, which implies $\gamma > 1$, an epidemic occurs, whereas if $S_0 < S_c$ it does not. The sensitivity of S_c to the parameters can only really be determined if r and γ are known with some confidence.

Exercises

1. The dynamics of a directly transmitted viral microparasite has been modelled by the system

$$\frac{dX}{dt} = bN - \beta XY - bX, \quad \frac{dY}{dt} = \beta XY - (b+r)Y, \quad \frac{dZ}{dt} = rY - bZ$$

where b, β and r are positive constants and X, Y and Z are the number of susceptibles, infectives and immune populations respectively. Here the population is kept constant by births and deaths (with a contribution from each class) balancing. Show that there is a threshold population size N_c such that if $N < N_c = (b+r)/\beta$ the parasite cannot maintain itself in the population and both the infectives and the immune class eventually die out. The quantity $\beta N/(b+r)$ is the *basic reproductive rate* of the infection.

2. In a criss-cross venereal infection model, with the removed class permanently immune, the infection dynamics is represented by

with the usual notation for the suceptibles, infectives and the removed class. Briefly describe the assumptions made for its model system to be

$$\frac{dS}{dt} = -rSI', \quad \frac{dS'}{dt} = -r'S'I,$$

$$\frac{dI}{dt} = rSI' - aI, \quad \frac{dI'}{dt} = r'S'I - a'I',$$

$$\frac{dR}{dt} = aI, \quad \frac{dR'}{dt} = a'I',$$

where the parameters are all positive. The intial values for S, I, R, S', I' and R' are S_0, I_0, 0 and S_0', I_0', 0 respectively.

Show that the female and male populations are constant. Hence show that $S(t) = S_0 \exp[-rR'/a']$, deduce that $S(\infty) > 0$ and $I(\infty) = 0$ with similar results for S' and I'. Obtain the transcendental equations which determine $S(\infty)$ and $S'(\infty)$.

Show that the threshold condition for an epidemic to occur is at least one of

$$\frac{S_0 I'_0}{I_0} > \frac{a}{r}, \quad \frac{S'_0 I_0}{I'_0} > \frac{a'}{r'}.$$

What single condition would ensure an epidemic?

3. For the drug use epidemic model in Section 19.6 show that the values given for the threshold parameter γ/S_0 in cases (iii) and (iv) in Table 19.1 are as given.

20. Geographic Spread of Epidemics

The geographic spread of epidemics is even less well understood and much less well studied than the temporal development and control of diseases and epidemics. The usefulness of realistic models for the geotemporal development of epidemics be they infectious disease, drug abuse fads or rumours or misinformation, is obvious. The key question is how to include and quantify spatial effects. In this chapter we shall describe a diffusion model for the geographic spread of a general epidemic which we shall then apply to a well known historical epidemic, namely the ever fascinating mediaeval Black Death of 1347–50. We shall then discuss practical models for the current rabies epidemic which has been sweeping through continental Europe and is now approaching the north coast of France. These type of models, of course, are not restricted to one disease.

20.1 Simple Model for the Spatial Spread of an Epidemic

We consider here a simpler version of the epidemic model discussed in detail in Section 19.1 in the last chapter. We assume the population consists of only two populations, infectives $I(\mathbf{x}, t)$ and susceptibles $S(\mathbf{x}, t)$ which interact. Now, however, I and S are functions of the space variable $\mathbf{x}$ as well as time. We model the spatial dispersal of I and S by simple diffusion and initially consider the infectives and susceptibles to have the same diffusion coefficient D. As before we consider the transition from susceptibles to infectives to be proportional to rSI, where r is a constant parameter. This form means that rS is the number of susceptibles who catch the disease from each infective. The parameter r is a measure of the transmission efficiency of the disease from infectives to susceptibles. We assume that the infectives have a disease induced mortality rate aI; $1/a$ is the life expectancy of an infective. With these assumptions the model mechanism for the development and spatial spread of the disease is then

$$\frac{\partial S}{\partial t} = -rIS + D\nabla^2 S \, ,$$

$$\frac{\partial I}{\partial t} = rIS - aI + D\nabla^2 I \, , \tag{20.1}$$

where a, r and D are positive constants. These equations are (19.1) and (19.2) in Section 19.1 with the addition of diffusion terms. The problem we are now interested in consists of introducing a number of infectives into a uniform population with initial homogeneous susceptible density S_0 and determining the *geotemporal* spread of the disease.

Here we consider only the one-dimensional problem, later in Section 20.4 we present the results of a two-dimensional study. We nondimensionalize the system by writing

$$I^* = \frac{I}{S_0}, \quad S^* = \frac{S}{S_0}, \quad x^* = \left(\frac{rS_0}{D}\right)^{1/2} x ,$$

$$t^* = rS_0 t, \quad \lambda = \frac{a}{rS_0} .$$

$$(20.2)$$

where S_0 is a representative population and the model (20.1) becomes, on dropping the asterisks for notational simplicity,

$$\frac{\partial S}{\partial t} = -IS + \frac{\partial^2 S}{\partial x^2} ,$$

$$\frac{\partial I}{\partial t} = IS - \lambda I + \frac{\partial^2 I}{\partial x^2} .$$

$$(20.3)$$

The three parameters r, a and D in the dimensional model (20.1) have been reduced to only one dimensionless grouping, λ. The basic *reproduction rate* (cf. Section 19.1) of the infection is $1/\lambda$: it has several equivalent meanings. For example, $1/\lambda$ is the number of secondary infections produced by one primary infective in a susceptible population. It is also a measure of the two relevant time scales, namely that associated with the contagious time of the disease, $1/(rS_0)$, and the life expectancy, $1/a$, of an infective.

The specific problem we investigate here is the spatial spread of an epidemic wave of infectiousness into a uniform population of susceptibles. We want to determine the conditions for the existence of such a travelling wave and, when it exists, its speed of propagation.

We look for travelling wave solutions, in the usual way (cf. Chapters 9, 11, 12 and 17) by setting

$$I(x,t) = I(z), \quad S(x,t) = S(z), \quad z = x - ct ,$$

$$(20.4)$$

where c is the wave speed, which we have to determine. This represents a wave of constant shape travelling in the positive x-direction. Substituting these into (20.3) gives the ordinary differential system

$$I'' + cI' + I(S - \lambda) = 0, \quad S'' + cS' - \lambda IS = 0 ,$$

$$(20.5)$$

where the prime denotes differentiation with respect to z. The eigenvalue problem consists of finding the range of values of λ such that a solution exists with positive wave speed c and non-negative I and S such that

$$I(-\infty) = I(\infty) = 0, \quad 0 \le S(-\infty) < S(\infty) = 1 . \tag{20.6}$$

The conditions on I imply a pulse wave of infectives which propagates into the uninfected population. Fig. 20.1 shows such a wave: Fig. 20.4 below, which is associated with the spread of a rabies epidemic wave, is another example, although there only the infectious population I diffuses.

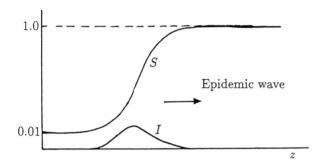

Fig. 20.1. Travelling epidemic wave of constant shape, calculated from the partial differential equation system (20.5) with $\lambda = 0.75$ and initial conditions with compact support compatible with (20.6). Here a pulse of infectives (I) moves into a population of susceptibles (S) with speed $c = 1$ which in dimensional terms from (20.2) is $(rS_0 D)^{1/2}$.

The system (20.5) is a fourth order phase space system. We can determine the lower bound on allowable wave speeds c by using the same technique we employed in Chapter 11, Section 11.2 in connection with wave solutions of the Fisher equation. Here we linearise the first of (20.5) near the leading edge of the wave where $S \to 1$ and $I \to 0$ to get

$$I'' + cI' + (1 - \lambda)I \approx 0 , \tag{20.7}$$

solutions of which are

$$I(z) \propto \exp\left[-c \pm \{c^2 - 4(1 - \lambda)\}^{1/2}\right] . \tag{20.8}$$

Since we require $I(z) \to 0$ with $I(z) > 0$ this solution cannot oscillate about $I = 0$, otherwise $I(z) < 0$ for some z. So, if a travelling wave solution exists, the wave speed c and λ must satisfy

$$c \ge 2(1 - \lambda)^{1/2}, \quad \lambda < 1 . \tag{20.9}$$

If $\lambda > 1$ no wave solution exists so this is the necessary threshold condition for the propagation of an epidemic wave. From (20.2), in dimensional terms the threshold condition is

$$\lambda = \frac{a}{rS_0} < 1 . \tag{20.10}$$

This is the same threshold conditon found in Section 19.1 for an epidemic to exist in the spatially homogeneous situation.

With our experience with the Fisher equation we expect such travelling waves to be unstable to initial conditions unless they have compact support (recall Sections 11.2 and 11.3, in Chapter 11) but, on the other hand, the computed wave solutions from the full nonlinear system will, except in exceptional conditions, evolve into a travelling wave form with the minimum wave speed $c = 2(1-\lambda)^{1/2}$. In dimensional terms, using (20.2), the wave velociy, V say, is given by

$$V = (rS_0D)^{1/2}c = 2(rS_0D)^{1/2}\left[1 - \frac{a}{rS_0}\right]^{1/2} , \qquad \frac{r}{aS_0} < 1 . \tag{20.11}$$

The travelling wave solution $S(z)$ cannot have a local maximum, since there $S' = 0$ and the second of (20.5) shows that $S'' = IS > 0$, which implies a local minimum. So $S(z)$ is a monotonic increasing function of z. By linearizing the second equation of (20.5) as $z \to \infty$, where $S = 1 - s$, s small, we have

$$s'' + cs' - I = 0 ,$$

which, with $I(z)$ from (20.8), shows that

$$S(z) \sim 1 - O\left(\exp\left[\{-3c \pm [c^2 - 4(1 - \lambda)]^{1/2}\}z/2\right]\right)$$

and so, as $z \to \infty$, $S(z) \to 1$ exponentially.

The threshold result (20.10) has some important implications. For example, we see that there is a minimum critical population density, $S_c = a/r$. for an epidemic wave to occur. On the other hand for a given population S_0 and mortality rate a, there is a critical transmission coefficient $r_c = a/S_0$ which, if not exceeded, prevents the spread of the infection. With a given transmission coefficient and susceptible population we also get a threshold mortality rate, $a_c = rS_0$, which, if exceeded, prevents an epidemic. So, the more rapidly fatal the disease is, the less chance there is of an epidemic wave moving through a population. All of these have implications for control strategies. The susceptible population can be reduced through vaccination or culling. For a given mortality and population density S_0, if we can, by isolation, medical intervention and so on, reduce the transmission factor r of the disease it may be possible to violate condition (20.10) and hence again prevent the spread of the epidemic. Finally with $a/(rS_0) < 1$ as the threshold criterion we note that a sudden influx of susceptible population can raise S_0 above S_c and hence initiate an epidemic.

Here we have considered only a two species epidemic model. We can extend the analysis to a three species SIR system. It becomes, of course, more complicated. In Sections 20.4-20.6 we discuss in some detail such a model for a current epidemic.

20.2 Spread of the Black Death in Europe 1347–1350

Historical Aside on the Black Death and Plague in the 20th Century

The fascination with the Black Death, the catastrophic plague pandemic that swept through Europe in the mid-14th century, has not abated with the passage of time. Albert Camus', *The Plague*, published in (1947) is one example, in a modern context. In the many accounts of the Black Death over the centuries, whether factual or romanticized, a vision has been conjured up of horrific carnage, wild debauchery, unbelievable acts of courage and altruism, and astonishing religious excesses.

The Black Death, principally bubonic plague, was caused by an organism (*Bacillus pestis*) and was transmitted by fleas, mainly from black rats, to man. It was generally fatal. The article by Langer (1964) gives a graphic description and some of the relevant statistics. The recent historical article by McEvedy (1988) discusses the pandemic's progress and surveys some of the current thinking on the periodic occurrences of bubonic plague. The plague was introduced to Italy in about December 1347, brought there by ship from the East where it had been raging for years. During the next few years it spread up through Europe at approximately 200–400 miles a year. About a quarter of the population died and approximately 80% of those who contracted the disease died within 2–3 days. Fig. 20.2 shows the geotemporal spread of the wave front of the disease.

After the Black Death had passed, around 1350, a second major outbreak of plague appeared in Germany in 1356. From then on periodic outbreaks seemed to occur every few years although none of them were in the same class as regards severity as the Black Death epidemic of 1347. In Section 20.3 we shall describe an obvious extension to the simple model in this section which takes into account the partial recovery of the population after the passage of an epidemic wave. Including this, results in periodic outbreaks, smaller ones, appearing behind the main front: see Fig. 20.6 below. Figs. 20.8 and 20.9 in Section 20.4, which considers a three-species model for the spatial spread of a rabies epidemic, exhibit even more dramatic periodic epidemic waves which follow the initial outbreak. There we can estimate the period of the recurring outbreaks analytically.

There was a great variety of reactions to plague in medieval Europe (just as there is to plague to-day and the current AIDS epidemic). Groups of penitents, vigorously flagellating their half naked bodies and preaching the coming of the end of the world, wandered about the countryside; some of the elegant and beautifully carved ivory handles of the flails of the richer flagellants survive. Cures for the plague abounded during this period. One late 15-th century cure

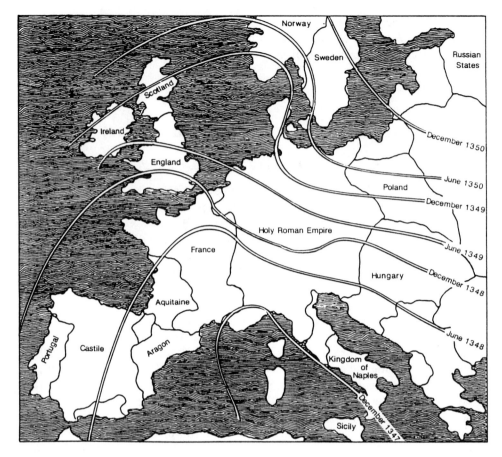

Fig. 20.2. Appriximate chronological spread of the Black Death in Europe from 1347–50. (Redrawn from Langer 1964)

was recently discovered in Westfalen-Lippe in northwest Germany and involved the following preparation. The tip of an almost hatched egg was cut off and the brood allowed to run out. The remaining egg yolk was mixed with raw saffron and the egg refilled and resealed with the shell pieces originally removed. The egg was afterwards fried until it turned brown. The recipe then called for the same amount of white mustard, some dill, a cranes beak and theriak (a popular quack medicine of the time). The mixture had to be swallowed by the victim who had to eat nothing more for 7 hours. There is no record of how effective this cure was!

The disease, of which there are three kinds, bubonic, pneumonic and septicemic, is caused by a bacillus carried primarily by fleas which are in turn carried by rats, mice and a host of other animals. Septicemic plague involves the bacilli multiplying extremely rapidly in the victim's blood and is almost invariably fatal, whether treated or not; the victim usually dies very quickly and often suddenly. Septicemic plague often develops from the pneumonic form and is extremely contagious. There are descriptions of plague victims who suddenly sat down and simply keeled over dead. These could well have been septicemic cases who contracted it from the coughs of pneumonic victims. Children at the time of the Great Plague of London from 1664–66, which peaked in 1665, used to sing the English nursery rhyme

> "Ring-a ring o'roses
> A pocket full of posies
> A-tishoo, A-tishoo
> We all fall down."

which is believed to date from that period. Onions and garlic were held to the nose, in 'posies' perhaps, to keep out the bad odours that were thought to be the cause of the disease.

There is considerably more data and information about the Great Plague of London in 1665 than is available about the Black Death. The people's reaction, however, seems not to have been dissimilar – fewer overt extreme penitents perhaps. The diarist Samuel Pepys describes the scene in my own university town of Oxford as one of 'lewd and dissolute behaviour'. Plus ca change...! Daniel Defoe's journal (see Brayley 1722) of the epidemic vividly conjures up a contemporary image and makes fascinating reading: 'It was then indeed, that man withered like the grass and that his brief earthly existence became a fleeting shadow. Contagion was rife in all our streets and so baleful were its effects, that the church-yards were not sufficiently capacious to receive the dead. It seemed for a while as though the brand of an avenging angel had been unloosed in judgement.'

There is a widely held belief that plague more or less ceased to be a problem after the Great Plague of London. This is far from the case, however, as clearly documented in the book by Gregg (1985). The last plague pandemic started in Yunnan in China about 1850 and only finished officially, according to the World Health Organisation, in 1959: more than 13 million deaths have been attributed to it, and it affected most parts of the world. The reported cases (and through ignorance or political expediency the figures must clearly be considered lower bounds) since 1959 makes it clear that plague epidemics are still with us. The thousands who died of it during the Vietnam war, particularly between 1965–75, is a dramatic case to point.

Plague was brought by ship to the northwest of America around 1900. About 200 deaths were recorded in the three year San Francisco epidemic which started just after the earthquake in 1906. As a result, the western part of the U.S.A., particularly New Mexico, is now one of the two largest residual foci of plague (in mice and voles particularly) in the world – the other is in Russia. The plague bacillus has spread steadily eastwards from the west coast and in 1984 was found among animals in the mid-west. The wave front has moved on average about 35 miles a year. The disease is carried by a large number of native wild animals. Rats are by no means the sole carrier: it has been found in nearly 30 different mammals including, for example, squirrels, chipmunks, coyotes, prairie dogs, mice, voles, domestic pets, and bats. The present complacency about the relatively small annual number of plague deaths is hardly justified. If, or rather *when*, plague reaches the east coast of the U.S.A. with its large urban areas, the potential for a serious epidemic will be considerable. New York, for example, has an estimated rat population of one rat per human; and mice – also effective disease carriers – probably number more. The prevailing lack of both concern and knowledge about the plague is dangerous. Plague symptoms are often not recognized or, at best, only belatedly diagnosed. Therefore the victim is free to expose a substantial number of people to the disease, particularly if it is pneumonic plague which is one of the most infectious diseases known.

To return now to our modelling, let us apply our simple epidemic model to the spread of the Black Death. We first have to estimate the relevant parameters, not a simple task with the paucity of hard facts about the social conditions of the time. Noble (1974) used such a model to investigate the spread of the plague and, after a study of the known facts, suggested approximate values for the parameters, some of which we use.

There were about 85,000,000 people in Europe in 1347 which gives a population density $S_0 \approx 50/\text{mile}^2$. It is particularly difficult to estimate the transmission coefficient r and the diffusion coefficient D. Let us suppose that the spread of news is governed by diffusion with a diffusion coefficient D. The time to cover a distance L miles purely by diffusion is then $O(L^2/D)$ years. Suppose, with the limited communications that existed at the time, that news and minor gossip, say, travelled at approximately 100 miles/year; this gives a value of $D \approx 10^4 \text{ miles}^2/\text{year}^2$. To transmit the disease the fleas have to jump from rats to humans and humans have to be close enough to infect other humans; this is reflected in the value for r. Noble (1974) estimated r to be $0.4 \, \text{mile}^2/\text{year}$. He took an average infectious period of two weeks (too long probably), which gives a mortality rate $a \approx 15/\text{year}$. These give $\lambda = a/(rS_0) \approx 0.75$. With the wave speed given by (20.11) in terms of the model parameters, we then get the speed of propagation, V, of the plague as

$$ V = 2(rS_0D)^{1/2} \left[1 - \frac{a}{rS_0}\right]^{1/2} \approx 140 \, \text{miles/year} . $$

Although this is somewhat lower than the speed of 200–400 miles/year, quoted by Langer (1964), it is not an unreasonable comparison in view of the gross estimates used for the unknown, and really, undeterminable parameters.

Of course, such a model is extremely simple and does not take into account a number of factors, such as the non-uniformity in population density, the stochastic element and so on. Nevertheless it does indicate certain global features of the geographic spread of an epidemic. As we noted in Section 19.1 in the last chapter, the stochastic model studied by Raggett (1982) for the plague epidemic of 1665–6 in the village of Eyam did not give as good comparison with the data as did the deterministic model. Stochastic elements however, are more important in spatial models, particularly when the numbers invovled are small.

If we now refer to Fig. 20.1 again, we see that, just as in the spatially uniform epidemic system situation discussed in Section 19.1 in the last chapter, after the epidemic has passed a proportion of the susceptibles have survived. It would be useful to be able to estimate this survival fraction analytically. This we can do in the following preliminary model for the spatial spread of rabies.

20.3 The Spatial Spread of Rabies Among Foxes I: Background and Simple Model

Rabies is widespread throughout the world and epidemics are quite common: Macdonald and Voigt (1985) discuss the global incidence of the disease and list the main animal carriers in the world scene. During the past few hundred years, Europe has been repeatedly subjected to rabies epidemics. It is not known why rabies died out some 50 or so years before the current epidemic started. The analysis of the models here, however, will provide a possible scenario.

The present epizootic (an epidemic in animals) seems to have started about 1939 in Poland and it has moved steadily westward at a rate of 30–60 km per year. It has been slowed down, only temporarily, by such barriers as rivers, high mountains and autobahns. The red fox is the main carrier and victim of rabies in the current European epidemic. A rabies epidemic is also moving rapidly up the east coast of America: the main vector here is the racoon. In this epidemic the progress was considerably enhanced by the importation into Virginia (by hunting clubs) of infected racoons from Georgia and Florida.

Rabies, a viral infection of the central nervous system, is transmitted by direct contact, and the dog is the principal transmitter of the disease to man. The incidence of rabies in man, at least in Europe and America, is now rare, with only very few deaths a year, but with considerably more in underdeveloped countries. It is a particularly horrifying disease for which there is no known case of a recovery once the disease has reached the clinical stage. During the incubation stage, vaccine seems to be 100% effective. The effect of rabies on other mammals, domestic and wild, however, is serious. In France, in 1980 alone, 314 cases of rabies in domestic animals were reported and 1280 cases in wild animals. The incidence of rabies in bats is now giving cause for concern in many countries. Rabies is a frightening disease which justifiably gives just cause for concern and warrants extensive study and development of control strategies, a subject we discuss later in Section 20.5.

Red foxes account for about 70% of the recorded cases in Western Europe. Although Britain has effectively been free from rabies since about 1900, the disease is likely to be re-introduced in the near future through the illegal importation of pets or even by infected bats, which are known to exist on the continent. The problem will be particularly serious in Britain because of the high rural and urban density of foxes, dogs and cats. In Bristol, for example, the density is of the order of 12 foxes/km^2 as compared with a rural population of 2–4 foxes/km^2. It is the comparatively high urban racoon density which is responsible for the current rapid spread of rabies up the east coast of America. The book on the fox and rabies by Macdonald (1980) provides many of the facts and data for Britain. General data on rabies in Europe is available from the *Centre National d'Etudes sur la Rage* in France (La Rage 1977). The books edited by Kaplan (1977) and Bacon (1985) are specifically concerned with the population dynamics of rabies and provide biological and ecological background together with useful data on the disease.

It is important to understand how the rabies epizootic wave front progresses into uninfected regions, what control methods might halt it and how the various parameters affect them. The remaining sections of this chapter will be concerned with these specific spatial problems. The material primarily comes from the model of Murray, Stanley and Brown (1986) and, in this section, from the much simpler, but less realistic, model of Källen, Arcuri and Murray (1985). The models and control strategies we propose in Section 20.5 are specifically related to the current European fox epizootic but the type of model is applicable to many other spatially propagating epidemics.

The spatial spread of epidemics is usually a very complex process, and rabies is no exception. In modelling such a complex process we can try to incorporate as many of the facts as possible, which necessarily involves many parameters, estimations of which are difficult to obtain with extant data. An alternative approach is to start with as simple a model as possible but which captures the key elements and for which it is possible to determine estimates for the fewer parameters. There is a trade-off between comprehensiveness and thus complexity, and the difficulty of estimating many parameters and a simpler approach in which parameter values can be reasonably assessed. For the models in this chapter we have opted for the latter strategy. In spite of their simplicity they nevertheless pose highly relevant practical questions and give estimates for various characteristics of importance in the spatial spread of diseases. Although in this section we describe and analyse a particularly simple model it is one for which we can obtain useful analytical results.

Although many animals are involved, a basic, and reasonable, assumption is that the ecology of foxes, the principal vectors, determines the dynamics of the spread of rabies. We further assume that the spatial spread of the epizootic is due primarily to the random erratic migration of rabid foxes. Uninfected foxes do not seem to wander far from their territory (Macdonald 1980). We divide the fox population into two groups – susceptible and rabid. Although the resulting model captures certain aspects of the spatial spread of the epizootic front, it leaves out a basic feature of rabies, namely the long incubation period of between 12 and 150 days from the time of an infected bite to the onset of the clinical infectious stage. We include this in the more realistic model discussed in Section 20.4.

To control, and ideally prevent, the spread of the disease it is important to have some understanding of how rabies spreads so as to assess the effects of possible control strategies. It is with this in mind that we first study a particularly simple modified version of the epidemic model system (20.1), which captures some of the key elements in the spread of rabies in the fox population. We shall then use it to derive some estimates of essential facts about the epizootic wave.

We consider the foxes to be divided into two groups, infectives I, and susceptibles S; the infectives consist of rabid foxes and those in the incubation stage. The principal assumptions are: (i) The rabies virus, contained in the saliva of the rabid fox, is transmitted from the infected fox to the susceptible fox. Foxes become infected at an average rate per head, rI, where r is the transmission coefficient which measures the rate of contact between the two groups. (ii) Rabies is invariably fatal and foxes die at a per capita rate a; that is the life expectancy of an infected fox is $1/a$. (iii) Foxes are territorial and divide the countryside into non-overlapping ranges. (iv) The rabies virus enters the central nervous system and induces behavioral changes in the fox. If the virus enters the spinal cord it induces paralysis whereas if it enters the limbic system it induces transient aggression during which it loses its sense of territory and the fox wanders about in a more or less random way. So, we assume that it is only the infectives which disperse with diffusion coefficient D km^2/year. With these assumptions our model is then (20.1) except that the susceptible foxes do not disperse. We exclude here the migration of cubs seeking their own territory. When they do move they try to stay as close to their original territory as possible. The model system in one dimension is then

$$\frac{\partial S}{\partial t} = -rIS \,,$$

$$\frac{\partial I}{\partial t} = rIS - aI + D\frac{\partial^2 I}{\partial x^2} \,. \tag{20.12}$$

From the analysis in the last section we expect this system to possess travelling wave solutions, whose speed of propagation depends crucially on the parameter values. The realistic estimation of these few parameters is important but still not easy.

Using the nondimensionalisation (20.2), the system (20.12) becomes (cf. (20.3))

$$\frac{\partial S}{\partial t} = -IS \,,$$

$$\frac{\partial I}{\partial t} = IS - \lambda I + \frac{\partial^2 I}{\partial x^2} \,, \tag{20.13}$$

where now S, I, x and t are dimensionless, and, as in the last section, $\lambda = a/rS_0$ is a measure of the mortality rate as compared with the contact rate. As before the contact rate is crucial and is not known with any confidence. We expect the threshold value to be again $\lambda = 1$ but we now verify this (see also Exercise 2).

Travelling wavefront solutions of (20.13) are of the form

$$S(x,t) = S(z), \quad I(x,t) = I(z), \quad z = x - ct \,. \tag{20.14}$$

where c is the wave speed and we look for solutions satisfying the boundary conditions

$$S(\infty) = 1, \quad S'(-\infty) = 0, \quad I(\infty) = I(-\infty) = 0 . \qquad (20.15)$$

Refer back to Fig. 20.1 for the type of wave anticipated. Note that it is the *derivative* of $S(z)$ which tends to zero as $z \to -\infty$ since we anticipate a residual number, as yet undetermined, of susceptible foxes to survive the epidemic. With (20.14) the system (20.13) becomes

$$cS' = IS ,$$
$$I'' + cI' + I(S - \lambda) = 0 . \qquad (20.16)$$

Linearizing about $I = 0$ and $S = 1$ exactly as we did in the last section and requiring I to be always non-negative, we find that this requires $\lambda < 1$, in which case the wave speed

$$c \geq 2(1 - \lambda)^{1/2}, \quad \lambda < 1 . \qquad (20.17)$$

With this specific model we are able to take the analysis further and find the actual fraction of susceptibles which survives the epidemic. From the first of (20.16), $I = cS'/S$, which on substituting into the second equation gives

$$I'' + cI' + \frac{cS'(S - \lambda)}{S} = 0 .$$

Integration gives

$$I' + cI + cS - c\lambda \ln S = \text{constant} .$$

Using the boundary conditions as $z \to \infty$ from (20.15), where $S = 1$, $I = 0$ and with $I' = 0$, we determine the constant to be c. If we now let $z \to -\infty$, again using (20.15) with $I = I' = 0$, we get the following transcendental equation for the surviving susceptible population, σ say, after the passage of the epizootic wavefront:

$$\sigma - \lambda \ln \sigma = 1, \quad \lambda < 1, \quad \sigma = S(-\infty) , \qquad (20.18)$$

which is independent of c. Writing this in the form

$$\frac{\sigma - 1}{\ln \sigma} = \lambda < 1 \quad \Rightarrow \quad 0 < \sigma < \lambda < 1 . \qquad (20.19)$$

From (20.19), with $\lambda = 0.4$, $\sigma = 0.1$ for example, whereas with $\lambda = 0.7$, $\sigma = 0.5$. λ is a measure of the severity of the epidemic. The smaller λ the fewer suceptibles survive; in other words, the worse the epidemic. Fig. 20.3 illustrates the surviving susceptible fraction σ as a function of λ obtained from (20.18): the curve was obtained by plotting λ as a function of σ.

The critical bifurcation value for λ is $\lambda = 1$, which in dimensional terms, from (20.2), means $a/(rS_0) = 1$. If $\lambda > 1$ no epidemic wave can propagate. This is to be expected since if $a > rS_0$ it means the mortality rate is greater than the rate of recruitment of new infectives. As before this bifurcation result says that

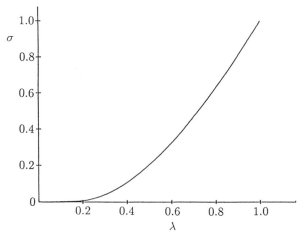

Fig. 20.3. The fraction, σ, of the original susceptible fox density which survive, after the passage of the epidemic wave, as a function of the epidemic severity: here, in terms of the original dimensional variables, $\sigma = S(-\infty)/S_0$ and $\lambda = a/(rS_0)$.

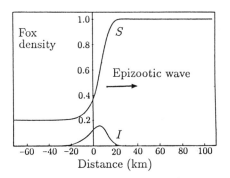

Fig. 20.4. Dimensionless epidemic wavefront solutions for the susceptible (S) and infected (I) fox populations computed from (20.13): here $\lambda = 0.5$. The wave speed is $c = \sqrt{2}$. Note the qualitative similarity with Fig. 20.1.

given r and a, there is a critical minumum fox density $S_c = a/r$ below which rabies cannot persist in the population and any infectives introduced will not cause an epidemic.

When rabies does persist, that is $\lambda < 1$, the computed speed of propagation of the epidemic wave is the minimum of the allowable speeds, namely $c = 2(1-\lambda)^{1/2}$, which in dimensional terms from (20.17) and (20.2) is

$$c = 2[D(rS_0 - a)]^{1/2} . \tag{20.20}$$

Fig. 20.4 shows an example of the computed travelling front solutions for S and I, from (20.13), for $\lambda = 0.5$. From Fig. 20.3 with $\lambda = 0.5$, The surviving fraction of susceptibles $\sigma \approx 0.2$.

Let us now compare the qualitative form of the susceptible fox population in the epidemic in Fig. 20.4 with that obtained from data from continental Europe

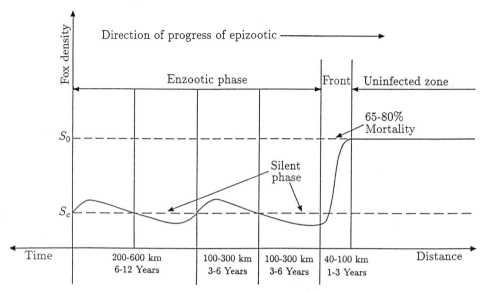

Fig. 20.5. Fluctuations in the susceptible fox population density as a function of the passage of the rabies epizootic obtained from data from Centre National d'Etudes sur la Rage 1977. In dimensional terms, S_0 is the uninfected susceptible population ahead of the epidemic wave. Note the periodic, but decreasing, fluctuations in S, which follow the main wavefront, as S tends to its steady state. (Redrawn from Macdonald 1980)

as illustrated in Fig. 20.5. There is a clear schematic difference in the behaviour behind the front in the two figures. The model (20.13) is only intended to cover the passage of an epidemic *front*. Clearly after the passage of the wavefront the suceptible population will start to increase again since the foxes find themselves in an environment which admits a larger carrying capacity. In other words, the time scale of the model (20.13) is considerably shorter than that associated with the oscillations in Fig. 20.5. To include in our model the situation which obtains after the front has passed we must include a term for the fox reproduction. If we model this by a simple logistic growth, the equation for the susceptibles in place of the first of (20.13) becomes

$$\frac{\partial S}{\partial t} = -rIS + BS\left(1 - \frac{S}{S_0}\right),\qquad(20.21)$$

where B is the linear growth rate. With the same non-dimensionalization (20.2) as before, the model now becomes

$$\frac{\partial S}{\partial t} = -IS + bS(1 - S),$$
$$\frac{\partial I}{\partial t} = I(S - \lambda) + \frac{\partial^2 I}{\partial x^2},\qquad(20.22)$$

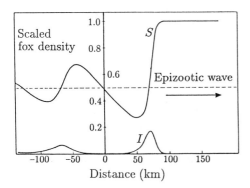

Fig. 20.6. Travelling epidemic wave solution for the susceptible (S) and infective (I) foxes from (20.22) when logistic growth is taken into account in the susceptible fox population: parameter values $b = 0.05$, $\lambda = 0.5$. Note the qualititative similarity with the data illustrated in Fig. 20.5. The initial front is succeeded by recurring, but smaller, outbreaks of the disease. (After Källén, Arcuri and Murray 1985)

where $b = B/rS_0$, that is the ratio of linear birth rate to the basic rate of infection per infective. Fig. 20.6 shows an example of the resulting epidemic wave of susceptibles and infectives, obtained by numerically solving (20.22): there is now good qualitative comparison between the results from this model and the data recorded in Fig. 20.5. The oscillations are decaying and eventually $S \to \lambda$ and $I \to b(1 - \lambda)$, the steady state solutions of (20.22), far behind the front.

Although the wavelength of the quasi-periodic outbreaks in both time and space are given by the numerical solutions, we can obtain some useful analytical results even for this more complex model (20.22). Let us start with the dimensional version of (20.22), namely

$$\frac{\partial S}{\partial t} = -rSI + BS\left(1 - \frac{S}{S_0}\right) ,$$

$$\frac{\partial I}{\partial t} = rSI - aI + D\frac{\partial^2 I}{\partial x^2} .$$

(20.23)

If we now introduce the nondimensional quantities

$$U = \frac{S}{S_0}, \quad V = \frac{rI}{BS_0}, \quad t^* = BT, \quad x^* = \left(\frac{B}{D}\right)^{1/2} x ,$$

$$\lambda = \frac{a}{rS_0}, \quad \alpha = \frac{rS_0}{B} ,$$

(20.24)

the dimensionless model becomes, on omitting the asterisks for notational simplicity,

$$U_t = U(1 - U - V) ,$$

$$V_t = \alpha V(U - \lambda) + V_{xx} .$$

(20.25)

These equations are exactly the same as equations (12.3) in Chapter 12, but with α and λ here in place of a and b there, and is the system we studied in detail in Section 12.2. The steady state solutions of (20.25) are $(0,0)$, $(1,0)$ and $(\lambda, 1 - \lambda)$ with the latter existing in the positive quadrant only if $\lambda < 1$. If we look for travelling wave solutions in the usual way (cf. equations (12.5)), the analysis of the three dimensional phase space (U, V, W), where $W = V'$, is given in Section 12.2. There we showed that with $\lambda < 1$ a travelling wave solution exists which joins the steady states $(1,0)$ and $(\lambda, 1 - \lambda)$. We also showed that a threshold $\alpha = \alpha^*$ exists such that if $\alpha > \alpha^*$ the approach to the steady state $(\lambda, 1 - \lambda)$ is oscillatory, whereas if $\alpha < \alpha^*$ it is monotonic (cf. Fig. 12.3). The computed solution in Fig. 20.6 is an example with $\alpha > \alpha^*$.

Let us now return to the observation in Section 20.2 about the subsequent outbreaks of plague which followed the initial Black Death epidemic. If we modify the susceptible equation in the model (20.1) to take into account the recovery of the population we again get subsequent periodic outbreaks of the disease following the initial epidemic similar to those shown in Fig. 20.5 and Fig. 20.6.

In spite of the simplicity of the model discussed here the results qualitatively capture some of the major phenomena observed. As with so many of the models we have discussed, even such a simple approach can elicit highly relevant questions.

20.4 The Spatial Spread of Rabies Among Foxes II: Three Species (SIR) Model

To be of practical use in developing control strategies to contain the spatial spread of an epidemic, we should consider more realistic and hence more complex models, which allow for quantitative comparison with known data and let us make practical predictions with more confidence. The model in the last section, although capturing certain aspects of the spread of an epizootic front, is rather too primitive for quantitative purposes. One of the major exclusions from the previous model is the long incubation period, which can be from 12 to 150 days, before the fox becomes rabid. In this section we consider a more realistic model which takes this, among other things, into account. With it we can obtain quantitative estimates for various times and distances of epidemiological and public health significance.

In this section we consider a three-species model where again the rabid foxes are considered the main cause of the spatial spread. The data on the movement of rabid foxes in the wild although rather scant is not zero – some of it will be used later when we estimate the crucial diffusion coefficient for rabid foxes.

The model we develop is still comparatively simple, but, even so, some of the parameters are difficult to estimate from the available data. Such parameter estimates will be required in any realistic models, so it is important to learn more about fox ecology and the impact of rabies on fox behaviour in order to improve upon the estimation of the more critical parameters.

The model, analysis and results we give here are based on the work of Murray et al. (1986) who give further details and results. It extends the work of Anderson et al. (1981), who considered the spatially homogeneous situation, by including spatial effects, specifically the crucial spatial dispersal of rabid foxes.

We consider a three-species SIR model in which we divide the fox population into susceptible foxes, S, infected, but non-infectious, foxes, I, and infectious, rabid foxes, R. The need for three species is primarily based on the long incubation period of from 12 to 150 days (and in some cases longer) that the rabies virus undergoes in the infected animal, during which time the animal appears to behave normally and does not seem to transmit the disease, and on the relatively short period (1 to 10 days) of clinical disease which follows.

The basic model assumptions are closely linked to those in the last section (we use a slightly different notation) but we reiterate them here so that this section can be read independently. The assumptions are:

(i) The dynamics of the fox population in the absence of rabies can be approximated by the simple logistic form

$$\frac{dS}{dT} = (a - b)S \left(1 - \frac{S}{K} \right) ,$$

where a is the linear birth rate, b is the intrinsic death rate, and K is the environmental carrying capacity. The parameters a, b and K may vary according to the habitat but at this stage we shall take them to be constants. Later, when we present the numerical results for the English 'experiment', we shall consider K to vary as it does in a major way in England.

(ii) Rabies is transmitted from rabid to susceptible fox by direct contact between foxes, usually by biting. Susceptible foxes become infected at an average per capita rate βR, which is proportional to the number of rabid foxes present, where the transmission coefficient β, taken to be constant, measures the rate of contact between the two species.

(iii) Infected foxes become infectious (rabid) at an average per capita rate, σ, where $1/\sigma$ is the average incubation time.

(iv) Rabies is invariably fatal, with rabid foxes dying at an average per capita rate α ($1/\alpha$ is the average duration of clinical disease).

(v) Rabid and infected foxes continue to put pressure on the environment, and die of causes other than rabies, but they have a negligible number of healthy offspring. These effects are small but are included for completeness.

To take into account the spatial effects we make the following further assumptions:

(vi) Foxes are territorial, and divide the countryside up into non-overlapping ranges.

(vii) Rabies acts on the central nervous system with about half of infected foxes having the so-called 'furious rabies', and exhibit the ferocious symptoms typically associated with the disease, while with the rest the virus affects the spinal cord and causes paralysis. Foxes with furious rabies may become aggressive and confused, losing their sense of direction and territorial behaviour, and wandering randomly. It is these we consider the main cause of the spatial spread of the disease.

These assumptions suggest the following model for the spatial and temporal evolution of the rabies epizootic:

$$\frac{\partial S}{\partial T} = aS - bS - \frac{(a-b)NS}{K} - \beta RS \, ,$$

$$\frac{\partial I}{\partial T} = -bI - \frac{(a-b)NI}{K} + \beta RS - \sigma I \, , \qquad (20.26)$$

$$\frac{\partial R}{\partial T} = -bR - \frac{(a-b)NR}{K} + \sigma I - \alpha R + D\frac{\partial^2 R}{\partial X^2} \, ,$$

where the total population

$$N = S + I + R \, . \qquad (20.27)$$

We have written the equations in this form to highlight what each term means. The only source term comes from the birth of susceptible foxes. All die naturally; the life expectancy is $1/b$ years. The term $(a-b)N/K$ in each equation represents the depletion of the food supply by all foxes. The transition from susceptible to infectious foxes is accounted for by the βRS term and from the infected to the infectious group by σI. Rabid foxes also die from rabies and thus is represented by the αR term; the life expectancy of a rabid fox is $1/\alpha$. Rabid foxes also diffuse with diffusion coefficient D. Typical parameter values, except for the crucially important D, are given in Table 20.1. If, in the absence of any spatial effects, we add equations (20.26) we get

$$\frac{dN}{dT} = aS - bN - \frac{(a-b)N^2}{K} - \alpha R \, , \qquad (20.28)$$

which is the equivalent logistic form for the total population.

Table 20.1. Parameter values for rabies among foxes (from Anderson et al. 1981)

Parameter	Symbol	Value
average birth rate	a	1 per year
average intrinsic death rate	b	0.5 per year
average duration of clinical disease	$1/\alpha$	5 days
average incubation time	$1/\sigma$	28 days
critical carrying capacity	K_T	1 fox km^{-2}
disease transmission coefficient	β	80 km^2 per year
carrying capacity	K	0.25 to 4.0 foxes km^{-2}

We have written the equations in one-dimensional form but we shall use the full two-dimensional form when we apply the model to the spread of the disease from a hypothetical outbreak in England, which we discuss later.

This model neglects the spatial dispersal of rabies by young, itinerant foxes, who may get bitten while in search of a territory and carry rabies with them before they become rabid. There is some justification for this since rabies is much less common in the young than in adults (Artois & Aubert 1982, Macdonald 1980).

The spatially homogeneous steady state solutions of (20.26), other than the zero steady state, is given, after some algebra, by

$$
\begin{aligned}
S_0 = {}& \beta^{-1}[\sigma\beta K - a(a-b)]^{-2}[(\alpha+b)\beta K \\
& + (a-b)(\alpha+a)][\sigma\beta K(\sigma+b) + \alpha(a-b)(\sigma+a)]\} , \\
I_0 = {}& [\sigma\beta K - a(a-b)]^{-1}[(\alpha+b)\beta K + (a-b)(\alpha+a)]R_0 , \\
R_0 = {}& \{\beta[\sigma\beta K - a(a-b)]\}^{-1}(a-b)[\sigma\beta K - (\sigma+a)(\alpha+a)] .
\end{aligned}
\tag{20.29}
$$

In the spatially uniform situation ($D = 0$), when rabies is introduced into a stable population of healthy foxes three possible behaviours are possible. Which behaviour occurs depends on the size of K relative to the critical carrying capacity K_T which is given by the condition for a nonzero value for the steady state R_0 in the last equation, namely

$$
K_T = \frac{(\sigma+a)(\alpha+a)}{\beta\sigma} .
\tag{20.30}
$$

If $K < K_T$, the epidemic threshold value of the carrying capacity, rabies eventually disappears ($R \to 0$, $I \to 0$), and the population returns to its initial value K. On the other hand, if K is larger than K_T, then the population oscillates about the steady state. From a standard linear stability analysis of the steady state (S_0, I_0, R_0), the equivalent of which we do below for $K > K_T$, it can be shown (after some algebra) that if K is not too much bigger than K_T the steady state is stable and perturbations die out in an oscillatory way. On the other hand, if K is sufficiently larger than K_T limit cycle solutions exist. There are thus 2 bifurcation values for K, namely K_T and the critical K between a limit cycle oscillation and a stable steady state.

From the epidemiological evidence, rabies seems to die out if the carrying capacity is somewhere between 0.2 and 1.0 foxes/km^2 (WHO Report 1973; Macdonald 1980; Steck & Wandeler 1980; Anderson et al. 1981; Boegel et al. 1981). β, which is a measure of the contact rate between rabid and healthy foxes, cannot be estimated directly given the difficulty involved in observing these contacts. Anderson et al. (1981) used the expression (20.30) as an indirect way to estimate β since we have estimates for K_T and all the other parameters except β. Parameter estimation is always an important aspect of any realistic modelling. Murray et al. (1986) discuss in some detail how they affect the spatial spread of rabies:

the model is quite robust to variations in many of the parameters within a band around the estimates used.

With $K > K_T$ the parameter choices listed in Table 20.1, give 3–5 year periods for the oscillations and 0–4% *equilibrium persistence*, p, of rabies, where p is defined by

$$p = \frac{R_0 + I_0}{S_0 + I_0 + R_0} . \tag{20.31}$$

These figures are in agreement with the available epidemiological evidence (Toma & Andral 1977; Macdonald 1980; Steck & Wandeler 1980; Jackson & Schneider 1984).

Travelling Epizootic Wavefronts and their Speed of Propagation

We introduce non-dimensional quantities by setting

$$s = \frac{S}{K}, \quad q = \frac{I}{K}, \quad r = \frac{R}{K}, \quad n = \frac{N}{K} ,$$

$$\varepsilon = \frac{a-b}{\beta K}, \quad \delta = \frac{b}{\beta K}, \quad \mu = \frac{\sigma}{\beta K}, \quad d = \frac{\alpha+b}{\beta K} , \tag{20.32}$$

$$x = \left(\frac{\beta K}{D}\right)^{1/2} X, \quad t = \beta K T ,$$

with which the model equations (20.26) with (20.27) become

$$\frac{\partial s}{\partial t} = \varepsilon(1-n)s - rs ,$$

$$\frac{\partial q}{\partial t} = rs - (\mu + \delta + \varepsilon n)q ,$$

$$\frac{\partial r}{\partial t} = \mu q - (d + \varepsilon n)r + \frac{\partial^2 r}{\partial x^2}, \tag{20.33}$$

$$n = s + q + r ,$$

which have a positive uniform steady state solution (s_0, q_0, r_0) given by equations (20.29) on dividing (S_0, I_0, R_0) by K. The condition (20.30) for an epidemic to occur, namely $K > K_T$, is then

$$0 < d < \left[1 + \frac{\delta+\varepsilon}{\mu}\right]^{-1} - \varepsilon . \tag{20.34}$$

The system (20.33) now depends on only 4 dimensionless parameters ε, δ, μ and d as compared with the original dimensional system's 7 parameters. Values for these dimensionless parameters are obtained from the parameter estimates of a, b, α, σ, K and β in Table 20.1. If we choose a representative carrying capacity of $K = 2$ foxes/km^2 we get $\varepsilon = \delta = 0.003$, $\mu = 0.08$ and $d = 0.46$. The fact that ε and δ are relatively small numbers compared to any of 1, μ, d and $1 - d$ can be used to simplify the analysis of the model system (20.33) and lets us derive useful *analytical* results: see below and Murray et al. (1986).

It is perhaps appropriate here to reiterate yet again the major benefit of non-dimensionalisation, namely that the parameter groupings show equivalent effects of variations in actual field parameters. For example, with ε and δ small this means that, during the epidemic, the infectious rate is relatively very much larger than the birth and death rates from causes other than rabies.

Let us now look for epizootic wave solutions to the system (20.33), which travel at a constant velocity v into an undisturbed, rabies-free region. (For algebraic simplicity we use a different notation from what we have used earlier in this chapter.) So, we look for solutions s, q and r as functions of the single variable $\xi = x + vt$, which thus satisfy

$$
\begin{aligned}
vs' &= \varepsilon(1 - n)s - rs \ , \\
vq' &= rs - (\mu + \delta + \varepsilon n)q \ , \\
vr' &= \mu q - (d + \varepsilon n)r + r'' \ , \\
n &= s + q + r \ ,
\end{aligned}
\tag{20.35}
$$

where prime denotes differentiation with respect to ξ and where $s \to 1$, $q \to 0$, $r \to 0$ as $\xi \to -\infty$, that is far ahead of the wave front. As usual, of course, we are interested only in non-negative solutions. We shall, in the following, use the fact that $\varepsilon \ll 1$ and $\delta \ll 1$.

The system (20.35) has 3 possible steady state solutions for (s, q, r) in the positive quadrant, namely $(1,0,0)$, $(0,0,0)$ and (s_0, q_0, r_0) where s_0, q_0 and r_0 are given by (20.29) on dividing by K. Since ε and δ are small, to first order in ε and δ,

$$
s_0 = d + \left[\varepsilon + \frac{\varepsilon d + \delta}{\mu}\right]d, \quad q_0 = \frac{\varepsilon d(1 - d)}{\mu}, \quad r_0 = \varepsilon(1 - d) \ .
\tag{20.36}
$$

From the full expressions for (s_0, q_0, r_0) all of s_0, q_0 and r_0 are non-negative only if the threshold condition (20.34) is satisfied.

A travelling wave solution to (20.35), with the required properties, is a trajectory in the 4-dimensional phase space of (20.35), which goes from the equilibrium at $s = 1$, $q = r = 0$ to one of the other two equilibrium points, $(0,0,0)$ or (s_0, q_0, r_0). We shall not carry out all the algebra since the proceedure is the customary one used throughout the book (cf. Chapters 3, 6, 11, 12): we shall just briefly describe the various steps and leave the details to be worked out as an exercise.

We write (20.35) as a 4-dimensional first order system in (s, q, r, r') and first linearize about the critical point $(s, q, r, r') = (1, 0, 0, 0)$. In the usual way, this gives a linear system whose solutions are linear combinations of the eigensolutions $\mathbf{x}_i \exp(\lambda_i \xi)$ where $\mathbf{x}_i$ and λ_i are the four eigenvectors and eigenvalues of the stability matrix. We can thus determine the solution behavior near the critical point by looking at all possible linear combinations of the eigensolutions. If

$\operatorname{Re}\lambda_i < 0$, then $x_i \exp[\lambda_i\xi] \to 0$ as $\xi \to \infty$ and the trajectory approaches the critical point, while if $\operatorname{Re}\lambda_i < 0$ the trajectory comes out of the critical point. Trajectories leaving the critical point thus correspond to linear combinations of those eigensolutions with $\operatorname{Re}\lambda_i > 0$. If an eigenvalue is complex, then its eigensolution is oscillatory. After some algebra we find that the four eigenvalues for the linear system near $(1,0,0,0)$ are $\lambda = -\varepsilon/v < 0$ and the roots of the cubic

$$f(\lambda) = \lambda^3 + \left(\frac{\mu+\delta+\varepsilon}{v} - v\right)\lambda^2 - (d+\mu+\delta+2\varepsilon)\lambda$$
$$+ \frac{\mu(1-d-\varepsilon)-(\delta+\varepsilon)(d+\varepsilon)}{v} \tag{20.37}$$

Note that $f(\lambda) \to \infty$ as $\lambda \to \infty$ and $f(\lambda) \to -\infty$ as $\lambda \to -\infty$. Further, if (20.36) holds, then $f(0) > 0$ and f has a negative slope at $\lambda = 0$. Depending on the values of the various parameters, $f(\lambda)$ can look like any of the forms illustrated in Fig. 20.7.

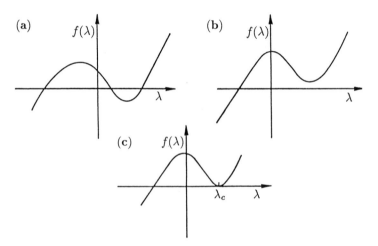

Fig. 20.7a-c. The eigenfunction cubic $f(\lambda)$ in (20.37), the zeros of which are eigenvalues of the linearized system about $(1,0,0)$. The solutions have either two positive roots as in (a), two complex roots with $\operatorname{Re}\lambda > 0$ as in (b), or a double real root at λ_c as in (c).

With all the parameters fixed, as the velocity v is varied $f(\lambda)$ sequentially looks like each of these shapes. Thus, as long as the threshold condition (20.34) holds, f has one negative real root and, depending on the value of the velocity of the wave, it has either two positive real roots, or two complex roots. When the velocity v is such that (20.37) has complex roots, these represent oscillatory solutions which imply negative populations and such waves are physical unrealistic. The bifurcation value for v, v_c say, between realistic and unrealistic solutions is the value when (20.37) has a double root as in Fig. 20.7 (c). Thus the range of allowable wave speeds of travelling waves is determined by v_c. This is given by setting $f = 0$ and $df/d\lambda = 0$ and eliminating λ to get an equation for v_c in terms of the parameters. After considerably more algebra we find that, to first order in ε and δ, v_c is given by the positive real roots of $g(v_c^2)$, where $g(z)$ is given by

$$g(z) = \left[4\mu + (d - \mu)^2\right]z^3 + 2\left[3\mu(1 - d)(3d + \mu) + (d + \mu)^2(2d + \mu)\right]z^2$$

$$+ \mu^2\left[(d + \mu)^2 - 6(1 - d)(3d + \mu) - 27(1 - d)^2\right]z - 4\mu^4(1 - d) . (20.38)$$

When the threshold criterion (20.34) holds, $g(z)$ is negative and d^2g/dz^2 is positive at $z = 0$. A rough sketch of $g(z)$ shows it has a unique positive root which corresponds to the minimum possible velocity for an epizootic wave.

We now show that it is not possible for a trajectory to go from the critical point at $s = 1$, $q = 0$, $r = 0$ to that at the origin where $s = q = r = 0$. On linearizing (20.35) about the origin we find (after more algebra) the eigensolutions

$$\begin{pmatrix} s \\ q \\ r \\ r' \end{pmatrix} = \mathbf{a}\exp\left[-\frac{(\mu + \delta)\xi}{v}\right], \quad \mathbf{b}\exp\left[\frac{v}{2} \pm \left(d + \frac{v^2}{4}\right)^{1/2}\right]\xi, \quad \mathbf{c}\exp\left[\frac{\varepsilon\xi}{v}\right],$$

where

$$\mathbf{a}^T = \left[0, \frac{d - \mu - \delta}{\mu} - \frac{(\mu + \delta)^2}{\mu v^2}, 1, -\frac{\mu + \delta}{v}\right],$$

$$\mathbf{b}^T = \left[0, 0, 1, \frac{v}{2} \pm \left(d + \frac{v^2}{4}\right)^{1/2}\right], \quad \mathbf{c}^T = [1, 0, 0, 0],$$

and the superscript T denotes the transpose. Sufficiently close to the origin, trajectories which approach the origin are linear combinations of the two eigensolutions with negative exponents, and so they approach the origin in the plane $s = 0$. For the system (20.35) "time" is reversible, in the sense that we can replace ξ by $-\xi$ and trace backwards along any trajectory. Setting $\tau = -\xi$ in (20.35) and taking $s = 0$ initially, we see that $s = 0$ for all positive τ irrespective of the initial values of r and q. This implies that a trajectory which has $s = 0$ for any ξ had $s = 0$ for all previous ξ, and has $s = 0$ for all subsequent ξ. So, a trajectory cannot come from $s = 1$, enter the $s = 0$ plane, and approach the origin.

This implies that a travelling wave can only occur if there is a trajectory from $s = 1$ to the critical point (s_0, q_0, r_0) and this requires, as we expected, that condition (20.34) must hold. To determine the behaviour of the wave as it approaches this critical point, we now linearize (20.35) about (s_0, q_0, r_0) to get (after more algebra) the eigenvalues

$$\lambda_1, \lambda_2 = \frac{1}{2}\left\{v - \frac{\mu}{v} \pm \left[\left(v - \frac{\mu}{v}\right)^2 + 4(\mu + d)\right]^{1/2}\right\} \qquad (20.39)$$

to first order in ε and δ, and

$$\lambda_3, \lambda_4 = \pm \frac{i}{v} \left[\frac{\varepsilon\mu d(1-d)}{\mu+d} \right]^{1/2}$$

$$- \varepsilon d \left[2v(\mu+d)^2 \right]^{-1} \left[\mu(1-d)\left(\frac{\mu}{v^2}-1\right) + (\mu+d)^2 \right]$$

(20.40)

to second order in ε and δ. λ_1 is positive and so, near the critical point, any solution which approaches (s_0, q_0, r_0) as $\xi \to \infty$ is a linear combination of the eigensolutions corresponding to λ_2, λ_3 and λ_4. Since $|\lambda_2| \gg |\mathrm{Re}\,(\lambda_3, \lambda_4)|$, the amplitude of its eigensolution decays much more rapidly than that of the eigensolutions of the complex eigenvalues. Thus, sufficiently far back in the tail of the wave (that is for sufficiently large ξ), the solutions corresponding to the complex eigenvalues govern the behavior of the travelling wave. The eigenvectors corresponding to these eigenvalues are given by

$$\begin{pmatrix} s - s_0 \\ q - q_0 \\ r - r_0 \\ r' \end{pmatrix} = \begin{pmatrix} 1 \\ \pm i \left[\dfrac{\varepsilon d(1-d)}{\mu(\mu+d)} \right]^{1/2} \\ \pm i \left[\dfrac{\varepsilon\mu(1-d)}{d(\mu+d)} \right]^{1/2} \\ \dfrac{\varepsilon\mu(1-d)}{v(\mu+d)} \end{pmatrix}$$

which, on taking an arbitrary real linear combination of the eigensolutions, gives, for sufficiently large ξ,

$$s - s_0 \sim [A \cos \omega\xi/v + B \sin \omega\xi/v] \exp\left(-\lambda\xi/v\right),$$

$$q - q_0 \sim \frac{\omega}{\mu}[A \sin \omega\xi/v - B \cos \omega\xi/v] \exp\left(-\lambda\xi/v\right),$$

(20.41)

$$r - r_0 \sim \frac{\omega}{d}[A \sin \omega\xi/v - B \cos \omega\xi/v] \exp\left(-\lambda\xi/v\right).$$

Here ω is the period of the waves, given by the imaginary part of the complex eigenvalues divided by v, and λ is the decay rate of the amplitude, given by the real part of these eigenvalues divided by v. A and B are constants, which depend on the way the trajectory approaches (s_0, q_0, r_0), which of course cannot be determined from a linear analysis.

Let us now exploit the smallness of ε and δ to obtain certain useful asymptotic analytical approximations (Murray et al. 1986). From the approximate steady state forms (20.36) we note that the rabid fox density $r_0 = \mu q_0/d$. From (20.41) we see that far back in the tail of the wave, that is ξ large, we also have $r - r_0 \sim \mu(q - q\infty)/d$. That is the profiles for the infected and rabid fox densities are similar, differing only in *scale*. In the simulations, such as given in Figs. 20.8 and 20.9 below, of the full nonlinear system, the striking profile similarity holds for the *entire* wave. This surprising fact suggests that, in view of the complexity of the three species model, it would be of considerable benefit if we could obtain analytically the conditions under which the travelling wave problem for the three species model could be modelled to a high degree of approximation with a two species model. That is we could replace, for example, the three-species SIR system of susceptible, infectious, rabid populations by a two species system of only susceptible and rabid populations. The infected, but not yet rabid, fox population is then given by a simple scaling of the rabid population, namely

$$q(\xi) \sim \frac{dr(\xi)}{\mu} .$$
(20.42)

Under certain conditions it is possible to give an analytical explanation as to why this phenomenon occurs. It is not obvious from the model system (20.35). The mathematical analysis is based on μ being small compared with both d and the non-dimensional wavespeed v, but large compared with ε and δ. The singular perturbation analysis is quite complicated and is given in detail by Murray et al. (1986).

To get the actual travelling wavefront solutions for the epizootic we must solve the partial differential equation system (20.33) numerically, starting with $s = 1$ (that is dimensionally the susceptible population is $S = K$, the undisturbed carrying capacity) everywhere and with a small concentration of rabid foxes at the origin. When the threshold criterion (20.34) is satisfied an epidemic wave forms and travels outward from the initial concentration of rabid foxes with near constant velocity. If, of course, the threshold inequality (20.34) is violated, then rabies dies out, and the fox population returns to the carrying capacity of the environment. Fig. 20.8 is an example of the traveling wavefront which evolves for parameter values appropriate to the current epizootic on continental Europe. The wave consists of the rabies front, in which the largest number of foxes die from the disease, followed by an oscillatory tail in which each successive outburst of rabies is smaller than the preceeding one. The oscillations gradually die out and the populations approach constant, nonzero values with the rabid and infected fox population zero. Fig. 20.9 illustrates the fluctuations in fox density for a travelling epizootic wave with parameters appropriate for England.

With the parameter values in Table 20.1, the threshold condition (20.34) is satisfied and

$$\varepsilon \text{ and } \delta \ll 1, \ d, \ \mu \text{ and } 1 - d .$$
(20.43)

Under these circumstances the wavefront is followed by an oscillatory tail. Analytically the minimum speed is given by $v = z^{1/2}$, where z is the unique positive root of the cubic (20.38). A contour plot for this root v is shown in Fig. 20.10 for $0 \leq d \leq 1$. All of the waves found numerically appear to travel at this minimum speed which, from (20.32), is given in dimensional form as

$$V = (D\beta K)^{1/2}v .$$
(20.44)

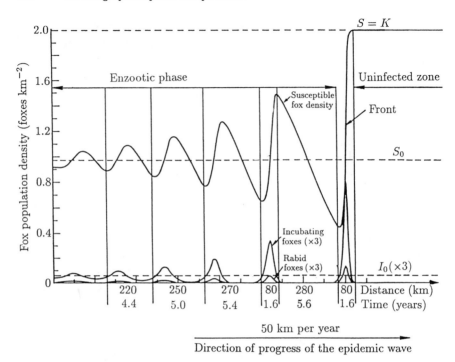

Fig. 20.8. The susceptible, infected and rabid fox populations due to the passage of a rabies epidemic wave from a numerical simulation of the model mechanism (20.37)–(20.40). The fox density in the uninfected region ahead of the front of the epidemic is taken to be at a carrying capacity of 2 foxes/km^2, a typical value (averaged over the yearly cycle) for much of continental Europe. The time and distance between the recurring outbreaks and the wave speed were obtained from the model using estimates for the field parameters given in Table 20.1 and a diffusion coefficient $D = 200$ km^2/yr. (From Murray, Stanley and Brown 1986)

For example, with the parameter values in Table 20.1, a diffusion coefficient of 200 km^2/yr and a carrying capacity of 2 foxes/km^2, we evaluate d and μ from (20.32) and then read off the appropriate v from Fig. 20.10: this gives the dimensional speed of propagation as $V = 51$ km/yr.

The linear analysis near the steady state (s_0, q_0, r_0), described above, shows that for sufficiently large times the wave tends to decaying oscillations given by (20.41). In terms of the original (x, t) variables these solutions can be written in the form

$$s(x, t) = s_0 + A \cos\left[\omega(t + x/v) + \psi\right] \exp\left[-\lambda(t + x/v)\right],$$
$$q(x, t) = q_0 + \mu^{-1}(s - s_0)' ,\qquad\qquad\qquad (20.45)$$
$$r(x, t) = r_0 + \mu(q - q_0)/d ,$$

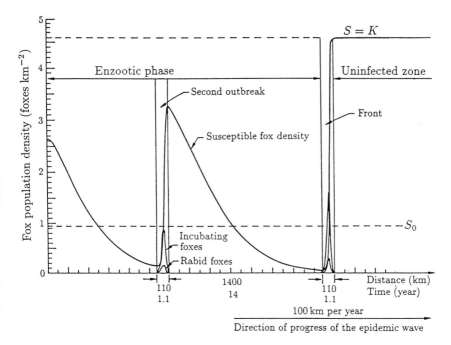

Fig. 20.9. Fox populations during the passage of the rabies epidemic wave when the fox density in front of the epidemic is at the carrying capacity of 4.6 foxes/km², which is common in parts of England. The diffusion coefficient $D = 200$ km²/yr, and the other parameters were taken from Table 20.1. Compare the different wavelengths and periods of the recurring epidemics following the front with those in Fig. 20.8. The epidemics for England are more severe. (From Murray et al. 1986)

to first order in ε and δ, where the prime denotes differentiation with respect to $(t + x/v)$ and the non-dimensional wavenumber ω is given by

$$\omega = \varepsilon^{1/2} \left[\frac{\mu d(1 - d)}{\mu + d} \right]^{1/2} + O(\varepsilon^{3/2}) , \qquad (20.46)$$

with the decay rate λ given by

$$\lambda = \varepsilon d \left[2(\mu + d)^2 \right]^{-1} \left[\mu \left(\frac{\mu}{v^2} - 1 \right) (1 - d) + (\mu + d)^2 \right] . \qquad (20.47)$$

A and ψ are constants. Note that the oscillations in the susceptible population are 90° out of phase with both the infected and, as noted above, the rabid populations. (This symmetry is broken if the oscillations are calculated to the next order in ε and δ.) As we also noted above, the $r - q$ proportionality relationship (20.42) seems to hold universally as indicated by the numerical simulations when physically reasonable parameters are used.

The singular perturbation analysis of Murray et al. (1986) yields several useful approximations regarding the epidemic. For example, the maximum density

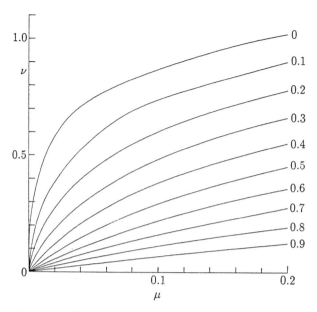

Fig. 20.10. The dimensionless velocity of propogation, v, of the epidemic front as a function of the dimensionless parameter μ for various values of d. Recall that μ is related to the incubation time for the rabies virus, and d is related to the duration time of the symptomatic, infectious, stage: see (20.32). The dimensional wave speed is given by $V = (\beta K D)^{1/2} v$. Note that $v = 0$ for $d \geq 1$, which corresponds to a carrying capacity less than the critical value. (From Murray et al. 1986)

of infected and rabid foxes in the first outbreak, are given by

$$
\begin{aligned}
r_{max} &\approx \mu \left[\ln d + \frac{1-d}{d} \right] , \\
q_{max} &\approx d \left[\ln d + \frac{1-d}{d} \right] ,
\end{aligned}
\tag{20.48}
$$

which in dimensional terms are

$$
\begin{aligned}
R_{max} &\approx \frac{\sigma K_T}{\alpha} \left(\ln \left[\frac{K_T}{K} \right] + \frac{K}{K_T} - 1 \right) , \\
Q_{max} &\approx K_T \left(\ln \left[\frac{K_T}{K} \right] + \frac{K}{K_t} - 1 \right) .
\end{aligned}
\tag{20.49}
$$

Since no epidemic ensues if $K \leq K_T$, the threshold carrying capacity, both R_{max} and Q_{max} are zero for $K = K_T$. Note that both R_{max} and Q_{max} increase as K increases above K_T.

Once we have the dimensionless wave speed $v(= z^{1/2})$ from (20.38) we can then determine the decay rate λ from (20.47). It turns out that λ is always positive. This means that the limit cycle behaviour which the diffusionless version of

(20.26), that is with $D = 0$, can exhibit for sufficiently large $K > K_T$ disappears when diffusion is taken into account: the oscillations always decay to the constant state (s_0, q_0, r_0). The dimensional decay rate is $\beta K \lambda$. The dimensional period of the recurring epidemics is

$$\tau = \frac{2\pi}{\beta K \omega} \, ,$$

where ω is given by (20.46): in terms of the original dimensional parameters, using (20.32), the period T is

$$T = 2\pi \left\{ (\alpha + \sigma + b) \left[(a - b)(\alpha + b)\sigma \left\{ 1 - \frac{\alpha + b}{\beta K} \right\} \right]^{-1} \right\}^{1/2} . \qquad (20.50)$$

Note that T decreases with K. So, in general, the greater the fox density before the appearance of rabies, the less frequently rabies outbreaks will appear far behind the front: this agrees with some observations (Macdonald 1980). However, numerically it was found that close to the front, where nonlinearities are important, the time between outbreaks may increase with K: see Figs. 20.8 and 20.9. Once we have the dimensional velocity V and period T we get the dimensional wavelength $L = VT$.

Estimate for the Diffusion Coefficient D and Sensitivity of Wavespeed and Epidemic Wavelength to Variations in D

To calculate the real dimensional speed V of the epizootic, and hence the period and wavelength of the recurring epidemics which follow the main front, we need an estimate for the diffusion coefficient, which is a measure of the rate at which a rabid fox covers ground in its wanderings. Little is known about the behaviour of rabid foxes in the wild, making it very difficult to estimate D.

Andral et al. (1982) tracked 3 rabid adult foxes in the wild. They accomplished this by innoculating captured foxes with rabies virus, equipping them with signal-emitting collars, and releasing them at the point of capture. They traced the fox movements first during the incubation period, to determine their home ranges and normal behaviour, and then during the rabid period, to observe the changes induced by the disease. Once the foxes became rabid their pattern of daily activity changed. Drawings showing, for each fox, the incubation period range and the principle displacements during the rabid period indicate that all three left their home range at some point during the rabid phase, but none travelled very far away.

Murray et al. (1986) used the results of Andral et al. (1982) to estimate the diffusion coefficient, in a rather primitive way, from the formula

$$D \approx \frac{1}{N} \sum_{j=1}^{N} \frac{\left(\text{straight line distance from the start} \right)^2}{4 \times \left(\text{time from the start} \right)} \, ,$$

where the sum is over the number of all foxes involved. Using the distance between the start of the rabid period and the point of death, along with the approximate length of the rabid period, gives an estimate of 50 km^2/yr for D. Since two of the three foxes happened to die much closer to their starting position than their mean distance away from it, this is most likely a lower bound on D. An extremely rough idea of an upper bound can be gained from the maximum distance that any one fox travelled away from its starting point. About halfway through the rabid phase, one fox got as far away from its starting point as 2.7 km, giving an estimate of 330 km^2/yr as an upper bound on D.

There are other ways of estimating diffusion coefficients. For example, D can be estimated as the product of the average territory size A and the average rate k at which a rabid fox leaves home. For their two-species model, Källén et al. (1985) supposed that infected foxes leave home at the end of the incubation period of one month, that is when they were assumed to become rabid. Taking an average territory size to be about 5 km^2, they obtained $D = 60$ km^2. To determine D for our 3-species model, we need an estimate for the average rate at which foxes leave their territories *after* the onset of clinical disease. If N infected foxes are observed, and the j-th one leaves its territory a time interval t_j after becoming rabid, then k can be estimated by

$$N^{-1} \sum_{j=1}^{N} t_j^{-1} .$$

Since roughly half of all infected foxes develop paralytic rabies and presumably never leave their home range, t_j is infinite for about $N/2$ foxes. For the furiously rabid foxes, if we suppose that half also never leave, and that the rest leave evenly spread out over the 6 days that the disease may take to run its course, then we can estimate

$$k \approx \frac{1}{N} \sum_{j=1}^{N/4} \frac{1}{t_j} = \frac{1}{24} \sum_{j=1}^{6} \frac{1}{j \text{ days}} = 40 \text{ yr}^{-1} .$$

Keeping the estimate of 5 km^2 for an average territory size (Toma & Andral 1977; Macdonald 1980), this gives $D = 190$ km^2/yr.

An alternative method is to estimate the mean free path and velocity of rabid foxes. The average total distance covered daily by the foxes observed by Andral et al. (1977) was 9 km during the rabid period. Suppose that this is not atypical and that, for example, a rabid fox goes 100 m at a stretch before becoming distracted and setting off in another direction. Then $D = $ (velocity)×(pathlength) gives a diffusion coefficient of 330 km^2/yr, the same as the upper bound that we estimated previously. All of these methods for estimating D should, in principle, be consistent if enough observations of rabid fox behaviour could be made. At this stage there is simply not enough known about fox behaviour to get much better estimates.

Since the speed of the wave is proportional to $D^{1/2}$, changing D from 50 to 330 km^2/yr increases V by a factor of 2.6. Table 20.2 shows the sensitivity of

Table 20.2. Dependence of the wavespeed and asymptotic wave length (that is the distance between recurring outbreaks) on the carrying capacity, calculated with $D = 200$ km^2/yr and other parameter values from Table 20.1.

K (foxes/km^2) or K/K_T* Carrying capacity	V (km/yr) velocity of the epidemic front	L (km) distance between successive outbreaks/peaks
1.5	35	150
2.	50	210
2.5	70	220
3.	80	250

* The parameters β and K only appear as the product βK in the calculations for the values in Tables 20.1 and 20.2. From equation (20.30), $\beta K = (K/K_T)(\sigma + a)(\alpha + a)/\sigma$, so that only a knowledge of the ratio of the actual carrying capacity to the critical value is necessary to obtain the results.

the wave speed and wavelength as a function of the carrying capacity for a given $D = 200$ km^2/yr.

Another difficult parameter to estimate is the disease transmission coefficient β. As we said above, this can be estimated by inverting the threshold expression (20.30). But, absolute values of fox population densities are in practice difficult to obtain; they are usually estimated from the numbers of foxes reported dead, shot or gassed, and some assumption on the percentage of the total population that this sample represents, or else by comparison of terrain with areas of known fox densities. This in turn means K_T is particularly difficult to estimate, and values of anywhere from 0.2 to 1.2 foxes/km^2 can be estimated from the values given in the literature (WHO Report 1973; Steck & Wandeler 1980; Macdonald et al. 1981; Gurtler & Zimen 1982). Since finding K/K_T only involves comparison of population sizes, this ratio might be easier to obtain than K and K_T separately.

A relevant question at this point is how sensitive the quantitative results are to the uncertainties in the parameters. This aspect and difficulties in estimating other parameter are given by Murray et al. (1986).

20.5 Control Strategy Based on Wave Propagation into a Non-epidemic Region: Estimate of Width of a Rabies Barrier

We discuss here one possible control strategy as developed by Murray et al. (1986), namely that of a possible protective barrier against the rabies epizootic which can be achieved by reducing the susceptible fox population below the critical density K_T in areas ahead of the advancing wave. This, for example, has been sucessful in Denmark, specifically Jutland. It has also been carried out in some regions of Italy and Switzerland, where it has been pursued with diligence, but it has had mixed results (Macdonald 1980; Westergaard 1982). Such a barrier can be created either by killing or vaccination. Since killing releases territories, there could be a more rapid colonization by young foxes which could in fact enhance the spread of the disease. Vaccination causes less disruption in the ecology and is also probably even more economic.

For a rabies "break" to be effective we must have reasonable estimates of both the width and the allowable susceptible fox density within it. Here we shall derive estimates analytically for how wide the protective break region needs to be to keep rabies from reaching the areas beyond. We shall also present some of the results from numerical simulations of the full equation system (20.33). In what follows, we use the term "infected fox" to refer to all foxes with rabies, whether infectious or not.

If we observe the passage of the rabies epizootic wave at a fixed place we note that each outbreak of the disease is followed by a long quiescent period, during which very few cases of rabies occur: refer to Figs. 20.8 and 20.9. The spatial and temporal dimensions are such that the secondary epidemic wave is sufficiently far behind so that the first wave will either have moved past the break, or have effectively died out by the time the second one arrives. Each succesive outbreak is weaker than the previous one. So, it seems reasonable to assume that the same population reduction schemes which eradicate the first outbreak will also be effective in stopping all subsequent outbreaks from passing through. We thus only need to consider how wide the break needs to be to stop the first outbreak. The width of the break is dependent on the size of the susceptible fox population density within it.

Since we model spatial dispersal by a deterministic diffusion mechanism it is, from a strict mathematical viewpoint, not possible for the density of infected foxes to vanish anywhere. This arises from treating the fox densities as continuous in space and time, rather than dealing with individual foxes, and from using classical diffusion to model the rabid fox dispersal. Thus we cannot simply have the epizootic wave move into a break of finite width and determine whether or not the density of infected foxes remains zero on the other side: it will always be positive, although exponentially small. Thus no matter how wide the break is, eventually enough infected foxes will in time leak through for the epizootic to start off again on the other side. Thus we must think instead of determining when the probability is acceptably small that an infected fox will reach the far side of the break.

Since the aim of any control scheme is to keep the density of foxes small, we treat the break region as one with a carrying capacity below K_T, the critical threshold value (20.30) for an epidemic, and we assume that the fox density has been reduced to this value well before the epizootic front arrives. To obtain estimates for the width of the break we investigate the behaviour of the model when the region of lowered susceptible fox density starts at $x = 0$ and extends to infinity. We first give here the results of the numerical simulations of the full system (20.33) and later in the section obtain approximate analytic results.

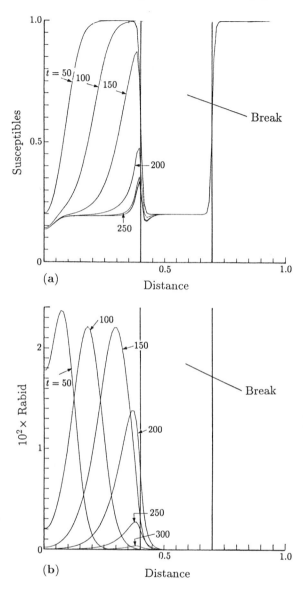

(a)

Susceptibles

$\bar{t} = 50$ / 100 / 150

200

250

0.5

Distance

1.0

Break

(b)

$10^2 \times$ Rabid

100

150

$\bar{t} = 50$

200

250

300

0.5

Distance

1.0

Break

Fig. 20.11a,b. The behavior of the travelling epizootic front when it encounters a break in the susceptible fox population. These plots show (a) the susceptible and (b) the rabid fox population densities for a sequence of times as the wave approaches the break region, stops and dissipates. They were obtained by solving equations (20.37)–(20.40) numerically with a carrying capacity of 2 foxes/km^2 in the region outside the vertical lines and of 0.4 foxes/km^2 in the region between them. Other parameters values were taken from Table 20.1. Note that the susceptible population just outside the break remains slightly higher than elsewhere, since few rabid foxes wander into this region from the right. The density of incubating foxes is proportional to the rabid population as we noted in Section 20.4: with the parameter values used, the incubating fox density is 5.6 times the rabid fox density. The times and distances are normalized values within the computer model. (From Murray et al. 1986)

Figs. 20.11 and 20.12 show what happens when the epizootic wave, coming in from the left, impinges on the break region. Remember that the epizootic wave cannot propagate when the carrying capacity is below the critical value K_T. Also, the point of maximum infected fox density will be at $x = 0$. As the infection wave moves into the region $x > 0$ it spreads out, decays in amplitude and the total number of infected foxes decreases. Eventually there will be less than p infected foxes/km^2 remaining, where p is some small number. Let $t_c(p)$ be the time at which this occurs. We now choose p sufficiently small that the probability of a rabid fox encountering a susceptible one after this critical time is

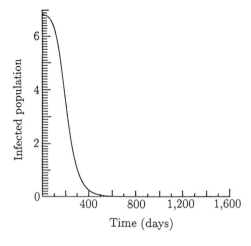

Fig. 20.12. This plot shows the total infected fox density per km (the integral over x of the infected foxes $I + R$) as a function of time for the case shown in Fig. 20.11 starting when the epidemic front first reaches the break. (From Murray et al. 1986)

neglible. Since the wave cannot propagate in the break region it simply decays, so, for all time the density of infected foxes is greatest at the edge of the break and decays with x – exponentially as x^2 in fact, as we show later. We choose the width of the break to be the point x_c where the infected fox density is a given (small) fraction m of the value at the origin, that is

$$I(x_c, t_c) + R(x_c, t_c) = m[I(0, t_c) + R(0, t_c)] \,.^{\dagger} \qquad (20.51)$$

Available evidence suggests that it has never been possible to eliminate all foxes from a region – a 70% reduction in population is about the best that can be achieved (Macdonald et al. 1981). Fig. 20.13 shows the dependence of the break width in terms of the percentage population reduction in the break, for different choices of the average duration time of clinical disease, $1/\alpha$.

In the numerical simulations for the curves in Fig. 20.13, the value of βK outside the break was held at 160 yr^{-1}, the number of infected foxes at the critical time was taken to be $p = 0.5$ foxes/km^2, the ratio m in (20.51) was arbitrarily chosen to be 10^{-4} and all other parameters except α are from Table 20.1. With these assumptions, for any given choice of α, the nondimensional forms in (20.32) give $d = (\alpha + 0.5\,\text{yr}^{-1})/(160\,\text{yr}^{-1})$ and (20.30) gives a carrying capacity outside the break region of $K = 149/(\alpha + 0.5\,\text{yr}^{-1})$ foxes km^{-2}yr^{-1}. For example, if we assume that the rabid period lasts an average of 3.8 days, then $d = 0.6$ and $K = 1.5$ foxes/km^2 outside of the break. If a reduction scheme can reduce the carrying capacity to 0.4 foxes/km^2 inside the break region well before the epidemic arrives, then $s_b = 0.26$ and Fig. 20.13 gives $x_b = 15$. Assuming a diffusion coefficient of 200 km^2/yr, (20.32) gives the predicted break width as 17 km. Of course, the choice of p and m depends on how cautious we want to be: Murray et al. (1986) discuss the sensitivity of the model to variations in these. The maximum value of $I + R$ at t_c for all of the calculations was less than 0.15 foxes/km^2. Even with m as large as $m = 10^{-2}$ there are fewer than 0.0015 infected foxes per square kilometre on the protected side of the break.

---

† Strictly the x_c and t_c are dimensionless here.

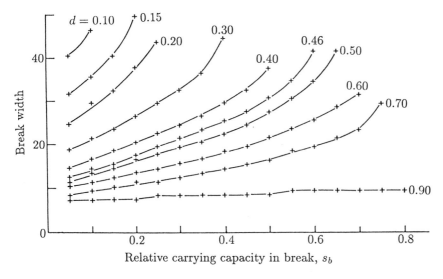

Fig. 20.13. The dependence of the break width on the initial susceptible population inside the break, as predicted by the model. The break width, in nondimensional terms, is plotted against the ratio of the carrying capacity in the break to the carrying capacity outside the break for various values of the duration time of clinical disease, $1/\alpha$ ($d \approx \alpha/\beta K$). The curves were obtained by solving (20.33) numerically until the total infected fox population in the first outbreak is 0.5 fox/km. As described in the text, we use these curves to calculate the break width, which can be put into dimensional form using relations (20.32). The dimensional break width X_c is given by $(D/\beta K)^{1/2} x_c$, where x_c is the nondimensional break width with $m = 10^{-4}$. βK was set at 160 yr^{-1}, and all other parameters, except α, were taken from Table 20.1. For example, if we assume $1/\alpha = 5$ days then $d = 0.46$ and the carrying capacity outside of the break is 2 foxes/km^2. If the carrying capacity inside the break is assumed to be 0.4 foxes/km^2 then $s_b = 0.2$ and this figure predicts $x_C = 18$. Assuming $D = 200$ km^2/yr, the predicted breakwidth X_C is then 20 km. (From Murray et al. 1986)

Analytic Approximation for the Width of the Rabies Control Break

We can determine analytically an approximate functional dependence of the break width on the parameters. The behaviour of the various fox population densities in the break region after the epizootic wave has reached it should be similiar to the situation in which a concentrated localized density of infected and rabid foxes at time $t = 0$ (with the same total number of I and R as for the epizootic wave) is introduced at $x = 0$ in a domain where the carrying capacity is everywhere equal to the initial fox density in the break. We can then obtain an estimate of the break width by looking at the following idealized problem. Suppose that the carrying capacity is zero for all x, which implies that the susceptible fox density $s = 0$. At time $t = 0$, take $r = r_0 \delta(x)$ and $q = q_0 \delta(x)$, where $\delta(x)$ is a Dirac delta function (that is we consider all of the r_0 rabid foxes are initially concentrated at $x = 0$).

We start by assuming that for $x \geq 0$, all of the susceptible foxes have been eliminated, for example by immunization or killing. In our analysis here we shall make the added approximation that the nonlinear terms in the equations for the incubating and rabid foxes can be neglected. Since ε and δ are small parameters, this should be a reasonable approximation. A further justification for these approximations comes from the numerical computations of the break width, where it was found that the computed break width did not change if these terms were neglected. With these assumptions, equations (20.33) reduce to the linear form

$$\frac{\partial q(x,t)}{\partial t} = -\mu q(x,t) ,$$

$$\frac{\partial r(x,t)}{\partial t} = \mu q(x,t) - dr(x,t) + \frac{\partial^2 r(x,t)}{\partial x^2} . \tag{20.52}$$

By symmetry, instead of considering the problem of a δ-function source of infected foxes at $x = 0$ and $t = 0$ which then move into the region $x \geq 0$, the initial conditions can be replaced by

$$q(x,0) = 2q_0\delta(x), \quad r(x,0) = 2r_0\delta(x) \tag{20.53}$$

and we then consider instead the region $-\infty < x < \infty$. The propagation of infected foxes into the break is described by equations (20.52) with initial conditions (20.53). The specific quantities of interest are the time t_c at which the population in the break has decayed to a given level, p, defined implicitly by the formula

$$\left(\frac{KD}{\beta}\right)^{1/2} \int_0^\infty [q(x,t_c) + r(x,t_c)]\, dx = p \tag{20.54}$$

and the breakwidth, x_c, which, as discussed above, is given implicitly by

$$q(x_c,t_c) + r(x_c,t_c) = m[q(0,t_c) + r(0,t_c)] . \tag{20.55}$$

We first estimate t_c. Integrating equations (20.52) with respect to x from 0 to ∞, we get the two ordinary differential equations

$$\frac{dQ^*(t)}{dt} = -\mu Q^*(t) ,$$

$$\frac{dF^*(t)}{dt} = -dF^*(t) + dQ^*(t) , \tag{20.56}$$

where

$$Q^*(t) = \int_0^\infty q(x,t)\, dx, \quad F^*(t) = \int_0^\infty [q(x,t) + r(x,t)]\, dx .$$

The initial conditions for (20.56) are $F^*(0) = q_0 + r_0$, $Q^*(0) = q_0$. The first of equations (20.56) is trivially solved for $Q^*(t)$ and Q^* can then be eliminated from the second equation, to obtain the following equation for F^*, namely the (scaled) total number of foxes present in the region $x > 0$:

$$\frac{dF^*}{dt} = -dF^* + dq_0 e^{-\mu t} . \tag{20.57}$$

With the given initial conditions, the solution to this equation is

$$F^*(t) = \left[q_0 + r_0 - \frac{dq_0}{d - \mu} \right] e^{-dt} + \frac{dq_0 e^{-\mu t}}{d - \mu} . \tag{20.58}$$

The critical time t_c can then be determined from (20.54) by solving the equation

$$F^*(t_c) = p \left(\frac{\beta}{KD} \right)^{1/2} . $$

Note that each of the two terms on the right-hand side of (20.58) involves an exponential factor. Since, for reasonable values of the field parameters, $d > \mu$ and $d - \mu = o(1/t_c)$, the first of those terms can be neglected in comparison with the second if t_c is sufficiently large. Let us assume this is the case, and verify it *a posteriori*. So, neglecting the first term, the resulting algebraic equation can be solved to give

$$t_c \approx \mu^{-1} \ln \left[\frac{d \left(\frac{KD}{\beta} \right)^{1/2} q_0}{p(d - \mu)} \right] . \tag{20.59}$$

Typical values for d and μ are 0.46 and 0.08, respectively. $(KD/\beta)^{1/2} q_0$ can be approximated from Fig. 20.12 and the fact that $q \approx dr/\mu$, so that the total number of infected foxes satisfies

$$\int_{-\infty}^{\infty} (I + R) \, dX = \left(\frac{KD}{\beta} \right)^{1/2} \left(1 + \frac{\mu}{d} \right) q_0 . $$

From Fig. 20.12,

$$\int_{-\infty}^{\infty} (I + R) \, dX \approx 6.9 \text{ foxes/km} , $$

giving $(KD/\beta)^{1/2} q_0 \approx 5.9$ foxes/km. For $p = 0.5$ fox/km, (20.59) gives an estimate of $t_c \approx 33$ for these values of the parameters, and so the ratio of the two exponentials $\exp[-dt_c]$ and $\exp[-\mu t_c]$ is approximately 3×10^{-6}, which justifies neglecting the smaller exponential in (20.58) in the above analysis.

We now derive an estimate for the break width x_c. This involves solving the problem posed by (20.52) with (20.53). The first of (20.52) gives

$$q(x, t) = 2q_0 \delta(x) e^{-\mu t} . \tag{20.60}$$

Substituting this into the second equation gives

$$\frac{\partial r}{\partial t} = -dr + \frac{\partial^2 r}{\partial x^2} + 2q_0 \mu \delta(x) e^{-\mu t} \tag{20.61}$$

the solution of which, with initial conditions (20.53), is of the form

$$r(x,t) = \frac{2r_0}{\sqrt{\pi t}} \exp\left[-\frac{x^2}{4t} - dt\right] + e^{-\mu t} r^*(x,t) \, ,$$

where $r^*(x,t)$ is the solution of

$$\frac{\partial r^*}{\partial t} = (\mu - d)r^* + \frac{\partial^2 r^*}{\partial x^2} + 2q_0\mu\delta(x) \tag{20.62}$$

with homogeneous initial data. This equation can be solved using Laplace transforms.

Denote the Laplace transform of r^* by ρ, that is

$$\rho(x,s) = \int_0^\infty r^*(x,t)\exp\left[-st\right]dt, \quad \mathrm{Re}\, s > 0 \, .$$

Then ρ satisfies the inhomogeneous ordinary differential equation

$$\frac{d^2\rho}{dx^2} + (\mu - d - s)\rho = -\frac{2q_0\mu\delta(x)}{s}, \quad -\infty < x < \infty, \quad \mathrm{Re}\, s > 0 \, . \tag{20.63}$$

We are only interested in the solution for $x \geq 0$; it is given by

$$\rho(x,s) = \mu q_0 \frac{\exp\left[-(s+d-\mu)^{1/2}x\right]}{s(s+d-\mu)^{1/2}} \, .$$

Thus, inverting the transform, we get

$$r^*(x,t) = \frac{\mu q_0}{2\pi i}\int_C \frac{\exp\left[-(s+d-\mu)^{1/2}x\right]\exp\left[st\right]}{s(s+d-\mu)^{1/2}}\,ds, \tag{20.64}$$

where C is the Bromwich contour. The singularities of the integrand are a pole at $s = 0$ and a branch point at $s = -(d-\mu)$. The branch cut can be taken along the negative real axis to the left of the branch point, and so the contour of integration can be deformed to lie above and below the negative real axis. Since it is only necessary to evaluate $r^*(x,t)$ for $t = t_c$, it can be assumed that $t \gg 1$ in the integral (20.64). If we now use the method of steepest descents (see, for example, Chapter 5 in the book by Murray 1984) the main contribution to the integral is given by the residue at the pole $s = 0$; the contribution from the branch cut is exponentially small in comparison, provided that

$$\left(\frac{x}{2t}\right)^2 \ll d - \mu \, . \tag{20.65}$$

This inequality will be shown to hold below. We thus arrive at the asymptotic solution for $r(x,t)$ given by

$$r(x,t) \sim \frac{r_0}{\sqrt{\pi t}}\exp\left[-\frac{x^2}{4t} - dt\right] + \frac{\mu q_0}{\sqrt{d-\mu}}\exp\left[-\mu t - (d-\mu)^{1/2}x\right] \, . \tag{20.66}$$

To estimate the break width, note that the formula (20.55) cannot be directly used since, with (20.60), $q(x,t)$ always involves a δ-function. Instead, we replace (20.55) by

$$r(x_c, t_c) = mr(0, t_c) . \tag{20.67}$$

The assumptions (20.65) and $t \gg 1$ can again be used to justify neglecting the first term in (20.66) as compared with the second. So, from (20.67) and (20.66), an estimate for the break width is given by

$$x_c \sim (d - \mu)^{-1/2} \ln \left[\frac{1}{m} \right] . \tag{20.68}$$

If we take $m = 10^{-4}$ together with the parameters used previously to estimate t_c, then assumption (20.65) is easily verified to be valid for $t = t_c$ and $x = x_c$ since $(x_c/2t_c)^2 \approx 0.05$ and $d - \mu \approx 0.38$.

Note that, at least to leading order, the formula for x_c is independent of the critical time t_c. The calculation of t_c was only necessary for the purpose of verifying the "t large" assumption that was made throughout the analysis.

In dimensional terms, (20.68) gives, using (20.32)

$$X_c \sim -(\beta K)^{-1} - \left[\frac{D}{\alpha + b - \sigma} \right]^{1/2} \ln m , \tag{20.69}$$

with typical values for these parameters given in Table 20.1.

In the expression (20.68), the dependence of x_c on d and m roughly agrees with Fig. 20.13. It also suggests that the break width should not be very sensitive to p, which, as shown by Murray et al. (1986) is the case when the carrying capacity in the break is not too close to the critical value.

20.6 Two-Dimensional Epizootic Fronts and Effects of Variable Fox Densities: Quantitative Predictions for a Rabies Outbreak in England

In general fox populations are not uniform, but instead vary according to the hospitality and carrying capacity of the local environment. This is very much the case in England, where interestingly some of the highest densities (by a factor of 2 to 3) are in the cities such as Bristol.

We first present the results of what happens when the epizootic wave encounters a localized region of different carrying capacity from the surrounding environment. The model system is still (20.26) except that in this two-dimensional situation, the diffusion term in the equation for the rabid population in (20.26) is replaced by $D\nabla^2 R$. Suppose that the carrying capacity, K, and the initial susceptible fox density are equal to a uniform value everywhere on a square region, except for a small patch in the center of the square, where they have different values. We now introduce a uniform distribution of rabid foxes along one edge of

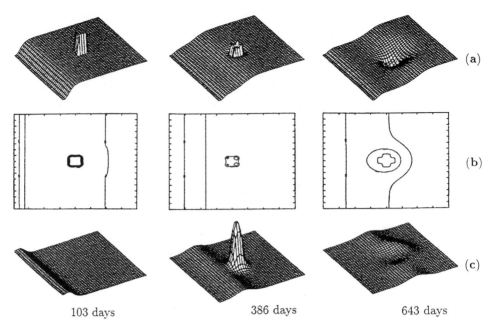

103 days   386 days   643 days

Fig. 20.14a-c. Effect on the epidemic front on encountering a pocket of higher initial susceptible fox density (and hence carrying capacity). Equations (20.37)–(20.40) were solved on a square, with the initial rabies-free fox density and the carrying capacity uniform everywhere except in a rectangular region in the center, where they were raised by a factor of 1.7. The results are shown for a sequence of three times, namely, as the wave comes in from one side, as it passes the higher density pocket and after it passes. (a) Three dimensional plot of the susceptible fox population density. (b) Contour plot of the susceptible fox density, with contour intervals of 0.1, where the density is normalized to have a maximum of 1. (c) Three dimensional plot of the rabid fox density at each point in the square. (After Murray et al. 1986)

the square, so that a one-dimensional epidemic front starts off across the square, and solve the model equations numerically. Fig. 20.14 shows the resulting rabid and susceptible fox population densities for the case of a higher initial susceptible density in the patch.

From Fig. 20.14 (b) we see that the front moves faster through the region of higher carrying capacity as we would expect heuristically. The residual fox population, once the first outbreak has moved past, is slightly lower in the pocket of higher K than in the surrounding region. The converse of these effects are obtained if the wave encounters a pocket of lower susceptible fox density. One interesting feature is that the pocket of lowered density provides a sort of protection to the region just adjoining it. There are never as many cases of rabies in a ring around the outside of this region, and the final susceptible population density is higher there than further away. The break region of the previous section also exhibits this feature, which arises because the region of lower density does not provide as many rabid foxes to diffuse into this area – there is, in effect, a preferential direction for the diffusion. The pocket of higher density has the

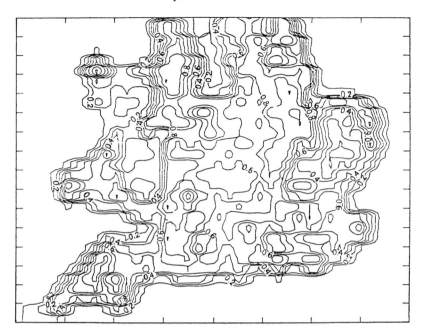

Fig. 20.15. Contour plot of fox densities in the southern half of England which were used in the numerical simulations. Values are scaled to lie between 0 and 1, with 1 corresponding to 2.4 adult foxes/km^2 in springtime, or to an average of 4.6 foxes/km^2 throughout the year. These values are based on Macdonald's (1980) estimates, who emphasizes that the density map is probably not very accurate but is based on educated estimates. (From Murray et al. 1986)

opposite effect. Here the epidemic moves ahead of the epidemic front into the pocket of higher density: see the central figure in Fig. 20.14 (c). This focusing effect could account for some of the cases when outbreaks of rabies appear in advance of the front.

As mentioned before, England has remained rabies-free (except for a minor epidemic after World War I) due mainly to the strict quarantine laws and high public awareness of the potential dangers. With the proximity of the disease in the north of France and the increased private boat traffic between continental Europe and Britain it seems inevitable that the disease will be brought into Britain in the near future. The appearance of rabies in Britain would be particularly serious, as we mentioned above, because of the high density of foxes, both urban and rural, in England. An additional cause for concern is the apparent compatibility of these urban foxes with cats (Macdonald 1980). If no control measures are applied, which admittedly would certainly not be the case, the epidemic would move quickly through England. We can use the model here to obtain a rough estimate for the position of the epidemic front after rabies is introduced into the fox population.

Macdonald (1980) gives a map of estimated fox densities in England (but excluding high urban pockets). Murray et al. (1986) covered the lower half of

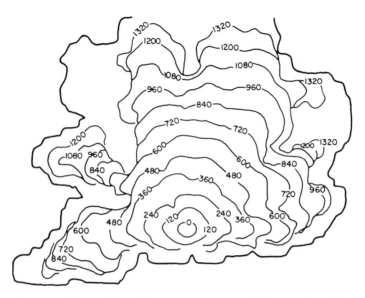

Fig. 20.16. The position of the wave front every 120 days predicted by the model (20.37)–(20.40) and the spatially heterogeneous fox densities in Fig. 20.15: that is some of the parameters are space dependent. Here a diffusion coefficient of 200 km^2/yr was taken with the other parameter values from Table 20.1. (From Murray et al. 1986)

England with a grid, and assigned a density to each square based on the values given on his map. Contour lines of these densities, normalized from 0 to 1, are shown in Fig. 20.15. A value of 1 corresponds to 2.4 adult foxes per square kilometre in springtime. The model we have been studying is, in fact, in terms of fox densities averaged over the yearly cycle. Prior to the introduction of rabies, the population increases to its yearly high just after whelping, then gradually returns to the adult springtime population. The average density is roughly the mean between the populations just before and just after whelping. The ratio of males to females is about 1.2 : 1, and females have an average of 3.7 to 4.2 cubs each year (Llyod et al. 1976). Thus the average population is about 1.9 times the springtime adult population, and 1 corresponds to a carrying capacity of 4.6 foxes/km^2 in Fig. 20.15.

Using the carrying capacities (and initial fox densities) shown in Fig. 20.15, and supposing, by way of illustration, that the rabies epidemic starts near Southampton, the two-dimensional form of (20.6) was solved numerically. The parameter values given in Table 20.1 were used, and the diffusion coefficient was taken to be 200 km^2/yr. The numerical simulations took about 120 minutes on a CRAY XMP-48 at the Los Alamos National Laboratory. The results are shown in Figs. 20.16 and 20.17. The position of the front every 120 days is shown in Fig. 20.16. We see that with such high fox densities the epidemic very quickly reaches most of the region studied. Within 4 years the front has effectively reached Manchester. The sequence in Fig. 20.17 shows that, just as in

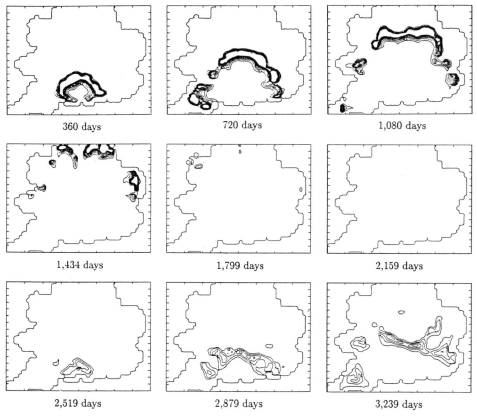

Fig. 20.17. The epidemic front as it moves through the southern part of England. This was obtained by numerically solving (20.37)–(20.40) with the local carrying capacities and initial susceptible fox densities shown in Fig. 20.15. A localized density of rabid foxes was initially introduced at Southampton on the south coast and allowed to spread. Contour plots of the rabid fox densities are given at a sequence of times, as the wave moves outward from its source. Note that, just as in the one dimensional case, there are few rabid foxes in the region behind the front. Note also the reappearance of the second epidemic wave of lower intensity, which starts about 7 years after the initial outbreak and moves outward at the same speed. Here $D = 200 \text{ km}^2/\text{yr}$ and other parameter values taken from Table 20.1. (From Murray et al. 1986)

the uniform density case, most of the cases of rabies are concentrated in a narrow band at the front; the susceptible population is effectively decimated by the epidemic and partially regenerates before another wave starts again. Fig. 20.17 shows the second outbreak starting off from Southhampton, about 7 years after the first one.

These quantitative predictions can, of course, only be rough estimates. Macdonald (1980) emphasizes that the fox densities in his map are only educated guesses, based on his knowledge of fox ecology. As we said above, not enough is known about the behaviour of rabid foxes to obtain a sharp estimate for the diffusion coefficient, which means that the speed of the wave may be anywhere from a half to four-thirds of our calculated result. We have also neglected such geographical factors as rivers, which tend to provide a channel for the epidemic, speeding its movement parallel to the banks and temporarily halting its direct passage. However, this relatively simple *SIR*-model provides a plausible quantitative first estimate for the progression of rabies in England if the epidemic was allowed to move unchecked. The model also provides a means of estimating realistic break widths which, at the very least, would seriously impede the spread of the disease.

The model we have investigated incorporates many of the salient features of the disease and the ecology of foxes. The model is sufficiently simple to enable us to obtain fairly reliable estimates for all of the parameters except the diffusion coefficient, for which we obtained a range of possible values. Analysis of the model produces certain predictions for the behaviour of the epidemic wave, in different environments, which provides some quantitative insight into the spatial spread of the epidemic and the transmission mechanisms responsible for its spread. For example, it is not known whether the primary reason for the spatial spread of the epidemic is the encroachment of confused rabid foxes onto their neighbour's territories, as we have assumed, or the migration of young foxes who carry the disease with them while healthy, or if both mechanisms are equally important. By isolating one of these mechanisms, we can determine how the epidemic wave behaves if that is the primary factor in its spatial spread, and compare the results with observation in continental Europe to see if it is possible for it to be the dominating factor. Our results indicate that the confused wandering of rabid foxes is indeed sufficient to account for much of the behaviour of the current epidemic. It would be interesting to investigate a model in which migrating young foxes are the primary cause for spatial spread of rabies. It is also known that a certain percentage of foxes are immune to rabies. Such effects as these can easily be incorporated into the model framework here.

The agreement of our model with the available epidemiological evidence is quite good, despite the uncertainty in the size of the diffusion coefficient. For an initial fox density of 2 foxes/km^2, which is similiar to densities reported for much of the continent, and for any reasonable choice of diffusion coefficient, the speed of the epidemic front, 25–65 km/yr, obtained from the model, encompasses the range of 30–60 km/yr usually observed. The speed of the wave increases with fox density, and drops to zero as the fox density decreases to the critical value. The model also predicts that rabies will essentially disappear for a period of about 5 years after the first outbreak, and then reappear, with the second outbreak weaker than the first. This correlates well with what has happened in many parts of Europe. Another interesting feature which emerges from the model is the enhanced movement of the rabies epidemic into regions of higher density *in advance of* the rest of the front. As we suggested, this may help to explain why outbreaks seemingly far in advance of the main epidemic occasionally occur.

It is possible for a strip of lowered susceptible fox population to check the progression of the epidemic, and protect an uninfected region ahead of the front. For this method of control to be efficiently applied, it is essential to have an indication of how wide an effective break region must be. For our model control scheme, Fig. 20.13 gives nondimensional estimates for this width. If there are 2 foxes/km^2 initially, and the reduction scheme is 80% effective, then Fig. 20.13 gives a break width of 10–25 km, depending on the diffusion coefficient. This is of the right order of magnitude when compared with the protective break which has proved effective in Denmark. There, intensive control measures were applied to a strip 20 km wide with less intensive measues used in an adjoining 20 km strip.

The question of what method should be used to contain an outbreak is interesting. The model here suggests that vaccination would be more effective than gassing or poisoning since the former would help to restrict the spread of infective foxes whereas the latter would enhance the spread. It seems that chicken heads impreganted with vaccine has proved reasonably effective in Ontario: it relies on efficient scavenging by the foxes. This is not necessarily the case with urban populations (personal communication from Dr. Stephen Harris, Bristol 1988). Another problem with vaccination in general is that sometimes the level of vaccination in one species may induce the disease in another as seems to be the case with the red and grey fox.

The probability that rabies will eventually reach England and other uninfected regions is high. It is clearly of considerable importance to understand as much as possible about the disease, its transmission and how it spreads well before it arrives. The density of foxes in England is much greater in many areas than on the continent, and the epidemic may proceed differently there. Figs. 20.16 and 20.17 summarizes some of the model's predictions for a particular choice of diffusion coefficient, and some estimates for the current fox populations in the southern half of England. Perhaps the most disturbing aspect of these results is the rapidity with which the epidemic would move through the central region, namely at speeds of around 100 km/yr. No less disturbing is the reappearance of the disease several years after the passage of the epidemic front: a relatively free rabies period is likely to give rise to complacency.

We have been primarily concerned here with the spatial propagation of an epidemic. There are important and interesting problems associated with control strategies when rabies, for example, is already in a community. An interesting and very practical model to deal with this situation was proposed by Frerichs and Prawda (1975) to deal with an urban area in Colombia: the model they proposed was for canine rabies.

The type of models we have discussed in this chapter have wider applicability, such as to the spatial spread of pests, killer bees (see Taylor 1977 for data on the South American spread), animals, plants and so on.

Exercises

1. Consider the dimensionless form of the epidemic model

    $$S_t = -IS + S_{xx}, \quad I_t = IS - \lambda I + I_{xx} ,$$

    where $\lambda > 0$ and look for travelling wave solutions $S(z)$ and $I(z)$, with $z = x - ct$, such that

    $$S'(-\infty) = 0, \quad S(\infty) = 1, \quad I(-\infty) = I(\infty) = 0 ,$$

    where prime denotes differentiation with respect to z.
    Prove that, for all finite z, $0 < S < 1$ by showing that $S'(z)$ is monotonic for all $-\infty < z < \infty$. Show also that $(S + I)' > 0$ and hence that for all $-\infty < z < \infty$, $S(z) + I(z) < 1$.
    Prove that

    $$\int_{-\infty}^{\infty} I(z')\, dz' > \int_{-\infty}^{\infty} I(z')S(z')\, dz' = \lambda \int_{-\infty}^{\infty} I(z')\, dz'$$

    and hence deduce that the threshold criterion for a travelling epidemic wave solution to exist is $\lambda < 1$.

2. A rabies model which includes a logistic growth for the susceptibles S and diffusive dispersal for the infectives is

    $$\frac{\partial S}{\partial t} = -rIS + bS\left(1 - \frac{S}{S_0}\right), \quad \frac{\partial I}{\partial t} = rIS - aI + D\frac{\partial^2 I}{\partial x^2} ,$$

    where r, b, a, D and S_0 are positive constant parameters. Nondimensionalize the system to give

    $$u_t = u_{xx} + uv - \lambda u, \quad v_t = -uv + bv(1 - v) ,$$

    where u relates to I and v to S. Look for travelling wave solutions with $u > 0$ and $v > 0$ and hence show, by linearizing far ahead of a wavefront where $v \to 1$ and $u \to 0$, that a wave may exist if $\lambda < 1$ and if so the minimum wave speed is $2(1 - \lambda)^{1/2}$. What is the steady state far behind the wave?

Appendix 1
Phase Plane Analysis

We discuss here, only very briefly, general autonomous second order ordinary differential equations of the form

$$\frac{dx}{dt} = f(x, y), \quad \frac{dy}{dt} = g(x, y) . \tag{A1.1}$$

We present the basic results which are required in the main text. There are many books which discuss phase plane analysis in varying depth, such as Jordan and Smith (1987) and Guckenheimer and Holmes (1983). A particularly good, short and practical exposition of the qualitative theory of ordinary differential equation systems, including phase plane techniques, is given by Odell (1980).

Phase curves or *phase trajectories* of (A1.1) are solutions of

$$\frac{dx}{dy} = \frac{f(x, y)}{g(x, y)} . \tag{A1.2}$$

Through any point (x_0, y_0) there is a unique curve except at *singular points* (x_s, y_s) where

$$f(x_s, y_s) = g(x_s, y_s) = 0 .$$

Let $x \rightarrow x - x_s$, $y \rightarrow y - y_s$, then $(0,0)$ is a singular point of the transformed equation. Thus without loss of generality we now consider (A1.2) to have a singular point at the origin, that is

$$f(x, y) = g(x, y) = 0 \quad \Rightarrow \quad x = 0, \quad y = 0 . \tag{A1.3}$$

If f and g are analytic near $(0,0)$ we can expand f and g in Taylor series and, retaining only the linear terms, we get

$$\frac{dx}{dy} = \frac{ax + by}{cx + dy}, \quad A = \begin{pmatrix} a & b \\ c & d \end{pmatrix} = \begin{pmatrix} f_x & f_y \\ g_x & g_y \end{pmatrix}_{(0,0)} \tag{A1.4}$$

which defines the matrix A and the constants a, b, c and d. The linear form is equivalent to the system

$$\frac{dx}{dt} = ax + by, \quad \frac{dy}{dt} = cx + dy . \tag{A1.5}$$

Solutions of (A1.5) give the parametric forms of the phase curves; t is the parametric parameter.

Let λ_1 and λ_2 be the eigenvalues of A defined in (A1.4), that is

$$\begin{vmatrix} a - \lambda & b \\ c & d - \lambda \end{vmatrix} = 0 \quad \Rightarrow \quad \lambda_1, \lambda_2 = \frac{(a+d) \pm [(a+d)^2 - 4\det A]^{1/2}}{2} \qquad \text{(A1.6)}$$

Solutions of (A1.5) are then

$$\begin{pmatrix} x \\ y \end{pmatrix} = c_1 \mathbf{v}_1 \exp[\lambda_1 t] + c_2 \mathbf{v}_2 \exp[\lambda_2 t] , \qquad \text{(A1.7)}$$

where c_1 and c_2 are arbitrary constants and $\mathbf{v}_1$, $\mathbf{v}_2$ are the eigenvectors of A corresponding to λ_1 and λ_2 respectively: they are given by

$$\mathbf{v}_i = (1 + p_i^2)^{-1/2} \begin{pmatrix} 1 \\ p_i \end{pmatrix}, \quad p_i = \frac{\lambda_i - a}{b}, \quad b \neq 0, \quad i = 1, 2 \qquad \text{(A1.8)}$$

Elimination of t in (A1.7) gives the phase curves in the (x, y) plane.

The form (A1.7) is for distinct eigenvalues. If the eigenvalues are equal the solutions are proportional to $(c_1 + c_2 t) \exp[\lambda t]$.

Catalogue of (Linear) Singularities in the Phase Plane

(i) λ_1, λ_2 real and distinct:

(a) λ_1 and λ_2 have the same sign. Typical eigenvectors $\mathbf{v}_1$ and $\mathbf{v}_2$ are illustrated in Fig. A1.1 (a). Suppose $\lambda_2 < \lambda_1 < 0$. Then, from (A1.7), for example for $c_2 = 0$, $c_1 \neq 0$

$$\begin{pmatrix} x \\ y \end{pmatrix} = c_1 \mathbf{v}_1 \exp[\lambda_1 t] ,$$

so the solution in the phase plane simply moves along $\mathbf{v}_1$ towards the origin as $t \to \infty$ in the direction shown in Fig. A1.1 (a) – along PO if $c_1 > 0$ and along QO if $c_1 < 0$.

From (A1.7) every solution tends to $(0, 0)$ as $t \to \infty$ since, with $\lambda_2 < \lambda_1 < 0$, $\exp[\lambda_2 t] = o(\exp[\lambda_1 t])$ as $t \to \infty$ and so

$$\begin{pmatrix} x \\ y \end{pmatrix} \sim c_1 \mathbf{v}_1 \exp[\lambda_1 t] \quad \text{as} \quad t \to \infty .$$

Thus close enough to the origin all solutions tend to zero along $\mathbf{v}_1$ as shown in Fig. A1.1 (a). This is called a *node* (Type I) singularity. With $\lambda_1 \le \lambda_2 \le 0$ it is a stable node since all trajectories tend to $(0, 0)$ as $t \to \infty$. If $\lambda_1 > \lambda_2 > 0$ it is an unstable node: here $(x, y) \to (0, 0)$ as $t \to -\infty$.

(b) λ_1 and λ_2 have different signs. Suppose, for example, $\lambda_1 < 0 < \lambda_2$ then $\mathbf{v}_1 \exp[\lambda_1 t] v_1 \to 0$ along $\mathbf{v}_1$ as $t \to \infty$ while $\mathbf{v}_2 \exp[\lambda_2] \to 0$ along $\mathbf{v}_2$ as $t \to -\infty$. There are thus different directions on $\mathbf{v}_1$ and $\mathbf{v}_2$: the solutions near $(0, 0)$ are as

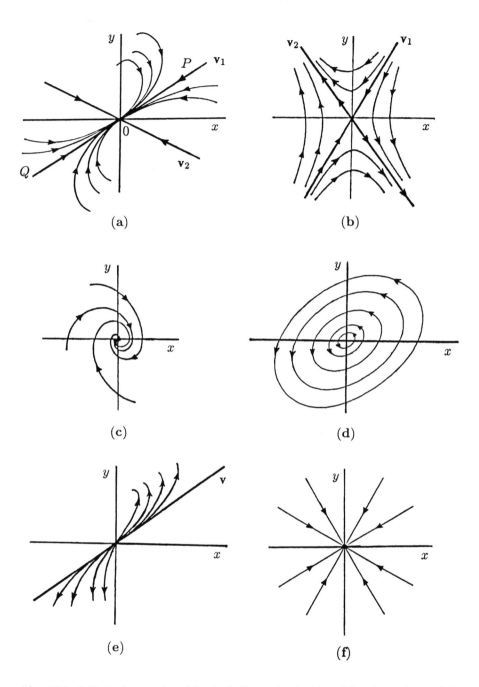

Fig. A1.1a-f. Typical examples of the basic linear singularities of the phase plane solutions of (A1.4). (a) Node (Type I): these can be stable (as shown) or unstable. (b) Saddle point. These are always unstable. (c) Spiral: these can be stable or unstable. (d) Centre. This is neutrally stable. (e) Node (Type II): these can be stable or unstable. (f) Star: these can be stable or unstable.

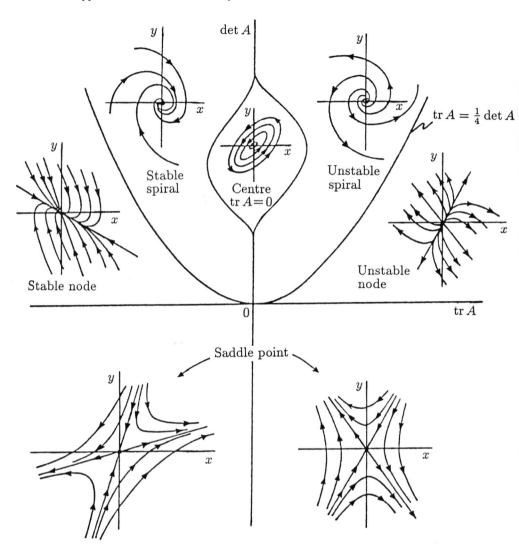

Fig. A1.2. Summary diagram showing how $\operatorname{tr} A$ and $\det A$, where A is the linearization matrix given by (A1.4), determine the type of phase plane singularity for (A1.1). Here $\det A = f_x g_y - f_y g_x$, $\operatorname{tr} A = f_x + g_y$, where the partial derivatives are evaluated at the singularities, the solutions of $f(x, y) = g(x, y) = 0$.

shown in Fig. A1.1 (b). This is a *saddle point* singularity. It is always *unstable*: except strictly along $\mathbf{v_1}$ any small perturbation from $(0, 0)$ grows exponentially.

(ii) λ_1, λ_2 complex: $\lambda_1, \lambda_2 = \alpha \pm i\beta$, $\beta \neq 0$. Solutions (A1.7) here involve $\exp[\alpha t] \exp[i \pm \beta t]$ which implies an oscillatory approach to $(0, 0)$.

(a) $\alpha \neq 0$. Here we have a *spiral*, which is stable if $\alpha < 0$ and unstable if $\alpha > 0$: Fig. A1.1 (c) illustrates a spiral singularity.

(b) $\alpha = 0$. In this case the phase curves are ellipses. This singularity is called a *centre* and is illustrated in Fig. A1.1 (d). Centres are not stable in the usual sense; a small perturbation from one phase curve does not die out in the sense of returning to the original unperturbed curve. The perturbation simply gives another solution. In the case of centre singularities, determined by the linear appoximation to $f(x,y)$ and $g(x,y)$, we must look at the higher order (than linear) terms to determine whether or not it is really a spiral and hence whether it is stable or unstable.

(iii) $\lambda_1 = \lambda_2 = \lambda$. Here the eigenvalues are *not* distinct.

(a) In general, solutions now involve terms like $t \exp[\lambda t]$ and there is only one eigenvector $\mathbf{v}$ along which the solutions tend to $(0,0)$. The t in $t \exp[\lambda t]$ modifies the solution away from $(0,0)$. It is called a *node* (Type II) singularity, an illustration of which is given in Fig. A1.1 (e).

(b) If the solutions do not contain the $t \exp[\lambda t]$ term we have a *star* singularity, which may be stable or unstable, depending on the sign of λ. Trajectories in the vicinity of a star singularity are shown in Fig. A1.1 (f).

The singularity depends on a, b, c and d in the matrix A in (A1.4). Fig. A1.2 summarizes the results in terms of the trace and determinant of A.

If the system (A1.1) possesses a confined set (that is a domain on the boundary ∂B of which the vector $(dx/dt, dy/dt)$ points into the domain) enclosing a single singular point which is an unstable spiral or node then any phase trajectory cannot tend to the singulariy with time, nor can it leave the confined set. The Poincaré-Bendixson theorem says that as $t \to \infty$ the trajectory will tend to a limit cycle solution. This is the simplest application of the theorem. If the sole singularity is a saddle point a limit cycle cannot exist. See, for example, Jordan and Smith (1987) for a proof of the theorem, its general application and some practical illustrations.

Appendix 2
Routh-Hurwitz Conditions, Jury Conditions, Descarte's Rule of Signs and Exact Solutions of a Cubic

A2.1 Characteristic Polynomials, Routh-Hurwitz Conditions and Jury Conditions

Linear stability of the systems of ordinary differential equations such as arise in interacting population models and reaction kinetics systems (cf. Chapters 3 and 5) is determined by the roots of a polynomial. The stability analysis we are concerned with, involves linear systems of the vector form

$$\frac{d\mathbf{x}}{dt} = A\mathbf{x} \qquad (A2.1)$$

where A is the matrix of the linearized nonlinear interaction/reaction terms: it is the Jacobian matrix about the steady state – the community matrix in ecological terms. Solutions are obtained by setting

$$\mathbf{x} = \mathbf{x}_0 e^{\lambda t} , \qquad (A2.2)$$

in (A2.1) where $\mathbf{x}_0$ is a constant vector and the eigenvalues λ are the roots of the *characteristic polynomial*

$$|A - \lambda I| = 0 , \qquad (A2.3)$$

where I is the identity matrix. The solution $\mathbf{x} = 0$ is stable if all the roots λ of the characteristic polynomial lie in the left-hand complex plane, that is $\operatorname{Re} \lambda < 0$ for all roots λ. If this holds then $\mathbf{x} \to 0$ exponentially as $t \to \infty$ and hence $\mathbf{x} = 0$ is stable to small (linear) perturbations.

If the system is of n-th order. the characteristic polynomial can be taken in the general form

$$P(\lambda) = \lambda^n + a_1 \lambda^{n-1} + \ldots + a_n = 0 , \qquad (A2.4)$$

where the coefficients a_i, $i = 0, 1, \ldots, n$ are all real. We tacitly assume $a_n \neq 0$ since otherwise $\lambda = 0$ is a solution, and the polynomial is then of order $n - 1$ with the equivalent $a_n \neq 0$. We require conditions on the a_i, $i = 0, 1, \ldots, n$ such that the zeros of $P(\lambda)$ have $\operatorname{Re} \lambda < 0$. The necessary and sufficient conditions for this to hold are the *Routh-Hurwitz conditions*. There are various equivalent forms of these one of which is, together with $a_n > 0$,

$$D_1 = a_1 > 0, \quad D_2 = \begin{vmatrix} a_1 & a_3 \\ 1 & a_2 \end{vmatrix} \quad D_3 = \begin{vmatrix} a_1 & a_3 & a_5 \\ 1 & a_2 & a_4 \\ 0 & a_1 & a_3 \end{vmatrix} > 0 ,$$

$$D_k = \begin{vmatrix} a_1 & a_3 & \cdot & \cdot & \cdot & \cdot \\ 1 & a_2 & a_4 & \cdot & \cdot & \cdot \\ 0 & a_1 & a_3 & \cdot & \cdot & \cdot \\ 0 & 1 & a_2 & \cdot & \cdot & \cdot \\ \cdot & \cdot & \cdot & \cdot & \cdot & \cdot \\ 0 & 0 & \cdot & \cdot & \cdot & a_k \end{vmatrix} > 0, \quad k = 1, 2, \dots, n . \tag{A2.5}$$

These conditions are derived, using complex variable methods, in standard texts on the theory of dynamical systems (see, for example, Willems 1970). As an example, for the cubic equation

$$\lambda^3 + a_1 \lambda^2 + a_2 \lambda + a_3 = 0$$

the conditions for $\operatorname{Re} \lambda < 0$ are

$$a_1 > 0, \quad a_3 > 0; \quad a_1 a_2 - a_3 > 0 .$$

Frankly it is hard to imagine anyone actually using the conditions for polynomials of order five or more.

Although (A2.5) are the necessary and sufficient conditions we need, the usual algebraic relations between the roots and the polynomial coefficients can often be very useful. If $\lambda_1, \dots, \lambda_n$ are the distinct non-zero roots of (A2.4) these are

$$\sum_{i=1}^{n} \lambda_i = -\frac{a_{n-1}}{a_n}, \quad \sum_{\substack{i,j \\ i \neq j}}^{n} \lambda_i \lambda_j = \frac{a_{n-2}}{a_n}, \dots \quad \lambda_1 \lambda_2 \dots \lambda_n = (-1)^n \frac{a_0}{a_n} . \tag{A2.6}$$

Lewis (1977) gives several useful ways of deriving qualitative results using the Routh-Hurwitz conditions together with network concepts directly on the matrices.

In the case of systems of discrete models for interacting populations (cf. Chapter 4), and with single population models with delay (cf. Chapter 2), stability is again determined by the roots of a characteristic polynomial (cf. Section 4.1 in Chapter 4). The linearised systems again give rise to matrix forms like (A2.3) and hence polynomials like (A2.4). With delay equations, with a delay of n time steps say, we have to solve linear difference equations typically of the form

$$u_{t+1} = b_1 u_t + \dots + b_n u_{t-n}$$

(equation (2.33) in Chapter 2 is an example). We solve this by setting $u_t \propto \lambda^t$ which again results in a polynomial in λ. Linear stability here however is determined by the magnitude of λ: stability requires $|\lambda| < 1$ since in this case $u_t \to 0$ as $t \to \infty$. So, for the linear stability analysis of discrete systems we require the conditions on the coefficients of the characteristic polynomial so that the solutions λ have magnitude less than 1. The *Jury conditions* are the conditions for this to the case.

The Jury conditions are given, for example, in the book by Lewis (1977), who describes and illustrates several useful, analytical and numerical, techniques concerning the size and signs of the roots of polynomials. For the polynomial $P(\lambda)$ in (A2.4), let

$$b_n = 1 - a_n^2, \quad b_{n-1} = a_1 - a_n a_{n-1}, \ldots, \quad b_{n-j} = a_j - a_n a_{n-j}, \ldots,$$

$$b_1 = a_{n-1} - a_n a_1;$$

$$c_n = b_n^2 - b_1^2, \quad c_{n-1} = b_n b_{n-1} - b_1 b_2, \ldots,$$

$$c_{n-j} = b_n b_{n-j} - b_1 b_{j+1}, \ldots, \quad c_2 = b_n b_2 - b_1 b_{n-1};$$

$$d_n = c_n^2 - c_2^2, \ldots, \quad d_{n-j} = c_n c_{n-j} - c_2 c_{j+2}, \ldots, \quad d_3 = c_n c_3 - c_2 c_{n-1};$$

and so on until we are left with only three elements of the type

$$s_n = r_n^2 - r_{n-3}^2, \quad s_{n-1} = r_n r_{n-1} - r_{n-3} r_{n-2}, \quad s_{n-2} = r_n r_{n-2} - r_{n-3} r_{n-1}.$$

The Jury conditions (necessary and sufficient) which ensure that the roots of the polynomial $P(\lambda)$ in (A2.4) all have magnitudes less than 1 are:

$$P(1) > 0, \quad (-1)_n P(-1) > 0,$$

$$|a_n| < 1, \quad |b_n| > |b_1|, \tag{A2.7}$$

$$|c_n| > |c_2|, \quad |d_n| > |d_3|, \ldots, \quad |s_n| > |s_{n-2}|.$$

A2.2 Descarte's Rule of Signs

Consider the polynomial (A2.4), and, as before, we take without loss of generality $a_n > 0$. Let N be the number of sign changes in the sequence of coefficients $\{a_n, a_{n-1}, \ldots, a_0\}$, ignoring any which are zero. Descarte's Rule of Signs says that there are at most N roots of (A2.4), which are real and positive, and further, that there are either N or $N - 2$ or $N - 4, \ldots$ real positive roots. By setting $\omega = -\lambda$ and again applying the rule, information is obtained about the possible real negative roots. Together these often give invaluable information on the sign of all the roots, which from a stability point of view is usually all we require.

As an example consider

$$\lambda^3 + a_2 \lambda^2 - a_1 \lambda + a_0 = 0, \quad a_i > 0 \quad \text{for all} \quad i = 0, 1, 2. \tag{A2.8}$$

There are 2 sign changes in the sequence of coefficients, and so there are either 2 or 0 real positive roots. If we now set $\lambda = -\omega$, the equation becomes

$$\omega^3 - a_2\omega^2 - a_1\omega - a_0 = 0 ,$$

which has 1 change of sign in the sequence, and so there is at most 1 real positive root ω. This means there is exactly one negative root λ of (A2.7).

A2.3 Roots of a General Cubic Polynomial

Sometimes it is helpful to have the actual roots of the characteristic polynomial, however complicated they may be. Although it is possible to find these for polynomials higher than order 3 the complexity is usually not worth the effort. The roots of a cubic are probably the most complicated that we would ever wish to have: a particularly simple derivation of these has been given by Namias (1985). The following Table A2.1 gives the explicit forms of the roots of a cubic.

Table A2.1. Explicit roots of the cubic polynomial $p(\lambda) = \lambda^3 + A\lambda^2 + B\lambda + C$, with A, B, and real

$\lambda^3 + A\lambda^2 + B\lambda + C = 0,$ $A \equiv 3a,$ $B \equiv 3b,$ $\alpha \equiv a^2 - b,$ $\beta \equiv 2a^3 - 3ab + C$

> 0	$\beta = 0$	$\lambda_1 = -a$	
		$\lambda_2 = (3\alpha)^{1/2} - a$	
		$\lambda_3 = -(3\alpha)^{1/2} - a$	
	$\lvert\beta\rvert \leq 2\alpha^{3/2}$	$\lambda_1 = 2\alpha^{1/2}\sin\phi - a$	$\phi = (1/3)\sin^{-1}\{\beta/[2\alpha^{3/2}]\}$
		$\lambda_2 = -2\alpha^{1/2}\sin(\pi/3 + \phi) - a$	$-\pi/6 \leq \phi \leq \pi/6$
		$\lambda_3 = 2\alpha^{1/2}\sin(\pi/3 - \phi) - a$	
	$\beta > 2\alpha^{3/2}$	$\lambda_1 = -2\alpha^{1/2}\cosh\psi - a$	$\psi = (1/3)\cosh^{-1}\{\lvert\beta\rvert/[2\alpha^{3/2}]\}$
		$\lambda_2 = \alpha^{1/2}\cosh\psi - a + i(3\alpha)^{1/2}\sinh\psi$	
		$\lambda_3 = \alpha^{1/2}\cosh\psi - a - i(3\alpha)^{1/2}\sinh\psi$	
	$\beta < -2\alpha^{3/2}$	$\lambda_1 = 2\alpha^{1/2}\cosh\psi - a$	
		$\lambda_2 = -\alpha^{1/2}\cosh\psi - a + i(3\alpha)^{1/2}\sinh\psi$	
		$\lambda_3 = -\alpha^{1/2}\cosh\psi - a - i(3\alpha)^{1/2}\sinh\psi$	
$= 0$	$-\infty < \beta < \infty$	$\lambda_1 = -\beta^{1/3} - a$	
		$\lambda_2 = \beta^{1/3}/2 - a + 3i\beta^{2/3}/4$	
		$\lambda_3 = \beta^{1/3}/2 - a - 3i\beta^{2/3}/4$	
< 0	$-\infty < \beta < \infty$	$\lambda_1 = -2(-\alpha)^{1/2}\sinh\theta - a$	$\theta = (1/3)\sinh^{-1}\{\beta/[2(-\alpha)^{3/2}]\}$
		$\lambda_2 = (-\alpha)^{1/2}\sinh\theta - a + i(-3\alpha)^{1/2}\cosh\theta$	
		$\lambda_3 = (-\alpha)^{1/2}\sinh\theta - a - i(-3\alpha)^{1/2}\cosh\theta$	

*Appendix 3
Hopf Bifurcation Theorem and Limit Cycles

This appendix discusses Hopf's (1942) theorem which is concerned with the conditions necessary for the existence of real periodic solutions of the real system of ordinary differential equations

$$\frac{d\mathbf{x}}{dt} = \mathbf{F}(\mathbf{x}, v) \ . \tag{A3.1}$$

where $\mathbf{F}$ and $\mathbf{x}(v, t)$ are n-dimensional vectors and v is a real parameter. We derive the conditions and give the proof for the existence of periodic solutions in the two-dimensional case of (A3.1): the proof is based on that given by Friedrichs (1971). Some illustrative examples are presented in detail at the end of the appendix. The book by Hassard, Kazarinoff and Wan (1979) disusses the theory fully as well as the practical techniques in its use.

Theorem: Let

$$\frac{d\mathbf{x}}{dt} = \mathbf{F}(\mathbf{x}, v) \ , \tag{A3.2}$$

be a second-order autonomous system of differential equations for each value of the parameter $v \in (-v_0, v_0)$ where v_0 is a positive number and the vector function $\mathbf{F} \in C^2(D \times (-v_0, v_0))$ where D is a domain in $\mathbb{R}^2$. Suppose that the system (A3.2) has a singular point for each v, that is

$$\mathbf{F}(\mathbf{x}, v) = 0 \quad \Rightarrow \quad \mathbf{x} = \mathbf{a}(v) \ , \tag{A3.3}$$

say. Let the matrix $\mathbf{A}(v)$ be the linearized matrix of (A3.2) about the singular point $\mathbf{a}(v)$, that is

$$\mathbf{A}(v) = [\nabla_{\mathbf{x}} \mathbf{F}(\mathbf{x}, v)]_{\mathbf{x}=\mathbf{a}(v)} \tag{A3.4}$$

Suppose that the matrix $\mathbf{A}(0)$ has purely imaginary eigenvalues $\pm iw$, $w \neq 0$, that is

$$\text{Tr}\,\mathbf{A}(0) = 0, \quad \det \mathbf{A}(0) > 0 \ . \tag{A3.5}$$

If the matrix $\mathbf{B}(v)$, defined by

$$\mathbf{A}(v) = \mathbf{A}(0) + v\mathbf{B}(v) \ , \tag{A3.6}$$

is such that

$$\mathrm{Tr}\,\mathbf{B}(0) \neq 0 \qquad (A3.7)$$

then there exists a periodic solution of (A3.2) for v in some neighbourhood of $v = 0$ and $\mathbf{x}$ in some neighbourhood of $\mathbf{a}$ with approximate period $T = 2\pi/w$ for small v.

Proof. The proof relies on showing that if the conditions of the theorem apply then, for any vector $\mathbf{b}$, there exist functions $c(u)$, $d(u)$, $\mathbf{y}(s,u)$, $v(u) = ud(u)$, $T(u) = T_0[1 + uc(u)]$, and

$$\mathbf{x}(u,t) = \mathbf{a}(v(u)) + u\mathbf{y}\left(\frac{T_0 t}{T(u)}, u\right) \qquad (A3.8)$$

satisfying

(i) $c(0) = 0 = d(0)$,   (A3.9)
(ii) $c(u)$ and $d(u)$ are $C^1[0, u_0)$ for some sufficiently small u_0 ,
(iii) $\mathbf{y}(0, u) = \mathbf{b}$,   (A3.10)
(iv) $\mathbf{y}(s, u)$ has period T_0 (in the variable s) ,
(v) $\mathbf{x}(u, t)$ is a solution of

$$\frac{d\mathbf{x}}{dt} = \mathbf{F}(\mathbf{x}, v(u))$$

with period $T(u)$: that is, the period varies with v.

Note that it is necessary to introduce u by $v = ud(u)$ so that it is possible to have a family of periodic solutions for only one value of v but several u. We come back to this point in the two illustrative examples below.

Introduce new variables by

$$s = \frac{T_0 t}{T(u)}, \quad v = ud(u), \quad T(u) = T_0[1 + uc(u)] , \qquad (A3.11)$$

and $\mathbf{y}(s, u)$ by (A3.8). With this independent variable s, if $\mathbf{x}(v, u, t)$ is periodic in t with period $T(u)$ then $\mathbf{y}(s, u)$ is periodic in s with period T_0. Thus, from (A3.8) and (A3.10)

$$\mathbf{x}(v, u, 0) = \mathbf{a}(v) + u\mathbf{b} . \qquad (A3.12)$$

Substituting (A3.8) and (A3.11) into (A3.2) gives

$$u\frac{d\mathbf{y}(s, u)}{ds} = \left[\frac{T(u)}{T_0}\right] \mathbf{F}(\mathbf{a}(v) + u\mathbf{y}, v) , \qquad (A3.13)$$

since $ds/dt = T_0/T(u)$. Now define the vector function $\mathbf{Q}(\mathbf{y}, v, u)$ by

$$\mathbf{F}(\mathbf{a} + u\mathbf{y}, v) = u\mathbf{A}(v)\mathbf{y} + u^2\mathbf{Q}(\mathbf{y}, v, u) , \qquad (A3.14)$$

and (A3.13) can be written in the form

$$\frac{d\mathbf{y}}{ds} = \mathbf{A}(0)\mathbf{y} + u\mathbf{G}(\mathbf{y}, u) \qquad (A3.15)$$

where, using (A3.6) and (A3.11),

$$\mathbf{G} = d(u)\mathbf{B}(ud(u))\mathbf{y} + c(u)\mathbf{A}(ud(u))\mathbf{y} + [1 + uc(u)]\mathbf{Q}(\mathbf{y}, ud(u), u) \ . \quad \text{(A3.16)}$$

Denote the principal matrix solution of (A3.15) by $\mathbf{Y}(s)$. (The principal matrix solution is the solution of $d\mathbf{y}/ds = \mathbf{A}(0)\mathbf{y}$.) So

$$\mathbf{Y}(s) = \cos{(ws)}\mathbf{I} + \frac{1}{w}\sin{(ws)}\mathbf{A}(0) \ , \quad \text{(A3.17)}$$

where $\mathbf{I}$ is the identity matrix. Recall that the eigenvalues of $\mathbf{A}(0)$ are $\pm iw$ (and $\operatorname{Tr}\mathbf{A}(0) = 0$ and $\det\mathbf{A}(0) > 0$) and so, since $\mathbf{A}^2(0) = -w^2\mathbf{I}$, we can verify that

$$\mathbf{Y}^{-1}(s) = \cos{(ws)}\mathbf{I} - \frac{1}{w}\sin{(ws)}\mathbf{A}(0) \ , \quad \text{(A3.18)}$$

and

$$\frac{d\mathbf{Y}^{-1}(s)}{ds} = -\mathbf{Y}^{-1}(s)\mathbf{A}(0) \ . \quad \text{(A3.19)}$$

Thus, with the last equation,

$$\begin{aligned}
\frac{d[\mathbf{Y}^{-1}(s)\mathbf{y}(s,u)]}{ds} &= \mathbf{Y}^{-1}\frac{d\mathbf{y}}{ds} - \mathbf{Y}^{-1}\mathbf{A}(0)\mathbf{y} \\
&= u\mathbf{Y}^{-1}(s)\mathbf{G}(\mathbf{y}(s,u),u) \ , \quad \text{(A3.20)}
\end{aligned}$$

on using (A3.15). Integrating (A3.20) gives

$$\mathbf{Y}^{-1}(s)\mathbf{y}(s,u) = u\int_0^s \mathbf{Y}^{-1}(z)\mathbf{G}(\mathbf{y}(z,u),u)\,dz + \text{constant} \ . \quad \text{(A3.21)}$$

Since, from (A3.8) and (A3.12), with s as in (A3.11),

$$u\mathbf{y}(0,u) = \mathbf{x}(u,0) - \mathbf{a}(v(u)) = u\mathbf{b} \ .$$

Thus, the constant of integration in (A3.21) is determined as $\mathbf{b}$, and so (since $\mathbf{Y}(0) = \mathbf{I}$)

$$\mathbf{y}(s,u) = \mathbf{Y}(s)\mathbf{b} + u\mathbf{Y}(s)\int_0^s \mathbf{Y}^{-1}(z)\mathbf{G}(\mathbf{y}(z,u))\,dz \ . \quad \text{(A3.22)}$$

with $\mathbf{G}$ a function of $\mathbf{y}$ from (A3.16); this is an integral equation for $\mathbf{y}$.

We must now determine the unknowns u, $c(u)$ and $d(u)$ such that $\mathbf{y}$ has period T_0 in s according to (iv) above. This must be so for all u. At this stage, we do not know that c and d are functions yet: they are simply unknowns to be determined. Putting $u = 0$ in (A3.22) gives, with $s = T_0$ and the periodicity T_0 of $\mathbf{y}$,

$$\mathbf{y}(T_0,0) = \mathbf{Y}(T_0)\mathbf{b} = \mathbf{y}(0,0) = \mathbf{b} \ .$$

Since **b** is arbitrary the last equation implies

$$\mathbf{Y}(T_0) = \mathbf{I} \quad \Rightarrow \quad \cos\,(wT_0)\mathbf{I} + \frac{1}{w}\sin\,(wT_0)\mathbf{A}(0) = \mathbf{I}$$

and so

$$T_0 = \frac{2\pi}{w}\,. \tag{A3.23}$$

Now, since $\mathbf{y}(s, u)$ is periodic with period T_0 for all u, $\mathbf{y}(T_0, u) = \mathbf{b}$ and so, putting $s = T_0$ in (A3.22), gives

$$\int_0^{T_0} \mathbf{Y}^{-1}(z)\mathbf{G}(z)\,dz = 0\,,$$

which with **G** from (A3.16) is

$$\int_0^{T_0} \mathbf{Y}^{-1}(z)[c(u)\mathbf{A}(ud(u))\mathbf{y}(z, u) + d(u)\mathbf{B}(ud(u))\mathbf{y}(z, u) \tag{A3.24}$$
$$+ (1 + uc(u))\mathbf{Q}(\mathbf{y}(z, u), ud(u), u)]\,dz = 0\,.$$

This equation represents two equations in the three variables u, c and d. We complete the proof if we can show that c and d can be expressed explicitly as functions of u when they are defined implicitly by (A3.24). We do this by using the implicit function theorem. We first show that $c(0) = 0 = d(0)$ and then that the appropriate Jacobian is non-zero and hence c and d can be determined.

Since, for small u, (A3.14) is essentially a Taylor series in $u\mathbf{y}$, with the $\mathbf{y}^2$ being included in **Q**, we see that $\mathbf{Q}(\mathbf{y}, v, 0)$ is a quadratic form in the components of **y**. Further, since $\mathbf{y}(s, 0) = \mathbf{Y}(s)\mathbf{b}$ from (A3.22), $\mathbf{Q}(\mathbf{y}, v, 0)$ is quadratic in the components of $\mathbf{Y}(s)$ and hence from (A3.17) is a linear combination of terms $\cos^n(ws)\sin^{2-n}(ws)$ with $n = 0, 1, 2$. $\mathbf{Y}^{-1}(z)\mathbf{Q}(\mathbf{y}(z, 0), 0, 0)$ is also a linear combination of such terms. Thus

$$\int_0^{T_0} \mathbf{Y}^{-1}(z)\mathbf{Q}(\mathbf{y}(z, 0), 0, 0)\,dz = 0\,. \tag{A3.25}$$

If we now put $u = 0$ in (A3.24) we have, using (A3.25),

$$\int_0^{T_0} [c(0)\mathbf{Y}^{-1}(z)\mathbf{A}(0)\mathbf{Y}(z)\mathbf{b} + d(0)\mathbf{Y}^{-1}(z)\mathbf{B}(0)\mathbf{Y}(z)\mathbf{b}]\,dz = 0\,. \tag{A3.26}$$

Let

$$\mathbf{B}^* = T_0^{-1}\int_0^{T_0} \mathbf{Y}^{-1}(z)\mathbf{B}(0)\mathbf{Y}(z)\,dz\,. \tag{A3.27}$$

With $\mathbf{A}(0)$ as in (A3.5) and **Y** and $\mathbf{Y}^{-1}$ as in (A3.17) and (A3.18), we see that

$$\int_0^{T_0} \mathbf{Y}^{-1}(z)\mathbf{A}(0)\mathbf{Y}(z)\,dz = T_0\mathbf{A}(0)\,, \tag{A3.28}$$

and so (A3.26) becomes

$$c(0)\mathbf{A}(0)\mathbf{b} + d(0)\mathbf{B}^*\mathbf{b} = 0 .$$

Since **b** is arbitrary the last equation means that

$$c(0)\mathbf{A}(0) + d(0)\mathbf{B}^* = 0 . \qquad (A3.29)$$

Now if we write, say

$$\mathbf{B}(0) = \begin{pmatrix} b_{11} & b_{12} \\ b_{21} & b_{22} \end{pmatrix} ,$$

(A3.27) gives, on integration of the right-hand side,

$$\mathbf{B}^* = \frac{1}{2} \begin{pmatrix} b_{11} + b_{22} & b_{12} - b_{21} \\ b_{21} - b_{12} & b_{11} + b_{22} \end{pmatrix} = \frac{1}{2}\{\mathbf{B}(0) - w^{-2}\mathbf{A}(0)\mathbf{B}(0)\mathbf{A}(0)\} . \qquad (A3.30)$$

Thus $\operatorname{Tr}\mathbf{B}^* = \operatorname{Tr}\mathbf{B}(0) \neq 0$ by the imposed condition in the statement of the theorem. Since $\operatorname{Tr}\mathbf{A}(0) = 0$ it follows that the determinant of the coefficients in the equation (A3.29) for $c(0)$ and $d(0)$ is not zero which implies that the only solution of (A3.29) is $c(0) = 0 = d(0)$ which proves (i) of the above requirements.

To get the $0(u)$ terms in $d(u)$ we differentiate (A3.24) with respect to d and then set $c = 0 = d = u$. The only term left in the integrand is from $\mathbf{B}(ud(u))\mathbf{y}$ and it gives

$$\int_0^{T_0} \mathbf{Y}^{-1}(z)\mathbf{B}(0)\mathbf{y}(z,0)\, dz$$

which, with $\mathbf{y}(z,0) = \mathbf{Y}(z)\mathbf{b}$ from (A3.22) and with the definition (A3.27) for $\mathbf{B}^*$, is

$$T_0\mathbf{B}^*\mathbf{b} . \qquad (A3.31)$$

In a similar way we differentiate (A3.24) with respect to c and set $c = 0 = d = u$ and the only term left is

$$\int_0^{T_0} \mathbf{Y}^{-1}(z)\mathbf{A}(0)\mathbf{y}(z,0)\, dz$$

which is

$$T_0\mathbf{A}(0)\mathbf{b} . \qquad (A3.32)$$

Thus, to complete the application of the implicit function theorem it remains to show that the determinant of the matrix $[\mathbf{B}^*\mathbf{b}, \mathbf{A}(0)\mathbf{b}]$ is non-zero for all non-zero **b**. (Here we are simply using the implicit function theorem for $F(x,y,z) = 0$, $G(x,y,z) = 0$. For these to have a solution $x = x(z)$, $y = y(z)$ the condition is that the Jacobian $\partial(F,G)/\partial(x,y) \neq 0$.) This is the non-zero Jacobian condition mentioned above.

First diagonalize $A(0)$ using the nonsingular principal axis transformation P for $A(0)$, that is we have

$$P^{-1}A(0)P = iw \begin{pmatrix} 1 & 0 \\ 0 & -1 \end{pmatrix}. \qquad (A3.33)$$

From (A3.30) and the last relation

$$2P^{-1}B^*P = P^{-1}B(0)P - w^{-2}P^{-1}A(0)B(0)A(0)P$$
$$= P^{-1}B(0)P + \begin{pmatrix} 1 & 0 \\ 0 & -1 \end{pmatrix} P^{-1}B(0)P \begin{pmatrix} 1 & 0 \\ 0 & -1 \end{pmatrix}.$$

If we write

$$P^{-1}B(0)P = \begin{pmatrix} \alpha & \beta \\ \gamma & \delta \end{pmatrix} \Rightarrow \begin{pmatrix} 1 & 0 \\ 0 & -1 \end{pmatrix} \begin{pmatrix} \alpha & \beta \\ \gamma & \delta \end{pmatrix} \begin{pmatrix} 1 & 0 \\ 0 & -1 \end{pmatrix} = \begin{pmatrix} \alpha & -\beta \\ -\gamma & \delta \end{pmatrix}$$

and so

$$P^{-1}B^*P = \begin{pmatrix} \alpha & 0 \\ 0 & \delta \end{pmatrix}. \qquad (A3.34)$$

Since P is nonsingular, it is sufficient to show that $\det\{P^{-1}[B^*b, A(0)b]\}$ is nonzero. For convenience define

$$k = \begin{pmatrix} k_1 \\ k_2 \end{pmatrix} = P^{-1}b$$

and so

$$P^{-1}[B^*b, A(0)b] = [P^{-1}B^*b, P^{-1}A(0)b] = \begin{pmatrix} \alpha k_1 & iwk_1 \\ \delta k_1 & -iwk_2 \end{pmatrix},$$

with determinant $-iwk_1k_2(\alpha + \delta)$ which is $-iwk_1k_2\,\mathrm{Tr}\,B^*$. Since $\mathrm{Tr}\,B^* \neq 0$, as proved above using (A3.30), we simply have to show that $k_1 \neq 0$ and $k_2 \neq 0$.

Let us write

$$b = \begin{pmatrix} b_1 \\ b_2 \end{pmatrix}, \quad P = \begin{pmatrix} P_{11} & P_{12} \\ P_{21} & P_{22} \end{pmatrix}, \quad A(0) = \begin{pmatrix} A_{11} & A_{12} \\ A_{21} & A_{22} \end{pmatrix}$$

and suppose $k_2 = 0$ and $k_1 \neq 0$. Then $b = Pk$ and

$$b_1 = P_{11}k_1, \quad b_2 = P_{22}k_1. \qquad (A3.35)$$

Since, from (A3.33),

$$A(0)P = iwP \begin{pmatrix} 1 & 0 \\ 0 & -1 \end{pmatrix}$$

we have

$$A_{11}P_{11} + A_{12}P_{21} = iwP_{11} , \qquad (A3.36)$$

$$A_{21}P_{11} + A_{22}P_{21} = iwP_{21} . \qquad (A3.37)$$

Multiplying (A3.36) by k_1 gives, using (A3.35),

$$A_{11}b_1 + A_{12}b_2 = iwb_1 .$$

Since $\mathbf{A}(0)$ and $\mathbf{b}$ are real, the last equation implies that $b_1 = 0$ and so (A3.35) implies that $P_{11} = 0$ and (A3.37) then implies that $A_{22} = iw$, which is a contradiction since $\mathbf{A}(0)$ is real. Thus $k_2 \neq 0$.

Similarly the supposition that $k_1 = 0$ and $k_2 \neq 0$ gives

$$b_1 = P_{12}k_2, \quad b_2 = P_{22}k_2 ,$$

$$A_{11}P_{12} + A_{12}P_{22} = -iwP_{12} ,$$

$$A_{21}P_{12} + A_{22}P_{22} = -iwP_{22} .$$

The same reasoning as above gives $b_1 = 0$, $P_{12} = 0$, $P_{22} \neq 0$, and hence $A_{22} = -iw$ which is again a contradiction.

Thus $\det[\mathbf{B}^*\mathbf{b}, \mathbf{A}(0)\mathbf{b}] \neq 0$ and hence $c(u)$ and $d(u)$ exist, for at least some $0 < u \leq u_0$ for a sufficiently small u, and satisfy (ii) above.

We have seen in the above analysis that (iii) is also satisfied, as is (iv). Thus, by virtue of the transformation (A3.8) and (A3.11) we have shown that (v) is true.

For small u, v is small and hence the period $T(u) \sim T_0$ and the frequency is approximately $2\pi/T_0$ with $T_0 = w$, the imaginary part of the imaginary eigenvalues of $\mathbf{A}(0)$.

Illustrative Examples

We now consider three examples which illustrate certain practical aspects of the bifurcation theorem. For simplicity we consider only two-dimensional examples.

Example 1. We consider first a simple linear case of (A3.1) namely

$$\frac{d\mathbf{x}}{dt} = \begin{pmatrix} \dfrac{dx_1}{dt} \\ \dfrac{dx_2}{dt} \end{pmatrix} = \begin{pmatrix} x_2 \\ -x_1 + vx_2 \end{pmatrix} = \mathbf{F}(\mathbf{x}, v) , \qquad (A3.38)$$

where v is a real parameter. The only equilibrium point is

$$x_1 = 0 = x_2 . \qquad (A3.39)$$

With this example, of course, we can solve (A3.38) exactly. We go through the analysis of the theorem for obvious pedagogical reasons.

From (A3.38) the single equation for x_1 is

$$\frac{d^2 x_1}{dt^2} - v\frac{dx_1}{dt} + x_1 = 0$$

with solution

$$x_2 = \frac{dx_1}{dt} ,$$

$$x_1 = e^{vt/2}\{A\cos\left[(1 - v^2/4)^{1/2}t\right] + B\sin\left[(1 - v^2/4)^{1/2}t\right]\} ,$$

$$(A3.40)$$

where A and B are constants and the form (A3.40) is for $v^2 < 4$: with $v^2 > 4$ there is no oscillatory behaviour at all. For $v \neq 0$, (A3.40) is oscillatory and the solution spirals into, or out of, the equilibrium point at the origin according to whether $v < 0$ or $v > 0$. The value $v = 0$ is a bifurcation point. This is just another way of saying that in the phase plane for (A3.38) the singular point (A3.39) of

$$\frac{dx_2}{dx_1} = -\frac{x_1 - vx_2}{x_2}$$

is a stable or unstable focus according to whether $v < 0$ and $v > 0$, but in either case $|v| < 2$.

For values of v in the vicinity of the equilibrium point, the *only* periodic solutions from (A3.40) are

$$v \equiv 0, \quad x_1 = A\cos t + B\sin t, \quad x_2 = \frac{dx_1}{dt} , \qquad (A3.41)$$

or, if $v \neq 0$, $x_1 = x_2 = 0$, the trivial solution. Thus, from the theorem point of view, a periodic solution exists but it is characterized by the function $v(u)$ in (A3.11) being *identically* zero and the periodic solutions of the system are then those of $d\mathbf{x}/dt = \mathbf{F}(\mathbf{x}, 0)$. This is exactly what the theorem would have produced had we carried through the detailed analysis.

Example 2. Consider now the non-trivial nonlinear system

$$\frac{d\mathbf{x}}{dt} = \begin{pmatrix} \dfrac{dx_1}{dt} \\ \dfrac{dx_2}{dt} \end{pmatrix} = \begin{pmatrix} x_2 \\ -x_1 + vx_2 - x_1^2 x_2 \end{pmatrix} = \mathbf{F}(\mathbf{x}, v) , \qquad (A3.42)$$

Before analysing (A3.42) using the analysis of the bifurcation theorem, note that on eliminating x_2 the system becomes

$$\frac{d^2 x_1}{dt^2} + (x_1^2 - v)\frac{dx_1}{dt} + x_1 = 0 , \qquad (A3.43)$$

which is a Liénard equation (see, for example, Jordan and Smith 1987). If $v > 0$ it has a unique stable limit cycle for each $v > 0$. For $v < 0$ the equation is not of Liénard type and no limit cycle solution exists in this case.

Let us now examine (A3.42) by the proceedure of the theorem. From (A3.3) and (A3.42),

$$\mathbf{F}(\mathbf{x}, v) = 0 \quad \Rightarrow \quad \mathbf{x} = 0 = \mathbf{a}(v) \tag{A3.44}$$

is the only singular point. If we linearize (A3.42) about $\mathbf{x} = 0$ we have (compare with (A3.4))

$$\mathbf{A}(v) = [\nabla_\mathbf{x}\mathbf{F}(\mathbf{x}, v)]_{\mathbf{x}=0} = \begin{pmatrix} 0 & 1 \\ -1 & v \end{pmatrix}, \tag{A3.45}$$

$$\mathbf{A}(0) = \begin{pmatrix} 0 & 1 \\ -1 & 0 \end{pmatrix}.$$

Thus (A3.5) is satisfied since

$$\operatorname{Tr}\mathbf{A}(0) = 0, \quad \det\mathbf{A}(0) = 1 > 0,$$

$$|\mathbf{A}(0) - \lambda\mathbf{I}| = 0 \quad \Rightarrow \quad \lambda = \pm i \quad \Rightarrow \quad w = 1 \quad \Rightarrow \quad T_0 = 2\pi. \tag{A3.46}$$

Since (compare with (A3.6))

$$\mathbf{A}(v) = \begin{pmatrix} 0 & 1 \\ -1 & 0 \end{pmatrix} + v\begin{pmatrix} 0 & 0 \\ 0 & 1 \end{pmatrix} = \mathbf{A}(0) + v\mathbf{B}(v)$$

$$\mathbf{B}(v) = \begin{pmatrix} 0 & 0 \\ 0 & 1 \end{pmatrix}, \quad \operatorname{Tr}\mathbf{B}(0) = 1 \neq 0. \tag{A3.47}$$

From (A3.30), with $w = 1$

$$\mathbf{B}^* = \frac{1}{2}\{\mathbf{B}(0) - \mathbf{A}(0)\mathbf{B}(0)\mathbf{A}(0)\} = \frac{1}{2}\begin{pmatrix} 1 & 0 \\ 0 & 1 \end{pmatrix}. \tag{A3.48}$$

The function $\mathbf{Q}$, defined by (A3.14), which, from (A3.42) with (A3.44) and (A3.45), is

$$\mathbf{F}(0 + u\mathbf{y}, v) = u\mathbf{A}(v)\mathbf{y} + u^2\mathbf{Q}(\mathbf{y}, v, u)$$

$$= u\begin{pmatrix} 0 & 1 \\ -1 & v \end{pmatrix}\mathbf{y} + u^2\begin{pmatrix} 0 \\ -uy_1^2 y_2 \end{pmatrix},$$

and so

$$\mathbf{Q}(\mathbf{y}, v, u) = \begin{pmatrix} 0 \\ -uy_1^2 y_2 \end{pmatrix}, \quad \mathbf{y} = \begin{pmatrix} y_1 \\ y_2 \end{pmatrix}. \tag{A3.49}$$

From (A3.17), with (A3.46) and (A3.47),

$$\mathbf{Y}(s) = \mathbf{I}\cos s + \mathbf{A}(0)\sin s = \begin{pmatrix} \cos s & \sin s \\ -\sin s & \cos s \end{pmatrix},$$

$$\mathbf{Y}^{-1}(s) = \begin{pmatrix} \cos s & -\sin s \\ \sin s & \cos s \end{pmatrix}. \tag{A3.50}$$

With (A3.10) and (A3.22)

$$\begin{pmatrix} y_1(s,0) \\ y_2(s,0) \end{pmatrix} = \mathbf{y}(s,0) = \mathbf{Y}(s)\mathbf{b}, \quad \mathbf{b} = \begin{pmatrix} b_1 \\ b_2 \end{pmatrix} . \tag{A3.51}$$

We know that $c(0) = 0 = d(0)$ as a consequence of the properties of (A3.29). The algebraic equations for $c'(0)$ and $d'(0)$ in the Taylor expansions

$$c(u) = c(0) + uc'(0) + \ldots = uc'(0) + \ldots ,$$
$$d(u) = d(0) + ud'(0) + \ldots = ud'(0) + \ldots$$

are now obtained from (A3.24) on expanding for small u and equating the coefficients of u using $\mathbf{A}(v)$, $\mathbf{B}(v)$ and $\mathbf{Q}$ from (A3.47) and (A3.49). Since, from (A3.51) $\mathbf{Y}^{-1}(s)\mathbf{y}(s,0) = \mathbf{b}$, the $O(u)$ terms in (A3.24) become, with (A3.27) and (A3.48), and $T_0 = 2\pi$ from (A3.46),

$$2\pi[c'(0)\mathbf{A}(0)\mathbf{b} + d'(0)\mathbf{B}^*\mathbf{b}]$$
$$+ \int_0^{2\pi} \mathbf{Y}^{-1}(z) \begin{pmatrix} 0 \\ -y_1^2(z,0)y_2(z,0) \end{pmatrix} dz = 0 . \tag{A3.52}$$

But, from (A3.51)

$$\begin{pmatrix} y_1(z,0) \\ y_2(z,0) \end{pmatrix} = \begin{pmatrix} b_1 \cos z + b_2 \sin z \\ -b_1 \sin z + b_2 \cos z \end{pmatrix} ,$$

and so

$$y_1^2(x,0)y_2(z,0) = (b_1 \cos z + b_2 \sin z)^2(-b_1 \sin z + b_2 \cos z) .$$

Substituting into the integral in (A3.52) and integrating the resulting functions $\sin^n z \cos^{4-n} z$, $n = 1, 2, 3, 4$ we have

$$\frac{1}{2\pi} \int_0^{2\pi} \mathbf{Y}^{-1}(z) \begin{pmatrix} 0 \\ -y_1^2(z,0)y_2(z,0) \end{pmatrix} dz$$
$$= -\frac{1}{2\pi} \int_0^{2\pi} \begin{pmatrix} -\sin z \\ \cos z \end{pmatrix} (b_1 \cos z + b_2 \sin z)^2(-b_1 \sin z + b_2 \cos z) dz$$
$$= -\frac{1}{8}(b_1^2 + b_2^2)\mathbf{b} = -\frac{|\mathbf{b}|^2 \mathbf{b}}{8} .$$

Thus, (A3.52) becomes, using (A3.45) for $\mathbf{A}(0)$ and (A3.48) for $\mathbf{B}^*(0)$,

$$c'(0) \begin{pmatrix} 0 & 1 \\ -1 & 0 \end{pmatrix} \mathbf{b} + \frac{1}{2}d'(0) \begin{pmatrix} 1 & 0 \\ 0 & 1 \end{pmatrix} \mathbf{b} = \frac{|\mathbf{b}|^2 \mathbf{b}}{8}$$

which gives

$$c'(0) = 0, \quad d'(0) = \frac{|\mathbf{b}|^2}{4}. \tag{A3.53}$$

Thus, for *small* u,

$$v = ud(u) \approx \frac{u^2|\mathbf{b}|^2}{4}$$

$$\Rightarrow \quad u = \frac{2v^{1/2}}{|\mathbf{b}|}, \quad T(u) = 2\pi[1 + o(u^2)] \tag{A3.54}$$

and

$$\mathbf{x}(u,t) = u\mathbf{y}(t,0) + O(u^2), \tag{A3.55}$$

where $\mathbf{y}(t,0)$ from (A3.22) is, with $\mathbf{Y}(t)$ from (A3.50),

$$\mathbf{y}(t,0) = \mathbf{Y}(t)\mathbf{b} = \begin{pmatrix} \cos t & \sin t \\ -\sin t & \cos t \end{pmatrix} \mathbf{b}. \tag{A3.56}$$

The bifurcation point of (A3.42) is at $v = v_c = 0$ and the bifurcation for periodic oscillatory solutions is into the parameter domain $v > 0$ since, from (A3.54), $v = u^2|\mathbf{b}|^2/4 > 0$ for periodic solutions. The period of the limit cycles for *small* positive v is $T_0 = 2\pi$ and the limit cycle solutions are, from (A3.55) and (A3.56).

$$\mathbf{x}(v,t) = \begin{pmatrix} x_1 \\ x_2 \end{pmatrix} = 2v^{1/2}|\mathbf{b}|^{-1} \begin{pmatrix} b_1\cos t + b_2\sin t \\ -b_1\sin t + b_2\cos t \end{pmatrix} + O(v)$$

$$= 2v^{1/2} \begin{pmatrix} \cos(t-\alpha) \\ -\sin(t-\alpha) \end{pmatrix} + O(v), \quad \alpha = \tan^{-1}(b_2/b_1), \quad 0 < v \ll 1. \tag{A3.57}$$

Since $\mathbf{b}$ is arbitrary, so is the angle α, which simply represents a phase change and reflects the fact that we can measure t from any base.

Note that the existence of limit cycles near the bifurcation point $v = 0$ requires $v > 0$. This is, of course, in agreement with the result stated at the beginning of this example about the single equation form (A3.43) for the original system (A3.42). Equation (A3.43) is an equation of fundamental importance in oscillator theory and is known as the *van der Pol equation*. It has limit cycle solutions in general for $v > 0$ and, as we have seen in (A3.57), the amplitude of the oscillations depend on the parameter v. (This equation is discussed in more detail in Murray's 1984 book.)

The final point is one of stability. In this case the equilibrium point at the origin is characterized by the linearized form of (A3.42) namely,

$$\begin{pmatrix} \dfrac{dx_1}{dt} \\ \dfrac{dx_2}{dt} \end{pmatrix} \approx \begin{pmatrix} x_2 \\ -x_1 + vx_2 \end{pmatrix} \quad \Rightarrow \quad \frac{dx_2}{dx_1} = \frac{-(x_1 - vx_2)}{x_2},$$

which is exactly the form discussed in Example 1 namely (A3.38) except here $v > 0$. As we saw there, as in this case, the singular point is an unstable focus if $0 < v < 2$. Thus, with the singular point an unstable focus and a limit cycle existing at least for small $v > 0$ the periodic solution (A3.57) is stable.

Example 3. Consider now

$$\frac{d\mathbf{x}}{dt} = \begin{pmatrix} \dfrac{dx_1}{dt} \\ \dfrac{dx_2}{dt} \end{pmatrix} = \begin{pmatrix} x_2 \\ -x_1 - vx_2 + \dfrac{x_2^3}{3} \end{pmatrix} = \mathbf{F}(\mathbf{x}, v) , \tag{A3.58}$$

As in Example 2 we can write the system as a single equation

$$\frac{d^2 x_2}{dt^2} + (v - x_2^2)\frac{dx_2}{dt} + x_2 = 0 . \tag{A3.59}$$

If we set $t = -t$ this equation becomes

$$\frac{d^2 x_2}{dt^2} + (x_2^2 - v)\frac{dx_2}{dt} + x_2 = 0 . \tag{A3.60}$$

which is a Liénard equation and hence has a unique limit cycle solution for each $v > 0$. This limit cycle solution is stable for $t \to \infty$ which implies that (A3.59) has a limit cycle only as t moves backwards and $t \to -\infty$. This means that the equation (A3.59) *cannot* have a stable limit cycle at $t \to +\infty$.

Let us now consider the results obtained from applying the bifurcation theorem. The analysis for (A3.55) parallels exactly that for Example 2 and we get

$$\mathbf{A}(v) = \begin{pmatrix} 0 & 1 \\ -1 & -v \end{pmatrix} , \quad \mathbf{A}(0) = \begin{pmatrix} 0 & 1 \\ -1 & 0 \end{pmatrix} , \quad T_0 = 2\pi ,$$
$$\mathbf{B}(v) = \begin{pmatrix} 0 & 0 \\ 0 & -1 \end{pmatrix} , \quad \mathbf{B}^* = -\frac{1}{2}\begin{pmatrix} 1 & 0 \\ 0 & 1 \end{pmatrix} , \quad \operatorname{Tr}\mathbf{B}(0) \neq 0 . \tag{A3.61}$$

$\mathbf{Y}(s)$ and $\mathbf{Y}^{-1}(s)$ are given as before by (A3.50) and

$$\mathbf{Q}(\mathbf{y}, v, u) = \begin{pmatrix} 0 \\ \dfrac{uy_2^3}{3} \end{pmatrix} . \tag{A3.62}$$

The equivalent of (A3.52) is now, with (A3.62),

$$2\pi[c'(0)\mathbf{A}(0)\mathbf{b} + d'(0)\mathbf{B}^*\mathbf{b}] + \int_0^{2\pi} \mathbf{Y}^{-1}(z) \begin{pmatrix} 0 \\ \dfrac{y_2^3(z, 0)}{3} \end{pmatrix} dz = 0 .$$

Since here

$$\frac{1}{2\pi} \int_0^{2\pi} \mathbf{Y}^{-1}(z) \begin{pmatrix} 0 \\ \dfrac{y_2^3(z,0)}{3} \end{pmatrix} dz$$

$$= \frac{1}{6\pi} \int_0^{2\pi} \begin{pmatrix} -\sin z \\ \cos z \end{pmatrix} (b_2 \cos z - b_1 \sin z)^3 \, dz$$

$$= \frac{|\mathbf{b}|^2 \mathbf{b}}{8}$$

we have

$$c'(0) \begin{pmatrix} 0 & 1 \\ -1 & 0 \end{pmatrix} \mathbf{b} - \frac{1}{2} d'(0) \begin{pmatrix} 1 & 0 \\ 0 & 1 \end{pmatrix} \mathbf{b} = -\frac{|\mathbf{b}|^2 \mathbf{b}}{8}$$

and hence

$$c'(0) = 0, \quad d'(0) = \frac{|\mathbf{b}|^2}{4} . \tag{A3.63}$$

This is the same as (A3.53). The rest of the analysis is now exactly the same as in Example 2 above and the result is (A3.57). As before the bifurcation value is $v = 0$ and the periodic solutions are for $v > 0$. The question now arises as to how this result relates to the analysis of the Liénard equation (A3.59).

In this case, in the vicinity of the singular point $\mathbf{x} = 0$ of (A3.58), the linearization gives

$$\frac{dx_1}{dt} = x_2, \quad \frac{dx_2}{dt} = -x_1 - vx_2, \quad v > 0 . \tag{A3.64}$$

and in the phase plane

$$\frac{dx_2}{dx_1} = -\frac{x_1 + vx_2}{x_2} .$$

The singular point is a *stable* focus if $0 < v < 2$. The stability is also immediately clear from the solution of (A3.64), namely

$$x_2 = \exp\left[-vt/2\right]\{A \cos\left[(1 - v^2/4)^{1/2}t\right] + B \sin\left[(1 - v^2/4)^{1/2}t\right]\} ,$$

$$x_1 = -\frac{dx_2}{dt} - vx_2 ,$$

since x_1 and x_2 both tend to zero, in an oscillatory manner, as $t \to \infty$ for small $v > 0$.

Thus, although the bifurcation theorem does indicate a limit cycle solution for (A3.58) for each $v > 0$ in the vicinity of the bifurcation point, it is *not* stable: it collapses into the single equilibrium point at the origin. This is the opposite of the situation in Example 2 where the equilibrium point is an unstable focus for $0 < v < 2$. It is still unstable there for all $v > 0$.

With the problem here, the Poincaré-Bendixson theorem, of course, immediately implies that there can be no limit cycle solutions for $v > 0$ since the equilibrium point is *stable*.

The bifurcation theorem is a very powerful general tool in establishing the existence of periodic solutions and, in the neighbourhood of the bifurcation points, it also gives their periods. However, it does not immediately provide any information as to their stability, which of course is as important as existence. It is of practical importance in treating systems of dimension higher than two: for two-dimensional systems the phase plane technique is almost always superior.

Appendix 4
General Results for the Laplacian Operator in Bounded Domains

In Chapter 14, Section 14.9, we used the result that a function $u(x)$ with $u_x = 0$ on $x = 0, 1$ satisfies

$$\int_0^1 u_{xx}^2 \, dx \geq \pi^2 \int_0^1 u_x^2 \, dx \tag{A4.1}$$

and the more general result

$$\int_B |\nabla^2 \mathbf{u}|^2 \, d\mathbf{r} \geq \mu \int_{\partial B} \|\nabla \mathbf{u}\|^2 \, d\mathbf{r} , \tag{A4.2}$$

where B is a finite domain enclosed by the simply connected surface ∂B on which zero-flux (Neumann) conditions hold, namely $\mathbf{n} \cdot \nabla \mathbf{u} = 0$ where $\mathbf{n}$ is the unit outward normal to ∂B. In (A4.2), μ is the least positive eigenvalue of $\nabla^2 + \mu$ for B with Neumann conditions on ∂B and where $\| \cdot \|$ denotes a Euclidean norm. By the Euclidean norm here we mean, for example,

$$\|\nabla \mathbf{u}\| = \max_{\mathbf{r} \in B} \left[\sum_{i,j} \left(\frac{\partial u_i}{\partial x_j} \right)^2 \right]^{1/2} \tag{A4.3}$$

$$\mathbf{r} = (x_j), \quad j = 1, 2, 3; \qquad \mathbf{u} = (u_i), \quad i = 1, 2, \ldots, n .$$

We prove these standard results in this section: (A4.1) is a special case of (A4.2) in which $\mathbf{u}$ is a single scalar and $\mathbf{r}$ a single space variable.

By way of illustration we first derive the one-dimensional result (A4.1) in detail and then prove the general result (A4.2).

Consider the equation for the scalar function $w(x)$, a function of the single space variable x, given by

$$w_{xx} + \mu w = 0 , \tag{A4.4}$$

where μ represents the general eigenvalue for solutions of this equation satisfying Neumann conditions on the boundaries, namely,

$$w_x(x) = 0 \quad \text{on} \quad x = 0, 1 . \tag{A4.5}$$

The orthonormal eigenfunctions $\{\phi_k(x)\}$ and eigenvalues $\{\mu_k\}$, $k = 0, 1, 2, \ldots$ for (A4.4) and (A4.5) are

$$\phi_k(x) = \cos \mu_k^{1/2} x, \quad \mu_k = k^2\pi^2, \quad k = 0, 1, \ldots . \tag{A4.6}$$

Any function $w(x)$, such as we are interested in, satisfying the zero-flux conditions (A4.5) can be written in terms of a series (Fourier) expansion of eigenfunctions $\phi_k(x)$ and so also can derivatives of $w(x)$, which we assume exist. Let

$$w_{xx}(x) = \sum_{k=0}^{\infty} a_k \phi_k(x) = \sum_{k=0}^{\infty} a_k \cos(k\pi x), \tag{A4.7}$$

where, in the usual way,

$$a_k = 2 \int_0^1 w_{xx}(x) \cos(k\pi x)\, dx, \quad k > 0$$

$$a_0 = \int_0^1 w_{xx}(x)\, dx = [w_x(x)]_0^1 = 0 .$$

Then, integrating (A4.7) twice and using conditions (A4.5) gives

$$w(x) = \sum_{k=1}^{\infty} -\frac{a_k}{\mu_k}\phi_k(x) + b_0\phi_0 ,$$

where b_0 and ϕ_0 are constants. Thus, since $a_0 = 0$,

$$\int_0^1 w_x^2(x)\, dx = [ww_x]_0^1 - \int_0^1 ww_{xx}\, dx$$

$$= -\int_0^1 ww_{xx}\, dx$$

$$= \int_0^1 \left[\sum_{k=1}^{\infty} \frac{a_k}{\mu_k} \cos(k\pi x)\right]\left[\sum_{k=1}^{\infty} a_k \cos(k\pi x)\right] dx$$

$$+ b_0\phi_0 \int_0^1 \left[\sum_{k=1}^{\infty} a_k \cos(k\pi x)\right] dx$$

$$= \frac{1}{2}\sum_{k=1}^{\infty} \frac{a_k^2}{\mu_k}$$

$$\leq \frac{1}{2\mu_1}\sum_{k=1}^{\infty} a_k^2$$

$$= \frac{1}{\mu_1}\int_0^1 w_{xx}^2\, dx = \frac{1}{\pi^2}\int_0^1 w_{xx}^2\, dx ,$$

which is (A4.1): μ_1 is the smallest positive eigenvalue μ_k for all k.

The proof of the general result (A4.2) simply mirrors the one-dimensional scalar version.

Again let the sequence $\{\boldsymbol{\phi}_k(\mathbf{r})\}$, $k = 0, 1, 2, \ldots$ be the orthonormal eigenvector functions of

$$\nabla^2 \mathbf{w} + \mu \mathbf{w} = 0$$

where $\mathbf{w}(\mathbf{r})$ is a vector function of the space variable $\mathbf{r}$ and μ is the general eigenvalue. Let the corresponding eigenvalues for the $\{\boldsymbol{\phi}_k\}$ be the sequence $\{\mu_k\}$, $k = 0, 1, \ldots$ where they are so ordered that $\mu_0 = 0$, $0 < \mu_1 < \mu_2 \ldots$. Note in this case also that $\boldsymbol{\phi}_0 = $ constant.

Let $\mathbf{w}(\mathbf{r})$ be a function defined for $\mathbf{r}$ in the domain B and satisfying the zero-flux conditions $\mathbf{n} \cdot \nabla \mathbf{w} = 0$ for $\mathbf{r}$ on ∂B. Then we can write

$$\nabla^2 \mathbf{w} = \sum_{k=0}^{\infty} a_k \boldsymbol{\phi}_k(\mathbf{r}) \, ,$$

$$a_k = \int_B \langle \nabla^2 \mathbf{w}, \boldsymbol{\phi}_k \rangle \, d\mathbf{r} \, , \tag{A4.8}$$

$$a_0 = \langle \boldsymbol{\phi}_0, \int_B \nabla^2 \mathbf{w} \, d\mathbf{r} \rangle = \langle \boldsymbol{\phi}_0, \int_{\partial B} \nabla \mathbf{w} \, d\mathbf{r} \rangle = 0 \, .$$

Here $\langle \cdot \rangle$ denotes the inner (scalar) product. Integrating $\nabla^2 \mathbf{w}$ twice we get

$$\mathbf{w}(\mathbf{r}) = \sum_{k=1}^{\infty} -\frac{a_k}{\mu_k} \boldsymbol{\phi}_k(\mathbf{r}) + b_0 \boldsymbol{\phi}_0$$

where b_0 and $\boldsymbol{\phi}_0$ are constants. With this expression together with that for $\nabla^2 \mathbf{w}$ we have, on integrating by parts,

$$\int_B \|\nabla \mathbf{w}\|^2 \, d\mathbf{r} = \int_{\partial B} \langle \mathbf{w}, \mathbf{n} \cdot \nabla \mathbf{w} \rangle \, d\mathbf{r} - \int_B \langle \mathbf{w}, \nabla^2 \mathbf{w} \rangle \, d\mathbf{r}$$

$$= \sum_{k=1}^{\infty} \frac{a_k^2}{\mu_k}$$

$$\leq \frac{1}{\mu_1} \sum_{k=1}^{\infty} a_k^2$$

$$= \frac{1}{\mu_1} \int_B |\nabla^2 \mathbf{w}|^2 \, d\mathbf{r} \, ,$$

which gives the result (A4.2) since μ_1 is the least positive eigenvalue.

Bibliography

Agladze, K.I., Krinsky, V.I.: Multi-armed vortices in an active chemical medium. Nature **286**, 424–426 (1982).

Aikman, D., Hewitt, G.: An experimental investigation of the rate and form of dispersal in grasshoppers. J. Appl. Ecol. **9**, 807–817 (1972).

Alberch, P.: Ontogenesis and morphological diversification. Amer. Zool. **20**, 653–667 (1980).

Alberch, P.: Developmental constraints in evolutionary processes. In: J.T. Bonner (ed.) Evolution and Development. Dahlem Conference Rep. **20**, 313–332. Berlin Heidelberg New York: Springer 1982

Alberch, P.: The logic of monsters: evidence for internal constraint in development and evolution. Geobios (1989) (in press).

Alberch, P., Gale, E.: Size dependency during the development of the amphibian foot. Colchicine induced digital loss and reduction. J. Embryol. exp. Morphol. **76**, 177–197 (1983).

Alberts, B., Bray, D., Lewis, J., Raff, M., Roberts, K., Watson, J.D.: Molecular Biology of The Cell. New York and London: Garland 1983.

Allessie, M.A., Bonke, F.I.M., Schopman, F.G.J.: Circus movement in rabbit atrial muscle as a mechanism of tachycardia. Circ. Res. **33**, 54–62 (1973).

Allessie, M.A., Bonke, F.I.M., Schopman, F.G.J.: Circus movement in rabbit atrial muscle as a mechanism of tachycardia. II. The role of nonuniform recovery of excitability in the occurrence of unidirectional block, as studied with multiple microelectrodes. Circ. Res. **39**, 168–177 (1976).

Allessie, M.A., Bonke, F.I.M., Schopman, F.G.J.: Circus movement in rabbit atrial muscle as a mechanism of tachycardia. III. The "leading circle" concept: a new model of circus movement in cardiac tissue without the involvement of an anatomical obstacle. Circ. Res. **41**, 9–18 (1977).

Alt, W., Lauffenburger, D.A.: Transient behaviour of a chemotaxis system modelling certain types of tissue inflammation. J. Math. Biol. **24**, 691–722 (1987).

Ammerman, A.J., Cavalli-Sforza, L.L.: Measuring the rate of spread of early farming. Man **6**, 674–688 (1971).

Ammerman, A.J., Cavalli-Sforza, L.L.: The Neolithic Transition and the Genetics of Populations in Europe. Princeton: Princeton University Press 1983.

Anderson, R.M.: Population Ecology of Infectious Disease Agents. In: R.M. May (ed.) Theoretical Ecology, 2nd edn., Oxford: Blackwell 1981, pp. 318–355.

Anderson, R.M., Jackson, H.C., May, R.M., Smith, A.M.: Population dynamics of fox rabies in Europe. Nature **289**, 765–771 (1981).

Anderson, R.M. (ed.): Population Dynamics of Infectious Diseases Theory and Applications. London: Chapman and Hall 1982.

Anderson, R.M.: The epidemiology of HIV infection: variable incubation plus infectious periods and heterogeneity in sexual behaviour. J. Roy. Statist. Soc. **A151**, 66–93 (1988).

Anderson, R.M., May, R.M.: Directly transmitted infectious diseases: control by vaccination. Science **215**, 1053–1060 (1982a).

Anderson, R.M., May, R.M.: Population dynamics of human helminth infections: control by chemotherapy. Nature **297**, 557–563 (1982b).

Anderson, R.M., May, R.M.: Vaccination and hers immunity to infectious diseases. Nature **318**, 323–329 (1985).

Anderson, R.M., May, R.M.: The invasion, persistence and spread of infectious diseases within animal and plant communities. Phil. Trans. Roy. Soc. (Lond.) **B314**, 533–570 (1986).

Anderson, R.M., Medley, G.F., May, R.M., Johnson, A.M.: A preliminary study of the transmission dynamics of the human immunodeficiency virus (HIV), the causitive agent of AIDS. IMA J. Maths. Appl. in Medicine and Biol. **3**, 229–263 (1986).

Andral, L., Artois, M., Aubert, M.F.A., Blancou, J.: Radio-tracking of rabid foxes. Comp. Immun. Microbiol. Infect. Dis. **5**, 285–291 (1982)

Aoki, K.: Gene-culture waves of advance. J. Math. Biol. **25**, 453–464 (1987).

Archer, C., Rooney, P., Wolpert, L.: The early growth and morphogenesis of lim cartilage. In: J. Fallon, A. Kaplan (eds.) Limb Development and Regeneration, Part A. New York: A.R. Liss 1983, pp. 267–276.

Arcuri. P., Murray, J.D.: Pattern sensitivity to boundary and initial conditions in reaction-diffusion models. J. Math. Biol. **24**, 141–165 (1986).

Arnold, R., Showalter, K., Tyson, J.J.: Propagation of chemical reactions in space. J. Chem. Educ. **64**, 740–742 (1987) [Translation of: Luther, R.-L: Räumliche Fortpflanzung Chemischer Reaktionen. Z. für Elektrochemie und angew. physikalische Chemie. **12**(32), 506–600 (1906).]

Aronson, D.G.: Density-dependent interaction-diffusion systems. In: W.E. Stewart, W.H. Ray, C.C. Conley (eds.) Dynamics and Modelling of Reactive Systems. New York: Academic Press 1980, pp. 161–176.

Artois, M., Aubert, N.F.A.: Structure des populations (age et sexe) de renard en zones indemnés ou atteintés de rage. Comp. Immun. Microbiol. infect. Dis. **5**, 237–245 (1982).

Babloyantz, A., Bellemans, A.: Pattern regulation in reaction diffusion systems – the problem of size invariance. Bull. Math. Biol. **47**, 475–487 (1985).

Babloyantz, A.: Molecules, Dynamics, and Life. An Introduction to Self-Organisation and Matter. New York: Wiley 1986.

Babloyantz, A., Hiernaux, J.: Models for cell differentiation and generation of polarity in diffusion-governed morphogenetic fields. Bull. math. Biol. **37**, 637–657 (1975).

Bacon, P.J. (ed.): Population Dynamics of Rabies in Wildlife. New York: Academic Press 1985.

Bailey, N.T.J.: The Mathematical Theory of Infectious Diseases, 2nd edn. London: Griffin 1975.

Banks, H.T., Kareiva, P.M., Lamm, P.K.: Modeling insect dispersal and estimating parameters when mark-release techniques may cause initial disturbances. J. Math. Biol. **22**, 259–277 (1985).

Bard, J.B.L.: A unity underlying the different zebra striping patterns. J. Zool. Lond. **183**, 527–539 (1977).

Bard, J.B.L.: A model for generating aspects of zebra and other mammalian coat patterns. J. theor. Biol. **93**, 363–385 (1981).

Barkley, D., Ringland, J., Turner, J.S.: Observations of a torus in a model for the Belousov-Zhabotinskii reaction. J. Chem. Phys. **87**, 3812–3820 (1987).

Barnes, R.D.: Invertebrate Zoology. Philapdelphia: Saunders 1980.

Beck, M.T., Váradi, Z.B.: One, two and three-dimensional spatially periodic chemical reactions. Nature (Phys. Sci.) **235**, 15–16 (1972).

Becker, P.K., Field, R.J.: Stationary patterns in the Oregonator model of the Belousov-Zhabotinskii reaction. J. Physical Chem. **89**, 118–128 (1985).

Beddington, J.R., Free, C.A., Lawton, J.H.: Dynamic complexity in predator-prey models framed in difference equations. Nature **255**, 58–60 (1975).

Beddington, J.R., May, R.M.: Harvesting Natural Populations in a Randomly Fluctuating Environment. Science **197**, 463–465 (1977).

Belousov, B.P.: A periodic reaction and its mechanism.: (1951, from his archives [Russian]). English translation: In: R.J. Field, M. Burger (eds.) Oscillations and Travelling Waves in Chemical Systems. New York: Wiley 1985, pp. 605–613.

Belousov, B.P.: An oscillating reaction and its mechanism. Sborn. referat. radiat. med., Medgiz, Moscow (Collection of abstracts on radiation medicine) (1959), p. 145.

Benchetrit, G., Baconnier, P., Demongeot, J.: Concepts and Formalizations in the Control of Breathing. Manchester: Manchester University Press 1987.

Ben-Yu, G., Mitchell, A.R., Sleeman, B.D.: Spatial effects in a two-dimensional model of the budworm-balsam fir ecosystem. Comp. and Maths. with Appls. **12B**, 1117–1132 (1986).

Berding, C., Keymer, A.E., Murray, J.D., Slater, A.F.G.: The population dynamics of acquired immunity to helminth infections. J. theor. Biol. **122**, 459–471 (1986).

Berding, C.: On the heterogeneity of reaction-diffusion generated patterns. Bull. math. Biol. **49**, 233–252 (1987).

Berding, C., Keymer, A.E., Murray, J.D., Slater, A.F.G.: The population dynamics of acquired immunity to helminth infections: experimental and natural infections. J. theor. Biol. **126**, 167–182 (1987).

Bergé, P., Pomeau, Y., Vidal, C.: Order within Chaos: Towards a deterministic approach to turbulence. New York: Wiley 1984.

Bernoulli, D.: Essai d'une nouvelle analyse de la mortalité causée par la petite vérole, et des avantages de l'inoculation pour la prévenir. Histoire de l'Acad. Roy. Sci. (Paris) avec Mém. des Math. and Phys., Mém., 1–45 (1760).

Best, E.N.: Null space in the Hodgkin-Huxley equations: a critical test. Biophys. J. **27**, 87–104 (1979).

Boegel, K., Moegle, H., Steck, F., Krocza, W., Andral, L.: Assessment of fox control in areas of wildlife rabies. Bull. WHO **59**, 269–279 (1981).

Bonner, J.T. (ed.): Evolution and Development. Report of the Dahlem Workshop on Evolution and Development, Berlin 1981. Berlin Heidelberg New York: Springer 1982.

Bonotto, S.: *Acetabularia* as a link in the marine food chain. In: S. Bonotto, F. Cinelli, R. Billiau (eds.) Proc. 6th Intern. Symp. on *Acetabularia*. Pisa 1984. Belgian Nuclear Center, C.E.N.-S.C.K. Mol, Belgium 1985, pp. 67–80.

Bonotto, S., Cinelli, F., Billiau, R. (eds.): Proc. 6th Intern. Symp. on *Acetabulria*. Pisa 1984. Belgian Nuclear Center, C.E.N.-S.C.K. Mol, Belgium 1985.

Born, W.: Monsters in art. CIBA Symp. **9**, 684–696 (1947).

Bray, W.C.: A periodic reaction in homogeneous solution and its relation to catalysis. J. Amer. Chem. Soc. **43**, 1262–1267 (1921).

Brayley, E.W.: A Journal of the Plague Year; or Memorials of the Great Pestilence in London, in 1665., by Daniel Defoe. London: Thomas Tegg 1722.

Bridge, J.F., Angrist, S.E.: An extended table of roots of $J'_n(x)Y'_n(bx) - J_n(bx)Y'_n(x) = 0$. Math. Comp. **16**, 198–204 (1962).

Britton, N.F., Murray, J.D.: Threshold wave and cell-cell avalanche behaviour in a class of substrate inhibition oscillators. J. theor. Biol. **77**, 317–332 (1979).

Britton, N.F.: Reaction-Diffusion Equations and their applications to Biology. New York: Academic Press 1986.

Buchanan, J.T., Cohen, A.H.: Activities of identified interneurons, motoneurons, and muscle fibers during fictive swimming in the lamprey and effects of reticulospinal and dorsal cell stimulation. J. of Neurophysiol. **47**, 948–960 (1982).

Bunow, B., Kernevez, J.-P., Joly, G., Thomas, D.: Pattern formation by reaction-diffusion instabilities: application to morphogenesis in *Drosophila*. J. theor. Biol. **84**, 629–649 (1980).

Butterworth, A.E., Kapron, M., Cordingley, J.S., Dalton, P.R., Dunne, D.W., Kariuki, H.C., Kimani, G., Koech, D., Mugambi, M., Ouma, J.H., Prentice, M.A., Richardson, B.A., Arap Siongok, T.K., Sturrock, R.F., Taylor, D.W.: Immunity after treatment of human schistomiasis mansoni II Identification of resistant individuals and analysis of their immune responses. Trans. Roy. Soc. Trop. Med. Hyg. **79**, 393–408 (1985).

Cahn, J.W., Hilliard, J.E.: Freee energy of a non-uniform system. I. Interfacial free energy. J. Chem. Phys. **28**, 258–267 (1958).

Cahn, J.W.: Free energy of a non-uniform system. II. Thermodynamic basis. J. Chem. Phys. **30**, 1121–1124 (1959).

Cahn, J.W., Hilliard, J.E.: Free energy of a non-uniform system. III. Nucleation in a two-component incompressible fluid. J. Chem. Phys. **31**, 688–699 (1959).

Canosa, J.: On a nonliear diffusion equation describing population growth. IBM J. Res. & Dev. **17**, 307–313 (1973).

Capasso, V., Paveri-Fontana, S.L.: A mathematical model for the 1973 cholera epidemic in the european mediterranean region. Rev. Epidém. et Santé Publ. **27**, 121–132 (1979).

Carelli, C., Audibert, F., Gaillard, J., Chedid, L.: Immunological castration of male mice by a totally synthetic vaccine administered in saline. Proc. Nat. Acad. Sci. U.S.A. **79**, 5392–5395 (1982).

Carpenter, G.A.: Bursting phenomena in excitable membranes. SIAM J. Appl. Math. **36**, 334–372 (1979).

Carslaw, H.S., Jaeger, J.C.: Conduction of Heat in Solids, 2nd edn. Oxford: Clarendon Press 1959.

Cartwright, M., Husain, M.A.: A model for the control of testosterone secretion. J. theor. Biol. **123**, 239–250 (1986).

Chance, B., Pye, E.K., Ghosh, A.K., Hess, B. (eds.): Biological and Biochemical Oscillators. New York: Academic Press 1973.

Charlesworth, B.: Evolution in age-structured populations. Cambridge: Cambridge University Press 1980.

Cheer, A., Nuccitelli, R., Oster, G.F., Vincent, J.-P.: Cortical activity in vertebrate eggs I: The activation waves. J. theor. Biol. **124**, 377–404 (1987).

Cherrill, F.R.: The Fingerprint System at Scotland Yard. London: H.M. Stationary Office 1954.

Choodnovsky, D.V., Choodnovsky, G.V.: Pole expansions of nonlinear partial differential equations. Nuovo Cim. **40**, 339–353 (1977).

Christopherson, D.G.: Note on the vibration of membranes. Quart. J. Math. (Oxford Ser.) **11**, 63–65 (1940).

Clark, C.W.: Mathematical Bioeconomics, the optimal control of renewable resources. New York: Wiley 1976a.

Clark, C.W.: A delayed-recruitment model of population dynamics with an application to baleen whale populations. J. Math. Biol. **3**, 381–391 (1976b).

Cocho, G., Pérez-Pascual, R., Rius, J.L., Soto, F.: Discrete systems, cell-cell interactions and color pattern of animals. I. Conflicting dynamics and pattern formation. II. Clonal theory and cellular automata. J. theor. Biol. **125**, 419–447 (1987).

Cohen, A.H., Wallén, P.: The neuronal correlate of locomotion in fish. Exp. Brain Res. **41**, 11–18 (1980).

Cohen, A.H., Harris-Warrick, R.M.: Strychnine eliminates alternating motor output during fictive locomotion in lamprey. Brain Res. **293** 164–167 (1984).

Cohen, A.H., Holmes, P.J., Rand, R.R.: The nature of coupling between segmental oscillators and the lamprey spinal generator for locomotion: a mathematical model. J. Math. Biol. **13**, 345–369 (1982).

Cohen, A.H., Rossignol, S., Grillner, S. (eds.): Neural Control of Rhythmic Movements in Vertebrates. New York: John Wiley 1988.

Cohen, D.S., Neu, J.C., Rosales, R.R.: Rotating spiral wave solutions of reaction-diffusion equations. SIAM J. Appl. Math. **35**, 536–547 (1978).

Cohen, D.S., Murray, J.D.: A Generalized Diffusion Model for Growth and Dispersal in a Population. J. Math. Biol. **12**, 237–249 (1981).

Cohen, Y. (ed.): Applications of Control Theory in Ecology. Lect. Notes in Biomathematics **73**. Berlin Heidelberg New York Tokyo: Springer 1987.

Courant, R., Hilbert, D.: Methods of Mathematical Physics. New York: Interscience 1962.

Cowan, J.D.: Some remarks on channel bandwidths for visual contrast detection. Neurosci. Res. Bull. **15**, 492–515 (1977).

Cowan, J.D.: Brain mechanisms underlying visual hallucinations. In: D. Paines (ed.) Emerging Syntheses in Science. New York: Addison-Wesley 1987.

Cowan, J.D.: Spontaneous symmetry breaking in large scale nervous activity. Intern. J. Quantum Chem. **22**, 1059–1082 (1982).

Crank, J.: The Mathematics of Diffusion. Oxford: Clarendon Press 1975.

Cummins, H., Midlo, C.: Finger Prints, Palms and Soles: An Introduction to Dermatoglyphics. Philadelphia: Blakiston 1943.

Cvitanović, P.: Universality in Chaos. Bristol: Adam Hilger 1984.

Dagg, A.I.: External features of giraffe. Extrait de Mammalia **32**, 657–669 (1968).

Darwin, C.: The origin of species, 6th edn. London: John Murray 1873.

Davidson, D.: The mechanism of feather pattern development in the chick. I. The time of determination of feather position. II. Control of the sequence of pattern formation. J. Embryol. exp. Morph. **74**, 245–273 (1983).

DeBach, P.: Biological Control by Natural Enemies. Cambridge: Cambridge University Press 1974.

Decroly, O., Goldbeter, A.: From a simple to complex oscillatory behaviour: analysis of bursting in a multipy regulated biochemical system. J. theor. Biol. **124**, 219–250 (1987).

Dee, G., Langer, J.S.: Propagating pattern selection. Phys. Rev. Letters **50**, 383–386 (1983).

Dellwo, D., Keller, H.B., Matkowsky, B.J., Reiss, E.L.: On the birth of isolas. SIAM J. Appl. Math. **42**, 956–963 (1982).

Dhouailly, D.: Formation of cutaneous appendages in dermoepidermal recombination between reptiles, birds and mammals. Wilhelm Roux Arch. EntwMech. Org. **177**, 323–340 (1975).

Dhouailly, D., Hardy, M., Sengel, P.: Formation of feathers on chick foot scales: a stage-dependent morphogenetic response to retinoic acid. J. Embryol. exp. Morphol. **58**, 63–78 (1980).

Dietz, K.: The population dynamics of onchocersiasis. In: R.M. Anderson (ed.) Population Dynamics of Infectious Diseases. London: Chapman and Hall 1982, pp. 209–241.

Dietz, K., Hadeler, K.P.: Epidemiological models for sexually transmitted diseases. J. Math. Biol. **26**, 1–25 (1988).

Driver, R.D.: Ordinary and Delay Differential Equations. Berlin Heidelberg New York: Springer 1977.

Duffy, M.R., Britton, N.F., Murray, J.D.: Spiral wave solutions of practical reaction-diffusion systems. SIAM J. Appl. Math. **39**, 8–13 (1980).

Dunbar, S.R.: Travelling wave solutions of diffusive Lotka-Volterra equations. J. Math. Biol. **17**, 11–32 (1983).

Dunbar, S.R.: Travelling wave solutions of diffusive Lotka-Volterra equations: a heteroclinic connection in R^4. Trans. Amer. Math. Soc. **268**, 557–594 (1984).

Edelstein, B.B.: The dynamics of cellular differentiation and associated pattern formation. J. theor. Biol. **37**, 221–243 (1972).

Edelstein-Keshet, L.: Mathematical Models in Biology. New York: Random House 1988.

Edmunds, L.N. (ed.): Cell Cycle Clocks. New York: Marcel Dekker 1984

Elsdale, T., Wasoff, F.: Fibroblast cultures and dermatoglyphics: the topology of two planar patterns. Wilhelm Roux Arch. **180**, 121–147 (1976).

Elton, C.S., Nicholson, M.: The ten-year cycle in numbers of lynx in Canada. J. Anim. Ecol. **191**, 215–244 (1942).

Ermentrout, G.B.: Stable small amplitude solutions in reaction-diffusion systems. Q. Appl. Math. **39**, 61–86 (1981).

Ermentrout, G.B.: $n{:}m$ phase-locking of weakly coupled oscillators. J. Math. Biol. **12**, 327–342 (1981).

Ermentrout, G.B., Cowan, J.: A mathematical theory of visual hallucination patterns. Biol. Cybern. **34**, 137–150 (1979).

Ermentrout, B., Campbell, J., Oster, G.: A model for shell patterns based on neural activity. The Veliger **28**, 369–388 (1986).

Feigenbaum, M.J.: Quantitative universality for a class of nonlinear transformations. J. Stat. Phys. **19**, 25–52 (1978).

Feroe, J.A.: Existence and stability of multiple impulse solutions of a nerve equation. SIAM J. Appl. Math. **42**, 235–246 (1982).

Field, R.J., Körös, E., Noyes, R.M.: Oscillations in chemical systems, Part 2. Thorough analysis of temporal oscillations in the bromate-cerium-malonic acid system. J. Am. Chem. Soc. **94**, 8649–8664 (1972).

Field, R.J., Noyes, R.M.: Oscillations in chemical systems, Part 4. Limit cycle behaviour in a model of a real chemical reaction. J. Chem. Phys. **60**, 1877–1844 (1974).

Field, R.J., Burger, M. (eds.).: Oscillations and Travelling Waves in Chemical Systems. New York: Wiley 1985.

Fife, P.C.: Mathematical Aspects of Reacting and Diffusing Systems. Lect. Notes in Biomathematics **28**. Berlin Heidelberg New York: Springer 1979.

Fife, P.C., McLeod, J.B.: The approach of solutions of nonlinear diffusion equations to travelling wave solutions. Archiv. Rat. Mech. Anal. **65**, 335–361 (1977).

Fisher, R.A.: The wave of advance of advantageous genes. Ann. Eugenics **7**, 353–369 (1937).

FitzHugh, R.: Impulses and physiological states in theoretical models of nerve membrane. Biophys. J. **1**, 445–466 (1961).

FitzHugh, R.: Mathematical models of excitation and propagation in nerve. In: H.P. Schwan (ed.) Biological Engineering. New York: McGraw-Hill (1969), pp. 1–85.

Folkman, J., Moscona, A.: Role of cell shape in growth control. Nature **273**, 345–349 (1978).

Frenzen, C.L., Maini, P.K.: Enzyme kinetics for a two-step enzymic reaction with comparable initial enzyme-substrate ratios. J. Math. Biol. **26**, 689–703 (1988).

Frerichs, R.R., Prawda, J.: A computer simulation model for the control of rabies in an urban area of Colombia. Management Science **22**, 411–421 (1975).

Friedrichs, K.O.: Advanced Ordinary Differential Equations. London: Nelson 1965.

Fujii, H., Nishiura, Y., Hosono, Y.: On the structure of multiple existence of stable stationary solutions in systems of reaction-diffusion equations. Studies in Maths. and its Applic. **18**. Patterns and Waves – Qualitative Analysis of Nonlinear Differential Equations (1986), pp. 157–219.

Ganapathisubramanian, N., Showalter, K.: Bistability, mushrooms and isolas. J. Chem. Phys. **80**, 4177–4184 (1984).

Gerber, A.: Die embryonale und postembryonale Pterylose der Alectromorphae. Rev. Suisse Zool. **46**, 161–324 (1939).

Gibbs, R.G.: Travelling waves in the Belousov-Zhabotinskii reaction. SIAM J. Appl. Math. **38**, 422–444 (1980).

Gierer, A., Meinhardt, H.: A theory of biological pattern formation. Kybernetik **12**, 30–39 (1972).

Gierer, A.: Generation of biological patterns and forms: some physical mathematical and logical aspects. Prog. Biophys. Molec. Biol. **7**, 1–47 (1981).

Gilkey, J.C., Jaffe, L.F., Ridgeway, E.B., Reynolds, G.T.: A free calcium wave traverses the activating egg of Oryzias latipes. J. Cell Biol. **76**, 448–466 (1978).

Gilpin, M.E.: Do hares eat lynx? Amer. Nat. **107**, 727–730 (1973).

Glass, L., Mackey, M.C.: Pathological conditions resulting from instabilities in physiological control control systems. Ann. N. Y. Acad. Sci. **316**, 214–235 (1979).

Goh, B.-S.: Management and Analysis of Biological Populations. Elsevier Sci. Pub.: Amsterdam 1982.

Goldbeter, A.: Models for oscillations and excitability in biochemical systems. In: L.A. Segel (ed.) Mathematical Models in Molecular and Cellular Biology. Cambridge: Cambridge University Press 1980, pp. 248–291.

Goldschmidt, R.: Die quantitativen Grundlagen von Vererbung und Artbildung. Berlin: Springer 1920.

Goldstein, S., Murray, J.D.: On the mathematics of exchange processes in fixed columns. III The solution for general entry conditions, and a method of obtaining asymptotic expressions. IV Limiting values, and correction terms, for the kinetic-theory solution with general entry conditions. V The equilibrium-theory and perturbation solutions, and their connexion with kinetic-theory solutions, for general entry conditions. Proc. R. Soc. Lond. **A257**, 334–375 (1959).

Goodwin, B.C.: Oscillatory behaviour in enzymatic control processes. Adv. in Enzyme Regulation **3**, 425–438 (1965).

Goodwin, B.C., Murray, J.D., Baldwin, D.: Calcium: the elusive morphogen in *Acetabularia*. In: S. Bonotto, F. Cinelli, R. Billiau (eds.) Proc. 6th Intern. Symp. on *Acetabulria*. Pisa 1984. Belgian Nuclear Center, C.E.N.-S.C.K. Mol, Belgium 1985, pp. 101–108.

Gray, P.: Instabilities and oscillations in chemical reactions in closed and open systems. Proc. Roy. Soc. **A 415**, 1–34 (1988).

Gray, P., Scott, S.K.: Autocatalytic reactions in the isothermal continuous stirred tank reactor. Chem. Eng. Sci. **38**, 29–43 (1983).

Gray, P., Scott, S.K.: A new model for oscillatory behaviour in closed systems: the autocatalator. Ber. Bunsenges. Phys. Chem. **90**, 985–996 (1986).

Green, H., Thomas, J.: Pattern formation by cultured human epidermal cells: development of curved ridges resembling dermatoglyphs. Science **200**, 1385–1388 (1978).

Greenberg, J.M.: Spiral waves for λ-ω systems. II. Adv. Appl. Math. **2**, 450–455 (1981).

Gregg, C.T.: Plague An Ancient Disease in the Twentieth Century. Albequerque: University of New Mexico Press 1985.

Griffith, J.S.: Mathematics of cellular control processes. I. Negative feedback to one gene. II. Positive feedback to one gene. J. theor. Biol. **20**, 202–216 (1968).

Grillner, S.: On the generation of locomotion in the spinal dogfish. Exp. Brain Res. **20**, 459–470 (1974).

Grillner, S., Kashin, S.: On the generation and performance of swimming fish. In: R.M. Herman, S. Grillner, P.S.G. Stein, D.G. Stuart (eds.) Neural Control of Locomotion. Plenum: New York 1976, pp. 181-202.

Grillner, S., Wallén, P.: On the peripheral control mechanisms acting on the central pattern generators for swimming in dogfish. J. exp. Biol. **98**, 1–22 (1982).

Guckenheimer, J., Holmes, P.J.: Nonlinear Oscillations, Dynamical Systems and Bifurcations of Vector Fields. New York Berlin Heidelberg: Springer 1983.

Gumowski, I., Mira, C.: Dynamique Chaotique. Toulouse: Collection Nabla, Cepadue Edition 1980.

Gurney, W.S.C., Blythe, S.P., Nisbet, R.M.: Nicholson's blowflies revisited. Nature **287**, 17–21 (1980).

Gurtler, W., Zimen, E.: The use of baits to estimate fox numbers. Comp. Immun. Microbiol. infect. Dis. **5**, 277–283 (1982).

Guttman, R., Lewis, S., Rinzel, J.: Control of repetitive firing in squid axon membrane as a model for a neurone oscillator. J. Physiol. (Lond.) **305**, 377–95 (1980).

Haeckel, E.: Die Radiolaren. Berlin: Georg von Reimer (Vol.I) 1862, (Vol.II) 1887.

Haeckel, E.: Art Forms in Nature. New York: Dover 1974.

Hagan, P.S.: Spiral waves in reaction diffusion equations. SIAM J. Appl. Math. **42**, 762–786 (1982).

Hamburger, V.: Monsters in nature. CIBA Sympos. **9**, 666–683 (1947).

Hanusse, P.: De l'éxistence d'un cycle limit dans l'évolution des systèmes chimique ouverts (On the existence of a limit cycle in the evolution of open chemical systems). Comptes Rendus, Acad. Sci. Paris, Ser. C **274**, 1245–1247 (1972).

Harris, A.K.: Traction, and its relations to contraction in tissue cell locomotion. In: R. Bellairs, A. Curtis, G. Dunn (eds.) Cell Behaviour. Cambridge: Cambridge University Press 1982, pp. 109–134.

Harris, A.K., Ward, P., Stopak, D.: Silicon rubber substrata: a new wrinkle in the study of cell locomotion. Science **208**, 177–179 (1980).

Harris, A.K., Stopak, D., Wild, D.: Fibroblast traction as a mechanism for collagen morphogenesis. Nature **290**, 249–251 (1981).

Harrison, L.G., Snell, J., Verdi, R., Vogt, D.E., Zeiss, G.D., Green, B.D.: Hair morphogenesis in *Acetabularia mediterranea*: temperature-dependent spacing and models of morphogen waves. Protoplasma **106**, 211–221 (1981).

Hasimoto, H.: Exact solution of a certain semi-linear system of partial differential equations related to a migrating predation problem. Proc. Japan Acad. **50**, 623–627 (1974).

Hassard, B.D., Kazarinoff, N.D., Wan, Y.-H.: Theory and Application of Hopf Bifurcation. Cambridge: Cambridge University Press 1981.

Hassell, M.P., May, R.M., Lawton, J.: Pattern of dynamic behaviour in single species populations. J. Anim. Ecol. **45**, 471–486 (1976).

Hassell, M.P.: The Dynamics of Arthropod Predator-Prey Systems. Princeton: Princeton University Press 1978.

Hastings, S.P., Murray, J.D.: The existence of oscillatory solutions in the Field-Noyes model for the Belousov-Zhabotinskii reaction. SIAM J. Appl. Math. **28**, 678–688 (1975).

Hastings, S.P., Tyson, J.J., Webster, D.: Existence of periodic solutions for negative feedback control systems. J. Differential Eqns. **25**, 39–64 (1977).

Hay, E.: Cell Biology of the Extracellular Matrix. New York: Plenum 1981.

Henke, K.: Vergleichende und experimentelle Untersuchungen an *Lymatria* zur Musterbildung auf dem Schmetterlingsflügel. Nachr. Akad. Wiss. Göttingen, Math.-Physik. Kl. (1943), pp. 1–48.

Herán, I.: Animal colouration: the nature and purpose of colours in invertebrates. London: Hamlyn 1976.

Hethcote, H.W., Stech, H.W., van den Driessche, P.: Nonlinear oscillations in epidemic models. SIAM J. Appl. Math. **40**, 1–9 (1981).

Hethcote, H.W., Yorke, J.A., Nold, A.: Gonorrhea modelling: a comparison of control methods. Math. Biosc. **58**, 93–109 (1982).

Hethcote, H.W., Yorke, J.A.: Gonorrhea Transmission Dynamics and Control. Lect. Notes in Biomaths. **56**. Berlin Heidelberg New York Tokyo: Springer 1984.

Hinchliffe, J.R., Johnson, D.R.: The Development of the Vertebrate Limb. Oxford: Clarendon Press 1980.

Hodgkin, A.L., Huxley, A.F.: A quantitative description of membrane current and its application to conduction and excitation in nerve. J. Physiol. (London) **117**, 500–544 (1952).

Holden, A.V. (ed.): Chaos. Manchester: Manchester University Press 1986.

Holder, N.: Developmental constraints and the evolution of the vertebrate digit patterns. J. theor. Biol. **104**, 451–471 (1983).

Hopf, E.: Abzweigung einer periodischen Lösung von einer stationären Lösung eines Differentialsystems (Bifurcation of a periodic solution from a stationary solution of a system of differential equations). Ber. math-phys. Kl. Sächs. Akad. Wiss. Leipzig **94**, 3–22 (1942).

Hopf, L.: Introduction to Differential Equations of Physics. New York: Dover 1948.

Hoppensteadt, F.C.: Mathematical Theories of Populations: Demographics, Genetics and Epidemics. CBMS Lectures Vol. **20**. Philadelphia: SIAM Publications 1975.

Hoppensteadt, F.C.: Mathematical Methods in Population Biology. Cambridge: Cambridge University Press 1982.

Hoppensteadt, F.C., Hyman, J.M.: Periodic solutions to a discrete logistic equation. SIAM J. Appl. Math. **32**, 985–992 (1977).

Hoppensteadt, F.C., Keller, J.B.: Synchronization of periodical cicada emergences. Science **194**, 335–337 (1976).

Hoppensteadt, F.C., Murray, J.D.: Threshold analysis of a drug use epidemic model. Math. Biosci. **53**, 79–87 (1981).

Hoppensteadt, F.C.: An Introduction to the Mathematics of Neurons. Cambridge: Cambridge University Press 1986.

Hosono, Y.: Travelling wave solutions for some density dependent diffusion equations. Japan J. Appl. Math. **3**, 163–196 (1986).

Hosono, Y.: Travelling wave for some biological systems with density dependent diffusion. Japan J. Appl. Math. **4**, 297–359 (1987).

Howard, L.N.: Nonlinear oscillations. Amer. Math. Soc. Lect. Notes in Appl. Math. **17**, 1–67 (1979).

Howard, L.N., Kopell, N.: Slowly varying waves and shock structures in reaction-diffusion equations. Studies in Appl. Math. **56**, 95–145 (1977).

Howe, A.H.: A theoretical Inquiry into the Physical Cause of Epidemic Diseases. London: J. Churchill and Son 1865.

Hsu, S.-B., Hubbell, S.P., Waltman, P.: A contribution to the theory of competing predators. Ecological Monographs **48**, 337–349 (1979).

Hubel, D.H., Wiesel, T.N.: Functional architecture of macaque monkey visual cortex. Proc. Roy. Soc. Lond. **B198**, 1–59 (1977).

Hubel, D.H., Wiesel, T.N., LeVay, S.: Plasticity of ocular dominance columns in monkey striate cortex. Phil. Trans. Roy. Soc. Lond. **B278**, 131–163 (1977).

Huberman, B.A.: Striations in chemical reactions. J. Chem. Phys. **65**, 2013–2019 (1976).

Huffaker, C.B. (ed.): Biological Control. New York: Plenum Press 1971.

Hunding, A.: Limit-cycles in enzyme systems with nonlinear feedback. Biophys. Struct. Mech. **1**, 47–54 (1974).

Hunding, A., Sørensen, P.G.: Size adaption of Turing prepatterns. J. Math. Biol. **26**, 27–39 (1988).

Hutchinson, G.E.: An Introduction to Population Biology. New Haven: Yale 1978.

Ikeda, N., Yamamoto, H., Sato, T.: Pathology of the pacemaker network. Math. Modelling **7**, 889–904 (1986).

International Whaling Commission.: Report No. **29**. International Whaling Commission, Cambridge 1979.

Isham, V.: Mathematical modelling of the transmission dynamics of HIV infection and AIDS: a review. J. Roy. Statist. Soc. **A151**, 5–30 (1988).

Ivanitsky, G.R., Krinsky, V.I., Zaikin, A.N., Zhabotinsky, A.M.: Autowave processes and their role in disturbing the stability of distributed excitable systems. Soviet Sci. Rev., Sect. D, Biol. Rev. **2**, 279–324 (1981).

Jackson, H.C., Schneider, L.G.: Rabies in the Federal Republic of Germany, 1950–81: the influence of landscape. Bull. WHO **62**, 99–106 (1984).

Jalife, J., Antzelevitch, C.: Phase resetting and annihilation of pacemaker activity in cardiac tissue. Science **206**, 695–697 (1979).

Jordan, D.W., Smith, P.: Nonlinear Ordinary Differential Equations, 2nd edn. Oxford: Clarendon Press 1987.

Källén, A., Arcuri, P., Murray, J.D.: A simple model for the spatial spread and control of rabies. J. theor. Biol. **116**, 377–393 (1985).

Kaplan, C. (ed.): Rabies: The Facts. Oxford: Oxford University Press 1977.

Kareiva, P.M.: Local movement in herbivorous insects: applying a passive diffusion model to mark-recapture field experiments. Oecologia (Berlin) **57**, 322–327 (1983).

Kath, W.L., Murray, J.D.: Analysis of a model biological switch. SIAM J. Appl. Math. 45, 943–955 (1986).

Kauffman, S.A.: Pattern formation in the *Drosophila* embryo. Phil. Trans. Roy. Soc. Lond. B295, 567–594 (1981).

Kauffman, S.A., Shymko, R., Trabert, K.: Control of sequential compartment in *Drosophila*. Science 199, 259–270 (1978).

Keener, J.P.: Waves in excitable media. SIAM J. Appl. Math. 39, 528–548 (1980).

Keener, J.P.: A geometrical theory for spiral waves in excitable media. SIAM J. Appl. Math. 46, 1039–1056 (1986).

Keener, J.P., Tyson, J.J.: Spiral waves in the Belousov-Zhabotinskii reaction. Physica 21D, 307–324 (1986).

Keller, E.F., Segel, L.A.: Travelling bands of chemotactic bacteria: a theoretical analysis. J. theor. Biol. 30, 235–248 (1971).

Keller, E.F., Odell, G.M.: Necessary and sufficient conditions for chemotactic bands. Math. Biosci. 27, 309–317 (1975).

Kermack, W.O., McKendrick, A.G.: Contributions to the mathematical theory of epidemics. Proc. Roy. Soc. A 115, 700–721 (1927); 138, 55–83 (1932); 141, 94–122 (1933).

Kingdon, J.: East African Mammals. An Atlas of Evolution in Africa. IIIA Carnivores. London: Academic Press 1978.

Kingdon, J.: East African Mammals. An Atlas of Evolution in Africa. IIIB Large mammals. London: Academic Press 1979.

Klüver, H.: Mescal and Mechanisms of Hallucinations. Chicago: University of Chicago Press 1967.

Koga, S.: Rotating spiral waves in reaction-diffusion systems – phase singularities of multi-armed spirals. Prog. Theor. Phys. 67, 164–178 (1982).

Kolmogoroff, A., Petrovsky, I., Piscounoff, N.: Étude de l'équation de la diffusion avec croissance de la quantité de matière et son application à un problème biologique. Moscow Univ. Bull. Math. 1, 1–25 (1937).

Kopell, N., Howard, L.N.: Horizontal bands in the Belousov reaction. Science 180, 1171–1173 (1973a).

Kopell, N., Howard, L.N.: Plane wave solutions to reaction-diffusion equations. Studies in Appl. Math. 42, 291–328 (1973b).

Kopell, N., Howard, L.N.: Target patterns and spiral solutions to reaction-diffusion equations with more than one space dimension. Adv. Appl. Math. 2, 417–449 (1981).

Kopell, N.: Toward a theory of modelling central pattern generators. In: A.H. Cohen, S. Rossignol, S. Grillner (eds.) Neural Control of Rhythmic Movements in Vertebrates. New York: John Wiley 1988, pp. 369–414.

Krebs, C.J., Myers, J.H.: Population cycles in small mammals. Adv. Ecol. Res. 8, 267–399 (1974).

Krinsky, V.I.: Mathematical models of cardiac arrhythmias (spiral waves). Pharmac. Ther. B 3, 539–555 (1978).

Krinsky, V.I., Medvinskii, A.B., Parfilov, A.V.: Evolutionary autonomous spiral waves (in the heart). Mathematical Cybernetics. Popular Ser. (Life Sciences) 8, 1–48 (1986). (In Russian)

Kühn, A., von Engelhardt, A.: Über die Determination des Symmetriesystems auf dem Vorderflügel von *Ephestia kühniella*. Z. Wilhelm Roux Arch. Entw. Mech. Org. 130, 660–703 (1933).

Kuramoto, K.: Instability and turbulence of wavefronts in reaction diffusion systems. Prog. Theor. Phys. 63, 1885–1903 (1980).

Kuramoto, K., Koga, S.: Turbulized rorating chemical waves. Prog. Theor. Phys. 66, 1081–1085 (1981).

Lajmanovich A., Yorke, J.A.: A deterministic model for gonorrhea in a nonhomogeneous population. Math. Biosci. 28, 221–236 (1976).

Landau, L., Lifshitz, E.: Theory of Elasticity, 2nd edn. New York: Pergamon 1970.

Lane, D.C., Murray, J.D., Manoranjan, V.S.: Analysis of wave phenomena in a morphogenetic mechanochemical model and an application to post-fertilisation waves on eggs. IMA J. Math. Applied in Medic. and Biol. 4, 309–331 (1987).

Langer, W.L.: The Black Death. Scientific American (February), 114–121 (1964).

La Rage: Informations Techniques des Service Veterinaires. Nos. 64–67. CNER: Paris 1977.

Lara Ochoa, F.: A generalized reaction diffusion model for spatial structure formed by mobile cells. Biosystems 17, 35–50 (1984).

Lara Ochoa, F., Murray, J.D.: A nonlinear analysis for spatial structure in a reaction-diffusion model. Bull. math. Biol. 45, 917–930 (1983).

Larson, D.A.: Transient bounds and time asymptotic behaviour of solutions of nonlinear equations of Fisher type. SIAM J. Appl. Math. 34, 93–103 (1978).

Lauffenburger, D.A., Keller, K.H.: Effects of leukocyte random motility and chemotaxis in tissue inflammatory response. J. theor. Biol. 81, 475–503 (1979).

Lauwerier, H.A.: Two-dimensional iterative maps. In: A.V. Holden (ed.) Chaos. Manchester: Manchester University Press 1986, pp. 58–95.

Leigh, E.: The ecological role of Volterra's equations. In: M. Gerstenhaber (ed.) Some mathematical problems in biology. Providence: Amer. Math. Soc. 1968, pp. 1–64.

Levin, S.A.: Dispersion and population interactions. Amer. Nat. 108, 207–228 (1974).

Levin, S.A.: Population dynamic models in heterogeneous environments. Ann. Rev. Ecol. 7, 287–310 (1976).

Levin, S.A.: The role of theoretical ecology in the description and understanding of populations in heterogeneous environments. Amer. Zool. 21, 865–875 (1981a).

Levin, S.A.: Models of population dispersal. In: S. Busenberg, K. Cooke (eds.) Differential Eqautions and Applications to Ecology, Epidemics and Population Problems. New York: Academic Press 1981b, pp. 1–18.

Levin, S.A., Segel, L.A.: Pattern generation in space and aspect. SIAM Rev. 27, 45–67 (1985).

Levinton, J.: Developmental constraints and evolutionary saltations: a discussion and critique. In: G. Stebbins, F. Ayala (eds.) Genetics and Evolution. New York: Plenum 1986.

Leviton, A.E., Anderson, S.C.: Description of a new species of *Cyrtodactylus* from Afghanistan with remarks on the status of *Gymnodactylus longpipes* and *Cyrtodactylus fedtschenkoi*. J. Herp. 18, 270–276 (1984).

Lewis, E.R.: Network Models in Population Biology. Berlin Heidelberg New York: Springer 1977.

Lewis, J., Slack, J.M.W., Wolpert, L.: Thresholds in development. J. theor. Biol. 65, 579–590 (1977).

Li, T.-Y., Yorke, J.A.: Period three implies chaos. Amer. Math. Monthly 82, 985–992 (1975).

Li, T.-Y., Misiurewicz, M., Pianigiani, G., Yorke, J.A.: Odd chaos. Phys. Letters 87A, 271–273 (1982).

Lions, P.L.: On the existence of positive solutions of semilinear elliptic equations. SIAM Rev. 24, 441–467 (1982).

Lloyd, H.G., Jensen, B., Van Haaften, J.L., Niewold, F.J.J., Wandeler, A., Boegel, K., Arata, A.A.: Annual turnover of fox populations in Europe. Zbl. Vet. Med. 23, 580–589 (1976).

Loesch, D.Z.: Quantitative dermatoglyphics: classification, genetics, and pathology. Oxford: Oxford University Press 1983.

Lorenz, E.N.: Deterministic nonperiodic flow. J. Atmos. Sci. 20, 131–141 (1963).

Lotka, A.J.: Contribution to the theory of periodic reactions. J. Phys. Chem. 14, 271–274 (1910).

Lotka, A.J.: Undaped oscillations derived from the law of mass action. J. Amer. Chem. Soc. 42, 1595–1599 (1920).

Lotka, A.J.: Elements of Physical Biology. Williams and Wilkins: Baltimore 1925.

Ludwig, D., Jones, D.D., Holling, C.S.: Qualitative analysis of insect outbreak systems: the spruce budworm and forest. J. Anim. Ecol. 47, 315–332 (1978).

Ludwig, D., Aronson, D.G., Weinberger, H.F.: Spatial patterning of the spruce budworm. J. Math. Biol. 8, 217–258 (1979).

Luther, R.-L: Räumliche Fortpflanzung Chemischer Reaktionen. Z. für Elektrochemie und angew. physikalische Chemie **12**(32), 506–600 (1906). [English translation: Arnold, R., Showalter, K., Tyson, J.J.: Propagation of chemical reactions in space. J. Chem. Educ. (1988) (in press)]

Macdonald, D.W.: Rabies and Wildlife: a biologist's perspective. Oxford Univ. Press: Oxford 1980.

Macdonald, D.W., Voight, D.R.: The biological basis of rabies models. In: P.J. Bacon (ed.) Population Dynamics of Rabies in Wildlife. London: Academic Press 1985, pp. 71–108.

MacDonald, N.: Bifurcation theory applied to a simple model of a biochemical oscillator. J. theor. Biol. **65**, 727–734 (1977).

MacDonald, N.: Time Lags in Biological Models. Lect. Notes in Biomath. **28**. Berlin Heidelberg New York: Springer 1979.

McEvedy, C.: The bubonic plague. Sci. Amer. (February), 74–79 (1988).

McKean, H.P.: Nagumo's equation. Adv. in Math. **4**, 209–223 (1970).

McKean, H.P.: Application of Brownian motion to the equation of Kolmogorov-Petrovskii-Piskunov. Comm. Pure Appl. Math. **28**, 323–331 (1975).

Mackey, M.C.: Periodic auto-immune hemolytic anemia: an induced dynamical disease. Bull. math. Biol. **41**, 829–834 (1979).

Mackey, M.C., Glass, L.: Oscillations and chaos in physiological control systems. Science **197**, 287–289 (1977).

Mackey, M.C., Dormer, P.: Enigmatic hemopoiesis. In: M. Rotenberg (ed.) Biomathematics and Cell Kinetics. Amsterdam: Elsevier North Holland 1981, pp. 87–103.

Mackey, M.C., Milton, J.G.: Dynamical diseases. Ann. N. Y. Acad. Sci. **504**, 16–32 (1988).

MacLulich, D.A.: Fluctuations in the Number of the Varying Hare (*Lepus americanus*). Univ. Toronto Stud. Biol. Ser. **43** (1937).

Maini, P.K., Murray, J.D.: A nonlinear analysis of a mechanical model for biological pattern formation. SIAM J. Appl. Math. **48**, 1064–1072 (1988).

Malthus, T.R.: An essay on the Principal of Population. 1798 [Penguin Books 1970].

Mandelbrot, B.B.: The Fractal Geometry of Nature. San Francisco: Freeman 1982.

Manoranjan, V.S., Mitchell, A.R.: A numerical study of the Belousov-Zhabotinskii reaction using Galerkin finite element methods. J. Math. Biol. **16**, 251–260 (1983).

Marek, M., Svobodová, E.: Nonlinear phenomena in oscillatory systems of homogeneous reactions – experimental observations. Biophys. Chem. **3**, 263–273 (1975).

Marek, M., Stuchl, I.: Synchronization in two interacting oscillatory systems. Biophys. Chem. **4**, 241–248 (1975).

Martiel, J.-L., Goldbeter, A.: A model based on receptor desensitization for cyclic AMP signalling in *Dictyostelium* cells. Biophys. J. **52**, 807–828 (1987).

Matano, H.: Asymptotic behaviour and stability of solutions of semilinear diffusion equations. Publ. Res. Inst. Math. Sci. Kyoto **15**, 401–454 (1979).

Maxwell, J.C.: Scientific Papers. New York: Dover 1952.

May, R.M.: Stability and Complexity in Model Ecosystems, 2nd edn. Princeton: Princeton Univ. Press 1975.

May, R.M.: Mathematical aspects of the dynamics of animal populations. In: S.A. Levin (ed.) MAA Studies in Mathematics **15**, **16**. Washington: Mathematical Association of America 1978, pp. 317–366.

May, R.M.: Periodical cicadas. Nature **277**, 347–349 (1979).

May, R.M., MacArthur, R.H.: Niche overlap as a function of environment variability. Proc. Nat. Acad. Sci. (U.S.A.) **69**, 1109–1113 (1972).

May, R.M., Oster, G.F.: Bifurcations and dynamic complexity in simple ecological models. Amer. Natur. **110**, 573–599 (1976).

May, R.M.: Mathematical models in whaling and fisheries management. In: R.C. Di Prima (ed.) Some Mathematical Questions in Biology. Lect. on Math. in Life Sci. **13**, 1–64. Providence: Amer. Math. Soc. 1980.

May, R.M.: Theoretical Ecology. Principles and Applications, 2nd edn. Oxford: Blackwell Scientific Publications 1981.

May, R.M.: Regulation of population with nonoverlapping generations by microparasites: a purely chaotic system. Amer. Nat. **125**, 573–584 (1985).

May, R.M.: When two and two do not make four: nonlinear phenomena in ecology. Proc. R. Soc. B**228**, 241–266 (1986).

Meinhardt, H.: Models of Biological Pattern Formation. London: Academic Press 1982.

Meinhardt, H.: Hierarchical inductions of cell states: a model for segmentation of *Drosophila*. J. Cell Sci. Suppl. **4**, 357–381 (1986).

Meinhardt, H., Klingler, M.: A model for pattern generation on the shells of molluscs. J. theor. Biol. **126**, 63–89 (1987).

Metz, J.A.J., Diekmann, O.: The Dynamics of Physiologically Structured Populations. Lect. Notes in Biomathematics **68**. Berlin Heidelberg New York Tokyo: Springer 1986.

Michaelis, L., Menten, M.I.: Die Kinetik der Invertinwirkung. Biochem. Z. **49**, 333–369 (1913).

Mikhailov, A.S., Krinsky, V.I.: Rotating spiral waves in excitable media: the analytical results. Physica **9D**, 346–371 (1983).

Mimura, M., Murray, J.D.: Spatial structures in a model substrate-inhibition reaction diffusion system. Z. Naturforsch. **33c**, 580–586 (1978).

Mimura, M., Murray, J.D.: On a diffusive prey-predator model which exhibits patchiness. J. theor. Biol. **75**, 249–262 (1978).

Mimura, M., Yamaguti, M.: Pattern formation in interacting and diffusing systems in population biology. Adv. Biophys. **15**, 19–65 (1982).

Mittenthal, J.E., Mazo, R.M.: A model for shape generation by strain and cell-cell adhesion in the epithelium of an arthopod leg segment. J. theor. Biol. **100**, 443–483 (1983).

Mollison, D.: Spatial contact models for ecological and epidemic spread. J. Roy. Stat. Soc. B**39**, 283–326 (1977).

Monk, A., Othmer, H. G.: Cyclic AMP oscillations in suspensions of *Dictyostelium discoideum*. Proc. Roy. Soc. B, (1988) (in press)

Monod, J., Jacob, F.: General conclusions: teleonomic mechanisms in cellular metabolism, growth and differentiation. Cold Spring Harbor Symp. Quant. Biol. **26**, 389–401 (1961).

Morse, P., Feshbach, H.: Methods of Theoretical Physics. Vol. 1. New York: McGraw Hill 1953.

Müller, S.C., Plesser, T., Hess, B.: The structure of the core of the spiral wave in the Belousov-Zhabotinskii reaction. Science **230**, 661–663 (1985).

Müller, S.C., Plesser, T., Hess, B.: Two-dimensional spectrophotometry and pseudo-color representation of chemical patterns. Naturwiss. **73**, 165–179 (1986).

Müller, S.C., Plesser, T., Hess, B.: Two-dimensional spectrophotometry of spiral wave propagation in the Belousov-Zhabotinskii reaction I. Experiments and digital representation. II. Geometric and kinematic parameters. Physica **24D**, 71–96 (1987).

Murray, J.D.: Singular perturbations of a class of nonlinear hyperbolic and parabolic equations. J. Maths. and Physics **47**, 111–133 (1968).

Murray, J.D.: Perturbation effects on the decay of discontinuous solutions of nonlinear first order wave equations. SIAM J. Appl. Math. **19**, 273–298 (1970a).

Murray, J.D.: On the Gunn effect and other physical examples of perturbed conservation equations. J. Fluid Mech. **44**, 315–346 (1970b).

Murray, J.D.: On Burgers' model equations for turbulence. J. Fluid Mech. **59**, 263–279 (1973).

Murray, J.D.: On a model for the temporal oscillations in the Belousov-Zhabotinsky reaction. J. Chem. Phys. **61**, 3610–3613 (1974).

Murray, J.D.: On travelling wave solutions in a model for the Belousov-Zhabotinskii reaction. J. theor. Biol. **52**, 329–353 (1976).

Murray, J.D.: Nonlinear Differential Equation Models in Biology. Oxford: Clarendon Press 1977.

Murray, J.D.: A pattern formation mechanism and its application to mammalian coat markings. 'Vito Volterra' Symposium on Mathematical Models in Biology. Accademia dei Lincei, Rome

Dec. 1979. Lect. Notes in Biomathematics **39**, 360–399. Berlin Heidelberg New York: Springer 1980.

Murray, J.D.: On pattern formation mechanisms for lepidopteran wing patterns and mammalian coat markings. Phil. Trans. Roy. Soc. (London) **B295**, 473–496 (1981a).

Murray, J.D.: A pre-pattern formation mechanism for animal coat markings. J. theor. Biol. **88**, 161–199 (1981b).

Murray, J.D.: Parameter space for Turing instability in reaction diffusion mechanisms: a comparison of models. J. theor. Biol. **98**, 143–163 (1982).

Murray, J.D.: Asymptotic Analysis, 2nd edn. Berlin Heidelberg New York Tokyo: Springer 1984.

Murray, J.D.: How the leopard gets its spots. Sci. Amer. **258**(3), 80–87 (1988).

Murray, J.D., Oster, G.F., Harris, A.K.: A mechanical model for mesenchymal morphogenesis. J. Math. Biol. **17**, 125–129 (1983).

Murray, J.D., Oster, G.F.: Generation of biological pattern and form. IMA J. Math. Appl. in Medic. & Biol. **1**, 51–75 (1984a).

Murray, J.D., Oster, G.F.: Cell traction models for generating pattern and form in morphogenesis. J. Math. Biol. **19**, 265–279 (1984b).

Murray, J.D., Sperb, R.P.: Minimum domains for spatial patterns in a class of reaction diffusion equations. J. Math. Biol. **18**, 169–184 (1983).

Murray, J.D., Maini, P.K.: A new approach to the generation of pattern and form in embryology. Sci. Prog. Oxf. **70**, 539–553 (1986).

Murray, J.D., Stanley, E.A., Brown, D.L.: On the spatial spread of rabies among foxes. Proc. Roy. Soc. (Lond.) **B229**, 111–150 (1986).

Murray, J.D., Maini, P.K., Tranquillo, R.T.: Mechanical models for generating biological pattern and form in development. Physics Reports **171**, 60–84 (1988).

Myers, J.H., Krebs, C.J.: Population cycles in rodents. Sci. Amer. (June), 38–46 (1974).

Nagawa, H., Nakanishi, Y.: Mechanical aspects of the mesenchymal influence on epithelial branching morphogenesis of mouse salivary gland. Development **101**, 491–500 (1987).

Nagorcka, B.N.: The role of a reaction-diffusion system in the initiation of skin organ primordia. I. The first wave of initiation. J. theor. Biol. **121**, 449–475 (1986).

Nagorcka, B.N., Mooney, J.R.: The role of a reaction-diffusion system in the formation of hair fibres. J. theor. Biol. **98**, 575–607 (1982).

Nagorcka, B.N., Mooney, J.R.: The role of a reaction-diffusion system in the initiation of primary hair follicles. J. theor. Biol. **114**, 243–272 (1985).

Nagorcka, B.N.: A pattern formation mechanism to control spatial organisation in the embryo of *Drosophila melanogaster*. J. theor. Biol. **132**, 277–306 (1988).

Nagorcka, B.N., Manoranjan, V.S., Murray, J.D.: Complex spatial patterns from tissue interactions – an illustrative model. J. theor. Biol. **128**, 359–374 (1987).

Nagumo, J.S., Arimoto, S., Yoshizawa, S.: An active pulse transmission line simulating nerve axon. Proc. IRE. **50**, 2061–2071 (1962).

Namias, V.: Simple derivation of the rots of a cubic equation. Am. J. Phys. **53**, 775 (1985).

Neu, J.C.: Chemical waves and the diffusive coupling of limit cycle oscillators. SIAM J. Appl. Math. **36**, 509–515 (1979a).

Neu, J.C.: Coupled chemical oscillators. SIAM J. Appl. Math. **37**, 307–315 (1979b).

Neu, J.C.: Large populations of coupled chemical oscillators. SIAM J. Appl. Math. **38**, 305–316 (1980).

Newell, P.C.: Attraction and adhesion in the slime mold *Dictyostelium*. In: J.E. Smith (ed.) Fungal Differentiation: A Contemporary Synthesis. Mycology Series **43**, pp. 43–71. New York: Marcel Dekker 1983.

Newman, W.I.: Some exact solutions to a nonlinear diffusion problem in population genetics and combustion. J. theor. Biol. **85**, 325–334 (1980).

Newman, W.I.: The long-timebehaviour of solutions to a nonlinear diffusion problem in population genetics and combustion. J. theor. Biol. **104**, 473–484 (1983).

Nicholson, A.J.: The self adjustment of populations to change. Cold Spring Harb. Symp. Quant. Biol. **22**, 153–173 (1957).

Nijhout, H.F.: Wing pattern formation in Lepidoptera: a model. J. exp. Zool. **206**, 119–136 (1978).

Nijhout, H.F.: Pattern formation in lepidopteran wings: determination of an eyespot. Devl. Biol. **80**, 267–274 (1980a).

Nijhout, H.F.: Ontogeny of the color pattern formation on the wings of *Precis coenia* (Lepidoptera: Nymphalidae). Devl. Biol. **80**, 275–288 (1980b).

Nijhout, H.F.: Colour pattern modification by coldshock in Lepodoptera. J. Embryol. exp. Morph. **81**, 287–305 (1984).

Nijhout, H.F.: The developmental physiology of colour patterns in Lepidoptera. Adv. Insect Physiol. **18**, 181–247 (1985a).

Nijhout, H.F.: Cautery-induced colour patterns in *Precis coenia* (Lepidoptera: Zool. Nymphalidae). J. Embryol. exp. Morph. **86**, 191–302 (1985b).

Nisbet, R.M., Gurney, W.S.C.: Modelling Fluctuating Populations. New York: Wiley 1982.

Noble, J.V.: Geographic and temporal development of plagues. Nature **250**, 726–728 (1974)

Odell, G.M.: Qualitative theory of systems of ordinary differential equations, including phase plane analysis and the use of the Hopf bifurcation theorem. In: L.A. Segel (ed.) Mathematical Models in Molecular and Cellular Biology. Cambridge: Cambridge University Press 1980, pp. 649–727.

Odell, G., Oster, G.F., Burnside, B., Alberch, P.: The mechanical basis for morphogenesis. Devel. Biol. **85**, 446–462 (1981).

Odum, E.P.: Fundamentals of Ecology. Philadelphia: Saunders 1953.

Okubo, A.: Diffusion and Ecological Problems: Mathematical Models. Berlin Heidelberg New York: Springer 1980.

Okubo, A., Chiang, H.C.: An analysis if the kinematics of swarming of *Anarete pritchardi* Kim (Diptera: Cecideomyiidae). Res. Popul. Ecol. **16**, 1–42 (1974).

Okubo, A.: Dynamical aspects of animal grouping: swarms, schools, flocks and herds. Adv. Biophys. **22**, 1–94 (1986).

Ortoleva, P.J., Ross, J.: On a variety of wave phenomena in chemical reactions. J. Chem. Phys. **60**, 5090–6107 (1974).

Ortoleva, P.J., Schmidt, S.L.: The structure and variety of chemical waves. In: R.J. Field, M. Burger (eds.) Oscillations and Travelling Waves in Chemical Systems. New York: Wiley 1985, pp. 333–418.

Oster, G.F.: Lectures in population dynamics. In: R.C. Di Prima (ed.) Modern Modelling of Continuum Phenomena. Amer. Math. Soc. Lect. in Appl. Math. **16**, 149–190 (1977).

Oster, G.F.: On the crawling of cells. J. Embryol. exp. Morphol. **83** Suppl., 329–364 (1984).

Oster, G.F., Alberch, P.: Evolution and bifurcation of developmental programs. Evolution **36**, 444–459 (1982).

Oster, G.F., Murray, J.D., Harris, A.K.: Mechanical aspects of mesenchymal morphogenesis. J. Embryol. exp. Morphol. **78**, 83–125 (1983).

Oster, G.F., Perelson, A.S.: A mathematical model for cell locomotion. J. Math. Biol. **21**, 383–388 (1984).

Oster, G.F., Murray, J.D., Odell, G.M.: The formation of microvilli. In: Molecular Determinants of Animal Form. New York: Alan R. Liss 1985, pp. 365–384.

Oster, G.F., Murray, J.D., Maini, P.K.: A model for chondrogenic condensations in the developing limb: the role of extracellular matrix and cell tractions. J. Embryol. exp. Morphol. **89**, 93–112 (1985).

Oster, G.F., Murray, J.D.: Pattern formation models and developmental constraints. In: J.P. Trinkaus Anniversary Volume 1989 (in press).

Oster, G.F., Shubin, N., Murray, J.D., Alberch, P.: Evolution and morphogenetic rules. The shape of the vertebrate limb in ontogeny and phylogeny. Evolution **45**, 862–884 (1988).

Oster, Gerold F.: 'Phosphenes' – The patterns we see when we close our eyes are clues to how the eye works. Sci. Amer. (February), 82–88 (1970).

Othmer, H.: Interactions of reaction and diffusion in open systems. Ph.D. thesis, Chem. Eng. Dept., Univ. Minnesota (1969).

Othmer, H.G.: Current problems in pattern formation. In: S.A. Levin (ed.) Amer. Math. Assoc. Lects. on Math. in the Life Sciences 9, 57–85 (1977).

Otto, H.: Die Beschuppung der Brevilinguir und Ascaleten. Jenaische Zeit. Wiss. 44, 193–252 (1908).

Pate, E., Othmer, H.G.: Applications of a model for scale-invariant pattern formation. Differentiation 28, 1–8 (1984).

Patou, M.: Analyse de la morphogenese du pied des oiseaux a l'aise de melanges cellulaires inter-specifiques. I. Étude morphologique. J. Embryol. exp. Morphol. 29, 175–196 (1973).

Peitgen, H.-O., Richter, P.H.: The Beauty of Fractals: Images of Complex Dynamical Systems. Berlin Heidelberg New York Tokyo: Springer 1986.

Penrose, R.: The topology of ridge systems. Ann. Hum. Genet., Lond. 42, 435–444 (1979).

Perelson, A.S., Maini, P.K., Murray, J.D., Hyman, J.M., Oster, G.F.: Nonlinear pattern selection in a mechanical model for morphogenesis. J. Math. Biol. 24, 525–541 (1986).

Perkel, D.H., Schulman, J.H., Bullock, T.H., Moore, C.P., Segundo, J.P.: Pacemaker neurons: effects of regularly spaced synaptic input. Science 145, 61–63 (1964).

Peterman, T.A., Drotman, D.P., Curran, J.W.: Epidemiology of the acquired immunodeficiency syndrome (AIDS). Epidemiology Reviews 7, 7–21 (1985).

Pianka, E.R.: Competition and niche theory. In: R.M. May (ed.) Theoretical Ecology: Principles and Applications. Oxford: Blackwells Scientific 1981, pp. 167–196.

Pielou, E.C.: An Introduction to Mathematical Ecology. New York: Wiley-Interscience 1969.

Pinsker, H.M.: Aplysia bursting neurons as endogenous oscillators. I Phase response curves for pulsed inhibitory synaptic input. II Synchronization and entrainment by pulsed inhibitory synaptic input. J. Neurophysiol. 40, 527–556 (1977).

Plant, R.E.: The effects of calcium^{++} on bursting neurons. Biophys. J. 21, 217–237 (1978).

Plant, R.E.: Bifurcation and resonance in a model for bursting nerve cells. J. Math. Biol. 11, 15–32 (1981).

Plant, R.E., Mangel, M.: Modelling and simulation in agricultural pest management. SIAM Rev. 29, 235–361 (1987).

Portmann, A.: Animal Forms and Patterns. A Study of the Appearance of Animals. (English translation) London: Faber and Faber 1952.

Prigogene, I., Lefever, R.: Symmetry breaking instabilities in dissipative systems. II. J. Chem. Phys. 48, 1665–1700 (1968).

Purcell, E.: Life at low Reynolds number. Amer. J. Phys. 45, 1–11 (1977).

Raggett, G.F.: Modelling the Eyam plague. Bull. Inst. Math. and its Applic. 18, 221–226 (1982).

Rand, A.H., Holmes, P.J.: Bifurcation of periodic motions in two weakly coupled van der Pol oscillators. Int. J. Nonlinear Mech. 15, 387–399 (1980).

Rand, A.H., Cohen, A.H., Holmes, P.J.: Systems of coupled oscillators as models of CPG's. In: A.H. Cohen, S. Rossignol, S. Grillner (eds.) Neural Control of Rhythmic Movements in Vertebrates. New York: John Wiley 1988, pp. 333–368.

Rapp, P.E.: Analysis of biochemical phase shift oscillators by a harmonic balancing technique. J. Math. Biol. 3, 203–224 (1976).

Rawles, M.: Tissue interactions in scale and feather development as studied in dermal-epidermal recombinations. J. Embryol. exp. Morph. 11, 765–789 (1963).

Rensing, L., an der Heiden, U., Mackey, M.C. (eds.): Temporal Disorder in Human Oscillatory Systems. Berlin Heidelberg New York Tokyo: Springer 1987.

Rinzel, J.: Models in neurobiology. In: R.H. Enns, B.L. Jones, R.M. Miura, S.S. Rangnekar (eds.) Nonlinear Phenomena in Physics and Biology. New York: Plenum Press 1981, pp. 345–367.

Rinzel, J.: On different mechanisms for membrane potential bursting. Proc. Sympos. on Non-linear Oscillations in Biology and Chemistry. Salt Lake City 1985. Lect. Notes in Biomath. **66**, 19–33. Berlin Heidelberg New York Tokyo: Springer 1986.

Rinzel, J., Keller, J.B.: Traveling wave solutions of a nerve conduction equation. Biophys. J. **13**, 1313–1337 (1973).

Rinzel, J., Terman, D.: Propagation phenomena in a bistable reaction-diffusion system. SIAM J. Appl. Math. **42**, 1111–1137 (1982).

Roberts, D.V.: Enzyme Kinetics. Cambridge: Cambridge University Press 1977.

Romer, A.S.: Vertebrate Paleontology. Chicago: University of Chicago Press 1962.

Rössler, O.E.: Chaotic behaviour in simple reaction systems. Z. Naturforsch. **31a**, 259–264 (1976).

Rössler, O.E.: Chemical turbulence: chaos in a simple reaction-diffusion system. Z. Naturforsch. **31a**, 1168–1172 (1976).

Rössler, O.E.: An equation for hyperchaos. Phys. Lett. **57A**, 155–157 (1979).

Rössler, O.E.: The chaotic hyerarchy. Z. Naturforsch. **38a**, 788–801 (1983).

Rotenberg, M.: Effect of certain stochastic parameters on extinction and harvested populations. J. theor. Biol. **124**, 455–472 (1987).

Rubin, R., Leuker, D.C., Flom, J.O., Andersen, S.: Immunity against *nematospiroides dubius* in CFW Swiss Webster mice protected by subcutaneous larval vaccination. J. Parasitol. **57**, 815–817 (1971).

Rubinow, S.I.: Introduction to Mathematical Biology. New York: Wiley 1975.

Sarkovskii, A.N.: Coexistence of cycles of a continuous map of a line into itself. (In Russian) Ukr. Mat. Z. **16**, 61–71 (1964).

Satsuma, J.: Explicit solutions of nonlinear equations with density dependent diffusion. J. Phys. Soc. Japan **56**, 1947–1950 (1987).

Schaffer, W.M.: Stretching and folding in lynx fur returns: evidence for a strange attractor in nature? Amer. Nat. **124**, 798–820 (1984).

Schmidt, S., Ortoleva, P.: Asymptotic solutions of the FKN chemical wave equation. J. Chem. Phys. **72**, 2733–2736 (1980).

Schnackenberg, J.: Simple chemical reaction systems with limit cycle behaviour. J. theor. Biol. **81**, 389–400 (1979).

Schwanwitsch, B.N.: On the ground-plan of wing-pattern in nymphalids and other families of rhopalocerous Lepidoptera. Proc. zool. Soc. Lond. **34**, 509–528 (1924).

Schwartz, V.: Neue Versuche zur Determination des zentralen Symmetriesystems bei *Plodia interpunctella*. Biol. Zentr. **81**, 19–44 (1962).

Searle, A.G.: Comparative Genetics of Coat Colour in Mammals. London: Academic Press 1968.

Segel, L.A.: Simplification and scaling. SIAM Rev. **14**, 547–571 (1972).

Segel, L.A. (ed.): Mathematical Models in Molecular and Cellular Biology. Cambridge: Cambridge University Press 1980.

Segel, L.A.: Modelling Dynamic Phenomena in Molecular and Cellular Biology. Cambridge: Cambridge University Press 1984.

Segel, L.A., Levin, S.A.: Application of nonlinear stability theory to the study of the effects of diffusion on predator prey interactions. In: R.A. Piccirelli (ed.) Amer. Inst. Phys. Conf. Proc. **27**: Topics in Statistical Mechanics and Biophysics 1976, pp. 123–152.

Segel, L.A., Slemrod, M.: The quasi-steady state assumption: a case study in perturbation. SIAM Rev. (1989) (in press)

Sengel, P.: Morphogenesis of Skin. Cambridge: Cambridge University Press 1976.

Sheldon, P.R.: Parallel gradualistic evolution of Ordovician trilobites. Nature **330**, 561–563 (1987).

Shibata, M., Bureš, J.: Reverberation of cortical spreading depression along closed pathways in rat cerebral cortex. J. Neurophysiol. **35**, 381–388 (1972).

Shibata, M., Bureš, J.: Optimum topographical conditions for reverberating cortical spreading depression in rats. J. Neurobiol. **5**, 107–118 (1974).

Shigesada, N.: Spatial distribution of dispersing animals. J. Math. Biol. **9**, 85–96 (1980).

Shigesada, N., Kawasaki, K., Teramoto, E.: Spatial segregation of interacting species. J. theor. Biol. **79**, 83–99 (1979).

Shigesada, N., Roughgarden, J.: The role of rapid dispersal in the population dynamics of competition. Theor. Popul. Biol. **21**, 353–372 (1982).

Showalter, K., Tyson, J.J.: Luther's 1906 discovery and analysis of chemical waves. J. Chem. Educ. **64**, 742–744 (1987).

Shubin, N., Alberch, P.: A morphogenetic approach to the origin and basic organisation of the tetrapod limb. In: M. Hecht, B. Wallace, W. Steere (eds.) Evolutionary Biology. Vol. **20**, 319–387. New York: Plenum 1986.

Sibatani, A.: Wing homeosis in Lepidoptera: a survey. Devl. Biol. **79**, 1–18 (1981).

Skellam, J.G.: The formulation and interpretation of mathematical models of diffusional processes in population biology. In: M.S. Bartlett, R.W. Hiorns (eds.) The Mathematical Theory of the Dynamics of Biological Populations. New York: Academic Press 1973, pp. 63–85.

Slack, J.M.W.: From Egg to Embryo. Determinative events in early development. Cambridge: Cambridge University Press 1983.

Slater, A.F.G., Keymer, A.E.: Heligmosomides polygyrus (Nematoda): the influence of dietary protein on the dynamics of repeated infection. Proc. Roy. Soc. Lond. **B229**, 69–83 (1986).

Smeets, J.L.R.M., Allessie, M.A., Lammers, W.J.E.P., Bonke, F.I.M., Hollen, J.: The wavelength of the cardiac impulse and the reentrant arrhythmias in isolated rabbit atrium. The role of heart rate, autonomic transmitters. temperature and potassium. Circ. Res. **73**, 96–108 (1986).

Smith, W.R.: Hypothalmic regulation of pituitary secretion of luteinizing hormone. II Feedback control of gonadotropin secretion. Bull. Math. Biol. **42**, 57–78 (1980).

Smith, J.C., Wolpert, L.: Pattern formation along the anteroposterior axis of the chick wing: the increase in width following a polarizing region graft and the effect of X-irradiation. J. Embryol. exp. Morph. **63**, 127–144 (1981).

Smoller, J.: Shock Waves and Reaction-Diffusion Equations. Berlin Heidelberg New York Tokyo: Springer 1983.

Southwood, T.R.E.: Bionomic strategies and population parameteers. In: R.M. May (ed.) Theoretical Ecology. Principles and Applications, 2nd edn. Oxford: Blackwell Scientific 1981, pp. 30–52.

Sparrow, C.: The Lorenz equations: Bifurcations, Chaos and Strange Attractors. Appl. Math. Sci. **41**. New York Berlin Heidelberg: Springer 1982.

Sparrow, C.: The Lorenz equations. In: A.V. Holden (ed.) Chaos. Manchester: Manchester University Press 1986, pp. 111–134.

Sperb, R.P.: Maximum principles and their applications. San Francisco: Academic Press 1981.

Steck, F., Wandeler, A.: The epidemiology of fox rabies in Europe. Epidem. Rev. **2**, 71–96 (1980).

Stefan, P.: A theorem of Sarkovskii on the existence of periodic orbits of continuous endomorphisms of the real line. Comm. Math. Phys. **54**, 237–248 (1977).

Stirzaker, D.: On a population model. Math. Biosc. **23**, 329–336 (1975).

Strogatz, S.H.: The Mathematical Structure of the Human Sleep-Wake Cycle. Lect. Notes in Biomath. **69**. Berlin Heidelberg New York Tokyo: Springer 1986.

Suffert, F.: Zur vergleichenden Analyse der Schmetterlingszeichnung. Bull. Zentr. **47**, 385–413 (1927).

Swindale, N.V.: A model for the formation of ocular dominance stripes. Proc. Roy. Soc. Lond. **B208**, 243–264 (1980).

Taddei-Ferretti, C., Cordella, L.: Modulation of *Hydra attenuata* rhythmic activity: phase response curve. J. Exp. Biol. **65**, 737–751 (1976).

Taylor, O.R.: The past and pssible future spread of Africanized honeybees in the Americas. Bee World **58**, 19–30 (1977).

Thoenes, D.: "Spatial oscillations" in the Zhabotinskii reaction. Nature (Phys. Sci.) **243**, 18–20 (1973).

Thomas, D.: Artificial enzyme membranes, transport, memory, and oscillatory phenomena. In: D. Thomas and J.-P. Kernevez (eds.) Analysis and Control of Immobilized Enzyme Systems. Berlin Heidelberg New York: Springer 1975, pp. 115–150.

Thompson, D'Arcy, W.: On Growth and Form. Cambridge: Cambridge University Press 1917.

Thorogood, P.: Morphogenesis of cartilage. In: B.K. Hall (ed.) Cartilage vol. 2: Development, Differentiation and Growth. New York: Academic Press 1983, pp. 223–254.

Tickle, C., Lee, J., Eichele, G.: A quantitative analysis of the effect of all-trans-retinoic acid on the pattern of chick wing development. Devl. Biol. **109**, 82–95 (1985).

Titchmarsh, E.C.: Eigenfunctions Expansions Associated with Second-Order Differential Equations. Oxford: Clarendon Press 1964.

Toma, B., Andral, L.: Epidemiology of fox rabies. Adv. Vir. Res. **21**, 15 (1977).

Tranquillo, R.T., Lauffenburger, D.A.: Consequences of chemosensory phenomena for leukocyte chemotactic orientation. Cell Biophys. **8**, 1–46 (1986).

Tranquillo, R.T., Lauffenburger, D.A.: Stochastic model of leukocyte chemosensory movement. J. Math. Biol. **25**, 229–262 (1987).

Tranquillo, R.T., Lauffenburger, D.A.: Analysis of leukocyte chemosensory movement. Adv. Biosci. **66**, 29–38 (1988).

Trinkaus, J.P.: Formation of protrusions of the cell surface during tissue cell movement. In: R.D. Hynes, C.E. Fox (eds.) Tumor Cell Surfaces and Malignancy. New York: Alan R. Liss 1980, pp. 887–906.

Trinkaus, J.P.: Cells into Organs. Englewood Cliffs: Prentice-Hall 1984.

Tsujikawa, T., Nagai, T., Mimura, M., Kobayashi, R., Ikeda, H.: C_0-semigroup approach to stability of travelling pulse solutions of the Fitzhugh-Nagumo equations. Japan J. Appl. Math. (1989) (in press).

Turing, A.M.: The chemical basis of morphogenesis. Phil. Trans. Roy. Soc. Lond. **B237**, 37–72 (1952).

Twitty, V.C.: Of Scientists and Salamanders. London: Freeman 1966.

Tyson, J.J.: The Belousov-Zhabotinskii Reaction. Lect. Notes in Biomath. **10**. Berlin Heidelberg New York: Springer 1976.

Tyson, J.J.: Analytical representation of oscillations, excitability and travelling waves in a realistic model of the Belousov-Zhabotinskii reaction. J. Chem. Phys. **66**, 905–915 (1977).

Tyson, J.J.: Periodic enzyme synthesis: reconsideration of the theory of oscillatory repression. J. theor. Biol. **80**, 27–38 (1979).

Tyson, J.J.: Scaling and reducing the Field-Körös-Noyes mechanism of the Belousov-Zhabotinskii reaction. J. Phys. Chem. **86**, 3006–3012 (1982).

Tyson, J.J.: Periodic enzyme synthesis and oscillatory suppression: why is the period of oscillation cloase to the cell cycle time? J. theor. Biol. **103**, 313–328 (1983).

Tyson, J.J.: A quantitative account of oscillations, bistability, and travelling waves in the Belousov-Zhabotinskii reaction. In: R.J. Field, M. Burger (eds.) Oscillations and Travelling Waves in Chemical Systems. New York: John Wiley 1985, pp. 92–144.

Tyson, J.J., Light, J.C.: Properties of two-component bimolecular and trimolecular chemical reaction systems. J. Chem. Phys. **59**, 4164–4172 (1973).

Tyson, J.J., Othmer, H.G.: The dynamics of feedback control circuits in biochemical pathways. Prog. Theor. Biol. **5**, 1–62 (1978).

Tyson, J.J., Fife, P.C.: Target patterns in a realistic model of the Belousov-Zhabotinskii reaction. J. Chem. Phys. **75**, 2224–2237 (1980).

Tyson, J.J., Keener, J.P.: Singular perturbation theory of travelling waves in excitable media (a review). Physica D **32**, 327–361 (1988).

Tyson, J.J., Alexander, K.A., Manoranjan, V.S., Murray, J.D.: Cyclic-AMP waves during aggregation of *Dictyostelium* amoebae. Physica D **34**, 193–207 (1989).

Tyson, J.J., Murray, J.D.: Cyclic-AMP waves during aggregation of *Dictyostelium* amoebae. Development **106**, 421–426 (1989).

Uppal, A., Ray, W.H., Poore, A.B.: The classification of the dynamic behaviour of continuous stirred tank reactors – influence of reactor residence time. Chem. Eng. Sci. **31**, 205–214 (1976).

Varek, O., Pospisil, P., Marek, M.: Transient and staionary spatial profiles in the Belousov-Zhabotinskii reaction. Scientific Papers, Prague Inst. Chem. Technol. **K14**, 179–202 (1979).
Verhulst, P.F.: Notice sur la loi que la population suit dans son accroissement. Corr. Math. et Phys. **10**, 113–121 (1838).
Vincent, J.L., Skowronski, J.M. (eds.): Renewable Resource Management. Lect. Notes in Biomath. **40**. Berlin Heidelberg New York: Springer 1981.
Volterra, V.: Variazionie fluttuazioni del numero d'individui in specie animali conviventi. Mem. Acad. Lincei. **2**, 31–113 (1926). (Variations and fluctuations of a number of individuals in animal species living together. Translation In: R.N. Chapman: Animal Ecology. New York: McGraw Hill 1931, pp. 409–448.)

Waddington, C.H., Cowe, J.: Computer simulations of a molluscan pigmentation pattern. J. theor. Biol. **25**, 219–225 (1969).
Walbot, V., Holder, N.: Developmental Biology. New York: Random House 1987.
Walsh, J.A., Warren, K.S.: Disease control in developing countries. New Eng. J. Med. **301**, 967–974 (1979).
Waltman, P.: Deterministic Threshold Models in the Theory of Epidemics. Lect. Notes in Biomath. **1**. Berlin Heidelberg New York: Springer 1974.
Waltman, P.: Competition Models in Population Biology. CMBS Lectures Vol. **45**. Philadelphia: SIAM Publications 1984.
Watt, F.M.: The extracellular matrix and cell shape. Trends in Biochem. Sci. **11**, 482–485 (1986).
Welsh, B.J., Gomatam, J., Burgess, A.E.: Three-dimensional chemical waves in the Belousov-Zhabotinskii reaction. Nature **304**, 611–614 (1983).
Wessells, N.: Tissue Interaction in Development. Menlo Park: W. A. Benjamin 1977.
Westergaard, J.M.: Measures applied in Denmark to control the rabies epizootic in 1977–1980. Comp. Immunol. Microbiol. Infect. Dis. **5**, 383–387 (1982).
Whitham, G.B.: Nonlinear Waves. New York: Academic Press 1974.
Whittaker, R.H.: Communities and Ecosystems, 2nd edn. New York: Macmillan 1975.
Wickwire, K.H.: Mathematical models for the control of pests and infectious diseases: a survey. Theor. Pop. Biol. **11**, 182–283 (1977).
Willems, J.L.: Stability Theory of Dynamical Systems. New York: Wiley 1970.
Williamson, M.: The Analysis of Biological Populations. London: Edward Arnold 1972.
Williamson, P.G.: Palaeontological documentation of speciation in Cenozoic molluscs from Turkana Basan. Nature **293**, 437–443 (1981).
Williamson, P.: Morphological stasis and developmental constraints: real problems for neo-Darwinism. Nature **294**, 214–215 (1981)
Winfree, A.T.: An integrated view of the resetting of a circadian clock. J. theor. Biol. **28**, 327–374 (1970).
Winfree, A.T.: Spiral waves of chemical activity. Science **175**, 634–636 (1972).
Winfree, A.T.: Rotating chemical reactions. Sci. Amer. **230**(6), 82–95 (1974).
Winfree, A.T.: Resetting biological clocks. Physics Today **28**, 34–39 (1975).
Winfree, A.T.: The Geometry of Biological Time. Berlin Heidelberg New York: Springer 1980.
Winfree, A.T.: The rotor in reaction-diffusion problems and in sudden cardiac death. In: M. Cosnard and J. Demongeot (eds.) Luminy Symposium on Oscillations. Lect. Notes in Biomath. **49**, 1983. Berlin Heidelberg New York: Springer 1981, pp. 201–207.
Winfree, A.T.: Sudden cardiac death: a problem in topology. Sci. Amer. **248**(5), 144–161 (1983).

Winfree, A.T.: The prehistory of the Belousov-Zhabotinskii oscillator. J. Chem. Educ. **61**, 661–663 (1984).

Winfree, A.T., Strogatz, S.H.: Organising centres for three-dimensional chemical waves. Nature **311**, 611–615 (1984).

Wolfram, S.: Cellular automata as models of complexity. Nature **311**, 419–424 (1984).

Wolpert, L.: Positional information and the spatial pattern of cellular differentiation. J. theor. Biol. **25**, 1–47 (1969).

Wolpert, L.: Positional information and pattern formation. Curr. Top. Dev. Biol. **6**, 183–224 (1971).

Wolpert, L.: The development of pattern and form in animals. Carolina Biol. Readers, No. **51** J.J. Head (ed.). Burlington, North Carolina: Scientific Publications Div., Carolina Supply Co. 1977, pp. 1–16.

Wolpert, L.: Positional information and pattern formation. Phil. Trans. Roy. Soc. (Lond.) **B295**, 441–450 (1981).

Wolpert, L., Stein, W.D.: Molecular aspects of early development. In: G.M. Malacinski, S.V. Bryant (eds.) Proc. Symp. on Molecular Aspects of Early Development (Annual Meeting Amer. Soc. Zoologists, 1982 Louisville). New York: Macmillan 1984, pp. 2–21.

Wolpert, L., Hornbruch, A.: Positional signalling and the development of the humerus in the chick limb bud. Development **100**, 333–338 (1987).

Wyatt, T.: The biology of *Oikopleara dioica* and *Fritillaria borealis* in the Southern Bight. Mar. Biol. **22**, 137–158 (1973).

Xu, Y., Vest, C.M., Murray, J.D.: Holographic interferometry used to demonstrate a theory of pattern formation in animal coats. Appl. Optics. **22**, 3479–3483 (1983).

Yagil, G., Yagil, E.: On the relation between effector concentration and the rate of induced enzyme synthesis. Biophys. J. **11**, 11–27 (1971).

Yoshikawa, A., Yamaguti, M.: On some further properties of solutions to a certain semi-linear system of partial differential equations. Publ. RIMS, Kyoto Univ. **9**, 577–595 (1974).

Young, D.A.: A local activator-inhibitor model of vertebrate skin patterns. Math. Biosciences **72**, 51–58 (1984).

Zaikin, A.N., Zhabotinskii, A.M.: Concentration wave propagation in two-dimensional liquid-phase self-organising system. Nature **225**, 535–537 (1970).

Zeeman, E.C.: Catastrophe Theory. Selected Papers 1972–77. Reading: Addison-Wesley 1977.

Zhabotinskii, A.M., Zaikin, A.N.: Autowave processes in a distributed chemical system. J. theor. Biol. **40**, 45–61 (1973).

Zhabotinskii, A.M.: Periodic processes of the oxidation of malonic acid in solution (Study of the kinetics of Belousov's reaction). Biofizika **9**, 306–311 (1964).

Zykov, V.S.: Modelling of Wave Processes in Excitable Media. Manchester: Manchester University Press 1988.

Index

Acetabularia 435, 468, 470
  hair patterning 458, 586
  whorl hairs 471
  whorl regeneration 469
  whorl regeneration mechanism 470
Activation 122
  waves 305
Activator 109, 119, 128, 375
Activator-inhibitor
  kernels 491
  kinetics 128, 376
  mechanism 128, 138, 177
  neural model 481, 490
  parameter space for periodic solutions
    177
  reaction diffusion system 379, 430, 431
  robustness 406
Actomyosin 566, 568
Age dependent model
  epidemic 640
  population 29
  similarity solution 31, 33
  threshold 32
Agladze, K.I. 313, 348, 349, 723
AIDS (acquired autoimmunodefficiency
  syndrome) 610, 618, 619, 624
  epidemic model 626
Aikman, D. 239, 723
Alberch, P. 563, 596, 597, 598, 599, 601, 602,
  605, 607, 608, 723, 737, 740
Alberts, B. 526, 528, 723
Alexander, K.A. 741
Algae 468
Allee effect 58, 88
Allessie, M.A. 344, 345, 723, 740
Allosteric
  effect 119
  enzyme 137
Alt, W. 241, 723
Ambystoma mexicanum (salamander) 602
  foreleg 604, 606, 607, 608

Ammerman, A.J. 281, 723
Amphibian eggs 302
  calcium waves 305
Ancyclostoma duodenale (hookworm) 630
an der Heiden, U. 738
Andersen, S.
Anderson, R.M. 610, 611, 625, 626, 629,
  630, 660, 668, 669, 723, 724
Anderson, S.C. 589, 733, 738
Andral, L. 670, 679, 680, 724, 725, 741
Angrist, S.E. 475, 725
Animal coat patterns 435
  computed patterns 441
  legs 439
  polymorphism 447
  size 438
  tail 439
  variation 447
Animal pole 305
Anteater (*Tamadua tetradactyl*) 446
Antzelevitch, C. 200, 212, 213, 214, 731
Aoki, K. 281, 724
Aperiodic solutions (discrete models) 46
Aphid (*Aphidicus zbeckistanicus*) 71
Apical ectodermal ridge 559
Aplysia 210
Arata, A.A. 733
Arcuri, P. 378, 405, 406, 407, 413, 414, 596,
  660, 665, 724, 731
Archer, C. 600, 724
Arimoto, S. 161, 736
Armadillo 589
Arnold, R. 277, 724
Aronson, D.G. 289, 291, 724, 733
Arrhenius temperature variation 472
Artificial joint rejection 587
Artois, M. 669, 724
Asaris lumbricoides 630
Atrial flutter 329
Aubert, M.F.A. 669, 724
Audibert, F. 726

Autocatalysis 376
Axial condensation (cells) 560
Axon 481

Babloyantz, A. 454, 471, 724
Bacconier, P. 19, 725
Bacilus pestis (plague) 665
Bacon, P.J. 660, 724
Bacterial
    colonies 346
    inflammation 241
Bailey, N.J.H. 611, 724
Baldwin, D. 468, 472, 479, 729
Baleen whale model 33, 53
    sustainable yield 53
Bankivia fasciata 517, 518, 519
Banks, H.T. 724
Bard, J.B.L. 436, 442, 724
Barkley, D. 198, 724
Barnes, R.D. 507, 724
Baron, A. 446
Basal lamina 566
Basic reproduction rate (epidemic) 652
Basin of attraction 395
Beck, M.T. 254, 724
Becker, P.K. 411, 724
Beddington, J.R. 24, 27, 95, 99, 100, 724
Begg, D. 580
Bellemans, A. 471, 724
Belousov, B.P. 142, 179, 725
Belousov reaction 179
    basic reaction 180
    bursting 198
    chaotic behaviour 198
    coupled systems 215
    Field-Noyes model 179
    hysteresis 198
    periodic-chaotic sequences 198
    relaxation model for limit cycle 192
    rhythm splitting 230
Belousov-Zhabotinskii reaction 179
    analytical approximation for period of
        oscillation 196
    kinematic waves 254, 287
    oscillations 142
    relaxation oscillation approximation
        190
    stationary spatial patterns 411
    travelling wavefront 322, 327
    wave speed 327
    weak diffusion 360
Benchetrit, G. 19, 725
Ben-Yu, G. 424, 725

Berding, C. 396, 397, 432, 630, 631, 632,
    633, 640, 725
Bergé, P. 50, 725
Bernoulli, D. 610, 640, 725
Best, E.N. 200, 210, 211, 212, 725
Bifurcating periodic solution 24
Bifurcation
    pitchfork 40, 44
    tangent 40, 45, 50
Bilharzia (schistomiasis) 88, 620
Billiau, R. 725
Binomial distribution 233
Biochemical
    reactions 109
    switch 456
Biological clock 141
    fruit fly 201
Biological oscillator 140, 148
    black holes 210
    breathing 140
    emergence of fruit flies 140
    emission of cAMP (*Dictyostelium* cells)
        141
    general results 148
    λ-ω system 160
    Lotka 124
    neural activity 140
    parameter domain determination 156,
        157
    two-species models 156
Biological pest control 106
Biological switch 129, 138, 140, 148
    general results 148
Biological time
    geometric theory 200
Biomass 86
Bistability 134
Black Death 624, 651
    spread 655, 656
Black holes (in oscillators) 200, 231
    cardiac oscillations 213
    real biological oscillator 210
    singularity 208, 212
Blancou, J. 724
Blythe, S.P. 729
Boegel, K. 669, 725, 733
Bombyx mori (silk moth) 241
Bonke, F.I.M. 345, 723, 740
Bonner, J.T. 725
Bonotto, S. 468, 725
Born, W. 605, 725
Bosch, Hieronymus 605
Brain hallucination patterns 494
Branching (or Y-) bifurcation 565, 600

Bray, D.   723
Bray, W.C.   141, 725
Brayley, E.W.   657, 725
Breathing   16, 140
    Cheyne-Stokes disease   15
    synchrony   200
Bridge, J.F.   475, 725
Britton, N.F.   274, 277, 335, 340, 725, 727
Brown, D.L.   660, 676, 736
Brusselator (reaction diffusion)   175
Bubonic plague   685, 656
Buchanan, J. T.   261, 267, 725
Budworm (spruce)
    critical domain sizes   418, 420
    maximum population   418
    model   4, 33
    outbreak   417
    outbreak spread   301
    refuge   417
    spatial pattern generation   414
    travelling waves   298
Bullock, T.H.   738
Bunow, B.   435, 725
Bureš, J.   329, 344, 345, 739, 740
Burger, M.   141, 179, 198, 336, 728
Burgers' equation   363, 366, 367
Burgess, A.E.   313, 348, 742
Burnside, B.   737
Bursting (periodic)   155
    Belousov-Zhabotinskii reaction   198
Butterfly (and moth) wing patterns   448, 451
    buckeye (*Precis coenia*)   452, 464, 466, 468
    cautery experiments   458, 459, 468
    Cethosia   469
    Crenidomimas cocordiae   469
    dependent patterns   461, 462
    Dichorragia nesimachus   451
    eyespots (see also ocelli)   450, 464
    Hamanumida daedalus   469
    Iterus zalmoxis   463
    Lymantria dispar   452
    Mycalesis maura   465
    ocelli   450, 464
    Papilionidae   461
    Precis coenia: see buckeye   452
    Psodos coracina   459
    Stichophthalma camadeva   451
    Taeneris domitilla   465
    temperature effects   466
    Troides haliphron   462
    Troides hypolitus   462
    Troides prattorum   463
    wing venation   453
Butterworth, A.E.   640, 725

Cahn, J.W.   251, 726
Calcium   469
    conservation equation (cytogel model)
        570
    effect on *Acetabularia* hair spacing   479
    stimulated calcium release mechanism
        302, 570
    threshold kinetics   569
    waves on amphibian eggs   305, 306, 573
California king snake   373
Campbell, J.   505, 517, 518, 728
Camus, A.   655
Canosa, J.   282, 284, 726
Capasso, V.   610, 726
Cardiac
    arrhythmias   200, 213
    black holes   214
    death   200, 213
    failure   213
    oscillator   213
    pacemaker cells (periodic beating)   212
Carelli, C.   175, 726
Carnivores   436
Carpenter, G.A.   329, 726
Carslaw, H.S.   370, 726
Cartilage
    abnormal patterns   603, 605
    condensations   558
    formation   411
    morphogenetic rules   563
Cartwright, M.   167, 170, 726
Catastrophe
    cusp   7
    Nile perch (Lake Victoria)   88
Cats (*felidae*)   440
    coat patterns   440
Cavalli-Sforza, L.L.   281, 723
Cell
    aggregations   559, 560
    -cell contact inhibition   535
    chemotactic flux   532
    chemotaxis model   600
    chondrocyte   559
    conservation equation   530
    convective flux   530
    dermal   526
    differentiation   374
    embryonic   526
    energy approach to diffusion   249
    energy density concept   250
    epidermal   526, 566

epthelial   507
excitatory   498
fibroblast   526
galvanotaxis flux   532
guidance cues   532
haptotactic flux   532
inhibitory   498
-matrix field equations   537
matrix mechanical interaction equation
   533
membrane   579
mesenchymal   526, 529
neuronal   481
pattern bifurcations   560
pigment   436, 461, 507
potential   249
proliferation rate   530
random dispersal   530
retinal   489
secretary   507
traction force   527, 529, 532, 535
transport   530
Cellular automata   436, 505, 506
Central ganglion   507
Central pattern generator   254, 258, 261
Central symmetry patterns   452, 453, 457,
   458, 459
   experiments   457, 458
   generating mechanism   454
   scale and geometry effects   457, 460
Centre National d'Études sur la Rage   660
Cerebral cortex   344
Cethosia   469
Chalcides ocellatus (lizard)   588
Chance, B.   726
Chaos   22, 46, 50, 155
   Belousov-Zhabotinsky (BZ) reaction
      198
   spatial   355
Chaotic solutions   41, 45
Characteristic polynomial   317, 433, 702
   neural activity (shell) model   511
   reaction diffusion system   383
Charlesworth, B.   29, 726
Chedid, L.   726
Cheer, A.   302, 306, 307, 309, 726
Cheetah (Acinonyx jubatis)   440, 441
Chemical prepattern   374, 525
   animal coat markings   438
   comparison with mechanochemical
      pattern generation   525
Chemoreceptors   16
Chemotaxis   232, 241
   cell   528

bacteria-nutrient system   357
   flux   242
   index   243
   induced movement (cells)   242
   log law   243
   parameter   434, 533
   reaction-diffusion system   243
   receptor law   243
Cherrill, F.R.   490, 726
Cheyne-Stokes respiration
   delay model   15
   periodic oscillations   19
Chiang, H.C.   240, 737
Chick limb chondrogenesis   559
Chlamydia (venereal disease)   619
Cholera   610
Chondrocyte   559
Chondrogenesis   559
   mechanical model   559
Chondrogenic focus   600
Choodnovsky, D.V.   367, 726
Choodnovsky, G.V.   367, 726
Christopherson, D.G.   393, 726
Cicadae   100
   life cycle   101
Cinelli, F.   725
Citarius picus   506
Clark, C.W.   24, 27, 28, 54, 726
Clerk Maxwell, James   244
Clostera cocina (black mountain moth)
   459, 460
Cobwebbing (discrete models)   38
Cocho, G.   436, 726
Cohen, A.H.   258, 259, 260, 261, 267, 273,
   725, 726, 738
Cohen, D.S.   249, 251, 321, 350, 353, 489,
   726
Cohen, Y.   54, 726
Colchicine   602, 608
Colour pattern   436
Community matrix (population)   66
Competition
   models   78
   ocular (visual)   490
   population   63
Confined set   76
Conservation equation (cell)   530, 532
Conservative system (population)   65
Contact guidance   528
Contractile mechanism (actomyosin)   566
Contraction waves   306
Control sytem (biological)   144
Conus episcopus   519
Conus marmareua   506

Conus textus 506
Convection 287
  cell 528, 530
  nonlinear 292
  population diffusion models 419
  predator-prey models 318, 419
Cooperativity (reaction kinetics) 118, 146
Cordella, L. 210, 740
Cortex 489
Cortical
  depression waves 329, 345
  magnification paramater 496
Coupled oscillators 200
  Belousov-Zhabotinsky reactions 227
  limit cycle coordinates 218
  model system 215
  phase locking 228
  singular perturbation analysis 217, 220
  synchronization 228
  two-time singular pertubation analysis
    223
Courant, R. 426, 726
Cowan, J.D. 495, 496, 497, 498, 503, 504,
  505, 727
Cowe, J. 505, 742
Crank, J. 235, 727
Crenidomimas cocordiae 469
Criss-cross
  disease 619
  SI (epidemic) model 621
Critical domain size 415
  analytical detemination 418, 419
Cross diffusion 321
CSCR (calcium-stimulated calcium release)
  569
Cummins, H. 490, 727
Curran, J.W. 626, 738
Cvitanović, P. 50, 727
Cyclic-AMP 314, 434, 480
Cyrtodactylus fedschenkoi (lizard) 589
Cytogel 566
  contractility 567
  force balance equation 569
  traction 568
Cytogel model
  equations 571
  linear analysis 571
  piece-wise linear caricature 573
  travelling waves 572
Cytoplasm 469, 480

Dagg, A.I. 444, 727
Darwin, C. 595, 727
Dasypus novemcinctus (armadillo) 589

Davidson, D. 554, 555, 557, 558, 727
DeBach, P. 107, 727
Decroly, O. 155, 727
Dee, G. 411, 727
Defoe, Daniel 657
Delay models
  physiological diseases 15
  testosterone control 167
Delay population model 8
  critical delay 13
  linear analysis 12
  periodic solution 10, 12, 14
Dellwo, D. 727
Demongeot, J. 19, 727
Dendrite 481
Dependent patterns 461, 462, 463
Dermal papillae 554
Descarte's rule of signs 702
Determination stream 453, 458, 459
  hypothesis 450
Developmental
  bifurcation programme 593
  biology 372
  laws 609
Development
  limb 411
  pattern and form 374
  sequence 406
Developmental constraint 564, 599, 605,
  606, 609
Dhouailly, D. 554, 587, 599, 727
Dictyostelium discoideum 242, 274, 314
  cell division 141
  kinetics models 242
  periodic emission, of cAMP 141, 155
  spatial patterns 434
  spiral patterns 346
  wave phenomena 141
Diekmann, O. 29, 735
Dietz, K. 620, 640, 727
Diffusing morphogen gene-activation system
  452
Diffusion 232
  cell potential 249
  density dependent diffusion model for
    insect dispersal 253
  energy approach 249
  Fickian 234, 244
  flux 234
  local 244
  long range 244
  short range 244
Diffusion coefficient 234
  anisotropic 468

biochemicals 275
chemicals 327
critical ratio 384
FHN (Fitzhugh-Nagumo) system 329
haemoglobin 234
insect dispersal 275
long range 521, 531
oxygen (in blood) 234
weak 360
Diffusion damping 428
Diffusion driven instability 375
   analogy 375
   boundary conditions 380, 385
   general conditions 380, 385
   initial conditions 386
Diffusion equation
   density dependent diffusion 238
   random walk derivation 232
   scalar 235, 275
Diffusion field 452, 467
Diffusivity: see diffusion coefficient
Digital arch 601
Dilation (matrix) 534
Discrete delay (population) models 51
   characteristic equation 52, 54
   crash-back 53
   extinction 53
Discrete population models
   aperiodic solutions 46
   chaotic solutions 45
   cobweb solution procedure 38
   critical bifurcation parameter 40
   densit dependent predator-prey 99
   eigenvalue 39
   extinction 58, 59
   graphical solution procedure 38
   harvesting 62
   logistic, 41
   m-periodic solutions 49
   maximum population 57
   minimum population 57
   odd periodic solutions 47
   orbit 49
   oscillations 39
   parasite epidemic 108
   period doubling 47
   periodic solutions 43, 47
   period-13 solution 105
   pitchfork bifurcation 40, 44
   predator-prey 96
   single species 36
   stability 47
   stability analysis 48
   synchronized emergence (locusts) 105

tangent bifurcation 40, 45, 50
trajectory 49
Dispersion relation 280, 293, 384, 397, 408, 501
   complex (mechanical models) 542
   cytogel model 571
   fast focusing 547
   infinite range of unstable wave numbers 487, 550
   long range diffusion 246
   mechanical (cell-matrix) model 538, 540
   mode selection 408
   neural activity (shell) model 512
   sol-gel 'reaction-diffusion'
      mechanochemical model 584
   spiral wave 351
   'vanilla' 408, 409
DNA (deoxyribonucleic acid) 143, 374
Dogfish 258
Domain of attraction 82
Dormer, P. 734
Driver, R.D. 15, 727
Drosophila melanogaster (fruit fly) 200, 435
Drotman, D.P. 626, 738
Drug
   hallucinogenic 495
   response 646, 647
   use epidemic model 646
Duffy, M.R. 340, 350, 353, 727
Dunbar, S.R. 315, 316, 318, 727
Dylan, Bob 100
Dynamic diseases (physiological) 15

ECM (extracellular matrix) 526
   adhesive sites 532
   displacement 530
Ecological
   Allee effect 58, 59
   caveats (modelling) 57
   control 419, 424
   extinction 53, 58, 59
   predation pit 59
   sterile insect control 61
Edelstein, B.B. 454, 727
Edelstein-Keshet, L. 1, 727
Edmunds, L.N. 141, 725
Eichele, G. 741
Eigenfunction
   axisymmetric 431
   linearly unstable 387
   1-dimensional 388
   plane tesselation 393, 551
   2-dimensional 389

reaction diffusion   387
Eigenvalue
  Acetabularia whorl regeneration problem
    474
  continuous spectrum   386
  discrete   382
  spatial reaction diffusion problem   382
Elastic
  body force   536
  parameters   536
  strain tensor   533
  stress tensor   534
Electric potential   261
Elephant   445
Elsdale, T.   490, 727
Elton, C.S.   67, 727
Embryology   372
Embryonic domain   386
Energy integral   426, 428, 432
  method for nonexistence of pattern   427
Enzyme   109
  basic reaction   109
  conservation   110
  hyaluronidase   564
  kinetics   109
  substrate reaction   109
  substrate complex   109
  uricase   376
Ephistia kuhniella: see moth
Epidemic   610, 612
  AIDS model   626
  age dependent threshold   642
  contact matrix   623
  contact rate   613, 622
  control strategy (rabies)   681
  critical population density   654
  critical transmission coefficient   654
  drug use   645
  drug use infectiousness   648
  equilibrium persistence   670
  fluctuations   664
  geographic spread   651
  immunity   630
  influenza   617
  modelling goals   639
  oscillatory behaviour   618, 628
  periodic outbreaks   655, 665
  plague   616, 617, 655
  rabies   659
  rabies model   660
  rabies outbreak frequency   677, 679
  rabies outbreak wave speed   678
  relative removal rate   613
  reproduction rate   613, 628

San Francisco plague   657
severity   614, 662
simple model for spatial spread   651
SIR model for spatial spread of rabies
  666
spatial spread among foxes   659
survival   645
threshold   613, 621, 640
threshold analysis   645
two-dimensional rabies wave front   689
two-dimensional rabies wave in a
  variable fox density   690
waves   618, 652, 665, 676
wave speed   653, 670
Epidermis
  mechanochemical model   566
Epithelium   452, 506
  model   566
Epizootic (rabies) wave   663
Ermentrout, G.B.   266, 338, 495, 498, 504,
  505, 506, 507, 508, 509, 510, 516, 517, 728
Exact solutions   286, 302
  excitable kinetics model   302
  cubic   702
Excitability
  definition   333
Excitable kinetics
  caricature   304
  Fitzhugh-Nagumo   328
  model   302, 303
Extracelluar matrix (see also ECM)   526
Evasion (predator-prey) model   318
Evolution   447
  backward   607
  morphological view   593
Evolutionary change
  morphogenetic view   596
Evolutionary homology   606
Eyespot patterns
  model mechanism   464

Feather germ
  formation   554
  hexagonal pattern   556, 557
Feedback
  control   122, 142, 176
  inhibition   125
Feedback mechanisms   28, 142, 144
  conditions for limit cycle solutions   146
  confined set   145
  frequency of oscillation   147
  inhibition   144
  negative   144, 176
  positive   144, 176

testosterone control   167
Feigenbaum, M.J.   47, 728
Feroe, J.A.   329, 340, 728
Feshbach, H.   244, 735
FHN: see Fitzhugh-Nagumo
Fibre alignment   535
Fibrillation (cardiac)   213, 329
    spiral waves   213, 344
Fictive swimming   258
Field, R.J.   142, 179, 180, 182, 198, 336, 411,
    724, 728
Fife, P.C.   274, 277, 299, 728, 741
Filopodia   526
Fingerprint   490
Firing
    threshold   308
    rate   498
Fisher, R.A.   237, 277, 728
Fisher equation   237, 277
    asymptotic solution   280
    axisymmetric   280
    exact solution   282, 283
    initial conditions   279
    minimum wave speed
    wave solution   278
    wave solution stability   284
Fishery management
    economic return   56
    model (discrete)   54
    optimisation problem   55
    stabilizing effect   57
    strategy   55
Fishing zone   432
Fitzhugh, R.   161, 163, 728
Fitzhugh-Nagumo model   161, 200, 210
    conditions for limit cycles   166
    equations   163
    piece-wise linear model   165, 178, 358
    space clamped model   178
FN (Field-Noyes) model for the BZ
    reaction   181
    confrined set   187
    comparison with experiment   189
    limit cycle solution   183
    linear stability analysis   183
    nondimensionalisation   182
    nonlocal stability   187
    periodic oscillations   185
Flom, J.O.   738
Focal condensation (cells)   565
Folkman. J.   527, 533, 728
Fox
    epizootic   660
    popoulation in England   691

Free, C.A.   724
Frenzen, C.L.   112, 728
Frerichs, R.R.   695, 728
Friedrichs, K.O.   706, 728
Frog (*Xenopus laevis*)   602, 604
Fruit fly (*Drosophila melanogaster*)   200,
    435
    biological clock   201
    emergence (eclosion)   201
Fulic atra (common coot)   589
Fujii, H.   728

Gaillard, J.   726
Gale, E.   602, 603, 723
Galvanotaxis   243, 528, 532
    flux   243
Ganapathisubrimanian, N.   135, 136, 137,
    728
Gaussian (normal) probability distribution
    233
Genes   525
Genet (*Genetta genetta*)   440, 441
Genetic mutation   606
Geographic spread of epidemics   651
Gerber, A.   589, 728
Ghosh, A.K.   726
Gibbs, R.J.   293, 328, 728
Gierer, A.   128, 376, 406, 728
Gilkey, J.C.   306, 307, 728
Gilpin, M.E.   67, 68, 728
Giraffe
    coat patterns   437, 443, 444
    embryo   444
    Giraffa camelopardalus   444
    Giraffa camelopardalus tippelskirchi
        437, 444
    Giraffa camelopardalus reticulata   444
    Giraffa camelopardalus rothschildi   444
Glass, L.   15, 16, 18, 20, 21, 728, 734
Glycolysis   140
Glyptadon (armadillo)   589
Goh, B.-S.   54, 728
Goldbeter, A.   141, 155, 242, 727, 729, 734
Goldschmidt, R.   452, 729
Goldstein, S.   292, 322, 729
Gomatam, J.   313, 348, 742
Gonorrhea   619
    control   622
    multi-group model   622
Goodwin, B.C.   143, 468, 469, 470, 472, 476,
    479, 480, 729
Goselerin (drug)   167
Grasshopper dispersal   239
Gray, P.   132, 133, 134, 729

Green, B.D.   730
Green, H.   490, 729
Greenberg, J.M.   351, 729
Gregg, C.T.   657, 729
Griffith, J.S.   144, 729
Grillner, S.   259, 726, 729
Guckenheimer, J.   266, 697, 729
Gumowski, I.   50, 729
Gurney, W.S.C.   1, 11, 729, 737
Gurtler, W.   681, 729
Guttman, R.   210, 211, 729

Hadeler, K.P.   620, 727
Haeckel, E.   373, 729
Haemoglobin   109, 118
Hagan, P.S.   351, 729
Hair
    Acetabularia whorl   468, 586
    colour   436
    follicle   436
    formation   558
    initiation in *Acetabularia*   469
    patterns   468
    spacing in *Acetabularia* whorl   471, 479
Hallucination patterns   494, 501, 504
    basic   495
    cortical images   497
    drug induced   494
    geometry   496
    polar form   496
Hamanumida daedulus   469
Hamburger, V.   605, 729
Hanusse, S.P.   143, 148, 187, 730
Haptotaxis   528, 532
Hardy, M.   727
Harris, A.K.   525, 528, 529, 729, 730, 736, 737
Harris, S.   695, 730
Harrison, L.G.   472, 730
Harris-Warrick, R. M.   259, 726
Harvesting strategy   27
Hasimoto, H.   321, 730
Hassard, B.D.   86, 706, 730
Hassell, M.P.   59, 95, 730
Hastings, S.P.   730
Hay, E.   528, 730
Heart muscle   329
    rotating waves   344
Heligmosoides polygurus   632
Helminth: see parasite
Hematopoiesis   19
Henke, K.   452, 458, 459, 730
Herán, I.   446, 730
Hess, B.   313, 344, 347, 726, 735

Heterogeneity integrals   426, 428, 432
Hethcote, H.W.   618, 619, 624, 730
Hewitt, G.   239, 723
Hiernaux, J.   454, 724
Hilbert, D.   426, 726
Hill
    coefficient   16, 144
    equation   122
    plot   122
Hill function   16
Hilliard, J.E.   251, 726
Hinchliffe, J.R.   537, 558, 564, 599, 601, 730
Hippocrates   610
Hippopotami   447
HIV (human immunodeficiency virus)   618, 619
    transmission dynamics   624
Hodgkin, A.L.   140, 161, 162, 210, 211, 212, 730
Hodgkin-Huxley
    Fitzhugh-Nagumo model   161, 210
    perturbed oscillations   210, 211
    piece-wise linear model   165
    space-clamped dynamics   161
    system excitability   163
    theory of nerve membranes   161
Holden, A.V.   50, 87, 730, 740
Holder, N.   526, 528, 561, 562, 599, 730, 742
Hollen, J.   740
Holling, C.S.   733
Holmes, P.J.   266, 697, 726, 729, 738
Holograph interferograms   449
Honey badger (*Mellivora capensis*)   445, 446
Hookworm (*Ancyclostoma duodenale*)   630
Hopf, E.   706, 730
Hopf, L.   244, 730
Hopf
    bifurcation theorem   143, 706
    limit cycles   143, 706
Hoppensteadt, F.C.   29, 50, 100, 105, 106, 286, 611, 618, 640, 645, 648, 730, 731
Hormone   166
Hornbruch, A.   560, 561, 562, 743
Hosono, Y.   297, 728, 731
Howard, L.N.   142, 255, 257, 335, 338, 350, 731, 732
Howe, A.H.   610, 731
Hsu, S.-B.   78, 731
Hubbell, S.P.   731
Hubel, D.H.   489, 490, 731
Huberman, B.A.   251, 731
Hudspeth, A.J.   580
Huffaker, A.F.   107, 731

Huffaker, C.B. 731
Humerus 560
Hunding, A. 146, 471, 731
Husain, M.A. 167, 170, 726
Hutchinson, G.E. 731
Huxley, A.F. 140, 161, 162, 210, 211, 212, 730
Hyaluronate 564
Hyaluronidase 564
Hydra attenuata 210
Hyman, J.M. 50, 730
Hypothalmus 167
Hysteresis 7, 153

Ikeda, H. 741
Ikeda, N. 329, 731
Infectious diseases
    control 611
    models 611
Inhibition 122
Inhibitor 109, 119, 128, 375
Insect
    aggregation 240
    break 423, 424
    control 106, 297
    dispersal model 238
    density dependent diffusion dispersal
        model 238, 253
    infestation break 302
    outbreak 4, 297
    pest control strategy 423
    population patterning 414
    population patterning with convection
        419
    population spread 297
    refuge 297
    synchronized emergence (locusts) 100
    swarm 241
Integument 436, 447
    mammalian embryo 438
    patterns: see animal coat patterns
Interacting populations
    characteristic polynomial (discrete
        model) 17
    community matrix 66
    competition 63, 238
    complexity and stability 68
    continuous models 63
    density dependent predator-prey
        (discrete) 99
    discrete growth models 95
    Lotka-Volterra 63
    lynx-snoeshoe hare 66
    mutualism 63

predator-prey 63
symbyosis 63
13-year (synchronized emergence)
    locusts model 101
International Whaling Commission (IWC)
    34, 53, 62
Iodate-arsenous acid reaction 134
Ion exchange 322
Isham, V. 629, 731
Isochronic lines 345
Isolas 130, 131, 134
Isotropy 535
Iterus zalmoxis 463
Ivanitsky, G.R. 731
IWC: see Internation Whaling Commission

Jacob, F. 143, 735
Jackson, H.C. 670, 723, 731
Jaeger, J.C. 370, 726
Jaffe, L.F. 728
Jaguar (*Panthera onca*) 440, 441
Jalife, J. 200, 212, 213, 214, 731
Jensen, B. 733
Johnson, A.M. 724
Johnson, D.R. 537, 558, 564, 599, 601, 730
Joly, G. 725
Jones, D.D. 733
Jordan, D.W. 316, 697, 701, 714, 731
Jury conditions 100, 702

Källén, A. 660, 665, 680, 731
Kaplan, C. 660, 731
Kareiva, P. 238, 275, 724, 731
Kashin, S. 259, 729
Kath, W.L. 303, 457, 462, 732
Kauffman, S.A. 435, 732
Kawasaki, K. 238, 740
Kazarinoff, N.D. 86, 706, 730
Keener, J.B. 329, 332, 334, 344, 346, 350,
    732, 744
Keller, E.F. 243, 335, 732
Keller, H.B. 727
Keller, J.B. 100, 105, 106, 329, 573, 731, 738
Keller, K.H. 241, 733
Kermack, W.O. 611, 612, 615, 616, 732
Kernel
    biharmonic contribution 248
    convolution 483, 498
    excitatory-inhibitory 248
    function 498, 499
    influence 482, 483
    local activation-long range inhibition
        483, 509
    moments 248, 487

symmetric exponential   485, 488, 498
Kernevez, J.-P.   725
Keymer, A.E.   631, 725, 740
Killer bees   424, 695
Kimani, G.   725
Kinematic waves   254
Kinetics
    delay   313
    excitable   302
    Gierer and Meinhardt   376
    marginal state   405
    multi-steady state   297
    Schnakenberg   376
    Thomas   376
Kingdon, J.   436, 443, 447, 732
Klingler, M.   505, 735
Klüver, H.   495, 732
Kobayashi, R.   741
Koga, S.   313, 351, 353, 354, 355, 732
Kolmogoroff equations   85
Kolmogoroff, A.   277, 279, 284, 732
Kopell, N.   255, 257, 258, 259, 267, 336, 338,
    350, 731, 732
Körös, E.   179, 180, 198, 728
Krebs, C.J.   11, 732, 736
Krinsky (Krinskii), V.I.   200, 313, 343, 344,
    348, 349, 350, 723, 731, 732, 735
Krocza, W.   725
Kruuk, H.   373, 437, 441
Kühn, A.   450, 452, 453, 457, 458, 459, 732
Kuramoto, K.   313, 351, 352, 354, 355, 732

Lajmanovich, A.   623, 732
Lake Victoria Nile perch catastrophe   88
λ-ω systems   337
    complex form   160
    oscillator   160
    polar form   351
    spiral waves   350, 353
    stability of wave train   340
    wavetrain solutions   337, 339
Lamm, P.K.   724
Lammelapodia   526
Lammers, W.J.E.P.   740
Lamprey   254, 259
Landau, L.   533, 534, 732
Lane, D.C.   302, 306, 307, 309, 573, 577, 578,
    579, 733
Langer, J.S.   411, 727
Langer, W.L.   655, 656, 658, 733
Laplacian operator
    general results   720
La Rage   733
Lara Ochoa, F.   251, 385, 501, 733

Larson, D.A.   280, 286, 733
Lateral inhibition: see long range inhibition
Latis niloticus (Nile perch)   88
Lauffenburger, D.A.   241, 243, 723, 733, 741
Lauwerier, H.A.   95, 733
Law of Mass Action   110
Lawton, J.H.   724, 730
Lee, J.   741
Lefever, R.   175, 738
Leigh, E.   68, 741
Lemke, L.   373
Lemming   11
Leopard (Panthera pardus)   373, 436, 437
    coat patterns   437
    pre-natal tail   440, 441
    tail patterns   440
Lepidoptera (see also butterfly, moth)
    generalized wing   453
    wing patterns   449
Leukemia   21
Leuker, D.C.   738
Le Vay, S.   731
Levin, S.A.   374, 385, 424, 733, 739
Levinton, J.   606, 733
Leviton, A.E.   589, 733
Lewis, E.R.   703, 704, 733
Lewis, J.   454, 723, 733
Lewis, S.   729
LH (luteinizing hormone)   167
LHRH (luteinizing hormone releasing
    hormone)   167
Li, T.-Y.   45, 47, 733
Lifshitz, E.   533, 534, 732
Light, J.C.   156, 741
Limantria dispar   458, 459
Limb bud   559, 562
Limit cycle   9
    analysis of BZ relaxation oscillation
        model   166
    conditions for FN model (BZ reaction)
        183
    coupled   215
    feedback control mechanisms   146
    Fitzhugh-Nagumo model   166
    λ-ω system   160
    local coordinate transformation   218
    period (delay model)   14
    phase locking   215, 228
    predator-prey model   77
    rhythm splitting   230
    simple example   203
    'simplest' kinetics   156
    stability condition   222
    tri-molecular reaction   156

variables 261
Lions, P.L. 418, 733
Lizard 588
    Chalcides ocellatus 588
    Cyrtodactylus fedschenkoi 589
    Scincus officinalis 588
    Terantola mauritanica 588
    Zonurus cordylus 588
Lloyd, H.G. 692, 733
Local (short range) activation 379, 483
Locusts 424
    Magicada 100
    predator-prey model 101
    13 year (synchronized emergence) 100, 105
Loesch, D.Z. 490, 733
Logistic
    discrete model 41, 60
    growth 2
Long range (lateral)
    diffusion 244, 521
    elastic parameters 535, 536
    inhibition 379, 483
    integral formulation 246
    interaction 481
    kernel function 246
Lorenz, E.N. 47, 87, 187, 733
Lorenz equations 87
Lotka, A.J. 64, 141, 733
Lotka reaction mechanism 124
Lotka-Volterra
    competition model 78
    multi-species model 69
    predator-prey model 63
    predator-prey system with dispersal 356, 434
    travelling wavefront 315
LSD 494
Lucila cuprina (sheep-blowfly) 10, 11
Ludwig, D. 4, 5, 417, 733
Luteinizing hormone (LH) 167
Luteinizing hormone releasing hormone (LHRH) 167
Luther, R.-L. 277, 734
Lynx-snowshoe hare interaction 66

MacArthur, R.H. 734
Macdonald, D.W. 659, 660, 664, 669, 679, 681, 682, 684, 691, 693, 734
MacDonald, N. 15, 146, 734
Mackey, M.C. 15, 16, 18, 20, 21, 22, 728, 734, 738
MacLulich, D.A. 734
Magicada (locust) 100

McEvedy, C. 655, 734
McKean, H.P. 280, 305, 329, 734
McKendrick, A.G. 611, 612, 615, 616, 732
McLeod, J.B. 299, 728
Maini, P.K. 525, 548, 553, 559, 564, 587, 600, 728, 734, 736, 737, 738
Malaria 620
Malthus, T.R. 2, 734
Mammalian coat patterns 436
Mammals
    East African 435
Mandelbrot, B.B. 95, 734
Mangel, M. 24, 738
Manoranjan, V.S. 280, 326, 558, 573, 577, 588, 589, 733, 734, 736, 741
Marek, M. 215, 216, 360, 368, 371, 734, 742
Martiel, J.-L. 141, 242, 734
Matano, H. 426, 734
Matkowsky, B.J. 727
Maximum
    economic yield (harvesting) 55
    growth rate 24
    sustainable yield 24, 55
Maxwell, J.C. 734
May, R.M. 1, 10, 11, 24, 27, 37, 50, 54, 84, 87, 95, 100, 106, 108, 610, 625, 629, 723, 724, 730, 734, 735
Mazo, R.M. 566, 735
Measles 618
Mechanical
    models 525
    shaping of form 526
Mechanical (cell-ECM)
    equation for cytogel contractility 567
    equilibrium equation 536
    field equations 537
Mechanical model
    bifurcation surface (for pattern) 543
    conceptual framework 538
    epidermis 566
    linear analysis 528
    matrix conservation equation 536
    simple models 542
    small strain approximation 552
    two-dimensional patterns 552
Mechanochemical model 398, 448, 564
    animal coat patterns 586
    cytogel sheet 583
    hair patterning in *Acetabularia* 586
    sol-gel-calcium cytogel sheet 580
Mechanotaxis 532
Medaka 274
    eggs 305
Medley, G.F. 724

Medvinskii, A.B.   732
Meinhardt, H.   128, 376, 406, 435, 458, 505, 728, 735
Melanin   436, 452
Melanoblast   436
Melanocyte   436, 443
Membrane potential   329
Menten, M.I.   109, 735
Mescaline   480
Mesenchymal cells   529
Metabolic control mechanism   144
Metabolic feedback   122
Metz, J.A.J.   29, 735
Michaelis, L.   109, 735
Michaelis constant   112, 117
Michaelis-Menten
    reaction   109
    theory   111
    uptake   117
Microcautery   458
Microvilli   579
    micrograph   580
Midlo, C.   727
Mikhailov, A.S.   250, 735
Milton, J.G.   15, 22, 734
Mimura, M.   350, 735, 744
Mira, C.   50, 95, 729
Misiurewicz, M.   733
Mitchell, A.R.   280, 326, 424, 725, 734
Mitchison, G.   454
Mitotic
    inhibitor   603, 606, 608
    rate   530
    time scale   537
Mittenthal, J.E.   566, 735
Mode
    fastest growing   385
    isolation   404
    polarity   406
    selection   404, 408
    unstable   388
Moegle, H.   725
Mollison, D.   279, 735
Mollusc: see shell
Monk, A.   141, 242, 735
Monkey   489
    macaque   489
Monocular
    deprivation   494
    vision   490
Monod, J.   143, 735
Monro, P.   446
Monsters   605
Mooney, J.D.   558, 587, 736

Moore, C.P.   738
Morphogen   374, 526
    calcium   435, 470
    map   374
    role in patterning   435
    switch mechanism   454, 456
Morphogenesis   372, 525, 593
    chemical theory   236, 374
    evolution and   592
    limb   558, 563
    mechanical models   525
    skin organ   554
Morphogenetic rules   563, 565, 599, 606
    basic bifurcations   600
    vertebrate limb   599, 600, 604
Morphoglogical divergence   606
Morse, P.   244, 735
Moscona, A.   527, 533, 728
Mosquito swarm   240
Moth (see also butterfly)
    antenna   373
    black mountain (*Clostera curtula*)   459, 460
    Chocolat chip (*Psodos coracina*)   459, 460
    Ephistia kuhniella   451, 452, 453, 458, 459
    Hyalophora cecropia   373
    simulated cautery experiments   459
    wing patterns   448
Motoneuron   259
Mouse (*Mus musculus*)   604
m-periodic solutions (discrete population models)   47
mRNA (messenger ribonucleic acid)   140, 143
Müller, S.C.   313, 343, 344, 347, 735
Murray, J.D.   15, 109, 112, 115, 119, 157, 169, 182, 183, 187, 190, 198, 217, 223, 233, 241, 249, 251, 252, 275, 281, 292, 293, 296, 302, 303, 321, 322, 326, 327, 334, 335, 340, 363, 370, 374, 376, 377, 378, 385, 398, 399, 400, 405, 406, 407, 410, 413, 414, 420, 423, 430, 431, 434, 436, 438, 441, 442, 443, 444, 445, 446, 448, 449, 450, 457, 458, 459, 460, 462, 468, 472, 479, 483, 487, 489, 501, 525, 529, 531, 548, 553, 554, 558, 559, 564, 566, 573, 577, 579, 587, 588, 589, 596, 599, 600, 645, 648, 660, 665, 666, 669, 671, 674, 675, 676, 677, 678, 679, 681, 683, 684, 685, 688, 689, 690, 691, 692, 693, 716, 724, 725, 727, 728, 729, 730, 731, 732, 733, 734, 735, 737, 738, 741, 742, 743
Mushroom (reaction kinetics)   130, 131, 134

Mus musculus (mouse) 604
Mutualism 63, 83
Mycalesis maura 465
Myers, J.H. 11, 732, 736
Myosin 566

Nagai, T. 741
Nagawa, H. 558, 566, 580, 588, 736
Nagorcka, B.N. 558, 587, 588, 589, 736
Nagumo, J.S. 161, 163, 736
Nagumo equation 305
Nakanishi, Y. 558, 566, 580, 588, 736
Namias, V. 705, 736
Natural selection 593
Nector americanus 630
Negative feedback loop (biological control)
    144
Neo-Darwinism 594
Nerrita turrita 517, 518
Nerve action potential 140. 305
Nerve cells 481
Nervous system 481
Neu, J. 215, 266, 360, 726, 736
Neural
    activity 495
    activity model for shell patterns 507
    activity oscillation 140
    Hodgkin-Huxley theory 161
    instability 498
    signalling 166
    stimulation 507
Neural model
    dispersion realtion 500, 501
    Hodgkin-Huxley (nerve membrane) 161
    pattern foramtion 481, 498
    shell pattern 507, 508
    stability analysis 499
Neuron 161, 258, 481, 489
    autonomous firing rate 481, 482
    mantle (shell/mollusc) 507
    periodic firing 164
Neuronal process 481
Neurotransmitter 261
Newell, P.C. 313, 314, 346, 736
Newman, W.I. 289, 736
Nicholson, A.J. 10, 11, 59, 737
Nicholson, M. 67, 737
Niewold, F.J.J. 733
Nijhout, H.F. 449, 450, 451, 452, 453, 737
Nile perch (*Lates niloticus*) 88
Nisbet, R.M. 1, 11, 729, 737
Nishiura, Y. 728
Noble, J.V. 658, 737
Nold, A. 624, 730

Nondimensionalisation 5
Nonlinear maps 36
Nonlocal
    dispersion (cells) 531
    effects 244
    elastic interactions 535
Normal (Gaussian) probability distribution
    233
Notochord 259
Noyes, R.M. 179, 180, 182, 198, 728
Nuccitelli, R. 726
Null clines
    excitable kinetics 328
    Fitzhugh-Nagumo system 328
    Gierer-Meinhardt kinetics 378
    Schnakenberg kinetics 378
    Thomas kinetics 378
Nymphalids 449, 451, 464
    ocelli 450
    wing pattern groundplan 449, 451

Ocelli patterns 464, 465
    model mechanism 464
    temporal growth 466, 467
    transplant experiments 464
Ocular dominance stripes 481, 489
    activation/inhibition kernels 491
    activation/inhibition domains 491
    effect of domain growth 494
    generating mechanism 491
Odell, G.M. 302, 335, 566, 572, 579, 599,
    697, 732, 737
Odum, E.P. 67, 737
Okubo, A. 236, 238, 240, 241, 737
Onchocersiasis (river blindness) 641
Open loop system 526
Optic nerve 489
Oregonator (BZ) model (oscillating
    reaction) 181
Ortoleva, P.J. 328, 358, 737, 739
Oscillator
    annihilation 214
    biological 140, 148
    black holes 200, 215
    BZ (Belousov-Zhabotinsky) model 181
    chemical (BZ reaction) 179
    coupled 200, 215
    determination of parameter space 157
    detuning 270
    independent 255
    λ-ω system 160
    neural 261
    Oregonator 181
    perturbed 200, 204

relaxation   154
  simple example   203
  stability   272
  phase-coupled   266, 268
  two-species models   156
  weak coupling   265
Osmotic pressure   564, 580
Osteoderm   588, 589
Oster, George F.   37, 50, 302, 398, 483, 505,
  525, 529, 554, 558, 559, 564, 566, 579, 596,
  597, 598, 599, 600, 603, 726, 728, 734, 736,
  737, 738
Oster, Gerold F.   495, 738
Othmer, H.G.   123, 141, 145, 146, 147, 242,
  429, 471, 531, 735, 738, 741
Otto, H.   588, 738

Pacemaker   313
  chaotic   313
  periodic (heart)   200
Paedomorphic form (limb)   606
Papilionidae (butterflies)
  wing patterns 462, 463
Papilla   554, 556, 598
Parameter space   399
  linear stability   74, 75
  parameteric method   157, 399
  two-species oscillations   156, 157
Parasite (helminth)
  acquired immunity   630
  Ancyclostoma duodenale (hookworm)
    630
  Asaris lumbricoides   630
  experiments   632
  Heligmosoides polygurus   632
  immune threshold   633
  immunological response   634, 640
  infection model   630
  Nector americanus   630
  population dynamics   630
  Schistosoma mansoni   640
  survival   633
  Trichurus trichiura (whipworm)   630
  Trichostrongylus retortaeformis   252
Parasite population dynamics model
  goals   639
  high protein diet   635
  low protein diet   635
Parfilov, A.V.   732
Pate, E.   471, 738
Patou, M.   603, 738
Pattern
  animal coat   435
  animal leg   439

bifurcation   441, 447
  butterfly wing   448
  cartilage (limb)   559
  chondrogenic   561
  complex   588
  computed   441
  critical domain size   415
  dependent   461
  dynamics in growing domains   411, 413
  developmental biology   372, 391
  doubly-periodic tesselation   496, 501
  ecological   414
  energy function   396
  finite amplitude   435
  formation   372
  generation in single species models   414
  hallucinogenic   494, 495, 504
  heterogeneity function   396, 426, 428
  hexagonal pattern of feather primordia
    557
  initiation   387
  initiation trigger   408
  leopard spot size   438
  lepidopteran wing   450
  microvilli   579
  neural firing   481
  nonexistence in reaction diffusion
    systems   424
  ocular dominance stripe   490, 494
  periodic feather germ   554
  periodic actin fibre   580
  polymorphism   447
  propagation   410
  retinal   496
  robustness   406
  scale and geometry effects   397, 443,
    445, 460
  shell   505
  size   398
  stripe preference   494
  superposition   587
  tail   439, 440, 441
  tapering cylinder   439
  tissue interaction   587
  travelling wave initiation   409
  variation   447
  visual cortex   496
  visual hallucination   494
  wave 'front'
Pattern formation mechanism
  animal coat markings   447
  cell(fibroblast)-matrix   537
  cytogel   566
  dependent (butterfly wing)   461

epidermal-dermal tissue interaction 587
    feather germ primordia 554
    initial conditions 406
    mechanical models 525
    microvilli 579
    mode selection 404
    neural 481, 498
    neural (shell) model 507
    ocular dominance stripe 492
    robustness 406
    sensitivity 406
    shell (mollusc) 507
    whorl (*Acetabularia*) 470
    wing pattern 454
Paveri-Fontana, S.L. 610, 726
Peitgen, H.-O. 46, 47, 95, 738
Penrose, R. 490, 553, 738
Pepys, Samuel 657
Perelson, A.S. 540, 555, 601, 737, 738
Pérez-Pascual, R. 726
Perichondrium 600
Period doubling 47, 155
Periodic
    bursting 155
    cell division 141, 143
    changes in enzyme synthesis 143
    chaotic sequences 198
    emergence (eclosion) of fruit flies 140
    emission of cyclic AMP (*Dictyostelium*)
        141
    neuron firing 163
    pacemaker 200
    pacemaker cells 212
    solutions of feedback control mechanisms
        146, 147
    testosterone (hormone) level 166
Perkel, D.H. 738
Peterman, T.A. 626, 738
Petrovsky, I. 277, 732
Phalange 560
Phase 200, 255
    critical 201
    indeterminate 209
    lag 260
    locked 261
    locking, 215, 228, 271
    resetting (in oscillators) 200, 202
    resetting curves 203
    shift 9, 201, 205
    shift equation 227
Phase plane
    analysis 697
    catalogue of singularities 698
    singulur points 697

Phase resetting in oscillators 200
    black hole (singularity) 208
    Type 1 curves 204
    Type 0 curves 206
Phenomenological pattern 606
Pheromone 241, 372
Phyletic gradualism 594
Physiological diseases 15
    Cheyne-Stokes respiration 15
    regulation of hematopoiesis 19
Pianigiani, G. 738
Pianka, E.R. 83, 738
PID (pelvic inflammatory disorders) 619
Pielou, E.C. 738
Pigment
    cells 436, 461, 507
    domain in wing (lepidopteran) patterns
        461
Pinsker, H.M. 738
Placode 554, 556, 598
Piscounoff, N. 277, 732
Pitchfork bifurcation 40, 44
Plankton-herbivore system 316
Plant, R.E. 24, 166, 738
Plague 616
    Bacilus pestes 656
    Black Death 655
    bubonic 618
    Great Plague (London) 657
    pneumonic 618
    residual foci 657
    San Francisco epidemic 657
    septicemic 656
    20th century 655
Plesser, T. 313, 344, 347, 735
Pomeau, Y. 50, 725
Poore, A.B. 742
Population
    birth rate 2
    crash-back 53
    carrying capacity 2
    competition models 78
    extinction 53
    harvesting 24
    hysterisis 7
    lemming 11
    logistic growth 2
    maximum growth rate 24
    predation 5
    predation threshold 5
    recovery, time 25
    self-regulation 37
    sigmoid growth 3
    synchronized emergence 100, 105

vole   11
world   2
Population interaction diffusion mechanism
   272
Population models
   age structured   29
   cautionary remarks   85
   conservation equation   1
   continuous interacting species   63
   continuous single species   1
   delay   8
   depensatory   102
   discrete (interacting species)   95
   discrete (single species)   36
   general   85
   harvesting   24
   insect outbreak   4
   interacting: see Interacting populations
   Kolmogoroff   85
   mutualism   83
   renewable resources   24
   symbiosis   83
   13 year locusts   100
Portmann, A.   436, 738
Porous media equation   239
Positional information   374, 392
Positive feedback loop (biological control)
   144
Pospisil, P.   742
Post-fertilization (egg) waves   306, 573
Prawda, J.   695, 728
Precis coenia (buckeye butterfly)   452, 464,
   468
Predator-prey
   blow-up   321
   convective model   310
   density dependent (discrete) model   99
   discrete growth model   96
   13 year (synchronized emergence)
      locusts model   101
   waves   315
   wolf-moose   321
   pursuit and evasion   318
   realistic models   70
Predator-prey model
   interacting populations   95
   limit cycle behaviour   72
   parameter domain of stability   72
Prepattern
   giraffe coat   443
   hair initiation (*Acetabularia*)   480
   morphogen   436
   theory   396
Prigogene, I.   175, 738

Primordia   528
   feather and scale   554
   hair   558
Principle of Competitive Exclusion   78
Proteus anquinus (salamander)   606, 607
Protozoa   373
Pseudo-steady state hypothesis   111, 117
Psodos coracina   459
Pteryla   554
Punctuated equilkibrium   594
Purcell, E.   534, 738
Pye, E.K.   726

Rabies (canine)   695
Rabies (fox)
   break   424, 681, 683
   break – analytical approximation   485
   break width   685, 695
   control strategies   660, 681, 695
   English 'experiment'   667
   epidemic   651
   epidemic fluctuations   664
   epidemic frequency   677
   epidemic wave   652, 662, 676
   epidemic wave speed   662, 678
   epizootic in England   692, 693
   estimate of model diffusion coefficient
      679
   fox density (England)   691
   furious   667
   outbreak in England   689
   outbreak predictions   689
   period of recurring epidemics   679, 694
   secondary outbreak   693
   simple model for spatial spread   659
   SIR model for spatial spread   666
   spread   424, 655, 656
   two-dimensional epizootic wave front
      680, 692, 693
   vaccination control   695
   wave speed dependence on carrying
      capacity   680
Radioloarian (*Trissocyclus spaeridium,
   Eucecryphalus genbouri*)   373
Radius (bone)   560
Raff, M.   723
Raggett, G.F.   617, 618, 658, 738
Rand, A.H.   258, 266, 726, 738
Rand, R.R.   726
Random walk (diffusion)   232
   biased   235, 252
Rapp, P.E.   146, 147, 738
Rate constants   110
Ratel: see honey badger

Rawles, M.   554, 587, 738
Ray, W.H.   742
Reaction
    Belousov   179
    bistability   134
    BZ (Belousov-Zhabotinsky)   179
    hydrogen peroxide – iodate ion   141
    hysteresis (steady steady)   130
    iodate-arsenous acid   134
    isolas   134
    kinetics   109
    Lotka   124
    matrix (stability)   125
    mushroom (steady state)   134
    rate   117
    rate constant   110
    rate limiting step   118
    uptake (velocity)   117, 122
Reaction diffusion equations   246
    computed patterns   441
    convection   419
    density-dependent diffusion   286
    exact solution   286, 302
    excitable kinetics   304
    λ-ω systems   337, 350
    limit cycle kinetics   336, 360
    linear stability analysis   380
    multi-species   311
    neiral activity (shell) analogue   520
    nonexistence of spatial patterns   424
    nonlinear convection   292
    oscillatory kinetics   336
    pattern robustenss   406
    scalar   275
    singular perturbation analysis   363
    weak diffusion   360
Reaction diffusion chemotaxis
    equation   242
    system   243
    mechanism   430, 434
Reaction diffusion mechanism
    analysis of pattern initiation   387
    chemotaxis   242
    convection   419
    practical applications   435
    Turing   375
Reaction kinetics   109
    activation   122
    activator-inhibitor   128
    autocatalysis   122, 123
    bistability   134
    Brusselator   175
    complex solution behaviour   153
    cooperative phenomena   118, 146

    fast dynamics   154
    first order   124
    gradient system   139, 176
    hysteresis (steady state)   130
    inhibition   122
    iodate-arsenous acid   134
    isolas   130, 134
    λ-ω model   160
    Lotka   124
    model autocatalysis   133
    multiple steady state   130
    mushrooms (steady state)   130, 134
    necessary and sufficient conditions for
        stability   148
    null clines – steady state local behaviour
        148
    periodic bursting   153
    rate limiting step   118
    slow dynamics   153
    'simplest' (limit cycle)   156
    singular perturbation analysis   112
    stability   127, 148
    threshold behaviour   129
    trimolecular (limit cycle)   156
Red spider mite   107
Refractory phase   336
Regeneration (*Acetabularia*)   472
    eigenvalue problem   474
    model mechanism   473
Regulation of hematopoiesis   19
    delay model   20
    oscillations   21
Reiss, E.L.   727
Rensing, L.   15, 738
Relaxation oscillator   154, 190, 198
    Belousov-Zhabotinsky (BZ)   180, 192
    model for Field-Noyes (FN) mechanism
        198
    period   190
    period of Field-Noyes (FN) model for
        BZ reaction   196
Retino-cortical magnification factor   496
Retinoic acid   562, 599
Reynolds, G.T.   728
Rhinoceri   447
Richter, P.H.   46, 47, 95, 738
Ridgeway, E.B.   728
Rius, J.L.   724
Ringland, J.   724
Rinzel, J.   140, 153, 155, 161, 166, 329, 333,
    335, 359, 573, 729, 738, 739
River blindness (onchocersiasis)   641
Roberts, D.V.   109, 739
Roberts, K.   726

Romer, A.S.   589, 739
Rooney, P.   600, 724
Rosales, R.R.   726
Ross, J.   737
Rossignol, S.   726
Rössler, O.E.   87, 739
Rotenberg, M.   24, 739
Roughgarden, J.   238, 740
Routh-Hurwitz conditions   702
Rubella (german measles)   619
Rubin, R.   637, 739
Rubinow, S.I.   109, 119, 739

Salamander: see *Ambystoma mexicanum*
Sarkovskii, A.N.   45, 50, 739
Sato, T.   731
Satsuma, J.   296, 739
Scale
    critical   391, 429
    effects   397, 440
    invariant mechanisms   471
    isolation of unstable modes   404
    parameter   386, 391
Scales   587
    epidermal   589
Schaffer, W.M.   68
Schistomiasis (bilharzia)   88, 620
Schistosoma mansoni   640
Schmidt, S.   328, 358, 737, 739
Schnakenberg, J.   156, 376, 406, 472, 739
Schnakenberg kinetics   376
    Turing space   401, 402
Schneider, L.G.   670, 731
Schopman, F.G.J.   345, 723
Schulman, J.H.   738
Schwanwitsch, B.N.   449, 451, 452, 463, 468, 739
Schwartz, V.   453, 739
Scincus officinalis (lizard)   588
Scott, S.K.   132, 133, 729
Searle, A.G.   436, 739
Segel, L.A.   5, 109, 112, 141, 242, 243, 274, 335, 374, 385, 732, 733, 739
Segmental condensation   565, 600
Segundo, J.P.   738
Self-organisation   380
Sengel, P.   554, 727, 739
Sex attractant   241
SHBG (sex hormone binding globulin)   167
Sheep-blowfly (*Lucila cuprina*)   10, 11
Sheldon, P.R.   595, 739
Shell (mollusc)
    Bankivia fasciata   517, 518, 519
    basic elements   506

basic structure   506
bifurcation to pattern   513
Conus episcopus   519
continuous time model   519
Citarius picus   506
Conus marmareua   506
Conus textus   506
Nerrita turrita   517, 518
neural activity model   505
pattern interaction with geometry   518
pattern polymorphism   505
stability analysis of neural model   510
Shibata, M.   329, 344, 345, 739, 740
Shigesada, N.   238, 239, 240, 740
Shihira-Ishikawa, I.   470
Shock solution   292, 296
Short range activation: see local activation
Showalter, K.   135, 136, 137, 277, 724, 728, 740
Shubin, N.   599, 601, 603, 607, 737, 740
Shymko, R.   732
Sibatani, A.   452, 740
Silk moth (*Bombyx mori*)   241
SIR (epidemic) models   611
    spatial spread of rabies among foxes   666
SIRM (sterile insect release method) pest control model   94
Skellam, J.G.   236, 740
Skin (organ) primordia   528, 590, 598
Skowronski, J.M.   24, 742
Slack, J.M.W.   372, 733, 740
Slater, A.F.G.   631, 725, 740
Sleeman, B.D.   424, 725
Slemrod, M.   112, 739
Smallpox   610
Smeets, J.L.   344, 740
Smith, A.M.   723
Smith, J.C.   561, 740
Smith, P.   316, 697, 701, 714, 731
Smith, W.R.   167, 168, 169, 740
Smoller, J.   274, 385, 740
Snell, J.   730
Sol-gel
    mechanochemical model   583
    simplified 'reaction diffusion' convection model   583
    stress tensor   582
    transition   580
Solitary pulse   330
    singular perturbation analysis   334
Sørensen, P.G.   471, 731
Soto, F.   726
Southwood, T.R.E.   59, 740

Sparrow, C. 47, 87, 740
Spatial pattern
    activation-inhibition neural model 481
    ecological 372, 414
    evolution 380
    formation 372
    neural (shell) model 510, 516
    stability 424
Speciation 594
Sperb, R.P. 420, 423, 426, 736, 740
Sperm entry point 305
Spinal
    cord 258
    transection 258
Spiral
    Archimedian 348
    logarithmic 348
    rotating 347
Spiral waves 313, 343, 346, 353
    Belousov-Zhabotinskii reaction 344
    brain tissue 344
    cardiac arrhythmias 345
    chaotic 354
    dispersion relation 351
    excitable systems 344
    heart muscle 345
    λ-ω systems 350
Spruce budworm: see budworm
Squid (giant) 161, 210
Stability
    necessary and sufficient conditions 74
    parameter domain 74, 75
Stanley, E.A. 660, 676, 736
STD (sexually transmitted disease) 619
Stech, H.W. 618, 730
Steck, F. 669, 670, 681, 725, 740
Stefan, P. 45, 740
Stefan problem 328
Stein, W.D. 561, 743
Stimulation function (shell) 508
Stimulation functional (shell) 509
Stimulus function (synapse) 491
Stimulus-timing-phase singularity 202
Stirzaker, D. 11, 740
Stopak, D. 528, 529, 730
Strain-gel 'reaction diffusion' model 584
Strain tensor 533
Stress-strain constitutive relation 534
Stress tensor 534
    cell (traction) 535
    elastic 534
    viscous 534
Stretch activation 570
Stripe formation mechanism 489

Strogatz, S.H. 313, 740, 743
Stuchl, I. 215, 216, 734
Substrate 109
    inhibition 126, 138, 376
Suffert, F. 449, 451, 452, 468, 740
Svobodová, E. 360, 368, 371, 734
Swimming pattern 260
Swindale, N.V. 490, 494, 740
Switch
    biological 148, 153
    hysteresis 153
    morphogen-gene mechanism 454
Symbiosis 63, 83
Synapse 481

Teddei-Feretti, C. 210, 740
Taeneris domitilla 465
Tangent bifurcation 40, 45, 50
Target patterns 312, 339
    Belousov-Zhabotinskii 314
Taylor, O.R. 424, 695, 741
Temperature shocks (wing patterns) 468
Teramoto, E. 238, 740
Terantola mauritanica (lizard) 588
Teratology 605
Terman, D. 329, 333, 335, 359
Tesselation patterns
    basic units 502, 551
    hexagon 394, 495, 503, 504, 551
    planar 393, 495
    polar form 395, 503, 551
    regular 393, 495, 551
    rhombus 395, 503, 505, 551
    roll/stripe 395, 502
    square 395, 503, 505, 551
Testosterone
    conditions for stability of model's steady
        state 170
    control model 166, 178
Tetrapod
    limb development 603
    vertebrate 599
Thoenes, D. 255, 741
Thomas, D. 15, 125, 376, 404, 405, 406, 407,
    438, 725, 741
Thomas, J. 490, 729
Thomas
    kinetics 151, 376
    mechanism 125, 438
Thompson, D'Arcy, W. 607, 741
Thorogood, P 558, 741
Threshold
    age structured population 32
    concentration 392

epidemic wave   654
Fitzhugh-Nagumo (piece-wise linear)
    model   165
functions   498, 499, 509
mantle (shell) activity   508
mortality rate (epidemic)   654
phenomena   88, 152
phenomenon (epidemic)   613
reaction kinetics   129, 152
switch mechanism   456
waves   335
Tickle, C.   562, 741
Tiger (*Felis tigris*)   437
coat patterns   437, 443
Tissue interaction   566, 587
mechanism   588
Titchmarsh, E.C.   286, 741
Toma, B.   670, 680, 741
Trabert, K.   732
Traction (cell) forces   535
Tranquillo, R.T.   241, 243, 525, 587, 736, 741
Travelling wave   274
Belousov-Zhabotinskii   322
form   276
general results   291
initiation of pattern   409
mechanical model   545, 551
microorganisms   357
polar form   337
pulse   330, 333
trains   312, 336, 367
Trichostrongylus retortaeformis (parasite
    worm)   252
dispersal model   252
Trifurcation   600, 603
Trinkaus, J.P.   529, 741
Troides
haliphron   462
hypolitus   462
prattorum   463
Trophic level   85
Trophic webb   63
Tsujikawa, T.   350, 741
Turing, A.M.   238, 374, 375, 435, 741
Turing
instability   380
mechanism   375
space   387, 397, 400, 478
Turner, J.S.   724
Twitty, V.C.   741
Tyson, J.J.   123, 143, 144, 146, 147, 148, 156,
    180, 182, 187, 189, 191, 198, 277, 313, 329,
    344, 345, 350, 724, 730, 732, 740, 741

Ulnus (bone)   560
Uppal, A.   132, 742

Vaccination   610
Valais goat (*Capra aegragrus hircus*)   445,
    446
van den Driessche, P.   618, 730
Van der Pol equation   190, 716
Van Haaften, J.L.   732
Váradi, Z.B.   254, 724
Varek, O.   411, 742
Vegetal pole   305
Venereal diseases   619
chlamydia   619
control model   623
gonorrhea   619, 623
syphilis   619
Ventral root   259
Verdi, R.   730
Verhulst, P.F.   2, 742
Verhulst process   37
Vertebrate
cartilage morphogenetic rules   599
limb construction scenario   609
limb development   564
skin   554
Vest, C.M.   448, 449, 743
Vibrations   448
plate   449
Vidal, C.   50, 725
Vincent, J.L.   24, 744
Vincent, J.-P.   726
Viscosity
bulk   534
shear   534
Vision
monocular   490
Visual cortex   481, 489
geometry of basic patterns   496
patterns   497, 502
stripe pattern formation mechanism
    489
Visuo-cortical transformation   496, 497
Voght, D.E.   730
Voight, D.R.   659, 734
Volterra, V.   63, 742
von Engelhardt, A.   450, 452, 453, 457, 458,
    459, 732
Von Foerster, equation   30
similarity solution   31

Waddington, C.H.   505, 742
Walbot, V.   526, 528, 561, 562, 742
Wallén, P.   259, 260, 726, 729

Walsh, J.A. 630, 742
Waltman, P. 613, 731, 742
Wan, Y.-H. 86, 706, 730
Wandeler, A. 669, 670, 681, 733, 740
Ward, P. 529, 730
Warren, K.S. 630, 742
Wasoff, F. 490, 727
Watson, J.D. 726
Watt, F.M. 533, 742
Wavefront solution 278
    asymptotic form 282
    Belousov-Zhabotinskii 322
    excitable kinetics 304
    Fisher equation 279
    stability 281
Wave length
    critical 386
    hair spacing (*Acetabularia*) 476
    variation with morphogen (calcium)
        concentration 479
Wavenumber 382
    critical 386, 398
    discrete 388
Wave
    activation 305
    back 575
    calcium 274
    chaotic 313
    contraction 306
    epidemic 653
    epizootic 661, 663, 670
    epizootic speed of propagation 670
    evasion 315
    excitable media 328
    front 256, 575
    gene-culture 281
    kinematic 254
    Lotka-Volterra 315
    multi-steady state kinetics 297
    muscle tissue 329
    plague 657
    post-fertilization 573
    pseudo 255
    propagating 277
    pulse 574
    pursuit 315, 319
    rabies epidemic 653, 659
    solation 573
    speed 276,
    speed dispersion relation 293
    spiral 313
    spread of farming 281
    steepness 283, 288
    three-dimensional 313

    trains 312, 336
    travelling 274
    two-dimensional epizootic (foxes) 689
    variable 276
Wave vector 386
Webster, D. 730
Weak solution 296
Weighting function 482
Weinberger, H.P. 733
Welsh, B.J. 313, 348, 742
Wessells, N. 554, 742
Westergaard, J.M. 682, 742
Whale
    baleen model 34
    discrete population model 53
Wickwire, K.H. 742
Whipworm (*Trichurus trichiura*) 630
Whitham, G.B. 367, 368, 742
Whittaker, R.H. 84, 742
WHO: see World Health Organisation
Whorl (*Acetabularia*) 468
    hairs 471
    regeneration 470
Wiesel, T.N. 489, 490, 731
Wild, D. 528, 730
Willems, J.L. 703, 742
Williamson, M. 68, 742
Williamson, P. 594, 606, 742
Winfree, A.T. 140, 142, 179, 200, 201, 202,
    204, 210, 212, 215, 274, 313, 314, 329, 343,
    344, 742, 743
Wolfram, S. 505, 517, 743
Wolpert, L. 374, 452, 525, 560, 561, 562,
    600, 724, 733, 740, 743
World Health Organisation (WHO) 657,
    669, 681
Wound
    scarring 587,
    healing 586
Wyatt, T. 315, 743

Xenopus laevis (frog) 602, 604
Xu, Y. 448, 449, 743

Yagil, E. 145, 743
Yagil, G. 145, 743
Yamaguti, M. 321, 735, 743
Yamamoto, H. 731
Y-(or branching) bifurcation 601
Yorke, J.A. 45, 619, 623, 624, 730, 732, 733
Yoshikawa, A. 321, 743
Yoshizawa, S. 161, 736
Young, D.A. 436, 743

Zaikin, A.N.   313, 731, 743
Zebra   436
    coat patterns   437, 440, 442
    embryo   442
    Equus burchelli   442
    Equus grevyi   437
    Equus zebra zebra   442
    gestation   436
    scapular stripes   442, 443
    stripe pattern   442, 490

Zeeman, E.C.   274, 743
Zeiss, G.D.   730
Zhabotinskii, A.M.   142, 179, 313, 731, 743
Zimen, E.   681, 729
Zone of influence (cells)   600
Zone of recruitment (cells)   600
Zonurus cordylus (lizard)   588
ZPA (zone of polarizing activity)   561
Zykov, V.S.   329, 346, 743

Bio-mathematics

Managing Editor: S. A. Levin

Editorial Board: M. Arbib, H. J. Bremermann, J. Cowan, W. M. Hirsch, S. Karlin, J. Keller, K. Krickeberg, R. C. Lewontin, R. M. May, J. D. Murray, A. Perelson, T. Poggio, L. A. Segel

In preparation

Volume 18

S. A. Levin, T. G. Hallam, L. J. Gross (Eds.)

Applied Mathematical Ecology

1989. 110 figures. Approx. 500 pages.
ISBN 3-540-19465-7

Contents: Part I. Introduction. *S. A. Levin:* Ecology in Theory and Application. – **Part II. Resource Management.** *C. W. Clark:* Bioeconomic Modeling and Resource Management. *R. McKelvey:* Common Property and the Conservation of Natural Resources. *M. Mangel:* Information and Area-wide Control in Agricultural Ecology. – **Part III. Epidemiology.** Fundamental Aspects of Epidemiology. *H. W. Hethcote:* Three Basic Epidemiological Models. **A. P. Dobson:** The Population Biology of Parasitic Helminths in Animal Populations. *J. L. Aron:* Simple Versus Complex Epidemiological Models. *H. W. Hethcote, S. A. Levin:* Periodicity in Epidemiological Models. – **Case Studies.** *H. W. Hethote:* Rubella. *W.-M. Liu, S. A. Levin:* Influenza and Some Related Mathematical Models. *C. Castillo-Chavez:* Review of Recent Models of HIV/AIDS Transmission. *R. M. May, R. M. Anderson:* The Transmission Dynamics of Human Immunodeficiency Virus (HIV). – **Part IV. Ecotoxicology.** *S. A. Levin:* Models in Ecotoxicology: Methodological Aspects. *R. V. Thomann:* Deterministic and Statistical Models of Chemical Fate in Aquatic Systems. *T. G. Hallam, R. R. Lassiter, S. A. L. M. Kooijman:* Effects of Toxicants in Aquatic Populations. – **Part V. Demography and Population Biology.** *L. J. Gross:* Mathematical Model in Plant Biology: An Overview. *J. Impagliazzo:* Stable Population Theory and Applications. *R. M. Nisbet, W. S. C. Burney, J. A. J. Metz:* Stage Structure Models Applied in Evolutionary Ecology. *C. Castillo-Chavez:* Some Applications of Structured Models in Population Dynamics. – Author Index. – Subject Index.

Volume 17

T. G. Hallam, S. A. Levin (Eds.)

Mathematical Ecology

An Introduction

1986. 84 figures. XII, 457 pages. ISBN 3-540-13631-2

Contents: Introduction. – Physiological and Behavioral Ecology. – Population Ecology. – Communities and Ecosystems. – Applied Mathematical Ecology. – Author Index. – Subject Index.

Volume 16

J. L. Casti, A. Karlqvist (Eds.)

Complexity, Language, and Life: Mathematical Approaches

1986. XIII, 281 pages. ISBN 3-540-16180-5

Contents: Allowing, forbidding, but not requiring: a mathematic for a human world. – A theory of stars in complex systems. – Pictures as complex systems. – A survey of replicator equations. – Darwinian evolution in ecosystems: a survey of some ideas and difficulties together with some possible solutions. – On system complexity: identification, measurement, and management. – On information and complexity. – Organs and tools: a common theory of morphogenesis. – The language of life. – Universal principles of measurement and language functions in evolving systems.

Volume 15

D. L. DeAngelis, W. M. Post, C. C. Travis

Positive Feedback in Natural Systems

1986. 90 figures. XII, 290 pages. ISBN 3-540-15942-8

Contents: Introduction. – The Mathematics of Positive Feedback. – Physical Systems. – Evolutionary Processes. – Organisms Physiology and Behavior. – Resource Utilization by Organisms. – Social Behavior. – Mutualistic and Competitive Systems. – Age-Structured Populations. – Spatially Heterogeneous Systems: Islands and Patchy Regions. – Spatially Heterogeneous Ecosystems: Pattern Formation. – Disease and Pest Outbreaks. – The Ecosystem and Succession. – Appendices. – References. – Subject Index. – Author Index.

Springer-Verlag Berlin Heidelberg New York London Paris Tokyo Hong Kong

Springer

Volume 14
C. J. Mode

Stochastic Processes in Demography and Their Computer Implementation
1985. 49 figures, 80 tables. XVII, 389 pages.
ISBN 3-540-13622-3

Volume 13
J. Impagliazzo

Deterministic Aspects of Mathematical Demography
An Investigation of the Stable Theory of Population including an Analysis of the Population Statistics of Denmark
1985. 52 figures. XI, 186 pages. ISBN 3-540-13616-9

Volume 12
R. Gittins

Canonical Analysis
A Review with Applications in Ecology
1985. 16 figures. XVI, 351 pages. ISBN 3-540-13617-7

Volume 11
B. G. Mirkin, S. N. Rodin

Graphs and Genes
Translated from the Russian by H. L. Beus
1984. 46 figures. XIV, 197 pages. ISBN 3-540-12657-0

Volume 10
A. Okubo

Diffusion and Ecological Problems: Mathematical Models
1980. 114 figures, 6 tables. XIII, 254 pages.
ISBN 3-540-09620-5

Volume 9
W. J. Ewens

Mathematical Population Genetics
1979. 4 figures, 17 tables. XII, 325 pages.
ISBN 3-540-09577-2

Volume 8
A. T. Winfree

The Geometry of Biological Time
1980. 290 figures. XIV, 530 pages. ISBN 3-540-09373-7

Volume 7
E. R. Lewis

Network Models in Population Biology
1977. 187 figures. XII, 402 pages. ISBN 3-540-08214-X

Volume 6
D. Smith, N. Keyfitz

Mathematical Demography
Selected Papers
1977. 31 figures. XI, 514 pages. ISBN 3-540-07899-1

Volume 5
A. Jacquard

The Genetic Structure of Populations
Translated by D. Charlesworth, B. Charlesworth
1974. 92 figures. XVIII, 569 pages.
ISBN 3-540-06329-3

Volume 4
M. Iosifescu, P. Tautu

Stochastic Processes and Applications in Biology and Medicine II
Models
1973. 337 pages. ISBN 3-540-06271-8

Volume 3
M. Iosifescu, P. Tautu

Stochastic Processes and Applications in Biology and Medicine I
Theory
1973. 331 pages. ISBN 3-540-06270-X

Volume 2
E. Batschelet

Introduction to Mathematics for Life Scientists
3rd edition. 1979. 227 figures, 62 tables.
XV, 643 pages. ISBN 3-540-09662-0

Springer-Verlag
Berlin Heidelberg New York
London Paris Tokyo Hong Kong